W9-CSF-004

THE

INTERNATIONAL SERIES
OF
MONOGRAPHS ON PHYSICS

PRINCIPLES OF ELECTRON TUNNELING SPECTROSCOPY

E. L. WOLF

Department of Physics
Ames Laboratory, U.S. Department of Energy
and
Microelectronics Research Center
Iowa State University

OXFORD UNIVERSITY PRESS • NEW YORK
CLARENDON PRESS • OXFORD
1985

Oxford University Press

London Glasgow New York Toronto
Delhi Bombay Calcutta Madras Karachi
Kuala Lumpur Singapore Hong Kong Tokyo
Nairobi Dar Es Salaam Cape Town
Melbourne Wellington

and associate companies in
Beirut Berlin Ibadan Mexico City

Published in the United States by
Oxford University Press, New York

Library of Congress Cataloging in Publication Data

Wolf, E. L.
Principles of electron tunneling spectroscopy.

(International series of monographs on physics; 71)
Includes index.
1. Tunneling spectroscopy. I. Title.
II. Series: International series of monographs on physics (Oxford, Oxfordshire); no. 71.
QC454.T75W64 1985 530.4′1 83-17470
ISBN 0-19-503417-1

British Library Cataloguing in Publication Data
Wolf, E. L.
Principles of electron tunneling spectroscopy.
(International series of monographs on physics; 71)
1. Tunneling spectroscopy
I. Title II. Series
530.4′1

ISBN 0-19-503417-1

Printed in Northern Ireland by The Universities Press (Belfast) Ltd.

In memory of

Norman and Harriet

Preface

Tunneling of electrons through classically forbidden barriers is a basic process in all matter at the atomic scale. Artificial tunneling structures, the topic of this book, contain barriers having an atomic or Angstrom length scale in one dimension, but, in the transverse directions have laboratory length scales of microns to millimeters. Overcoming the technical difficulties in preparing such uniform barriers in solid state structures and using these to reproducibly demonstrate the predictions of elementary quantum mechanics has been relatively recent, dating from the work of Esaki on semiconductor junctions in 1958, quickly followed by Giaever's work on metal-insulator-metal thin film structures. The field of research opened up by these reproducible "macroscopic" tunneling barrier structures has had a major impact on condensed matter physics and on technology. Contributions to both basic physics and technology were hastened by the verification of the Bardeen, Cooper and Schrieffer superconducting energy gap by Giaever and by the brilliant theoretical prediction in 1962 by B. D. Josephson of dc and ac supercurrents in tunneling barrier diodes with both electrodes in the superconducting state. The Josephson effects, verified in the early 1960s, have led to major advances in detection of magnetic flux, current and voltage, and also to promising new families of logic devices for computer application. The importance of this new area of research to the advancement of physics and technology was recognized in 1973 by award of the Nobel Prize in Physics to Esaki and Giaever, and to Josephson.

The present book is intended to review in a comprehensive fashion the various techniques, principally tunneling energy spectroscopies, for study of condensed matter, based on tunnel junctions, and also to review the solid state physics and materials properties that have been thus revealed. An effort has been made to provide an overall conceptual basis in which superconducting effects and the various normal state effects, including the vibrational excitation spectroscopy of organic molecules in tunnel barriers (IETS), can be uniformly treated. A rather large number of figures has been included, to illustrate all forms of energy spectroscopy that evolve from barrier tunneling, and to provide examples from as many as possible of the different physical situations or systems which have been studied. This material has been taken from what is by now an extensive and rather widely scattered literature, and has been quite rigidly organized, usually according to the physical mechanism by which the spectroscopic data are thought to arise. The conceptual organization will be evident from the

Table of Contents, especially with reference to the introductory first chapter, particularly the discussion at the end of Chapter 1.

This book has been made possible by assistance from many sources. Granting of leave with financial support from Iowa State University during 1981 and 1982, and hospitality extended to me as a visitor at the University of Pennsylvania, and later at the Yorktown Heights Research Center of IBM, were essential to getting the project started. The assistance, particularly, of Elias Burstein and E. Ward Plummer at Penn and of Charles Kircher and William Gallagher at IBM is gratefully acknowledged. Thanks are due also to Douglas Finnemore, Margaret Avery, Jerome Ostenson and Tar-Pin Chen for their assistance during my absence from Ames. Part of the manuscript was written at the Aspen Center for Physics, and all was typed with great efficiency and good humor by Ms. Lesley Swope of the Ames Laboratory—USDOE at Iowa State University. I am indebted to many Ames colleagues, and especially to John R. Clem, Douglas K. Finnemore and Bruce N. Harmon for discussions and for their encouragement and support during this project.

Permission to reproduce figures from (at least one of) the authors listed in the figure captions is most gratefully acknowledged. Particular thanks for permission to include previously unpublished material go to G. B. Arnold, W. J. Gallagher, F. Gompf, Z. G. Khim, B. N. Taylor and D. G. Walmsley. I am also grateful to G. B. Arnold, F. Gompf, K. E. Gray, R. C. Jaklevic, D. G. Walmsley and others for providing original figures, enlargements, or other material. G. B. Arnold is particularly acknowledged as the source of many of the figures in Chapter 5. Special thanks are also due to B. D. Josephson for assistance in obtaining a copy of his Fellowship dissertation at Trinity College, University of Cambridge.

Present and past members of our Tunneling Group have assisted in improving the manuscript: specific thanks are due to Hong-Jie Tao, Tar-Pin Chen, Mark Albers, Patricia Allen, Siyuan Han, Bret Hess, Quiming Li, Kwok-Wai Ng, Danny Shum, Anne Thomas, and Youwen Xu.

Finally, I am most grateful to my family for their support during this project: specifically to Douglas for helping to establish an orderly relation between the innumerable figures and their captions, to David for searching and sorting among the unending illegible references, and to Carol for her patient and cheerful accommodation of the many inconveniences, which, happily, are now over.

Ames, Iowa E. L. W.
October, 1984

Contents

1
Introduction

The several forms of energy spectroscopy of solids made possible by quantum mechanical tunneling of electrons between two metallic electrodes separated by a thin barrier are the subject matter of this book. In addition to discussing techniques of tunneling spectroscopy which have evolved over the past twenty-five years or so, we also attempt to collect and organize the wealth of information, primarily concerning the electronic and vibrational properties of metals, and notably superconductors and semiconductors, that tunneling has made available.

Tunneling spectroscopies may be classified by the range of energies probed. These are, at most, on the order of metallic Fermi energies, or about 5 eV, in studies relating to aspects of electronic band structure. More typically, one measures in the range of a few to a few hundred millielectronvolts in studies of phonons and molecular vibrations. In the important applications of tunneling to determination of the excitation spectra of superconductors, a further characteristic scale is that of the superconducting energy gap, typically a few millielectronvolts. In the majority of cases, the object of study is the tunneling electrode, which may be a single crystal of metal or semiconductor but more typically is an evaporated polycrystalline or possibly amorphous metallic film. The energy resolution of tunneling spectroscopy is a few times the thermal energy kT, so that, for energy resolution of millivolts or less, measurements are usually made at liquid helium temperature. In the study of electrodes, one may be primarily interested in the normal-state properties of the given electrode material, in its particular superconducting properties, or in universal features of the superconducting state which are relatively insensitive to the choice of material. Indeed, a particularly fruitful application of tunneling has been to elucidate the properties of coupled superconductors, as first predicted by Josephson (1962*a*), in the superconductor-insulator-superconductor (SIS) configuration. An additional important application of tunneling is to quantitative determination of the electron-phonon spectral function $\alpha^2F(\omega)$ of many superconductors, from which a rather complete specification of their properties has been obtained. In other instances, the object of spectroscopic study is the insulating barrier, necessarily limited in thickness to the order of 100 Å to allow a measurable tunnel current. In one special case of this type of study, one measures the vibrational spectrum of a monolayer of molecules present at one edge of the insulator. Such spectra are obtained in a form similar to conventional infrared and Raman spectra. An important advantage of *inelastic electron tunneling spectroscopy* (IETS) is that comparable spectra may be obtained from samples containing orders-of-magnitude fewer molecules than are required in conventional optical techniques.

1.1 Concepts of Quantum Mechanical Tunneling

The basis for the several forms of electron tunneling spectroscopy is the inherently smooth behavior of the probability density $\psi^*\psi$ even at points of finite discontinuity in the potential energy $V(x)$. At a metal-vacuum interface, in particular, the wavefunction for an electron at the Fermi energy does not vanish outside the metal but rather decays exponentially as $\exp(-\kappa x)$, where $\kappa \cong (2m/\hbar^2)^{1/2}\phi^{1/2}$ and x is the distance into the vacuum; at metal-*insulator* interfaces, the electronic mass m and metal work function ϕ must be altered to reflect the properties of the solid-state insulator I. If, then, a metal-insulator-metal (MIM) junction is biased by an external voltage V, electrons in an energy range eV between the two shifted Fermi energies may elastically tunnel from one side to empty states opposite. Note that the most energetic of the injected electrons appear at an energy precisely eV above the Fermi energy of the positively biased electrode, in the limit $T=0$. Similarly, eV is precisely the maximum energy (at $T=0$) which can be given up in an inelastic process and yet allow the tunneling electron to appear in an unoccupied state of the positively biased electrode.

Tunneling processes are fundamental to quantum mechanics, following directly from the nature of the solutions $\psi(x)$ of Schrödinger's equation and the probability interpretation of $\psi^*\psi$. The rate at which such processes occur is dominated by the exponential decay in the classically forbidden barrier region, where the *potential energy* $V(x)$ or $U(x)$ exceeds E_x. The symbol $U(x)$ for potential energy will often be used in cases where the potential barrier may be changed by an applied voltage: $U(x, V)$. For many purposes the rate can be adequately estimated from the Wentzel-Kramers-Brillouin (WKB) approximation

$$D = \exp(-2K) \tag{1.1}$$

$$K = \int_{x_1(E_x)}^{x_2(E_x)} \kappa(x, E_x)\,\mathrm{d}x \tag{1.2}$$

$$\kappa(x, E_x) = \left\{\frac{2m^*[V(x) - E_x]}{\hbar^2}\right\}^{1/2} \tag{1.3}$$

In Eqns. (1.1)–(1.3), E_x is the kinetic energy; x_1, x_2 are the classical turning points; and D is properly regarded as the fraction of the probability current $\hbar k/m$ carried by an incident wave e^{ikx} which passes the barrier $V(x) > E_x$ entering (1.3). The probability current is defined as

$$j = \frac{i\hbar}{2m}\left(\psi\frac{\partial\psi^*}{\partial x} - \psi^*\frac{\partial\psi}{\partial x}\right) \tag{1.4}$$

1.2 Occurrence of Tunneling Phenomena

The fundamental nature of the tunneling phenomenon makes it a pervasive feature in the behavior of small-mass particles in rapidly spatially varying potentials. These conditions apply to electrons in atoms, molecules, and solids. Tunneling phenomena are thus inescapably involved in the structure and behavior of matter viewed on an atomic scale.

The first phenomena to be identified and convincingly explained in terms of tunneling allowed unequivocal detection of particles escaped into the vacuum by barrier tunneling from metastable states in matter. Three such phenomena were already identified in 1928: the natural decay of certain heavy nuclei by α-particle emission, the ionization of atomic hydrogen in a strong electric field, and, similarly, the emission of electrons from a cold, clean metal surface under the application of a strong electric field. As sketched in Fig. 1.1, the α particle which is subsequently observed outside the nucleus with a kinetic energy E_k, typically a few MeV, was assumed (Gamow, 1928; Gurney and Condon, 1929) to occupy a metastable state at energy E inside the nucleus prior to decay. If we take the tunneling barrier as $V(r)=(Z-2)2e^2/r$ for $r>R$, with R the nuclear radius, application of the WKB formula (1.2) gives

$$K=\left[\frac{2m_\alpha}{\hbar^2}(Z-2)2e^2\right]^{1/2}\int_R^b\left[\frac{1}{r}-\frac{E}{(Z-2)2e^2}\right]^{1/2}dr$$

with $b=(Z-2)2e^2/E$, m_α the α-particle mass, and Z the atomic number.

Fig. 1.1. Decay of radioactive nucleus by tunneling of alpha particle through Coulomb barrier. Regularity between energy E_k of emerging alpha particle and lifetime τ_α against decay substantiates tunneling model.

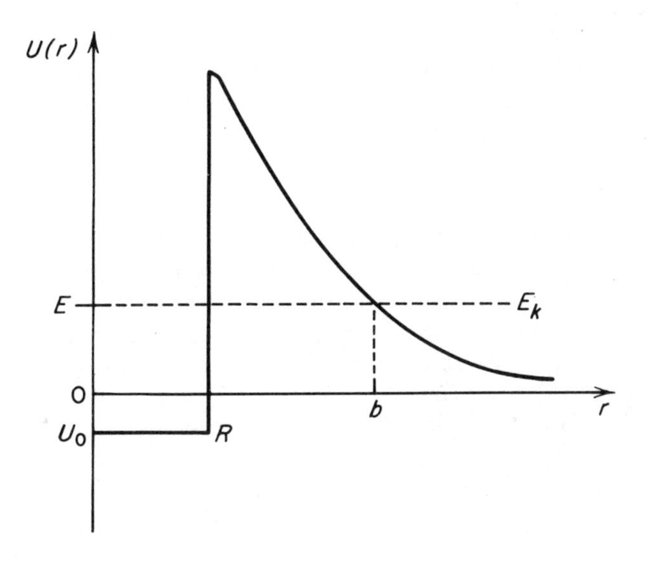

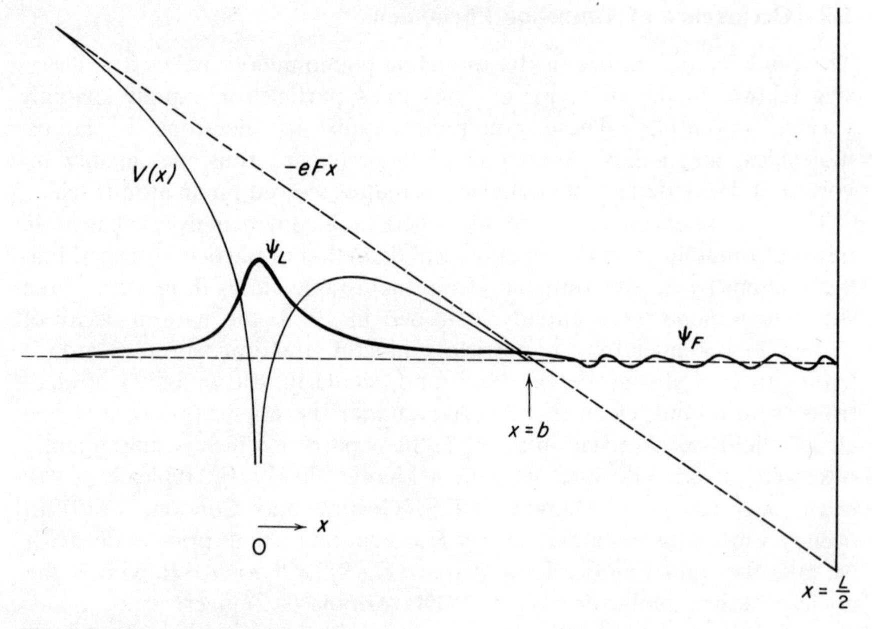

Fig. 1.2. Hydrogen atom in an applied electric field decays as the electron tunnels through the Coulomb barrier to the classically allowed region at positive x.

Evaluation gives an inverse lifetime $(1/\tau_\alpha)$ of

$$\frac{1}{\tau_\alpha} = A\,\exp(-2K) = A\,\exp\left\{-BE^{-1/2}\cos^{-1}\left[\left(\frac{R}{b}\right)^{1/2} - \left(\frac{R}{b} - \frac{R^2}{b^2}\right)^{1/2}\right]\right\}$$

with $B = 2\hbar^{-1}(2m_\alpha)^{1/2}2(Z-2)e^2$ (Gamow's formula). For heavy nuclei, $R/b = E/V(R)$ is reasonably small, so the $\cos^{-1}$ term is $\pi/2$, with slowly varying corrections. Neglecting the latter gives

$$\ln \tau_\alpha = -\ln A + \tfrac{1}{2}\pi B E_k^{-1/2}$$

which approximately describes τ_α values spanning 15 orders of magnitude. This expression provided a great success of the early quantum theory.

A second application of tunneling concepts in 1928 was Oppenheimer's treatment of the field ionization of hydrogen. A one-dimensional hydrogen atom at the center $(x = 0)$ of a box with a field $F = V/L$ is sketched in Fig. 1.2. For small values of F, there are two states at energy E_0 for the electron: the "localized atomic" state ψ_L is peaked near the "nucleus," while in the "free" states, ψ_F, the electron oscillates between the classical turning point $x = b$ and the boundary $x = L/2$. In Oppenheimer's treatment for weak fields F, the atomic wavefunction ψ_L is calculated by neglecting the field, and ψ_F can be calculated as the eigenfunction at energy E_0 of the uniform field alone.

Oppenheimer (1928) showed that the initial rate of transition w from state ψ_L to the continuum of final states ψ_F (in the limit of large L) can be calculated to good accuracy from the Golden Rule as

$$w = \frac{2\pi}{\hbar} |M|^2 \rho_F(E) \tag{1.5}$$

where $|M|^2 = |\langle\psi_F| H'(x) |\psi_L\rangle|^2$ and $H'(x) = -eFx$ is the perturbing Hamiltonian. Oppenheimer estimated the lifetime of the hydrogen atom against ionization in a field F of 10^3 V/m as $10^{10^{10}}$ sec.

The third early example of tunneling, which is closer to the subject of this book, is field emission of electrons from metals, treated in 1928 by Fowler and Nordheim. Again using (1.2) to estimate the decaying exponential behavior, assuming a triangular barrier as in Fig. 1.3, one finds for electrons at the Fermi energy $E_x = \mu_F$

$$2K = 2\int_0^b \left(\frac{2m}{\hbar^2}\right)^{1/2} (\phi - eFx)^{1/2}\, dx = \frac{8\pi\sqrt{2m}\,\phi^{3/2}}{3heF}$$

Here, ϕ is the work function, F is the field strength, and $b = \phi/eF$ is the classical turning point. Thus, the transmitted fraction of the electron current incident upon the surface barrier at the Fermi energy is $D = e^{-2K}$. The total emitted current is obtained by integration over the available electron energy states weighted by their thermal occupation, given by the Fermi function defined as

$$f(E) = \left\{1 + \exp\left[\frac{(E - \mu_F)}{kT}\right]\right\}^{-1} \tag{1.6}$$

Hence, since there are two electron states per k-space volume $(2\pi)^3$, with group velocity $v_x = \hbar^{-1}\,\partial E/\partial k_x$, the current density J_x is

$$J_x = \frac{2e}{(2\pi)^3} \iiint D(E_x) f(E) \left(\frac{1}{\hbar}\right)\left(\frac{\partial E}{\partial k_x}\right) dk_x\, dk_y\, dk_z \tag{1.7}$$

leading to the original result of Fowler and Nordheim (1928),

$$J_x = \frac{4\sqrt{\mu_F \phi}}{(\mu_F + \phi)} \frac{e^3 F^2}{8\pi h\phi} \exp\left(-\frac{8\pi\sqrt{2m}\,\phi^{3/2}}{3heF}\right) \tag{1.8}$$

which successfully describes the measured field emission current, typically yielding straight-line plots of $\ln(J/F^2)$ vs. $1/F$.

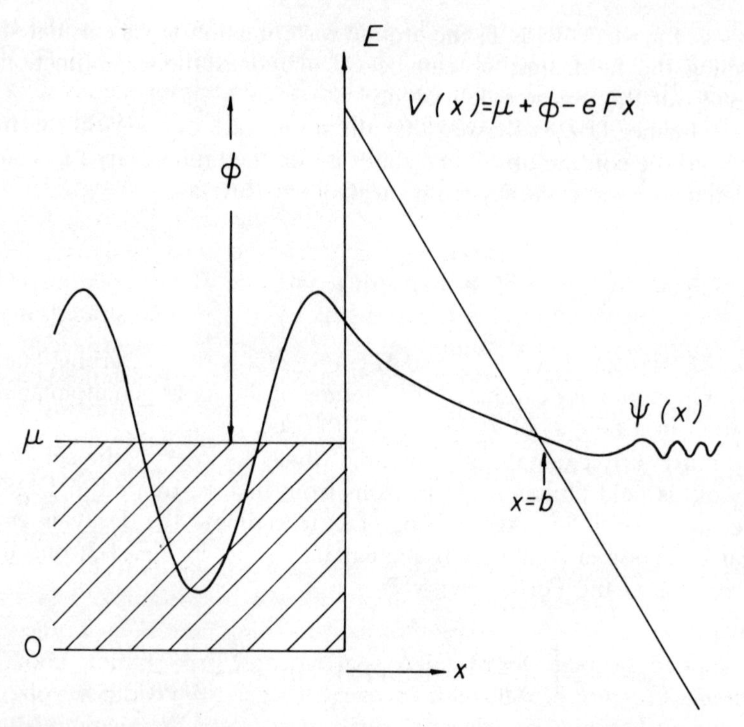

Fig. 1.3. Field emission of electrons from cold metal surface under the application of the strong electric field, as proposed by Fowler and Nordheim (1928), occurs by electron tunneling through the potential barrier formed by the work function and the electric field.

1.3 Electron Tunneling in Solid-State Structures

Electron tunneling within a crystalline solid, where the Bloch states are of the form

$$\psi_{\vec{k}}(\vec{r}) = u_{\vec{k}}(\vec{r})e^{i\vec{k}\cdot\vec{r}} \tag{1.9}$$

with $u_{\vec{k}}(\vec{r})$ a periodic function of $\vec{r}$, occurs in the band gap, where $k = i\kappa$ is imaginary, resulting in exponentially damped waves.

The magnitude of the decay constant $\kappa(E)$ at energy E depends upon the location of E within the band gap $E_g = E_c - E_v$ and the effective masses m_c^*, m_v^* which characterize the conduction and valence bands.

In principle, the application of a uniform electric field to a homogeneous solid, as, for example, an intrinsic semiconductor at $T = 0$, should permit interband tunneling, or "internal field emission," as illustrated in Fig. 1.4a. This effect was treated by Zener (1934) by a WKB method, yielding a transmission factor $D = \exp(-maE_g w/4\hbar^2)$, with a the lattice constant, F the electric field, and the "barrier" width $w = E_g/|eF|$. While this effect has probably never been observed in a homogeneous solid, it is closely related to tunneling effects observed by Esaki (1958) in narrow semiconductor *pn* junctions.

The schematic diagram of Fig. 1.4b shows Esaki's structure, in which the width w is of order 100 Å, corresponding to a junction field $F \simeq 3 \times 10^7$ V/m. The Esaki diode is degenerately doped p type on the left and n type on the right such that the metallic Fermi energies μ_F are ~0.05 eV. Application of a forward bias $eV \simeq \mu_F$, as illustrated, may be expected to allow a peak tunnel current, as all of the donated electrons in n can tunnel elastically into empty electron states (filled holes states) in the p-type region. One might expect further increase in forward bias to reduce the current, as there will become fewer available final electron states. This effect produces the characteristic and technologically useful negative-resistance region, first observed by Leo Esaki in 1958. The Esaki diode is an important invention, permitting construction of oscillators and many other circuits with wide applications. A further scientific importance of Esaki's work was to provide the first example in which, as shown in Fig. 1.4c, the variation of the tunneling current as a function of the applied voltage V revealed spectroscopic information about the *electronic structure* of the solids involved. Clearly, the width in voltage of the current "hump" is at least a rough measure of the energy widths of the filled states on the n side and the empty states on the p side, i.e., the local Fermi energies as measured from the respective band edges. Any such spectroscopic aspect depends upon having highly conducting electrode regions such that the applied potential drops entirely across the thin barrier. One assumes that the tunneling process is elastic, leading to the injection of electrons on the other side whose energies extend precisely to the applied bias energy eV and that the tunneling process depends on the densities of states involved. A second type of spectroscopy is also evident in the results of Esaki (Fig. 1.4c) in the lower curve measured at 4.2 K. In this case, the points of maximum curvature of the I–V characteristic labeled as 18, 55, and 83 mV are simply related to phonon energies (arrows) of the silicon lattice. This case is an example of a tunneling *threshold spectroscopy* in which the tunneling process acquires a higher probability (and is said to be *assisted*) if some other mode of the system can be excited. As we shall see, the phonons are required in this case because the electronic transition in silicon from the lowest conduction band to the highest valence band involves a change in wavevector from the zone boundary to the zone center; conservation of crystal momentum within the single-crystal structure hence requires phonon emission or absorption to balance the change in wavevector incurred in the electronic transition. Many other types of excitations have been observed in this general fashion as onsets of new processes which, even to a small degree,

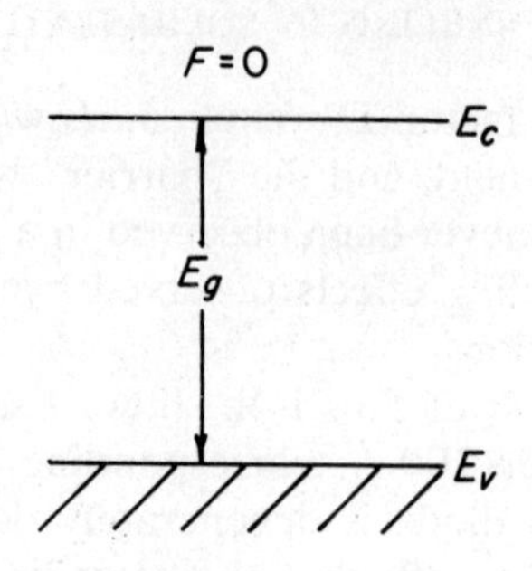

F=0
Ec
Eg
Ev

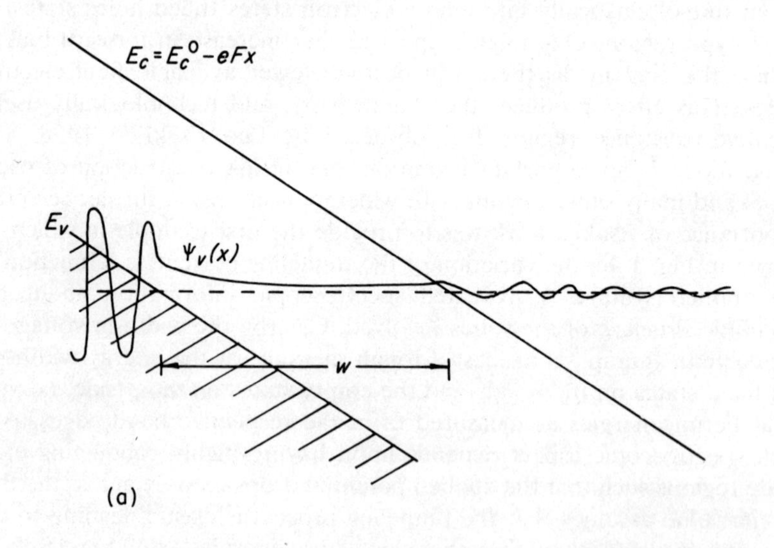

Ec = Ec0 − eFx
Ev
ψv(x)
w
(a)

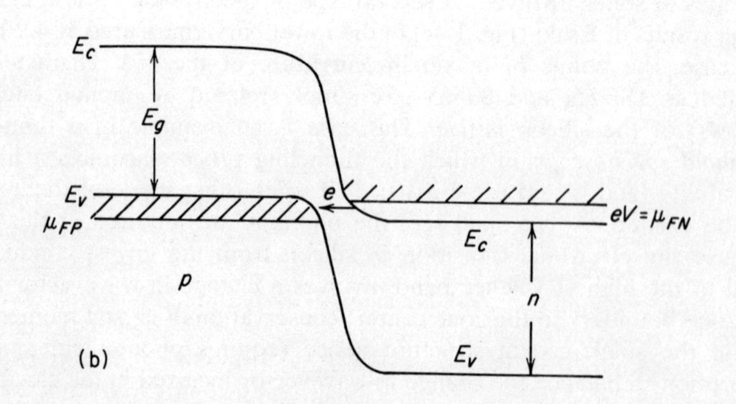

Ec
Eg
Ev
μFP
e
eV = μFN
Ec
p
n
(b)
Ev

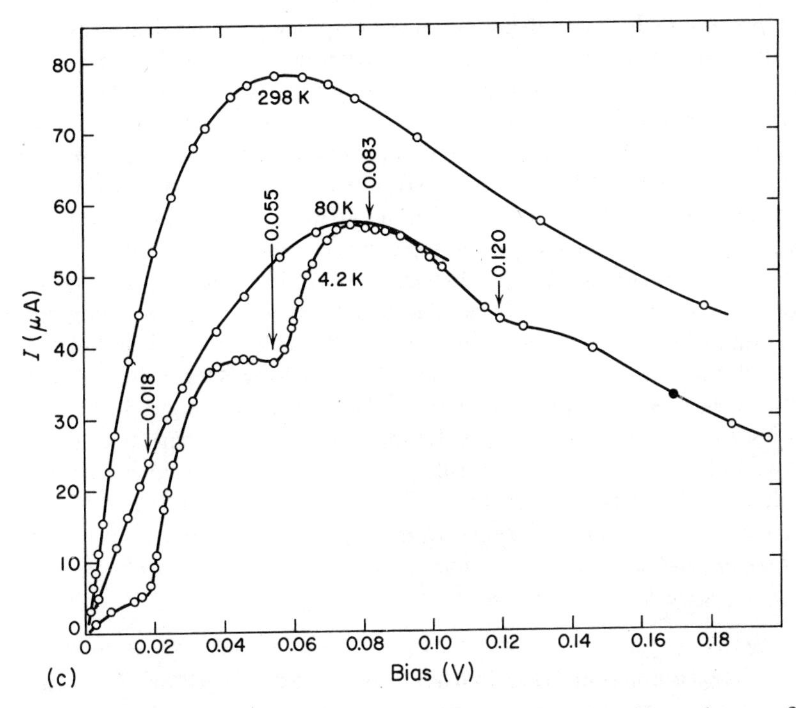

Fig. 1.4. (a) Schematic diagram illustrating intrinsic semiconductor in zero field (top) and at high electric field $F = E_g/ew$ (bottom). Zener interband tunneling occurs from valence band maximum to conduction band minimum through the semiconductor band gap, forming a barrier of width w. (b) Schematic of narrow tunneling (Esaki) *pn* junction between degenerately doped regions of a semiconductor single crystal. (c) Currents observed in narrow *pn* junction or tunnel diode by Esaki and Miyahara (1960) at three different temperatures. The arrows in the figure at 18, 55, 83, and 120 mV locate simple multiples of the zone boundary phonon energies in silicon, indicating that phonon emission is required in the tunneling process.

alter the measured tunneling current; examples include excitation of plasmons, molecular vibrations, and Zeeman transitions of impurities in tunnel junctions. It is difficult to overemphasize the third importance of Esaki's discovery: that for the first time a solid-state structure could be fabricated *reproducibly* in which the parameters of importance for the tunneling probability could be estimated, putting the calculation of a tunnel current on a semiquantitative basis. Earlier attempts to fabricate MIM structures, with insulators thin enough to allow tunneling to occur, had suffered from lack of reproducibility and difficulty in reliably estimating such relevant parameters as the thickness and the barrier height of the insulating layer. Esaki was awarded the Nobel Prize in 1973 for his invention and other contributions in this area of research.

It also seems likely that Esaki's success spurred renewed interest in the question of fabricating MIM thin-film structures in a similarly reproducible and quantifiable fashion. Such an effort was undertaken at the General Electric laboratory by Fisher and Giaever, who by early 1959 (Giaever, 1974) had demonstrated reproducibility of currents measured in new thin-film MIM tunneling structures. These were fabricated by allowing a thin, natural oxide layer to grow on an initially evaporated film, followed by evaporation of a second, crossing counterelectrode film (Fisher and Giaever, 1961). It was found that the results could be explained semiquantitatively within the framework of the elementary tunneling theory which had been developed much earlier (Frenkel, 1930; Frenkel and Joffe, 1932; Nordheim, 1932; Wilson, 1932), in part stimulated by early work on metal contacts (Holm and Meissner, 1932, 1933). An essentially unexpected development of great importance in 1960 allowed Giaever and Fisher to rule out appreciable current contributions in their MIM structures from mechanisms other than tunneling. This discovery was to initiate a long and fruitful connection between tunneling and superconductivity, and to demonstrate the simplicity and elegance of tunneling as a spectroscopic tool.

1.4 Superconducting (Quasiparticle) and Josephson (Pair) Tunneling

Giaever (1960*a*, 1960*b*) discovered that MIM structures, in which the tunneling barrier was formed by exposing an evaporated film of Al or Pb to the atmosphere, leaving an oxide layer typically 20 Å thick (Fisher and Giaever, 1961), developed a pronounced nonlinearity in the I–V curve when cooled well below the superconducting transition temperature T_c of one of the electrodes. The limiting behavior at very low temperature was zero current below an onset bias $V=\Delta/e$, where Δ is half the energy gap in the superconducting electrode. Further, Giaever (1960*a*) (see Fig. 1.5) and Nicol, Shapiro, and Smith (1960) established that the slope of the I–V characteristic $G(V)=\mathrm{d}I/\mathrm{d}V$, in the region of the nonlinearity, was directly related to the "density of states," or of quasiparticle excitations, of the superconductor, as predicted by the 1957 theory of Bardeen, Cooper, and Schrieffer (BCS). The many important consequences of this discovery, for which Giaever was awarded the Nobel Prize in 1973 (Giaever, 1974), can be regarded as being of two types. First, as we have indicated, this observation for the first time unequivocally demonstrated that tunneling could be the origin of the entire current (to better than one part per thousand) in properly made MIM structures. This result spurred further experimental and theoretical work, which have led to a considerably improved technology of tunneling structures and to much greater sophistication in understanding related phenomena. Second, a high-resolution energy spectroscopy based on tunneling was made available for detailed study of the phenomena of superconductivity. Giaever's discovery of Fig. 1.5 led to a great deal of further experimental and theoretical

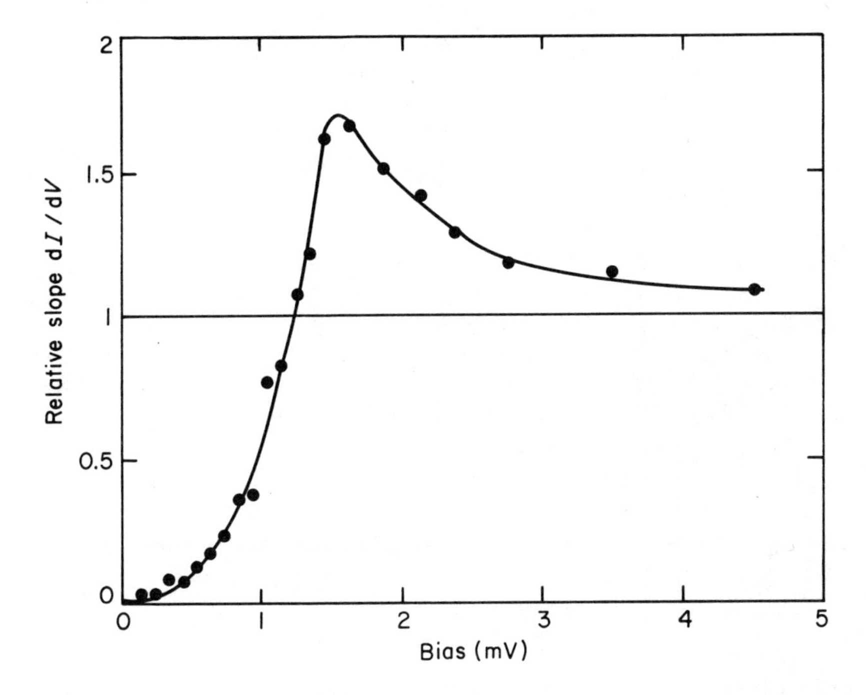

Fig. 1.5. Original observation of the superconducting energy gap in the tunnel conductance dI/dV of an aluminum–oxide–lead junction at 1.6 K (after Giaever, 1960a). The plotted points are slopes of I–V characteristics. Such measurements were soon shown to be quantitatively in agreement with the BCS theory of superconductivity if one included the effect of thermal smearing in the counterelectrode.

work on superconductors, providing detailed verification of many features of the BCS theory, as well as providing evidence of anomalous behavior (Fig. 1.6a) in soft metals such as Pb and Hg, which could not be explained by the BCS theory. The anomalous appearance of structure in the normalized conductance $\sigma(V) = G(V)_S/G(V)_N$ in the bias range corresponding to phonon energies (visible from energy 4Δ to 8Δ in Fig. 1.6a) was rather quickly understood by extensions of the theory of superconductivity to strong electron-phonon coupling, due notably to Migdal (1958), Eliashberg (1960), Nambu (1960), and Schrieffer, Scalapino, and Wilkins (1963). In the hands of McMillan and Rowell (1965), this phonon-related structure became the basis for extraction of the effective phonon spectrum $\alpha^2F(\omega)$ (Eliashberg function) of the superconductor (Fig. 1.6b). This technique, involving an elegant combination of experiment and the strong-coupling theory, has provided the most sensitive probe of the superconducting state.

The theoretical possibility of a supercurrent at zero voltage tunneling through a barrier between two superconductors was described in 1962 by

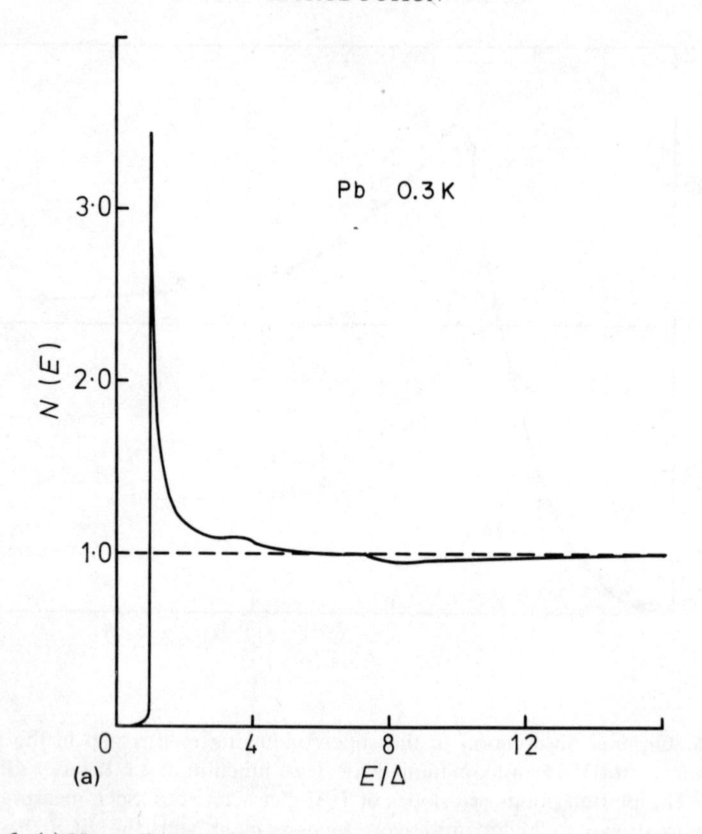

Fig. 1.6. (a) Normalized conductance of a tunnel junction involving lead at 0.3 K (after Giaever, Hart, and Megerle, 1962). Note the extremely sharp energy gap. The small deviations of the density of states from unity in the 4–10 mV range are due to the phonons of lead. (b) Illustration of the use of tunneling to determine the effective phonon spectrum $\alpha^2F(\omega)$ of a strong-coupling superconductor. The Pb phonons are revealed in detail by the analysis of McMillan and Rowell (1965). Curves *A*, *B*, and *C*, respectively, show the second derivative, first derivative, and effective phonon spectrum for lead.

B. D. Josephson. This further inherent feature of SIS tunneling is shown at $V=0$ in the I–V curve of Nicol, Shapiro, and Smith (1960) (Fig. 1.7) but remained unexplained before Josephson's work. Identifying features of this *dc Josephson effect,* which corresponds to tunneling of Cooper pairs of electrons, are a well-defined upper limit J_c of the $V=0$ current density with a sensitive dependence on magnetic field, which, without shielding, may destroy the effect. The characteristic magnetic field dependence of the effect was confirmed by Anderson and Rowell in 1963. A second prediction of Josephson (1962*a*) was that the SIS structure exhibiting pair tunneling, when driven out of the zero-voltage state to a bias voltage *V*, should emit radiation of frequency $\nu=2eV/h$. This *ac Josephson effect* was first directly observed by Giaever (1965) and by Yanson, Svistunov,

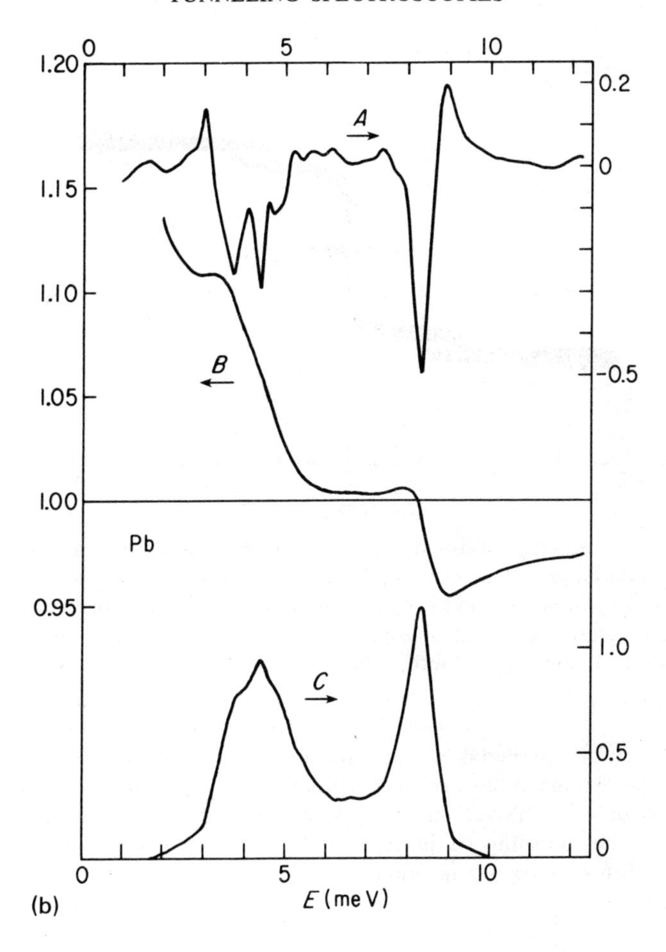

and Dmitrenko (1965). Josephson also was awarded the Nobel Prize in 1973 for these theoretical discoveries, which have had many consequences.

1.5 Tunneling Spectroscopies

A common feature of the solid-state-barrier tunneling structures mentioned in the two preceding sections, and collected in Fig. 1.8, is that application of a bias voltage V leads to tunneling of electrons with a well-defined range of energies $0 \leqslant E \leqslant eV$. This feature makes possible several forms of spectroscopy of the solid electrodes and of the barrier, with energy resolution set by kT. (Better resolution is possible when both electrodes are superconductors.) These forms include a spectroscopy of the superconducting state, which probes both the details of the energy gap structure and of the phonon spectrum which produces the paired-electron system.

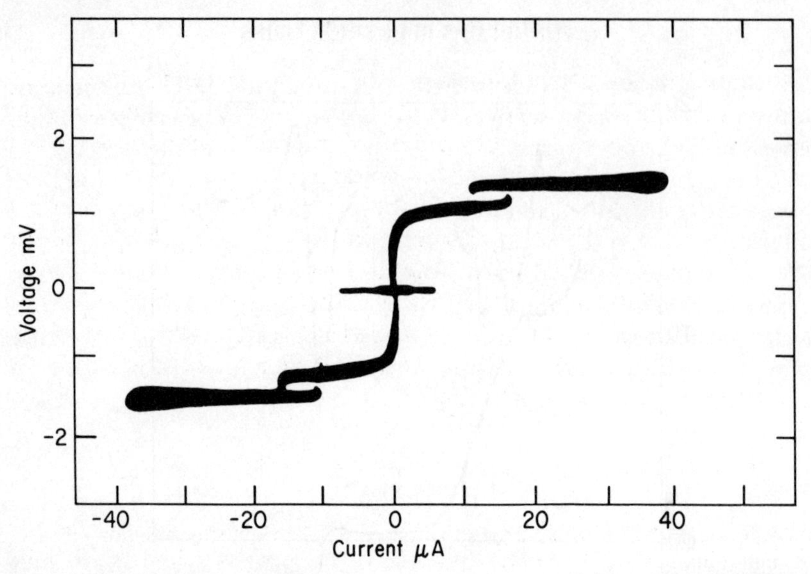

Fig. 1.7. The original published report of the current-voltage relation between two superconductors. The current, here plotted horizontally, rises at the sum energy gap of aluminum and lead, and also at $V = 0$, showing for the first time the Josephson supercurrent and quasiparticle current in the same tunnel junction. (After Nicol, Shapiro, and Smith, 1960.)

Fig. 1.8. (a) In the metal-insulator-metal tunnel junction, the barrier $U(x)$ between the two metal films is usually produced by oxidation of a film deposited on a glass slide. (b) Typical cross-stripe junction and elementary measurement circuit. (c) The tunneling *pn* junction or Esaki diode. (d) Metal-semiconductor contact or Schottky barrier junction.

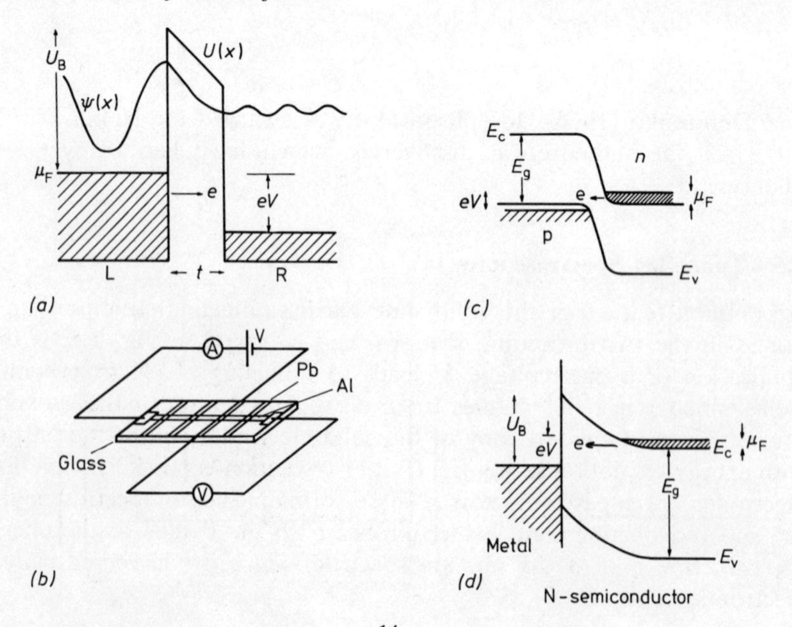

A second major area of spectroscopy, known as IETS for inelastic electron tunneling spectroscopy, involves measurement of inelastic excitations of the electrodes (usually in the normal state) and the barrier. An example is a threshold for phonon generation in an Esaki diode at a bias $V = \hbar\omega_{ph}/e$, which is detected by an accompanying step increase in $G(V)$ and thus a peak in d^2I/dV^2. In similar fashion, energies of plasmons, of spin waves, of spin-flip (Zeeman) transitions of paramagnetic ions, and of a variety of other excitations occurring in or near the barrier in various tunneling structures have been measured. A related area of particular activity in IETS is the measurement of vibrational frequencies of

Fig. 1.9. (a) Curve of d^2V/dI^2 reveals vibrational spectrum of phenolate anion at 2 K on aluminum oxide. (After McMorris, Brown, and Walmsley, 1977.) (b) Schematic diagram of circuitry used for obtaining the second derivative d^2V/dI^2.

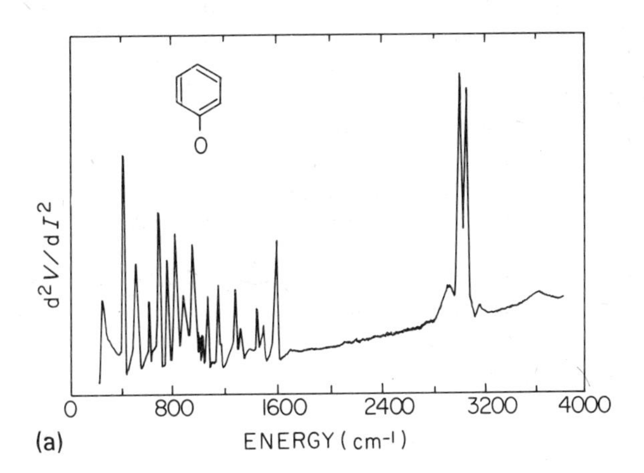

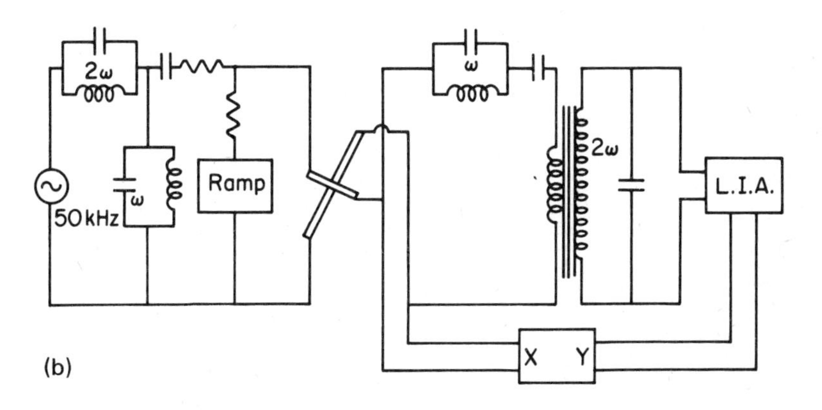

molecules (Jaklevic and Lambe, 1966), including large organic molecules, incorporated in an MIM tunnel junction by adsorption to the barrier oxide. A recent example of such a tunneling spectrum is shown in Fig. 1.9a, while a typical circuit for obtaining such spectra is shown in Fig. 1.9b. This work provides information about the bonding of molecules to solid surfaces and about the interactions or reactions which adsorbed molecules may undergo. Interest in mechanisms of catalysis is a motivating factor in much of this work.

A third class of spectroscopy in junction tunneling relates to the distribution of electron energy states either in the final electrode or, occasionally, in the barrier. Perhaps surprisingly (although theoretical and experimental reasons will be discussed), it has turned out that tunneling is usually not a particularly good diagnostic tool for measuring the normal density of states in a metallic solid. Nevertheless, spectroscopic observation of Landau levels and of size-quantized electronic states have been made, and energy band positions in semiconductors and metals have been measured. In addition, normal metal tunneling has recently provided useful spectroscopic information about fundamental electronic correlation effects in the metallic phase near the localization (Mott) transition.

1.6 Perspective, Scope, and Organization

These forms of spectroscopy and the physics that they reveal are the central themes of the present work, which is intended further to provide a reasonably comprehensive introduction to the rather wide range of topics in solid-state physics, and especially superconductivity, which have been probed by the tunneling spectroscopies. We do not take a strictly historical approach but rather emphasize developments since the most recent full survey of tunneling by Solymar in 1972. The reader is referred to Duke (1969) for a thorough history of tunneling in solids prior to 1969. Of several edited works, including the treatise "Superconductivity" (Parks, 1969) and the proceedings of the International Conferences on Low Temperature Physics and of the recent conferences on superconductivity in d- and f-band metals, the volume *Tunneling Phenomena in Solids* (Burstein and Lundqvist, 1969) and the recent volume on molecular spectroscopy (IETS) edited by P. K. Hansma (1982) are particularly useful. With regard to the Josephson effects and their consequences, which are not emphasized here, other sources include the monograph of Kulik and Yanson (1970) and the careful research-level review of Waldram (1976). Recent texts (van Duzer and Turner, 1981; Barone and Paterno, 1982), with strong emphasis on applications, are also recommended.

Some comments on the organization of the book may be useful. The treatment of tunneling in normal-state structures is divided into two parts. Chapter 2 introduces the subject in some detail from both the theortical and the experimental points of view and is intended to provide enough coverage of tunneling in real structures to prepare for Chaps. 3 through 7, which deal with tunneling applied primarily to the study of superconductivity. Chapter 2 thus stops short of treating the subject of many-body effects in normal-state tunneling, although it does include accounts of other theoretical advances. An example is the work of Feuchtwang in extending the WKB transmission calculation to the case of a barrier coupling periodic potentials (Feuchtwang, 1970) and further in treating the barrier itself as a periodic structure (Leipold and Feuchtwang, 1974). On the experimental side, Chap. 2 describes enough of the real life pathology of tunneling structures, including the effects of various types of defects in the barrier, to forewarn the reader, who may be interested primarily in superconductivity, of various extraneous effects which may occur in practice. In the same vein, some emphasis is given to techniques for checking the degree of perfection (homogeneity, sharpness of interfaces, etc.) of tunneling barriers, although the systematic discussion of experimental techniques is deferred to Appendix A. The treatment of superconducting tunneling appears in Chap. 3, which introduces the theory of superconductivity and of superconductive tunneling in such a way as to emphasize the close relationships between quasiparticle and Josephson tunneling. This chapter summarizes the tunneling study of the superconducting energy gap, including cases of strong coupling and the effects of anisotropy, multigap structure, and external perturbations. Chapter 4 describes the basic method of McMillan and Rowell for determining the phonon spectral function $\alpha^2F(\omega)$ of superconductors and illustrates the method with the classic results for lead (McMillan and Rowell, 1965, 1969). This chapter also discusses the strong-coupling theory of superconductivity, i.e., the Eliashberg equations, allowing generalization to cases of anisotropy, spin fluctuations, and nonconstant density of electronic states, and finally notes the somewhat limited applications of the conventional phonon spectroscopy of superconductors. Chapter 5 is devoted to proximity tunneling methods as a means to extend the superconducting phonon spectroscopy to those metals, notably in the transition series, which have not been accessible to the conventional methods. Chapter 5 first surveys phenomena in bilayer junctions and traces the development of a quantitative spectroscopic technique for the study of both the N and S layers of C–I–NS proximity junctions. Chapter 6 then surveys strong-coupling superconductivity results, primarily the $\alpha^2F(\omega)$ functions, obtained to date by the various methods of tunneling spectroscopy. In this chapter, superconductors are separated into classes of s–p-band elements, alloys, and amorphous alloys; transition-metal–related superconductors; and extremely weak coupling metals (studied by the proximity techniques); finally, the chapter provides a survey of what has been learned about the systematics of superconductivity from tunneling. Tunneling studies of nonequilibrium superconductors are reviewed in Chap. 7, which begins with a brief discussion of nonequilibrium phenomena in normal structures (e.g., ballistic propagation of injected quasiparticles and phonons, detected using tunnel junctions) and proceeds through measurements of characteristic lifetimes in superconductors to discussion of gap perturbations, including gap enhancement induced by irradiation or by tunneling injection in double-tunnel junctions. Chapter 8 returns to the study of normal-state structures, with emphasis on the possible spectroscopies. These can be broken down roughly into final-state effects and threshold spectroscopies, where the former subject again falls into two parts, roughly the simple one-body effects and the many-body effects. A separate survey of zero-bias anomalies is included in Chap. 8. Chapter 9 introduces recent work on artificial superlattices and nonconventional conductors, including lower-dimensional metals as well as ultrathin films and small-particle superconductors. Chapter 10 is devoted to the field of molecular spectroscopy (IETS); Chap. 11 describes some applications of tunneling phenomena. Appendix A on experimental topics emphasizes recent advances in preparation of electrodes and barriers which promise to extend the application of tunneling spectroscopy to a wider range of materials. Appendix B is devoted to discussion of the specialized numerical methods primarily as required in the treatment of superconducting tunneling data and especially of proximity tunneling data. Appendix C provides a compendium of material properties, including a good deal of information about superconductors as obtained by tunneling methods.

2

Tunneling in Normal-State Structures. I

2.1 Introduction

In this chapter we are concerned with calculating the current $I(V)$ and conductance $dI/dV = G(V)$ in structures, as shown in Fig. 2.1, consisting of two metallic electrodes separated by a narrow potential barrier $U(x)$. There are two basic approaches to such a calculation: a steady-state approach, as used originally by Gamow and by Condon and Gurney in treating the alpha decay problem mentioned in Chap. 1, and a transfer Hamiltonian approach indicated in the lower portion of Fig. 2.1, traceable to Oppenheimer's original treatment of ionization of the hydrogen atom. Making simple assumptions about metals and the barrier, we will develop expressions for the current density $J(V)$ by using the two methods and show their equivalence. The consequences of making the model more realistic in various ways will be discussed, and the theoretical expectations for basic cases will be described and compared with some experimental results, including metal-insulator-metal (MIM), metal-insulator-semimetal, Schottky barrier, and direct *pn* junction cases. Within the framework of a one-electron model, the influence of electronic band structure and density of states on the current and conductance will be discussed, and the consequences of various types of barrier defects will be considered in some detail.

2.2 Calculational Methods and Models

An extended wavefunction Ψ_k, as indicated in Fig. 2.1, can be constructed from solutions of Schrödinger's equation:

$$-\frac{\hbar^2}{2m}\frac{\partial^2\psi(x)}{\partial x^2} + U(x)\psi(x) = E_x\psi(x) \tag{2.1}$$

where

$$\Psi = \psi(x)\exp(ik_y y + ik_z z)$$

within each of the three regions by matching the wavefunctions together at the electrode-insulator boundaries $x = 0$ and t such that the wavefunction ψ and its derivative $d\psi/dx$ are continuous at these points. For simplicity, we assume free-electron metals, i.e., $U(x) = 0$ in metals 1 and 2 and $U(x) > 0$ in the barrier region. A wave incident from the left, defined as e^{ikx}, suffers partial reflection with amplitude R at $x = 0$, is

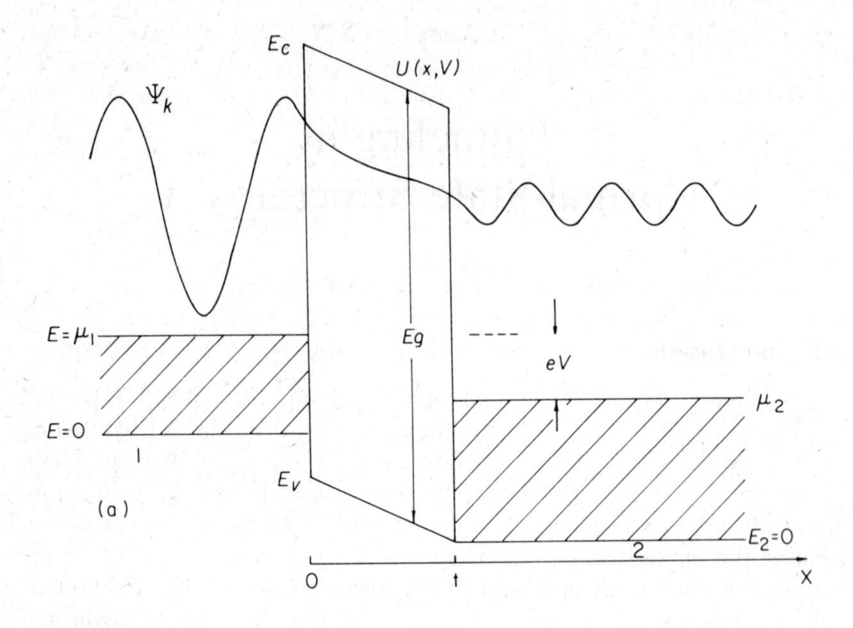

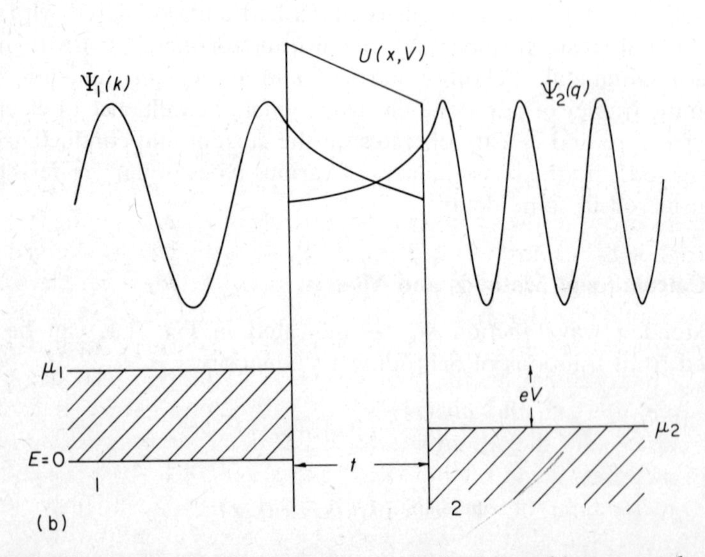

Fig. 2.1. (a) Steady-state wavefunction extending smoothly across the entire tunneling structure. (b) Separate wavefunctions ψ_1 and ψ_2 are, respectively, standing waves in the left and right regions of the structure. These approximate wavefunctions decay exponentially into the barrier and beyond into the opposite electrode. Such wavefunctions are regarded as initial and final states, respectively, in the tunneling Hamiltonian picture.

exponentially decaying, $e^{-\kappa x}$, in the interval $0<x<t$, and emerges for $x>t$ as Te^{iqx}, where T^2 measures the probability of the particle penetrating the barrier. The probability current operator, (1.1), applied to the wavefunction in metals 1 and 2, respectively, will give $\hbar k/m_1 \times (1-R^2)$ and $T^2 \hbar q/m_2$, respectively, which must be equal. The same probability current must also be obtained within the barrier region. Using (1.1), one sees that this result is possible only if the wave reflected from $x=t$ is included.

The quantity of importance in calculating the tunneling current is D, (Eq. 1.1), defined as the fraction of the incident probability current $\hbar k/m_1$ which is transmitted; this fraction is $T^2 \times m_1 q/m_2 k$. The assumption is made that the tunneling process is elastic, i.e., E, but not E_x separately, is invariant. We also assume that components of a wavevector in the transverse directions k_z, k_y, referred to simply as k_t, are conserved. i.e. $k_t = q_t$. In the important case of small transmission, the exact expression for D, the transmission factor, has been given (Harrison, 1958) by (2.2), where κ is the decay constant in the barrier region, defined in (2.3):

$$D(E_x) = g \exp(-2K) \tag{2.2a}$$

$$g = \frac{16kq\kappa^2}{(k^2+\kappa^2)(q^2+\kappa^2)} \tag{2.2b}$$

$$K = \int_0^t \kappa(x, E_x)\, dx \tag{2.2c}$$

$$\kappa(x, E_x) = \left(\frac{2m}{\hbar^2}\right)^{1/2} [U(x)-E_x]^{1/2} \tag{2.3}$$

The last expression, (2.3), for a particle of fixed mass m in a potential $U(x)$ may not be a good approximation to κ if the barrier, as in Fig. 2.1a, is regarded as the forbidden gap of an insulator, unless the energy of the particle lies close to the conduction or valence band edge of the insulator so that the usual electron or hole mass will be appropriate. The generalization of this approach to a more realistic three dimensional picture has been discussed by Harrison (1961), who finds that Eqs. (2.2) are still valid. The prefactor g (2.2b), exact for a square barrier, results from the matching requirement on the wavefunctions at the two interfaces. This prefactor is unity in the usual WKB approximation, which is formally valid if variation of the potential $U(x)$ is sufficiently slow in the vicinity of the turning points 0 and t. However, the exponent K is correctly given by (2.2c), independent of the steepness of the potential variation at the turning point. Note that the expression for g varies linearly with k and q or, equivalently, with $(E_x)^{1/2}$, predicting $D(E_x)$ to vanish linearly in $(E_x)^{1/2}$ at the bottom of a conduction band. This behavior will obviously be noticeably altered depending upon whether g is taken as expression (2.2b) or as unity. However, if the electron energies in question are far removed from band edges, then the factor g has a slow variation with E, and its presence may be scarcely detectable in comparison with the typically exponential variation of D.

2.2.1 Stationary-State Calculation

To obtain the current density $J(V)$ in a junction between metals 1 and 2, we assume each is described by an equilibrium Fermi function. In notation suited to the case $\mu_1 - \mu_2 = eV$, we define

$$
\begin{aligned}
f_1 &= \left[\exp\left(\frac{E_1 - \mu_1}{kT}\right) + 1\right]^{-1} \equiv f(E) \\
f_2 &= \left[\exp\left(\frac{E_2 - \mu_2}{kT}\right) + 1\right]^{-1} \equiv f(E + eV)
\end{aligned}
\tag{2.4}
$$

We conventionally measure energy $E_1 = E$ from the bottom of the left-hand conduction band. Here, μ_1 and μ_2 are the Fermi energies measured from the bottom of the respective bands, and E_1 and E_2 are the total energies on the left and right, similarly measured. We take the convention that positive bias lowers by eV the Fermi level of the right-hand electrode, as shown in Fig. 2.1, and calculate the net current density $J = J_{12} - J_{21}$:

$$
\begin{aligned}
J_{12} &= \frac{2e}{(2\pi)^3} \iiint d^2k_t \, dk_x \left(\frac{1}{\hbar}\frac{\partial E}{\partial k_x}\right) Df(E)[1 - f(E + eV)] \\
J_{21} &= \frac{2e}{(2\pi)^3} \iiint d^2k_t \, dk_x \left(\frac{1}{\hbar}\frac{\partial E}{\partial k_x}\right) Df(E + eV)[1 - f(E)]
\end{aligned}
\tag{2.5}
$$

$$
J = \frac{2e}{(2\pi)^2 h} \int_0^\infty dE_x [f(E) - f(E + eV)] \iint d^2k_t \, D(E_x, V) \tag{2.6}
$$

To obtain J_{12}, one must integrate over all available k states in metal 1, weighting each k by the corresponding group velocity,

$$
v_x = \hbar^{-1} \frac{\partial E}{\partial k_x} \tag{2.7}
$$

multiplying by the transmission factor $D(E_x)$ and by the Fermi functions, guaranteeing that the initial state 1 is occupied and that the final state 2 is empty. In (2.5), the factor 2 represents the spin degeneracy, and the factor $(2\pi)^3$ in the denominator gives the number of states per unit volume in k space. In arriving at (2.6) for the current density J under influence of applied bias voltage V, we employed (2.7) to obtain the integration variable E_x. Other forms of the k-space integral (2.6) for J are possible, and will be discussed in Sec. 2.3.

In connection with (2.5), it is appropriate to discuss some of the properties of $D(E_x)$, which is explicitly a function of $E_x = E - E_t = E - \hbar^2 k_t^2 / 2m$ and implicitly a function of bias V, which distorts the barrier and hence influences the potential function $U(x) = U(x, V)$. Regarding D

as calculated via (2.3) reveals that a maximum D, and strong preference in transmission, is given for electrons of a given total E which have a *minimum* value of k_t, i.e., for $k_t = 0$, $E_t = 0$, and hence $E_x = E$. This result can be given in terms of an angular dependence.

Consider a Fermi surface electron, $E = \mu_1$, and let ϕ be the angle between its wavevector of magnitude k_F and the barrier normal, so that $k_x = k_F \cos\phi$ and $E_x \simeq \mu_1(1-\sin^2\phi)$, and assume g = 1, which is strictly appropriate only to a slowly varying $U(x)$. Consider the simple case of a square barrier $U(x) = U_0$, whose height from μ_1 is defined as $U_B = U_0 - \mu_1$. The WKB result is, approximately, for fixed total energy $E = \mu_1$ and varying tunneling angle ϕ,

$$D \simeq \exp(-2\kappa t)\exp\left(-\kappa t \frac{\mu_1 \sin^2\phi}{U_B}\right) \tag{2.8}$$

Taking values $U_B = 1$ eV, $t = 20$ Å, $\mu_1 = 5$ eV, and $m = m_e$, for example, one finds

$$D \simeq D_0 \exp(-\beta\phi^2)$$

with $\beta = 51$ corresponding to transmission reduced to e^{-1} at an angle $\phi = 8°$.

The effect of bias potential V on an ideal dielectric barrier of thickness t can be written as

$$U(x, V) = U(x, 0) - \frac{eVx}{t} \tag{2.9}$$

which produces a trapezoidally shaped tunneling barrier.

The dependence of D on bias voltage V can be treated by (2.9) in cases in which (2.3) is employed. As may be seen by inspecting Fig. 2.1, the effect of bias upon the barrier is such as to reduce the average barrier height for electrons tunneling in the forward direction (i.e., from the higher Fermi energy level μ_1), and to reduce the probability for reverse flow from μ_2. The net effect is to increase the flow.

2.2.2 *Transfer Hamiltonian Calculations*

The transfer Hamiltonian approach, originating in the work of Oppenheimer in 1928 and extended to solid-state structures by Bardeen (1961), is motivated by the fact that a near-unity probability of reflection by the barrier occurs in a typical tunneling structure such as shown in Fig. 2.1a. In a real sense, then, the waves on the left side of the structure, rather than being traveling waves e^{ikx} are standing waves $\cos(kx)$, and the same is true on the right-hand side. The barrier can be regarded as separating the system into two nearly independent portions, and the weak residual coupling can be treated by a perturbing Hamiltonian H^T,

$$H = H_1 + H_2 + H^T \tag{2.10}$$

which is then regarded as driving electron transitions from one side to the other in the usual Golden Rule calculation given by (2.11),

$$w_{12} = \left(\frac{2\pi}{\hbar}\right) |\langle\psi_2| H^T |\psi_1\rangle|^2 \rho(E_2)\delta(E_2 - E_1) \tag{2.11}$$

Equation (2.11) describes the transition rate w_{12} from a given state 1 (as on the left side in Fig. 2.1b) to a set of states of equal energy and density $\rho(E_2)$ on the right. The explicit appearance in this formula of the density of states $\rho(E)$ was useful to Bardeen (1961) in providing a mathematical explanation for the discovery of Giaever (1960*a,b*) that tunneling measured the density of excitations in the superconducting case. We first show that this formulation leads to the same results as the steady-state calculation. We will later see that the transfer Hamiltonian calculation can be generalized to cases where the tunneling transition is assisted, or proceeds by interaction of the tunneling electron with the barrier and also another mode of the system. From the treatment of Bardeen (1961) and Harrison (1961) in a WKB approximation, the wavefunction

$$\psi_1 = \begin{cases} w^{-1/2} \exp[i(k_y y + k_z z)]\cos(k_x x + \delta), & x < 0 \\ \frac{1}{2} w^{-1/2} \exp[i(k_y y + k_z z)]g \exp(-\kappa x), & 0 < x < t \\ 0, & x > t \end{cases} \tag{2.12}$$

is taken as a standing wave normalized in the left electrode of width w. This is matched to an exponentially decaying wave in the barrier, which is assumed then to continue smoothly to zero in the right-hand region rather than oscillating. A similar expression applies for side 2, which again is a standing wave on the right and assumed to be zero on the left side of the structure. In order to identify the matrix element (T) for the operator H^T, one substitutes an initial time-dependent wavefunction

$$\psi(t) = a(t)\psi_{10} e^{-iE_{10}t/\hbar} + \sum_n b_n(t)\psi_{2n} e^{-iE_{2n}t/\hbar} \tag{2.13}$$

into Schrödinger's time-dependent equation

$$H\psi = i\hbar \frac{\partial \psi}{\partial t} \tag{2.14}$$

This gives

$$T = \int \psi_1^*(H - E_{2n})\psi_{2n} \, d\tau, \tag{2.15}$$

after manipulations making use of the fact that the wavefunctions ψ_1 and ψ_2 overlap only in the barrier region. It is established that the appropriate matrix element is essentially that of the quantum mechanical current density operator j,

$$T_{q,k} = -i\hbar\langle\psi_q| j |\psi_k\rangle \tag{2.16}$$

which must be evaluated within the barrier region, and in which we maintain our convention of q, k for wavevectors on the right and left. This expression was evaluated by Harrison (1961) to give for $|T_{q,k}|^2$

$$|T_{q,k}|^2 = \delta(k_t, q_t) g \exp(-2\kappa)(4\pi\rho_{1x}\rho_{2x})^{-1} \tag{2.17}$$

within the WKB approximation, assuming conservation of the transverse wavevector. Here, ρ_{1x} is a one-dimensional density of states on the left, assuming box normalization within the width w of the electrode:

$$\rho_{1x} = \frac{w}{\pi}\left(\frac{\partial E}{\partial k_x}\right)^{-1} \tag{2.18}$$

In (2.17), recall that g is the prefactor required in the expression for D if wavefunction matching is employed as, for example, given by (2.2b); if $g = 1$, then the WKB approximation appropriate to a smoothly varying potential $U(x)$ is assumed. Now that we have obtained the appropriate matrix element, it is straightforward to calculate the current J by summing transition rates from all filled states to all empty states and multiplying by $2e$:

$$J = 2e\sum_{k,q}\frac{2\pi}{\hbar}|T_{q,k}|^2\,[f_k(1-f_q) - f_q(1-f_k)] \tag{2.19}$$

This expression is equivalent to (2.6) as obtained in the steady-state method, with all of the density-of-states factors, excepting those contained in the prefactor g, having disappeared in the transformation from the sum to the integral over k space. Thus, apart from k dependence in the prefactor g, there is no explicit dependence of the tunnel current on the one-dimensional density of states. The generalization of this method to accommodate assisted tunneling processes is given in Sec. 2.6.

2.2.3 *Ideal Barrier Transmission*

Clearly, the barrier is the crucial element in the tunneling experiment, providing the required decoupling between the electrodes so that carriers from one can be injected at a precise energy in the other. The central feature of the barrier tunneling process is the decaying wavefunction $e^{-\kappa x}$ in the classically forbidden region, which experimentally implies an extremely rapid variation of the resistance of a tunnel junction with thickness of the barrier. This feature is illustrated in Fig. 2.2, in which the characteristic exponential dependence of the resistance on the barrier thickness is obtained. Indeed, this dependence is one means of testing for the presence of a tunneling process. The purpose of the present section is to gain more detailed insight into transmission of particles through the most nearly ideal of the experimentally available barriers, which are typically thermally grown oxides and less frequently single crystals, as in Schottky barrier tunneling or in special cases involving layer compounds such as GaSe. The questions involved are of two types. The first is a quantum mechanical problem: given a barrier potential $U(x)$, what

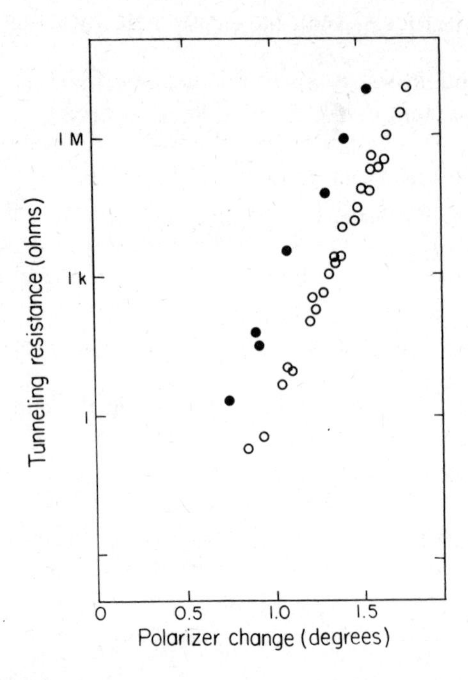

Fig. 2.2. The dominant role of the exponential decay of the wavefunction with increasing barrier thickness in setting the resistance of the junction is demonstrated in a logarithmic plot of the junction resistance, in ohms, at $V=0$ vs. insulator thickness, measured ellipsometrically, with conversion $0.53° = 10$ Å. In both cases, the first metal film is aluminum and the insulator is aluminum oxide grown in a plasma discharge. The thickness range lies between 15 and 30 Å. The solid circles represent measurements made with a lead second electrode; the open circles represent measurements made with aluminum as the second electrode. (After Knorr and Leslie, 1973.)

practical means are there for calculating the transmission of an electron at a given energy E? The second questions lie behind the first and involve more realistic methods to approximately describe the interaction and transmission of a tunneling electron through the barrier regarded as a periodic insulating solid. Finally, we investigate some of the practical dependences of transmission upon variables such as the energy of the tunneling particle, its tunneling angle ϕ with respect to the normal to the barrier, and the bias voltage V.

We begin with a convenient method (Kane, 1969) for solving tunneling problems involving stepwise constant potentials $V(x)$. Solutions of the one-dimensional, time-independent Schrödinger equation, Eq. (2.1), for constant potential V are the form

$$\psi = \alpha e^{ikx} + \beta e^{-ikx} \tag{2.20}$$

where, if $E > V$, k is real, while for $E < V$, $k = i\kappa$, with real $\kappa = [(2m/\hbar^2)(V-E)]^{1/2}$.

In the frequently encountered case of a constant potential with steps at positions x_p, taking barrier potential values V_p for $x_{p-1} \leq x \leq x_p$, specification of α_p, β_p, and k_p is obtained by requiring that ψ and $d\psi/dx$ be continuous at each x_p.

Setting $2m/\hbar^2=1$, one finds that the two equations resulting from continuity of ψ and $d\psi/dx$ at x_p are

$$\alpha_p e^{ik_p x_p}+\beta_p e^{-ik_p x_p}=\alpha_{p+1}e^{ik_{p+1}x_p}+\beta_{p+1}e^{-ik_{p+1}x_p} \quad (2.21)$$

$$\alpha_p k_p e^{ik_p x_p}-\beta_p k_p e^{-ik_p x_p} = \alpha_{p+1}k_{p+1}e^{ik_{p+1}x_p}-\beta_{p+1}k_{p+1}e^{-ik_{p+1}x_p} \quad (2.22)$$

Multiplying the first equation by k_p and adding and subtracting the two, one identifies the matrix equation

$$\begin{pmatrix}\alpha_p\\ \beta_p\end{pmatrix}=\mathbf{R}^p\begin{pmatrix}\alpha_{p+1}\\ \beta_{p+1}\end{pmatrix} \quad (2.23)$$

with $\mathbf{R}^p$ given as the 2×2 matrix:

$$\mathbf{R}^p=\begin{pmatrix}\dfrac{k_p+k_{p+1}}{2k_p}e^{i(k_{p+1}-k_p)x_p} & \dfrac{k_p-k_{p+1}}{2k_p}e^{-i(k_{p+1}+k_p)x_p}\\ \dfrac{k_p-k_{p+1}}{2k_p}e^{i(k_{p+1}+k_p)x_p} & \dfrac{k_p+k_{p+1}}{2k_p}e^{-i(k_{p+1}-k_p)x_p}\end{pmatrix} \quad (2.24)$$

(Kane, 1969).

The problem of a plane wave $k>0$ incident upon a single-step potential $V_2>E_1$ at $x_1=0$ such that $k_2=i\kappa_2=i(V_2-E)^{1/2}$ is stated as

$$\begin{pmatrix}1\\ \beta_1\end{pmatrix}=\mathbf{R}^1\begin{pmatrix}\alpha_2\\ 0\end{pmatrix}$$

where β_1 and α_2, respectively, are the amplitudes of reflected and transmitted (decaying) waves, as indicated by (2.20). Hence, one has

$$1=\mathbf{R}^1_{11}\alpha_2=\frac{k_1+i\kappa_2}{2k_1\alpha_2} \qquad \beta_1=\mathbf{R}^1_{21}\alpha_2=\frac{k_1-i\kappa_2}{2k_1\alpha_2}$$

Thus, in the forbidden region $x>0$, one recovers

$$\psi(x)=2k_1(k_1+i\kappa_2)^{-1}e^{-\kappa_2 x} \quad (2.25)$$

The same problem with a square barrier of width t, as in Fig. 2.1, assuming $0<k_3<k_1$, as if the Fermi energy on the right were less than that on the left, is stated as

$$\begin{pmatrix}1\\ \beta_1\end{pmatrix}=\mathbf{R}^1\mathbf{R}^2\begin{pmatrix}\alpha_3\\ 0\end{pmatrix} \quad (2.26)$$

To obtain the transmitted wave $\psi_3(x)$, one evaluates

$$\alpha_3=[(\mathbf{R}^1\mathbf{R}^2)_{11}]^{-1} \qquad (\mathbf{R}^1\mathbf{R}^2)_{11}=\mathbf{R}^1_{11}\mathbf{R}^2_{11}+\mathbf{R}^1_{12}\mathbf{R}^2_{21} \quad (2.27)$$

In general, then, there are two terms, so that a single exponential decay is not obtained. Supposing $\kappa t\gg 1$, however; the contribution $\mathbf{R}^1_{12}\mathbf{R}^2_{21}$ is exponentially small compared with $\mathbf{R}^1_{11}\mathbf{R}^2_{11}$, and

$$\psi_3(x)=\frac{4ik_1\kappa_2 e^{-\kappa_2 t}e^{ik_3(x-t)}}{(k_1+i\kappa_2)(i\kappa_2+k_3)} \quad (2.28)$$

In the symmetrical case $k_1 = k_3 = k$, the exact result (2.27) for α_3, retaining both terms of $(\mathbf{R}^1\mathbf{R}^2)_{11}$, is

$$\alpha_3 = \frac{2ik\kappa e^{ikt}}{(k^2 - \kappa^2)\sinh \kappa t + 2ik\kappa \cosh \kappa t} \tag{2.29}$$

(Baym, 1969). The fraction of the incident probability current $\hbar k_1/m$ transmitted is

$$\frac{\alpha_3^2 k_3}{k_1} = T^2 = \frac{16 k_1 k_3 \kappa_2^2 e^{-2\kappa_2 t}}{(k_1^2 + \kappa_2^2)(k_3^2 + \kappa_2^2)} \tag{2.30}$$

which provides a derivation of (2.2b). The prefactor of the exponential, designated as g, is seen to go to zero linearly in k_1 and k_3. As has been noted before, this feature is absent in the WKB treatment, where $g = 1$, and suggests a sensitivity of g to details of the assumed interface potential. A generalization of this expression for the case of a trapezoidal barrier is given by Brinkman, Dynes, and Rowell (1970). Note that the incident wave e^{ik_1x} is transmitted as $e^{ik_3(x-t)}$. The loss in phase, k_3t, with respect to the wave that would occur if the barrier were absent, when the problem is cast in a wave packet form, can make it appear that the particle spends no time in the barrier.

A variety of one-dimensional well problems can be handled with Kane's method. For example, the finite square well is described by $k_1 = k_3 = i\kappa$, $k_2 > 0$, and the condition for a bound state is $\alpha_1 = 0$, $\beta_3 = 0$, corresponding to only $e^{+\kappa x}$ for $x < 0$ and only $e^{-\kappa x}$ for $x > t$. The condition is then $(R^1R^2)_{11} = 0$, which reduces to

$$\frac{k + i\kappa}{k - i\kappa} = e^{ikt} = e^{2i\phi} \tag{2.31}$$

where $\phi = \tan^{-1}(\kappa/k)$. Thus, $\tan(kt/2) = \kappa/k$, which is the condition for the first bound state. Extensions of matching calculations to more realistic three-dimensional cases are discussed by Harrison (1961), Conley, Duke, Mahan, and Tiemann (1966), Dowman, MacVicar, and Waldram (1969), Feuchtwang (1970), and Leipold and Feuchtwang (1974).

Unfortunately, such exact matching calculations can rarely be accomplished in realistic cases, except by numerical means, which is the reason for the great practical value of the WKB method. There is a considerable literature devoted to methods of approximating solutions of the one-dimensional Schrödinger equation following the original work by Wentzel, Kramers, and Brillouin (for a review of the WKB approximation, see Schiff, 1955). Notable contributions to this literature have been made by Langer (1937) and by Miller and Good (1953), which have been commented on by Franz (1969). A central feature of WKB methods is evident in the expression for the approximate wavefunction

$$\psi(x) = \exp[i\hbar^{-1}S(x)] \tag{2.32a}$$

$$\simeq p(x)^{-1/2} \exp\left[\pm i\hbar^{-1} \int^x p(\xi)\, d\xi\right] \tag{2.32b}$$

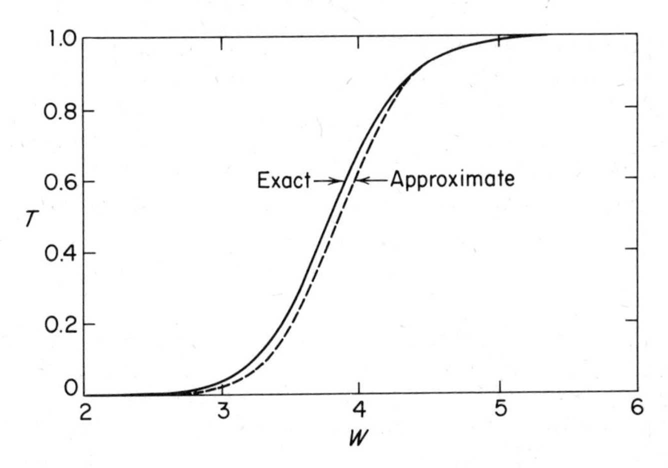

Fig. 2.3. Tests of the accuracy of the WKB approximation for the transmission coefficient for a particular potential $V(x)$. The abscissa in this figure is a measure of the energy of the particle. (After Miller and Good, 1953.)

as the appearance of the momentum to the inverse $\frac{1}{2}$ power. This has the awkward consequence that the wavefunction becomes infinite at the edge of the barrier, i.e., at the classical turning point. The usual WKB wavefunction, Eq. (2.32b), is obtained by substituting the initial wavefunction (2.32a) into the one-dimensional Schrödinger equation

$$-S'^2 + i\hbar S'' + p^2 = 0 \tag{2.33}$$

(in units such that $\hbar^2/2m = 1$), and then expanding the unknown function $S(x)$ as powers of $\hbar$, which is assumed to be a small quantity:

$$S(x) = S_0(x) + S(x)\hbar + \cdots \tag{2.34}$$

The method of Miller and Good can be paraphrased in the following way. Consider a soluble problem such as the harmonic oscillator for which the exact eigenfunctions corresponding to (2.32) are known. Of course, these functions also become large at the classical turning points for large quantum number n, as is familiar from elementary quantum mechanics, but the functions are finite at the classical turning points. Miller and Good (1953) established a mathematical transformation between the eigenfunctions of a soluble model system, such as the harmonic oscillator problem, and the problem at hand, and in this fashion obtained approximate wavefunctions well behaved at the two separate turning points in the typical well or barrier problem. The accuracy of their method for the calculation of the transmission coefficient

$$|T|^2_{\mathrm{MG}} = \left\{1 + \exp\left[+2\int_{x_1}^{x_2} \kappa(x)\,\mathrm{d}x\right]\right\}^{-1} \tag{2.35}$$

for an exactly soluble model barrier problem is illustrated in Fig. 2.3. Note that (2.35) agrees with (2.2) in the limit of an opaque barrier. The paper of Miller and Good provides expressions from which the error in applying the approximate solutions to a given situation can be estimated: as with the usual WKB method, the approximations are best for a slowly varying potential. If, however, one blindly applies (2.35) to the case of the square barrier, which was treated exactly (thick-barrier case) in (2.29), then the exponential factor is reproduced correctly but the prefactor g is missing. This result is believed to be the general result in applying the WKB method beyond its range of applicability (Landau and Lifshitz, 1958).

It has been stated (Harrison, 1961) that application of the WKB method to determination of T^2 in solid-state structures generally gives such a result with $g(E)=1$, i.e., devoid of any prefactor. However, an interesting extension of the WKB method to a barrier between regions in which the potential is specifically periodic, as in a true solid, leads to the reappearance of the prefactor (Feuchtwang, 1970), as indicated in

$$|T|_F^2=\frac{v_r}{a_r q_r}\frac{v_\ell}{a_\ell q_\ell}|T_B|^2 \tag{2.36}$$

Here, v_r and v_ℓ, respectively, are the group velocities $\partial E/\partial q$ on the right and left, a_r and a_ℓ are the periods of the potential on the right and left, and the q's are wavevectors. The remaining barrier factor T_B is determined by the properties of the insulator and reduces to a single exponential dependence only if the barrier is thick.

A different but related extension of the treatment of barrier transmission betwen two free-electron metals was given by Schnupp (1967a,b), who treated the periodicity in a real crystalline barrier by an array of delta functions. This approach has been extended by Leipold and Feuchtwang (1974), who treat the barrier as a repetitive array of one-dimensional square well potentials. The motivation for this work is indicated in Fig. 2.4, which sketches the $E(k)$ one would expect for a tunneling barrier assumed to be a crystalline solid. The straight-line portions of E vs. k^2 correspond to effective masses in the conduction and valence bands. The smooth connection of these sections through the middle of the energy gap suggests that the decay constant κ in the center of the gap will not be given reasonably by any formula such as (2.3) involving a single mass. In the work of Leipold and Feuchtwang, summarized in Fig. 2.5, the transmission factor as a function of energy for a finite array of square wells has been calculated and compared with that obtained on WKB-type approximations in which the function $\kappa(E)$ was taken as that characteristic of the infinitely extended periodic array of square wells. An advance in the work of Leipold and Feuchtwang was in an application of an electric field, simulating a bias, to their periodic structure. This application makes possible a connection to the problem of interband tunneling first treated by Zener (1934) and later by Kane (1959), although the biases used by Leipold and Feuchtwang are never so great as to permit tunneling through the upper (conduction) band. The

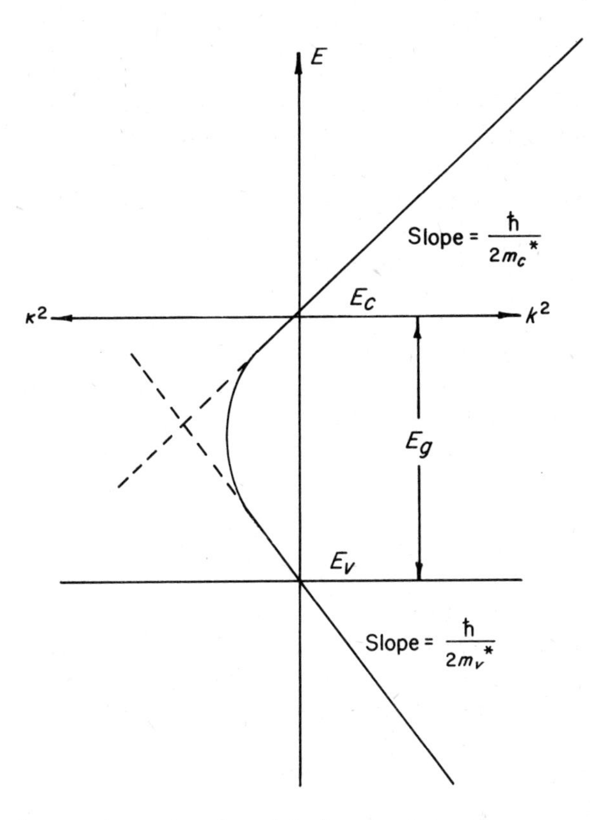

Fig. 2.4. In the simple two-band model of an insulator, the wavevector squared k^2 becomes negative in the energy range between E_c and E_v, i.e., in the forbidden band gap, corresponding to decaying exponential waves. One can see from this plot that estimates of the decay constant κ can be in error in the middle of the gap where the solid line departs noticeably from the dashed portions, obtained by using the effective masses m_c and m_v for the conduction and valence bands, respectively.

formula of Zener which is tested is

$$|T|_z^2 = \exp\left\{-2\int_{x_1}^{x_2} \mathrm{Im}[k(E, x)]\,\mathrm{d}x\right\} \tag{2.37}$$

and k is determined from an infinite repeating array of square wells. The dimensions of the square well potential, shown as it would appear under a bias eV in Fig. 2.5a, are well widths of 4 Å, a barrier width of 2 Å, and a well depth of 5.5 eV. The individual isolated wells would support bound states at 1.15 eV and 4.16 eV above the well bottom. In a repetitive array, the ground-state levels, as one would expect, broaden into bands which are identified as the valence and conduction bands of the solid

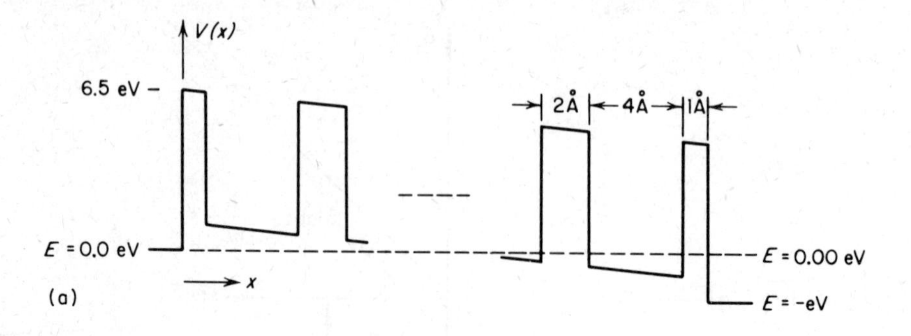

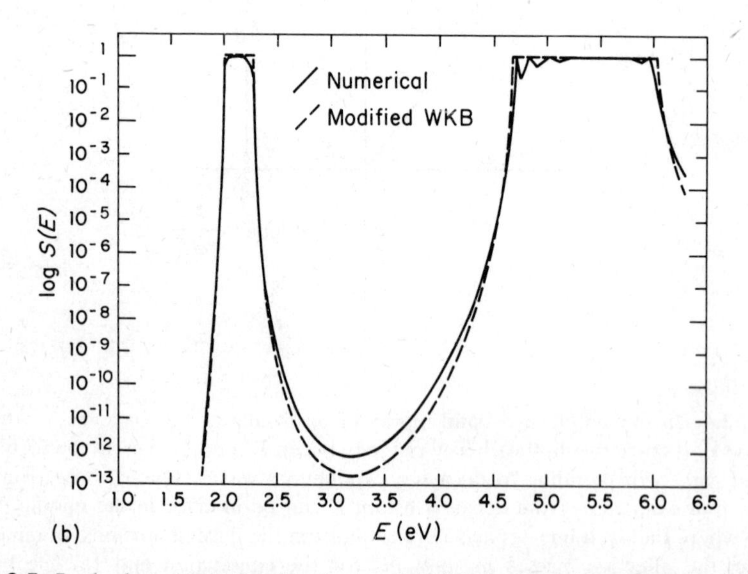

Fig. 2.5. In the heterojunction barrier model of Leipold and Feuchtwang (1974), the tunneling barrier is treated as a finite, repetitive array of cells. Each cell is a square well potential supporting two bound states (presumed empty) and lying, respectively, 2 and 5 eV above the zero of energy. (a) The potential $V(x)$ resulting from application of bias is shown. (b) The transmission coefficient of the structure at zero bias is displayed as a function of incident electron energy. The two regions of large transmission correspond approximately to coincidence with the bound-state levels within the individual wells. The transmission is a slowly varying function of energy only for energies well away from the bound states. (c) The transmission as a function of energy for the structure under an applied bias. As before, the structure has 8 unit cells of width 6 Å. The bias assumed here is about 2.5 V. Here, as in (b), the smooth curve is obtained by exact calculation, while the dotted curve is obtained by Zener's modified WKB approximation, Eq. (2.37). The sharp structures in the solid curve arise from the resonant transmission effects. (After Leipold and Feuchtwang, 1974.)

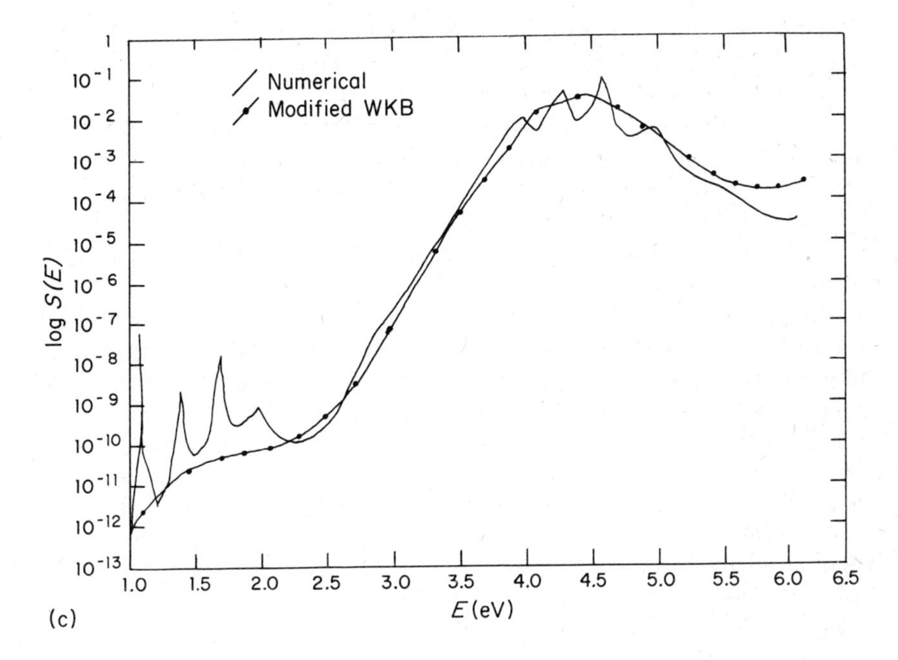

represented by the array. The corresponding widths of the valence and conduction bands as determined by Leipold and Feuchtwang are 0.26 and 1.36 eV, respectively. Referring to Fig. 2.5b, these bands lead to regions of unity transmission coefficient separated by a region of low transmission coefficient which is interpreted as the gap of the corresponding repetitive array. In Fig. 2.5c, corresponding to an applied bias of about 2.5 V, great changes occur in the transmission coefficient as a function of energy. The increase by orders of magnitude of the transmission toward the center of the gap is interpreted as the lowering of the average barrier by the applied field. The sharp features in the low- and high-energy regions (corresponding to the valence and conduction bands, determined by the numerical exact calculation shown by the smooth curve) are interpreted as the splitting by the electric field of the individual resonance levels of the cells in the barrier. In comparison of Figs. 2.5b and 2.5c, it is clear that the Zener transmission factor (2.37), follows the trend of the exact calculation in the band gap reasonably well. This observation suggests that working backwards from an experimentally determined transmission factor to infer the decay constant, making use of (2.37), might not lead to an unreasonably large error. The results of Fig. 2.5 have been obtained from a barrier which is only eight cells thick, corresponding to an overall barrier width of about 50 Å. A close approximation to the E vs. k relationship of the infinite solid is obtained: for this purpose, a very thin sample on the order of 10 to 15 wells is entirely adequate. As Leipold and

Feuchtwang (1974) pointed out, while significant errors occur, even in the center of the gap, between the exact and the WKB results based upon the infinite-sample $k(E)$ relationship, the trend of the data is predicted remarkably well. Further, the major change in the transmission evaluated at 3.2 V in Figs. 2.5b and 2.5c upon application of 2.5 V bias is consistent with a reduction of the average barrier height. Finally, however, the authors point out that the presence of a disordered, rather than crystalline, barrier drastically alters the transmission coefficient as a function of energy.

Armed with this information, we will adopt the point of view that the transmission in practical cases will be well approximated by a WKB factor in which the decay constant is that of the solid forming the insulator. We remain skeptical of the WKB result that there are no prefactors but repeat that prefactors, which are generally slowly varying functions of wavevector, or $E^{1/2}$, are likely to play an appreciable role only if k or $E^{1/2}$ goes to zero, i.e., if one considers tunneling from the bottom of a band in one of the materials. In adopting this practical point of view, we shall go further and point out in a very rough way the dependences of D upon E_x, the kinetic energy normal to the barrier, and the bias voltage V. In doing so, we find it convenient to introduce $\phi(x)$ as the potential barrier measured from the Fermi energy,

$$\begin{aligned} \phi(x) &= U(x) - \mu \\ W &= E_x - \mu \end{aligned} \tag{2.38}$$

and similarly to measure the x component of kinetic energy in excess of μ by the symbol W in (2.38). For rough estimates, the average value $\bar{\phi}$ of the barrier

$$\bar{\phi} = \frac{1}{t}\int_{x_1}^{x_2} \phi(x)\,dx \tag{2.39}$$

can be used to observe that the application of positive bias V, which lowers the right-hand Fermi energy, will decrease the average value of the potential barrier by $eV/2$,

$$\phi(V) = \bar{\phi} - \frac{eV}{2} \tag{2.40}$$

a result due to Sommerfeld and Bethe (1933). When this result is inserted into (2.2) and (2.3), the transmission factor for $g = 1$ becomes

$$D = \exp\left[-2\sqrt{\frac{2m}{\hbar^2}}\,t\left(\bar{\phi} - \frac{eV}{2} - W\right)^{1/2}\right] \tag{2.41}$$

The lowest-order expansion of the square root term involving the energies gives

$$D \simeq \exp(-2\kappa_0 t)\exp\left(\frac{eV}{2E_0}\right)\exp\left(\frac{W}{E_0}\right), \qquad eV, W \ll \bar{\phi} \tag{2.42}$$

$$\kappa_0 = \sqrt{\frac{2m\bar{\phi}}{\hbar^2}} \qquad E_0 = \frac{\bar{\phi}}{\kappa_0 t} = \frac{(\hbar^2\bar{\phi}/2m)^{1/2}}{t} \tag{2.43}$$

which we emphasize is valid only for $eV \ll \bar{\phi}$, $W \ll \bar{\phi}$. For small variations in V and W, D then is seen to vary exponentially on a scale set by the parameter E_0, which is typically a tenth of a volt if $\bar{\phi}$ is 1 or 2 eV.

2.3 Basic Junction Types

We will typically consider junctions in which at least one electrode consists of a free-electron metal with a well-developed Fermi surface corresponding to Fermi energy μ. Supposing, in Fig. 2.1a, the left member of the junction to be such a free-electron metal, biased negatively by voltage V with respect to the electrode on the right, we may take the current density J as a function of V to be given by (2.6) and (2.7). Referring to Fig. 2.1a: at $T = 0$, from simply an energy point of view, electrons lying between μ and $\mu - eV$ are eligible to tunnel in that only these face empty states in the right-hand electrode. Referring now to a k-space picture of the left electrode in Fig. 2.6, the k-vectors for the electrons eligible to tunnel to the right lie between spherical shells characterized by energies μ and $\mu - eV$. Since we consider only electrons moving to the positive x direction, only half of the volume included between the spherical shells need be integrated over in computing the current density. Note that the transmission factor D is constant for constant values of E_x and is maximum for $E_x = \mu$ (or $W = 0$). It is thus convenient to integrate over disc-shaped regions in performing the k-space integration, as indicated by shading in Fig. 2.6. For $E_x < \mu - eV$, however, the surface of constant E_x becomes an annulus, Fig. 2.6a,

Fig. 2.6. Schematic shows k-space location of electrons eligible to tunnel from left to right in a junction under bias such as shown in Fig. 2.1. (a) States of constant tunneling probability for $k_x < k_{min} = [(2m/\hbar^2)(\mu_1 - eV)]^{1/2}$. (b) States of constant tunneling probability for $k_{min} < k_x < k_{max}$, where $k_{max} = [(2m/\hbar^2)\mu_1]^{1/2}$. (After Floyd and Walmsley, 1978.)

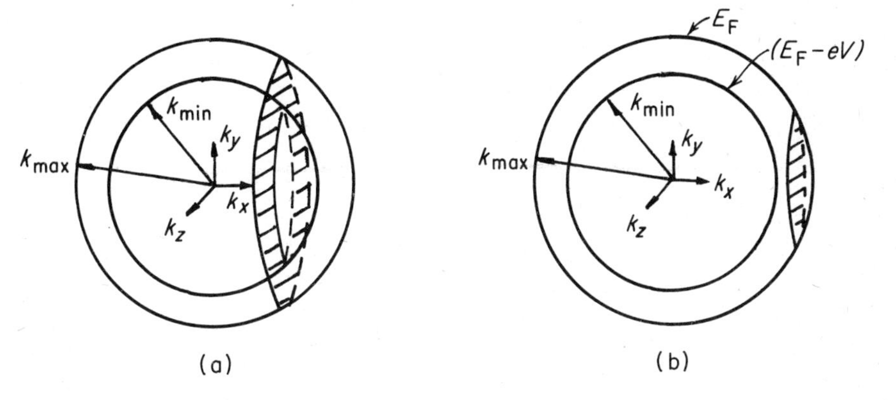

corresponding to $k_x < k_{\min} = [(2m/\hbar^2)(\mu_1 - eV)]^{1/2}$. This breakup of the region of integration into two parts is reflected in (2.44) for the current density as a function of V:

$$J(V) = \frac{2e\rho_t}{h}\left[\int_{\mu-eV}^{\mu} D(E_x, V)\,dE_x \int_0^{\mu-E_x} dE_t + \int_0^{\mu-eV} D(E_x, V)\,dE_x \int_{\mu-eV-E_x}^{\mu-E_x} dE_t\right] \quad (2.44)$$

where one assumes $T = 0$ and has defined a two-dimensional density of states ρ_t by

$$d^2k_t = 2\pi k_t\,dk_t = \frac{2\pi m_t}{\hbar^2}\,dE_t = (2\pi)^2\rho_t\,dE_t \quad (2.45)$$

In obtaining this expression from (2.6) and (2.7), we have used the group velocity factor to express k_x in terms of E_x, and the integral over transverse wavevector k_t has been transformed into an integration over the corresponding kinetic energy E_t in (2.45) by adoption of a density of states ρ_t for transverse motion. The transverse integrals are performed, leading to

$$J(V) = \frac{2e\rho_t}{h}\left[eV\int_0^{\mu-eV} D(E_x, V)\,dE_x + \int_{\mu-eV}^{\mu} D(E_x, V)(\mu - E_x)\,dE_x\right] \quad (2.46)$$

valid for $T = 0$. Inspection of this equation, in which the first bracketed term is proportional to the voltage V, indicates that the current-voltage relationship will be linear for small V, a result first obtained by Sommerfeld and Bethe (1933).

2.3.1 *Metal-Insulator-Metal Junctions*

Evolution from the very early work of Frenkel (1930), Sommerfeld and Bethe (1933), and Holm and Kirschstein (1935, 1939) from the free-electron model represented by (2.46) to the quantitative treatment of MIM tunneling has required careful consideration of the transmission factors D in (2.46), even within the WKB approximation, neglecting the image force, and consideration of the effects of temperature, which have been neglected in (2.46). Quantitative developments which we will mention are due to Stratton (1962), drawing upon earlier work of Murphy and Good (1956), and to Simmons (1963*a*, 1963*b*, 1963*c*, 1964), making use of earlier work by Holm (1951). The developments of Simmons and Stratton applied to the case of a symmetrical barrier, in extension of (2.46), both arrive at a cubic V^3 correction term to the basic linear dependence of current density on voltage:

$$J(V) = \alpha V + \gamma V^3 + \cdots \quad (2.47a)$$

$$G(V) = \alpha + 3\gamma V^2 + \cdots \quad (2.47b)$$

Equation (2.47b) corresponds in the symmetric junction case to a parabolic conductance $G(V)$ symmetric about $V=0$. We shall find later that a frequently observed modification of this background conductance $G(V)$ is a parabola displaced from $V=0$ as a result of a barrier profile which is not symmetric. The offset parabolic dependence of $G(V)$ is also obtained by numerical methods working from (2.46).

The approach of Simmons to obtaining (2.47) is based on the use of the average barrier $\bar{\phi}$ given by (2.39). The approximate transmission factor of Simmons (1963*b*) is given as

$$D(E_x) \simeq \exp[-A(\bar{\phi}-W)^{1/2}]$$
$$A = \left(\frac{4\pi\beta t}{h}\right)(2m)^{1/2} = (16\pi\beta^2 t^2 \rho_t)^{1/2} \tag{2.48}$$

which is defined apart from the correction factor β, which is usually near unity. The use of (2.48) in (2.46) leads to Simmons' basic result for the current density:

$$J = J_0\{\bar{\phi}\exp(-A\bar{\phi}^{1/2}) - (\bar{\phi}+eV)\exp[-A(\bar{\phi}+eV)^{1/2}]\}$$
$$J_0 = \frac{e}{2\pi h(\beta t)^2} \tag{2.49}$$

The dependence of (2.49) comes entirely from the limits of integration in (2.46); when (2.40) is used in (2.49), Simmons' expression becomes

$$J(V) = \frac{e}{2\pi h(\beta t)^2}\left\{\left(\bar{\phi}-\frac{eV}{2}\right)\exp\left[-\frac{4\pi\beta t}{h}(2m)^{1/2}\left(\bar{\phi}-\frac{eV}{2}\right)^{1/2}\right]\right.$$
$$\left. - \left(\bar{\phi}+\frac{eV}{2}\right)\exp\left[-\frac{4\pi\beta t}{h}(2m)^{1/2}\left(\bar{\phi}+\frac{eV}{2}\right)^{1/2}\right]\right\} \tag{2.50}$$

which, for low voltages, reduces to the form (2.47), where α and γ are given by

$$\alpha = \frac{(2m)^{1/2}}{t}\left(\frac{e}{h}\right)^2 \bar{\phi}^{1/2}\exp(-A\bar{\phi}^{1/2})$$
$$\frac{\gamma}{\alpha} = \frac{(Ae)^2}{96\bar{\phi}} - \frac{Ae^2}{32\bar{\phi}^{1/2}} \tag{2.51}$$

in the case that $\beta=1$ (Simmons, 1963*a*, 1962*b*). Although the analysis of Simmons presented thus far is intended to accommodate asymmetric barriers, it incorrectly predicts for these cases a parabolic dependence of G upon V centered at $V=0$. A typical observed conductance spectrum is illustrated in Fig. 2.7, showing a small offset from $V=0$.

The method of Stratton (1962) differs in its treatment of the penetration factor D in the evaluation of (2.46):

$$\ln D(W) = -2\left(\frac{2m}{\hbar^2}\right)^{1/2}\int_{x_2}^{x_1}[\phi_1(x, V)-W]^{1/2}\,dx \tag{2.52}$$
$$= -[b_1 + c_1(-W) + f_1 W^2 + \cdots] \tag{2.53}$$

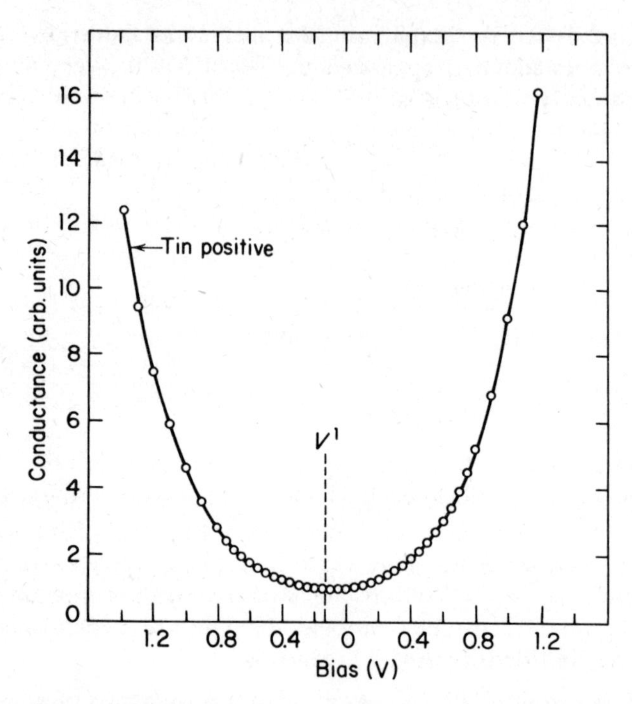

Fig. 2.7. Typical experimental conductance of an Al–I–Sn tunnel junction at 4.2 K. For voltages greater than 0.4 V, the plot is symmetric about $V = -0.125$ V. Note that the conductance is relatively flat in the region between ±400 mV. The conductance has an offset in its minimum, and the variation is generally parabolic at low biases and becomes more rapid at higher biases. (After Rowell, 1969*b*.)

with

$$\phi_1(x, V) = \phi(x) - \frac{eVx}{t} \tag{2.54}$$

Equation (2.52) is a restatement of (2.2) and (2.3) for an arbitrary barrier function ϕ_1 which includes the effect of the applied field by (2.54). Motivated by the fact that the largest contributions to the tunneling current come from electrons whose E_x value is about μ, i.e., $W = 0$, Stratton, following Murphy and Good (1956), expanded the integral in (2.53) in powers of W. The voltage-dependent expansion coefficients b_1 and c_1 of (2.53) are given by

$$b_1(V) = \sqrt{\frac{8m}{\hbar}} \int_{x_{11}}^{x_{21}} (\phi_1)^{1/2}\, dx = b_{10} - b_{11}V + b_{12}V^2 + \cdots \tag{2.55}$$

$$c_1(V) = \sqrt{\frac{2m}{\hbar}} \int_{x_{11}}^{x_{21}} (\phi_1)^{-1/2}\, dx = c_{10} - c_{11}V + c_{12}V^2 + \cdots \tag{2.56}$$

These coefficients in principle must be evaluated by integration of the indicated powers of ϕ_1, a function of x and V. However, useful results have been obtained by expanding, e.g., $b_1(V)$ as $b_{10}-b_{11}V$, etc. The result of Stratton's analysis is

$$J(V)=\frac{4\pi me\exp(-b_1)}{h^3c_1^2}[1-\exp(-c_1V)] \tag{2.57}$$

written for $T=0$ and further expanded as

$$J(V)\simeq\frac{4\pi me}{h^3c_{10}^2}\exp(b_{11}V-b_{12}V^2)[1-\exp(-c_{10}V)] \tag{2.58}$$

and

$$J(V)\simeq\frac{4\pi me}{h^3c_{10}}[V+(b_{11}-\tfrac{1}{2}c_{10})V^2+(\tfrac{1}{2}b_{11}^2-\tfrac{1}{2}b_{11}c_{10}+\tfrac{1}{6}c_{10}^2-b_{12})V^3+\cdots] \tag{2.59}$$

Contact between (2.59) and the expected behavior is made since the coefficient of the quadratic term in V in (2.59) is zero for a symmetric barrier, leaving, as required, linear and cubic terms in the eV dependence of the current density J. An advantage of the Stratton method is that it leads more directly to the expected temperature dependence of the tunneling current density $J(V, T)$, which is

$$J(V,T)=e\frac{4\pi mkT}{h^3}\int_0^\infty D(E_x)\ln\left\{\frac{1+\exp(-W/kT)}{1+\exp[(-W-V)/kT]}\right\}\mathrm{d}E_x \tag{2.60}$$

The logarithmic function in the integrand replaces the limits of integration in (2.46), thus generalizing that result for finite temperature T. This expression leads to a useful prediction,

$$\begin{aligned}\frac{J(V,T)}{J(V,0)}&=\frac{\pi c_1kT}{\sin(\pi c_1kT)}\\&\simeq1+\tfrac{1}{6}(\pi c_1kT)^2+\cdots\end{aligned} \tag{2.61}$$

for the ratio of the current at a temperature T normalized to the current at the same voltage at $T=0$; namely unity plus a small term in T^2. This fact, which was first found by Murphy and Good (1956), has also been derived by Simmons (1964). Some refinements and comparisons between these two methods have been offered by Hartman (1964).

To summarize, briefly, the results of the approximate analytic analyses of Simmons, Stratton and others: in the WKB model of tunneling between normal metal films, the current-voltage relationship is linear, with small cubic corrections at low voltage, and becomes exponential at high voltage. In the low-voltage region, the observed offset to the parabolic dependence of the conductance of $G(V)$ is not obtained (it is obtained by exact numerical methods, as we will see), but the temperature dependence appears to be correctly predicted. Turning to the experimental side, one can say generally, as was observed by Fisher and Giaever (1961), that the I–V relationship is linear at low voltage and exponential at high voltage in tunneling through thermally grown oxides. On a quantitative basis, however, the early experiments were based upon thermally oxidized films which are known to be amorphous and rather sensitive to preparation conditions. One straightforward source of difficulty is simply obtaining a good measurement of the thickness t of the oxide film which may be only 20 Å. The use of too large a value of t then could lead the researcher to an unreasonably small value of the mass (Stratton, 1962). Still, the general features, and particularly the temperature dependence, obtained by early workers (Hartman and Chivian, 1964; Pollack and Morris, 1965) left no doubt that tunneling was being observed. Because more recent results involving single-crystal barriers, as in metal-semiconductor Schottky barriers and tunneling through thin-film crystal films, have since been put in quantitative agreement with the theory, it seems most reasonable to attribute discrepancies in the early thermally oxidized tunnel barrier literature to variations in the preparation of the films involved.

The work of Giaever on superconductive tunneling, through the observation of the superconducting gap in tunnel junctions involving oxides of aluminum, tin, and lead made clear that, at least at 4 K, tunneling could be the only appreciable mechanism of charge transfer. In an accumulating body of evidence in the conductance spectra $G(V)$ of such junctions, the lower-voltage ranges, less than ± 0.1 V, indicate the weak parabolic behavior of (2.47), as shown in Fig. 2.7. The important work of Brinkman, Dynes, and Rowell (1970) examines, within the framework of (2.46) but including trapezoidal barriers, the influence of the prefactor g, which we have thus far ignored; also, the image force corrections. This work has answered many of the questions left by the early analyses, most notably the origin of the offset minimum in the conductance. Figure 2.8, calculated numerically by using the WKB approximation with the prefactor ignored for a barrier of thickness 15 Å and average height $\bar{\phi}$ of 2 V but with varying degrees of asymmetry in the barrier parameters, demonstrates the offset of the parabolic minimum in conductance. These calculations have assumed the free electron mass and have ignored the image correction. The curves as plotted do not show the expected changes in conductance level which has here been normalized to the same value at $V = 0$. The barrier asymmetry has been included in the calculation by adopting (2.54) in which $\phi(x)$ is originally taken as a trapezoidal barrier with heights ϕ_1 and ϕ_2 on the two sides. Numerical calculations were extended to include an exact prefactor g, a generalization of (2.2b) to unequal barrier heights; comparison of the conductance curves is shown in Fig. 2.9, in which the lower curve is obtained with the prefactor. In both cases, the thickness is 15 Å, the trapezoidal heights ϕ_1, ϕ_2 on the two sides are 1 and 3 V respectively, and the Fermi energies μ are 10 V. It can be seen here that the exact prefactor (see Eq. 9 of Brinkman,

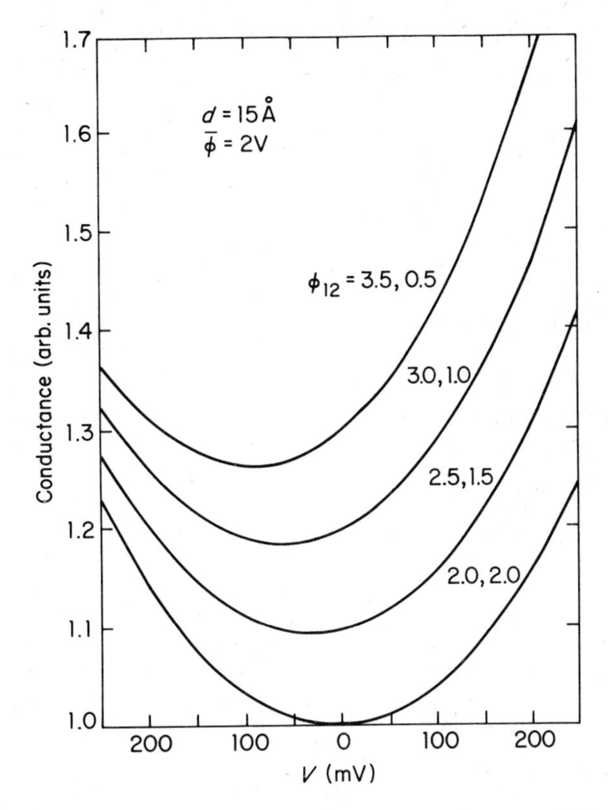

Fig. 2.8. Calculated conductance vs. voltage curves for a metal-insulator-metal junction, illustrating the effect of different barrier heights on the two sides in producing the offset in the conductance minimum. The successive curves shown have been normalized to fixed conductance at $V = 0$ and have been shifted vertically for clarity. (After Brinkman, Dynes, and Rowell, 1970.)

Dynes, and Rowell, 1970) has very little influence on the appearance of the conductance.

While the conductance curves calculated in the previous two figures have the right shapes and magnitudes of offset to agree reasonably with experiment, the results have been obtained by using rather small values of thickness t, quoted as 15 Å, less than experimental estimates which range typically from 20 to 30 Å. Two plausible origins of this discrepancy are the utilization of the full electron mass rather than the possibly reduced effective mass in the barrier oxide (see Fig. 2.4) and neglect of the image force correction to the barrier potential, which might be appreciable for a thickness of only 15 Å.

Inclusion of the image force for an electron tunneling into vacuum from a pure metal surface is essential to calculate the correct probability of field emission; and this effect was treated carefully in the early work (Sommerfeld and Bethe, 1933). The modification of the trapezoidal

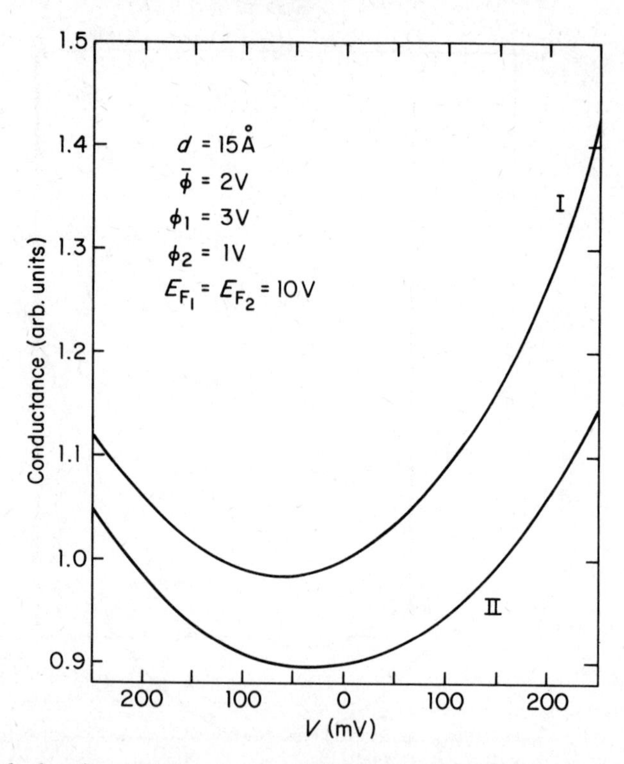

Fig. 2.9. Calculated conductance vs. voltage curves, using barrier parameters $d = 15$ Å, $\phi_1 = 3$ V, $\phi_2 = 1$ V, with differing assumptions concerning the boundaries of the junction. In model I (upper curve), the boundaries are assumed to be diffuse (g = 1), whereas in model II, the boundaries are assumed to be sharp and the exact prefactor included in the calculation. The conductances are normalized at $V = 0$ and have been offset to show the curves more clearly. (After Brinkman, Dynes, and Rowell, 1970.)

barrier $\phi(x)$ by the image force is a strong function of the assumed dielectric constant, as indicated in Fig. 2.10 (after Simmons, 1963). The effect of the image force is to round the edges of the barrier and to slightly lower the average barrier height. The two $\phi(x)$ curves shown in Fig. 2.10 include, respectively, the exact image term and an approximation shown in the last term in (2.62):

$$\phi(x, V) = \phi_1 + \frac{x}{t}(\phi_2 - eV - \phi_1) - \frac{1.15\lambda t^2}{x(t-x)} \tag{2.62}$$

$$\lambda = e^2 \ln \frac{2}{8\pi\varepsilon\varepsilon_0 t} \tag{2.63}$$

This form was used by Brinkman, Dynes, and Rowell (1970) to obtain the conductance curve shown in Fig. 2.11, assuming a dielectric constant ε of 4.0. The effect of the image term is judged to be minor by these authors.

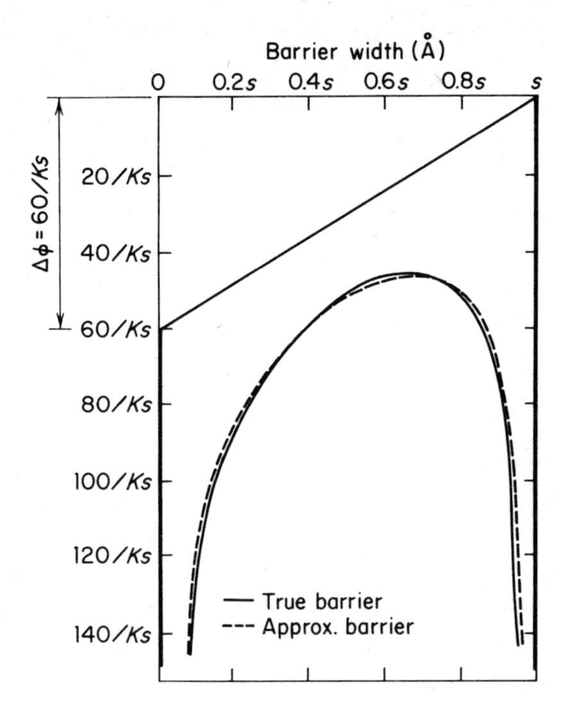

Fig. 2.10. The effect of the image force on a typical trapezoidal barrier profile. The rounding of the barrier which occurs very close to its edges is frequently ignored, its effects being absorbed in effective barrier height and thickness parameters. (After Simmons, 1963.)

Fig. 2.11. Calculated conductance for metal-insulator-metal junction having thickness $d = 20$ Å, with $\phi_1 = 1$ V, $\phi_2 = 3$ V, and including the effect of the image force. The image force tends to round the edges of the barrier, increasing the overall transmission. Inclusion of the image force slightly reduces the asymmetry of the conductance and leads to a conductance which increases more rapidly at higher bias. (After Brinkman, Dynes, and Rowell, 1970.)

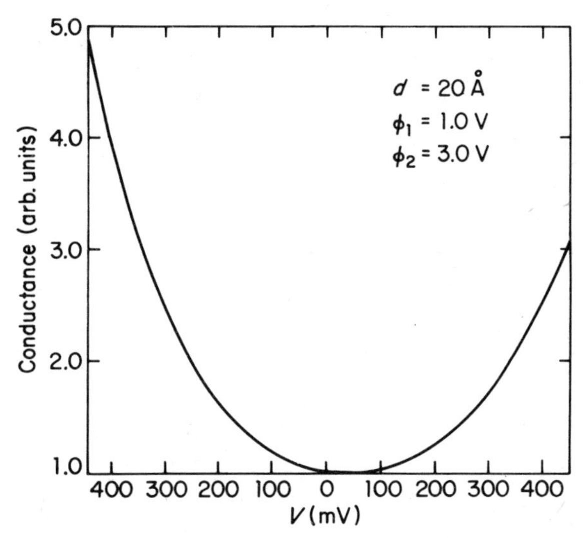

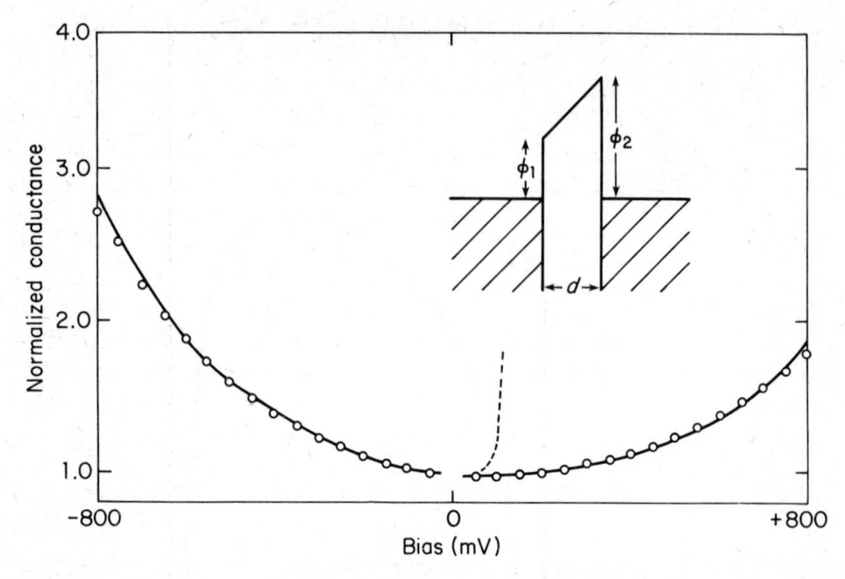

Fig. 2.12. The solid line is the measured conductance of an Al–Al_2O_3–Pb tunnel junction, while the points are obtained from calculation, assuming sharp interfaces and the following parameters: $\phi_1 = 1.57$ V, $\phi_2 = 4.07$ V, and $d = 11.6$ Å. The calculation accurately reproduces the numerical value of the junction conductance at $V = 0$. (After Walmsley, Floyd, and Timms, 1977.)

The main changes are that the conductance offset is slightly reduced and that the rate of increase of conductance with voltage is made somewhat more steep. Hence, in most of their work and in the work of later researchers, the image term has been omitted. The use of WKB relations based on (2.46) is illustrated finally in Fig. 2.12, after Walmsley, Floyd, and Timms (1977), who have actually fit data from Al–I–Pb tunnel junctions (solid line) with calculated values (dots) corresponding to a trapezoidal barrier, as indicated, with parameter values $\phi_1 = 1.57$ V, $\phi_2 = 4.7$ V, and $t = 11.6$ Å; these parameters rather closely match the measured zero-bias conductance, given as 42.4 mmho mm^{-2}.

The first definitive measurement of the energy-dependent $\kappa(E)$ from tunneling through a crystalline insulator was reported by Kurtin, McGill, and Mead (1970) (Fig. 2.13) in their study of GaSe. The barrier transmission was treated within the WKB approximation with no prefactors, as given in (2.12). The temperature dependence of the Fermi function was included, and the integration over transverse k was done by a method of steepest descents (Kurtin, McGill, and Mead, 1971). The key to this work was development of a technique for peeling single-crystal films of the layer compound GaSe down to thicknesses on the order of 100 Å. A series of independent measurements established not only the thickness but also the individual barrier heights ϕ_1 and ϕ_2, specifying the trapezoidal barrier profile as in (2.62), neglecting the image correction. Figure

2.13a shows a configuration of a Cu–GaSe–Au tunnel junction whose thickness is known to be 83 Å. These measurements established that the band gap of the GaSe is 2.0 eV, the low-frequency dielectric constant $\varepsilon = 8$, and the optical dielectric constant $\varepsilon = 7$, while internal photoemission measurements determined separately the interfacial barrier heights on the copper and gold sides, as measured from the valence band of the GaSe, to be 0.68 and 0.52 eV, respectively. A further important assumption is necessary to obtain the purely trapezoidal profile, namely the lack of band bending, as could result from Debye charge layers at the interfaces. This assumption is confirmed by a low value of about $3 \times 10^{14}\,\text{cm}^{-3}$ of trapped- and free-carrier density in the specimen. The dielectric constant and the rather large thickness may be taken as justification of the neglect of the image correction and adoption of the simple trapezoidal barrier shown. Figure 2.13a shows the dependence upon bias of both signs of the current measured at 77 K, verified to differ only slightly from the values at room temperature. As tunneling occurs, a trajectory through the forbidden gap of the GaSe is followed, which corresponds to a constant energy relative to the external Fermi levels but a changing energy with respect to the band edge of the GaSe. This trajectory is obviously substantially changed by the application of biases ranging from ±0.6 V. Thus, a substantial portion of the GaSe energy gap from near the valence band edge to approximately 1.5 V above the

Fig. 2.13. (a) Current density J as a function of applied voltage V for a tunnel junction having a single-crystal GaSe barrier layer of 83 Å thickness. The electrodes in this case are copper and gold. The plotted points are experimentally determined, while the solid line is calculated from the theory, using the deduced E vs. k relationship. (b) Inferred E vs. k^2 relation for the band gap of GaSe. The evident agreement between theory and experiment confirms that the displayed $E(k)$ relation is an accurately determined property of GaSe. Note that the $E(\kappa)$ is determined in the gap from the valence band edge to 1.5 V above this energy. In the region near the valence band edge, the result is consistent with an effective mass $m^*/m_e = 0.07$. (After Kurtin, McGill, and Mead, 1971.)

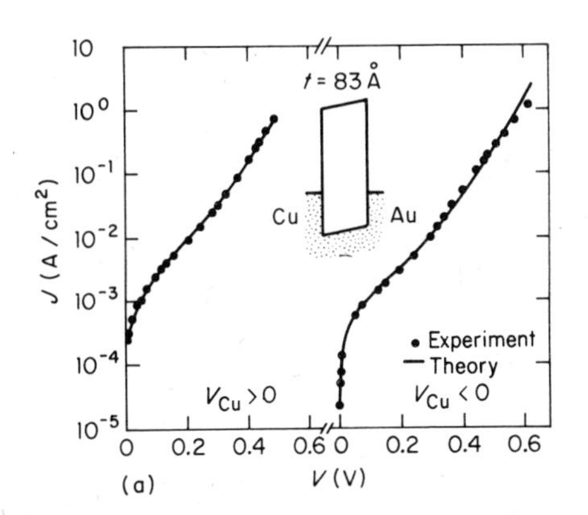

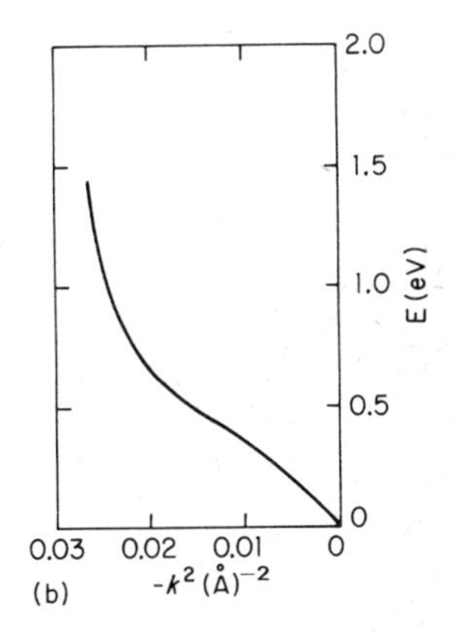

valence band edge has been sampled in measurements interpreted in terms of (2.2a) and (2.2c) and leading finally to the dependence shown in Fig. 2.13b for the variation of $-k^2 = \kappa^2$ as a function of energy in GaSe at low energies (i.e., near the valence band edge). The limiting slope of this plot is constant and corresponds to the valence band effective mass $m^* = 0.07$.

2.3.2 *Metal-Insulator-Semiconductor Junctions*

The characteristic feature of a semimetal, a small value of the Fermi degeneracy μ, should bring into an easily measurable range of bias voltages the condition when the bottom of the conduction band crosses the Fermi energy of the opposite electrode. At $eV = \mu$, one would expect to see a noticeable change in the tunneling conductance, because beyond this bias no further change occurs in the number of electrons able to tunnel. Further change in the conductance for $eV > \mu$ could be attributable to the change in the barrier shape alone. An estimate of the expected behavior can be obtained by the use of (2.46), which we have rewritten as

$$J = \frac{2e\rho_t}{h}\left[eV\int_0^{\mu} gD(E_x)\,dE_x + \int_{\mu}^{\mu-eV} D(E_x)g(E_x + eV - \mu)\,dE_x\right] \tag{2.64}$$

If we neglect the voltage dependence of D, then the first term in (2.64) will vary smoothly through the condition $eV = \mu$ and hence will contribute only a constant to the derivative dJ/dV. Therefore, we concentrate on the second term in (2.64) and examine dJ/dV near $\mu = eV$, which gives

$$\frac{dJ}{dV} = \frac{2e^2\rho_t}{h}\int_{\mu}^{\mu-eV} gD(E_x)\,dE_x + \text{smaller terms} \tag{2.65}$$

because the integrand evaluated at the upper limit vanishes, and we are neglecting the voltage dependence of D. The band edge behavior in the conductance $g(E)$ then is estimated by assuming a simple form (2.66)

$$gD(E_x) = D_0 E_x^n \exp\frac{E_x}{E_0} \tag{2.66}$$

for the transmission factor. In this equation the possibility of a prefactor is included by the factor E_x^n. Examination of the V-dependence of (2.65) reveals, in addition to smoothly varying terms, a term which disappears at the band edge $eV = \mu$ as the first power of $\mu - eV$ (taking $g = 1$) and quadratically in $\mu - eV$, assuming g is proportional to E_x. Thus, for $eV \simeq \mu$ and $n = 0, 1$, one has

$$G(V) \simeq \frac{2e^2\rho_t}{h} D_0(\mu - eV)^{n+1}\theta(\mu - eV) + \text{other terms} \tag{2.67}$$

The presence of the prefactor, linear in E_x, will make the variation of the conductance rather smooth and the $eV = \mu$ effect difficult to observe in the presence of the other smoothly varying terms. This result is consistent with the early conclusion of Harrison (1961) that band structure effects in normal metal tunneling would be difficult to observe and is also consistent with the rather weak effects which have been seen in tunneling studies of bismuth and Bi–Sb alloys (Chu, Eib, and Henriksen, 1975). Stronger effects related to band structure are observable when different bands have significantly different transmission factors, as in semiconductors in comparison of zone center and zone boundary conduction bands, or in transition metals comparing wavefunctions of predominantly s vs. predominantly d character.

2.3.3 Schottky Barrier Junctions

The direct metal contact to a heavily doped semiconductor produces a tunnel junction, as illustrated in Fig. 2.14, which is both rather easy to make and susceptible to detailed characterization. The barrier in this case is the single-crystal semiconductor depletion region. Charge from this region transfers to the metal or to the interface states, leaving a uniform space charge, i.e., a region in which the electrostatic potential varies quadratically with distance. The thickness of this region can be controlled by the concentration of donor or acceptor impurities in the semiconductor. Schottky barrier junctions can be made by cleavage of a semiconductor crystal in the presence of an evaporating metal (e.g., Wolf and Compton, 1969), producing an abrupt and contamination-free interface between metal and crystalline semiconductor. In tunneling studies made at low

Fig. 2.14. Schematic band diagram of Schottky barrier contact to a degenerate (metallic) n-type semiconductor. Conventional forward bias in metal-semiconductor contacts is that sign which reduces the height of the barrier seen from the semiconductor.

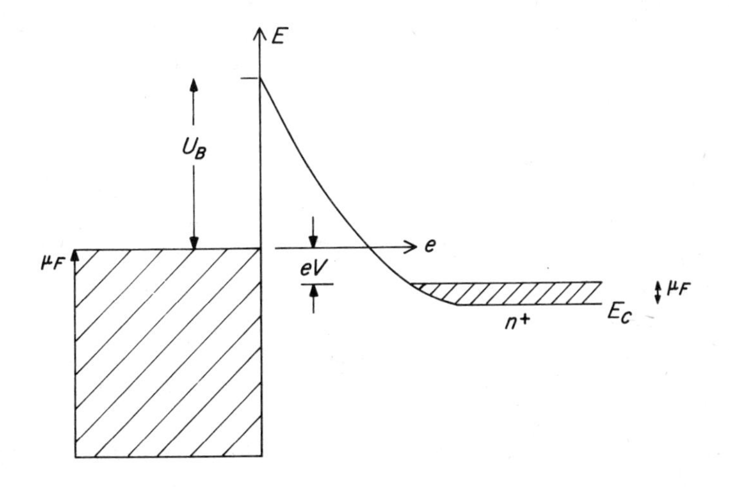

temperatures, it is important that the semiconductor have sufficient doping density to lie on the metallic side of the Mott transition, so that, as indicated in Fig. 2.14, a metallic Fermi energy μ typically in the range 10–100 mV is obtained. This condition is required in practice to ensure that the applied voltage drops across the depletion region barrier and not across the bulk of the semiconductor crystal, especially at 4 K where the resistance of a nonmetallic semiconductor becomes exponentially large. A complicating feature of Schottky barrier contacts is that as the free-carrier concentration in the semiconductor is increased with impurity concentration N_D, the thickness of the barrier is reduced proportionally to $N_D^{-1/2}$ (n-type case), with a consequent decrease in the resistance level of experimental devices. The resistance can also be adjusted to some extent by choosing different contact metals to vary the barrier height U_B, which also affects the zero-bias thickness of the barrier. A second complication is that the thickness of the Schottky barrier is a function of bias voltage. In the n-type case shown in Fig. 2.14, at a bias V equal to the barrier height U_B, the thickness of the barrier is zero.

The tunneling problem posed by the parabolic depletion layer barrier has been solved exactly (Conley, Duke, Mahan, and Tiemann, 1966), including the effect on the conductance of the bias variation of barrier thickness. The conductance spectra of Schottky barrier contacts also permit determination of the band edge (i.e., the Fermi degeneracy μ) referred to in the previous section. The conductance $G(V)$ falls to a minimum at a bias eV equal to μ, as shown (in resistance) in the calculated dV/dI curves for Schottky barrier contact on n-type germanium (Fig. 2.15). In this figure, the arrows locate the μ values calculated independently from the known properties of the semiconductor crystals.

The calculation of Conley, Duke, Mahan, and Tiemann (1966) (CDMT) is based upon matching of exact eigenfunctions in the metal, the semiconductor barrier, and the semiconductor bulk, i.e., following (2.7) in which D is obtained from the exact eigenfunctions and hence includes the proper prefactor. The barrier potential $\phi(x)$ used by Conley, Duke, Mahan, and Tiemann is

$$\phi(x) = \frac{e^2 N_D (d-x)^2}{2\varepsilon} + eV - \mu, \qquad 0 < x < d \tag{2.68}$$

$$d = \left[\frac{2\varepsilon(U_B + \mu - V)}{eN_D}\right]^{1/2} \tag{2.69}$$

Here, ϕ is understood to be measured from the Fermi energy μ of the semiconductor and the coordinate x is measured from the metal-semiconductor interface into the semiconductor; N_D and ε are, respectively, the donor concentration and the dielectric constant. Positive bias V raises the semiconductor conduction band. The parameter d is the bias-voltage-dependent width of the depletion layer as expressed in (2.69), where U_B (indicated in Fig. 2.14) is the barrier height at the metal-

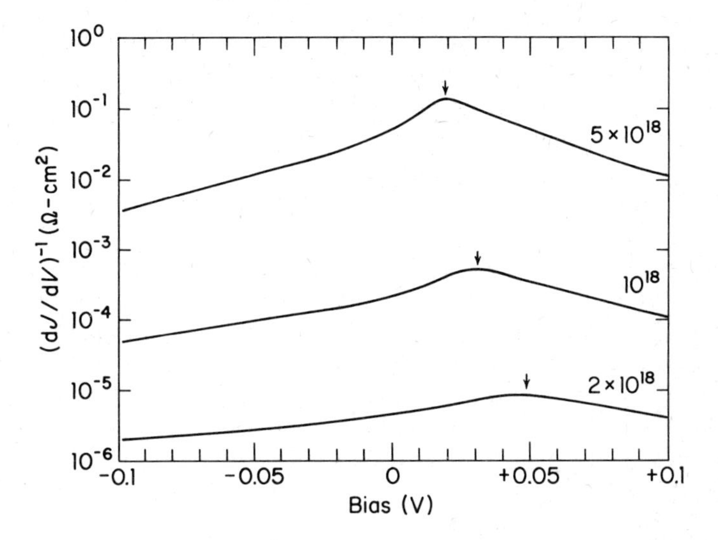

Fig. 2.15. Calculated incremental resistance dV/dI, plotted logarithmically, for Schottky barrier contact of indium on degenerate n-type germanium. Positive bias corresponds to raising the semiconductor Fermi level with respect to that of the metal. The arrows denote the Fermi energies of the degenerate semiconductors in the three cases shown, indicating the utility of such measurements in determining the Fermi energy. (After Conley, Duke, Mahan, and Tiemann, 1966.)

semiconductor interface. The value of U_B is determined, at least partially, by location of interface and surface states on the semiconductor surface. The barrier potential (2.68) is expected to be accurate except in the so-called reserve region of the semiconductor, where the free-carrier density is rising from zero (in the depletion region) to its bulk value. A further approximation in (2.68) is neglect of the image force. However, this approximation is not expected to be important because of the dielectric constant of the semiconductor and the typically large thicknesses of the depletion layer. Exact solution of this problem is possible because the Hamiltonian in the barrier region $0<x<d$,

$$H_b = -\frac{\hbar^2}{2m_s}\frac{d^2}{dx^2}+\phi(x), \qquad 0<x<d \tag{2.70}$$

is a form of a standard parabolic cylinder equation (Miller, 1965). The eigenfunctions are thus determined to be

$$\psi(x) = AU(-\kappa_s^2, L)+BV(-\kappa_s^2, L) \tag{2.71}$$

with

$$\begin{aligned} \kappa_s &= \lambda k_s \\ \lambda &= \left(\frac{h^2\varepsilon_s}{4m_s e^2 N_D}\right)^{1/4} \\ L &= \frac{d-x}{\lambda} \end{aligned} \tag{2.72}$$

which are expressed as a linear combination of the parabolic cylinder functions U and V, with auxiliary relations defined in (2.72), where subscript s denotes values in the semiconductor. The exact transmission factor is found by imposing continuity conditions

$$\psi(x), \qquad m(x)^{-1}\frac{d\psi(x)}{dx} \qquad \text{continuous at } x=0, d \tag{2.73}$$

at the interface $x=0$ and the edge of the depletion region $x=d$. The transmission factor is evaluated by taking a constant effective mass, i.e., neglecting variation of the decay constant κ through the energy gap of the germanium. The calculation is successful in predicting both the general shape of the conductance and the location of its minimum at $eV=\mu$. In several cases, the correct magnitude of the conductance, taking the appropriate parameters from experimental measurements rather than as fitting parameters, has been predicted. Such a comparison is shown in Fig. 2.16 for Pb contacts to Ge:Sb at $N_D=7.5\times10^{18}\ \text{cm}^{-3}$ measured at 4.2 K. The three solid curves indicate a range of conductance levels obtained in three different contacts, which may reflect local doping variations. The dashed curve is obtained numerically on the model of Conley, Duke, Mahan and Tiemann (CDMT) taking the barrier height parameter U_B from capacitance measurements and the parameter N_D as inferred from resistivity or Hall measurements, and making use of the measured junction area and other known parameters of Ge. The calculated Fermi energy position (arrow) agrees with the minimum in the conductance curve. Generalization of the Conley, Duke, Mahan, and Tiemann calculation to the ellipsoidal energy surfaces appropriate to the conduction band of Ge or Si has been carried out by Christopher, Darley, Lehman, and Tripathi (1975). Wimmers and Christopher (1981) have noted the possible experimental importance of dopant striations, i.e., periodic variations in the concentration of dopant along the growth direction of semiconductor crystals. These occur as a consequence of the Czochralski method of crystal growth, which involves a combination of rotation and linear translation. The spacing of striations is typically a few tenths of a millimeter, possibly larger than a single junction dimension (Wimmers and Christopher, 1981). This result may possibly account for the variations in conductance level noted in Fig. 2.16. An additional feature which may be important in estimating the resistance level of Schottky barriers in the high-dopant concentration (thin-barrier) limit has been discussed by Wolf and Losee (1970) in a study of silicon Schottky barrier junctions. The assumption of a uniform space charge density becomes inadequate, and random fluctuations in the distribution of donor impurities in the depletion layer, which may, at very high dopant concentration, be only tens of angstroms thick, are shown to lead to corresponding local fluctuations in the transmission. An appreciable increase in overall transmission may occur, as this is dominated by regions of higher local donor density. The same effect can be shown to be unimportant in

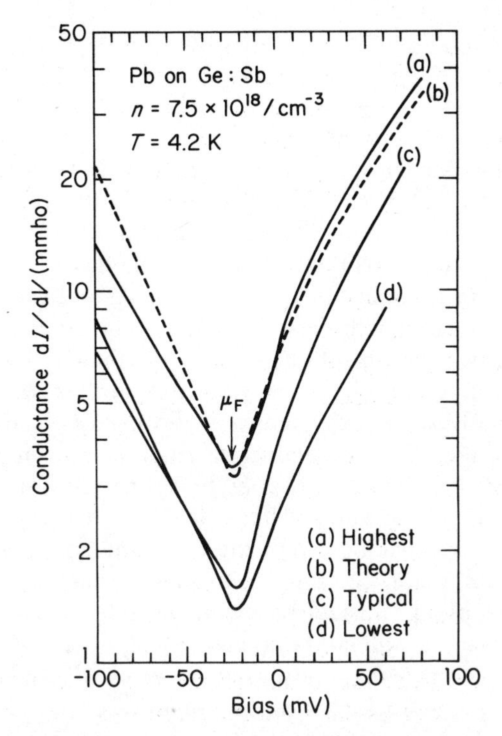

Fig. 2.16. Experimental conductance of Schottky barrier contacts to germanium doped with antimony, measured at 4.2 K. The dashed line is the conductance calculated from the CDMT theory, making use of the independently measured barrier height $U_B = 0.63$ eV and the experimentally measured antimony concentration, $7.5 \times 10^{18}\,\text{cm}^{-3}$. Note that both the theory and the experimental curves have conductance minima at the correct Fermi energy, 25 mV. The variation between the measured curves is presumably due to fluctuations in the concentration of antimony dopant in the sample of germanium. (After Steinrisser, Davis, and Duke, 1968.)

the case of the less heavily doped germanium samples mentioned previously.

Tunneling measurements on Schottky barriers of GaAs have been reported by Conley and Mahan (1967) and Mahan and Conley (1967). Conley and Mahan (1967) clarify the relationship between the exact CDMT calculation described above and the calculation based upon (2.7) using the WKB approximation. According to Conley and Mahan (1967), a prefactor proportional to $E^{1/2}$ is required to agree with the exact calculation. These authors also incorporate into the WKB calculation a realistic treatment of the decay constant variation $\kappa(E)$ through the energy gap, neglected by CDMT. This effect is more important in the case of gallium arsenide because the barrier height U_B is greater. These authors discuss the parameter E_0, which is defined as

$$E_0 = \left(\frac{\pi e^2 N_D \hbar^2}{\varepsilon m_s}\right)^{1/2} \tag{2.74}$$

for the Schottky barrier case. Conley and Mahan point out that a minimum precisely at $eV = \mu$ (Fermi energy) in the conductance spectrum is neither expected nor observed if μ is much greater than E_0. A further restriction of the CDMT analysis is that U_B be equal to or less than $\frac{1}{2}E_g$. Mahan and Conley (1967) confirm that a prefactor $g(E) \simeq E^{1/2}$ is required to obtain agreement between a WKB and an exact treatment of the Schottky barrier. This result suggested to Mahan and Conley that the prefactor $g(E)$ may be identified with the density of states in the semiconductor and may be expected to depart from $E^{1/2}$ in a region close to the bottom of the effective conduction band, i.e., in the impurity band region. In analysis of Schottky barrier tunneling data on p-type GaAs, shown in Fig. 2.17, these authors have determined $g(E)$ (the density of states) from the data, making use of the two-band approximation for the WKB transmission factor (see Fig. 2.17b). The results shown as circles in the figure indeed agree satisfactorily with the $E^{1/2}$ density of states that one might expect in the valence band and also show features near positive bias of 50 mV, which were interpreted in terms of a metallic impurity band formed from interacting acceptor impurities. The latter effects are shown to become unimportant at higher concentrations, in agreement with the observations cited for germanium.

Schottky barrier I–V curves have been analyzed to yield the E vs. κ relationship in the forbidden gap of the semiconductor InAs by Parker and Mead (1968). In the study of InAs, indicated in Fig. 2.18, the energy-wavevector plot in the band gap InAs was obtained through the equation

$$-k^2(E) = \frac{N_A e^2}{2\varepsilon}(U_B - eV)\left(\frac{d \ln J}{dV}\right)^2 \tag{2.75}$$

This result applies to a Schottky barrier on a p-type semiconductor of acceptor density N_A under forward-bias conditions (i.e., bias sign is that which reduces barrier). This formula assigns all of the change in the current to change in the barrier transmission factor, which is appropriate once the bias has exceeded the Fermi degeneracy. As may be seen from the figure, rather good agreement is obtained between the various sets of points and a theoretical curve

$$E = \left(\frac{E_g^2}{4} + \frac{\hbar^2 k^2 E_g}{2m^*}\right)^{1/2} - \frac{E_g}{2} \tag{2.76}$$

after Franz (1956), which predicts the E vs. κ relationship assuming the effective masses of the conduction and valence band electrons are the same; here, $m^* = 0.02$. Similar results have been obtained by a slightly different method for GaAs by Conley and Mahan (1967).

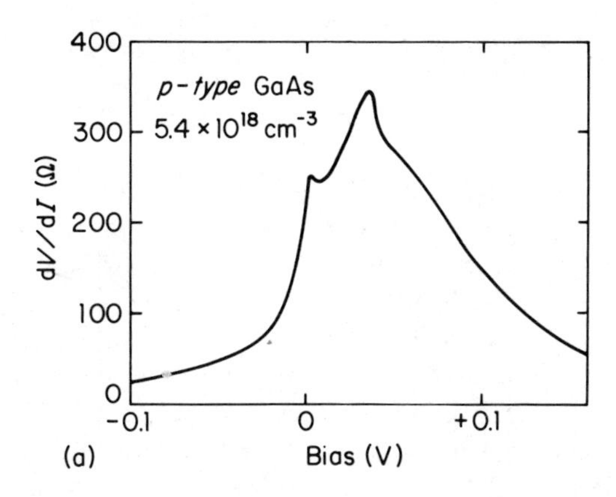

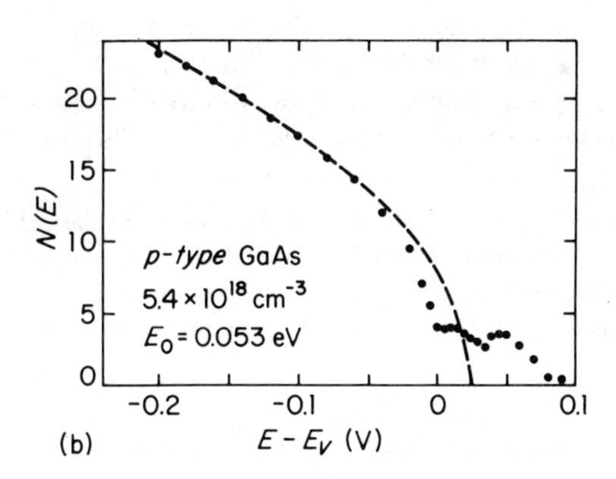

Fig. 2.17. (a) Differential resistance dV/dI spectrum of Schottky barrier contact of gold on p-type GaAs. (b) Analysis of the data shown in (a) to deduce the density of states $g(E)$ in the p-type GaAs below the valence band edge, making use of the theory of Conley and Mahan (1967). The dashed line is an $E^{1/2}$ density of states, which is seen to approximate the results fairly well deep in the valence band. The small features in the vicinity of the valence band edge were interpreted as an impurity band of the zinc acceptor impurities. (After Mahan and Conley, 1967.)

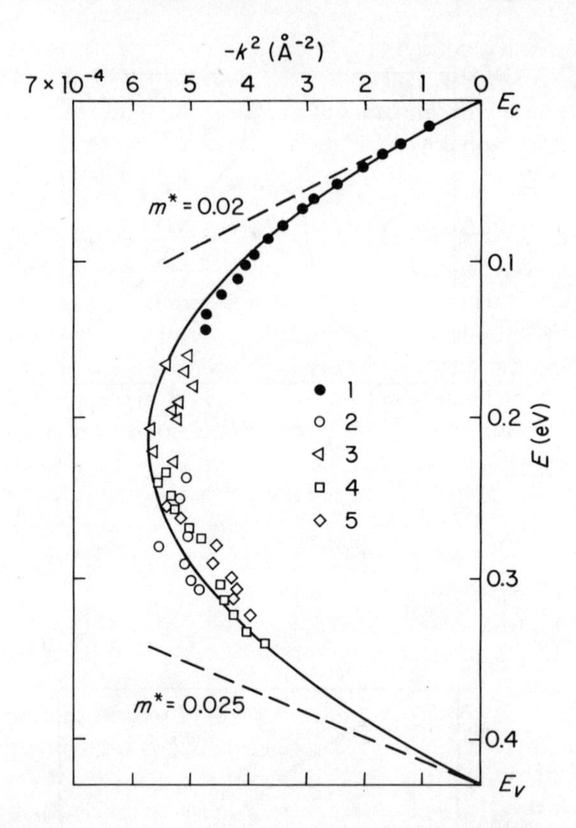

Fig. 2.18. Determination of the energy-wavevector relationship through the forbidden gap of InAs from analysis of Schottky barrier tunneling data. The data points are in reasonable agreement with the two-band model of InAs for which both effective masses are 0.02. (After Parker and Mead, 1968.)

There are several other aspects of tunneling in Schottky barrier junctions to which we will return later. These include zero-bias anomalies associated with localized donor states in the reserve region, effects associated with localization of the wavefunctions in the semiconductor at the Mott transition, and phonon effects appearing both as simple, inelastic assisted tunneling and also as many-body, density-of-states effects. Finally, Schottky barriers have been used as a prepared tunneling-barrier-electrode combination in order to tunnel into substances of interest, e.g., metallic polyacetylene. In all that we have described so far, it is assumed that the electron transition from one electrode to another can occur at constant k_t, i.e., that assistance in the form of a phonon absorbed or emitted is not necessary to transfer the carrier from one part of the Brillouin zone to the other. While the tunneling need not be strictly specular in the sense of preserving k_t exactly, no large changes in k_t have been necessary. In the case of metal contacts to n-type germanium, for example, where the conduction electron states occur near the boundary of the Brillouin zone, it is assumed that the projections of the Fermi surfaces of the semiconductor and metal overlap, and only in these regions of overlap are the tunneling transitions allowed.

2.3.4 pn *Junction (Esaki Diode)—Direct Case*

We have already seen, in Fig. 1.4, the I–V curves observed in *pn* junctions by Esaki (1958), where we pointed out that the kinks observed at the low temperature occur at zone boundary phonon energies in silicon. Thus, this case represents a complicated situation of assisted tunneling, in which the electronic transition is accompanied by emission of a phonon. In continuing the line of discussion of previous sections, we will initially ignore the complications implied by phonon assistance. Hence, we will consider direct *pn* junctions in which the conduction and valence bands have their minima and maxima at the same point in the Brillouin zone. This result occurs, for example, in GaAs but not in Ge or Si. The direct tunneling problem which is left is essentially the same as the one treated by Kurtin, McGill, and Mead (1971) in their study of tunneling currents through single-crystal GaSe. That is, the transition rate is dominated by the decay of the wavefunction through the forbidden gap of the semiconductor on a trajectory $E(k)$ which carries it from the valence band on one side to the conduction band on the other side. The *pn* junction problem might appear to be simpler than the MIM problem of Kurtin, McGill, and Mead, for the ranges of occupied electron and hole states in the p and n regions of semiconductors forming the Esaki diode are at most tenths of an electronvolt in width, while in the case cited, much wider ranges of states are filled in the metal electrodes on both sides of the junction. This difference, of course, accounts for the qualitative difference in the I–V curves in the two cases and, in particular, leads to the negative-resistance region in the I–V characteristic of the Esaki diode.

The basic problem in the calculation of the $J(V)$ characteristic of an Esaki diode is that first addressed by Zener (1934), namely the rate of tunneling from the valence to the conduction band in an insulator with a uniform applied electric field. In the uniform-field approximation to an Esaki diode, the "tilted band" picture of Zener occurs in the junction region between the degenerate n and p regions. Zener's approach to this problem was one-dimensional and was not adaptable to including realistic details of the band structure. The most refined result available through extensions of Zener's method yields, for the number of electrons n passing into the conduction band per unit volume per unit time,

$$n = N_v \frac{eFa}{h} \exp \frac{-\pi\sqrt{2m^*}\, E_g^{3/2}}{2ehF} \tag{2.77}$$

More realistic calculations of the interband tunneling current have been given by Kane (1959, 1961), building upon work of Keldysh (1958*a*, 1958*b*). An important point in these more rigorous calculations is the use of energy eigenfunctions $q_n(k)$, given by

$$q_n(\vec{k}) = N_n^{1/2} \exp\left\{\frac{i}{F}\int_0^{k_x} [E - E_n(k')]\,dk'_x\right\} \delta(k_y - k_{y0})\,\delta(k_z - k_{z0}) \tag{2.78}$$

where n is an index corresponding to energy band $E_n(k)$ and N_n is a normalization constant. Interband tunneling corresponds to transitions between q's of differing band index n, under the influence of a perturbing Hamiltonian depending upon the field. The corresponding transition rate is evaluated by Kane, whose results may be expressed in terms of the transmission coefficient

$$T = \frac{\pi^2}{9} \exp\left(-2\int_{x_2}^{x_1} \kappa\,dx\right) \tag{2.79}$$

which differs only by the prefactor, nearly unity, from the WKB expression. Following Kane (1961), the transmission coefficient is evaluated as

$$T = \frac{\pi^2}{9} \exp\left[\frac{-\pi m^{*1/2} E_g^{3/2}}{2\sqrt{2}\,\hbar F}\right] \exp\frac{-2E}{\bar{E}_t} \tag{2.80}$$

in which E_t is the kinetic energy parallel to the barrier and the parameter $\bar{E}_t$ is defined by

$$\bar{E}_t = \frac{\sqrt{2}\,\hbar F}{\pi m^{*1/2} E_g^{1/2}} \tag{2.81}$$

where F and E_g are, respectively, the junction field and band gap of the semiconductor. The corresponding expression for the current density is

$$J = \frac{em^*}{18\hbar^3} \exp\left[\frac{-\pi m^{*1/2} E_g^{3/2}}{2\sqrt{2}\,\hbar F}\right] \iint [f(E) - f(E + eV)] \exp\left(\frac{-2E_t}{\bar{E}_t}\right) dE\,dE_t \tag{2.82}$$

The evaluation of this integral to obtain the forward-bias current hump of the Esaki diode is discussed by Kane (1961), who also gives extensions for the case of indirect or phonon assisted tunneling. In application of these expressions which show only one effective mass, m^* should be chosen as the smaller of the m_c^* and m_v^* values. Fredkin and Wannier (1962) give a slightly different result for J as a consequence of a different treatment of the E vs. κ relationship in the semiconductor band gap. A comparison of the two expressions is given by the latter authors. A more detailed discussion of *pn* tunnel diode calculations and comparison with experimental results is given by Duke (1969, especially in Sections 8 and 13).

2.3.5 *Vacuum Tunneling*

The usual meaning of vacuum tunneling is field emission (Fig. 1.3), first quantitatively treated by Fowler and Nordheim (1928). By application of a voltage to a finely pointed metal specimen, it is relatively easy to achieve a sufficiently narrow barrier to allow measurable tunneling electron emission currents. The alternative planar geometry, analogous to Fig. 2.1, is more difficult to achieve, from the necessity of maintaining metal surfaces separated by some 10 Å and particularly in avoiding vibrations which would modulate the spacing. Modeling the metal by a step work function barrier, the elementary quantum mechanics of these situations is straightforward, even if the image force correction is included. The latter is important in a vacuum tunneling case both because the barrier height (work function) is usually relatively large, making the thickness of the equivalent vacuum barrier small, and because the vacuum offers no dielectric screening. A detailed review of the field emission case has been given by Good and Müller (1956).

On a second look, also, both vacuum tunneling configurations are still interesting. The closely spaced vacuum gap, from a technical point of view, offers the possibility of mapping surface topography on an angstrom scale (Binnig and Rohrer, 1982) as well as a scheme, at least in principle, for performing tunneling spectroscopy on an arbitrary clean electrode, free of disturbance by an attached barrier layer. From a different point of view, vacuum tunneling, particularly the field emission case, has given additional insight into the questions of the effects of the normal density of states and the wavefunction character on the tunneling conductance. Sketched in Fig. 2.19 is a simple d-band model for a ferromagnet: note that the density of spin-up (majority-spin) d electrons at the Fermi energy vanishes. However, the barrier for extraction into the vacuum of electrons (at equal energy) of either spin projection is presumably the same: the work function ϕ. Assuming that one can separately detect spin-up and spin-down electrons emitted into the vacuum, measurement of the polarization of field emission current from a ferromagnetic metal point offers tests of the dependence of the tunneling current on the underlying density of states. In the WKB approximation, with $g=1$, the current is independent of the density of states in a band but vanishes beyond the band edges. This result only partially explains (accepting the picture of Fig. 2.19, with $\Delta E > 0$) the observations of Gleich, Regenfus, and Sizman (1971) of spin-polarized field emission from the (100) plane of ferromagnetic Ni in which the polarization P is found to be only -10%, with the preferred emission of electrons of minority-spin direction. This result is qualitatively consistent with Fig. 2.19 but quantitatively inconsistent with the expected $P = -100\%$. The spin polarization P is defined in terms of the spin-polarized current densities $J^{\pm}$ as

$$P = \frac{J^{+} - J^{-}}{J^{+} + J^{-}} \tag{2.83}$$

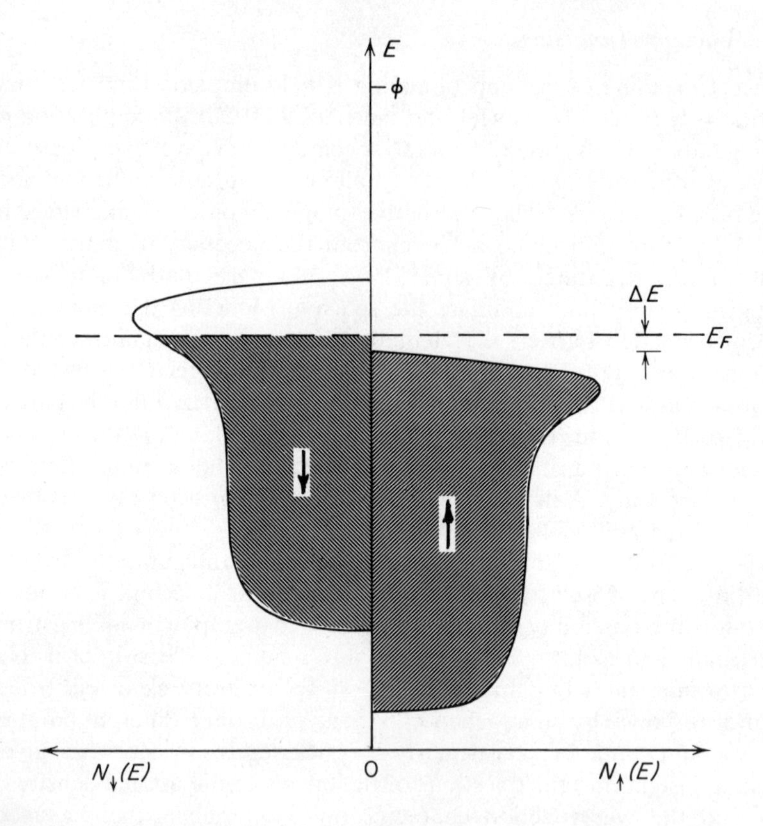

Fig. 2.19. Simple d-band model of a ferromagnet. The majority spins reside in a filled band slightly below the Fermi energy.

where the superscript + indicates an emitted electron of the same polarization as the majority spins within the ferromagnet. The interpretation of these and related results in light of an earlier calculation by Politzer and Cutler (1970*a*, 1970*b*) provides insight into the basic nature of tunneling from a metal. The work of Politzer and Cutler concerns both the magnitude of the predicted d-band spin polarization, much greater than observed, and also the relative tunneling probabilities of s- and d-band electrons. The calculations of Politzer and Cutler predict that the polarization due to the 3d bands alone is −80%, to be compared with the −10% observed. The implication is that a larger fraction of the tunnel current must come from 4s–p electrons (not shown in Fig. 2.19) than from the 3d electrons, which are spin-polarized. Analysis of this and similar situations then leads (Plummer, 1975) to two definite conclusions: (1) the emitted current sensitively depends, through prefactors, upon the density of states at the Fermi level to provide the observed partial polarization, and (2) the character of the wavefunction, by its effect on the tunneling matrix element, has a strong influence, in addition to the density of states, on the probability of tunneling. The additional fact that the polarization magnitude, and even its sign, varies with crystalline face of the ferromagnetic tip indicates that tunneling can be a sensitive probe of the electronic states in a normal metal. We have initially discussed single-crystal field emitter tips, rather than experiments involving oxide barrier junctions, because of the simplicity of the field emitter. We will return below to extensive measurements on tunnel junctions, which have given clear evidence of spin polarization, and to the further questions involved in comparing results to those on field emitters.

2.4 Dependence of J(V) and G(V) on Band Structure and Density of States

Because of the interest in measurements of "the density of states" in normal metal tunneling, we briefly summarize what has been said on this question. Returning to (2.5), (2.6), and (2.44)–(2.46), giving the basic expressions for the tunneling current, recall that the factor v_x in (2.5), which is the inverse of a one-dimensional density of states, disappeared from the current in (2.6) in changing the integration variable from k_x to the energy E_x. Viewing expression (2.6), one sees that there are then two ways in which the band structure and density of states can enter: through prefactors of D and through the influence of the shape of the Fermi surface on the integrations to obtain $J(V)$.

2.4.1 Fermi Surface Integrals

Consider the integration at fixed k_x over transverse wavevectors k_t, which can also be recast as (2.44) in terms of integration over the transverse component E_t of kinetic energy. In connection with (2.44), recall that, at a bias V, the k space contributing to the current is a shell of width eV below the Fermi surface for a nominal positive bias as shown in Fig. 2.6 or on the outside of the Fermi surface for nominal reverse bias. If, for the moment, we overlook the effect of the transmission factor $D(E_x)$, which, in fact, strongly weights the contributions of largest longitudinal wavevector k_x, then $J(V)$ can be regarded as proportional to an average transverse cross-sectional area (i.e., πk_t^2, the shaded regions in Fig. 2.6) over the shell of varying width eV above or below the Fermi surface. The reason that the average should characterize only the shell of thickness eV, rather than the full Fermi surface, is that we assume tunneling into a metal counterelectrode, which can occur only from those k states matching the k_t and E of empty states in the counterelectrode. This restriction does not apply to vacuum tunneling, where the average would be over the whole Fermi surface. In the same approximation, neglect of the transmission factor, the incremental current, i.e., the *conductance* at voltage V, is a measure of an average transverse *radius* k_t of an energy shell displaced by eV from the Fermi surface, since the increment in the k-space area in the direction transverse to the x or tunneling direction varies as $2\pi k_t\, dk_t$. These considerations are usually strongly modified by the E_x dependence of the transmission factor, typically such as to increase by a factor of e in an increase of $E_0 \simeq 100$ mV. If, on the other hand, the variation of the transmission factor is taken as the *dominant* influence, selecting only states of *maximum* E_x at $E_x = eV$, then the conductance at voltage $V(V<0$ in Fig. 2.6) will measure the rate of increase of cross-sectional area of a slice through the Fermi surface perpendicular to the x or tunneling direction at the maximal energy eV above the Fermi surface. In this case, the variation of D evaluated at $E_x = eV$ would have to be divided out of the measured conductance in order to extract the variation of the cross-sectional area in question. It is evident from these considerations that the spectroscopy involved is rather weak, revealing only averages of the Fermi surface properties.

We now turn briefly to a special case in which the spectroscopy available through the transverse integration becomes sharper and the result is not distorted by the variation $D(E_x)$: the case of *two-dimensional* final states in which the allowed k_x values are discrete. Such states are bound states for motion in the x direction. The tunneling electrons in this case come only from disclike areas transverse to the k_x direction and at specific allowed values of k_x. This result occurs, for example, if the electrode is thin, with ideal specular boundary conditions such that the states within the electrode are space-quantized, in effect, standing wave bound states across the electrode. In a free-electron picture, the total kinetic energy of such electrons specified by a given bound-state wavevector, say $k_x = k_B$, and a variable transverse k_t will be

$$E = E_t + E_B \tag{2.84}$$

where in the simplest case

$$\begin{aligned} E_t &= \frac{\hbar^2 k_t^2}{2m_t} \\ E_x = E_B &= \frac{\hbar^2 k_B^2}{2m} \end{aligned} \tag{2.85}$$

In this case, the transmission factor is identical for all E_t, if one neglects any V-dependence of $D(E_B)$. Alternatively, one may say that the transmission is determined by the *bound-state* wavefunction $\psi(x)$, which is independent of E_t. Hence, the conductance dJ/dV is a direct measure of the density of states $\rho_t(E_t)$ for transverse motion and, in the configuration we have been discussing, will be given by

$$\frac{dJ}{dV} \propto \rho_t(\mu - eV - E_b)D(E_B) \tag{2.86}$$

This expression can also be arrived at by (2.44), with the restriction that E_x has only one value, namely E_B, so that integrations over E_x disappear.

This interesting case has been observed in thin metal films, as mentioned, and also in two-dimensional states occurring at a semiconductor surface under so-called accumulation conditions. We will return to these cases in Chap. 8.

2.4.2 *Prefactors: Wavefunction Matching at Boundaries*

In addition to the density-of-states information which comes through the transverse integral in (2.7), information may be available through prefactors dependent upon k in the transmission factor D. While such prefactors are absent in the simplest WKB approximation which we have assumed in the discussion of Sec. 2.4.1, comparison with exact matching calculations suggests that a prefactor proportional to k_x may more generally belong in the transmission factor D, even in the WKB limit, as is argued by Mahan and Conley (1967) and supported by the considerations of Feuchtwang (1970). There is some evidence for the view that proportionality to k_x is a general property of the transmission factor, even in the WKB limit. The possibility of further modification of the dependence beyond simply k_x as a consequence of detailed interface conditions also exists. A series of model calculations stimulated primarily by observation of transmission resonances in field emission experiments (see also Appelbaum and Brinkman, 1970) has led to the view that the effect of the transmission resonances can be formally recast as observation of a local or surface density of states (Plummer and Young, 1970; Penn and Plummer, 1974). The physical origin of the effects was first discussed by Duke and Alferieff (1967).

In the work of Plummer and Young (see also Gadzuk, 1970), a model calculation based on field emission through a triangular surface barrier supports the identification

$$\frac{J'(E)}{J_0'(E)} \propto |\psi(z=0)|^2$$

where $J'(E) = \mathrm{d}J/\mathrm{d}E$, with $J(E)$ the field-emitted current density. The subscript in $J_0(E)$ denotes the result of a free-electron calculation, while $J(E)$ incorporates the effect of an attractive square well potential at the surface. The resultant enhancement of J can be correlated with enhancement of the wavefunction in the potential well in the manner indicated. In a more complete discussion of this topic, Plummer and Young (1974) find a formal expression for the field emission current per unit energy at energy E:

$$J'(E) \simeq \frac{2\hbar}{m} \lambda^{-2}(E) \sum_m D_0^2(E_x^m)\, |\psi_m(x_m)|^2\, \delta(E-\varepsilon_m)$$

Here, D_0^2 is the barrier penetration probability including image corrections, $E_x = E - E_t$, $\psi_m(x_m)$ is the metal wavefunction at the classical turning point x_m, and $\lambda(E)$ is a slowly varying function of E. These authors state that the D_0^2 factor strongly weights electron states with small k_t, so that $J'(E)$ "measures the density of states at x_m arising from the component of the bulk band structure normal to the surface." The product $|\psi|^2/\lambda^2$ here is equivalent to the preexponential factor $P(E, k_t)$ appearing in the paper of Politzer and Cutler (1970*b*). A summarizing discussion of this area is given by Plummer (1975).

Extensions of junction tunneling to single-crystal oriented electrodes are found in recent work of Meservey, Tedrow, Kalvey, and Paraskevopoulos (1979), who have measured the spin polarization from oriented nickel electrodes and made comparison with results from field emission.

2.5 Nonideal Barrier Transmission

In the context of the conventional, two-electrode, Giaever tunnel junction configuration, typically applied to the study of superconductors, it may be useful to discuss the various effects that an imperfect barrier may have upon tunneling spectra, specifically confusing their interpretation in terms of electrode properties alone. Considering the requirements on the tunneling barrier—i.e., that it be of uniform thickness in the 10–20-Å range, be free of pinholes, and be able to withstand at least 0.1-V bias, which, however, may correspond to an electric field of 10^8 V/m—it is remarkable that nearly ideal behavior is ever achieved.

2.5.1 Approach to Ideal Behavior

Perhaps the simplest test for ideal behavior is afforded in the Giaever experiment under biases less than the energy gap of the superconductor making up one of the electrodes. In this situation, the current should be zero for $V<\Delta/e$ and should rise (at very low temperatures $T \ll \Delta/k_B$) rather abruptly to an ohmic current at $V=\Delta/e$. The most obvious source of *leakage* current, i.e., nontunneling current observed at voltages $eV<\Delta$, is a microshort, possibly caused by a pinhole in the oxide film, filled in during the evaporation of the counterelectrode. Such defects can easily result from small foreign particles, for example, on the film of metal being oxidized to form the barrier. In practice, reduction of leakage currents to less than one part in 10^3 is common, and there have been reports of leakages as low as 10^{-6}, as in Hubin and Ginsberg (1969) using aluminum oxide barriers no more than 30 Å thick. While this test is the best single test for desirable barrier behavior, as we will see, the proof of tunneling as the only mechanism of current flow is only established rigorously for voltages less than Δ, typically a millivolt, while even in low-energy phonon spectroscopy of superconductors, one will require biases of at least 20 mV. How, one may ask, can we be sure that tunneling is still the only transfer mechanism at 20 mV, given that it is the only mechanism operative below 1 mV? Unfortunately, there is no definitive answer to this question. Ideal tunneling behavior provides a slow parabolic increase (possibly offset from $V=0$) in conductance with bias. Thus, if the junction in the normal state shows such a conductance in addition to showing low leakage below $eV=\Delta$ when the electrode is superconducting, then tunneling is reasonably established as the mechanism, even to considerably higher biases.

A rather frequent observation is that of a conductance which has a minimum at zero bias but rises much more steeply, for bias of either sign, than the behavior we have considered earlier. It is not, in this case, obvious that the barrier is in fact defective; the barrier may simply have a much lower barrier height, perhaps 0.2 V rather than 2 V. In this case, in measurements at biases up to 20 mV, common in superconducting tunneling, no error would be made in reducing the data in the usual fashion. However, a conductance rising sharply from $V=0$ may also be the result of microscopic defects, e.g., magnetic impurities at one edge of the barrier, and the effects in this case on the superconducting tunneling measurements may be more serious.

An obvious question about an oxide tunneling barrier concerns its uniformity in thickness. An elegant answer to this question in the case of oxides grown on tin has been provided by Dynes and Fulton (1971) through measurements of the magnetic field dependence of the critical Josephson supercurrent, which ideally varies as the Fraunhofer diffraction function vs. applied field. The results of their analysis, which will be described later, are shown in Fig. 2.20, in which the current density $J(x)$ as a function of position in the junction is shown. Within the spatial resolution of the method, which is estimated as $0.04a$, where a, the lateral dimension, was 0.22 mm, the current density is constant to within about 20% over the junction, falling off at its edges. The latter behavior may be a consequence of plasma discharge growth of the oxide in this case. The main point, however, is the relative constancy of the tunneling current density J. Since J is expected to be an exponential function of the local thickness, $J=J_0e^{-2\kappa t}$, the fractional variation is $\Delta t/t=(\Delta J/J)/2\kappa t$. This result indicates a variation in thickness of the order of a few percent, if $2\kappa t$ is on the order of 10. Related measurements of the tunnel current density J as a function of the oxide thickness, measured ellipsometrically, are shown in Fig. 2.21 (after Eldridge and Matisoo, 1971). In these measurements on Pb–PbO–Pb junctions, the current varies exponentially with thickness, as expected, and can be fit by $2\kappa=0.67\ \text{Å}^{-1}$. Such an observation is itself a good indication of ideal barrier behavior. Such a law would not be obeyed by a leakage current or by thermionic current above a barrier.

Another important question about an oxide barrier is the sharpness or abruptness of the interface between the underlying metal and the growing oxide. The second interface, namely that between the grown oxide and the deposited counterelectrode, would be expected to be relatively sharp as long as the grown oxide is not subjected to undue heat and the

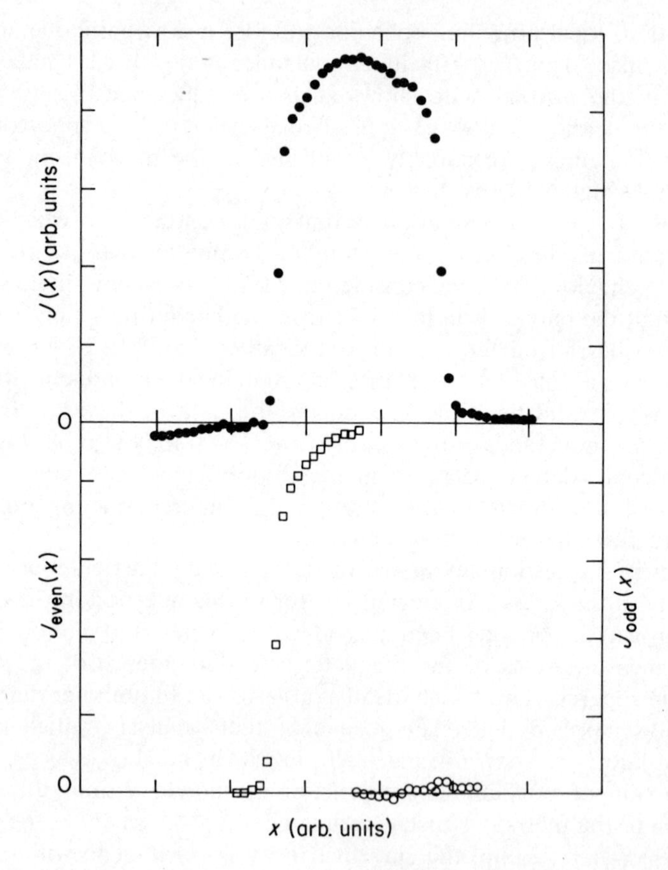

Fig. 2.20. Current density profile $J(x)$ deduced from Josephson tunneling measurements of Sn–oxide–Sn junction at 1.46 K. (After Dynes and Fulton, 1971.)

evaporating metal atoms have only typical thermal energies, to avoid diffusion or penetration of metal into the oxide. There is no definite answer to the question of interfacial sharpness, and indeed, a variety of degraded tunneling characteristics, especially in transition metal systems, may well be due to gradations at the metal-oxide interface. In the early literature of aluminum–aluminum oxide–aluminum tunneling, Fisher and Giaever (1961) discuss this possibility. Examples of sharp interfaces are also known, and we will describe two cases in which sharpness of the metal-oxide interface on the order of a typical Fermi wavelength, $\lambda_F = 2\pi/k_F$, can be inferred from tunneling characteristics. The first case involves a technologically important oxide interface, that occurring between silicon and thermally grown silicon dioxide in a metal-oxide-semiconductor (MOS) junction, shown in Fig. 2.22 (after Maserjian, 1974). Here, the tunneling structure is Cr–SiO_2–Si (p type), indicated under bias eV leading to electron flow into the p-type Si. It can be seen

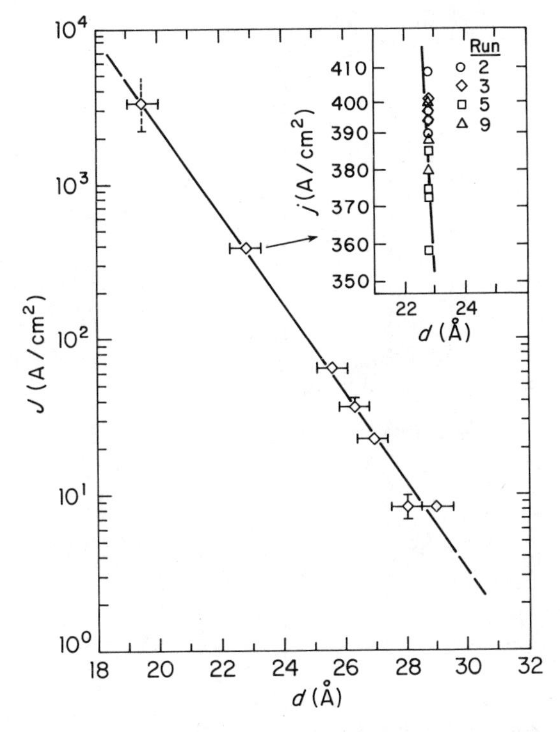

Fig. 2.21. Logarithmic plot of the Josephson current density J_c vs. barrier thickness t in a Pb–PbO–Pb tunnel junction. (After Eldridge and Matisoo, 1971.)

from the parameters in the figure that at metal positive bias V exceeding 4.2 V, electrons from the Fermi energy of the Cr will enter the conduction band of the SiO_2; the conditions then exist for Fowler-Nordheim tunneling, sketched in the lower portion of Fig. 2.22. Under these conditions, an oscillatory wavefunction occurs in the SiO_2 conduction band over a (bias-dependent) portion of the SiO_2 width. A portion of this wave will be reflected at the interface between the SiO_2 and the Si. This situation will lead, if the Si–SiO_2 interface is suitably abrupt, to interference of the reflected and incident waves, which, in turn, can modify the current-voltage relationship of the overall structure. From Maserjian (1974), the effect of such coherent reflections from the SiO_2–Si interface, which occurs only in an appropriate voltage range, is to multiply the transmission factor D by an oscillatory factor b,

$$b = \left[Ai^2(-\alpha x_1) + \left(\frac{\alpha}{k}\right)^2 Ai'^2(-\alpha x_1) \right]^{-1} \tag{2.87}$$

where Ai and Ai' are the Airy function and its derivative, the field- (voltage) dependent parameter x_1 is given by

$$x_1 = w - (\phi_m + E')/eF \tag{2.88}$$

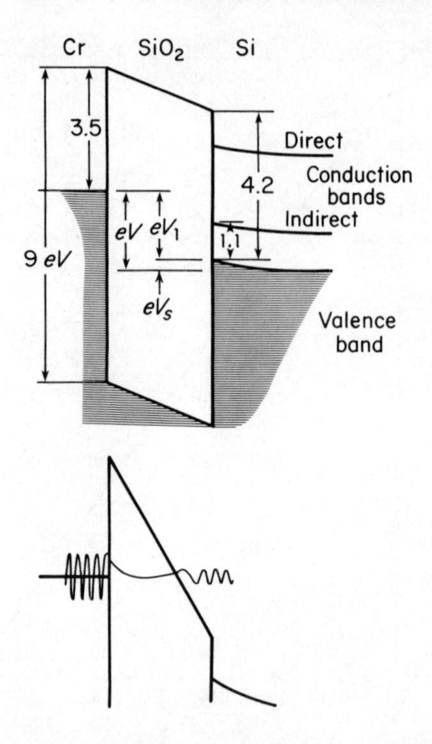

Fig. 2.22. Diagram of Cr–SiO_2–Si diode structure and typical electron wavefunction under bias conditions for Fowler–Nordheim tunneling. Interference effects in the tunneling current are possible in this configuration if the silicon dioxide–silicon interface is sufficiently sharply defined. (After Maserjian, 1974.)

and α by

$$\alpha = \left(\frac{2m^*eF}{h^2}\right)^{1/2} \tag{2.89}$$

Here, m^*, F, and w, respectively, are the effective mass, electric field, and overall width in the SiO_2, and ϕ_m is the metal work function. Term E' is the energy of the electron in question measured below the metal Fermi energy, and k is the electron wavevector in the silicon direct conduction band. These equations, when incorporated into the usual Fowler-Nordheim tunneling expression,

$$J = \bar{b}B_0F^2 \exp \frac{-2\kappa\phi_m}{eF} \tag{2.90}$$

provide a modulating factor $\bar{b}$, an oscillatory function of the field strength F, obtained from (2.87) by integrating over E', representing the initial state energies in the Cr. Finally, the effect is most conveniently displayed by dividing J by b_0F^2 to isolate the oscillatory term $\bar{b}$. The log of this quantity is shown in Fig. 2.23 plotted against electric field F. A more complete account of the theory involved has been given by Gundlach (1966). The important point here is that good agreement is obtained between the experimentally observed oscillations and a theory assuming a sharp oxide-silicon interface. The voltage range is such that, at the largest field shown, the energy of the electron in the SiO_2 conduction band is about 2 V, corresponding, with an effective mass of 0.85, to an electron wavelength of about 9 Å. Since the electron of 9-Å wavelength is reflected perfectly (i.e., its reflection is accurately predicted on the assumption of a perfectly sharp interface), the width of the interface may be assumed to be much less than 9 Å.

A second example, which we will mention only rather briefly here and return to in Chap. 8, involves the observation of space-quantized standing

Fig. 2.23. Oscillatory behavior in the current density vs. applied field, in volts per angstrom, in the Cr–SiO_2–Si structure shown in Fig. 2.22. Observation of such oscillations is a stringent test of sharpness of the silicon–silicon dioxide interface. (After Maserjian, 1974.)

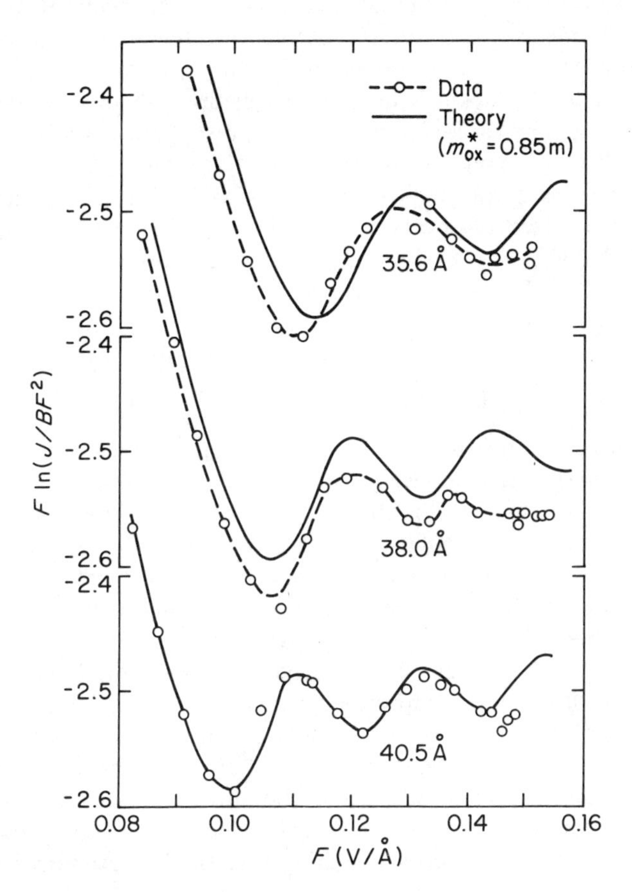

waves across thin films (Jaklevic and Lambe, 1975). Well-defined oscillatory effects, implying electron waves specularly reflected by sharp metal film boundaries, are observed for Pb, Ag, Au, and Mg metal films. It is clear, at least in the case of Mg, that such a film will have an oxide layer; hence, observation of the oscillatory effects implies that the Mg–MgO interface must be sharp—again, on the scale of the Fermi wavelength, on the order of 10 Å or less.

We have seen two examples both involving oxides grown thermally on a deposited metal film in which evidence of uniform oxide thickness and a well-defined metal-oxide boundary can be given. Certainly, in many cases, such ideal behavior is not present. For example, when fluctuations in barrier thickness exist, the exponential decrease of the transmission factor with thickness leads to a heavy weighting of the overall current to regions of minimum barrier thickness. Such an effect has been identified in Schottky barriers at heavy doping, where statistical fluctuations in the number of ionized impurities lead to significant local variations in the thickness and transmission of the barrier (Wolf and Losee, 1970). Fluctuations in the thickness of barriers produced by deposition are generally expected to be more substantial than those of thermally grown oxides, because no means exist for smoothing out fluctuations arising from random nucleation of depositing atoms into islands on the substrate. Reduction of such thickness fluctuations in the case of *deposited* Al_2O_3 barriers, using electron beam evaporation of sapphire, is achieved in the recent work of Moodera, Meservey, and Tedrow (1982) by cooling the substrate to 77 K before deposition of Al_2O_3. Such artificial barriers are reported to be free of shorts only in average thicknesses estimated as between 14 and 24 Å. Cooling the substrate presumably improves the uniformity of thickness by increasing the density of nucleation sites and by inhibiting the growth of grains.

2.5.2 *Resonant Barrier Levels*

One circumstance in which the barrier can have a profound effect upon measured energy dependences, rather than in simply establishing the scale of measured current, is that in which the barrier has one or more internal resonant levels. An example is indicated in Fig. 2.24a in which a one-dimensional barrier includes an internal, square well attractive potential of width W. The meaning of the two resonance levels, indicated by the dashed lines, is that if the thickness B of the barriers in the figure were made infinitely large, then the dashed lines would represent true bound states of an electron in the attractive square well. For finite values of the barrier width B, however, an electron placed in one of the states of the square well can tunnel out in a finite lifetime. If this time is short one finds only transmission resonances, rather than bound states internal to the barrier. When the bias is such that the electron tunneling from one of the electrodes has an energy matching one of the resonant levels, the

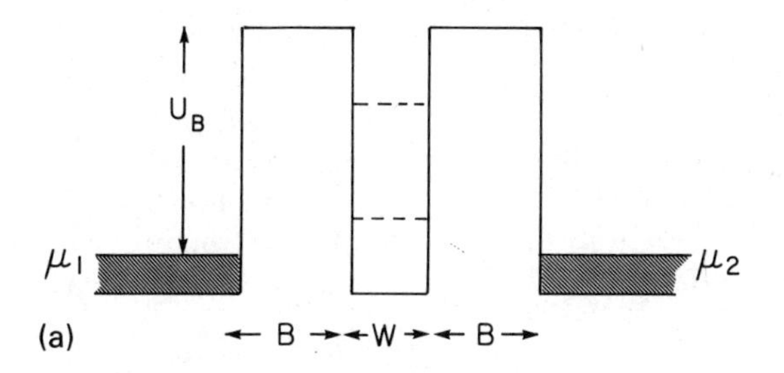

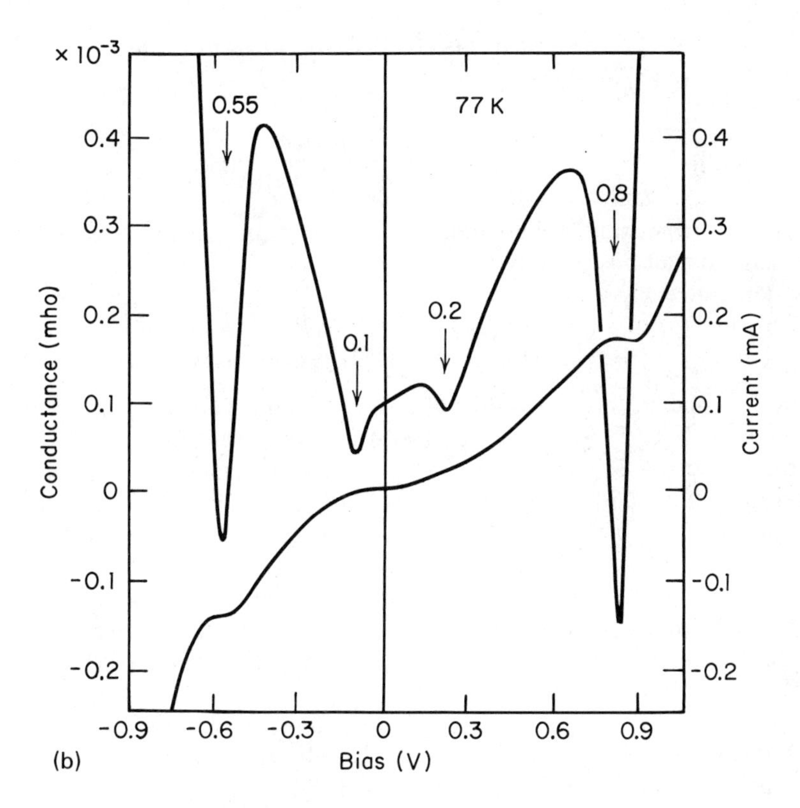

Fig. 2.24. (a) Diagram shows barrier with internal potential well, of width W, capable of supporting two bound states in limit of large B. (b) Tunneling conductance and current of an artificial three-layer barrier structure supporting internal resonances, or partially bound states, at 0.1 and 0.5 eV and leading to strong structure in the measured conductance at these voltages. The barrier is composed of 50-Å layers of GaAs and $Ga_{0.3}Al_{0.7}As$. (After Chang, Esaki, and Tsu, 1974.)

transmission of the barrier can be increased by orders of magnitude. The transmission may even become unity, as in the Ramsauer-Townsend effect, for example, or in the passband (conduction band) of a Kronig-Penney lattice. From Kane (1969), the ratio of transmitted to incident current through a potential barrier such as shown in Fig. 2.24a, with regions labeled 1–5 from the left side of the structure, can be calculated, assuming an initial wave $\alpha_1 e^{ik_1x}$ and final wave $\alpha_5 e^{ik_5x}$, as

$$\frac{j_t}{j_i}=\frac{|\alpha_5|^2 k_5}{|\alpha_1|^2 k_1} \tag{2.91}$$

where

$$\alpha_1=(\mathbf{R}^1\mathbf{R}^2\mathbf{R}^3\mathbf{R}^4)_{11}\alpha_5 \tag{2.92}$$

The matrix **R** has been defined in (2.24). The result of this analysis is

$$\frac{j_t}{j_i}=\frac{2^8 k_1\kappa_2^2 k_3^2\kappa_4^2 k_5/K}{(k_1^2+\kappa_2^2)(\kappa_2^2+k_3^2)(k_3^2+\kappa_4^2)(\kappa_4^2+k_5^2)} \tag{2.93}$$

where the dominant exponential terms occur in K, which is in the denominator and is given by

$$K=f_1\exp(\kappa_2 W_2+\kappa_4 W_4)+f_2\exp(\kappa_2 W_2-\kappa_4 W_4)$$
$$+f_3\exp(-\kappa_2 W_2+\kappa_4 W_4)+f_4\exp(-\kappa_2 W_2-\kappa_4 W_4) \tag{2.94}$$

Here, W_2 and W_4 are the widths of barrier regions 2 and 4, denoted B in Fig. 2.24a. It is shown by Kane (1969) that the prefactor f_1 of the first (largest) term in K goes to zero when the bound-state condition in the interior well is fulfilled. Under these conditions, the next largest terms in K, the second and third, limit the transmission. However, both of these terms represent ratios of the transmission of the individual barriers (regions 2 and 4) and hence can be adjusted to unity by making $\kappa_2 W_2=\kappa_4 W_4$, i.e., making the structure symmetric. Under these conditions of resonance in the interior attractive well and of symmetry of the outer barrier, unit transmission through the whole structure should occur. The treatment is idealized in being strictly one-dimensional and also neglects distortion of the barrier by the applied bias voltage. Nevertheless, vestiges of this effect have been observed in the I–V and conductance characteristics of multilayer barrier structures such as shown in Fig. 2.24b (after Chang, Esaki, and Tsu, 1974). The arrows above the upper dI/dV curve locate the biases for transmission through the two resonant levels indicated in the inset.

An important example of transmission resonance occurs in field emission from single-crystal tips with adsorbed atoms which can provide a resonant level for electrons and enhance tunneling through the triangular work function barrier (Duke and Alferieff, 1967; Clark and Young, 1968; and Plummer and Young, 1970). Large enhancements have been reported in the latter two references with adsorption of Sr atoms on W points; it has been demonstrated that a single Sr atom can be detected. In general, metallic adsorbates lead to resonances in the transmission probability whose widths are on the order of 1 eV. The peak enhancements of the emission current are factors ranging from 10^1 to 10^4. Neutral adsorbate potentials, on the other hand, lacking bound states, lead to reductions in both the emission probability and the current. This feature affords a useful means of locating levels associated with adsorbed atoms on field emitter tips. For a review of this work, the reader is referred to Plummer (1975).

One expects an important connection between the *dimensionality* of the potential in which the bound state exists and even qualitative features of the energy dependence of the enhancement. In a strictly one-dimensional case, the enhancement of the transmission factor at a resonance is approximated by a Lorentzian peak at the energy of the resonance:

$$|T|^2 \simeq \frac{\Gamma^2}{(E_x - E_{x0})^2 + \Gamma^2} \tag{2.95}$$

where Γ, the half-width of the peak in the transmission factor, is associated with the rate of decay of a wave packet prepared in the well. In a hypothetical, purely one-dimensional tunneling experiment involving a symmetrical barrier containing a bound state at energy E_{x0}, a true peak in conductance would occur at a bias $eV = 2E_{x0}$, the factor of 2 appearing because the potential well occurs at the center of the barrier, where the energy is shifted by only half the applied potential. Of course, one-dimensional experiments are not possible, and a real experiment would be more like that illustrated in Fig. 2.24, in which motion in the two dimensions transverse to the tunneling direction would ideally be unaffected by the potential. Thus, it is the maximum value of the *total* energy $E = E_x + E_t$ which is set by the bias eV. In the symmetrical case, when the total energy is $E = eV/2$, those tunneling electrons directed in the x direction—i.e., with $E_t = 0$, $E_x = E_{x0} = eV/2$—will first satisfy the resonance condition. However, for larger values of bias, an increasing number of electrons, propagating slightly off normal incidence, i.e., with increasing values of E_t, will have E_x values matching the condition for resonance. Since the transmission factor depends only upon E_x, the resonance contribution in the conductance for biases above the threshold will not fall off with E_t. This result should have a profound effect upon the line shape observed in the conductance. To a first approximation, this effect would convert the Lorentzian peak mentioned earlier into an upward step or threshold behavior, whose details depend upon the transverse k integration.

This discussion does not apply to the adsorbed atoms on the field emitter tip, for the atomic potential wells are not extended in the transverse directions but are inherently three-dimensional, although embedded in a work function barrier which, in the first approximation, is two-dimensional. The current enhancements produced by adatoms on field emitter tips have definite peak energies and widths, as if the transmission through the adatom were so great as to make negligible contributions from other portions of the barrier. Indeed, it has been mentioned that a single Sr atom can be observed by its increase in the current. This limit may essentially reduce the problem to the simplified case treated by Duke and Alferieff (1967). Still, it would be desirable to extend this treatment to fully include the three-dimensional aspects of tunneling.

2.5.3 Two-Step Tunneling

In junction devices, the tunneling barrier is an oxide or other insulator. A typical defect of such real materials is a trap, i.e., a state in which an electron can be localized. The effect of such states on tunneling spectra was first identified by Parker and Mead (1969) in studies of Schottky barrier tunneling on n-type CdTe. Their observations are shown in Fig. 2.25 in a logarithmic plot of current density J as a function of forward-bias voltage V. As can be seen from Fig. 2.14, under forward bias of the Schottky barrier, the number of electrons eligible to tunnel into the metal

Fig. 2.25. Logarithm of current density vs. applied voltage for Schottky barrier contacts on n-type CdTe of varying donor concentrations. The change in slope in these curves is interpreted as a change from tunneling through the entire barrier (upper portion) to tunneling in two steps via an intermediate impurity state (lower portion). (After Parker and Mead, 1969.)

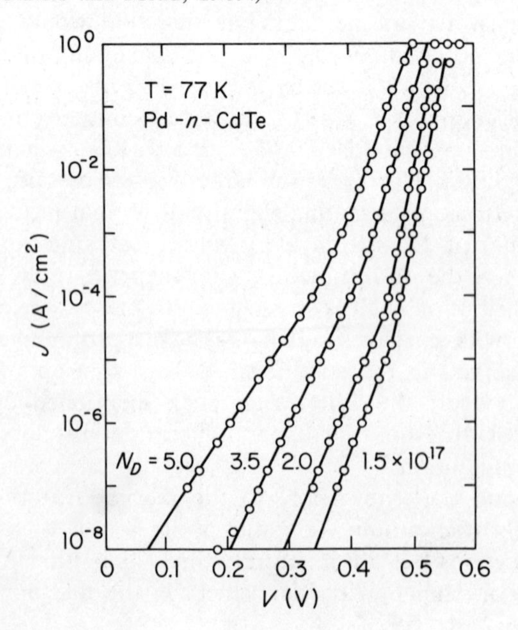

becomes constant as soon as the semiconductor band edge is raised above the Fermi level of the counterelectrode metal. Beyond this point, increase in current is due only to increase in the barrier transmission factor. In the Schottky barrier case, the height and width of the barrier seen from the semiconductor side both decrease with forward-bias eV. The curves shown in Fig. 2.25, thus, are basically plots of the transmission factor of the Schottky barrier as a function of bias. The interesting feature of these particular curves is the change in the logarithmic slope (by a factor of 2), which occurs at voltages which are dependent upon the doping level (and hence, barrier width) in the semiconductor. The higher-bias portions of these curves with the steeper slopes are shown by Parker and Mead (1969) to correspond to the usual Schottky barrier transmission, Eq. (2.96),

$$D \simeq \exp\left[-2\int_0^t \kappa(x)\,dx\right] \tag{2.96}$$

characterized by a parameter E_0 given by (2.74). The lower-bias portions of the curves are found to be characterized by a parameter which is just *twice* E_0. These measurements are interpreted in terms of a process in which electrons tunnel through the barrier in two steps, forming a real intermediate state on a trap inside the barrier at coordinate x_1. Assuming this process, the relevant transmission factors for steps 1 (into the trap) and 2 (out of the trap) are given, respectively, by Eqs. (2.97) and (2.98):

$$D_1 = \exp\left[-2\int_0^{x_1} \kappa(x)\,dx\right] \tag{2.97}$$

$$D_2 = \exp\left[-2\int_{x_1}^t \kappa(x)\,dx\right] \tag{2.98}$$

Following Parker and Mead (1969), one can write the corresponding rates w_1 and w_2 for tunneling in and out of the trap states; these must depend upon the occupation probability f_1 of the trap:

$$w_1 \simeq N_t(1-f_1)D_1 \qquad w_2 \simeq N_t f_1 D_2 \tag{2.99}$$

Here, N_t represents the concentration of trap states, assumed uniformly distributed in position and energy. For a steady state, w_1 must equal w_2, achieved by adjustment of the occupation probability f_1 to the value

$$f_1 = \frac{D_1}{D_1 + D_2} \tag{2.100}$$

The rate of the two-step process is then

$$w = w_1 = w_2 = \frac{N_t D_1 D_2}{D_1 + D_2} = \frac{N_t D}{D_1 + D_2}$$

making use of $D_1 D_2 = D$ from (2.96)–(2.98). The resulting two-step tunneling rate w is maximum when the transmission probabilities D_1 and

D_2 are equal, so that

$$w_{\max} \leftrightarrow \sim\tfrac{1}{2}N_t D^{1/2} \tag{2.101}$$

Hence, the maximum rate is determined by the square root of the original rate, (2.96), which explains the change in logarithmic slope noted by Parker and Mead and evident in Fig. 2.25. Summarizing the situation treated by Parker and Mead, assuming that the distribution of trap states is uniform in energy and position, we find that the most important of these traps are situated in the barrier such that the tunneling probability to either side is equal. The contribution to the tunneling current, proportional to the number of traps, is governed by a transmission factor approximately equal to that for the individual transition from the trap to either side. The relative importance of two-step tunneling thus increases with the thickness of the overall barrier.

From the point of view of tunneling spectroscopy, such processes reduce the energy resolution. The electrons which tunnel from the traps into the opposite electrode may have an energy distribution which is influenced by the energy distribution of the traps. For example, if the important traps occur near the Fermi energy at the center of the barrier, the most energetic tunneling electrons will give up an energy $eV/2$ in the trapping process. After the second tunneling step to the opposite electrode, the maximum energy of such carriers is $eV/2$, rather than eV, for direct tunneling. Tunneling spectroscopy is not possible under these conditions.

We now consider the connection between real intermediate state or two-step tunneling and the resonant tunneling discussed in the previous section. To clarify the difference between these cases, consider, first, the decay of an electron bound in a trap which lies above the Fermi energy in an unbiased junction. The probability of finding this electron in the trap will decay exponentially with time according to a mean lifetime τ, after which the electron can be found, with roughly equal probabilities, in either of the two electrodes. In the resonant tunneling situation, on the other hand, the wavefunction is oscillatory on one side of the junction, decays exponentially into the barrier, peaks up in the vicinity of the potential well, and then continues to decay towards the distant side of the barrier. A discussion of this distinction in a one-dimensional context has been given by Baym (1969), who considers the effect of a resonant level, of energy width Γ, upon the propagation of a wave packet incident on the barrier from the left. An expression for such a packet is

$$\psi_{\text{inc}}(x, t) = \frac{1}{2\pi}\int_0^\infty e^{i(kx-\omega t)} f(k)\, dk \tag{2.102}$$

where the function $f(k)$ is sharply peaked at $k = k_0$, which is related to energy or frequency ω_0 by

$$k_0 = \frac{\sqrt{2m\omega_0}}{\hbar} \tag{2.103}$$

The function $f(k)$ is characterized by a width $\Delta k \simeq 1/\Delta x_p$, which is inversely proportional to the extent of the wave packet in real space. Equation (2.103) makes clear that the plane waves making up the packet correspond to a spread of energies $\Delta\omega$. In Baym's discussion, the equivalent energy width of the packet, in comparison with the energy width Γ of the resonance, is a determining factor. The transmitted wave packet is given by

$$\psi_{\text{trans}}(x, t) = \frac{1}{2\pi} \int e^{i(kx - \omega t)} T(\omega) f(k)\, dk \tag{2.104}$$

The transmission function as a function of energy near a resonance is of the form

$$T(\omega) \simeq \frac{i\Gamma/2}{\hbar(\omega - \omega_0) + i\Gamma/2} \tag{2.105}$$

The form of the transmitted packet, (2.104), depends upon the relative sharpness of T and f considered as functions of ω. If T is a broad function, then the emerging packet has much the same shape as the incident packet. On the other hand, if T is sharp, and Γ much less than the equivalent energy width in the incoming packet, then a different behavior occurs, which can be estimated by taking the function of $f(k)$ out of the integral. This manipulation leads to

$$\psi_{\text{trans}}(x, t) \simeq \begin{cases} \exp\left[-\dfrac{\Gamma}{2\hbar}\left(t - \dfrac{x}{v_0}\right)\right] e^{ik_0(x - \omega_0 t)}, & x < v_0 t \\ 0, \quad x > v_0 t,\ v_0 = \dfrac{\hbar k_0}{m} \end{cases} \tag{2.106}$$

which is written for time t after the arrival time of the packet at the well, and the coordinate x is measured from the edge of the well. In this expression, an oscillating wave, typical of the center of the packet, is multiplied by an exponential horn function which opens towards positive x and, for fixed $x = 0$, reveals an exponentially decaying amplitude at the well. Hence, the probability of finding the particle at the well decays as $e^{-\Gamma t/\hbar}$, which corresponds exactly to an exponential decay law for a trapped particle. While it is unfortunately not entirely clear how to apply this criterion in the context of junction tunneling, it would appear reasonable to take the width Δk of the wave packet, (2.103), to correspond approximately to the inverse of the mean free path ℓ in the metal electrode: $\Delta k \simeq 1/\ell$.

A slightly different approach, based on the properties of localized states in the barrier, may also be useful. The wavefunction of a localized state in the barrier, assumed to be infinitely extended, will also have a form similar to (2.102) (with $t = 0$); $f(k)$ specifies the range of plane wave states necessary to compose the localized state. The range of k involved can again be expressed in terms of the energies; the more localized the state, the larger is the range of energies involved. Consider now the effect of reducing the width of the barrier such that the localized electron can tunnel out into either electrode in a time $\tau = h/\Gamma$ [(with Γ given by an expression similar to (2.97) and (2.98)]. The effect on the localized wavefunction may be unimportant until Γ is of the same order as the equivalent energy width of $f(k)$: this will destroy the localized state, leading to resonant tunneling. This criterion can be restated as

$$\Gamma a^{*2} \ll \frac{\hbar^2}{2m^*} \tag{2.107}$$

(Wolf, 1975) from which it is seen that the formation of real intermediate states is favored by a small value of a^*, the wavefunction radius characteristic of a local state. Real intermediate states can strongly influence the voltage and temperature dependences of tunneling spectra, especially near zero bias. In important cases, it is known that there is a minimum energy U, required in order to place an electron into the real intermediate state. Thus, increased barrier transmission is available only for biases V such that $eV \gtrsim U$ or at temperatures T such that $kT \gtrsim U$. This result can lead to a minimum conductance at $V = 0$, the minimum value of conductance being dependent upon temperature. The result is a zero-bias resistance peak anomaly, and a carefully studied example will be discussed in Chap. 8. In the work of Giaever and Zeller (1969), real intermediate states form on small metal particles embedded in an oxide barrier; the minimum energy U required to localize the tunneling electron on the particle in this case can be estimated simply from electrostatic considerations.

2.5.4 *Barrier Interactions*

Of other forms of interaction of tunneling electrons with the barrier, probably the most important is interaction with the lattice vibrations. As we will see, this form can manifest itself as features in the tunneling spectra at characteristic barrier phonon energies. A tunneling electron may, if it has sufficient energy, excite a phonon of the barrier, losing energy $\hbar\omega$, and enter the opposite electrode at an energy $eV - \hbar\omega$. An additional case, similar in principle, can occur if the barrier happens to be magnetic oxide so that tunneling can occur with excitation of a spin wave. Variations on this basic inelastic process occur if the barrier contains impurity molecules with low-energy vibrational excitations. Examples are organic molecules at surfaces of the barrier. A further example is afforded by magnetic ions, in the barrier or at its edge, with which the tunneling electron can interact via its spin magnetic moment. These are examples of assisted tunneling processes, of which a more complete discussion is necessary.

2.6 Assisted Tunneling Processes

In direct elastic tunneling, as sketched in Fig. 2.1, the final state of an electron tunneling from the Fermi surface of the left-hand electrode is a conduction band state at energy eV above the Fermi surface of the right-hand electrode. In real systems, the tunneling electron may be able to interact with other modes, such as phonons or localized vibrations of the barrier or electrode. When sufficient energy is available to excite these modes, more complicated final states may have an appreciable probability. For example, if $eV = \hbar\omega_p$, where ω_p represents an oxide or interface phonon, a final state is possible in which one phonon is generated and the electron appears just at the Fermi surface in the right-hand electrode. From one point of view, the condition $eV = \hbar\omega_p$ represents the threshold for opening a new inelastic channel of tunneling. This condition represents an increased phase space of final states; hence, an abrupt increase in the tunnel conductance will occur at the threshold $eV = \hbar\omega_p$, beyond which the elastic and inelastic processes can both proceed. Since the threshold condition corresponds to an upward step in conductance, the derivative of the conductance, d^2I/dV^2, has a peak. Hence, one obtains in d^2I/dV^2 a spectrum of peaks at the various thresholds which resembles typical infrared absorption spectra for the vibrations.

While a steady-state theory of such processes is possible (Brailsford and Davis, 1970) and will be mentioned in Chap. 8, the most direct and convenient route to quantitative treatment of assisted tunneling lies in the transfer Hamiltonian approach. It is necessary to include in the Hamiltonian the interactions between the electrons and the various modes of the system which are important. A more general development of the transfer Hamiltonian method sketched briefly following (2.10) is necessary in order to include the assisted processes. It is appropriate to express the Hamiltonian involved in the second quantization notation, including all two-electron interactions. In this approximation,

$$H = \sum_i \frac{p_i^2}{2m} + \sum_i V(\vec{x}_i) + \frac{1}{2} \sum_{i \neq j} W(\vec{x}_i - \vec{x}_j) \tag{2.108}$$

$$H = H_0 + H_I \tag{2.109}$$

is the Hamiltonian for the entire tunnel junction. In second quantization, the Hamiltonian breaks up into two parts,

$$H = \int \psi^*(\vec{x}) \left[\frac{p^2}{2m} + V(\vec{x}) \right] \psi(\vec{x}) d^3\vec{x} + \frac{1}{2} \int \psi^*(\vec{x}) \psi^*(\vec{x}') W(\vec{x} - \vec{x}') \times \psi(\vec{x}') \psi(\vec{x})\, d^3\vec{x}\, d^3\vec{x}' \tag{2.110}$$

representing, respectively, one-electron and two-electron interactions, the latter represented by $W(\vec{x}_i - \vec{x}_j)$. The destruction operator $\psi(\vec{x})$ is expanded in terms of states ψ_i^a and ψ_i^b, which represent, respectively, complete sets of states on the left and right sides of the junction and overlap in the barrier region itself, as described in Sec. 2.2. Thus,

$$\psi(\vec{x}) = \sum_i a_i \psi_i^a(\vec{x}) + \sum_i b_i \psi_i^b(\vec{x}) \qquad (2.111)$$

with an analogous expression for $\psi^*(x)$. Note that the states created by $\psi^*(\vec{x})$ are not restricted to plane wave states but may be localized states, as are required in a theory of Appelbaum (1967) for tunneling assisted by magnetic scattering. In this case, which is probably the best example of the application of the transfer Hamiltonian theory to assisted processes, a single localized state ψ_d is assumed to exist near one edge of the barrier; the other electron states are taken to be plane waves in the two electrodes, represented by indices k and q, respectively. The destruction operator is appropriately rewritten as

$$\psi(\vec{x}) = \sum_{k\sigma} a_{k\sigma} \phi_{k\sigma}(\vec{x}) + \sum_{q\sigma} b_{q\sigma} \phi_{q\sigma}(\vec{x}) + \sum_{\sigma} \mathrm{d}_\sigma \phi_{d\sigma}(\vec{x}) \qquad (2.112)$$

where $a_{k\sigma}$ destroys an electron of wavevector k and spin projection σ on the left, $b_{q\sigma}$ destroys an electron of wavevector q and spin projection σ on the right, and d_σ destroys the localized state of spin projection σ. An important role is played in this problem by spin-flip scattering processes in which the spin projection of one electron increases and the spin projection of the localized moment or d state decreases. Inserting (2.112) into the Hamiltonian, (2.110), Appelbaum finds that the terms are grouped into those describing the electrode separately and terms which describe transfer and interaction:

$$H = H_1 + H_2 + H_3 + \cdots \qquad (2.113)$$

The first term

$$H_1 = \sum_k \varepsilon_k a_k^* a_k + \sum_q \varepsilon_q b_q^* b_q \qquad (2.114)$$

represents the energies of electrons on the left and right sides of the barrier, while

$$H_2 = \sum_{k,q} (T_{kq} a_k^* b_q + T_{qk} b_q^* a_k) + \cdots \qquad (2.115)$$

describes transfer by tunneling from one side to the other. The matrix element of T_{kq} as derived by Bardeen was given earlier as (2.16). The first of the terms involving interaction of the localized state, assumed to be on the left-hand side of the barrier, is given by

$$H_5 = \sum_{k,k',\sigma} W_{kk'}^d d_\sigma^* a_{k\sigma'}^* d_{\sigma'} a_{k'\sigma} + \text{H.c.} \qquad (2.116)$$

where H.c. indicates the Hermitian conjugate expression. The process described is one in which an electron in the left-hand metal interacts with the localized state by changing its spin projection: a spin-flip scattering event between the local moment (represented by d_σ) and an electron in the nearby electrode. The term H_7 given by

$$H_7 = \sum_{k,q,\sigma} W^d_{kq} a^*_{k\sigma} d^*_{\sigma'} b_{q\sigma} d_\sigma + \text{H.c.} + \cdots \qquad (2.117)$$

is analogous to the previous term, except that, in the interaction with the localized moment, an electron is simultaneously transferred across the barrier: a spin-flip tunneling process. For example, an up-spin electron on the left-hand side of the barrier may tunnel, appearing with down-spin on the other side and, simultaneously, changing the spin projection of the local moment. It is natural to recast processes involved in (2.116) and (2.117) in terms of raising and lowering operators S^+ and S^- for the localized moment; this recasting is accomplished in

$$\begin{aligned} H^T = T_J \sum_{k,q} \{ & S_z[(a^*_{k^+} b_{q^+} - a^*_{k^-} b_{q^-}) + (b^*_{q^+} a_{k^+} - b^*_{q^-} a_{k^-})] \\ & + S^+(a^*_{k^-} b_{q^+} + b^*_{q^-} a_{k^+}) + S^-(a^*_{k^+} b_{q^-} + b^*_{q^+} a_{k^-})\} \\ & + T \sum_{k,q,\sigma} (a^*_{k\sigma} b_{q\sigma} + b^*_{q\sigma} a_{k\sigma}) + \cdots \end{aligned} \qquad (2.118)$$

Here, the first group of terms with prefactor T_J describes the spin-flip tunneling, while the final terms in this operator describe direct tunneling not involving the impurity. Similarly, the interaction term H^I, as expressed in

$$H^I = J \sum_{k'} [S_z (a^*_{k^+} a_{k'^+} - a^*_{k^-} a_{k'^-}) + S^+ a^*_{k^-} a_{k'^+} + S^- a^*_{k^+} a_{k'^-}] \qquad (2.119)$$

with coefficient J, is the second quantization expression for the "s–d" exchange interaction,

$$H^{s-d} = -2J\vec{S} \cdot \vec{\sigma}$$

Finally, the Hamiltonian H' used to calculate the rate of transition according to the Golden Rule is

$$H' = H^T + H^I \qquad (2.120)$$

Equation (2.120) is then used in the Fermi Golden Rule calculation for the transition rate, given by

$$w_{i,j} = \frac{2\pi}{\hbar} \left\{ |H'_{ij}|^2 + \sum_{k \neq i} \frac{H'_{ik} H'_{kj} H'_{ij}}{E_i - E_k} + \text{H.c.} \right\} \cdot \delta(E_i - E_j) \qquad (2.121)$$

It is necessary to go beyond second order in this particular case, as indicated, to include a term involving a sum over intermediate states. In the large number of terms that result from combining e.g. (2.120) and (2.121), important terms are those which contribute to the conductance as

$$g_1 \simeq T^2 \qquad g_2 \simeq T_J^2 \qquad g_3 \simeq T_J^2 J \qquad (2.122)$$

Here, g_1 is the ordinary elastic conductance through the barrier without interaction with the impurity, and g_2 represents contributions in which interaction with a magnetic impurity takes place and is proportional to T_J^2. The presence of this term in experiment can be identified by steps in conductance at $eV = \pm g\mu H$ (H is an applied magnetic field) which represent the thresholds for exciting a Zeeman transition of the magnetic moment. The third term, which appears to order $g_3 \simeq T_J^2 J$, can be described most simply as a Kondo tunneling process, analogous to the Kondo scattering process (Kondo, 1964), known to lead to a resistance minimum in metals containing magnetic impurities. The term g_3 is forward Kondo scattering across the barrier. We will return to discuss this topic in Chap. 8. In general, the extra contributions to the tunneling conductance can be the result of either inelastic or elastic processes, the latter being more difficult to separate from the usual elastic conductance. Calculations of the general sort which we have sketched, following Appelbaum (1967), are applicable to tunneling assisted by phonon emission, plasmon emission, and excitation of localized vibrational modes, among others. One of the most important applications of the transfer Hamiltonian formalism is in the treatment of superconductive tunneling.

3

Spectroscopy of the Superconducting Energy Gap: Quasiparticle and Pair Tunneling

3.1 Basic Experiments of Giaever and Josephson Tunneling

Fisher and Giaever undertook, in 1959, research on mechanisms of conduction between thin films of metals separated by a thermal oxide, with particular interest in tunneling mechanisms (Fisher and Giaever, 1961). Giaever recognized that the theory of superconductivity put forth by Bardeen, Cooper, and Schrieffer (BCS) (1957) predicting a characteristic density-of-states structure

$$N_T(E) = \mathrm{Re}\,\frac{|E|}{(E^2 - \Delta^2)^{1/2}} \tag{3.1}$$

offered a means of identifying currents arising from elastic tunneling. As recounted by Giaever (1974), the low-temperature measurements found early success. The nonlinearity in the I–V characteristic of Fig. 1.5 (Giaever, 1960*a*, 1960*b*), when one electrode was superconducting, was such that the slope $\mathrm{d}I/\mathrm{d}V$, when normalized by the normal state $\mathrm{d}I/\mathrm{d}V$, indeed appeared to follow (3.1). Interpretations of Giaever's experiment by Bardeen (1961) and by Cohen, Falicov, and Phillips (1962) confirmed Giaever's assumption that the conductance in this case must measure the density of states. The first published report in which the experimental tunneling current was quantitatively compared with a calculation based upon (3.1) was that of Nicol, Shapiro, and Smith (1960), whose curves and calculated points are shown in Fig. 3.1a. It is equivalent to say that the slope of these curves, the normalized conductance, can be quantitatively calculated on the basis of

$$\sigma = \frac{(\mathrm{d}J/\mathrm{d}V)_{\mathrm{S}}}{(\mathrm{d}J/\mathrm{d}V)_{\mathrm{N}}} = \int_{-\infty}^{\infty} N_T(E')\left[\frac{-\partial}{\partial E'} f(E' + eV)\right] \mathrm{d}E' \tag{3.2}$$

taking $N_T(E)$ from (3.1). The bracketed function in (3.2) is sharply peaked and positive at $E' = -eV$, with half-width $3.5\,kT$. Hence, in the limit of $T = 0$, $\sigma = G_s/G_N \to N_T(-eV) = N_T(eV)$, a simple and beautiful result. These demonstrations opened the way for detailed studies of the superconducting energy gap structure in various situations, which will be the subject of this chapter.

An extremely close relationship exists between Giaever quasiparticle

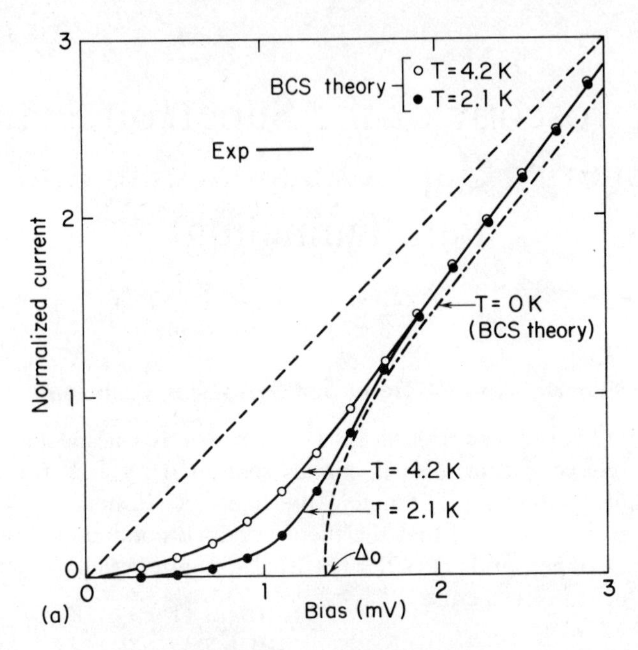

Fig. 3.1. (a) First published report of tunneling between a normal metal and a superconductor in which the nonlinearity of the I–V curve was quantitatively treated in terms of the BCS theory. In this figure, the solid lines represent experimental measurements at 4.2 and 2.1 K between aluminum and lead, while the points have been calculated from the BCS theory. The dashed and dot-dashed curves, respectively, are the normal-state current and the expected current at absolute zero. (After Nicol, Shapiro, and Smith, 1960.) (b) The close relationship between the Josephson supercurrent and the quasiparticle current is illustrated in this plot of the I–V curves of an Sn–CdS–Sn junction measured at 2.2 K. The tunneling transmission coefficient of the CdS insulator is evidently increased by exposure to light: the upper curves, which clearly show the supercurrent, are obtained from a quasiparticle current junction simply by increasing the transmission factor. (After Giaever, 1965.) (c) Characteristic magnetic field dependence of the maximum Josephson current displayed by a narrow Pb–I–Pb junction at 1.3 K. (After Rowell, 1963.)

tunneling, revealing the density of quasiparticle excitations, and the Josephson tunneling *supercurrent* between two superconductors separated by a narrow oxide barrier. This result was not at first appreciated, even though such a supercurrent clearly accompanied the quasiparticle current in the figure of Nicol, Shapiro, and Smith (1960), which was shown in Fig. 1.7. Josephson (1962*a*) extended the transfer Hamiltonian formalism, previously applied to explain Giaever's results, to predict that such a supercurrent should indeed occur and that one characteristic feature would be a dependence upon the difference in superconducting phase, ϕ, across the junction. This, in turn, leads to a striking magnetic-field

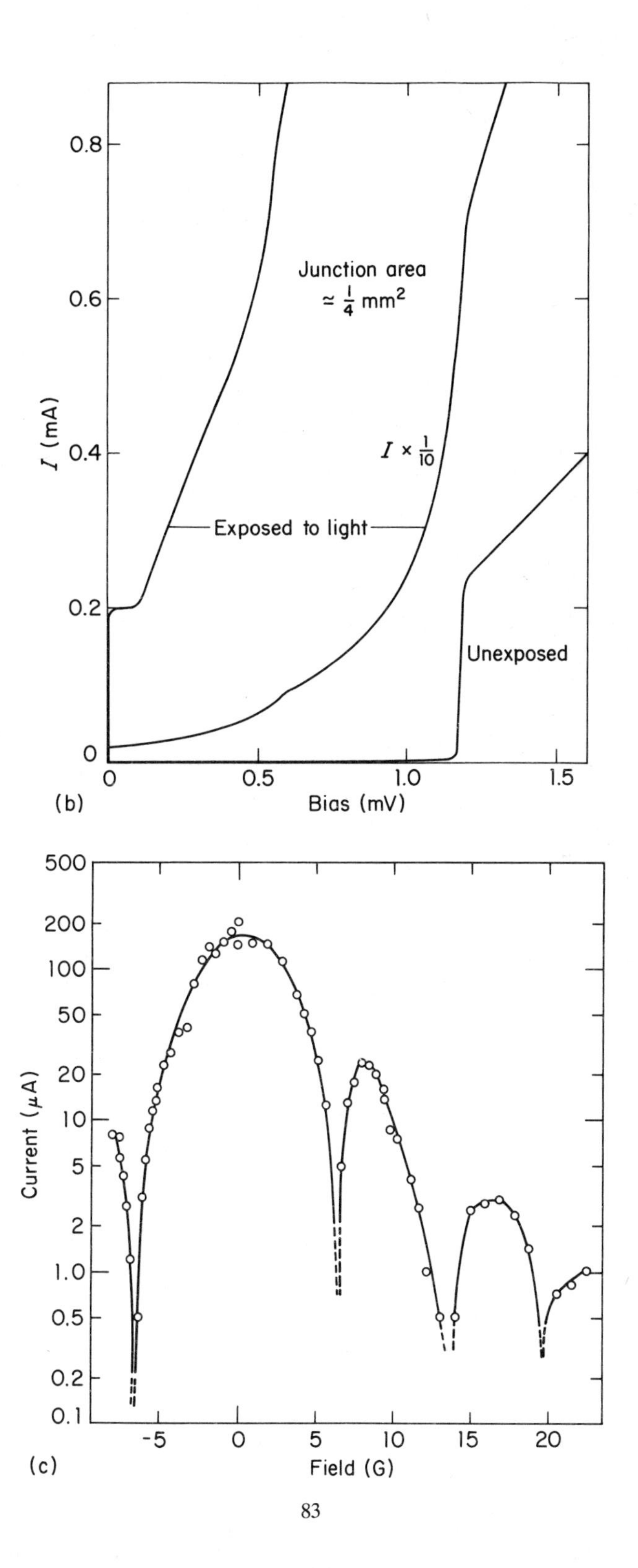

0.8
0.6
0.4
0.2
0
I (mA)
Junction area
≈ 1/4 mm²
I × 1/10
Exposed to light
Unexposed
0
0.5
1.0
1.5
(b)
Bias (mV)
500
200
100
50
20
10
5
2
1.0
0.5
0.2
0.1
Current (μA)
-5
0
5
10
15
20
(c)
Field (G)

dependence. Specifically, the supercurrent density J has a critical upper value J_c dependent on the phase difference ϕ as

$$J_c = J_0 \sin \phi \tag{3.3}$$

The phase difference ϕ between the two superconductors varies according to

$$\phi = \theta_1 - \theta_2 - \frac{2\pi}{\Phi_0} \int_{\vec{r}_1}^{\vec{r}_2} \vec{A} \cdot d\vec{r}$$

where

$$\Phi_0 = \frac{h}{2e} = 2.07 \times 10^{-15}\ \text{Wb} = 2.07 \times 10^{-7}\ \text{gauss-cm}^2 \tag{3.4}$$

is the flux quantum unit and $\vec{A}$ is the magnetic vector potential. Since the vector potential $\vec{A}$ can be introduced and easily varied by a suitable magnetic field, confirmation of Josephson's prediction required demonstrations of a close connection between the supercurrent and the quasiparticle current and of the characteristic dependence upon magnetic field. These two features are shown, respectively, in Figs. 3.1b and 3.1c. The data of Giaever (1965) shown in Fig. 3.1b, which appeared later and thus did not affect the historical development of the subject, show unequivocally that the tunneling supercurrent arises from the same mechanism as the quasiparticle current and is observable if the tunneling matrix elements are large enough. In the curves of Fig. 3.1b, the only change which is made in the junction to make the supercurrent appear is the absorption (at 2.2 K) of photons, which, through trapping of charge in the barrier, lower its height or width, thereby increasing the tunneling matrix element and demonstrating the intimate connection between the two currents. The first demonstrations of the key magnetic field dependence were provided by Anderson and Rowell (1963) and Rowell (1963), whose data are shown in Fig. 3.1c in the form of a Fraunhofer diffraction pattern as a function of magnetic field. This observation provided convincing evidence for the Josephson mechanism.

If the supercurrent of (3.3) can be regarded as arising by tunneling of pairs of electrons, then in such a junction biased to a voltage V, the final pair state should have energy $2eV$ above the ground state of the second superconductor. Josephson (1962*a*) predicted radiation of quantum energy $2eV$ in this situation:

$$\frac{d\phi}{dt} = \frac{2eV}{\hbar}$$
$$\nu = (2\pi)^{-1} \frac{d\phi}{dt} = 483.6\ \text{MHz}/\mu\text{V} \tag{3.5}$$

This prediction has also been confirmed by tunneling and other means, as we shall see.

In order to appreciate the close relationships between these results and to prepare for extensions of these basic experiments to provide a very detailed experimental probe of superconductivity, we summarize some of the basic features of the superconducting state.

3.2 Superconductivity

Two basic experimental features characterizing superconductivity are the disappearance of electrical resistance (Onnes, 1911) and a complete expulsion of magnetic flux below a critical field H_c (Meissner and Ochsenfeld, 1933). An alternative and perhaps more fundamental characterization of the superconducting state is as a macroscopic quantum state, described by an order parameter $\psi(r, t)$. This function, which has many of the properties of a ground-state wavefunction, describes the condensate of pairs in a superconductor, known to exhibit phase coherence over macroscopic distances. These universal superconducting features can be summarized as

$$\rho = 0 \qquad \text{(Onnes)} \tag{3.6a}$$

$$B = 0 \qquad \text{(Meissner)} \tag{3.6b}$$

$$\psi(\vec{r}, t) = \sqrt{n_s} e^{i\theta}$$

$$\vec{J} = e^* \left[\frac{i\hbar}{2m} (\psi \vec{\nabla} \psi^* - \psi^* \vec{\nabla} \psi) - \frac{e^*}{mc} |\psi|^2 \vec{A} \right] \tag{3.6c}$$

Here, we understand that $e^* = 2e$, and n_s is a local density of pairs. Equation (3.6c) includes, in addition, the relationship between the current density J and the wavefunction ψ, which has the form expected in quantum mechanics. Thus, the statements we have listed are redundant. The third statement, which cannot be directly proven experimentally, enables one to infer statements a and b and further to explain other phenomena characteristic of the superconducting state, which, in fact, are also closely related to the Josephson tunneling effects.

The experiment most clearly implying the macroscopically occupied ground-state wavefunction, (3.6c), is the observation of quantized magnetic flux through a superconducting loop (Deaver and Fairbank, 1961; Doll and Nabauer, 1961), as illustrated in Fig. 3.2a (after Goodman, Willis, Vincent, and Deaver 1971). The physical requirement that the wavefunction (3.6c) be single-valued on the loop implies that the phase difference (3.4) should be a multiple of 2π or, equivalently, that the flux enclosed, given by the line integral of the vector potential around the loop, should be a multiple of the basic flux quantum $\Phi_0 = h/2e$.

The measurements, of which the data in Fig. 3.2a is a more recent example, establish within the available precision that the effective charge e^* is indeed $2e$, as required by the pairing model. Coherence of the pair phase in a superconductor can apparently extend over truly macroscopic

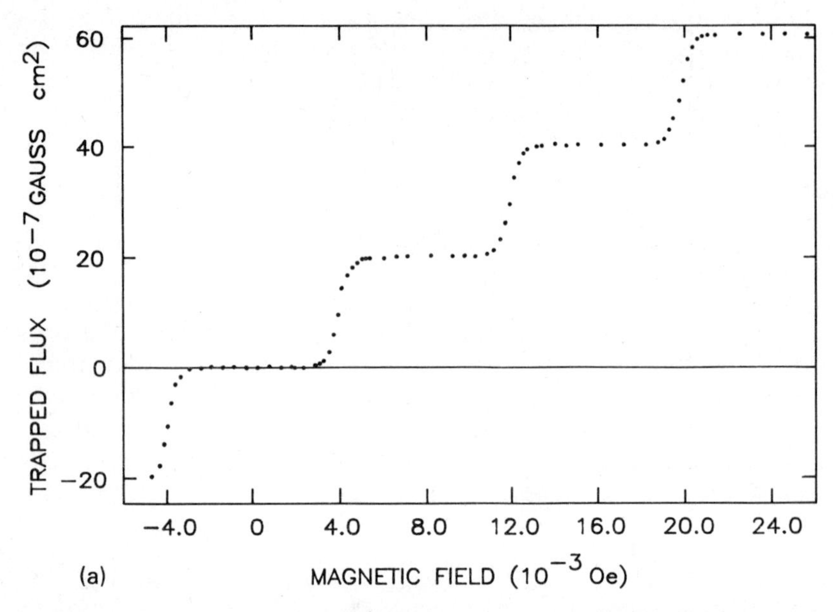

Fig. 3.2. (a) Measurements which clearly show quantization of magnetic flux trapped in Sn cylinder cooled below T_c in ambient magnetic field (shown on abscissa). Such results verify the expression for the flux quantum $\Phi_0 = h/e^*$ with $e^* = 2e$. The Sn cylinder has 56 μm i.d., length 24 mm, and wall thickness near 5000 Å. (After Goodman, Willis, Vincent, and Deaver, 1971.) (b) Specific heat data on Ga in superconducting (solid line) and normal states (dashed) show jump at T_c corresponding to second-order phase transition. (After N. E. Phillips, 1964.) (c) Reduced entropy S in the superconducting state can be inferred from specific heat measurements such as shown in (b).

distances. The physical origin of this remarkable effect within the microscopic pairing model comes from the large spatial extent ξ_0 of the individual pairs, with a consequence that enormous numbers of pairs overlap. This result, in turn, implies a high degree of correlation in their motions.

The macroscopic quantum-state picture of superconductivity implied by flux quantization and required for the Josephson effects also implies the more directly observable infinite conductivity and flux exclusion or Meissner effect. Josephson (1965), pointed out that the order parameter ψ transforms under gauge transformations as

$$\vec{A} = \vec{A}' + \vec{\nabla}\chi$$

$$U = U' - \frac{1}{c}\frac{\partial \chi}{\partial t},$$

where $\vec{A}$ and U, respectively, are vector and scalar potentials, and χ a scalar function. ψ thus transforms the same fashion as an ordinary

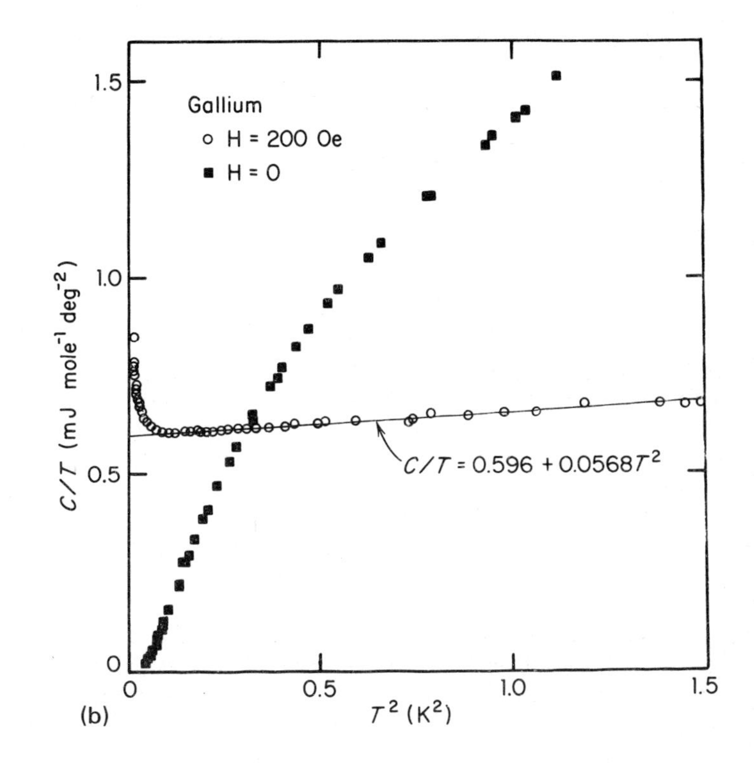

Gallium
H = 200 Oe
H = 0
C/T (mJ mole⁻¹ deg⁻²)
C/T = 0.596 + 0.0568T²
T² (K²)
(b)

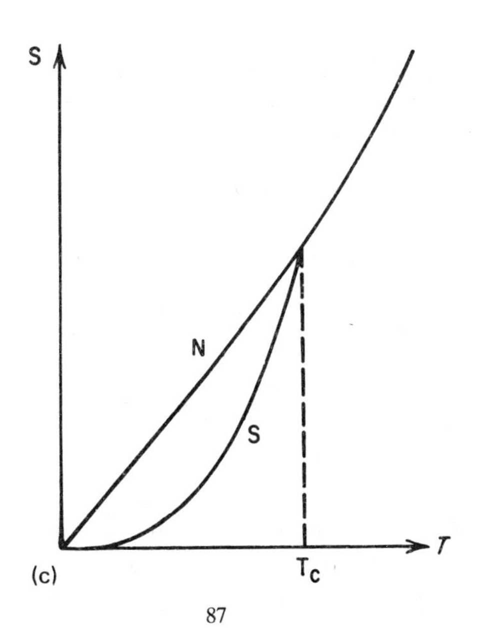

S
N
S
T
Tc
(c)

wavefunction;

$$\psi = \psi' \exp \frac{i\chi e^*}{\hbar c}$$

and this allows $\vec{J}$ to remain gauge-invariant. The form of $\vec{J}$, which includes a term proportional to $\vec{A}$, leads to the Meissner effect (London and London, 1935*a*,*b*).

In order to satisfy gauge invariance, the time dependence of ψ, for a superconductor in equilibrium, must be $\psi(\vec{r}, t) = \psi(\vec{r})e^{-iEt/\hbar}$, just that of an ordinary wavefunction. If one borrows from later theory the existence of a condensed phase or reservoir of pairs, $e^* = 2e$, with which individual pairs can be freely exchanged, then $E = 2\mu$, where μ is the chemical potential (Fermi energy), so that $\arg\psi = \theta(\vec{r}, t)$ varies as $(\partial/\partial t)\theta = -2\mu(\vec{r}, t)/\hbar$. If a current is established, one may expect a local correction μ', depending on the degree of departure from equilibrium:

$$\frac{\partial}{\partial t}\theta(\vec{r}, t) = \frac{-2[\mu(\vec{r}, t) + \mu']}{\hbar}$$

Fig. 3.3. (a) Type I superconductor (dashed line) exhibits perfect diamagnetism up to thermodynamic critical field H_c. Type II materials (solid line) admit flux in vortex form at $H = H_{c1} < H_c$ and become normal, except for surface regions, at $H = H_{c2} > H_c$. (b) Low-temperature critical field H_c data for some elements and simple alloys exhibit approximate $T_c^{1.3}$ dependence. Transition elements and alloys lie slightly above the line characterizing s–p elements and alloys. (Data compiled by Roberts, 1976.) (c) The temperature dependence of H_c is close to parabolic: $h = H_c/H_c(0) = 1 - t^2$. Deviations $D(t) = h - (1 - t^2)$ are usually less than 5% and, as seen, are characteristically negative for weak-coupling and positive for strong-coupling superconductors. (After Swihart, Scalapino, and Wada, 1965.)

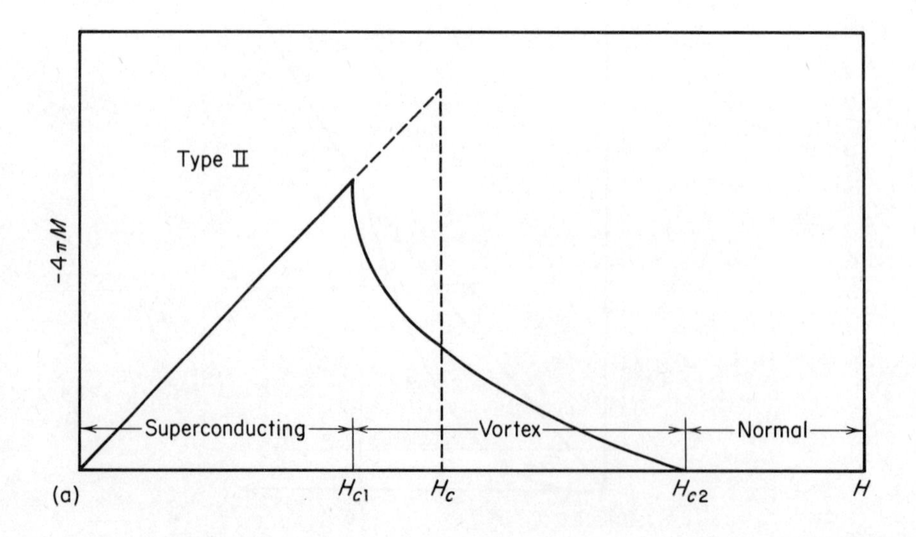

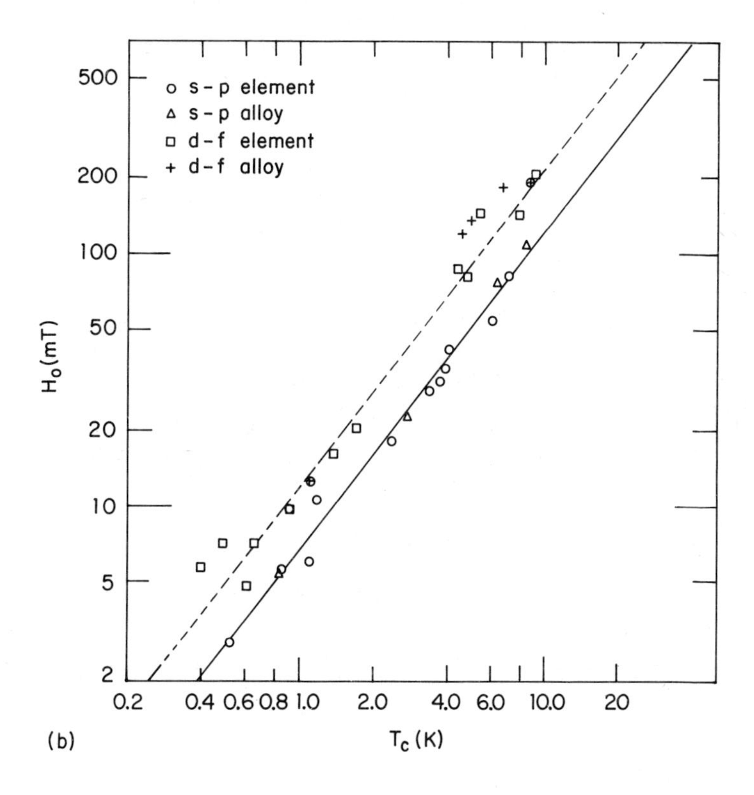

o s-p element
Δ s-p alloy
□ d-f element
+ d-f alloy
500
200
100
50
20
10
5
2
H_0(mT)
0.2
0.4 0.6 0.8 1.0
2.0
4.0 6.0
10.0
20
T_c (K)

(b)

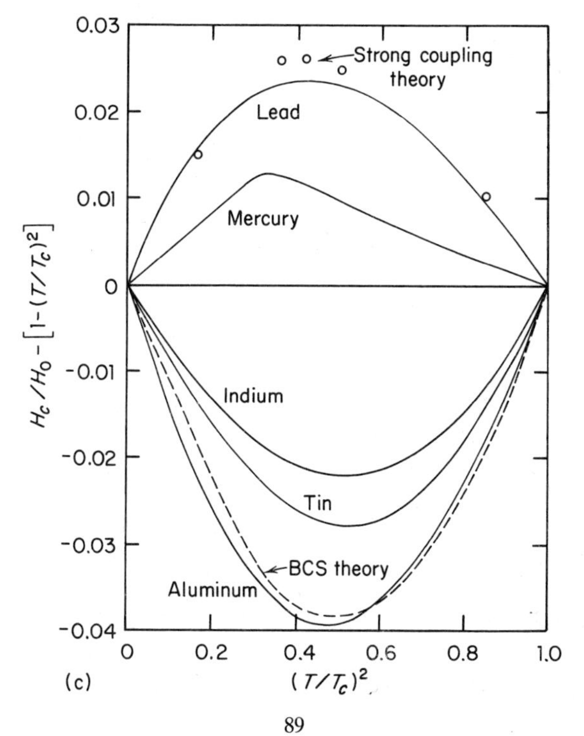

Strong coupling theory
Lead
Mercury
Indium
Tin
BCS theory
Aluminum
0.03
0.02
0.01
0
-0.01
-0.02
-0.03
-0.04
$H_c/H_0 - [1-(T/T_c)^2]$
0
0.2
0.4
0.6
0.8
1.0
$(T/T_c)^2$

(c)

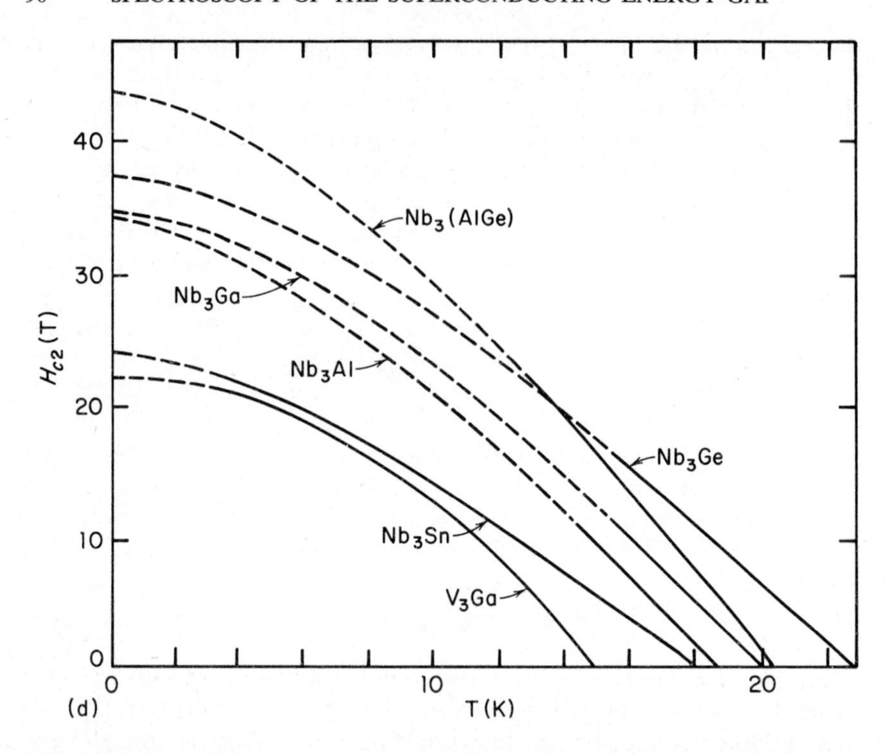

Fig. 3.3. (cont.) (d) Examples of extremely high critical fields in the A-15 class of superconductors, which exhibit T_c values near 20 K. (After Foner et al., 1970; Foner, McNiff, Gavaler, and Janocko, 1974; compiled by Dew-Hughes, 1975.)

In a steady-state flow, the wavefunction can vary only with a phase factor:

$$\psi(\vec{r}, t) = \psi(\vec{r}, 0)\exp[i\alpha(t)]$$

where α is not a function of $\vec{r}$. Hence, the difference of the potential $\delta[\mu(\vec{r}, t) + \mu']$ between two points displaced by $\delta\vec{r}$ depends on $\delta\mu'$, the difference of the local departures from equilibrium. Since this potential difference does not scale with the displacement $\delta\vec{r}$, Josephson (1965) concludes that the conductivity must be infinite.

Another effect which is consistent with the notion of a highly ordered quantum state below the superconducting transition temperature T_c is the jump in the specific heat at the second-order phase transition, which leads, with further reduction in temperature, to a state of substantially reduced entropy. These features are illustrated in Figs. 3.2b and 3.2c.

The occurrence of the superconducting state is limited also by critical values of magnetic field and current density. Two types of superconductor can be distinguished by their behavior in a magnetic field, as indicated in Fig. 3.3a. Type I materials exhibit perfect diamagnetism (Meissner flux expulsion) up to the thermodynamic critical field H_c sketched by the dashed line in the figure. Type II superconductors completely expel flux for H below a lower critical field H_{c1}. Above H_{c1}, flux threads through the superconductor in an array of vortex lines with a normal core surrounded by screening currents. With increasing field, the density of vortex lines increases until at H_{c2} the whole structure becomes normal. The transition between Type I and Type II behavior was elegantly explained by Abrikosov (1957). The latter behavior can frequently be induced by reducing the mean free path, as by forming an alloy. Figure 3.3b illustrates the regularity between the zero-temperature values of H_c and the observed transition temperature T_c for most of the elemental superconductors and a few of their alloys. The straight line drawn in this figure through the lowermost set of points representing s–p metals, which are Type I superconductors, corresponds to an approximate $T_c^{1.3}$ dependence for $H_c(0)$. The points in this figure which scatter about the dashed line are primarily transition metal superconductors, some of which are of Type II. The temperature dependence of the thermodynamic critical field H_c is typically described in relation to the simple parabolic law:

$$H_c(t) = H_{c0}\left[1-\left(\frac{T}{T_c}\right)^2\right] \tag{3.7}$$

Weak (approximately 5%) deviations from this equation are characteristic of weak-coupling (negative deviations) and strong-coupling (positive deviations) superconductors, as shown in Fig. 3.3c. While the critical temperatures T_c of known superconductors are rather narrowly limited, the highest presently being 23.2 K in Nb_3Ge (Gavaler, 1973), materials with critical field values H_c greatly enhanced over those typical of elemental superconductors have been found, as indicated in Fig. 3.3d. Since the critical values of current density J_c are found to be similarly enhanced, these materials are of practical use in making magnets.

A first evidence for an energy gap 2Δ at the Fermi surface as a typical feature of superconductors came from a specific heat measurements (Corak, Goodman, Satterthwaite, and Wexler, 1956) and far-infrared absorption (Glover and Tinkham, 1956) before the development of tunneling methods. The ratio $2\Delta/kT_c$, given by 3.53 in the BCS theory, lies between the BCS value and about 4.0, over a remarkably wide range of superconductors. The dependence of the energy gap on T is close to a universal function of temperature. These and further regularities in the behavior of all superconductors are generally described as the "law of corresponding states."

The diversity of materials and circumstances in which superconductivity occurs is remarkable: many elements exhibit the effect, with the highest T_c, 9.25 K, occurring in Nb. A great many metallic alloys and intermetallic compounds are also superconductors. Regularities in these classes of materials have been noted by Matthias (1955). Superconductivity also appears in a few doped semiconductors, such as $SrTiO_3$ and semimetals (GeTe, SnTe, . . .), and has recently been observed in the polymeric "metal" $(SN)_x$, the organic metal $(TMTSF)_2PF_6$ (at high pressure), and at zero pressure in the similarly quasi-one-dimensional metal $Hg_{3-\delta}AsF_6$. As a further example of an unusual diversity, in materials such as the ternary compound $ErRh_4B_4$, the phenomena of superconductivity and ferromagnetism or antiferromagnetism compete or coexist in ordered microdomain structures.

The existence of a single macroscopic quantum state, and its consequences on the behavior of weakly interacting superconductors, makes possible a remarkable series of applications, including ultrasensitive sensors of magnetic flux, electrical current, and voltage, and related devices arrayed in high-performance logic circuits for computers. These quantum superfluid devices all followed the theoretical discoveries of Josephson (1962*a*) regarding the behavior of weakly coupled specimens of any superconductor. The useful sensitivity to magnetic field H arises from the resultant variations of phase θ governed by (3.3) and (3.4), taking note of the small value of the magnetic flux quantum

$$\Phi_0 = \frac{h}{2e} = 2.07 \times 10^{-15}\ \mathrm{Wb}$$

Although an energy gap $2\Delta = 3.53kT_c$, as predicted by the BCS theory, and typically 1 or 2 meV, is a characteristic feature of superconductivity, it is not essential: certain gapless superconductors display all of the effects mentioned above. The more fundamental microscopic property leading to (3.6) is the ordered electronic ground state in which electrons near the Fermi surface exhibit pair correlations $(\vec{k}\uparrow, -\vec{k}\downarrow)$. While this state normally leads to a gap, there is a fairly narrow range of shortened pair-correlation lifetimes which obscures the gap without destroying the superconducting phase. The spatial extent of the pair, essentially the coherence length $\xi_0 = \hbar v_F/\pi\Delta$ in the BCS theory, is large on an atomic scale (even at $\xi_0 \simeq 50$ Å, for large-gap–low-v_F metals like Nb_3Sn), which implies extreme overlap and hence correlation of the motion of nearby electron pairs. A supercurrent shifts the whole paired Fermi surface. Nonmagnetic impurity scattering does not break pairs but simply rotates them around the Fermi surface, maintaining their center of mass velocity and hence their contribution to the current. As long as the single quantum state exists, there is essentially zero dissipation and no observable decay of persistent currents.

A further demonstration of a coherence property of the superconducting state is provided by the proximity effect (Holm and Meissner, 1932), in which a layer of ordinarily normal metal (such as gold or silver), placed in proximity to a superconductor, can itself support a supercurrent. The experimental requirement is that electrons can pass freely between the two metals. The proximity coherence length (of electronic pairing correlations into the "normal" metal) may be as great as 1 or even 10 μm. Specific heat measurements have demonstrated that the entropy of the normal region is reduced by its proximity to the superconductor. We will see that tunneling spectroscopy can reveal details of the superconducting pair potential induced in the normal metal by the proximity effect.

While, in principle, the electron pairing could be driven by various mechanisms, the only mechanism fully and convincingly documented at present is that provided by electron-phonon coupling. The first strong hint of the importance of electron-phonon coupling was the discovery (Maxwell, 1950; Reynolds, Serin, Wright, and Nesbitt, 1950) of the isotope effect,

$$T_c M^a = \text{constant}$$

where M is the isotopic mass and a is typically near 0.5, as for Hg, Pb, Sn, Tl, and Zn. As the isotopic mass value has an appreciable influence only on the lattice vibrations, and metallic conductivity is electronic, this observation was interpreted as strong evidence that the electron-phonon interaction must be involved in the superconducting state. This feature has since been fully confirmed by results of tunneling spectroscopy, which support the electron-phonon mechanism even for transition metal superconductors (Ta, Nb, V) for which a magnetic pairing interaction had earlier been mentioned (Matthias, 1960; Matthias, Geballe, and Compton, 1963).

3.3 Electron-Phonon Coupling and the BCS Theory

The essential physical idea of the attractive electron-electron interaction is simple: lattice polarization by one conduction electron lowers the energy of a second electron appearing at the same position at a later time. This notion already makes plausible the large spatial extent of the electron Cooper pair. In the interval $\tau \simeq \omega_D^{-1} \simeq 10^{-13}$ sec for the maximum amplitude (overscreening) of local lattice distortion to develop, the polarizing electron typically moves a distance of order $\ell \simeq v_F \tau \simeq 1000$ Å, taking $v_F = 10^8$ cm/sec. This value is on the order of a typical BCS coherence length; $\xi_0 = \hbar v_F / \pi \Delta$. In a quantum mechanical picture (Fig. 3.4a), the initial electron $\vec{k}$ scatters to $\vec{k} - \vec{q}$ by emitting a phonon, of wavevector $\vec{q}$ and energy $\hbar\omega_q$, which is absorbed as the second electron scatters from $-\vec{k}$ to $-\vec{k} + \vec{q}$. Any phonon $\vec{q}$ of the system can participate to the extent that the matrix element $M_{\vec{q}}$ is nonzero, because the energy is reabsorbed by the second electron. This process scatters pairs around the Fermi surface, as in Fig. 3.4b.

The difficulties in isolating and treating this rather weak effect in realistic quantum mechanical terms were addressed by Fröhlich (1952) and, including electronic screening, by Bardeen and Pines (1955). The extension to include retardation was later given by Eliashberg (1960). The Bardeen–Pines interaction Hamiltonian H_I is

$$H_I = \sum_{k,k',q} \frac{\hbar\omega_q \, |M_q|^2 \, a^*_{k'-q} a_{k'} a^*_{k+q} a_q}{(\varepsilon_k - \varepsilon_{k+q})^2 - (\hbar\omega_q)^2} + H_{\text{coul}} \tag{3.8}$$

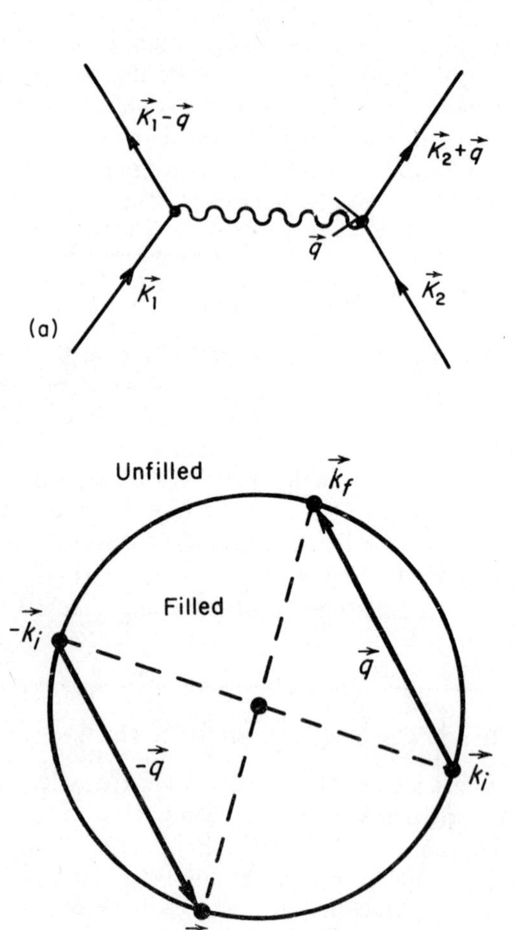

Fig. 3.4. (a) Scattering of two electrons by emission and absorption of a phonon. This process lowers the energy if the two electron energies differ from the Fermi energy by less than the phonon energy. (b) Scattering of pairs of electrons about the Fermi surface by the phonon process indicated in (a). (c) The BCS model potential assumes a constant attractive energy V for pairs of electrons whose energies are less than the cutoff energy from the Fermi surface. (d) The pairing self-energy function $\Phi(\omega)$ as determined by electron tunneling spectroscopy for Pb. The real part of this function is regarded as a generalization of the BCS potential. (After McMillan and Rowell, 1969.)

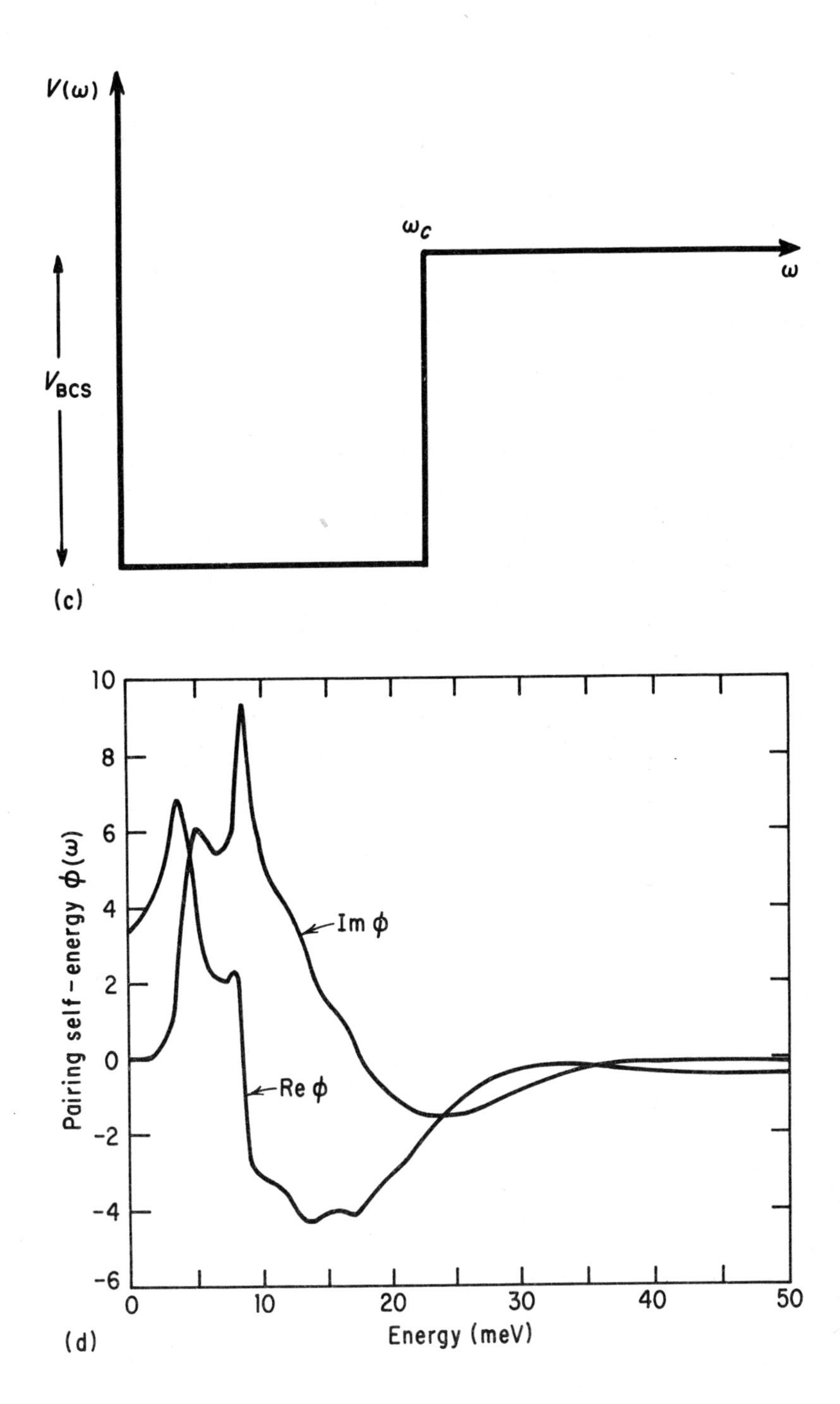
$V(\omega)$
ω_c
ω
V_{BCS}
(c)
10
8
6
4
2
0
-2
-4
-6
Pairing self-energy $\Phi(\omega)$
Im ϕ
Re ϕ
0
10
20
30
40
50
Energy (meV)
(d)

Here, $a_{\vec{k}}^*$ creates an electron $\vec{k}$ (spin indices understood), $\varepsilon_{\vec{k}}$ is the energy of the normal quasiparticle at $\vec{k}$ measured from the Fermi surface, and H_{coul} represents the residual, screened Coulomb interaction between quasiparticles. The phonon interaction term is seen to be negative for $|\varepsilon_k - \varepsilon_{k+q}| < \hbar\omega_q$ and may dominate the Coulomb repulsion U so that $V \simeq H_I + U$ is negative, corresponding to an attractive interaction.

3.3.1 The Pair Ground State

Cooper (1956) showed that two electrons ($\vec{k}\uparrow, -\vec{k}\downarrow$) interacting via any attractive V above an otherwise noninteracting filled Fermi sea at $T=0$ form a bound state. Hence, for a net attractive V at $T=0$, the N-electron system also appeared unstable against transition to a state in which all electrons pair: ($\vec{k}\uparrow, -\vec{k}\downarrow$) are both occupied or both empty, for all $\vec{k}$.

Bardeen, Cooper, and Schrieffer (1957) (BCS) assumed such a ground state and found, by a variational calculation, the occupation function h_k of the pair states which produced the minimum energy.

To obtain h_k, BCS drastically simplified the interaction $V_{kk'}$, which had already been truncated to include only the electron-phonon coupling and screened Coulomb repulsion (taken as $4\pi e^2/(q^2+k_s^2)$, where k_s is a screening constant), involved in transitions between pair states. The actual interaction used by BCS was taken as constant below a cutoff (Fig. 3.4c). The function $\Phi(\omega)$, a generalization of $V_{kk'}$ known as the pairing self-energy, as inferred from tunneling results on Pb, is shown in Fig. 3.4d.

We now follow BCS in taking an attractive interaction $V_{kk'}$ as positive and constant below cutoff. Then, the energy W_0 of the paired vs. unpaired systems (at $T=0$) is

$$W_0 = \sum_{k>k_F} 2\varepsilon_k h_k + \sum_{k<k_F} 2|\varepsilon_k|(1-h_k) - \sum_{k,k'} V_{kk'}[h_{k'}(1-h_{k'})]^{1/2}[h_k(1-h_k)]^{1/2} \tag{3.9}$$

In the interaction term, the occupation factors specify the probability for a pair transition $k \to k'$: that initially k be occupied, k' be unoccupied, and, finally, k' be occupied while k is empty. The desired h_k minimizing W_0 satisfies $\partial W_0/\partial h_k = 0$, or from (3.9),

$$2\varepsilon_k = \frac{\Delta_k(1-2h_k)}{\sqrt{h_k(1-h_k)}} \tag{3.10}$$

where the *gap function* Δ_k is defined as

$$\Delta_k \equiv \sum_{k'} V_{kk'}\sqrt{h_{k'}(1-h_{k'})} \tag{3.11}$$

Solution of (3.10), which is quadratic in h_k, gives (Fig. 3.5a)

$$h_k = \frac{1}{2}\left(1 - \frac{\varepsilon_k}{E_k}\right) \equiv v_k^2 \qquad 1-h_k = \frac{1}{2}\left(1 + \frac{\varepsilon_k}{E_k}\right) \equiv u_k^2 \tag{3.12}$$

where

$$E_k \equiv (\varepsilon_k^2 + \Delta_k^2)^{1/2} \tag{3.13}$$

as sketched in Fig. 3.5b.

The quasiparticle excitation spectrum of the superconductor is of central importance to tunneling measurements. From the definition $E_k = \sqrt{\varepsilon_k^2 + \Delta_k^2}$, the density of superconducting excitations is (Fig. 3.5c)

$$\frac{dn}{dE_k} = \frac{dn}{d\varepsilon_k}\frac{d\varepsilon_k}{dE_k} = \frac{N(0)\,|E|}{\sqrt{E_k^2 - \Delta_k^2}} \equiv N(0)N_T(E)$$

Term $N_T(E)$, the "tunneling density of states," is rather directly measured by tunneling (Giaever, 1960*a*,*b*) according to (3.1). The elementary excitation $E_{\vec{k}}$ of the fully paired state is defined as $\vec{k}\uparrow$ occupied, $-\vec{k}\downarrow$ unoccupied. The meaning of E_k changes smoothly from holelike ($\varepsilon_k < 0$, $h_k = v_k^2 \simeq 1$) for $k < k_F$ to electronlike ($\varepsilon_k > 0$, $1 - h_k \simeq u_k^2 \simeq 1$) for $k > k_F$. This result may be seen by referring to Figs. 3.5a,b, in the sketches of v_k^2 vs. k and E_k vs. k.

The gap function Δ_k is seen from (3.12) and (3.13) to set the energy range above the $T = 0$ Fermi surface where the pair-state occupation probability h_k remains appreciable, as indicated in Fig. 3.5a. Similarly, $1 - h_k$, the probability of finding an empty pair state at $\vec{k}$, is nonzero for a range of order Δ_k below the Fermi surface. The pair density function $[h_k(1 - h_k)]^{1/2} = u_k v_k$ is sketched in Fig. 3.5d.

The central equation of BCS theory (1957), the gap equation, determines Δ_k by combining (3.10)–(3.13). This equation, when generalized to finite T, is

$$\Delta_k = \sum_{k'} V_{k,k'} \frac{\Delta_{k'}}{2E_{k'}}[1 - 2f(E_{k'})] \tag{3.14}$$

Strictly, to obtain (3.14), BCS minimized the free energy $F = U - TS$ against both h_k and the single-particle occupation function f_k. For f_k, the usual Fermi function of argument E_k resulted, $f_k = [1 + \exp(E_k/kT)]^{-1}$. This function reveals E_k as the true quasiparticle excitation of the condensed phase, whose minimum value, Δ_k, from (3.13), then represents a gap in the excitation spectrum.

If the sum on k' is converted to an integral over E_k, one finds,

$$\sum_{k'} \rightarrow \int 2N(0)\, d\varepsilon_k = \int 2N(0) \frac{E_k}{\sqrt{E_k^2 - \Delta_k^2}}\, dE_k \tag{3.15}$$

using (3.13) to determine $d\varepsilon_k/dE_k$. If V is also taken as constant below a cutoff $\hbar\omega_c$, (3.14) becomes

$$\Delta = N(0)V \int_{\Delta}^{\hbar\omega_c} \frac{\Delta}{\sqrt{E^2 - \Delta^2}}[1 - 2f(E)]\, dE \tag{3.16}$$

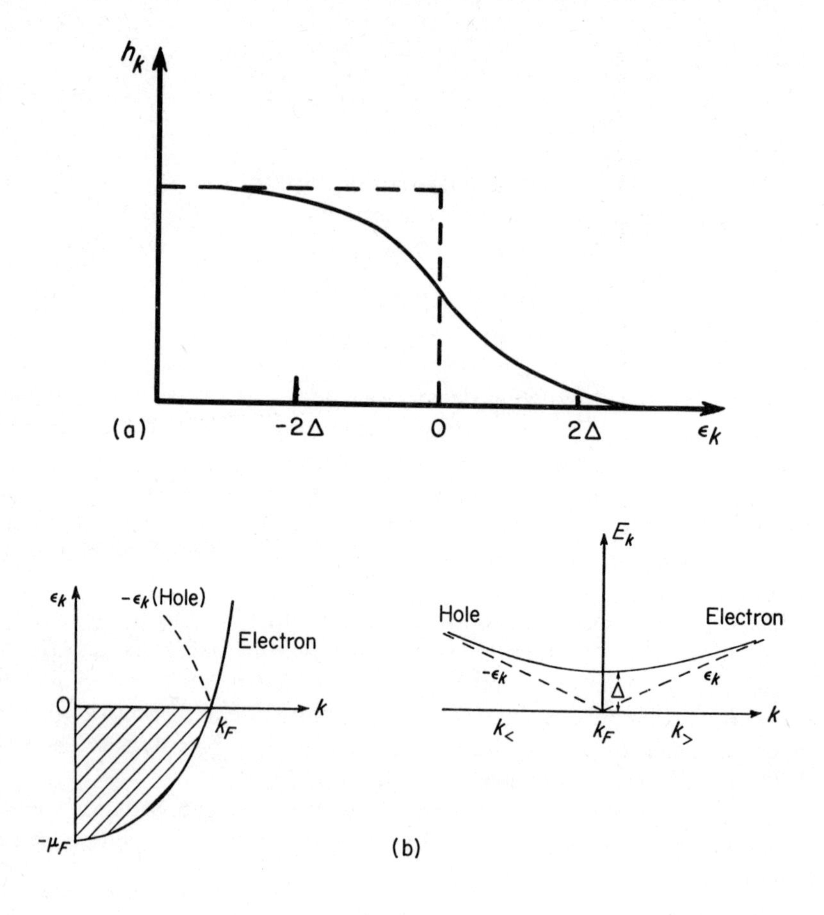

Fig. 3.5. (a) Pair-state occupation function $h_k = v_k^2$ for BCS superconductor at $T=0$ (solid), and $T=0$ Fermi function (dashed). Single-particle states above normal Fermi surface by $E \simeq \Delta$ are substantially occupied in the $T=0$ paired state, almost as in the normal state at $T=T_c$. (b) The excitation spectrum of a superconductor in a small range of wavevector close to k_F. In contrast to the normal state, there is now a minimum excitation energy Δ, with the excitation spectrum smoothly connecting to the electron branch of the normal phase for $k \gg k_F$ and smoothly connecting to the hole branch of the normal-phase excitation relation for $k \ll k_F$.

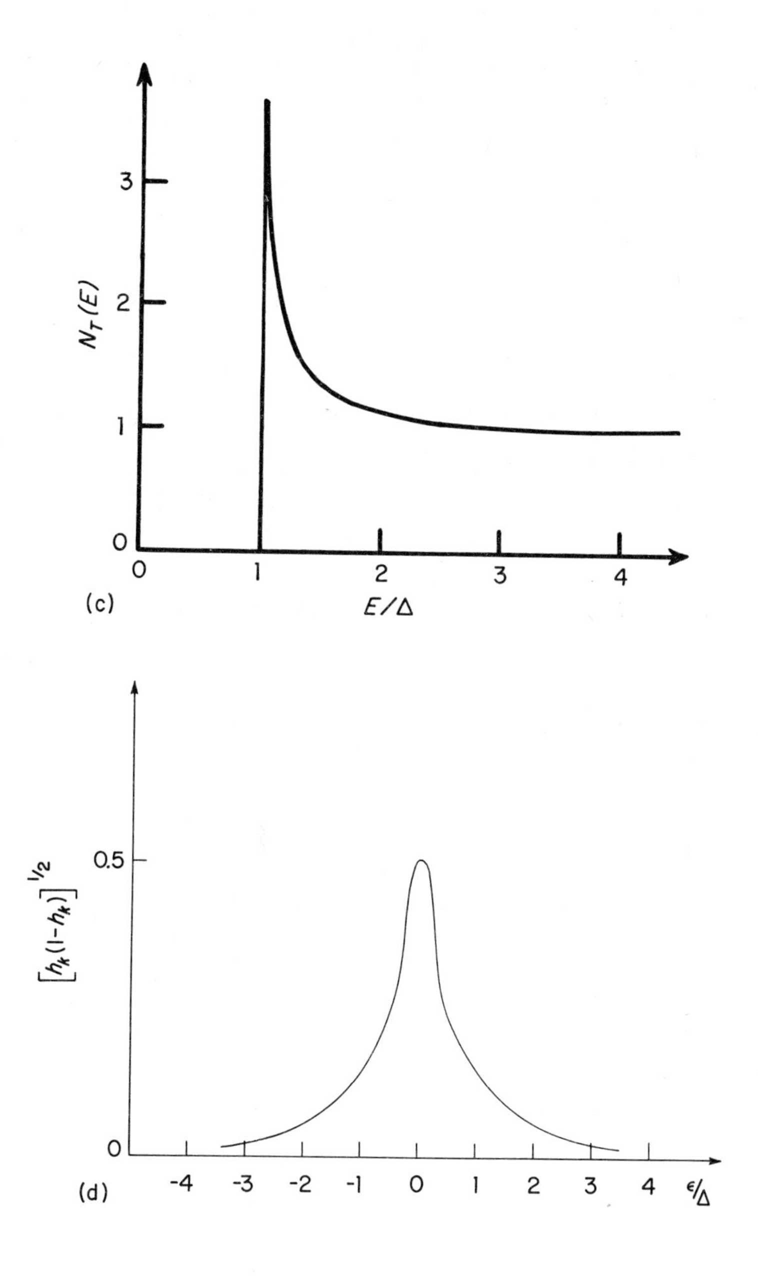

(c) The energy density of excitations $N_T(E)$ corresponding to the spectrum of (b). A square root singularity results from pushing the low energy excitations away from the gap. (d) The pair density function of a superconductor. At the Fermi energy, the probability of occupation of a given pair state is 0.5, maximizing the rate of pair-scattering as shown in Fig. 3.4b.

This equation yields, for $T=0$, $f(E)=0$:

$$\Delta(0)=\frac{\hbar\omega_c}{\cosh[1/N(0)V]}$$

$$\simeq 2\hbar\omega_c \exp\frac{-1}{N(0)V} \qquad \text{for} \qquad N(0)V \ll 1 \tag{3.17}$$

The critical temperature T_c from (3.16) corresponds to setting $\Delta=0$ (after canceling Δ's on right and left):

$$\frac{1}{N(0)V}=\int_0^{\hbar\omega_c}\frac{\tanh(E/2kT)\,dE}{E} \tag{3.18}$$

This result gives the estimate, for $1/N(0)V \ll 1$,

$$kT_c \simeq 1.13\hbar\omega \exp\frac{-1}{N(0)V} \tag{3.19}$$

and [comparing with (3.17)] the more accurate ratio

$$2\Delta(0)=3.53kT_c \tag{3.20}$$

The BCS function $\Delta(T)$, shown later (Fig. 3.14) in comparison with data, varies steeply near T_c as

$$\Delta(T)\simeq 1.8\Delta(0)\left(1-\frac{T}{T_c}\right)^{1/2} \qquad (T\simeq T_c) \tag{3.21}$$

and is quite close to its $T=0$ value for $T/T_c<0.5$. These results, valid for weak coupling, $1/N(0)V \ll 1$, do not differ greatly from results in the strong-coupling limit of the same model (Thouless, 1960). In the Thouless theory, the exact condition

$$\frac{\Delta(T)}{\Delta(0)}=\tanh\left[\frac{T}{T_c}\frac{\Delta(T)}{\Delta(0)}\right] \tag{3.22}$$

is found, while $2\Delta(0)=4.0kT_c$ and $\Delta(T)\simeq 1.73\Delta(0)[1-(T/T_c)]^{1/2}$ for $T\simeq T_c$.

That occupation of the quasiparticle states E_k is detrimental to superconductivity may be seen directly from (3.14), where $f_k>0$ reduces the summand. Physically, excitations block both empty and filled pair states, reducing the overall scattering rate, which produces the lowered energy. It follows from (3.14) that the gap edge states ($E_k=\Delta$) are most important in this regard. It was pointed out by Eliashberg (1970) that absorption of radiation of energy $\hbar\omega$, to induce transitions from Δ to $\Delta+\hbar\omega$ (restricting $\hbar\omega<2\Delta$, to avoid generating quasiparticles), should tend to empty gap edge excitations and thus increase Δ. Such effects have been observed, using both microwave and phonon radiation of $\hbar\omega<2\Delta$.

As we will see, it is also possible to similarly change the quasiparticle populations by use of a tunnel junction biased at $eV < 2\Delta$.

3.3.2 *Elementary Excitations of Superconductors*

To add an *electron*, as distinct from a quasiparticle, to the superconductor at fixed energy E requires a linear combination of elementary excitations and thus *two* k values, $k'>$ and $k''<$, such that $u_{k'>}^2 + u_{k''<}^2 = 1$. For $E \gg \Delta_k$, the electron goes, with high probability, in the branch $k > k_F$, the electronlike branch.

The appropriate creation operators α_k^* for the elementary excitation $E_{\vec{k}}$, defined independently by Bogoliubov (1958a,b) and by Valatin (1958), may be written

$$\begin{aligned} \alpha_{k\uparrow}^* &= u_k a_{k\uparrow}^* - v_k a_{-k\downarrow} \\ \alpha_{-k\downarrow}^* &= u_k a_{-k\downarrow}^* + v_k a_{k\downarrow} \end{aligned} \tag{3.23}$$

(Here, the notation follows Waldram (1976); the notation γ_{k0}^*, γ_{k1}^* for $\alpha_{k\uparrow}^*$, $\alpha_{-k\downarrow}^*$ is also in use. We take u, v as real.)

These expressions make particularly clear that the elementary excitation in a superconductor is a mixture of electron and hole. The strength of this mixing is controlled by the anomalous potential Δ, which may be said to scatter electrons into holes. The inverse transformation is

$$\begin{aligned} a_{k\uparrow}^* &= u_k \alpha_{k\uparrow}^* + v_k \alpha_{-k\downarrow} \\ a_{-k\downarrow}^* &= u_k \alpha_{-k\downarrow}^* - v_k \alpha_{k\uparrow} \end{aligned} \tag{3.24}$$

Some discussion is worthwhile regarding the effect of these operators, say $\alpha_{k\uparrow}^*$, on the total number of electrons, including those in pairs, in the sample. If the pair state $\vec{k}\uparrow$ is unoccupied, with probability amplitude $u_{k\uparrow}$, then creation of the $\vec{k}\uparrow$ electron does not change the number of pairs. However, generation of the quasiparticle excitation, if the pair state k is occupied with probability amplitude v_k, proceeds by destroying the $-\vec{k}\downarrow$ member of the k pair, thus reducing the number of pairs by one. For most purposes, it is correct to allow the pair number to change, assuming a large reservoir of pairs. Indeed, that this assumption must be made in order to calculate the supercurrent between weakly coupled superconductors can be viewed as the central point of Josephson's theoretical discovery (Josephson, 1962*a*,*b*; Waldram, 1976) that the pair current is of the same order in T^2 as the quasiparticle tunneling current. For purposes of discussion and for specific applications, Josephson (1962*a*) and, independently, Bardeen (1961) modified the Bogoliubov-Valatin operators to specify the creation (annihilation) of definitely one electron (hole). Thus, adopting subscripts 0, 1 for spin-up and spin-down, respectively, on the quasiparticle operators α, and understanding $k = \vec{k}\uparrow$, $-k = -\vec{k}\downarrow$ on electron operators, one can define a complete set of operators as

$$\begin{aligned}
\alpha^*_{ek0} &= u_k a^*_k - v_k S^* a_{-k} \\
\alpha^*_{hk0} &= u_k S a^*_k - v_k a_{-k} \\
\alpha^*_{ek1} &= u_k a^*_{-k} + v_k S^* a_k \\
\alpha^*_{hk1} &= u_k S a^*_{-k} + v_k a_k \\
\alpha_{ek0} &= u_k a_k - v_k S a^*_{-k} \\
\alpha_{hk0} &= u_k S^* a_k - v_k a^*_{-k} \\
\alpha_{ek1} &= u_k a_{-k} + v_k S a^*_k \\
\alpha_{hk1} &= u_k S^* a_{-k} + v_k a^*_k
\end{aligned} \tag{3.25}$$

where S^* creates and S removes a pair of electrons.

Thus, α^*_{ek0} (α^*_{ek1}) creates an up-spin (down-spin) quasiparticle excitation definitely adding an electron to the system; similarly, α^*_{hk0} and α^*_{hk1} create quasiparticles while definitely removing an electron. The inverse relations are

$$\begin{aligned}
a^*_k &= u_k \alpha^*_{ek0} + v_k \alpha_{hk1} \\
a^*_{-k} &= u_k \alpha^*_{ek1} - v_k \alpha_{hk0} \\
a_k &= u_k \alpha_{ek0} + v_k \alpha^*_{hk1} \\
a_{-k} &= u_k \alpha_{ek1} - v_k \alpha^*_{hk0}
\end{aligned} \tag{3.26}$$

3.3.3 *Generalizations of BCS Theory*

It may be useful to clarify the relation of the BCS theory to the well-known Hartree and Hartree-Fock treatments of many electrons interacting via a two-body potential in a metal. The general approach (Anderson, 1961) is to extract from the full two-body interaction Hamiltonian,

$$H_{\text{int}} = \int \psi^*(\vec{x})\psi^*(\vec{y})V(\vec{x}, \vec{y})\psi(\vec{y})\psi(\vec{x})\,d\vec{x}\,d\vec{y} \tag{3.27}$$

groups of operators which can be identified with the average potential, the Fock exchange potential, and the anomalous BCS potential.

Following Anderson (1961), the factorization of the interaction Hamiltonian is viewed from the equation of motion,

$$\begin{aligned}
i\hbar\frac{\partial\psi^*(x)}{\partial t} &= [H, \psi^*(\vec{x})] \\
&= \left(\frac{p^2}{2m} + U - \mu\right)\psi^*(x) + \int \psi^*(\vec{x})\psi^*(\vec{y})V(\vec{y}-\vec{x})\psi(\vec{y})\,d\vec{y}
\end{aligned} \tag{3.28}$$

In (3.28) the interaction (integrand) term may be reduced to a one-body potential by factorizing and averaging the product of the particle operators, in three possible ways:

$$\int \psi^*(\vec{x})[\langle\psi^*(\vec{y})\psi(\vec{y})\rangle_{av} V(\vec{y}-\vec{x})]\, d\vec{y} = V_{av}\psi^*(\vec{x}) \tag{3.29a}$$

$$\int [\langle\psi^*(\vec{x})\psi(\vec{y})\rangle_{av} V(\vec{y}-\vec{x})]\psi^*(\vec{y})\, d\vec{y} = -A\psi^*(\vec{x}) \tag{3.29b}$$

$$\int [\langle\psi^*(x)\psi^*(y)\rangle_{av} V(\vec{y}-\vec{x})]\psi(\vec{y})\, d\vec{y} \tag{3.29c}$$

The first two factorizations give the usual Hartree potential and the Fock exchange term. Concentrating on the third term, (3.29c), by absorbing the others in the usual Hartree–Fock Hamiltonian H_{HF}, one obtains

$$i\hbar\frac{\partial\psi}{\partial t} = H_{HF}\psi^*(\vec{x}) + \int \Delta^*(\vec{x}, \vec{y})\psi(\vec{y})\, d\vec{y} \tag{3.30}$$

where

$$\Delta^*(\vec{x}, \vec{y}) = F^*(\vec{x}, \vec{y})V(\vec{x}-\vec{y}) \qquad \text{and} \qquad F^*(\vec{x}, \vec{y}) = \langle\psi^*(\vec{x})\psi^*(\vec{y})\rangle_{av}$$

Anderson interprets Δ^* as a mean potential for pair formation: a particular electron under consideration pairs with another electron, leaving a hole, as represented by the annihilation operator $\psi(\vec{y})$ in the integrand of (3.30). The cooperative nature of the pair potential is seen in its proportionality to F^* and thus to the strength of the existing pair correlations, as well as to the actual coupling V. Anderson showed that this picture leads to the usual BCS results.

The first importance of this generalized approach, beyond its conceptual aspect, is that it allows the pair potential Δ to be $\vec{k}$-dependent, to treat gap anisotropy, or dependent on position, as in the important cases of inhomogeneous superconductivity. Anisotropy in Δ_k can enter either through the interaction $V_{kk'}$ itself or through the sum on states k' over an anisotropic Fermi surface. Further, there is the possibility, if the Fermi surface has multiple sheets, that these can lead to different solutions for Δ, corresponding to multiple gaps.

A generalization relevant to many real superconductors was made by Anderson (1959) to include "dirty superconductors" of short mean free path. Pointing out that many nearly amorphous metals are good superconductors, Anderson generalized the pairing to time-reversed states. The states which now interact via $V_{kk'}$ are the true scattered states in the sample. Pairing is preserved with potential scattering because spins are unaffected and the orbits are changed symmetrically. All results of the BCS theory come out as before, with matrix elements now averaged over the Fermi surface. In a sense, dirty superconductors are those most accurately fitting the BCS model, for all anisotropies are averaged out in the limit of rapid scattering.

Generalization to strong coupling was made by Thouless (1960), who found only minor changes, noted above, in the limit $N(0)V > 1$ of the

BCS model ($V_{kk'} = V$ for $\omega < \omega_c$). This situation is not precisely the same case as the strong-coupling case treated by using the Eliashberg electron-phonon coupling. However, in both cases, the ratio $2\Delta/kT_c$ exceeds the BCS value: Thouless found 4.0 for this ratio, while in Eliashberg theory, the value varies with coupling and may exceed 4.0.

While Coulomb interaction of a pair of electrons with a fixed charge, for example, will be expected to scatter $\vec{k}\uparrow$ and $-\vec{k}\downarrow$ through the same angle, preserving the pair, a magnetic interaction will in general affect the $\uparrow$ and $\downarrow$ electrons differently and may destroy the pairing. It is thus not surprising that magnetic impurities have a stronger effect in depressing T_c of a superconductor than do nonmagnetic impurities. In fact, several pair-breaking mechanisms are known to act in this way and introduce a lifetime for the pair and a consequent broadening Γ of energy levels. One might guess that the condition $\Gamma > \Delta$ would destroy superconductivity, but this guess is not always correct, as was discovered theoretically by Abrikosov and Gor'kov (1960) in an analysis of the effects of magnetic impurities. While magnetic impurities do destroy superconductivity above a critical concentration, there is a range of concentrations, extending, in the Abrikosov-Gorkov theory, 10% below the critical concentration, in which gapless superconductivity, $\Gamma > \Delta$, exists. A rather similar situation occurs with a thin sample in a parallel magnetic field. This observation makes clear that the essential feature required for superconductivity is the electron pair correlation and not the gap in itself.

3.4 Theory of Quasiparticle and Pair Tunneling

The basic experiments of Giaever (1960) are summarized by (3.1) and (3.2) for $\sigma = (\mathrm{d}J/\mathrm{d}V)_S/(\mathrm{d}J/\mathrm{d}V)_N$ and by the corresponding expression for the current I generalized to allow both electrodes to be superconducting:

$$I_0(V) = G_{NN}\int_{-\infty}^{\infty} \rho_L(E)\rho_R(E+eV)[f(E)-f(E+eV)]\,\mathrm{d}E \qquad (3.31)$$

Here, $\rho_{L,R}$ represents $N_T(E)$ in (3.1) or (3.14), and $G_{NN} = R_{NN}^{-1}$ is the junction conductance when both electrodes are in the normal state. A further inherent feature of the SIS configuration is the Josephson supercurrent $I_J(V=0)$. Since this current was originally difficult to distinguish from a microshort current, its close relation to the quasiparticle current I_0, made theoretically clear in 1962–63, was at first not widely understood.

The experimental relationship (3.31) for the quasiparticle current I_0 is surprisingly simple (Fig. 3.6), depending only upon the total densities of quasiparticle excitations, $\rho(E)$, and, in particular, devoid of the coherence factors u, v. This simplicity, whose origin we will further discuss, prompted description of the tunneling results by Bardeen in the "semiconductor model." As sketched in Fig. 3.6 for the NIN, NIS, and SIS cases, the $I-V$ features can all be described assuming a density of states ρ symmetrically split about the Fermi level, so that at low temperatures the upper quasiparticle states have occupation probability less than $f(\Delta) \simeq e^{-\Delta/kT}$ for

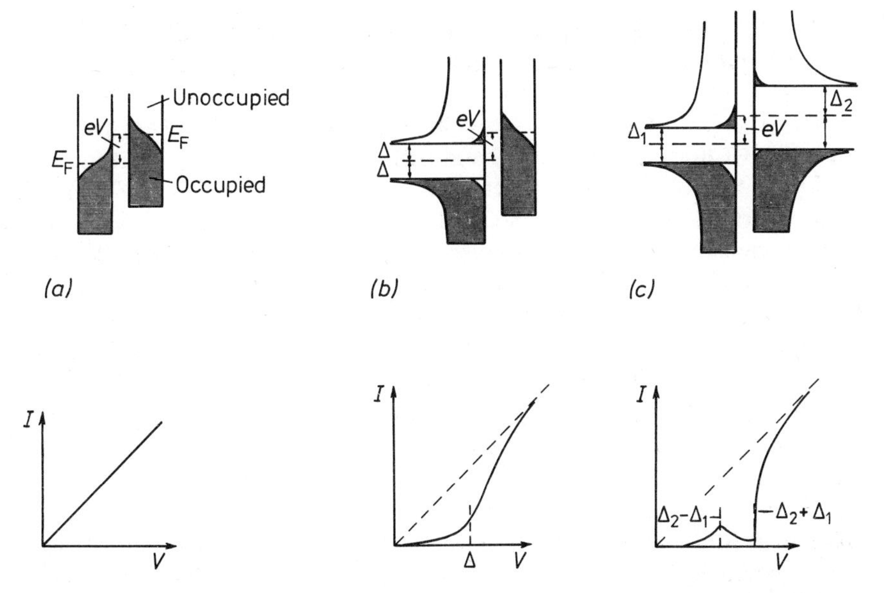

Fig. 3.6. Sketches indicate the quasiparticle current–voltage characteristics expected in the three basic cases: (a) two normal metals, (b) a normal metal and a single superconductor, and (c) two superconductors. In the latter case one may expect to determine both superconducting energy gaps from structures in $I(V)$ arising at the sum and difference of the two gap energies.

$kT \ll \Delta$, while the states below μ are full, except for the equal numbers of "holes" left by the thermal excitation. Application of a bias V shifts the densities of states rigidly past each other with, for example, the electron current to the right obtained by integrating the product $(1-f_R)\rho_R(E+eV)\times f_L\rho_L(E)$ over energies, where $f_R=f(E+eV)$, $f_L=f(E)$. Subtraction of a similar current to the left gives the integrand of (3.31).

In the NIS case, the current below $eV=\Delta$ arises only to the extent of thermal occupation of electrons above the Fermi energy in N and of holes in the filled states below μ in the superconductor. At $eV=\Delta$, a rise in current occurs, broadened by the $\sim 3.5kT$ width of the Fermi edge. At low temperature, $T \ll \Delta/k$, this edge is sharp and locates Δ accurately. The corresponding derivative in this low-T limit directly measures $\rho(E)$, via (3.2), as can also be seen from the diagram, making the usual assumption that the barrier transmission is unchanged for small bias voltages $V \simeq \Delta/e \ll U_B$.

The SIS case produces, at $T=0$, no current below $eV=\Delta_L+\Delta_R$, where an abrupt rise occurs, providing a means for precise determination of $\Delta_L+\Delta_R$. At finite temperature, this sharp rise persists unbroadened, because of the crossing singularities in ρ_L and ρ_R. An additional thermally activated current has a similarly unbroadened cusp which occurs at $eV=\Delta_R-\Delta_L$ (assuming $\Delta_R>\Delta_L$), with a negative-resistance region above this point. This feature can be used to locate the value $\Delta_R - \Delta_L$ if the observation temperature is high enough that $\exp(-\Delta_L/kT)$ is appreciable.

The basic transfer processes which must account for (3.31) are sketched in Figs. 3.7 and 3.8 in the NIS and SIS cases, respectively. Considering, first, the NIS case with an electron at a definite energy $E \leq eV$ incident from N, there are two possible final states of $E = E_k$: at $k'>$, where the chance that the excitation is an electron is $u_{k'}^2$, and at $k''<$, corresponding to $u_{k''}^2$. Following (3.12), and since $E_{k''} = E_{k'}$ implies $\varepsilon_{k'} \simeq -\varepsilon_{k''}$ (for $k', k'' \ll k_F$, ε_k is the energy of the single-particle state measured from the Fermi energy), the sum of the probabilities of finding an electronlike excitation at $k'>$ and at $k''<$ is unity:

$$u_{k'>}^2 + u_{k''<}^2 = \frac{1}{2}\left(1 + \frac{\varepsilon_{k'>}}{E_k}\right) + \frac{1}{2}\left(1 + \frac{-\varepsilon_{k'>}}{E_k}\right) = 1$$

Since for $E_k \gg \Delta$, $u^2 \simeq 1$, $v^2 \simeq 0$, in this case the electron is accommodated with high probability on the electronlike branch $k>$. Similarly, for bias of the opposite sign, the resulting excitation is, with high probability, a hole.

More formal justification of the use of the semiconductor model in the simple formulas (3.2) and (3.31) was given by Bardeen (1961) and Cohen, Falicov, and Phillips (1962), using the tunneling Hamiltonian. Extension and further clarification of this method by Josephson (1962) predicted the tunneling supercurrent. Subsequently, the results have been confirmed by more general methods, e.g., by Ambegaokar and Baratoff (1963). The tunneling Hamiltonian approach requires adapting the transfer Hamiltonian operator,

$$H^T = \sum_{k,q} (T_{kq} a_k^* a_q + T_{qk} a_k a_q^*) \tag{3.31a}$$

Fig. 3.7. Tunneling from a normal metal to a superconductor in the excitation representation. The electron which tunnels into the superconductor is accommodated with probability $u_{k'}^2$ on the electron branch and with probability $u_{k''}^2 = 1 - u_{k'}^2$ on the holelike branch. The two particular k values are specified by the conservation of energy in the tunneling process.

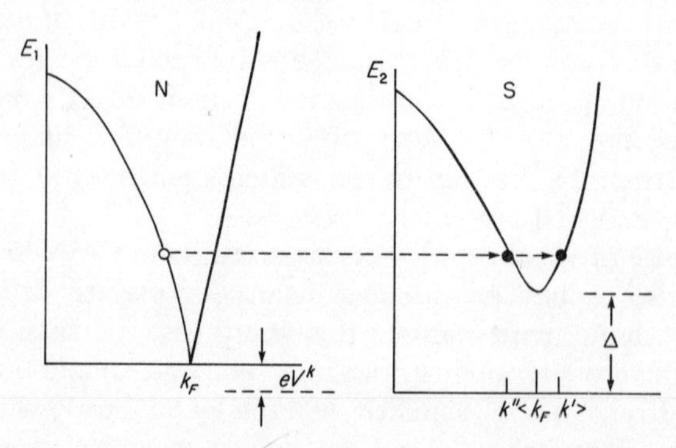

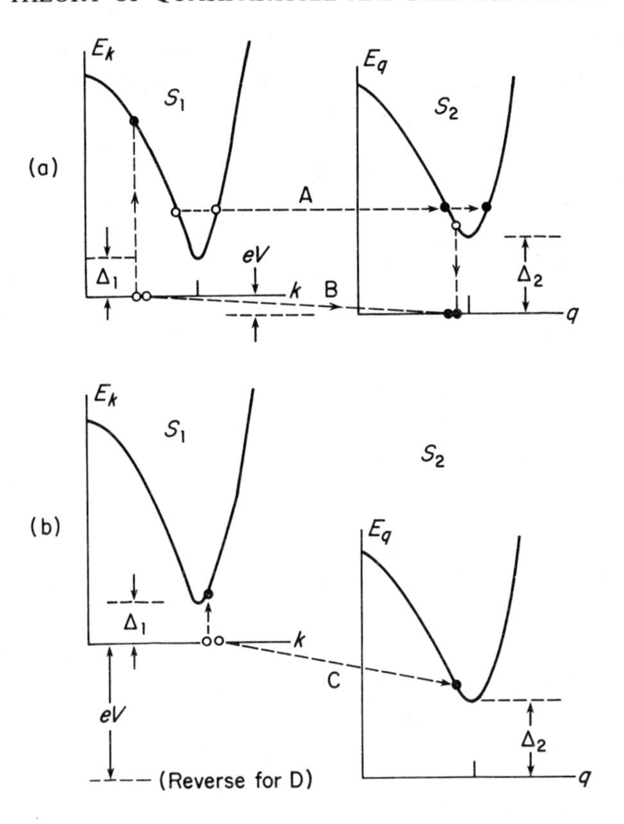

Fig. 3.8. Basic processes or channels in S_1IS_2 quasiparticle tunneling in the excitation representation. (a) In Channel A, an existing quasiparticle excitation in the first superconductor tunnels into S_2. Channel B is similar, with the exchange of a pair. (b) In Channel C, a pair in the condensate of S_1 is broken up, creating quasiparticle excitations in both superconductors. Channel D, the reverse of C, is energetically possible only for bias of reversed sign.

to the superconducting case, by expressing the electron operators a_k, a_q in terms of the superconductor quasiparticle excitation operators α_k, α_q:

$$H^{\mathrm{T}} = \sum_{k,q} [T_{kq}(u_k\alpha^*_{ek0} + v_k\alpha_{hk1})(u_q\alpha_{eq0} + v_q\alpha^*_{hq1}) + T^*_{qk}(u_q\alpha^*_{eq0} + v_q\alpha_{hq1})(u_k\alpha_{ek0} + v_k\alpha^*_{hk1})] \tag{3.31b}$$

Calculating the current brings in the coherence factors u_k, v_k and u_q, v_q, which, in the end, fall out of the final result.

The method of Cohen, Falicov, and Phillips (1962), elaborated by Josephson (1962), starts by calculating the rate of change of the expectation number of electrons d/dt $\langle N_L \rangle$, where

$$N_L = \sum_k a^*_k a_k \tag{3.32}$$

from the equation of motion of the operator N_L, which is given by its commutator with the Hamiltonian $H=H^L+H^R+H^T$:

$$\dot{N}_L=\frac{1}{i\hbar}[N_L, H^T]=\frac{1}{i\hbar}\sum_{k,q}(T_{kq}a_k^*a_q-T_{qk}a_q^*a_k) \tag{3.33}$$

We have used (3.31) and the fact that H^L and H^R commute with N_L. The current operator J_{op} is then

$$J_{op}=e\dot{N}_L=\frac{e}{i\hbar}[N_L, H^T] \tag{3.34}$$

It is necessary to evaluate the expectation $\langle\psi|J_{op}|\psi\rangle$ in the perturbed state ψ resulting from the application of a bias voltage. Josephson (1962*b*) carried out this evaluation in the interaction representation, in which operators have a simple exponential time dependence which is factored out of the wavefunction. The wavefunction follows from a Schrödinger equation involving only the perturbation:

$$\dot{\psi}=-\frac{i}{\hbar}H^T(t)\psi \tag{3.35}$$

It is assumed that the coupling $H^T(t)$ is turned on slowly at $t=-\infty$, mathematically accomplished by multiplying by $e^{\varepsilon t}$, where the small positive ε will later be set to zero. If ψ_0 represents the wavefunction at $t=-\infty$, then, to first order,

$$\psi(t)\simeq\left[1-\frac{iE(t)}{\hbar}\right]\psi_0 \tag{3.36}$$

where

$$E(t)=\int_{-\infty}^{t} e^{\varepsilon t'}H^T(t')\,dt' \tag{3.37}$$

From the definition of J_{op}, this equation gives

$$\langle\psi(t)|J_{op}|\psi(t)\rangle=\frac{e}{i\hbar}\left\langle\psi_0\left|\left[1+\frac{iE(t)}{\hbar}\right][N_L, H^T]\left[1-\frac{iE(t)}{\hbar}\right]\right|\psi_0\right\rangle \tag{3.38}$$

from which Josephson obtains the compact expression

$$J(t)=\frac{e}{\hbar^2}\langle[[N_L, H^T], E]\rangle_{th} \tag{3.39}$$

where a thermal average has also been indicated.

Josephson's calculation proceeds by evaluating $J(t)$ using (3.33) and (3.37), in which all indicated electron operators as in (3.31) are expanded into quasiparticle operators according to (3.26) (assuming the electrode is superconducting).

For example, in the expansion of H^T in the quasiparticle operators, the first term is $T_{kq}u_k u_q \alpha^*_{ek0}\alpha_{eq0}$. In the calculation of the contribution of this term to $E(t)$ via (3.37), the integral

$$\int_{-\infty}^{t} e^{\varepsilon t'}\alpha^*_{ek0}(t')\alpha_{eq0}(t')\,dt' = \alpha^*_{ek0}(0)\alpha_{eq0}(0)\int_{-\infty}^{t} \exp\left[\left(\varepsilon + \frac{i\,\Delta E}{\hbar}\,t'\right)\right]dt' \tag{3.40a}$$

occurs, with $\Delta E_1 = \mu_k + E_k - \mu_q - E_q$ representing the difference in energy of the two excitations, including the shift of Fermi energies. This contribution to $E(t)$, found to be

$$\frac{i\hbar T_{kq}u_k u_q \alpha^*_{ek0}(t)\alpha_{eq0}(t)}{\Delta E_1 - i\hbar\varepsilon} \tag{3.40b}$$

has a characteristic energy denominator $\Delta E_1 = eV + E_k - E_q$, since $\mu_k - \mu_q = eV$. Altogether, four distinct denominators ΔE are found.

In following the outlined procedure, Josephson found two types of terms of nonzero expectation value appearing in $[N_L, H^T]$, which enters (3.39). These are, first, terms like

$$\begin{aligned}[\alpha^*_{ek0}\alpha_{eq0}, \alpha^*_{eq0}\alpha_{ek0}] = \cdots &= n_{k0}(1-n_{q0}) - (1-n_{k0})n_{q0} \\ &= n_{k0} - n_{q0}\end{aligned} \tag{3.41}$$

where occupation numbers n are defined as

$$n_{ks} = \alpha^*_{eks}\alpha_{eks} = \alpha^*_{hks}\alpha_{hks}, \qquad s = 0, 1 \tag{3.42}$$

and, second, terms like

$$\begin{aligned}[\alpha^*_{ek0}\alpha_{eq0}, \alpha_{hk0}\alpha^*_{hq0}] = \cdots &= n_{k0}S^*_k(1-n_{q0})S_q - (1-n_{k0})S^*_k n_{q0}S_a \\ &= -(n_{k0} - n_{q0})S^*_k S_q\end{aligned} \tag{3.43}$$

using relations such as

$$\alpha^*_{eks}\alpha_{hks} = S^* n_{ks} \tag{3.44}$$

After thermal averaging, the difference $(n_{k0} - n_{q0})$ brings in the familiar Fermi factor difference $(f_k - f_q)$. The particular significance of the $S^*_k S_q$ combination is that it depends upon the phase difference between the pairs in the two electrodes. Further, following the procedure of evaluating (3.39), the result

$$J = J_0 + \tfrac{1}{2}J_1 S^*_k S_q + \tfrac{1}{2}J^*_1 S^*_q S_k = J_0 + J_J \tag{3.45}$$

was obtained by Josephson (1962), in which the phase difference $\phi = \theta_q - \theta_k$ entering through the S operators affects the contributions of the last two terms.

In (3.45), the two closely related contributions to J, found in the same order $|T|^2$, in the transfer matrix element, are from quasiparticle tunneling (J_0) (Bardeen, 1961; Cohen, Falicov, and Phillips, 1962) and from pair tunneling (J_J) (Josephson, 1962; Ambegaokar and Baratoff, 1963).

The two contributions, in more complete form, to $J = J_0 + J_J$ are

$$J_0 = \frac{-eT^2}{i\hbar} \sum_{k,q} \left[\frac{u_k^2 u_q^2 (f_q - f_k)}{E_q - E_k + eV + i\eta} + \frac{v_k^2 v_q^2 (f_k - f_q)}{-E_q + E_k + eV + i\eta} + \frac{u_k^2 v_q^2 (1 - f_q - f_k)}{-E_q - E_k + eV + i\eta} + \frac{v_k^2 u_q^2 (f_q + f_k - 1)}{E_q + E_k + eV + i\eta} + \text{c.c.} \right] \tag{3.46}$$

and

$$J_J = \frac{-eT^2}{i\hbar} \sum_{k>, q>} \left[\frac{4 u_k u_q v_k v_q (f_q - f_k) e^{i\phi}}{E'_q - E'_k + eV + i\eta} + \text{c.c.} \right] \tag{3.47}$$

where $\eta = \hbar\varepsilon$. Here, $\phi = \theta_R - \theta_L$, where $\theta_R(\theta_L)$ is the phase of the pairs on the right (left). In (3.47), compactness has been achieved by restricting k, q to electronlike branches $k>$ using the symmetry properties of u, v: exchange of branches turns u_k^2 into $1 - u_k^2$ and v_k^2 into $1 - v_k^2$ and is assumed to leave f_k unchanged. Further, the primes on E'_q, etc. now indicate that these may take negative values, as in the semiconductor model of Bardeen.

Reduction of these formulae to give (3.31) (from J_0) and a corresponding simplification of J_J involves changing sums on k to integrals on energy, bringing in densities of states $\rho(E)$. In performance of the first integration, the energy denominators ΔE lead to δ-functions on energy, according to the relation

$$\frac{1}{\Delta E - i\eta} = P\left(\frac{1}{\Delta E}\right) + \pi i \delta(\Delta E) \tag{3.48}$$

where P indicates the principal part in taking the integral and πi is the residue at the pole. In all terms in J_0, the principal parts cancel. In the four terms of (3.46), further, the u, v factors drop out, as a consequence of the symmetry with regard to exchange of branches $k>$, $k<$. In the four terms remaining, the first two combine if E is allowed to take both positive and negative values, while the third and fourth cancel, leaving the simple result

$$J_0 = \frac{2\pi e\, |T|^2 N_{0L} N_{0R}}{\hbar} \int_{-\infty}^{\infty} [f(E - eV) - f(E)] \rho_R(E - eV) \rho_L(E)\, dE \tag{3.49}$$

equivalent to (3.31), with reversed sign of V. The normal-state densities of states $N_{0L,R}$ have been factored out of $\rho_{L,R}$, as has an averaged matrix element, assumed to show negligible energy variation on the scale of the superconducting gap. (In practice, as we have noted above, a weak parabolic increase in dJ/dV results from T^2 variation with V.) This expression reduces to the normal-state result with $\rho_{L,R} = 1$; such a result could have been obtained through the analysis if the electron operators had been carried through to the end, leading to a single term like the first of (3.46), with $u_k = 1$ and with E replaced by $|\varepsilon_k|$.

The term J_J of (3.47) can be further reduced (Waldram, 1976) and includes a principal part contribution (because of the $e^{i\phi}$ factor) as well as a contribution from the residue. The result is

$$J_J = J_1(V)\sin\phi + J_2(V)\cos\phi \tag{3.50}$$

$$J_1(V) = \frac{2eT^2}{\hbar} N_{0L}N_{0R}P\int_{-\infty}^{\infty} d\varepsilon \frac{\Delta_1\Delta_2[f(E_1)-f(E_2)]}{E_1E_2(E_1-E_2+eV)} \tag{3.51}$$

$$= \frac{\pi\,\Delta(T)}{2eR_{NN}}\tanh\frac{\Delta(T)}{2kT} \quad \text{if} \quad V=0, \Delta_1=\Delta_2=\Delta \tag{3.52}$$

$$J_2(V) = \frac{2\pi eT^2}{\hbar} N_{0L}N_{0R}\int_{-\infty}^{\infty} d\varepsilon \frac{\Delta_1\Delta_2}{(E-eV)E}[f(E-eV)-f(E)] \tag{3.53}$$

Note that the maximum supercurrent at $T=0$ corresponds to the current in the normal-state junction, biased at $eV=\pi\,\Delta/2$.

The supercurrent J_J is of the same order (T^2) in the barrier factor as is the quasiparticle current J_0, even though pairs of electrons are involved, because of the coherence of initial and final states in the transfer process, represented by $e^{i\phi}$, as compared to the incoherence of contributions to J_0. The coherence is stabilized by a coupling energy of order T^2 (Anderson, 1964);

$$\Delta E = \frac{-\hbar J_1(0)}{2e} = \frac{-\Phi_0 J_1(0)}{2\pi} \tag{3.54}$$

which is small and often overcome by fluctuating signals in experimental situations. Without the phase locking, the contributions J_J are not seen, leaving only the quasiparticle contributions.

Having carried through the analysis following Josephson (1962) at the level of detail required to establish the intimate connection between the quasiparticle- and super- currents, we now focus on the four quasiparticle contributions in (3.46). These have become identified as the four S_1IS_2 quasiparticle tunneling channels, labeled in order A, B, C, and D and further detailed in Table 3.1 and Fig. 3.8. The eight terms, grouped according to common energy constraints to give the four channels, can be generated by carrying out the indicated multiplications in the expression for H^T, (3.31b). Channels A and B, as may be seen from Table 3.1, always subtract an existing excitation from either 1 (E_k) or 2 (E_q) and hence occur only at $T>0$. This result is clear from the energy constraint of the form $eV=\pm(E_q-E_k)$. The two remaining channels are identical except for the direction of transfer, and have an energy constraint $eV=\pm(E_q+E_k)$. This implies that only one of C and D can occur for a given sign of V (since the excitation energies E_q, E_k are inherently positive). In the open channel, the process simply corresponds to generation of excitations on both sides by destruction of a pair. This basic superconductive tunneling process (labeled C in Fig. 3.8b) has threshold $eV=\Delta_1+\Delta_2$ and accounts for the familiar current onset at the sum-gap

Table 3.1. S_1IS_2 Quasiparticle Tunneling Channels (after Tinkham, 1972)

Channel	Terms in H^T	Probability	Electrons (Excitations) Added $S_1(k)$, $S_2(q)$	Excitation Energy E_q in S_2
A	$u_k u_q \alpha^*_{ek0} \alpha_{eq0}$	$u_k^2 u_q^2 (1-f_k) f_q$	$-1(-1)$, $1(1)$	$E_k - eV$
	$u_q u_k \alpha^*_{eq0} \alpha_{ek0}$	$u_q^2 u_k^2 (1-f_q) f_k$	$1(1)$, $-1(-1)$	
B	$v_k v_q \alpha_{hk1} \alpha^*_{hq1}$	$v_k^2 v_q^2 f_k (1-f_q)$	$-1(1)$, $1(-1)$	$E_k + eV$
	$v_q v_k \alpha_{hq1} \alpha^*_{hk1}$	$v_q^2 v_k^2 f_q (1-f_k)$	$1(-1)$, $-1(1)$	
C	$v_k u_q \alpha_{hk1} \alpha_{eq0}$	$v_k^2 u_q^2 f_k f_q$	$-1(-1)$, $1(-1)$	$-E_k - eV$
	$u_q v_k \alpha^*_{eq0} \alpha^*_{hk1}$	$u_q^2 v_k^2 (1-f_q)(1-f_k)$	$1(1)$, $-1(1)$	
D	$u_k v_q \alpha^*_{ek0} \alpha^*_{hq1}$	$u_k^2 v_q^2 (1-f_k)(1-f_q)$	$-1(1)$, $1(1)$	$-E_k + eV$
	$v_q u_k \alpha_{hq1} \alpha_{ek0}$	$v_q^2 u_k^2 f_q f_k$	$1(-1)$, $-1(-1)$	

voltage. In the semiconductor model, this process is represented by tunneling from "filled" quasiparticle states to "empty" quasiparticle states at $eV = \Delta_1 + \Delta_2$. The channel *A* process, elastic transfer of an existing excitation from one side to the other, is represented in the semiconductor picture by a thermally excited quasiparticle in the unfilled band transferring to the opposite unfilled band. Channel *B* is reminiscent of an Auger process, as illustrated in Fig. 3.8b. An existing quasiparticle condenses with an electron from the other electrode to form a pair, releasing energy $E_q + eV$, which then is the energy E_k of the final quasiparticle. Note that the resultant quasiparticle E_k in the process labeled *B* in Fig. 3.8b could transfer elastically to the right in channel *A*, arriving back on the right with additional energy eV. A sequence of n (B, A) processes increases the energy of the quasiparticle by neV. The semiconductor description of the channel *B* process would involve a quasiparticle from the filled band tunneling into a vacant position in the filled band of the adjacent electrode.

The Josephson tunneling supercurrent in the same representation arises from a sum over intermediate states of the sort described in (3.47) and (3.51): a schematic representation is given in Fig. 3.9.

Fig. 3.9. Josephson tunneling of pairs between two superconductors.

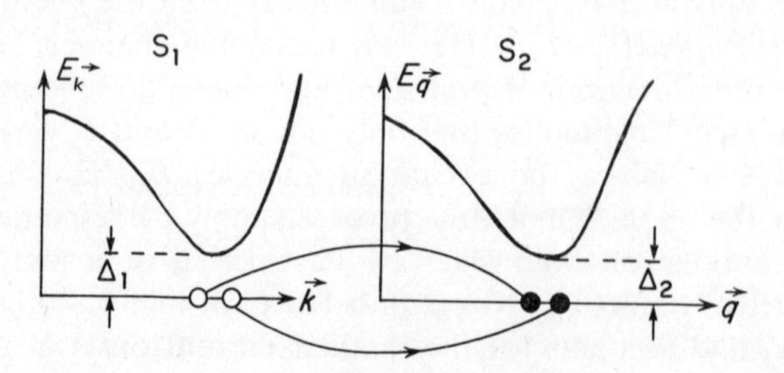

3.5 Gap Spectra of Equilibrium BCS Superconductors

The expression for the quasiparticle current density for the case of unequal gaps Δ_1, Δ_2 is given by (3.31), or, equivalently,

$$J_0(V) = G_{NN}e^{-1}\int_{-\infty}^{\infty} dE \frac{|E-U|}{[(E-U)^2-\Delta_1^2]^{1/2}} \frac{|E|}{(E^2-\Delta_2^2)^{1/2}}[f(E-U)-f(E)] \tag{3.55}$$

where $U = eV$ and $f(E) = [1+\exp \beta E]^{-1}$, $\beta = 1/kT$, and

$$G_{NN} = \frac{2\pi e}{\hbar}|T|^2 N_1(0)N_2(0) \tag{3.56}$$

In general, (3.55) must be evaluated numerically. At $T=0$, as shown by Douglass and Falicov (1964), this equation can be expressed in terms of κ and ε, which are, respectively, complete elliptic integrals of the first and second kind. (The current is zero, at $T=0$, for $eV < \Delta_1 + \Delta_2$, which corresponds to the threshold for channel C, D processes.) Thus,

$$\frac{J_0(V)}{G_{NN}} = \begin{cases} 0, & 0 \leq U \leq \Delta_1 + \Delta_2 \\ [U^2-(\Delta_2-\Delta_1)^2]^{1/2}\varepsilon(x) - 2\Delta_1\Delta_2[U^2-(\Delta_2-\Delta_1)^2]^{-1/2}\kappa(x), & U > \Delta_1 + \Delta_2 \end{cases} \tag{3.57}$$

where

$$x^2 = \frac{U^2-(\Delta_1+\Delta_2)^2}{U^2-(\Delta_2-\Delta_1)^2} \tag{3.58}$$

Since the limit as $x \to 0$ ($eV = \Delta_1 + \Delta_2$) of both $\kappa(x)$ and $\varepsilon(x)$ is $\pi/2$, the magnitude of the discontinuity in the current density at the threshold $eV = \Delta_1 + \Delta_2$ is

$$\Delta J = G_{NN}\frac{\pi}{2e}\sqrt{\Delta_1\Delta_2} = J_{NN}(\Delta_1+\Delta_2)\frac{\pi}{2}\frac{\sqrt{\Delta_1\Delta_2}}{\Delta_1+\Delta_2} \tag{3.59}$$

at $T=0$. This current jump thus precisely equals the $T=0$ Josephson current, given above in (3.52), in the case $\Delta_1 = \Delta_2$. The physical origin of the discontinuity is the crossing of the gap edge singularities in ρ_L and ρ_R at $eV = \Delta_L + \Delta_R$ shown in Figs. 3.6c and 3.6d, which graphically indicate the terms in (3.55). For this reason, the corner discontinuity in the J–V curve may be expected to remain sharp at finite temperature. This behavior is shown in Fig. 3.10 in a sequence of curves obtained from Al–Al_2O_3–Al junctions (Blackford and March, 1968). The magnitude of the current jump as a function of T was calculated by Shapiro et al. (1962) and can be expressed as

$$\Delta J = G_{NN}\frac{\pi}{4e}\sqrt{\Delta_1\Delta_2}\frac{\sinh[(\Delta_1+\Delta_2)/2kT]}{\cosh(\Delta_1/2kT)\cosh(\Delta_2/2kT)} \tag{3.60}$$

where an added factor 2 has appeared from expressing exponential functions in hyperbolic functions, and the $T=0$ limit agrees with (3.59).

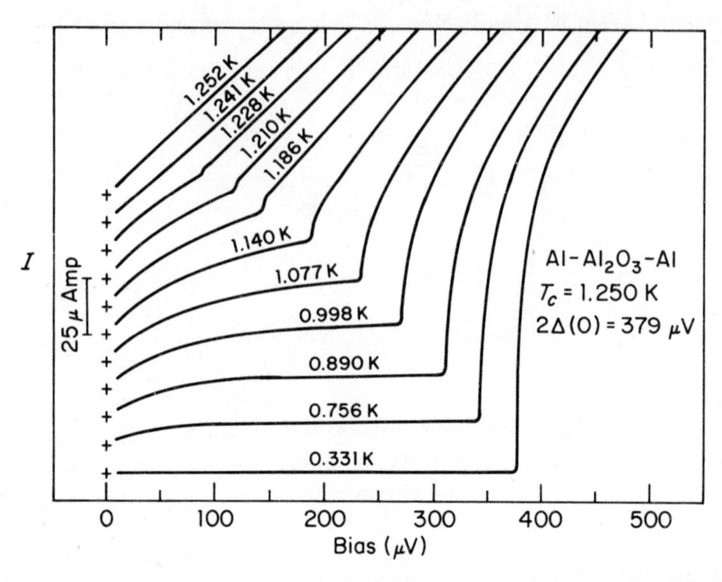

Fig. 3.10. Experimental results for quasiparticle tunneling between identical superconductors, in this case Al, measured at a sequence of temperatures rising from 0.331 K. The sharp corner at $eV = 2\Delta(\mathrm{T})$ is retained as the temperature rises because it arises from the crossing of singularities in the densities of states in the opposing electrodes. (After Blackford and March, 1968.)

The second point of singularity occurs at $eV = |\Delta_1 - \Delta_2|$, at which the A and B channels, dependent upon thermally excited quasiparticles, also experience a crossing of square root singularities in $\rho(E)$, as suggested by Fig. 3.6c. Analysis of (3.55) in the BCS limit of real Δ_1, Δ_2 (Shapiro et al., 1962) predicts a logarithmic infinite peak in J_0 centered at $eV = \Delta_2 - \Delta_1$:

$$J_0 \propto \ln |eV - |\Delta_1 - \Delta_2|| \cdot \left(\exp \frac{-\Delta_1}{kT} - \exp \frac{-\Delta_2}{kT}\right) \tag{3.61}$$

Although lifetime broadening (but not thermal smearing) will remove the infinity, careful observation, such as those in Fig. 3.11, after Tsang and Ginsberg (1980*b*), indeed reveals a sharp peak, even at the relatively high measurement temperature 1.08 K. The symmetry of the current peak at about $eV = \Delta_2 = \Delta_1$ implies a negative-resistance region. Calculations of the cusp feature in several cases are shown in Fig. 3.12 (Taylor, 1963).

As is consistent with the factor $[\exp(-\Delta_1/kT) - \exp(-\Delta_2/kT)]$ in (3.61), the logarithmic peak disappears as $eV = \Delta_2 - \Delta_1 \to 0$. In this limit, an approximate expression for J_0 for $0 \leqslant eV < 2\Delta$ is (Taylor, 1963)

$$J_0 = 2G_{\mathrm{NN}} \exp\left(\frac{-\Delta}{kT}\right) \sqrt{\frac{2\Delta}{eV + 2\Delta}}\,(eV + \Delta) \sinh\left(\frac{eV}{2kT}\right) K_0\left(\frac{eV}{2kT}\right) \tag{3.62}$$

where K_0 is a modified Bessel function of the second kind. This function displays a weak negative-resistance behavior when $kT < 0.3\Delta$.

This feature is shown in numerical calculations by Taylor (1963) in Fig. 3.13 for four temperatures $T/T_c = 0$, 0.07, 0.17, and 0.85. Note that, at $T = 0$, this current is entirely absent and that the pair breaking current at threshold $eV = 2\Delta$ is just $\pi/4 = 0.785$ of the normal-state current.

In the NIS case at $T = 0$ (Fig. 3.1a), the current [Eq. (3.55)] becomes

$$J_0(V) = \begin{cases} 0, & eV < \Delta \\ \dfrac{G_{NN}}{e} \displaystyle\int_{\Delta}^{eV} \dfrac{E}{\sqrt{E^2 - \Delta^2}} \, dE = \dfrac{G_{NN}}{e} [(eV)^2 - \Delta^2]^{1/2}, & eV \geq \Delta \end{cases} \tag{3.63}$$

For finite T, an expansion of the Fermi functions in (3.55) was shown by Giaever and Megerle (1961) to give the rapidly convergent series

$$J_0(V) = 2G_{NN} \frac{\Delta}{e} \sum_{m=1}^{\infty} (-1)^{m+1} K_1 \frac{m\Delta}{kT} \sinh \frac{meV}{kT} \tag{3.64}$$

valid for $eV < \Delta$, where $K_1(x)$ is the first-order, modified Bessel function of the second kind.

In practice, the NIS gap structure is usually studied in the derivative

Fig. 3.11. The tunneling current-voltage relationship for two dissimilar superconductors, in this case Pb and Al (after Tsang and Ginsberg, 1980). In the upper curve, a sharp cusp is observed at the energy corresponding to the difference of the two gaps. In the lower curve, the sharpness of the feature has been destroyed, presumably by addition of 69 atomic parts per million of manganese, which is a magnetic impurity in the Pb.

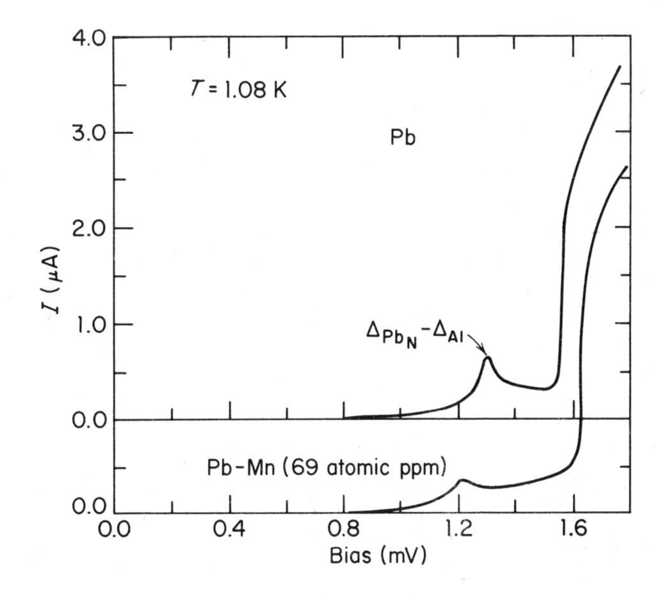

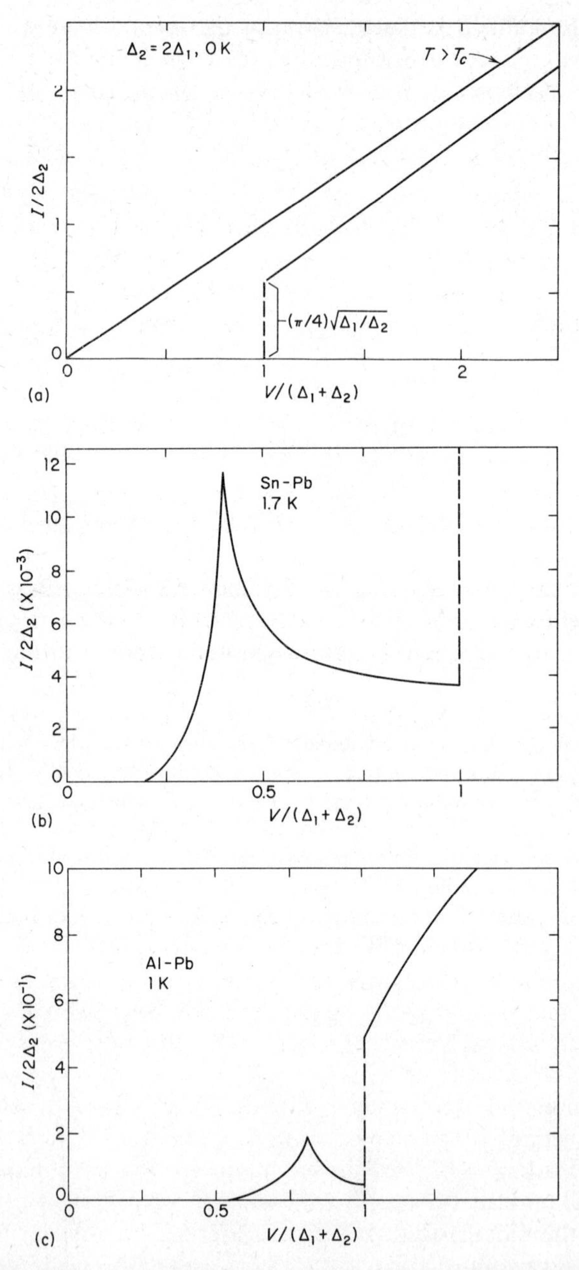

Fig. 3.12. Theoretical *I–V* curves for tunneling between dissimilar superconductors. (a) Current-voltage relationship at $T = 0$ shows an abrupt rise at the sum gap. (b) Modeling the current-voltage relationship for a tin-lead junction at 1.7 K. Note the changes of scale in viewing the cusp structure at the difference gap. (c) Current-voltage relationship modeled for an aluminum-lead junction at 1 K. Note that the cusp structure at the difference gap is relatively larger in this case. This curve may be compared with the upper curve in Fig. 3.11. (After Taylor, 1963.)

$dJ/dV = G(V)$, which is given, from (3.55) with $\Delta_1 = 0$, by

$$G(V) = G_{NN} \int_{-\infty}^{\infty} \frac{|E|}{(E^2 - \Delta^2)^{1/2}} \left[-\frac{\partial f}{\partial E}(E + eV) \right] dE \tag{3.65}$$

The function $\sigma(V, T, \Delta) = G(V)/G_{NN}$, which has been calculated numerically and tabulated by Bermon (1964) (see also Mühlschlegel, 1959; Douglass and Falicov, 1964), is frequently used to fit measurements and thereby determine Δ. Since $-f'(E)$ is a sharply peaked function at $E = 0$ of width $3.5kT$, in cases $3.5kT \ll \Delta \simeq 1.76kT_c$ or $T/T_c \ll 1/2$, $-f'$ is effectively a δ function, so that, to a good approximation,

$$\sigma(V) \simeq \begin{cases} 0, & eV < \Delta \\ \dfrac{|eV|}{[(eV)^2 - \Delta^2]^{1/2}}, & eV > \Delta \text{ and } T = 0 \end{cases} \tag{3.66}$$

The measurement of σ for Pb at 0.3 K, shown in Fig. 1.6a, is well in this limit, at $T/T_c \simeq 0.04$, and Δ can essentially be read from the peak position in σ.

An approximate determination of Δ can also be obtained from the T dependence at $T \ll T_c$ of $\sigma(0)$, as derived from (3.64), by taking only the lowest term, $m = 1$, and using the asymptotic form

$$K_1(x) \simeq \left(\frac{\pi}{2x}\right)^{1/2} e^{-x}; \qquad x \gg 1 \tag{3.67}$$

Thus, one obtains

$$\sigma(0) = \left(\frac{2\pi\Delta}{kT}\right)^{1/2} \exp\frac{-\Delta}{kT} \tag{3.68}$$

It may be pointed out that this expression is similar in form to the expression for the specific heat at low temperature.

Several methods of determining gap parameter values are either evident from the preceding discussion or have been used for reasons of simplicity. In a roughly decreasing order of reliability, these methods are as follows:

1. Measurement of the point of discontinuity in the I–V relation of SIS or S_1IS_2 junctions is the method used to determine 2Δ or $\Delta_1 + \Delta_2$. In the event of broadening of the current jump, the point of maximum slope dI/dV should preferably be used to locate $\Delta_1 + \Delta_2$. This point will be well defined if the singularities Δ_1, Δ_2 are lifetime-broadened, for example (Dynes, Narayanamurti and Garno, 1977), and in this case, the method of Gasparovic, Taylor, and Eck (1966), extrapolating the rise in I down to its intersection with the subgap current, may underestimate $\Delta_1 + \Delta_2$. In case of an I–V rise of nearly constant slope, the center point of the rise is chosen (McMillan and Rowell, 1969), presumably representing the average of a distribution of gap values. In the S_1IS_2 case, the cusp at $\Delta_2 - \Delta_1$ must also be located to get Δ_1 and Δ_2.

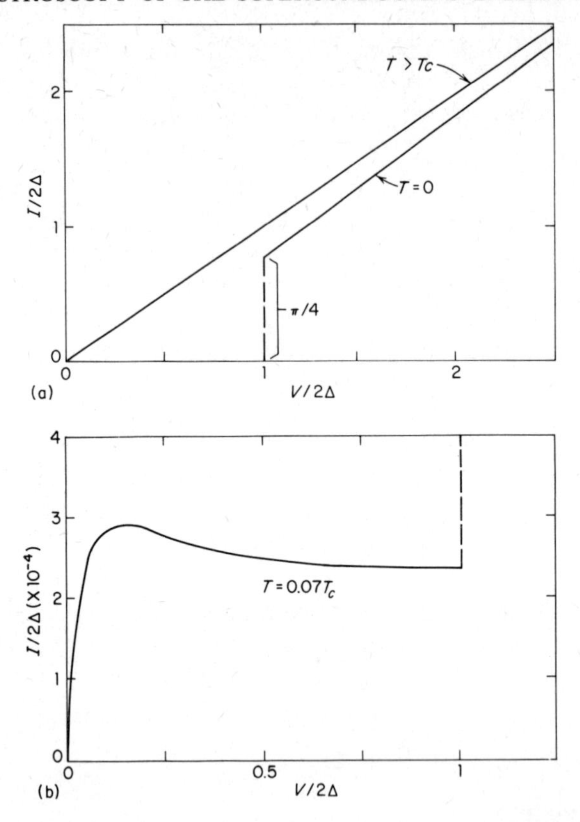

Fig. 3.13. Theoretical I–V curves for tunneling between two identical superconductors: (a) $T=0$, (b) $T=0.07T_c$, (c) $T=0.17T_c$, and (d) $T=0.85T_c$. (After Taylor, 1963.) Note the scale changes in parts (b) and (c) of this figure.

2. Accurate fitting of normalized NIS conductance data to the integral of (3.65) calculated numerically, is the best method in NIS cases where the BCS calculation is expected to be valid. An accurate measurement of the junction temperature should be used in the calculation. Matching the $\sigma(V)$ peak voltage and magnitude by inspection of the Bermon table is an alternative procedure, less accurate and less likely to reveal systematic deviation of the data from the BCS model, but still useful. A rough measure of Δ may be obtained from an I–V plot by finding the voltage at which $I(V)$ is parallel to the normal-state curve. The temperature dependence of $\sigma(0)$ measured over a wide range extending to $T \ll T_c$ can also be used to estimate Δ. A plot of (3.68) is given by Douglass and Falicov (1964).

The most direct test of the BCS theory made possible by the tunneling technique is quantitative verification of the form of $\rho(E)$, accomplished in the earliest papers (Nicol, Shapiro, and Smith, 1960; Giaever and Megerle, 1962). A further test of theory afforded by the unambiguous measurement of the dependence of Δ on T is illustrated in Fig. 3.14. The agreement with the BCS functional form of $\Delta(T)/\Delta(0)$ vs. T/T_c for

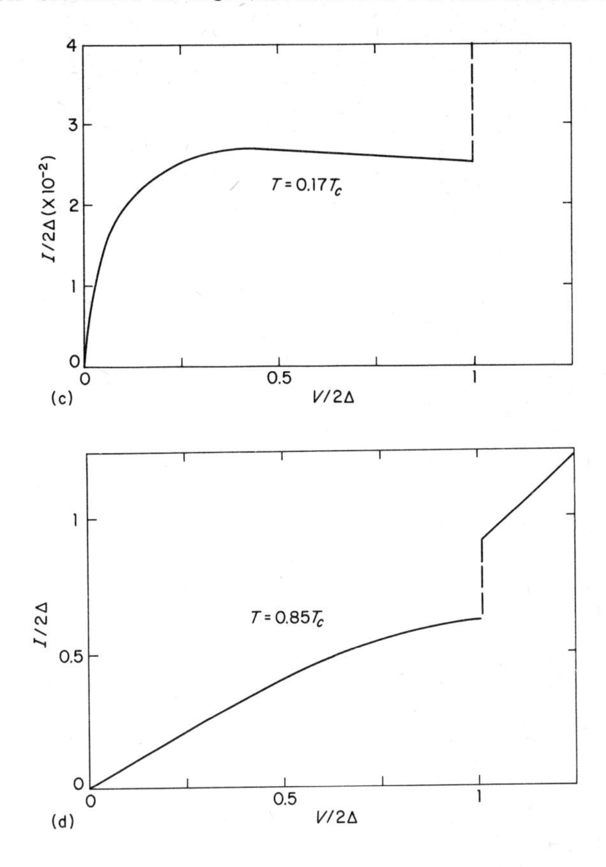

weak-coupling superconductors is essentially perfect and has historically been limited only by the accuracy of the measurements.

The accumulated results for $\Delta(0)$ and $2\Delta(0)/kT_c$ for the variety of superconductors studied by tunneling are collected in Appendix C.

The maximum Josephson current, in simplest terms, can be compared in its magnitude at $T=0$ with the expectation

$$J_1(0)=\frac{\pi}{4}\frac{2\Delta}{eR_{\mathrm{NN}}}=\Delta J|_{2\Delta} \tag{3.69}$$

in the case $\Delta_1=\Delta_2$, and also in its functional dependence on $t=T/T_c$. At a second level, the voltage dependence of the Josephson currents, depending on $\sin\phi$ and $\cos\phi$, (3.50), can be tested, at least in principle.

An early confirmation (Fiske, 1964) of the normalized T dependence of J_1 (at $V=0$) is shown in Fig. 3.15 for Sn–Sn and Sn–Pb junctions. The theoretical curve for $\Delta_1\neq\Delta_2$ has been calculated following the generalization of (3.52) (Ambegaokar and Baratoff, 1963):

$$J_1(0)=\frac{\Delta_1(T)}{eR_{\mathrm{NN}}}K\left\{\left[1-\frac{\Delta_1^2(T)}{\Delta_2^2(T)}\right]^{1/2}\right\} \tag{3.70}$$

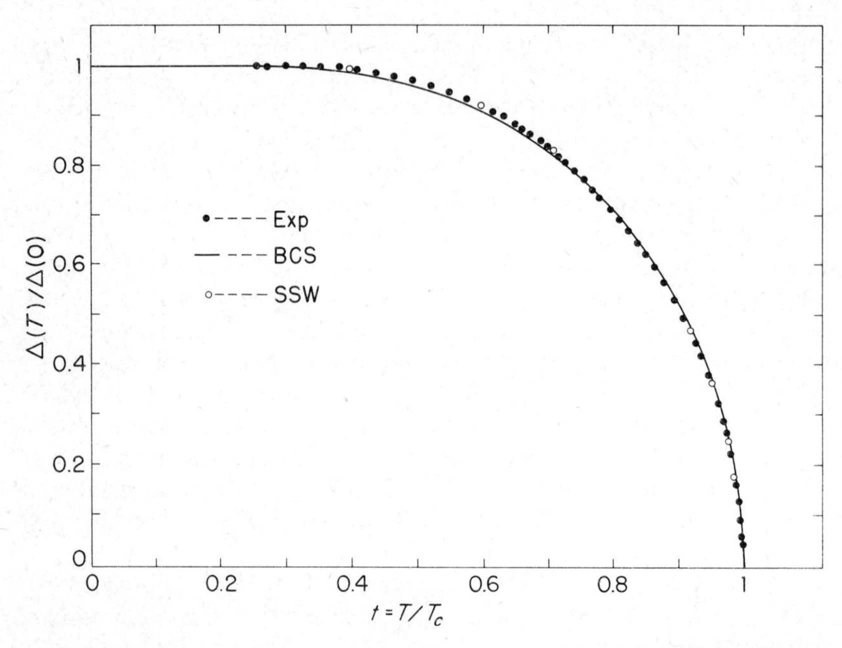

Fig. 3.14. Temperature dependence of the superconducting gap 2Δ in aluminum (dotted points are data obtained from the curves of Fig. 3.10, while the curve represents the temperature dependence of the BCS theory) for aluminum samples which have a T_c of 1.25 K and an energy gap 2Δ of 0.379 mV. (After Blackford and March, 1968.)

Fig. 3.15. Temperature dependence of normalized, zero-voltage Josephson supercurrents of Sn–I–Sn and Pb–I–Sn junctions, shown, respectively, as circles and squares, in comparison with the calculation of Ambegaokar and Baratoff (1963). (After Fiske, 1964.)

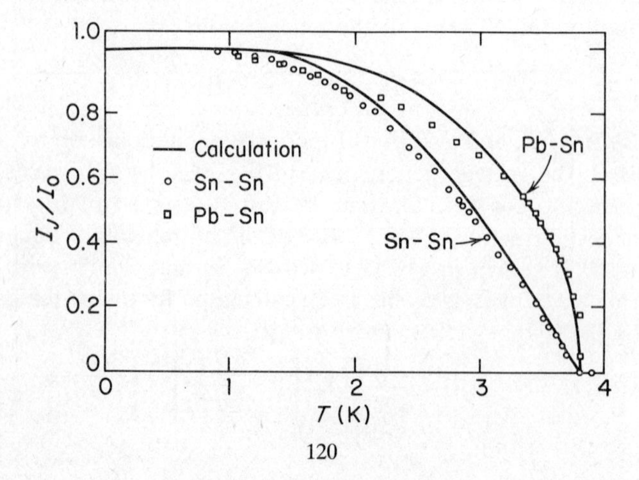

where $K(x)$ is the complete elliptic integral of the first kind, which reduces to $\pi/2$ at $x=0$. Improved agreement has been subsequently obtained in more detailed measurements and, as we will see later, by incorporating strong-coupling corrections which are clearly appropriate for Pb and not negligible for Sn. It has been found by Hauser (1967) that similar results persist when the electrodes are gapless superconductors. The observed magnitude of J_1 typically falls somewhat below the weak-coupling theoretical values (3.69, 3.70), possibly from failure to obtain perfectly uniform junctions of low impedance and the necessity in measurement to avoid electromagnetic noise pickup, which can easily disturb the required phase locking $\phi=\theta_2-\theta_1$ between the two electrodes. A good early I–V curve for the Sn–Sn case, shown by Jaklevic, Lambe, Mercereau, and Silver (1965), corresponds to $J_1/\Delta J \simeq 0.7$, while a corrected theoretical expectation in the case of Sn is 0.91 and for Pb is 0.755 (Harris, Dynes, and Ginsberg, 1976). However, reports of measurement of the theoretically predicted $J_1/\Delta J$ relation for Pb (Eldridge and Matisoo, 1971; Scott, 1970) remove serious doubt that the theoretical relation is correct, however difficult it may be to achieve in practice. We will return to the details of the T dependence below in consideration of strong-coupling effects.

Fig. 3.16. The three terms in the S_1IS_2 tunnel current at $T=0$—$J_0(V)$ (quasiparticle) and the Josephson $J_1(V)\sin\phi$ and $J_2(V)\cos\phi$—are shown as solid lines. The dashed line is the normal-state current. (After Harris, 1974.)

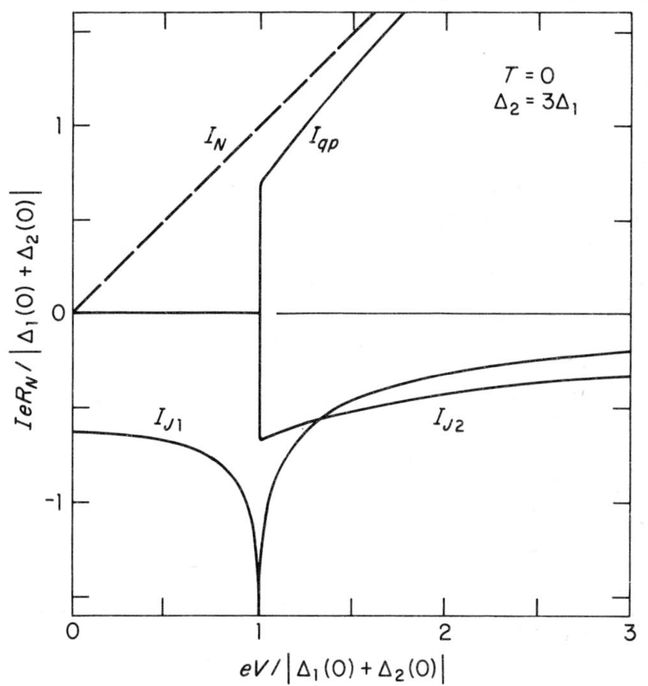

The situation with regard to the voltage dependence of the two, separate, pair tunneling terms and particularly the second, dependent upon $\cos\phi$, is much less clear, quite apart from strong-coupling effects. The theoretical relationships in the SIS case of the three terms $J = J_0(V) + J_1(V)\sin\phi + J_2(V)\cos\phi$ are plotted (at $T=0$) in Fig. 3.16. At $V>0$, the $\sin\phi$ term corresponds to an ac current at $\omega = 2eV/\hbar$, implying electromagnetic radiation, as observed by Giaever (1965) and by Yanson, Svistunov, and Dmitrenko (1965). The singularity in this term at the gap edge was predicted by Riedel (1964) and observed by Hamilton and Shapiro (1971). Initial experiments attempting to measure the $\cos\phi$ term by Pederson, Finnegan, and Langenberg (1972) implied the right magnitude but, apparently, the opposite sign, as discussed by Langenberg (1974). Since this current oscillates as $\cos(2eVt/\hbar)$, it is expected to average to zero in the typical low-frequency, I–V measurements. There is no evidence that this term has any effect on SIS quasiparticle tunneling spectroscopy as commonly practiced, and, of course, it is eliminated altogether by reverting to the NIS configuration.

3.6 Gap Spectra in More General Homogeneous Equilibrium Superconductor Cases

Effects due to nonequilibrium and to proximity layers will be treated separately in Chaps. 7 and 5, respectively.

3.6.1 Strong-Coupling Superconductors

The effects of strong-coupling superconductivity upon gap region spectra are relatively minor, involving ~10% corrections to parameters but no qualitatively new phenomena. In this section, we will use several generalized, strong-coupling expressions but will defer all but a minimum discussion of the theoretical treatment of strong coupling to Chap. 4.

Briefly, strong electron-phonon coupling leads to breakdown of the concept of quasiparticles of long lifetime and sharply defined energy. A real electron is dressed by a cloud of electronic and lattice polarization (quasiparticle), but in strong coupling, its lifetime may be short if its energy is great enough to create, e.g., plasmons or phonons. The conventional treatment of strong coupling is in terms of the energy-dependent complex quantities

$$\Delta(\omega) = \Delta_1(\omega) + i\Delta_2(\omega) \tag{3.71a}$$

$$Z(\omega) = Z_1(\omega) + iZ_2(\omega) \tag{3.71b}$$

whose imaginary parts are inversely proportional to the lifetimes. The corresponding generalizations of the densities of states for quasiparticles $[\rho(\omega)]$ and pairs $[p(\omega)]$ are

$$\rho(\omega) = \mathrm{Re}\left\{\frac{|\omega|}{[\omega^2 - \Delta^2(\omega)]^{1/2}}\right\} \tag{3.72}$$

$$p(\omega) = \mathrm{Re}\left\{\frac{\Delta(\omega)}{[\omega^2 - \Delta^2(\omega)]^{1/2}}\right\} \tag{3.73}$$

The quasiparticle tunneling expressions derived earlier remain valid, with the understanding that (3.72) is to be used, evaluated with the complex value of $\Delta(\omega)$. The first effects of this substitution in strong-coupling cases are $\sim$10% increases in the NIS conductance peak at Δ and the SIS current jump at $eV=2\Delta$. The reason for these minor changes is that $\Delta_1(\omega)$ is an increasing function of ω near the gap energy Δ_0, which is defined as $\Delta(\Delta_0)=\Delta_0$. Thus, $\Delta_1(\omega)\simeq\Delta_0+\partial\Delta/\partial\omega(\Delta-\Delta_0)$ and

$$\rho(\omega)=\rho_0(\omega)\left[1+\frac{\partial\Delta}{\partial\omega}\bigg|_{\omega=\Delta_0}\right] \tag{3.74}$$

(McMillan and Rowell, 1969), where

$$\rho_0(\omega)=\frac{|\omega|}{(\omega^2-\Delta_0^2)^{1/2}}$$

The correction factor in (3.74) enters the expression for the strong-coupling current jump ΔJ_s as the square,

$$\Delta J_s(2\Delta)=\frac{2\Delta_0}{eR_{NN}}\frac{\pi}{4}\left(1+\frac{1}{2}\frac{\partial\Delta}{\partial\omega}\right)^2 \tag{3.75}$$

because, as we have noted, this discontinuity arises from the crossing of the two singularities in the opposite electrodes. As may be seen in Table 3.2 (after Harris, Dynes, and Ginsberg, 1976), the correction to $\Delta J_s/\Delta J$ is about 1% for Sn, Tl, and In, is $\sim$5% for Pb and, at maximum, is 14% for a strong-coupling PbBi alloy.

The variation of $\Delta_0(T)/\Delta_0(0)$, shown in Fig. 3.17 for Pb and a strong-coupling PbBi alloy (Adler and Chen, 1971), is seen to deviate somewhat from the BCS result (solid curve), which, as we have seen in Fig. 3.14, is extremely accurate for weak coupling. As originally shown by Gasparovic, Taylor, and Eck (1966) and more recently by Lim et al. (1970) for Pb, the deviations can be calculated by using the full Eliashberg theory for $T>0$ to obtain the complex functions $\Delta(\omega, T)$, $Z(\omega, T)$ of the particular material; such calculations were first performed for Pb by Swihart, Scalapino, and Wada (1965). Calculations of this sort, which will be taken up in Chapter 4, also provide a quantitative understanding of the observed values of $2\Delta_0(0)/kT_c$, several of which are tabulated in Table 3.2 and which exceed the BCS value, 3.53, by as much as $\sim$35%. Strong-coupling corrections to $J_1(T=0)$ were first calculated for Pb by Fulton and McCumber (1968), making use of a generalization of the Ambegaokar and Baratoff calculation by Nam (1967*a*, 1967*b*). As may be seen from Table 3.2, the resultant $T=0$ strong-coupling ratio $J_{1s}/\Delta J_s$ of the Josephson current to the quasiparticle current jump is reduced from its weak-coupling value, 1.0, to 0.91 for Sn–Sn, to 0.755 for Pb–Pb, and to only 0.65 for PbBi junctions. The ratio $J_{1s}/J_1=J_{1s}/\Delta J$ is obtained from the product of the second and third columns, and, e.g., is 0.797 for

Table 3.2. Strong-Coupling Corrections for ΔJ and J_1 at $T=0$

Material[a]	$\frac{2\Delta}{kT_c}$	$\frac{\Delta J_s}{\Delta J}$	$\frac{J_{1s}}{\Delta J_s}$
Sn	3.7[b]	1.009	0.907
Tl	3.6[c]	1.011	0.895
In	3.68[c]	1.014	0.888
Pb	4.67[d]	1.056	0.755
Hg	4.61[e]	1.075	0.735
β Ga		1.027	0.844
Amorphous Ga	4.5[f]	1.078	0.743
Amorphous Bi		1.102	0.691
$In_{0.9}Tl_{0.1}$[c]		1.016	0.880
$In_{0.73}Tl_{0.27}$[c]		1.021	0.861
$In_{0.67}Tl_{0.33}$[c]		1.019	0.869
$In_{0.5}Tl_{0.5}$[c]		1.014	0.887
$Tl_{0.9}Bi_{0.1}$	3.58[c]	1.012	0.880
$Pb_{0.8}Tl_{0.2}$	4.37[g]	1.064	0.744
$Pb_{0.8}Bi_{0.2}$	4.7	1.093	0.694
$Pb_{0.7}Bi_{0.3}$	4.86[g]	1.103	0.678
Amorphous $Pb_{0.45}Bi_{0.55}$[h]		1.135	0.649

[a] Except as noted, these data are from J. M. Rowell, W. L. McMillan, and R. C. Dynes (unpublished).
[b] Rowell, McMillan, and Feldmann (1969).
[c] Dynes (1970).
[d] McMillan and Rowell (1965, 1969).
[e] Hubin and Ginsberg (1969).
[f] Chen, Chen, Leslie, and Smith (1969).
[g] Dynes and Rowell (1975).
[h] Allen and Dynes (1975*a*, 1975*b*).

Pb. The expression for J_{1s} derived by Nam (1967) is

$$J_{1s}=\frac{2}{\pi}\frac{1}{eR_{NN}}\int_0^\infty\int_0^\infty d\omega_1\, d\omega_2\left[\frac{f(\omega_1)-f(\omega_2)}{\omega_1-\omega_2}+\frac{1-f(\omega_1)-f(\omega_2)}{\omega_1+\omega_2}\right]\rho_1(\omega_1)\rho_2(\omega_2) \tag{3.76}$$

where 1, 2 refer to the two superconductors and $\rho_{1,2}$ are given by (3.72). Equation (3.76) applies also to $T>0$ if, in ρ, the appropriate value $\Delta(\omega, T)$ is used. Extensive measurements of $J_{1s}(T)/J_{1s}(0)$ for Pb–Pb Josephson junctions have been reported by Hauser (1967) and by Lim et al. (1970). The latter authors employed the full theoretical approach mentioned above and found good agreement with their experimental results, significantly improved over that obtained with weak-coupling theory.

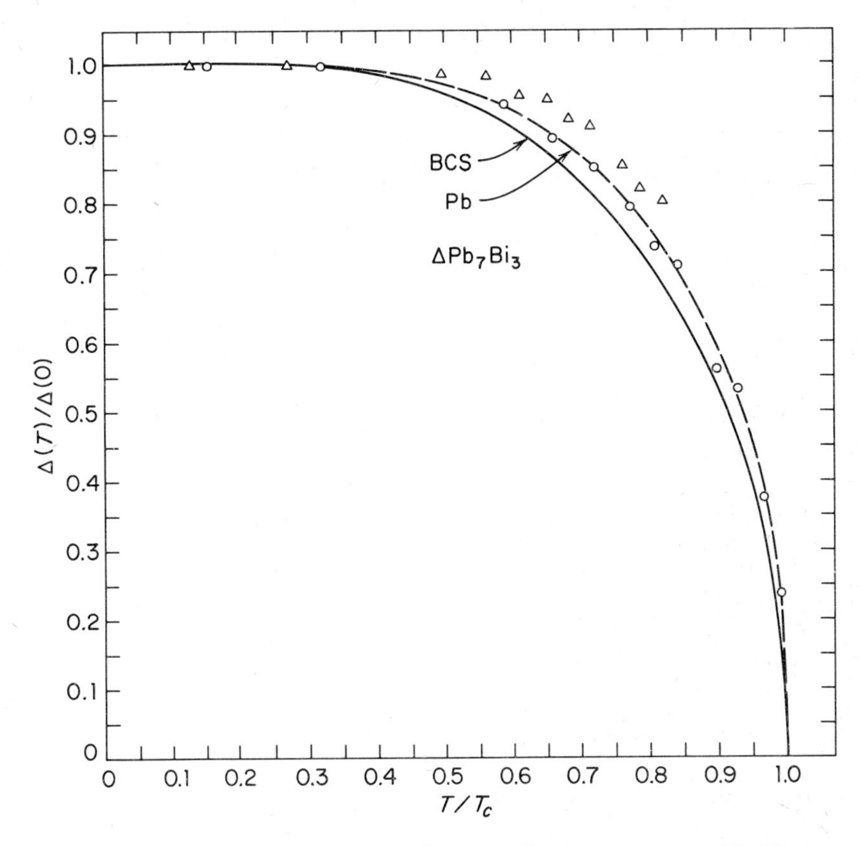

Fig. 3.17. Temperature variations of the normalized energy gaps $\Delta(T)/\Delta_0$ for a strong-coupling $Pb_{0.7}Bi_{0.3}$ alloy (shown as triangles) and for Pb (shown as dots) can be seen to depart slightly from the BCS theory (the solid line). (After Adler and Chen, 1971.)

3.6.2 Gap Anisotropy

The possibility of Δ_k being, in fact, dependent upon the direction of $\vec{k}$, i.e., the position of the electron on the Fermi surface in a superconductor free of scattering, is implicit in the BCS gap equation, (3.15),

$$\Delta_k = \sum_{k'} V_{k,k'} \frac{\Delta_{k'}}{2E_{k'}} [1 - 2f(E_{k'})]$$

if the interaction $V_{k,k'}$ is k-dependent or if the Fermi surface (sum over k') is anisotropic. There are two main sources of anisotropy in $V_{k,k'}$: the electron-phonon matrix element and the phonon spectrum. A complete discussion of this topic will be deferred to Chapter 4 in connection with the Eliashberg theory. For the purpose of the present chapter, we mention a simple model introduced by Markowitz and Kadanoff (1963);

elaborated and further applied by Clem (1966*a*, 1966*b*):

$$V_{k,k'} = (1 + a_k)V(1 + a_{k'}) \tag{3.77}$$

where a_k is an anisotropy factor chosen to average to zero over the Fermi surface. Substitution of (3.77) into (3.15) leads to a variational solution which differs from BCS only in that Δ takes the angular dependence

$$\Delta_k = \Delta_0(1 + a_{\vec{k}}) \tag{3.78}$$

Hence, the mean square variation of Δ over the Fermi surface is simply

$$\langle a^2 \rangle \equiv \frac{\langle (\Delta_k - \langle \Delta_k \rangle)^2 \rangle}{\langle \Delta_k \rangle^2} \tag{3.79}$$

The elementary theory of tunneling through a potential barrier strongly favors transitions with $\vec{k}$ perpendicular to the surface, into a cone of opening angle 5–10°. In the clean limit, if the angular variation of Δ_k is large enough, and if the barrier growth is well behaved (Dowman, MacVicar, and Waldram, 1969), measurement of $\Delta_{\vec{k}}$ should be possible, with angular resolution on the order of this cone angle, by a sequence of junctions formed on different faces of the superconductor.

A brief summary of results from the more extensively studied single-crystal cases is given in Table 3.3. In the selection of data for this table, preference is given to cases in which a free crystal serves as substrate, with an evaporated film, preferably of the same metal, as counterelectrode. This configuration minimizes the opportunity for strains to occur in cooling from the fabrication temperature to the measurement temperature of typically 1 K, which can produce comparable gap energy shifts unrelated to anisotropy. Thin films of the superconductor under study are undesirable because microcrystals of many different orientations may be present and also because scattering at the film boundaries and at internal grain boundaries can lead to the dirty superconductor situation, with an averaged isotropic gap. However, thick, epitaxial, single-crystal films, in which strain is minimized by reasonable matching of thermal expansion coefficients (Lykken, Geiger, Dy, and Mitchell, 1971), have been used in the case of Pb and give results in good agreement with those obtained from freestanding single crystals.

Figure 3.18, after Blackford and Hill (1981), summarizes the variation of $2\Delta/kT_c$ ($T_c = 3.73$ K) for freestanding Sn single crystals, with Sn and Pb film counterelectrodes. The overall variation is ~17%, if some averaging of the quoted values near the maximum ($\theta \simeq 25°$, $\phi \simeq 15°$) and minimum (75°, 30°) is done, consistent with a tunneling cone width of 5–10°. Agreement between nearby values inferred with Sn counterelectrode films (triangles) and with Pb counterelectrode films (circles) seems to average about 3.5%, or ~20% of the total variation, with Sn counterelectrode points for $\theta > \sim 70°$ giving systematically higher Δ. The results are not highly precise, but a clear trend of variation well above the local fluctuation is evident. When compared with the earlier results of

Table 3.3. Gap Anisotropy in Crystalline Superconductors

Metal	$\langle a^2 \rangle^a$	Δ_{min}	Δ_{max}	Comment	Reference
Al	0.009	0.15 ± 0.01^b	0.195 ± 0.01^b	Al film on free-standing Al crystal	Blackford, 1976
	0.0084	0.143	0.200	Theory	Leavens and Carbotte 1972
Ga	—	0.199 ± 0.002^b	0.209 ± 0.003^b	Ag film on free Ga crystal	Kalvey and Gregory, 1979
Pb		1.18 ± 0.01	1.24 ± 0.01	(Lower gap)[c] Pb films on free Pb crystals	Blackford and March, 1969
	0.007				
		1.37 ± 0.01	1.39 ± 0.01	(Upper gap)[c]	
		1.18 ± 0.01	1.24 ± 0.01	(Lower gap)[c,d] Pb film on epitaxial Pb crystal	Lykken, Geiger, Dy, and Mitchell, 1971
	0.009				
		1.37 ± 0.01	1.40 ± 0.01	(upper gap)[c]	
	—	1.27	1.43	Theory	Bennett, 1965
	0.009	1.20	1.60	Theory	Carbotte, 1977
Sn	0.026	0.434	0.691	Sn crystallized against glass	Zavaritskii, 1964, 1965
	—	0.518	0.619	Freestanding Sn crystals; Sn and Pb films	Blackford and Hill, 1981

[a] Several of these values are taken from Bostock and MacVicar (1977).
[b] Averaged values with uncertainty estimates added.
[c] Two distinct gaps are observed, each of which varies with orientation.
[d] In the case of [100] normal orientation, a third distinct gap at 2.32 meV was observed. 2.36 meV is the smallest value of the lower gap $2\Delta_{min}$ in those cases in which two gaps were seen.

Zavaritskii (1964, 1965) obtained from Sn droplets frozen against glass slides, the angular variation (mapped by the dashed line boundaries; *H*, *M*, and *L* indicating high, medium, and low gap values) is roughly coincident, but the numerical range of the recent values (Table 3.3) is significantly smaller than that reported earlier.

The situation is complicated by the appearance of double gaps observed only with Sn film counterelectrodes. Observation of double gaps, well established in only a few cases (Section 3.6.3), is presumably the result of contributions from different sheets of Fermi surface in the Sn crystals at the same angle $\vec{k}/|\vec{k}|$. The double gap is not expected in the thin-film electrode because of gap averaging. To explain the appearance of two gaps with Sn film counterelectrodes only, one is led to consider either an inherently broader gap edge in the Pb versus Sn films (incomplete averaging), obscuring the detailed structure in the Sn, or possibly an inhomogeneous gap variation in the Pb film, due either to a spatial distribution of strains or to differently oriented Pb crystallites responding differently to a uniform strain. This effect might well be smaller with the

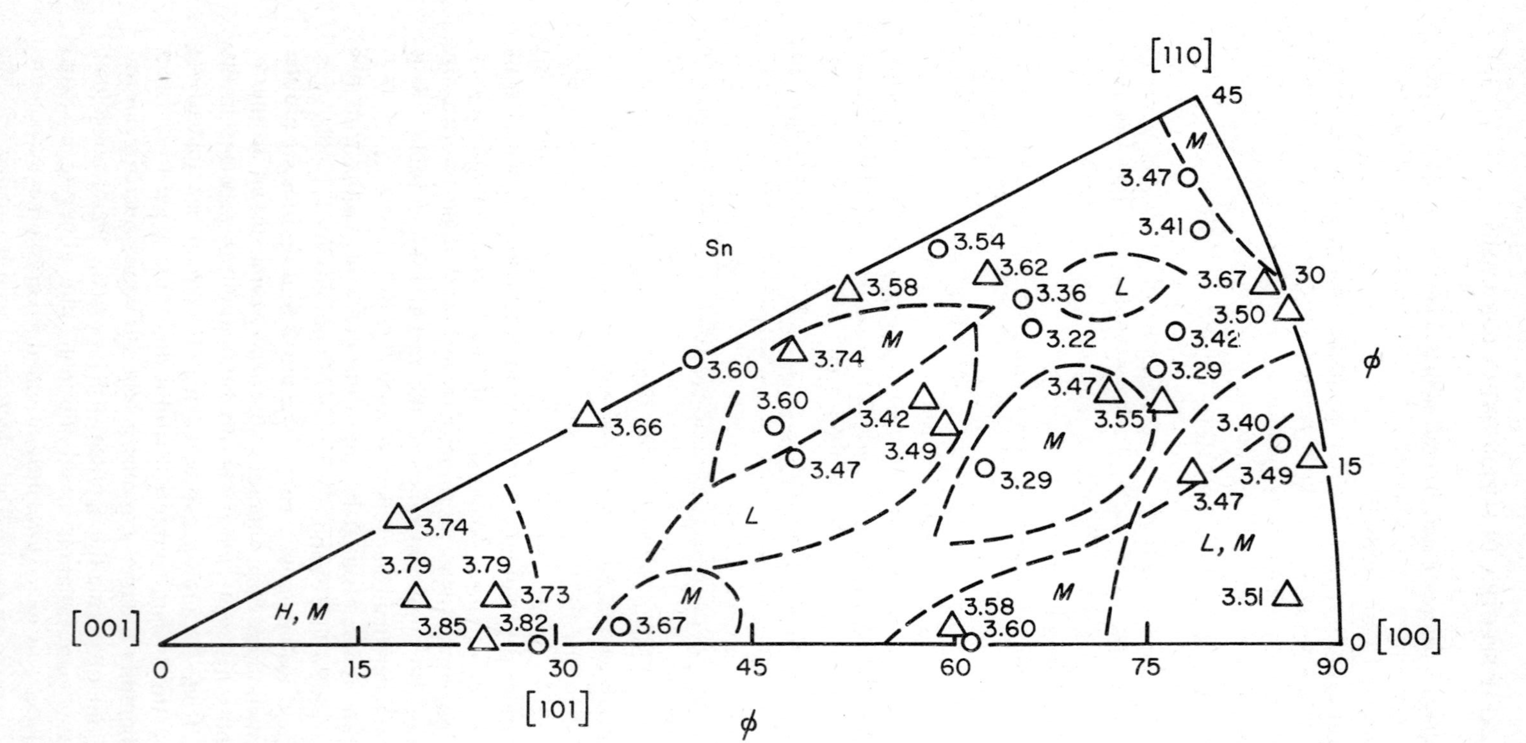

Fig. 3.18. Crystalline anisotropy of the reduced-gap parameter $2\Delta/kT_c$ for tin plotted vs. the orientation of the tunnel junction fabricated on single-crystal tin. Here, θ and ϕ are the usual spherical polar angles, with the polar axis chosen along the [001] direction. The data points represented by circles correspond to junctions with lead top films, while the triangles are obtained from junctions which have tin top films. These data represent approximately a 17% deviation in the reduced-gap parameter. (After Blackford and Hill, 1981.)

Sn films because of expected matching of thermal contraction against the Sn crystal. Comments on the importance of strain in films have been made by Banks and Blackford (1973).

The situation with Pb crystals is generally more consistent and has been summarized by Lykken et al. (1971). However, the extra (third) gap seen in thick [100] oriented epitaxial Pb films by Lykken et al., strong enough in some cases to be seen directly in the *I–V* curves, is not clearly observed in junctions on (100) facets of freestanding Pb crystal junctions (Blackford and March, 1969). Interestingly, the third peak is not seen, either, in Pb epitaxial films growing 5° off the [100] direction. While there is no definite resolution of this point, generally there are many ways that a sharp feature, seen under optimum conditions, can be obscured under less-than-optimum conditions.

An extensive review of experimental studies of gap anisotropy, including data from techniques of ultrasonic attenuation, specific heat, critical magnetic field, and others, in addition to tunneling, has been given by Bostock and MacVicar (1977).

As is emphasized in this review, tunneling measurements of anisotropy are difficult and subject to reduction and distortion by several extraneous factors. Reduced anisotropy, as we have said, is a result of scattering rapid enough to produce an isotropic average gap. Since tunneling is an inherently surface-sensitive measurement, the condition $\ell/\xi_0 \gg 1$ may not be satisfied in the sampled region of thickness $\sim\xi$, even though measurement of the residual resistance ratio of the bulk specimen may imply clean conditions in the bulk. This effect may account for the lack of tunneling evidence for anisotropy in Nb (Bostock, Agyeman, Frommer, and MacVicar, 1973) and other transition metals. Distortion of measurements by strain, either uniform or random and obscuring features, is also a severe problem in an experiment where a sandwich of dissimilar materials is required. As in any tunneling measurement, it is essential to make proper four-terminal connections to the electrode films and to use a voltmeter of sufficiently high input impedance that there are no voltage drops along leads. The possibility of measurable voltage drops along normal-state electrode films, consequent to the tunneling current flow (Osmun, 1980), should be borne in mind: this problem is avoided by use of two superconducting electrodes.

The theory of superconductivity is well developed and implies, taken with other real properties known to be anisotropic (band structure, phonon structure), anisotropy of various superconducting properties, as reviewed by Carbotte (1977). The surprising aspect is not that a few anisotropic features have been observed but that there are not more and better-documented examples. Progress is likely to be made with wider use of epitaxial single-crystal films, whose surfaces are typically of higher purity, with tunneling barriers formed *in situ*, e.g., by Al or Mg-deposited films, subsequently oxidized, leaving the upper layers of the metal electrode clean.

3.6.3 Multiple Gaps, Two-Band Superconductivity

The Fermi surface in the reduced-zone scheme consists of several portions (sheets) which have been translated by reciprocal lattice vectors from second, third, etc. Brillouin zones in the extended-zone scheme. The electron properties on these portions may differ significantly. In a given direction, there generally will be more than one sheet of Fermi surface, on which the gap parameter solutions Δ_k will probably be different. In the absence of scattering, the two or more gaps should be resolvable by tunneling, leading to multiple-gap structure.

This explanation is indeed the conventional one for the double gap seen in single-crystal Pb in the theory of Bennett (1965). In general, Bennett finds three values of Δ, from electrons in the first zone, holes in

Fig. 3.19. (a) Normalized conductance spectra in the gap region of In Schottky barrier tunnel junctions on superconducting (n-type) $SrTiO_3$:Nb, measured at 0.1 K. The sum gap peak is observed to split as the Nb concentration, increasing in progression from junctions 6 through 8, becomes sufficient to fill a second conduction band in the $SrTiO_3$. Evidence is given that this situation corresponds to a case of two-band superconductivity. (b) The separate gap parameters Δ_1 and Δ_2 inferred from the spectra of (a) displayed vs. the Nb doping concentration in the $SrTiO_3$ (upper scale) and the Fermi degeneracy μ_F in meV (lower scale). The second gap Δ_2 appears at a Fermi degeneracy of about 32 meV, which is in modest agreement with a band splitting of 20 meV in from the band structure of Mattheiss (1972).

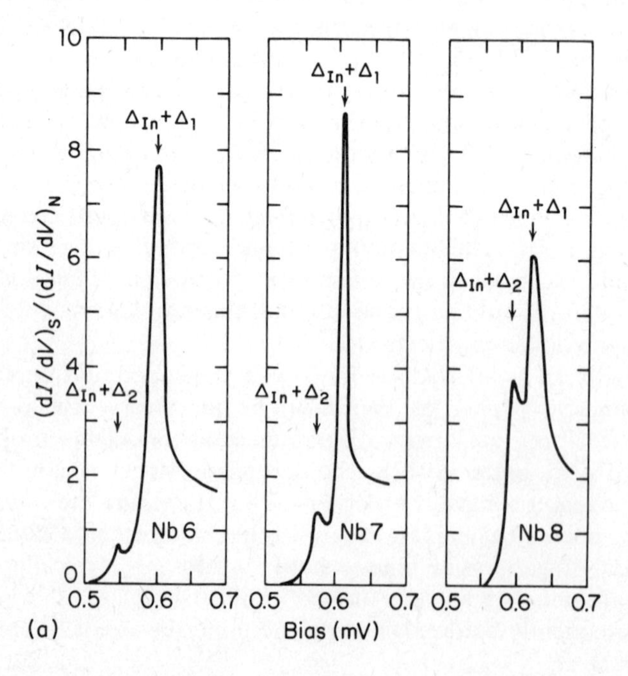

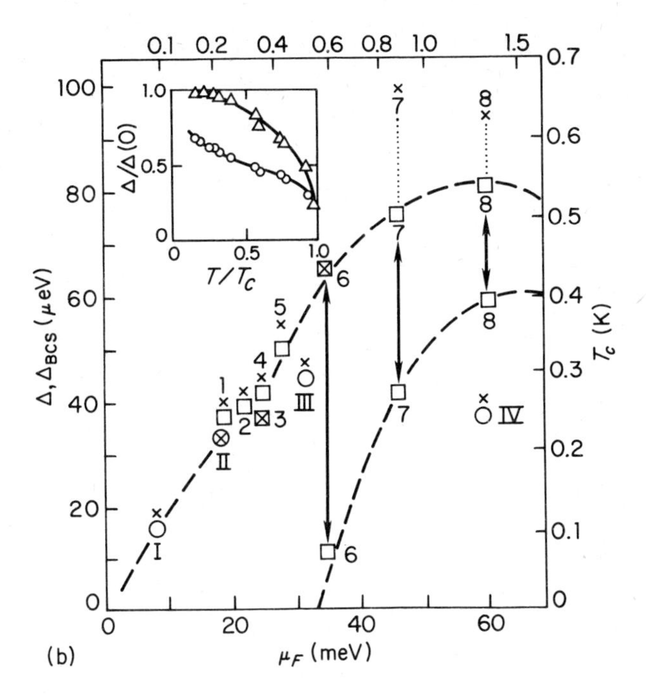

the second zone, and electrons in the third zone. In the [100] direction, the gap values of Bennett are 2.60, 2.75, and 2.86 meV, which agree in their number, if not with much precision in their values, with the 2.32, 2.56, and 2.80 meV observed by Lykken et al. for [100] epitaxial Pb films. In the [110] and [111] directions, approximate coincidences of calculated gap values occur, consistent with observation of only double gaps in these directions. The analysis of Léger and Klein (1971) appears to reasonably explain the absence of correspondingly doubled phonon peaks in Pb.

A rather different example of double-gap structure is shown in Fig. 3.19, after Binnig, Baratoff, Hoenig, and Bednorz (1980), in the n-type degenerate semiconductor $SrTiO_3$ doped with variable amounts of Nb donors. The second gap appears at a minimum concentration $\sim 5\times 10^{19}$ cm^{-3} Nb, corresponding to filling of the lowest conduction band to $\mu \simeq 32$ meV, inferred from Schottky barrier tunneling characteristics. This result presumably corresponds to the onset of electron occupation of a second band, predicted to lie 20 meV higher in the band structure of $SrTiO_3$ of Mattheiss (1972). The systematic variation of the superconducting gaps with doping, shown in Fig. 3.19b, provides strong evidence for the correctness of the interpretation in two-band superconductivity.

It seems, at first, surprising that clean conditions could occur in a doped

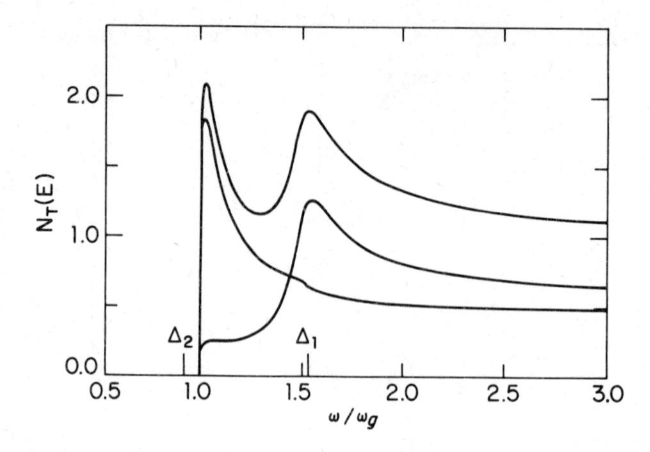

Fig. 3.20. Partial and total densities of states at $T=0$ for a particular model of a two-band superconductor. (After Schopohl and Scharnberg, 1977.)

material, with randomly distributed ionized donor centers. In this particular case, however, dielectric screening is known to be very effective, with dielectric constant values exceeding 10^4 below 10 K.

A second feature of this work, namely that measurements are made on [100] single-crystal faces produced by cleavage of bulk doped $SrTiO_3$, makes these results especially compelling. Theoretical curves assuming a two-band model and qualitatively resembling those observed for $SrTiO_3$ have been presented by Schopohl and Scharnberg (1977). This calculation, in fact, was applied to Nb, for which detailed band structure calculations predict three separate sheets of Fermi surface. The absence of experimentally observed splittings in Nb is presumably explained by interband scattering, which, in the model, is shown to progressively reduce and eliminate the double-gap structure. This feature is shown in Fig. 3.20. Further discussion of the unusual nature of superconductivity in $SrTiO_3$ is given by Baratoff et al. (1982).

3.6.4 *Excess Currents, Subharmonic Structure*

Excess currents in the subgap range $eV<\Delta_1+\Delta_2$ may be defined as those in excess of the prediction of BCS theory. The first reports of such currents, believed due to additional tunneling mechanisms, were made by Taylor and Burstein (1963) (whose results are shown in Fig. 3.21) and, independently, by Adkins (1963). Characteristic features are onsets of additional current at gap submultiples $2\Delta_j/m$, $m=1,2,4,\ldots$ and $j=1,2$, of the gap energies $2\Delta_1$, $2\Delta_2$ in the junction (termed the even series), as indicated for $m=2$ ($m=1$ corresponding to the gap edge) in Fig. 3.21 and to higher m in the semilogarithmic plot of Fig. 3.22, after Rowell and Feldmann (1968). The *absence* of the steps in the Al data of Fig. 3.21 is an early indication that these depend in some way on a particular feature of

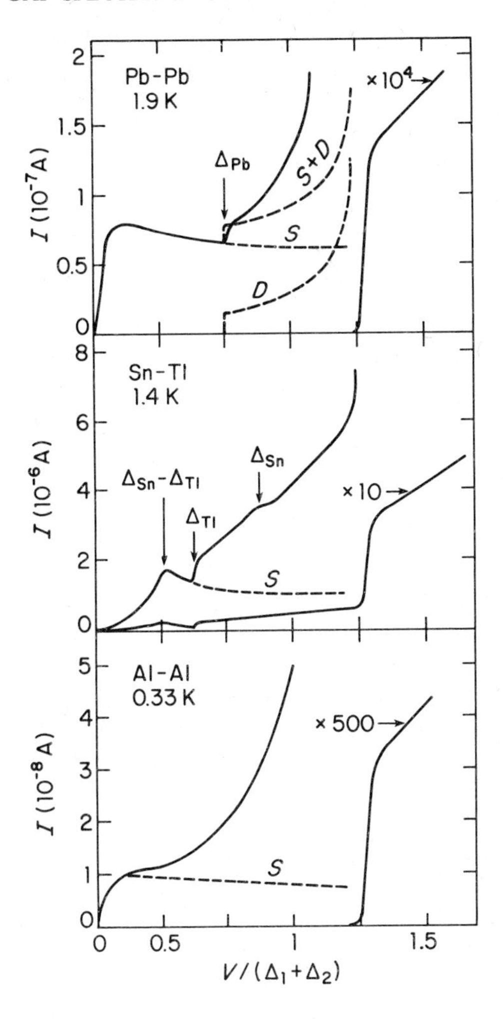

Fig. 3.21. The original published report of excess currents in electron tunneling between two superconductors, after Taylor and Burstein (1963). The excess currents may be seen as the difference between the dashed curves (marked *S*) and the solid data curves rising above them. The dashed curve (marked *D*) in the upper panel of the figure is calculated on the basis of a double-particle tunneling process, shown in Fig. 3.24.

the barrier. Higher-order current steps in an asymmetric junction can also appear at odd submultiples $eV = (\Delta_1 + \Delta_2)/m$, $m = 3$, 5 (odd series), according to theoretical models. Recent observations (Epperlein, 1980, 1981) show (Fig. 3.23) the $m = 3$ member of this series. Note the clearly steplike nature of the $2\Delta/2$ increase at $\Delta_1 = 1.48$ meV (Nb) and $\Delta_2 = 1.70$ meV (PbBi) in this figure, and note also that the $m = 2$ step heights are proportional to Δ_j in size. A generally observed feature is lack of temperature or magnetic field dependence of the step structure.

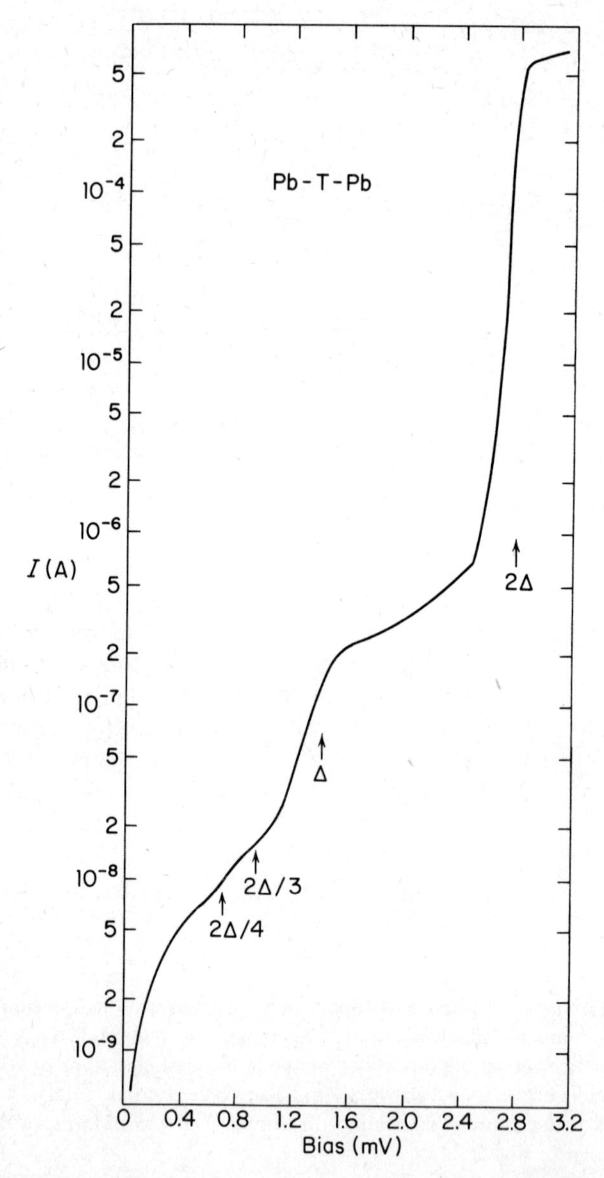

Fig. 3.22. Tunneling current on a logarithmic scale vs. voltage for a Pb–I–Pb junction measured at 1 K. Further features occurring at submultiples of the gap energy are pointed out. Note the extremely low values of current which are obtained relative to the current at the voltage 2Δ, and note that the current drops by three orders of magnitude quite sharply at the energy 2Δ. (After Rowell and Feldmann, 1968.)

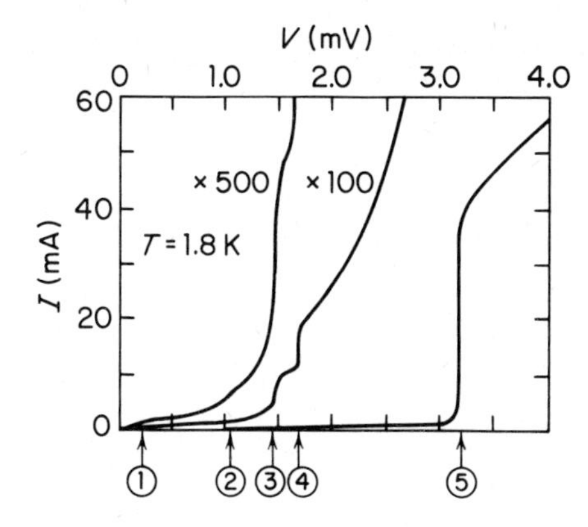

Fig. 3.23. Recent I–V measurements of an Nb–I–PbBi tunnel junction measured at 1.8 K reveal a structure that is believed to be due to the multiple-particle tunneling processes shown in Fig. 3.24. The onset biases indicated by arrows 1 through 5 are identified as $\Delta_2 - \Delta_1$, $(\Delta_1 + \Delta_2)/3$, $\Delta_1/2$, $\Delta_2/2$ and $\Delta_1 + \Delta_2$. (After Epperlein, 1981.)

A fundamental explanation of the step structure, offered by Schrieffer and Wilkins (1963) (see also Wilkins, 1969), is simply simultaneous transfer of two or more (m) particles (MPT), in a higher-order process of expected rate $|T|^{2m}$, where T is the tunneling matrix element. Two possible processes of $m = 2$ are shown in Fig. 3.24, corresponding to an energy balance at thresholds $2eV = 2\Delta_1$, $2\Delta_2$, or $V = 2\Delta_1/2e$, $2\Delta_2/2e$. Following Wilkins (1969), the rate for an initial state I to a set of final states F,

$$w = \frac{2\pi}{\hbar} \sum_F \left| \sum_M \frac{\langle F|H^T|M\rangle\langle M|H^T|I\rangle}{E_M - E_I} \right| \delta(E_F - E_I) \tag{3.80}$$

Fig. 3.24. Schematic indication of two processes which transfer two particles simultaneously through the tunneling barrier. Note that if single-particle tunneling is of order T^2 in the tunneling matrix element T, then processes such as shown here are of order T^4. (After Schrieffer and Wilkins, 1963.)

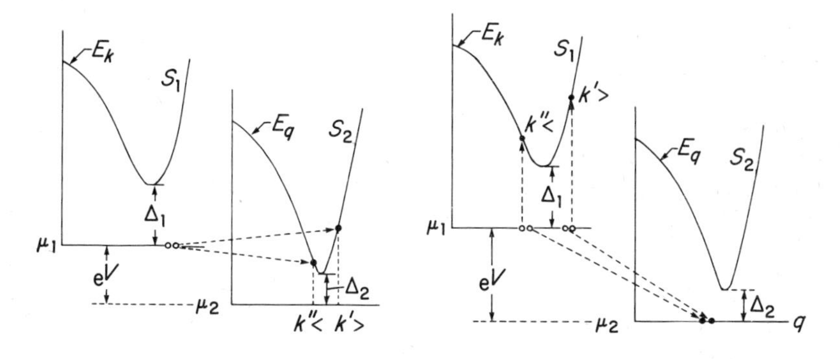

proceeds through all possible intermediate states M. This result leads to a current contribution proportional ($T=0$) to

$$|T|^4 \int_{\Delta_1}^{2eV-\Delta_1} \frac{\rho_1(E)\rho_1(2eV-E)\,\mathrm{d}E}{\Delta_2^2-(E-eV)^2} \tag{3.81}$$

where ρ is the BCS density of states. This expression gives a discontinuity at the lower limit of E for $eV=\Delta_1$, for both numerator terms are singular with a finite denominator. The current jump is followed by a linear rise, becoming steeper and again jumping at $eV=\Delta_2$. A final rapid increase occurs at $\Delta_1+\Delta_2$. The current jump at the m-particle threshold $eV=2\Delta/m$ in a symmetric junction can be written

$$\Delta J_m = \frac{e\Delta}{4\pi\hbar t^2} \tanh\left(\frac{\Delta}{2kT}\right)\left(\frac{e^{-s}}{16}\right)^m (s+m^{-1})\left[\left(\frac{2}{m!}\right)\left(\frac{m}{2}\right)^m\right]^2 \tag{3.82}$$

(Wilkins, 1969; Hasselberg, Levinsen, and Samuelson, 1974), where $s=2\kappa t$, with t the oxide thickness. As expected, the step heights, in case of a uniform oxide of width t, fall off rapidly with order m: $\Delta J_m \simeq e^{-ms}$. A further useful result is the ratio, for $m=2$, of ΔJ at Δ_1 and Δ_2:

$$\frac{\Delta J(\Delta_1)}{\Delta J(\Delta_2)} = \Delta_1 \frac{\tanh(\Delta_1/2kT)}{\Delta_2} \tanh\frac{\Delta_2}{2kT} \tag{3.83}$$

The additional assumption of an inhomogeneous or patchy oxide having a small fraction $A_2/(A_1+A_2)$ of its area with decreased t or κ and hence increased $|T|$ has allowed semiquantitative fitting to the multiparticle model in several cases beyond the early work of Schrieffer and Wilkins. Measurements of PbIn–I–Pb junctions, where the insulator I is actually the oxide of the PbIn alloy, which has been reported to be inhomogeneous (e.g., Broom and Mohr, 1977), have been successfully interpreted by Mukhopadhyay (1979). This result is confirmed and extended by Epperlein (1980, 1981) in Nb–NbO_x–PbBi, Sn–SnO_x–Sn, and other junctions.

A second fundamental explanation for the subharmonic structure, giving nearly the same features (e.g., the same even and odd series, temperature independence), is envisaged from self-absorption (or self-coupling) of Josephson radiation, JSC. Recall that since $I_J=I_1 \sin\phi$, one has, at a fixed bias V_0, an oscillating current, since

$$\frac{\mathrm{d}\phi}{\mathrm{d}t} = 2e\frac{V_0}{\hbar}$$

One expects a corresponding oscillating voltage δV, supposing a junction capacitance C (since $I=C\,\mathrm{d}V/\mathrm{d}t$), whose fundamental component (at $\omega_0=2eV_0/\hbar$) is

$$\delta V = \frac{I_1}{\omega_0 C}\cos(\omega_0 t+\alpha)$$

This voltage will correspondingly produce electromagnetic radiation, both at the fundamental ω_0 *and* at its overtones $n\omega_0$, the latter due to the inherent self-coupling and nonlinearity in the problem.

The junction is treated as having capacitance only, and its electromagnetic modes are ignored (the resonant frequency and damping set equal to zero). The *RF* current in the external circuit is zero (no external driving current); hence (Werthamer, 1966; Giaever and Zeller, 1969; Strässler and Zeller, 1971; Hasselberg, Levinsen, and Samuelson, 1974, 1975), $I_1 \sin \phi + C\, dV/dt = 0$. This expression reduces, since $V = \hbar\dot{\phi}/2e$, to the simple nonlinear pendulum equation

$$\ddot{\phi} + \omega_J^2 \sin \phi = 0 \tag{3.84}$$

where

$$\omega_J^2 = \frac{2eI_1}{\hbar C} \tag{3.85}$$

For small values of the characteristic coupling $\Gamma = (\omega_J/\omega_0)^2$, the current and voltage are purely sinusoidal at ω_0. Higher harmonics $n\omega_0$ come in, as with the pendulum for large Γ. The solution in the presence of impressed dc voltage V_0 is

$$\phi = \omega_0 t + \sum_{\ell=1}^{\infty} 2\alpha_\ell \sin(\ell\omega_0 t) \tag{3.86}$$

where, for small $\Gamma \ll 1$ (large voltage),

$$\alpha_\ell = \left(\frac{2}{\ell}\right)\left(\frac{\Gamma}{4}\right)^\ell \tag{3.87}$$

and for large Γ, the α_ℓ saturate to values

$$\alpha_\ell = \frac{1}{\ell} \tag{3.88}$$

(Hasselberg, Levinsen, and Samuelson, 1974). The self-generated photons of energy $\hbar\omega = n2eV$ can be absorbed by the individual superconductors when $2\Delta_i = n2eV$. Since the energy absorbed from this radiation at threshold $V = 2\Delta_i/2ne$ (even series) comes from the battery in the tunneling circuit, a structure appears in the current at this voltage. A second process involving absorption of photons is photon-assisted tunneling (Dayem and Martin, 1962), for which the energy balance is $eV + \hbar\omega = \Delta_1 + \Delta_2$. In the present case, we have $eV + n2eV = \Delta_1 + \Delta_2$, or

$$V = \frac{\Delta_1 + \Delta_2}{(2n+1)e} \qquad \text{(odd series)}$$

Comparisons of details of the multiparticle (MPT) and Josephson self-coupling (JSC) mechanisms have been given by Hasselberg, Levinsen, and Samuelson, (1974, 1975) and by Mukhopadhyay (1979). Summarizing, the two mechanisms are similar in prediction of the voltages at which features appear and also in prediction of substantially temperature-dependent and magnetic field–independent behavior. Both theories predict (Hasselberg, Levinsen, and Samuelson, 1975) microwave-induced satellite structure at voltages $eV = (2\Delta \pm nh\nu)/m$, with m and n integers, ν the frequency of the radiation, and h Planck's constant. The case $m = 1$ corresponds to simple microwave-assisted tunneling (Dayem and Martin, 1962), to be discussed further in Section 3.6.8. The microwave-induced structures have the same dependence on RF power.

The primary differences occur in the predicted line shapes, which are truly steplike in MPT but can be either steps or peaks in JSC; and in the dependence on the tunneling matrix element. While MPT with a truly uniform barrier varies at $|T|^{2m}$ (where m is the number of electrons transferred), the JSC predicts all features to vary simply as $|T|^2$. The higher-order structures fall off more rapidly with m in MPT than in JSC, even if the $(e^{-s})^m$ dependence is ignored, assuming suitable fluctuations in local barrier thickness or potential height. This result has been expressed in terms of maximum amplitude ratios of the successive step structures. Mukhopadhyay (1979) finds, for example,

$$\left|\frac{\Delta J(2\Delta/3)}{\Delta J(2\Delta)}\right|_{\max} = \begin{cases} 0.0017, & \text{MPT} \\ 0.18, & \text{JSC} \end{cases}$$
$$\left|\frac{\Delta J(2\Delta/4)}{\Delta J(2\Delta/2)}\right|_{\max} = \begin{cases} 0.0032, & \text{MPT} \\ 0.59, & \text{JSC} \end{cases} \tag{3.89}$$

A further difference between the two mechanisms is that, in JSC, contrary

Fig. 3.25. Differential resistance dV/dI spectrum [data after Rowell and Feldmann (1968)] showing more pronounced subharmonic structure. The computed curve is a model calculation based on the assumption that the junction contains metallic microshorts. (After Klapwijk, Blonder, and Tinkham, 1982.)

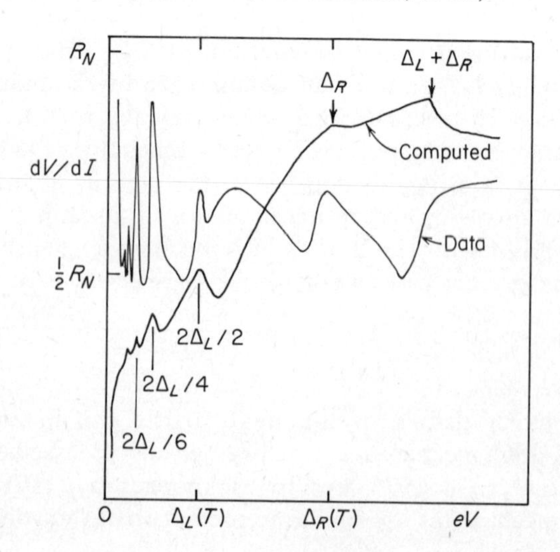

to (3.83), qualitative arguments indicate that the structure arising from the smaller of Δ_1, Δ_2 will be the *larger*. This result is contrary to the observations in Fig. 3.23.

Figure 3.25 gives an example of a much leakier junction with extended subharmonic structure. Since the observed features index up to $m = 12$, such data cannot be accounted for in terms of MPT. A strong case for interpretation of similar data in terms of JSC alone was made by Giaever and Zeller (1970). These authors make a connection between strong JSC structure and microshorts. The data of Fig. 3.25 have been successfully modeled in terms of a microshort model based on Andreev scattering, to be taken up in Chap. 6.

3.6.5 *Effects of Magnetic Field*

A typical effect of a pair-breaking perturbation is illustrated in Fig. 3.26, in the response of an Al–I–Pb junction to parallel magnetic field. The

Fig. 3.26. Measurements of the tunneling density of states of an Al–Al_2O_3–Pb junction at 1.74 K with applied magnetic fields parallel to the junction. Here, the dashed curves are calculated on the basis of pair breaking by the magnetic field, in addition to the usual thermal smearing from a normal counterelectrode, with the assumption of a small mean free path in the lead. (After Räsing, Salemink, Wyder, and Strässler, 1981.)

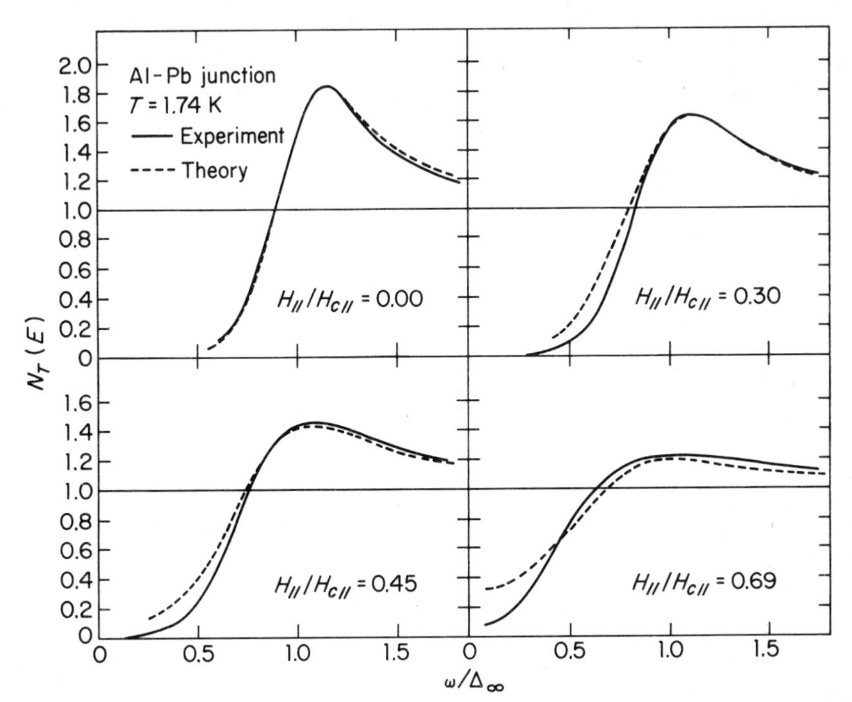

simplest description of the physics is that any perturbation acting unequally on the members of the pair, breaking up the exactly time-reversed states $\vec{k}\uparrow, -\vec{k}\downarrow$ in a time τ, will lead to an energy lifetime broadening

$$2\alpha = \frac{\hbar}{\tau_K} \tag{3.90}$$

where α is called the pair breaker strength. This will eventually fill in the energy gap (when $2\alpha = \Delta$). It is an interesting fact, discovered theoretically by Abrikosov and Gor'kov (1960) in the closely analogous case of paramagnetic impurity additions, that the gap in general may become zero before the normal state reappears. In clean situations, or when the perturbation is unaffected by scattering, 2α is simply defined as the energy difference produced between states $k\uparrow, -k\downarrow$.

The basic effects of a parallel magnetic field H on a thin superconductor ($d \ll \sqrt{\ell\xi_e}$) enter the Hamiltonian via terms $(\vec{p} - e\vec{A}/c)^2/2m$, where $\vec{A}$ is the vector potential, and $\pm\mu_B H$ from the electron spin, where μ_B is the Bohr magneton. The leading orbital contribution is $-(e/mc)\vec{p}\cdot\vec{A}$. Other parameters of importance in the terms involving H, are the electron mean free path ℓ, often expressed by the diffusivity $D = (1/3)v_F\ell = (1/3)v_F^2\tau_{tr}$, and the spin-orbit scattering time τ_{s0}, conventionally expressed as a dimensionless parameter $b = \hbar/3\Delta\tau_{s0}$. The dominant effect in Fig. 3.27 is orbital

Fig. 3.27. Illustration of the spin splitting of the excitation spectrum of a superconductor in extremely high magnetic fields in the case of small spin orbit scattering. In the upper portion of the figure, spin orbit scattering is taken to be zero, whereas in the more realistic calculation in the lower portion of the figure, the spin-orbit parameter b is taken as 0.2. (After Engler and Fulde, 1971; Tedrow and Meservey, 1971.)

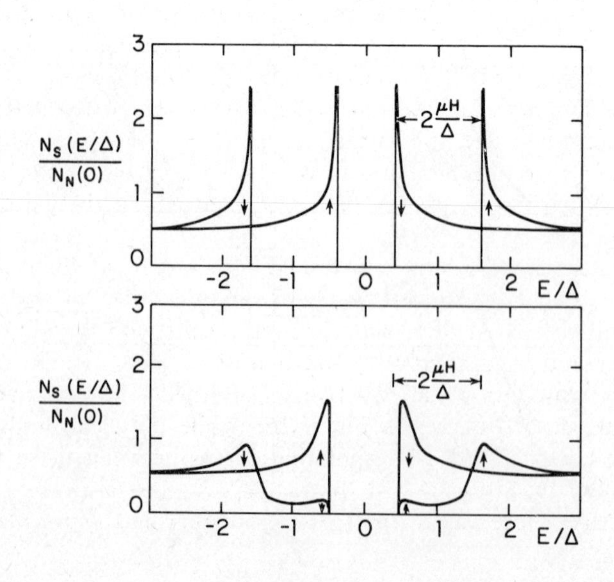

pair breaking, for which the relation

$$\alpha = \frac{\tau_{tr}}{18\hbar}(v_F\, deH)^2 \tag{3.91}$$

applies and the dirty limit $\ell/\xi \ll 1$ has been assumed. In the dirty limit, the pair-breaking time τ_K is taken (de Gennes, 1964*a*, 1964*b*, 1966) as the time for unit phase difference $\phi_\uparrow - \phi_\downarrow \simeq 1$ to accumulate, as a result of many random increments $\Delta\phi$ occurring in intervals of the transport free time $\tau_{tr} \ll \tau_K$ between members of the pair, described by $e^{-iEt/\hbar}$. As the energy difference $\Delta E = 2(e/mc)\vec{p} \cdot \vec{A} = (2e/c)\vec{v}_k \cdot \vec{A}(\vec{r})$ fluctuates in sign after τ_{tr}, a random walk in phase difference occurs. Hence, since $\Delta\phi = \Delta E\tau_{tr}/\hbar$, the condition $\Delta\phi_{rms}(\tau_K/\tau_{tr})^{1/2} = 1$ becomes

$$\frac{(2e/c)^2 v_F^2 \langle A^2\rangle \tau_{tr}^2}{\hbar^2(\tau_K/\tau_{tr})} = 1$$

or

$$\alpha = \frac{\hbar}{2\tau_K} = \frac{2\tau_{tr} e^2 v_F^2}{3\hbar}\langle \vec{A}^2(\vec{r})\rangle \tag{3.92}$$

This expression leads to Eq. (3.91) for a thin film if appropriate conditions $\langle A(\vec{r})\rangle = 0$ and $\vec{\nabla} \cdot \vec{A}(\vec{r}) = 0$ are imposed on $\vec{A}$ (Maki, 1964). In contrast, α for the $\vec{k}$-independent spin splitting is simply μH, from the energy difference, if spin-orbit scattering is negligible.

Under special conditions, the spin-splitting $2\mu_B H$ and spin-orbit effects can be seen clearly in the tunneling conductance, as illustrated in Figs. 3.27 and 3.28, after Tedrow and Meservey (1971). The orbital effects can be made negligible by use of extremely thin (~40 Å) Al metal films in an accurately parallel H field. Under these conditions, even though the bulk H_c of Al is only about 100 Oe, parallel critical fields as great as 55 kOe have been observed. The role of the spin-orbit parameter b is illustrated in Fig. 3.27b for $b = 0.2$. Actual fitting of the Al data in Fig. 3.28, using theory developed by Engler and Fulde (1971) and Bruno and Schwartz (1973), determines the spin-orbit parameter b of Al as 0.07. A much larger value, $b = 0.6$, is similarly inferred for Ga in Fig. 3.29. Extensions of this work testing theories of high-field superconductivity are given by Tedrow and Meservey (1979).

A careful study of orbital pair breaking in which the finite mean free path ℓ was explicitly treated was reported by Millstein and Tinkham (1967), who compared their experimental results with theoretical results of Strässler and Wyder (1967). The density of states in the dirty limit $\ell = 0$ was treated in detail by Skalski, Betbeder-Matibet, and Weiss (1964) (e.g., dashed curves in Fig. 3.26), while the clean limit was first treated by Larkin (1965); interpolating curves between those two cases are presented by Strässler and Wyder (1967). Qualitatively, the major difference is a more rapid onset of broadening and gaplessness in the

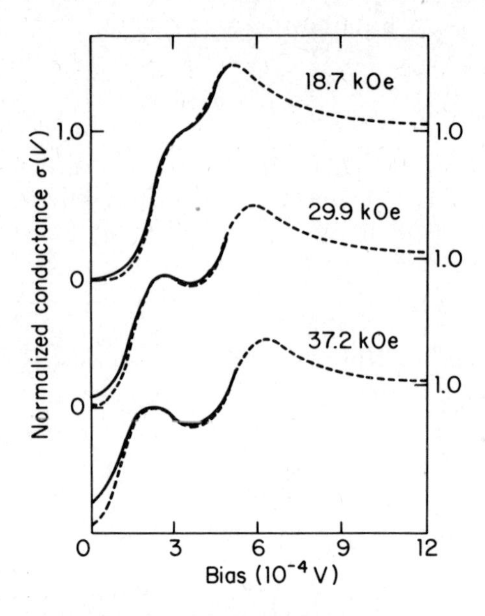

Fig. 3.28. Normalized conductance measurements on thin films of aluminum in various magnetic fields reveal spin-split superconducting density-of-state spectra as predicted in the previous figure. The solid curves are measured conductances, and the dashed curves are obtained from the theory of Bruno and Schwartz (1973). The parameter b for these data for aluminum is determined to be 0.07. (After Meservey, 1978.)

clean case. The extreme clean limit appears not to have been tested experimentally. A related clean limit case, that of uniform current flow, has been treated theoretically by Fulde (1965). Reviews of these results have been given by Maki (1969) and Fulde (1969).

Effects of flux penetration in the parallel configuration (Type II superconductors) have been observed by Sutton (1966) and by Moore and Beasley (1977). Changes in conductance associated with vortices normal to the tunneling electrodes have been observed by Fulton, Hebard, Dunkleberger, and Eick (1977), who compare their results with those obtained earlier by Band and Donaldson (1973) and Gray (1974).

3.6.6 *Magnetic Impurities*

It has long been known that magnetic impurities have a more pronounced effect in depressing T_c than do nonmagnetic impurities, with ΔT_c linearly dependent on concentration c; e.g., see Schwidtal (1960). A microscopic treatment of this effect based on the exchange interaction $-J\vec{S}\cdot\vec{\sigma}$ between localized moments $\vec{S}$ and conduction electrons $\vec{\sigma}$ (Abrikosov and Gor'kov, 1960) predicted, further, a linear decrease in Δ with c, more

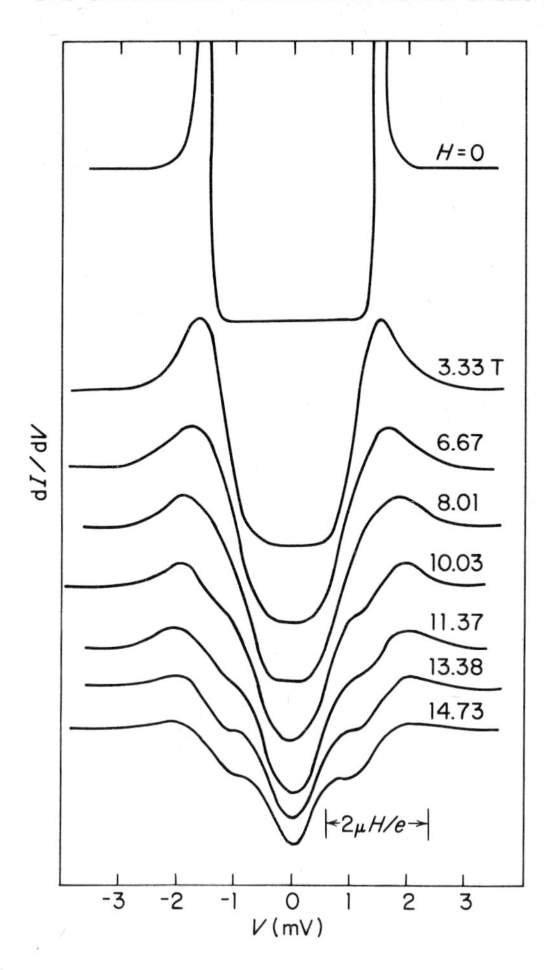

Fig. 3.29. Traces of the tunneling conductance curves for Al–Al_2O_3–Ga junctions in several values of magnetic field running up to 14.7 T reveal again the spin-split superconducting density-of-states peaks, but indicate a stronger spin-orbit scattering than in the case of aluminum. The value of the spin-orbit parameter b is determined to be 0.6. (After Tedrow and Meservey, 1975.)

rapid than that of T_c, and, in fact, that $\Delta = 0$ should occur for the dirty limit at $c/c_{cr} = 2\exp(-\pi/4) = 0.912$, with c_{cr} corresponding to $T_c = 0$. Tunneling observations (Reif and Woolf, 1962; Woolf and Reif, 1965) convincingly confirmed this remarkable prediction, establishing the first example of gapless superconductivity. Some of the conductance data are shown in Fig. 3.30, in comparison with curves calculated from an extension of the Abrikosov–Gor'kov (AG) theory of Skalski, Betbeder-Matibet, and Weiss (1964). These curves are seen to be similar in nature to those obtained for thin dirty films in a parallel field (Fig. 3.25). In fact,

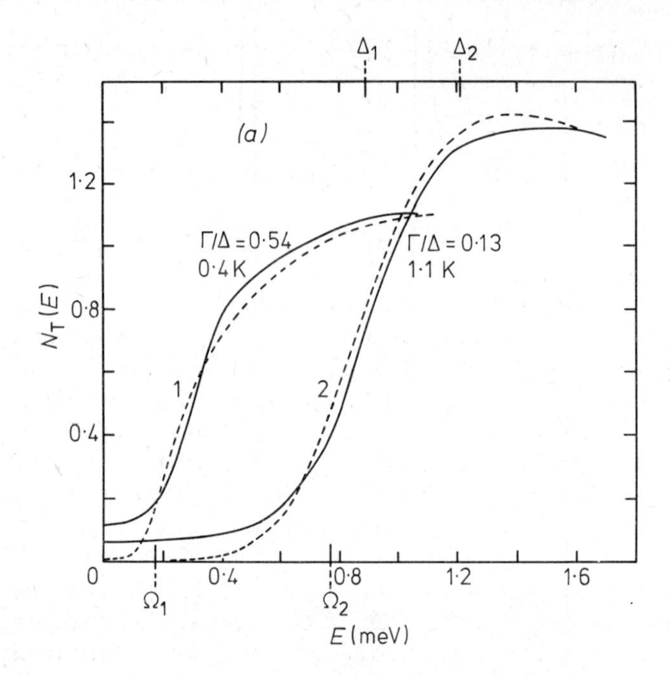

Fig. 3.30. The effect of magnetic impurities and lifetime broadening on the superconducting density-of-states structure is shown in measurements of the tunneling conductance into lead containing gadolinium impurities. The measurements are shown as the solid curves, while calculations based upon the theory of Skalski, Betbeder-Matibet, and Weiss (1964) are shown as dashed, from which the depairing parameters can be deduced. (After Reif and Woolf, 1962.)

since the paramagnetic pair-breaking parameter,

$$\alpha \simeq \frac{c}{4} N(0)S(S+1)J^2$$

is proportional to c, in analogy to the case when $\alpha \simeq H^2$ in the magnetic field, one correctly infers $H/H_{c\|} = \sqrt{2}\exp(-\pi/8) = 0.955$ (dirty limit) for the onset of gaplessness in parallel fields. Because of the universal aspect of the pair-breaking situation described by the parameter α, the result of AG for the T_c depression,

$$\ln(T_{c0}/T_c) = \psi\left(\frac{1}{2} + \frac{\alpha}{2\pi kT_c}\right) - \psi\left(\frac{1}{2}\right) \tag{3.93}$$

where $\psi(x) = \Gamma'(x)/\Gamma(x)$, is the digamma function and T_{c0} is the value of T_c with $\alpha = 0$, applies in a variety of dirty superconductor cases. Expansion of $\psi(x)$ about $x = \frac{1}{2}$ leads to the linear relation $\Delta T_c = -\pi\alpha/4$, while the condition for the normal-state transition $T_c = 0$ is simply $2\alpha = \Delta_0$ (the condition for gaplessness being $2\alpha = \Delta$).

In the gapless regime (again, for $\ell \ll \xi$), a rather simple expression for the tunneling density $N_T(E)$ has been given by de Gennes (1964):

$$N_T(E) \simeq 1 + \frac{|\Delta|^2}{2\alpha^2} \frac{(E/\alpha)^2 - 1}{[(E/\alpha)^2 + 1]^2} \tag{3.94}$$

where $2\alpha = \hbar/\tau_K$. This regime has been examined in experiment by Guyon, Martinet, Matricon, and Pincus (1965).

The AG approach treated a single magnetic impurity (independent scattering) within a Born approximation. That this approximation of weak scattering is too rough to predict localized impurity states, occurring even at low concentration and physically corresponding to excited states of Cooper pairs near the impurity, is now clear theoretically (Shiba, 1968; Rusinov, 1969; Zittartz and Müller-Hartmann, 1970; Zittartz, Bringer, and Müller-Hartmann, 1972) and experimentally (Levin, Selisky, Heim and Buckel, 1976; Dumoulin, Guyon and Nédellec, 1977; Tsang and Ginsberg, 1980*a*, 1980*b*; Bauriedl, Ziemann and Buckel, 1981). The results of Bauriedl, Ziemann, and Buckel (1981) for Mn^+ ions implanted into Pb (through the tunnel barrier) at cryogenic temperature are shown in Fig. 3.31a,b. Two bands of localized states are evident in the gap, observed at 0.2 K. One suspects that temperature smearing may have obscured such features in some earlier work. These results have been fitted to an extension (addition of a second band) of the theory of Zittartz, Bringer, and Müller-Hartmann (1972), which is a Kondo effect theory. The possibility may still be open that a fit could also be obtained within the simpler classical spin Shiba theory, shown by Ginsberg (1979) to yield additional bands when higher-order scattering (beyond S waves) is incorporated.

A further theoretical possibility, of including the full Eliashberg theory in treatment of the paramagnetic impurities, has been explored by Schachinger, Daams, and Carbotte (1980). Even though the situation of dilute ($c \leqslant 0.1\%$) paramagnetic additions to superconductors apparently remains unsettled, several bold attempts in the high-concentration, spin-glass regime have been reported in the past few years. The experimental study of the spin-glass AgMn at ~1% concentration, backed by Pb to induce superconductivity, of Schuller, Orbach, and Chaikin (1978) has indicated a marked increase in the spin-flip scattering time as T is reduced through the temperature at which the susceptibility shows a cusp (T_G). An extensive theoretical attack on the spin-glass problem, including calculation of the tunneling density of states, has been launched by Nass, Levin, and Grest (1980, 1981), who derive and solve Eliashberg-like equations for a nondilute magnetically ordered superconductor. A qualitative comparison of the derivative of their $N_T(E)$ is shown in Fig. 3.32, with further tunneling data from a 400-Å AgMn film backed by Pb and measured at 0.05 K.

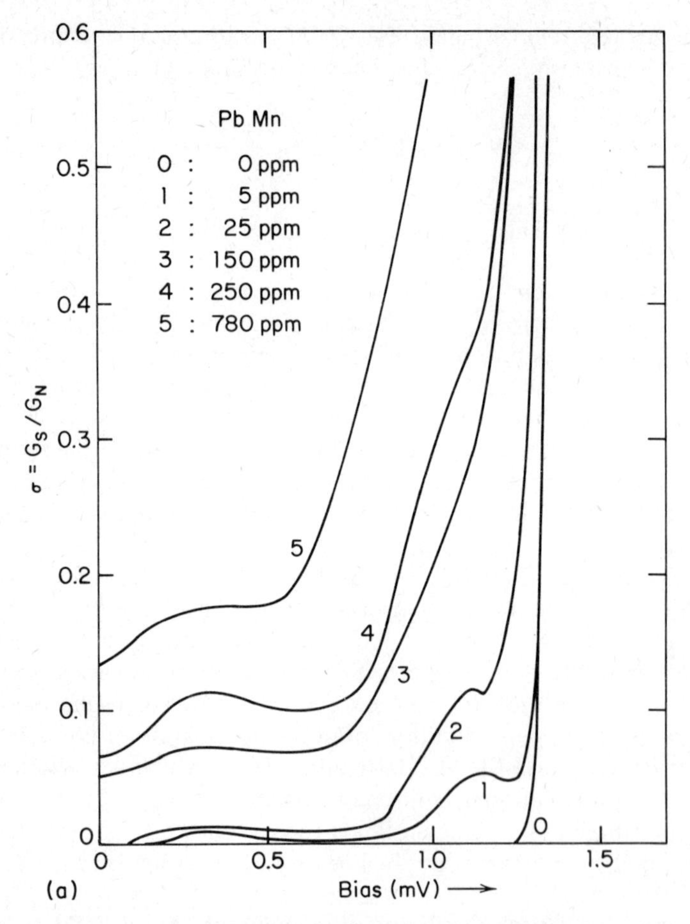

Fig. 3.31. (a) Observation of magnetic impurity bands in the gap of PbMn alloys of varying Mn concentration in normalized conductance measurements of Mg–MgO–PbMn junctions at 0.2 K. The lead films were prepared by implantation of Mn^+ ions into annealed lead films held at low temperature, and the tunneling conductance was measured *in situ* after each implantation step. (b) Comparison of normalized conductance measurements, here shown over a wider voltage range and at three concentrations of manganese impurity in lead, with theory (shown dashed). (After Zittartz and Müller-Hartmann, 1970; Bauriedl, Ziemann, and Buckel, 1981.)

3.6.7 *Pressure Effects*

The effects of hydrostatic pressure on the superconducting energy gap, as well as on the phonon spectra, can be studied by tunneling. A recent example of such experimental work is shown in Fig. 3.33 (after Svistunov, Chernyak, Belogolovskii, and D'yachenko, 1981). At 12 kbar in a pure Pb–Al counterelectrode junction, a 19% decrease in sum gap occurs, as well as a decrease in the gap of the granular Al electrode. Comparison of

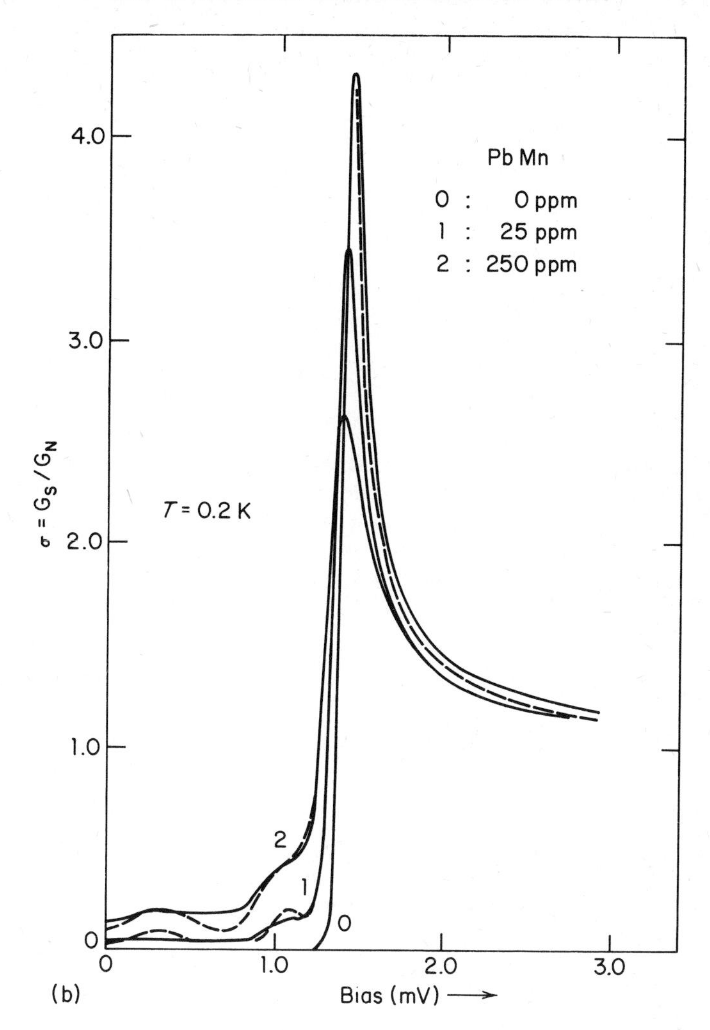

these substantial effects in Δ with corresponding changes in T_c within the $PbIn_x$ alloy system shows that the fractional decrease in Δ with pressure P exceeds that of T_c:

$$\left|\frac{d(\ln \Delta_0)}{dP}\right| > \left|\frac{d(\ln T_c)}{dP}\right|$$

A related result is a decrease in the ratio $2\Delta_0/kT_c$ from 4.53 (pure Pb, atmospheric pressure) to 4.35 at 12.2-kbar pressure. A qualitatively similar behavior had earlier been observed in In, but no change in $2\Delta/kT_c$ with pressure was found with Tl (Galkin, Svistunov, and Dikii, 1969). Svistunov, Chernyak, Belogolovskii, and D'yachenko (1981) employ the

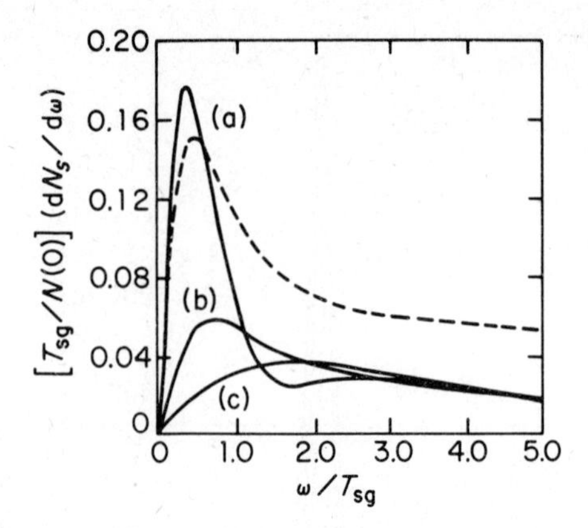

Fig. 3.32. Calculations of the derivative of the density of states vs. energy for a concentrated magnetic impurity system, or spin-glass. The dashed curve in this figure is due to I. Schuller and P. M. Chaikin and represents measurement at 0.05 K of tunneling into a 400-Å AgMn film backed by 3000 Å of Pb with Mn concentration of 0.2%. (After Nass, Levin, and Grest, 1980.)

relation

$$\frac{d(\ln \Delta_0)}{dP} = k\frac{d(\ln T_c)}{dP} - (k-1)\gamma^*$$

in correlating these results, where $k > 1$ is an adjustable parameter and $\gamma^* = d(\ln \bar{\omega})/dP$, with $\bar{\omega}$ a suitable average phonon frequency. The origin of changes in T_c and Δ with P lies in changes in the normal-state properties—the phonon spectrum, electron-phonon coupling, the electronic density of states, and the Coulomb repulsion—all of which enter the Eliashberg equations. Further discussion of these comprehensive studies, such as that of Svistunov, Chernyak, Belogolovskii, and D'yachenko (1981), including changes in phonon spectra, etc., will be given in Chapter 6.

3.6.8 Interactions with Electromagnetic Radiation

There are many effects which one can observe in SIS junctions which arise from interactions of electromagnetic radiation with Josephson and quasiparticle tunneling currents. Several of these, in the order of their discovery, are photon-assisted (quasiparticle) tunneling at $eV = 2\Delta \pm n\hbar\omega$, $n = 1, 2, 3$ (Dayem and Martin, 1962); photon-assisted pair tunneling, or frequency-modulated supercurrents; current steps, or zero-slope regions in $I(V)$ (Shapiro steps), symmetric about $V = 0$, $V = \pm n\hbar\omega/2e$, $n = 1, 2, 3, \ldots$, predicted by Josephson (1962*a*) (Shapiro, 1963); self-detection of Josephson radiation via interaction with magnetic field–tuned, slow-wave (Swihart, 1961) modes (Eck, Scalapino, and Taylor,

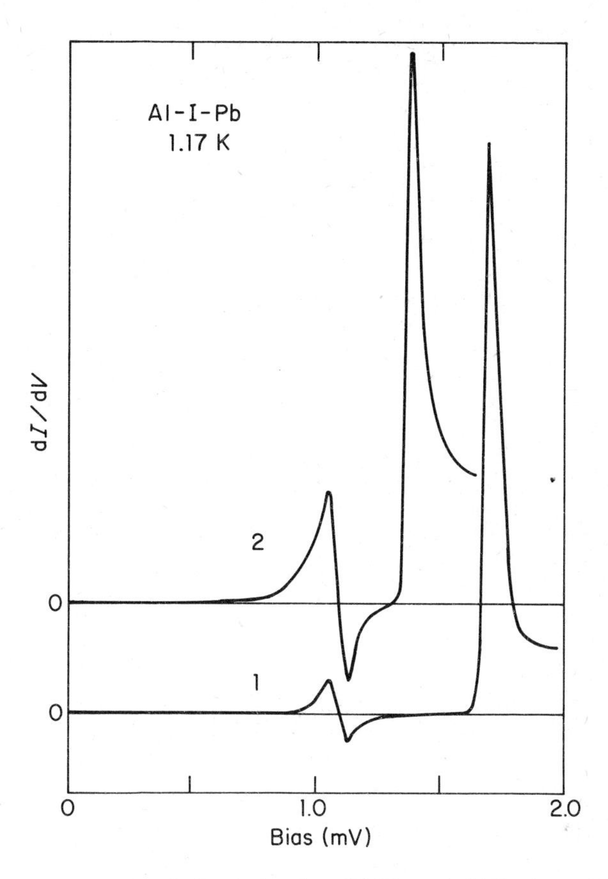

Fig. 3.33. Measurements of the derivative dI/dV at 1.17 K of an Al–Al_2O_3–Pb junction at atmospheric pressure in curve 1 and at 12.2 kbar in curve 2. Note that the position of the sum-gap structure has changed appreciably under the application of pressure, whereas the structure representing the difference gap is less affected, although its amplitude is increased. (After Svistunov, Chernyak, Belogolovskii, and D'yachenko, 1981.)

1964) and via high-Q cavity modes, or Fiske steps (Fiske, 1964); and, finally, external detection of the Josephson radiation in a coupled tunnel junction by Giaever (1965) and also directly in a microwave cavity Langenberg, Scalapino, Taylor, and Eck, 1965; Yanson, Svistunov, and Dmitrenko, 1965.

Multiphoton quasiparticle tunneling (Dayem-Martin effect) is illustrated in Fig. 3.34, taken from Tien and Gordon (1963). The splitting of the sum-gap edge can be semiquantitatively described, as these authors and Cohen, Falicov, and Phillips (1962) showed, through an effective quasiparticle density of states in one of the electrodes, composed of a sum of BCS peaks uniformly spaced by $\pm n\hbar\omega$ around $E = \Delta_0$.

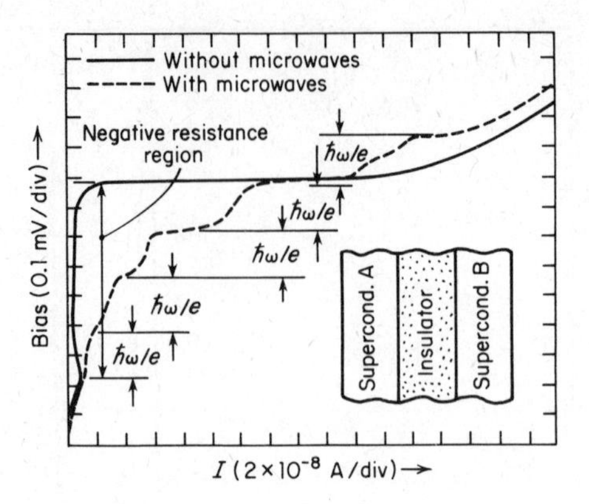

Fig. 3.34. The effects of an applied microwave electromagnetic field upon the I–V characteristics of an Al–Al_2O_3–In diode (both electrodes superconducting) reveals a breakup of the sum-gap edge into a set of rising portions spaced exactly by the photon energy (solid curve). These data are due to Dayem and Martin (1962), as displayed by Tien and Gordon (1963).

Tien and Gordon assumed that the externally applied microwave field leads to a perturbation $H' = e\delta \cos \omega t$ experienced by a quasiparticle at the edge of one of the electrodes. Suppose that $\Psi_0 = \psi(\vec{r})e^{-iEt/\hbar}$ describes the quasiparticle in the absence of the perturbation. The important effect of the uniform microwave field is not a possible change in the space part of Ψ_0 but, rather, an alteration of the time dependence, by introducing frequency modulation. Tien and Gordon thus assumed a corrected wavefunction

$$\Psi = \Psi_0\left(\sum_n B_n e^{-in\omega t}\right)$$

which, upon direct substitution in Schrödinger's equation, $H\Psi = i\hbar\, \partial\Psi/\partial t$, leads to a Bessel function recursion relation for the B_n. Thus,

$$\Psi = \Psi_0 \sum_{n=-\infty}^{\infty} J_n(\alpha)e^{-in\omega t}, \qquad \alpha = \frac{e\delta}{\hbar\omega} \tag{3.95}$$

where J_n is the nth-order Bessel function of the first kind. The effective split density of states is

$$\rho_{\text{eff}}(E) = \sum_{n=-\infty}^{\infty} \rho(E + n\hbar\omega)J_n^2(\alpha) \tag{3.96}$$

(Tien and Gordon, 1963), a result also obtained from a different calculation by Werthamer in 1966. The experimental observations of multiphoton absorption were confirmed and extended by Cook and Everett (1967); Sweet and Rochlin (1970); and Hamilton and Shapiro (1970); among others. A thorough discussion of this development has been given by Solymar (1972).

The Shapiro step structure, a splitting of the supercurrent I–V relation under microwave irradiation, follows directly from the Josephson relations. The microwave voltage $\delta\cos(\omega t+\theta)$, as above, taken with the Josephson equation for the relative pair phase ϕ, $\partial\phi/\partial t = 2eV(t)/\hbar$, yields

$$\phi(t) = \omega_0 t + \frac{\omega_0 \delta}{\omega V_0}\sin(\omega t+\theta)+\phi_0 \tag{3.97}$$

Here, $\omega_0 = 2eV_0/h$, where V_0 is the dc voltage across the junction and θ and ϕ_0 are the initial values of the microwave and relative pair phases, respectively. Substituting in the Josephson current relation, one finds

$$J = J_1 \sin\left[\omega_0 t + \frac{\omega_0\delta}{\omega V_0}\sin(\omega t+\theta)+\phi_0\right] \tag{3.98}$$

which, when expanded in a Fourier series, becomes

$$J(t) = J_1 \sum_{n=-\infty}^{\infty} J_n\left(\frac{n\delta}{V_0}\right)\sin[(\omega_0+n\omega)t+n\theta+\phi_0] \tag{3.99}$$

where J_n is the usual Bessel function of order n. Evidently, this expression gives a dc current whenever $\omega_0+n\omega=0$ or $V=n\hbar\omega/2e$, $\pm n = 1, 2, 3, \ldots,$ of magnitude

$$J_{\text{step}}^{(n)} = J_1(-1)^n J_n \frac{n\delta}{V_0}\sin(\phi_0 - n\theta) \tag{3.100}$$

Observation and interpretation of this step structure by Shapiro (1963) constituted the first experimental confirmation of the ac Josephson effect.

Self-detection of the Josephson radiation in a single junction is also possible, under certain circumstances, leading also to distinct features in the I–V characteristic. It was pointed out by Josephson (1965) that resonant behavior of the type required cannot occur in a perfectly homogeneous barrier in zero magnetic field. In the first report, by Eck, Scalapino, and Taylor (1964), a small magnetic field was applied in the plane of the junction and was shown to control the bias ($0<eV<\Delta$) at which a weak resonant current peak of width ~0.2 mV appeared in the I–V characteristic of a Pb–PbO–Pb Josephson junction at 1.2 K. Following a clear tutorial discussion by Langenberg, Scalapino, and Taylor (1966), we assume a small junction of width t to lie in the y, z plane, with normal vector $\hat{n}=\hat{x}$, and the magnetic field to be uniform, $\vec{H}=H_0\hat{y}$, and we allow the phase ϕ and current density J to vary spatially in the y, z plane. In this geometry, the variation $\phi(y, z)$ is governed by

$$\vec{\nabla}\phi = \frac{2ed}{\hbar c}(\vec{H}\times\hat{n}) \tag{3.101}$$

where $d = 2\lambda + t$, with λ the penetration depth, which reduces to

$$\frac{\partial\phi}{\partial z} = -\frac{2ed}{\hbar c}H_0, \qquad \frac{\partial\phi}{\partial y}=0 \tag{3.102}$$

Integration of (3.102) and substitution into the current equation gives

$$\vec{J}(z)=J_1 \sin\left(\phi_0-\frac{2ed}{hc}H_0 z\right)\hat{n} \tag{3.103}$$

Digressing, if the junction width in the z direction is L, integration over this width predicts a maximum current density

$$J_m=J_1\left|\frac{\sin(\pi\Phi/\Phi_0)}{\pi\Phi/\Phi_0}\right| \tag{3.104}$$

where the flux enclosed by the junction is $\Phi=H_0\,dL$ and the flux quantum is $\Phi_0=hc/2e=2.07\ \mathrm{G}\cdot\mathrm{cm}^2$. This variation, (3.104), was demonstrated by Rowell (1963), whose data were shown in Fig. 3.1c. Observation of this dependence requires $L\ll\lambda_J$, where the Josephson penetration depth

$$\lambda_J=\left(\frac{hc^2}{8\pi eJ_1 d}\right)^{1/2} \tag{3.105}$$

arises from the $\phi(z)$ phase difference variation produced by the magnetic fields of the Josephson current itself (Ferrell and Prange, 1963). This effect limits the Josephson current flow to the edges of junctions of large dimensions.

Returning to (3.103) (for $L\ll\lambda_J$) and including the time dependence, we note that this equation is of the form $J(z,t)=J_1\sin(\omega_0 t-kz)$, with magnetic field–dependent wavevector $k=2edH_0/\hbar c$, which represents a current density wave traveling across the z dimension of the junction. The phase velocity $\omega_0/k=V_0c/H_0d$ is proportional to the *ratio* of the applied bias V_0 to magnetic field. The resonant current peak of Eck, Scalapino, and Taylor (1963), therefore, occurs when the phase velocity of the current density wave matches that of electromagnetic waves propagating between the superconducting electrode plates. Such transverse magnetic TM waves, considered by Swihart (1961), are substantially slowed by their penetration into the superconductors and propagate parallel to the barrier plane, with phase velocity

$$\bar{c}=c\left(\frac{t}{\varepsilon d}\right)^{1/2}$$

Here t and d, respectively, are the oxide thickness, and $t+2\lambda_L$ (λ_L is the London penetration depth), and ε is the dielectric constant. A typical value of $\bar{c}$ is $c/20$. When the resonant condition occurs, the voltage δ of the TM wave at $\omega=\omega_0$ increases, and analysis similar to (3.98) and (3.99) leads to a first-order result corresponding to experiment.

Much stronger effects (Fiske, 1964) occur when high-Q electromagnetic resonances of the open-ended tunnel junction cavity match the Josephson frequency $\omega_0=2eV_0/\hbar$. Assuming antinodes occur at the junction edges,

$$L = \frac{n\lambda}{2} = \frac{n\bar{c}}{2\nu}$$

hence, $V_0 = \pm nh\bar{c}/4Le$ gives the locations of the uniformly spaced steps. A magnetic field is necessary (unless the junction is inhomogeneous; see Matisoo, 1969) to couple to these modes. The intensity of the steps is controlled by the magnetic field through considerations of the overlap of the current density waves with the electromagnetic modes, discussed by Coon and Fiske (1965) and others; see Langenberg, Scalapino, and Taylor (1966). Further examples of these effects are given by Dmitrenko, Yanson, and Svistunov (1965); Langenberg, Scalapino, Taylor, and Eck (1965); and Matisoo (1969).

The detection of Josephson radiation with a second electromagnetically coupled tunnel junction was reported by Giaever (1965), whose junction structure and results are illustrated in Fig. 3.35. The three-electrode Sn structure incorporates a low-impedance Josephson generator junction, biased at voltage V_{23}, and a higher-impedance detection junction whose quasiparticle current I_{12} is plotted vs. V_{12} in Fig. 3.35b. The oxide barriers of the two junctions connect and are regarded as coupled electromagnetic cavities. Giaever's observation of Dayem-Martin steps (see Fig. 3.34) corresponding to radiation of energy $\hbar\omega = 2eV_{23}$ is illustrated. While the spectral resolution afforded by the Dayem-Martin effect measurement is limited, it is certainly adequate to establish the crucial factor of 2 in the Josephson frequency condition.

Josephson radiation of $\sim 10^{-14}$ W was detected in a microwave cavity by Yanson, Svistunov, and Dmitrenko (1965); soon thereafter measurement of 10^{-12} W at 9.2 GHz was reported by Langenberg, Scalapino, Taylor, and Eck (1965), who confirmed the Josephson frequency to 10^{-4} accuracy. The coupling between the ac Josephson current and the electromagnetic modes was improved by application of a 19-G magnetic field in the plane of the junction, as above. This work was extended by Parker, Taylor, and Langenberg (1967) and Parker, Langenberg, Denenstein, and Taylor (1969) to determine the ratio e/h to an absolute accuracy of 2.4×10^{-6}. Interpretation of this measurement, working *from* the Josephson relation $2eV = \hbar\omega$, to obtain fundamental constants involves an assumption of exact pairing in the superconducting state. This assumption was checked to $\sim 2 \times 10^{-6}$ by Parker, Taylor, and Langenberg (1967) by use of different superconductors. The techniques are now applied to provide standards of voltage.

The question of exactness of pairing in different materials was addressed to higher precision (1 in 10^8) by Clarke (1968), as illustrated in Fig. 3.36. This work made use of pairs of SNS junctions, where N is copper ($\sim 1\ \mu$m in thickness) and with S = Pb, Sn, In, and it exploited the extremely sharp, characteristic radiation-induced steps at $V_n = n\hbar\omega/2e$, with n an integer or a half-integer (Anderson and Dayem, 1964). As

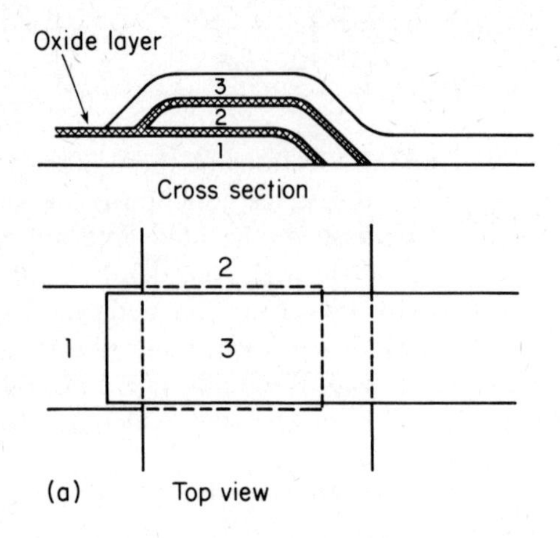

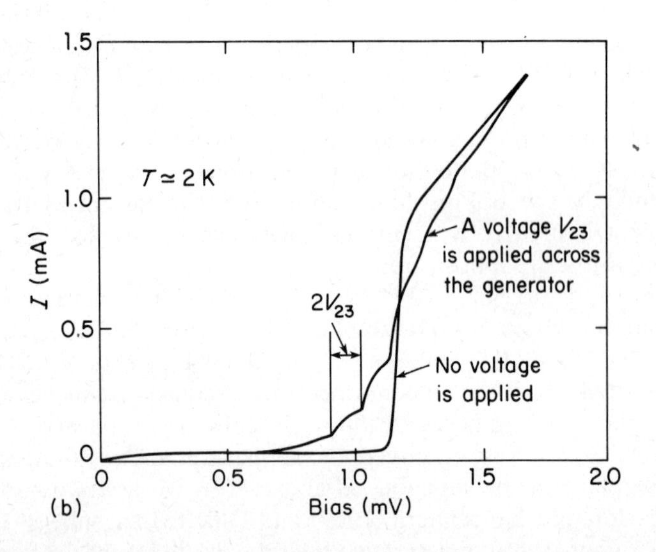

Fig. 3.35. (a) Schematic drawing of the structure employed by Giaever (1965) for detection of the Josephson radiation. Note that the structure contains two separate junctions, all films are made of Sn, and that the oxide layers between the tin films are regarded as microwave cavities. (b) The resulting $I-V$ characteristics. The curve rising smoothly above the sum gap is measured with no voltage supplied across the generator junction, whereas the curve with cusp-like structures indicates the response when a voltage is applied across the generator junction. Note the similarity of these results to those of fig. 3.34 in which microwave radiation breaks up the I–V characteristic of a tunnel junction. (After Giaever, 1965.)

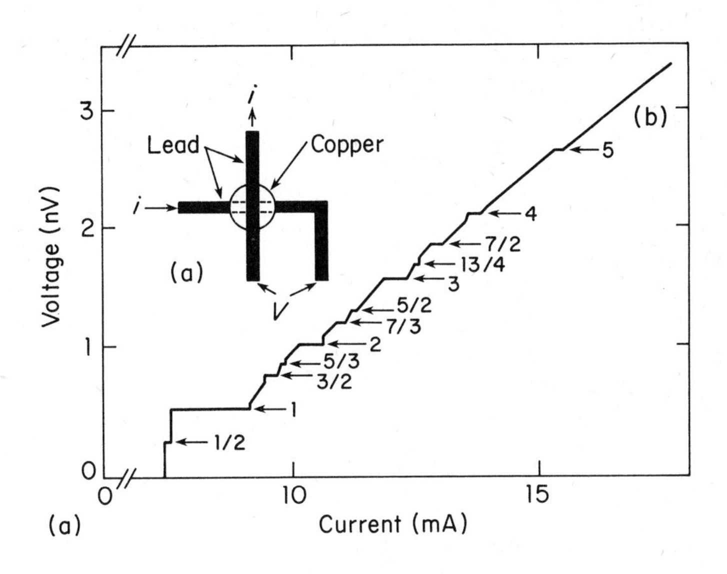

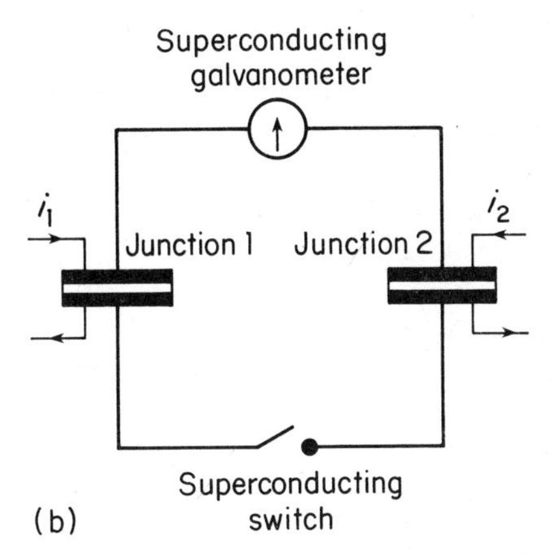

Fig. 3.36. (a) The current-voltage relationship of the Pb–Cu–Pb structure shown in the inset is dominated by constant-voltage-current Josephson steps, which occur at voltages $V_n = n \times h\nu/2e$, where n can be an integer or a simple ratio of integers. Here $h\nu$ is the energy of the radiation quantum falling upon the structure, and $2e$ is the charge of the electron pair. In the case shown, the structure is held at 4.2 K and irradiated at 250 kHz. (After Clarke, 1968.) (b) The circuit used for comparing the voltages developed across two junctions made from different superconductors. Separate bias currents i_1 and i_2 were supplied so that each structure corresponds to the same multiple n as shown in (a). In such a way, a very precise measure of the ratio $e^*/2e$ was obtained. (After Clarke, 1968.)

shown in Fig. 3.36b, dissimilar junctions biased to the same quantum index n were compared, using the same radiation. The nanovoltmeter should read

$$\Delta V_n = n\hbar\omega[(2q_1)^{-1} - (2q_2)^{-1}]$$

where q_1, q_2 are values of the pair charge appropriate to the differently composed junctions and n is the index of the step. This quantity was always observed to be zero, independent of the combination of metals, temperature, and values of several other parameters. Actually, the maximum value of ΔV was set at the extremely small value $\Delta V \leqslant 1.7 \times 10^{-17}$ V, implying the constancy of e/h to one part in 10^8 by use of a superconducting galvanometer, as shown. First, Clarke (1968) showed that a current induced around the loop (Fig. 3.36b) decayed with a time constant $\tau = L/R \geqslant 3 \times 10^6$ sec, where the loop inductance was previously measured as $(1.0 \pm 0.1) \times 10^{-7}$ H. Hence, $R = L/\tau$ was less than $3.3 \times 10^{-14}\ \Omega$. This result implies that the junction dV/dI biased on the step structure cannot exceed this value. Further, with the junction current at a typical value, 0.5 mA, then the voltage width of the step is less than 10^{-17} V, less than in tunnel junctions, making the SNS device superior in this experiment. Second, with the superconductor galvanometer of resolution 0.3 μA, the circuit of Fig. 3.36b was closed, and the galvanometer current was observed to remain zero for $t = 1.8 \times 10^3$ sec. If a voltage ΔV had existed, it would have led to a growing current according to $\Delta L = L\, dI/dt$. Using the measured values $\Delta I < 3 \times 10^{-6}$ A, $\Delta t = 1.8 \times 10^3$ sec, and $L = 1.0 \times 10^{-7}$ H, we see $\Delta V \leqslant 1.7 \times 10^{-17}$ V.

These experiments show that the basic current-carrying entity in superconductors is exactly twice the electron charge. However, if *small* contributions from quartets or higher-order groups of electrons were present, these would lead to subharmonic or lower-order harmonic steps and would not affect the voltage at which the steps occur.

3.6.9 *Superconducting Fluctuations*

The density of states in a fluctuation state of superconductivity at $T > T_c$ in granular Al has been measured as shown in Fig. 3.37. Other, more exhaustively studied features of fluctuation superconductivity are the dc conductivity at $T \geqslant T_c$, addressed experimentally by Glover (1967) and theoretically by Aslamazov and Larkin (1968), and the diamagnetic susceptibility (Gollub, Beasley, and Tinkham, 1970; Prange, 1970; Lee and Payne, 1971).

The existence of thermodynamic fluctuations into a paired state above T_c is a general property of superconductors. However, the effects are made more easily observable in dirty thin-film specimens of small coherence length ($\xi \sim \sqrt{\xi_0 \ell}$, $\ell \ll \xi_0$) in which the competing normal-state effect (diamagnetism or conductivity) is reduced. This effect is illustrated by the formula of Aslamazov and Larkin (1968) for the conductivity of a film of

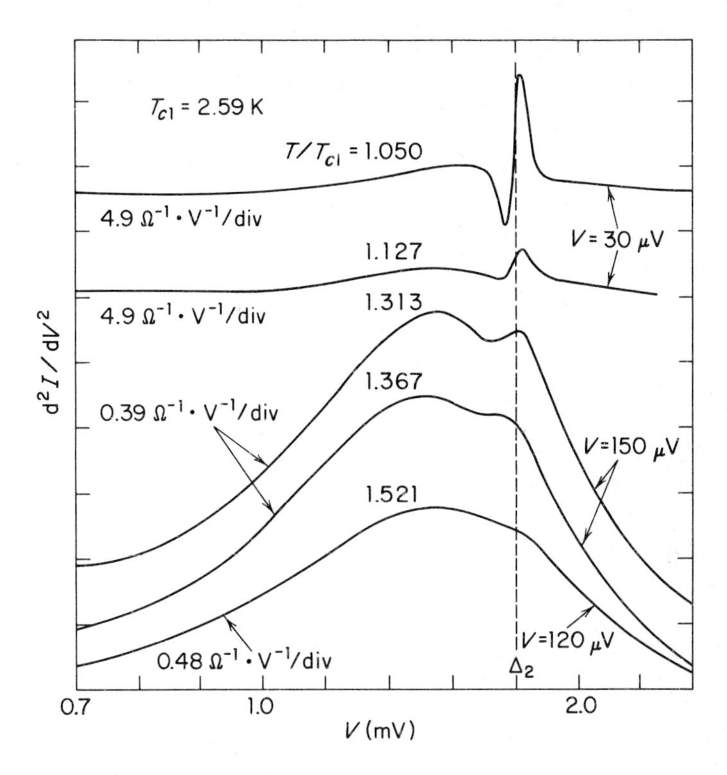

Fig. 3.37. Second-derivative d^2I/dV^2 measurements of an $Al–Al_2O_3–Pb_{0.7}Bi_{0.3}$ junction reveal structure at the sum gap above the T_c of the granular aluminum film, attributed to superconducting fluctuations. The lowest curve in the figure corresponds to the NIS case: even at a temperature $1.5 \times T_c$ of the aluminum film, the superconducting characteristic can be seen clearly. The uppermost second derivative characteristic corresponds to the observation of the corner which occurs in the SIS $I–V$ curve at the sum-gap energy. The lower portion of the figure shows the tunneling density of states inferred for the amorphous aluminum film as it fluctuates into the superconducting state. (After Cohen, Abeles, and Fuselier, 1969.)

thickness d with normal-state conductivity σ_0:

$$\sigma = \sigma_0 + \frac{e^2}{16d\hbar}\frac{T - T_c}{T_c}$$

in which the relative contribution of the universal fluctuation term increases with decreasing d and σ_0.

In the tunneling study of Cohen, Abeles, and Fuselier (1969), the greatest effects were observed in the films having the smallest mean free paths and, consequently, the smallest coherence lengths. The structure in the tunneling conductance and d^2I/dV^2 (as shown in Fig. 3.37 for a film

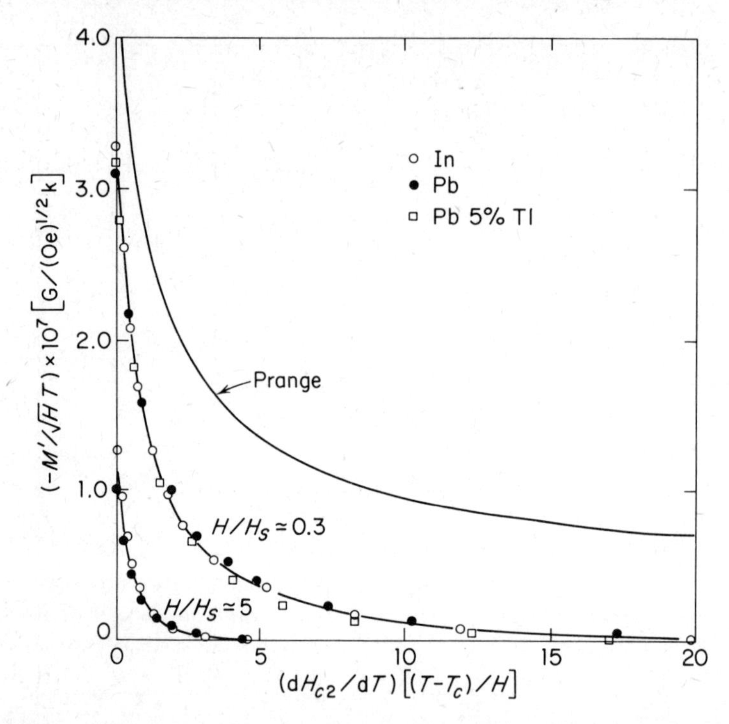

Fig. 3.38. Universal behavior in the temperature dependence of diamagnetic fluctuations observed in superconductors above their critical temperature T_c. The quantity M', which is plotted in a normalized fashion on the ordinate, represents the magnetization of the sample in a given parallel magnetic field (at the specified temperature) minus the magnetization at the same field at considerably higher temperature; M' then eliminates the temperature-independent magnetization in the normal state of the metal. The crosses, solid dots, and squares represent measurements on indium, lead, and a lead-thallium alloy, respectively. (After Gollub, Beasley, and Tinkham, 1970.)

with $\xi = 120$ Å) occurred in the same temperature range in which fluctuation effects were evident in the resistive transitions. The density of states above T_c for the Al films inferred from the tunneling curves, shown in the figure, is strongly gapless in nature and becomes BCS-like at T_c. From these data, Cohen, Abeles, and Fuselier inferred values of a pair-fluctuation lifetime $\hbar/\alpha$ associated with the density of states given by (3.94).

4

Conventional Tunneling Spectroscopy of Strong-Coupling Superconductors

4.1. Introduction

The power of the tunneling method is enhanced by the fundamental role of the normalized conductance $\sigma = (dI/dV)_S/(dI/dV)_N$ as a nearly exact measure of the density of excitations of the superconductor. Thus, any weak variations in the particular barrier's normal-state conductance $G(V)$ almost completely cancel, leaving a fundamental quantity which can be analyzed precisely in terms only of superconductive properties, as shown by Schrieffer, Scalapino, and Wilkins (1963). The full quantitative power implied by these circumstances was first demonstrated by McMillan and Rowell (1965) in providing a comprehensive and detailed picture of the strong-coupling superconductivity of Pb working from spectra such as had earlier been recorded by Giaever (see Fig. 1.6a). Even though the deviations in $\sigma(V)$ from the BCS prediction in the phonon energy range $V \simeq \hbar\omega_{ph}/e$ were only a few percent, possibly of the same order as the variation of the normal-state conductance $G(V)$, because of the fundamental role of the normalized quantity, the former deviations could be interpreted as arising only from energy dependence in $\Delta(E)$. Working backward from the normalized conductance, McMillan and Rowell in fact determined $\Delta(E)$ and the underlying electron-phonon spectral function $\alpha^2F(\omega)$, also to an accuracy of a few percent. On this basis, McMillan and Rowell (1969), in a full exposition of their quantitative methods, have suggested that tunneling provides the most detailed picture of the superconducting state.

We outline the methods of conventional McMillan–Rowell spectroscopy, starting with a brief account of the Eliashberg–Nambu theory of strong-coupling superconductors. Inversion of tunneling data in this method is accomplished on the assumption of isotropic superconducting properties, in both the original McMillan variational program and, more recently, by Galkin, D'yachenko, and Svistunov (1974) in a more direct method. The use of such results to establish the accuracy of the isotropic $T = 0$ strong-coupling theory and extensions of the method to more general situations are considered in turn. Finally, some comments are offered on the limitations of the conventional spectroscopy.

4.2 Eliashberg-Nambu Strong-Coupling Theory of Superconductivity

The electron-electron coupling provided by the final Eliashberg (1960) electron-phonon function

$$V_{k,k+q} \propto \frac{E_{k+q} + \hbar\omega_q}{(E_{k+q} + \hbar\omega_q)^2 - E_k^2} \tag{4.1}$$

is local in space and retarded in time, reflecting the delay $\tau \simeq 1/\omega_{\text{ph}}$ in the development of lattice overscreening. This result is in contrast to the nonlocal instantaneous nature of the BCS model interaction, attractive for any pair of electrons both within $\hbar\omega_D$ of the Fermi surface. A realistic and accurate theory of superconductivity evolved through the contributions of several authors. For example, (4.1) was preceded by the work of Fröhlich (1950, 1952), Bardeen and Pines (1955), and Bogoliubov (1959). The possibility of accurately treating even so strong an electron-phonon coupling that the quasiparticle concept is invalid (i.e., the damping $\hbar/\tau$ may even exceed the quasiparticle energy E_k) was established by Migdal in 1958. Migdal's calculation in the normal state of the one-electron self-energy benefited by the fact that the ion mass M greatly exceeds the electron mass m; an accuracy of order $(m/M)^{1/2} \simeq 10^{-2}$ was demonstrated. Extension of this self-energy calculation to the superconducting state was accomplished by Eliashberg (1960) and by Nambu (1960). The residual (screened) Coulomb repulsive interaction between Coulomb quasiparticles was considered by Bogoliubov, Tolmachev, and Shirkov (1959) and by Morel and Anderson (1962). The Coulomb potential V_c, usually expressed as the dimensionless $\mu = N(0)V_c$, is reduced by consideration of the difference in time scales of the electronic and ionic motion (Allen and Dynes, 1975*b*) to

$$\mu^* = \frac{\mu_c}{1 + \mu_c \ln(\omega_{p1}/\omega_D)} \tag{4.2}$$

where ω_{p1} and ω_D are, respectively, the electron plasma frequency and the lattice Debye frequency. This typical reduction from $\mu \simeq 1.0$ to $\mu^* \simeq 0.1$ arises from the increased spacing between paired electrons that results from the relatively slow response of the ions to the electronic motion.

The unique property of a superconductor is the existence of a self-consistent pairing potential

$$\Phi = Z\Delta \tag{4.3}$$

which allows for condensation of quasiparticles into pairs. Acting on a single electron, the action of Φ or Δ is to produce a pair and also a hole, resulting from removal of a second electron to generate the pair. This aspect of superconductivity was made explicit by Bogoliubov (1959), who found Schrödinger-like equations for the electron-hole amplitudes u, v of the quasiparticle wavefunction $\psi=\begin{pmatrix} u \\ v \end{pmatrix}$. For time-independent potentials V, Δ, the Bogoliubov equations are

$$\left(-\frac{\hbar^2}{2m}\nabla^2-\mu+V\right)u+\Delta v=Eu \tag{4.4}$$

$$\left(\frac{\hbar^2}{2m}\nabla^2+\mu-V\right)v+\Delta u=Ev \tag{4.5}$$

where μ is the chemical potential. These weak-coupling equations are equivalent to a matrix potential acting on a vector wavefunction. Generalization of these equations to strong coupling is discussed by Arnold (1978).

The generalization to strong coupling of this theory by Nambu (1960) leads to the isotropic Eliashberg equations, conventionally used in connection with tunneling spectroscopy. In Nambu's theory, the electron annihilation operator is

$$\psi_k=\begin{pmatrix} a_{k\uparrow} \\ a_{-k\downarrow} \end{pmatrix} \tag{4.6}$$

and the Hamiltonian is

$$\begin{aligned} H=&\sum_k \varepsilon_k\psi_k^*\hat{\tau}_3\psi_k+\sum_q \Omega_q p_q^* p_q \hat{1}+\sum_{k,k'} g_{kk'}\phi_{k-k'}\psi_{k'}^*\hat{\tau}_3\psi_k \\ &+\frac{1}{2}\sum_{k_1k_2k_3k_4}\langle k_3k_4|\,V_c\,|k_1k_2\rangle(\psi_{k_3}^*\hat{\tau}_3\psi_{k_1})(\psi_{k_4}^*\hat{\tau}_3\psi_{k_2}) \end{aligned} \tag{4.7}$$

where $\phi_q=p_q+p_{-q}^*$, with p_q a phonon annihilation operator, $\hat{1}$ the unit matrix, $\hat{\tau}_i$ $(i=1, 2, 3)$ the Pauli spin matrices, and phonon polarization indices have been absorbed. In this expression, Ω_q, $g_{k,k'}$, and V_c are unrenormalized phonon frequencies, electron-phonon, and electron-electron couplings, respectively. The Fourier-transformed electron Green's function is found to be a 2×2 matrix, whose elements G_{11}, G_{22} relate to electrons and holes separately, while the off-diagonal entries G_{12} and G_{21} are the functions F and F^* (Gor'kov, 1958) describing pair correlations. The central result of the theory is an electron matrix self-energy $\sum(\vec{k},\omega)$ given as

$$\sum(\vec{k},\omega)=[1-Z(\vec{k},\omega)]\omega_1+Z(\vec{k},\omega)\Delta(\vec{k},\omega)\hat{\tau}_1+\delta\varepsilon(k)\hat{\tau}_3 \tag{4.8}$$

Here, Z is the renormalization function, Δ the pair potential or gap function, and $\delta\varepsilon(\vec{k})$, which arises from the Coulomb interaction, can usually be treated as a simple scale change of $\varepsilon(\vec{k})$, which is the same in the normal and superconducting states. The matrix self-energy equation (4.8) represents a set of four coupled integral equations, known as the Eliashberg equations. Averaged over the Fermi surface, as is appropriate for dirty superconductors, at $T=0$, and neglecting the weak k dependence (on the scale of k_F) of Δ and Z, these equations become (Scalapino,

Schrieffer, and Wilkins, 1966)

$$\Delta(\omega) = [Z(\omega)]^{-1} \int_0^{\omega_c} \mathrm{d}\omega' P(\omega')[K_+(\omega, \omega') - \mu^*] \tag{4.9}$$

$$[1 - Z(\omega)]\omega = \int_0^{\infty} \mathrm{d}\omega' N(\omega') K_-(\omega, \omega') \tag{4.10}$$

where

$$P(\omega) = \mathrm{Re}\left\{\frac{\Delta(\omega)}{[\omega^2 - \Delta^2(\omega)]^{1/2}}\right\} \tag{4.11}$$

$$N(\omega) = \mathrm{Re}\left\{\frac{|\omega|}{[\omega^2 - \Delta^2(\omega)]^{1/2}}\right\} \tag{4.12}$$

and

$$K_\pm(\omega, \omega') = \int_0^{\infty} \mathrm{d}\Omega\, \alpha^2(\Omega) F(\Omega) \left(\frac{1}{\omega' + \omega + \Omega + i\delta} \pm \frac{1}{\omega' - \omega + \Omega - i\delta}\right) \tag{4.13}$$

In these equations, ω_c is a cutoff typically taken as five to ten times the maximum phonon frequency. The only material-dependent quantities here are μ^* and $\alpha^2(\Omega)F(\Omega)$. The latter is given by

$$\alpha^2(\Omega)F(\Omega) = \frac{\displaystyle\int \frac{\mathrm{d}S_{k'}}{|\vec{v}_{k'}|} \int \frac{\mathrm{d}S_k}{|\vec{v}_k|} \frac{1}{(2\pi)^3 \hbar} \sum_\lambda |g_{k',k,\lambda}|^2 \delta[\Omega - \omega_{\lambda,k'-k}]}{\displaystyle\int \frac{\mathrm{d}S_k}{|\vec{v}_k|}} \tag{4.14}$$

where $\vec{v}_k$ is the group velocity of state $\vec{k}$ on the Fermi surface, $\mathrm{d}S_k$ is an element of Fermi surface area, and $g_{k',k,\lambda}$ describes electron scattering from $\vec{k}$ to $\vec{k}'$ on the Fermi surface with creation of a phonon of energy $\hbar\omega_{\lambda,k'-k}$ and polarization λ.

A useful measure of the strength of electron-phonon coupling is the McMillan parameter λ,

$$\lambda = 2 \int \alpha^2(\omega) F(\omega) \omega^{-1}\, \mathrm{d}\omega \tag{4.15}$$

shown (McMillan, 1968a) to be approximately factorizable as

$$\lambda = \frac{N(0)\langle I^2 \rangle}{M\langle \omega^2 \rangle} \tag{4.16}$$

in an elemental metal of ionic mass M, Fermi level density of states (single spin) $N(0)$, and mean square phonon frequency $\langle \omega^2 \rangle$. Term $\langle I^2 \rangle$ is the Fermi surface average of the squared electron-phonon matrix element. The renormalization function at $E = 0$, further, obeys the simple relation

$$Z(0) = 1 + \lambda \tag{4.17}$$

which, e.g., describes the renormalization of the electron mass by the electron-phonon coupling. The Eliashberg function $\alpha^2(\omega)F(\omega)$ is fundamental, further, in providing the appropriate weighting function to compute averages, defined as

$$\langle f(\omega) \rangle \equiv \frac{2}{\lambda} \int_0^\infty \alpha^2 F(\omega) f(\omega) \omega^{-1} \, d\omega \tag{4.18}$$

Within Eliashberg theory, the transition temperature T_c is the lowest temperature for which $T \neq 0$ versions of (4.9) and (4.10), generalized to describe thermal populations of quasiparticles via Fermi functions and phonons via Bose-Einstein functions, have solutions only for $\Delta = 0$. The pioneering work of McMillan (1968*a*) in finding a simplified formula for T_c in terms of λ and μ^* has been extended by several authors, notably Allen and Dynes (1975*b*), who find

$$T_c = \frac{f_1 f_2 \omega_{\log}}{1.2} \exp\left[-\frac{1.04(1+\lambda)}{\lambda - \mu^*(1+0.62\lambda)} \right] \tag{4.19}$$

Here $\omega_{\log}$, to be expressed in kelvins, is defined as

$$\omega_{\log} = \exp\langle \ln \omega \rangle \tag{4.20}$$

and the expectation value implies use of Eq. (4.18). The correction prefactors f_1 and f_2 are near unity for $\lambda \leq 1.5$ and are defined as

$$\begin{aligned} f_1 &= \left[1 + \left(\frac{\lambda}{\Lambda_1} \right)^{3/2} \right]^{1/3} \\ f_2 &= 1 + \frac{(\bar{\omega}_2/\omega_{\log} - 1)\lambda^2}{\lambda^2 + \Lambda_2^2} \\ \Lambda_1 &= 2.46(1 + 3.8\mu^*) \\ \Lambda_2 &= 1.82(1 + 6.3\mu^*)\left(\frac{\bar{\omega}_2}{\omega_{\log}} \right) \end{aligned} \tag{4.21}$$

It is understood in connection with (4.18)–(4.20) that the value of $\mu^* = \mu^*_{\text{ph}}$ be further corrected to the phonon cutoff frequency ω_{ph} by the relation

$$\frac{1}{\mu^*(\omega_{\text{ph}})} = \frac{1}{\mu^*(\omega_c)} + \ln \frac{\omega_c}{\omega_{\text{ph}}} \tag{4.22}$$

4.3 Tunneling Density of States

It is implicit in the strong-coupling theory that a definite relationship between energy and wavevector is lost, because of the finite lifetime of a state. The spectral function $A(k, \omega)$, such that

$$N_T(\omega) = \frac{1}{(2\pi)^3} \int A(k, \omega)\, d^3k \qquad (4.23)$$

describes the resulting situation. The spectral function for the strong-coupling superconductor,

$$A(k, \omega) = \frac{1}{\pi} |\mathrm{Im}\, G_{11}(k, \omega)| \qquad (4.24)$$

and $N_T(\omega)$, making use of the weak k dependence of Δ and Z, are:

$$A(k, \omega) = \frac{1}{\pi} \left| \mathrm{Im} \left\{ \frac{Z(k, \omega)\omega + \varepsilon_k}{Z^2(k, \omega)\omega^2 - \varepsilon_k^2 - [Z(k, \omega)\Delta(k, \omega)]^2} \right\} \right| \qquad (4.25a)$$

$$N_T(\omega) = \mathrm{Re} \left\{ \frac{|\omega|}{[\omega^2 - \Delta^2(\omega)]^{1/2}} \right\} \qquad (4.25b)$$

$N_T(\omega)$ is precisely the same as (3.1), but with the understanding of a complex, energy-dependent $\Delta(\omega)$. As shown by Scalapino, Schrieffer, and Wilkins (1966), the same expressions for the tunneling current and conductance are recovered. In general, the expression for the current in the strong-coupling case is

$$J = \frac{4\pi e}{h} \sum_{k_1, k_2} |T_{k_1, k_2}|^2 \int_{-\infty}^{\infty} \frac{dE_1}{2\pi} A_1(k_1, E_1) \int_{-\infty}^{\infty} \frac{dE_2}{2\pi} A_2(k_2, E_2) \times \delta(E_1 - E_2 - eV)[f(E_1) - f(E_2)] \qquad (4.26)$$

4.4 Quantitative Inversion for $\alpha^2 F(\omega)$: Test of Eliashberg Theory

A practical method of determining the strong-coupling superconductor functions from tunneling data was demonstrated by McMillan and Rowell (1965, 1969) using Pb. The input information required is the normalized phonon range conductance $\sigma(V)$, typically presented in its ratio to the structureless BCS conductance as the BCS-reduced conductance $\sigma/\sigma_{BCS} - 1$ (see Fig. 4.1 and also Fig. 1.6), and the gap edge value $\Delta(\Delta_0) = \Delta_0$. The outputs are, first, the Eliashberg function $\alpha^2 F(\omega)$ and Coulomb pseudopotential μ^* and then the derived superconductor functions [using (4.9) and (4.10)] $\Delta(\omega)$ and $Z_S(\omega)$. The normal-state renormalization function $Z_N(\omega)$ is also derived from $\alpha^2 F(\omega)$ via the same equations by setting $\Delta(\omega) = 0$. Figure 4.2 sketches the relationship between phonon features in $\alpha^2 F(\omega)$, in the pair potential Δ, and in the tunneling density of states. It may be noted that the imaginary (lifetime) part of $\Delta(\omega)$ (dashes) becomes nonzero as sufficient energy becomes available for real phonon emission. The peak position in $\alpha^2 F(\omega)$ corresponds accurately to the point of the most negative slope in the conductance; hence, in d^2I/dV^2 data, a negative peak locates a peak in the phonon spectrum. [This result

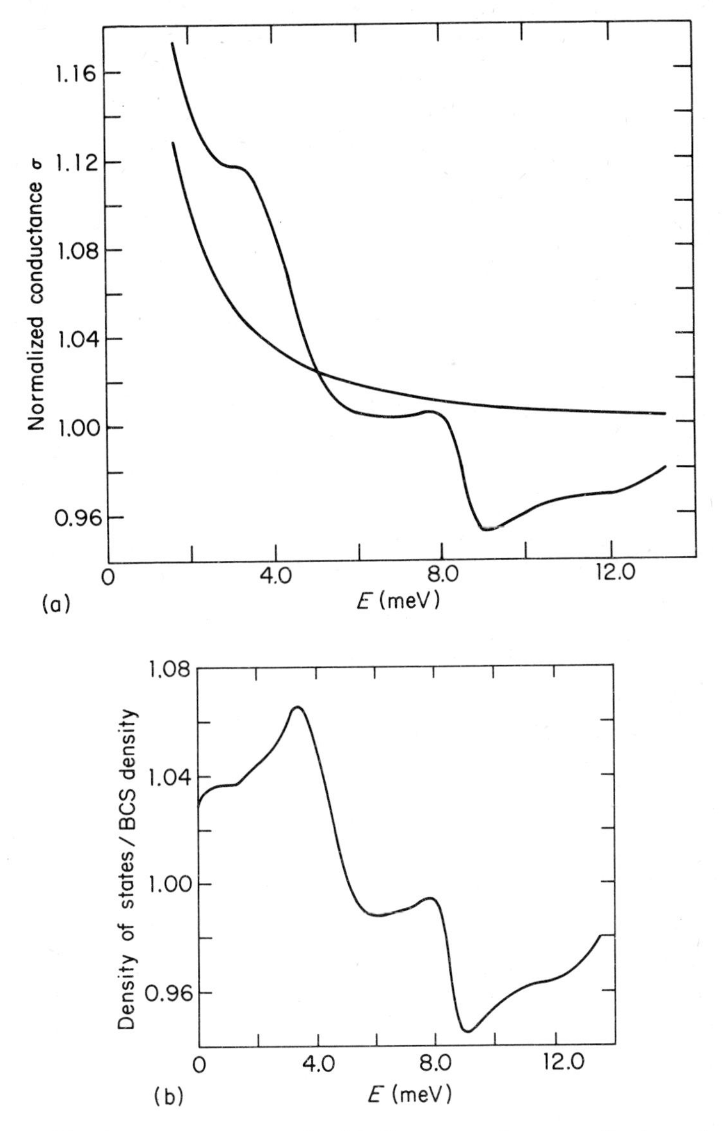

Fig. 4.1. (a) The separate curves in this figure are the normalized conductance (tunneling density of states) measured for lead in the phonon range of energies and the corresponding prediction (smooth curve) of the BCS theory. (b) The *ratio* of the tunneling density of states measured for Pb to the corresponding BCS density of states. This latter BCS-reduced density of states is used for determination of the phonon spectrum of lead, which was shown in Fig. 1.6b. (After McMillan and Rowell, 1969.)

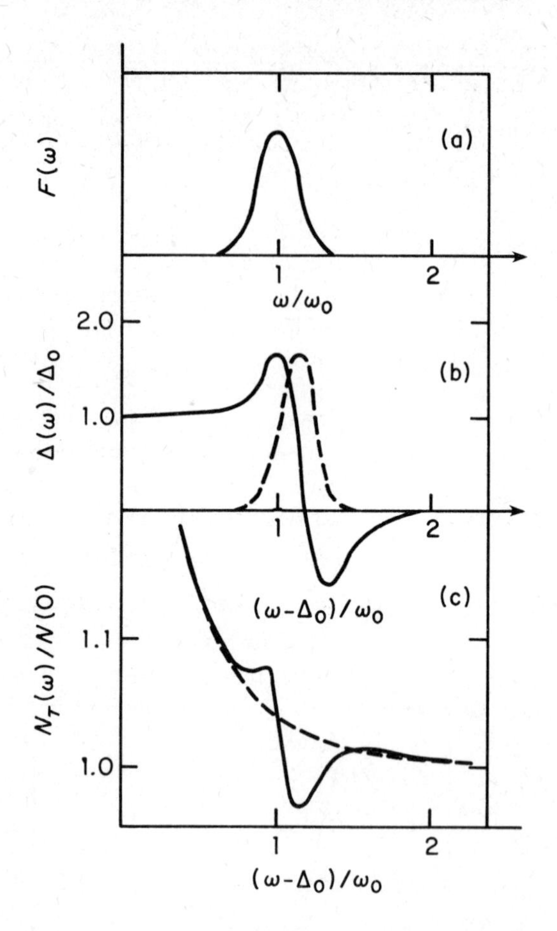

Fig. 4.2. The influence of a single peak in the phonon spectrum of a superconductor (a) on (b), the energy-dependent gap function $\Delta(\omega)$. Here, the solid curve is the real part and the dashed curve is the imaginary part of $\Delta(\omega)$. (c) The contribution of the peak in the phonon spectrum in the tunneling density of states. Note that the energy position of the most negative slope of dI/dV, i.e., the negative peak in d^2I/dV^2, corresponds to the phonon peak. (After Scalapino, Schrieffer, and Wilkins, 1966.)

was also shown in Fig. 1.6; the d^2I/dV^2 data were, in fact, integrated by McMillan and Rowell (1965, 1969) to obtain an improved input function $\sigma/\sigma_{BCS}-1$ for the inversion.]

Since the Eliashberg equations are nonlinear in $\Delta(\omega)$, inversion was accomplished by McMillan variationally in two nested iterative loops [a description of the programs has been given by Hubin (1970)]. An initial $\alpha^2F^0(\omega)$ function, a starting value μ^{*0} and an arbitrary initial $\Delta(\omega)$ are integrated numerically in (4.9) and (4.10) to get a first-improved $\Delta(\omega)$;

this process (the inner loop) is repeated until $\Delta(\omega)$ converges to the function $\Delta(\omega)^0$ required by the initial α^2F^0 and μ^{*1}. An improved μ^* is then obtained by requiring that $\Delta(\omega)^0$ agree with the measured gap value. At this point, the corresponding conductance (density of states)

$$\sigma_c = \mathrm{Re}\left\{\frac{|\omega|}{[\omega^2 - \Delta^0(\omega)^2]^{1/2}}\right\}.$$

is calculated, and its difference $\delta\sigma^0 = \sigma_c^0 - \sigma_{\mathrm{exp}}$ from the data is used with a calculated functional derivative $\delta\sigma(\omega)/\delta\alpha^2F(\omega)$ to get a correction to $\alpha^2F(\omega)^0$ and, hence, a first-improved $\alpha^2F(\omega)^1$. The outer-loop iteration is continued until $\alpha^2F(\omega)$ and all other functions converge.

An important feature of the analysis of Pb, as shown in Fig. 4.3, is that the variation of $\alpha^2F(\omega)$ occurs only over the phonon range of energies, while the tunneling data and the predicted conductance corresponding to the $\alpha^2F(\omega)$ extend to much higher energies. The remarkably good agreement over the extended energy range, including details of multiphonon-like structure generated by the nonlinear Eliashberg equations, is, as McMillan and Rowell (1969) have emphasized, a stringent test not only of the inversion scheme but also of the underlying Eliashberg (1960) theory. It was concluded that this theory must be accurate to within a few percent.

More recently, Galkin, D'yachenko, and Svistunov (1974) have demonstrated an inversion method which, in fact, does not require iterations. This method is based on the identity

$$\mathrm{Im}\left\{\frac{\omega}{[\omega^2 - \Delta^2(\omega)]^{1/2}}\right\} = \frac{2}{\pi}\int_{\Delta_0}^{\infty} \frac{\sigma(\omega') - \sigma_{\mathrm{BCS}}(\omega')}{\omega^2 - \omega'^2}\,\omega'\,d\omega' \tag{4.27}$$

which, taken together with

$$\sigma(\omega) = \mathrm{Re}\left\{\frac{|\omega|}{[\omega^2 - \Delta^2(\omega)]^{1/2}}\right\}$$

provides two equations from which the two unknowns, $\mathrm{Re}\,\Delta(\omega)$ and $\mathrm{Im}\,\Delta(\omega)$, can be determined directly. While this is only the first step toward determining $\alpha^2F(\omega)$, the remaining problem is a straightforward linear integral equation for $\alpha^2F(\omega)$ in which μ^* does not appear and for which iterations are not needed.

Figures 4.3b and 4.3c show recent Pb data of Svistunov, Chernyak, Belogolovskii, and D'yachenko (1981) in which the Galkin, D'yachenko, and Svistunov method has been employed. A direct comparison of the nearly identical $\alpha^2F(\omega)$ functions generated by the two methods, as applied to the Pb tunneling data of Rowell, is shown in Fig. 4.4, while Fig. 4.5 contains all of the additional functions describing Pb as inferred from the Rowell data.

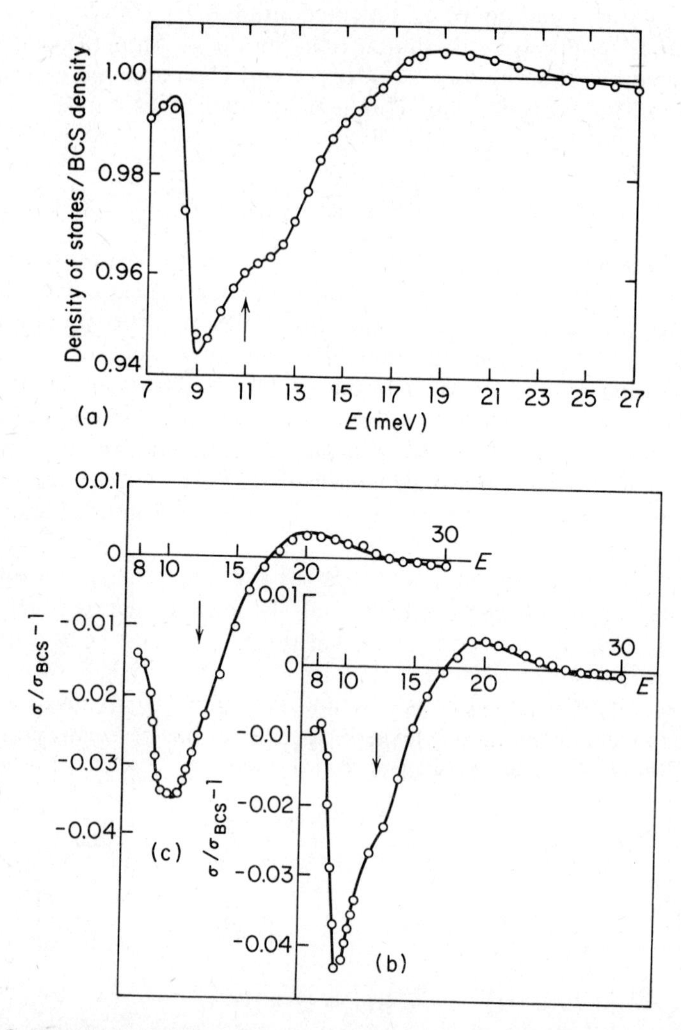

Fig. 4.3. Details of the fit obtained by inversion of tunneling data on lead in the phonon energy range provide a test of the accuracy of the Eliashberg equations. In these figures, the arrow at 11 meV above the gap-edge locates the upper energy to which the variational process is carried out; the quality of the fit beyond this energy is a test of the accuracy of the theory. (a) $N_T(E)$ (after McMillan and Rowell, 1969) which corresponds to the reduced conductance data shown in Fig. 4.1. (b) The $N_T(E)$ curve obtained by Svistunov, Chernyak, Belogolovskii, and D'yachenko, (1981) for pure lead. (c) Results for lead under pressure. These two curves (b) and (c), are obtained by an inversion scheme which does not involve a variational step. (From Svistunov, Chernyak, Belogolovskii, and D'yachenko, 1981.)

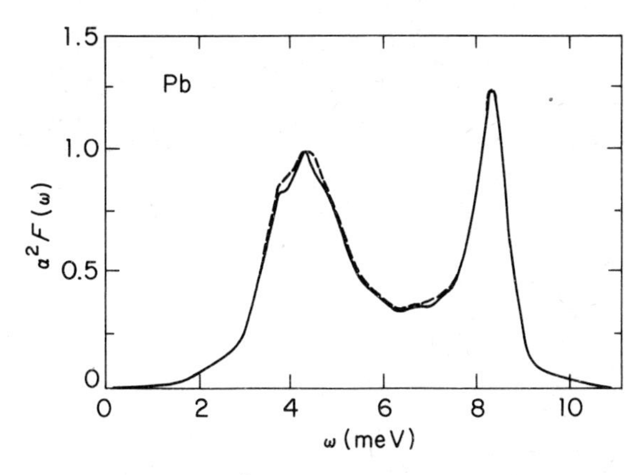

Fig. 4.4. A comparison of the $\alpha^2F(\omega)$ functions for lead obtained from the data of McMillan and Rowell (1969) as reduced using the variational scheme (dashed curve) and using the nonvariational scheme of Galkin, D'yachenko, and Svistunov (1974) (solid curve). (After Galkin, D'yachenko, and Svistunov, 1974.)

Fig. 4.5. (a) The real and imaginary parts of the computed gap function $\Delta(\omega)$ for lead obtained from the data of McMillan and Rowell (1969). In this figure, the dashed curve is the imaginary part and the solid curve is the real part of the gap function.

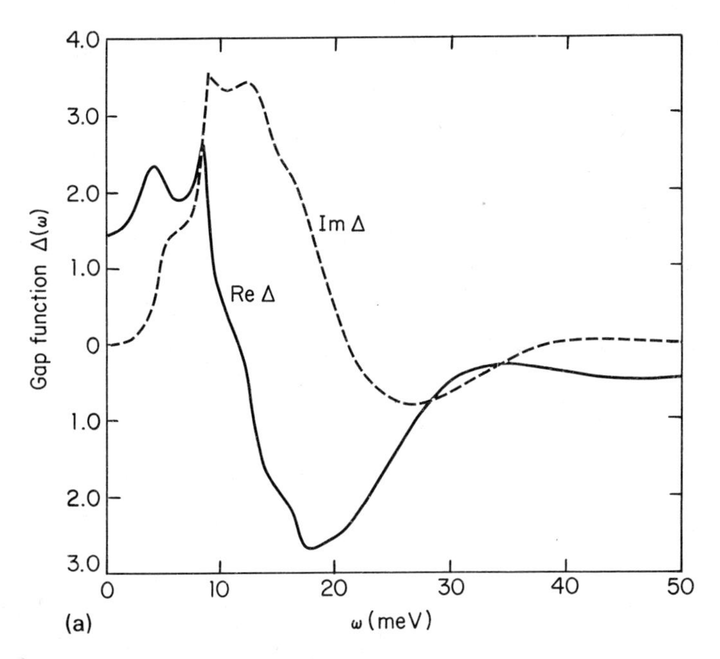

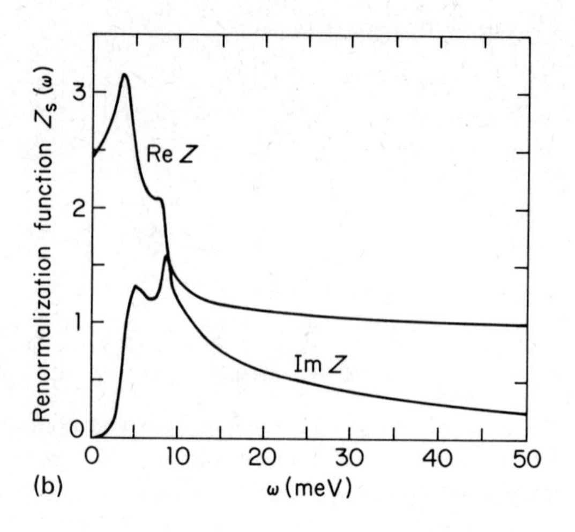

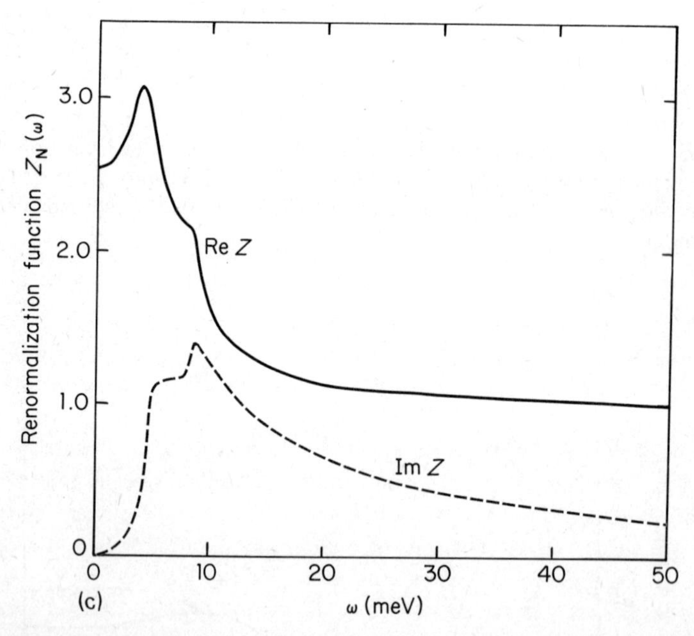

Fig. 4.5. (cont.) (b) Real and imaginary parts of the renormalization function $Z_S(\omega)$ for the superconducting state of lead. (c) The renormalization function $Z_N(\omega)$ for lead in the normal state; Z_N is obtained from the Eliashberg equations by setting the gap function $\Delta(\omega)$ equal to zero. Note its similarity to the renormalization function in the superconducting state. (After McMillan and Rowell, 1969.)

4.5 Extension to More General Cases

The isotropic $T=0$ theory sketched above has been employed in the majority of applications of tunneling spectroscopy. We briefly mention a few of the more important generalizations of the theory.

4.5.1 *Finite Temperature*

The equilibrium nonzero T case can be incorporated in (4.9)–(4.13) by replacing (4.13) with

$$K_{\pm}(\omega, \omega') = \int_0^{\infty} d\Omega\, \alpha^2(\Omega)F(\Omega)\left\{\left[\frac{f(\omega')+n(\Omega)}{\omega'+\omega+\Omega+i\delta} \pm \frac{f(-\omega')+n(\Omega)}{\omega'-\omega+\Omega-i\delta}\right]\right.$$
$$\left. \mp\left[\frac{f(\omega')+n(\Omega)}{-\omega'+\omega+\Omega+i\delta} \mp \frac{f(\omega')+n(\Omega)}{-\omega'-\omega+\Omega-i\delta}\right]\right\} \tag{4.28}$$

where

$$f(\omega) = \left[1+\exp\left(\frac{\omega}{kT}\right)\right]^{-1} \tag{4.29}$$

and

$$n(\omega) = \left[1-\exp\left(\frac{\omega}{kT}\right)\right]^{-1} \tag{4.30}$$

and by replacing μ^* in (4.19) by

$$\mu^* \tanh\frac{\omega}{2kT_c} \tag{4.31}$$

(Ambegaokar and Tewordt, 1964; Scalapino, Wada and Swihart, 1965). Comparisons of the results of these integral equations with temperature-dependent tunneling data are shown in Fig. 4.6, after Franck and Keeler (1967), whose spectra (near T_c) are shown as solid lines at various reduced temperatures. A more refined calculation with a better $\alpha^2F(\omega)$ function for Pb was later given by Vashishta and Carbotte, substantially removing the departures between theoretical and experimental curves. Figure 4.7 shows the gap variation of Pb (strong coupling), with T (near T_c) due to Franck and Keeler (1967).

It should be mentioned that there is a separate and extensive literature on solution of the $T\neq 0$ Eliashberg equations for T_c by summing residues at poles along the imaginary axis. For a discussion of this topic and earlier references, see Allen and Dynes (1975*b*).

4.5.2 *Anisotropy*

To the extent that barrier tunneling is regarded as a directional process, with most electrons injected in a cone of angle $\sim 10°$ about the barrier normal direction (see, however, Dowman, MacVicar and Waldram,

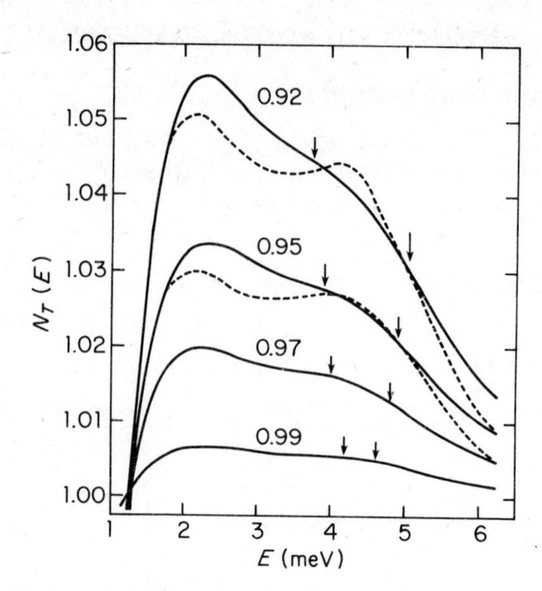

Fig. 4.6. Plots of the experimental normalized conductance for Pb–Al_2O_3–Al junctions for reduced temperatures varying from 0.92 to 0.99 (solid curves). (After Franck and Keeler, 1967.) The dashed $N_T(\omega)$ curves are obtained by solution of temperature-dependent Eliashberg equations by Scalapino, Wada, and Swihart (1965), whose early calculation necessarily made use of an only approximate lead phonon spectrum. More recent work by Vashishta and Carbotte (1970) confirms that the discrepancies here are resolved by the use of an accurate Pb phonon spectrum.

Fig. 4.7. Plot of the energy gap squared versus reduced temperature for lead are shown as the solid symbols connected by the solid line; the open circles are from the calculation of Scalapino, Wada, and Swihart (1965); the dashed line is the variation expected on the BCS theory. (After Franck and Keeler, 1967.)

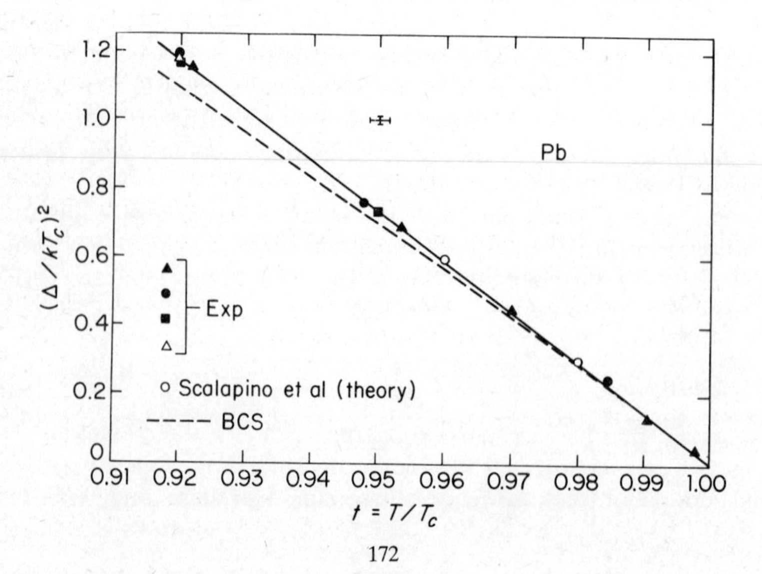

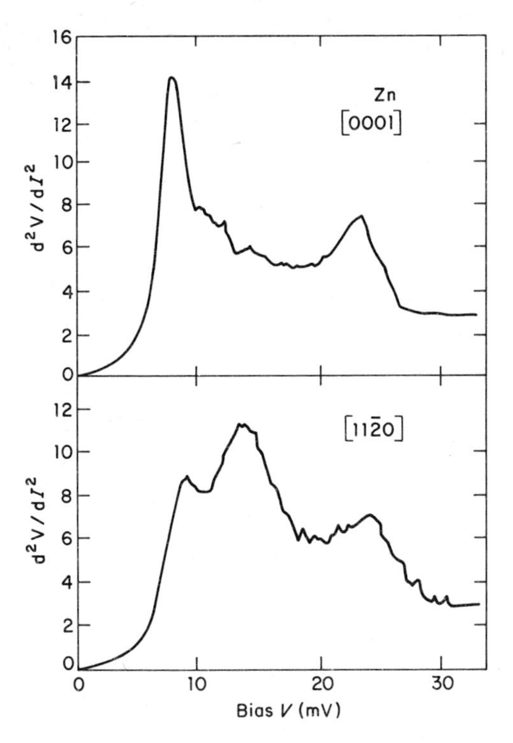

Fig. 4.8. Measured point-contact spectra of zinc at 1.5 K for two perpendicular orientations indicate the main features of the α^2F functions. Substantial anisotropy is indicated in these results. (After Yanson and Batrak, 1978.)

1969), the confirmed anisotropy of Fermi surfaces and phonons in real, clean materials would be expected to produce anisotropy in the $\alpha^2F(\omega)$ functions deduced from tunneling into different crystalline faces. Surprisingly, there seems at present to be no good example of this expected effect from barrier tunneling, although such effects have been observed in a related experiment, that of point-contact spectroscopy, as shown in Fig. 4.8. for Zn observed in two mutually perpendicular directions. The theory of anisotropy is familiar, e.g., from the work of Carbotte and collaborators [Carbotte, 1977; see Truant and Carbotte 1972 for the case of Zn], while the most complete and detailed calculations to date are probably those of Tomlinson and Swihart (1979), also for Zn. Some of the latter results are shown in Fig. 4.9.

The understanding of such an anisotropic $\alpha_k^2(\omega)F_k(\omega)$ is basically in terms of (4.14) before Fermi surface averaging:

$$\alpha_k^2(\omega)F_k(\omega)=\frac{1}{(2\pi)^3}\int\frac{dS_{k'}}{\hbar\,|\vec{v}_{k'}|}\sum_{\lambda}|g_{k',k\lambda}|^2\,\delta[\omega-\omega_{\lambda,k'-k}] \qquad (4.32)$$

Here the $\vec{k}$ subscripted to $F_k(\omega)$ is identified with the cone of states into

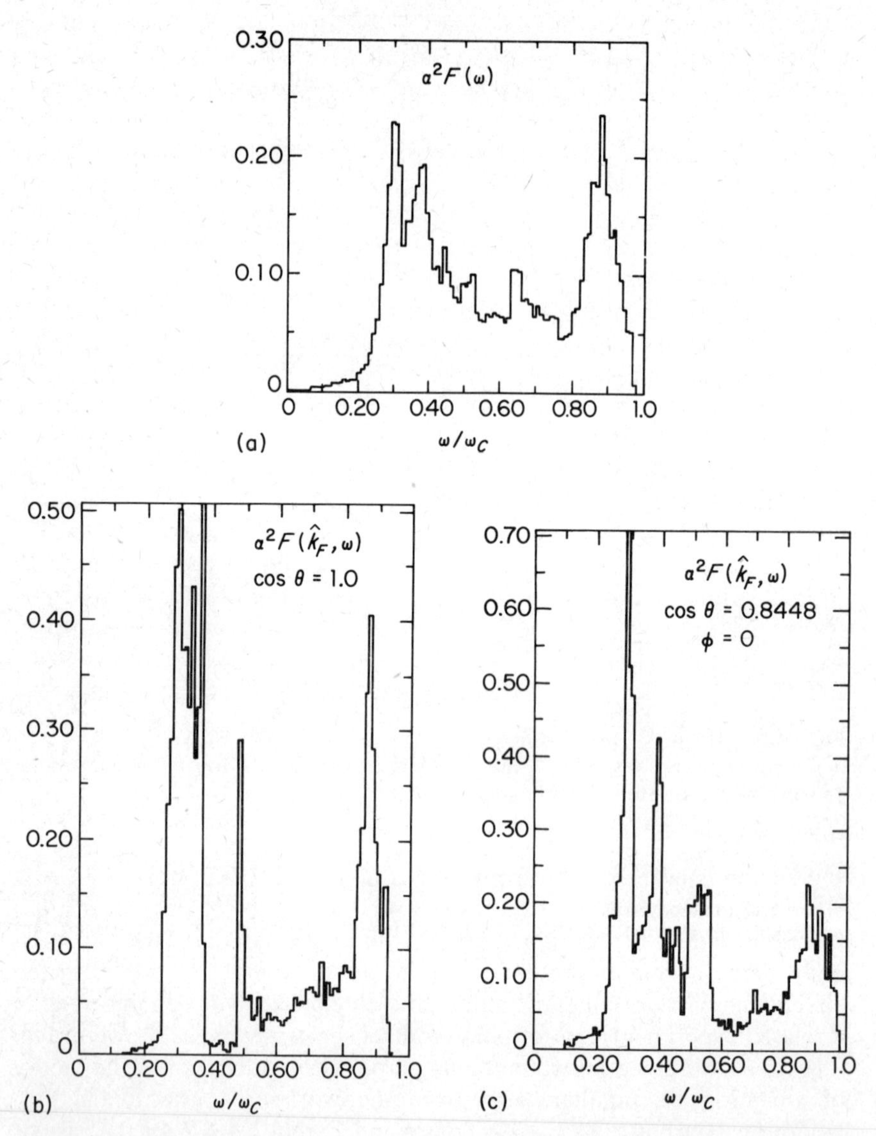

Fig. 4.9. (a) Total $\alpha^2F(\omega)$ function calculated for zinc by Tomlinson and Swihart (1979). (b) The directional $\alpha^2F_k(\omega)$ calculated for $\vec{k}$ along the (0001) (c) axis. (c) Calculated $\alpha^2F_k(\omega)$ function for $\vec{k}$ at an angle 32° from the c axis. (After Tomlinson and Swihart, 1979.)

which the tunnelling transitions occur. A corresponding λ_k can be defined in analogy with (4.14). As emphasized by Carbotte (1977) and by Tomlinson and Swihart (1979), there are several sources of $\vec{k}$ dependence in (4.31). These are the phonon properties ($\omega_{k,\lambda}$) (and also polarization vectors of the normal modes), the Fermi surface geometry, the electronic wavefunctions, the group velocities $\vec{v}_k$; and the electron-phonon interaction.

In the face of a well-established theoretical background and demonstration of anisotropy in point-contact spectra, one may ask why similar effects have not yet been seen in barrier tunneling spectroscopy. There appears to be no consensus on this point. The requirement of little scattering, i.e., that the single-crystal electrodes be in the clean limit, is often attained in the usual bulk sense, although it is probably more pertinent and at the same time more difficult to establish clean conditions specifically in the surface region of the electrode actually being sampled. The detailed nature of the oxide-metal interface is also possibly relevant as a source of local strains or dissolved gases which may produce locally dirty conditions. Some of these possible undesirable features may be avoided in the proximity junction geometry to be discussed in Chap. 5. In this case, a thin epitaxial metallic N layer on the single-crystal S electrode may be less disturbing to S than a grown oxide interface, and at the same time, it may act as a barrier against undesired gases diffusing into the sampled region of S.

4.5.3 Spin Fluctuations

The tendency of some metals to form a magnetic ground state can suppress the superconducting T_c, as was originally pointed out by Berk and Schrieffer (1966) and by Doniach and Engelsberg (1966). An extreme example is Pd, where the superconductivity is totally suppressed. In such cases, the large paramagnetic susceptibility indicates ferromagnetic spin correlations (spin fluctuations, or paramagnons) which oppose the singlet-state pairing leading to superconductivity. As Berk and Schrieffer point out, if the exchange interaction is nearly strong enough to give ferromagnetic alignment, a given up-spin electron will tend to be surrounded by other up-spin electrons. A second down-spin electron, attempting to lower its energy via local phonon coupling to the first electron, will have to pass through a barrier region produced by its unfavorably oriented (up-spin) neighbors. If the size and lifetime of the spin fluctuations are sufficiently large, singlet pairing can be prevented.

The physics of this situation can be incorporated, at least approximately, into the Eliashberg equations by the addition of the paramagnon spectral density $p(\omega)$ in (4.13), which becomes

$$K\pm(\omega,\omega')=\int d\Omega\,[\alpha^2F(\Omega)\pm p(\Omega)]\left(\frac{1}{\omega'+\omega+\Omega+i\delta}\pm\frac{1}{\omega'-\omega+\Omega-i\delta}\right) \tag{4.33}$$

Following Berk and Schrieffer (1966) and Schrieffer (1968),

$$p(\omega)=\frac{3N(0)}{2\pi}\int_0^{2k_F}\frac{q\,dq}{2k_F^2}\,\mathrm{Im}\,t(q,\omega) \tag{4.34}$$

with

$$t(q,\omega)=\frac{1}{3}\frac{U+J(q)}{1-[U-J(q)]N(0)u(q,\omega)}+\frac{2}{3}\frac{U}{1-UN(0)u(q,\omega)} \tag{4.35}$$

and

$$J(q)=J_0(1-C^2q^2)$$

In these expressions, t represents the particle-hole t matrix, in which U and $J(q)$, respectively, are a contact potential and a q-dependent interatomic interaction, $u(q, \omega)$ is the Lindhard dielectric function, and $C=1/\sqrt{2}k_F$, with k_F the Fermi wavevector. In the more recent work of Rietschel, Winter, and Reichardt (1980) [whose calculated $p(\omega)$ curves are shown in Fig. 4.10)], the Stoner factor

$$S=\frac{1}{1-N(0)I} \tag{4.36}$$

and related quantities

$$I=U+J_0 \qquad K=\frac{J_0}{I} \tag{4.37}$$

are introduced, where K is a measure of the nonlocality of the electron-electron Coulomb interaction and is at present treated as a parameter. The main point to be made is that for reasonable parameter choices, the major spectral weight in $p(\omega)$ occurs at energies on the order of E_F, much larger than the typical phonon energy $\hbar\omega_D$. In view of this result, it has been demonstrated by Daams, Mitrović, and Carbotte (1981) (see also Mitrović and Carbotte, 1981b; Leavens and MacDonald, 1983) that the low-energy ($\omega \simeq \omega_{ph}$) behaviour in the system containing paramagnons can be simulated approximately by simple rescaling of quantities in terms of the spectral weight

$$\lambda_S=2\int p(\omega)\omega^{-1}\,d\omega \tag{4.38}$$

of the paramagnon spectrum (see Fig. 4.11). The effective values are found to be

$$\alpha^2F(\omega)_e=\frac{\alpha^2F(\omega)_{ph}}{1+\lambda_S} \qquad \lambda_e=\frac{\lambda_{ph}}{1+\lambda_S} \tag{4.39}$$

$$\mu_e^*\simeq\frac{\mu_{ph}^*+\lambda_S}{1+\lambda_S} \tag{4.40}$$

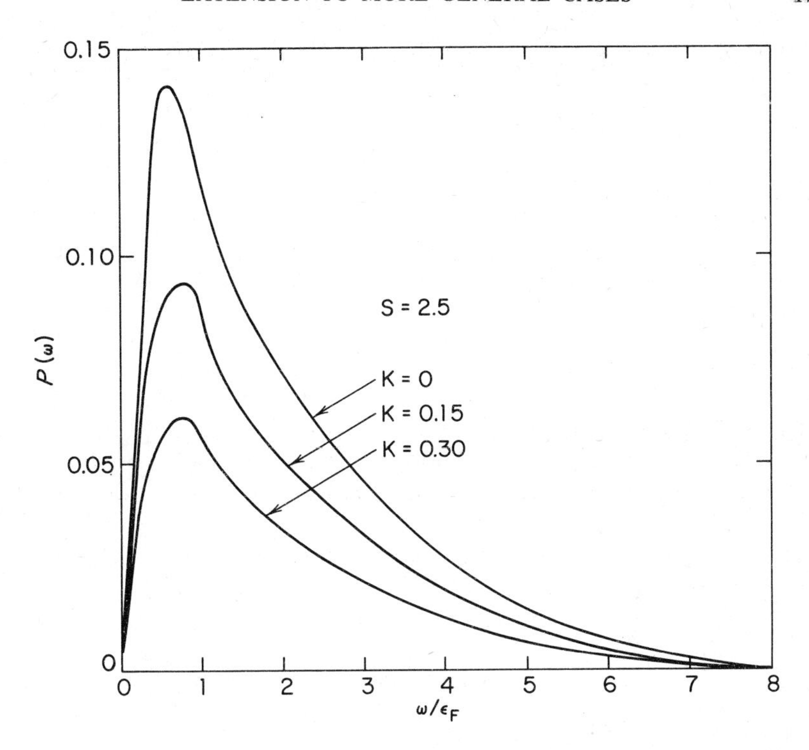

Fig. 4.10. Calculations of the paramagnon spectral weight function $p(\omega)$ plotted against energies normalized by the Fermi energy, assuming a Stoner factor S of 2.5 and several values of K, which is a measure of the nonlocality of the Coulomb interaction. (After Rietschel, Winter, and Reichardt, 1980.)

Refinement of (4.40) is discussed by Leavens and MacDonald (1983). The suppression of the low-energy pair potential $\Delta(E)$ and tunneling spectrum $N_T(E)$ by the spin fluctuations, within the analysis of Daams, Mitrović, and Carbotte (1981), can thus be calculated from the effective $\alpha^2F_e(\omega)$, whose scale is reduced from that in the absence of paramagnons $[\alpha^2F_{ph}(\omega)]$ by the factor $(1+\lambda_S)^{-1}$, as shown in Fig. 4.11.

4.5.4 Electronic Density-of-States Variation

The range of energies involved in superconducting pairing is rather small, a few multiples of a typical phonon energy or $\hbar\omega_D$ on either side of the Fermi surface. In most cases, then, the standard assumption of a constant density $N(0)$ of electronic states is a good one. In several important cases, such as the A–15 compounds V_3Si, Nb_3Ge, and Nb_3Sn, however, it is now reasonably clear, both from anomalous strong temperature dependences of experimental quantities such as the magnetic susceptibility (for a review, see Testardi, 1975) and from high-resolution band structure

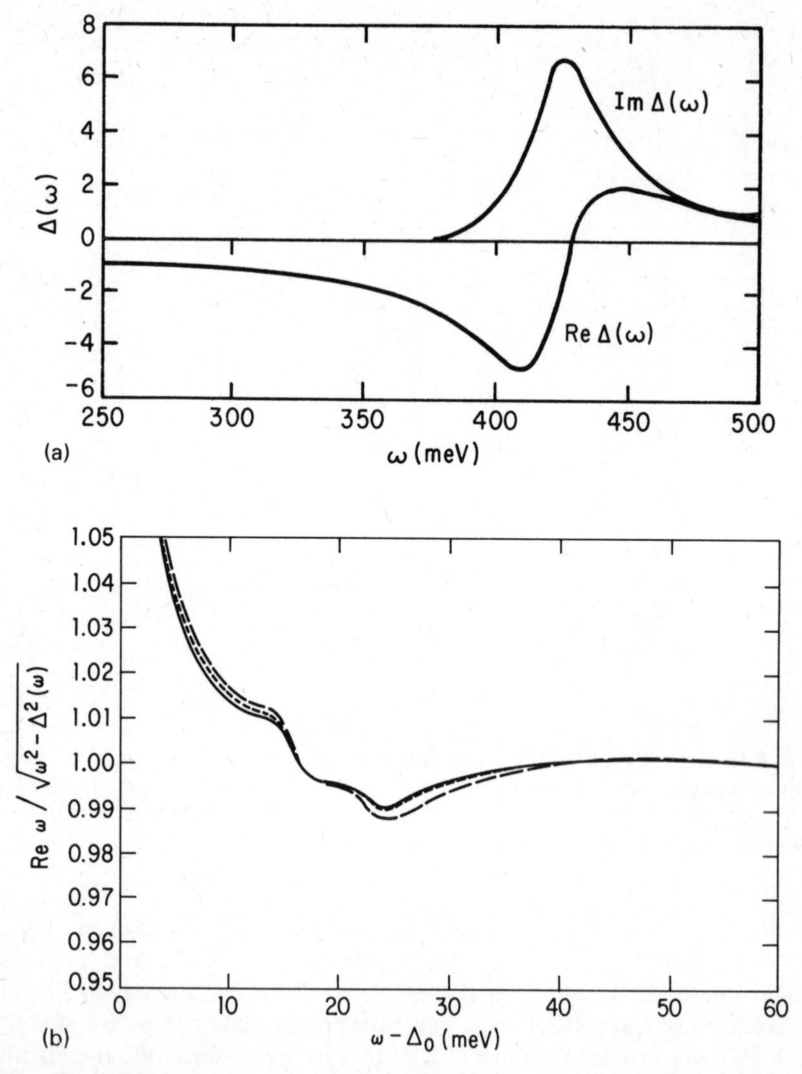

Fig. 4.11. (a) Repulsive pair-potential contribution in Nb calculated assuming a paramagnon spectrum consisting of a δ function at 400 mV. (b) Influence of the paramagnon spectrum [whose pair potential was shown in (a)] on the tunneling density of states, assuming the experimental $\alpha^2F(\omega)$ of niobium. The dashed curve is the tunneling of states in the absence of the paramagnon spectrum. The solid curve is the tunneling density of states obtained from solution of the Eliashberg equations, including the paramagnon peak, while the dotted curve is that obtained by using the scaling equations defined in the text and restricting the solution of the Eliashberg equations to the region of the phonon spectrum. (After Daams, Mitrovic, and Carbotte, 1981.)

calculations (see, e.g., Ho, Cohen, and Pickett, 1978; Jarlborg, 1979; Klein, Boyer, Papaconstantopoulos, and Mattheiss, 1978), that significant variations in $N(E)$ on a scale of $\hbar\omega_D$ can indeed occur, which must be considered for their possible influence on the superconducting properties. The calculated variations relevant to Nb_3Sn are shown in Fig. 4.12.

Incorporation of $N(E)$ variation into the strong-coupling Eliashberg theory has been discussed by several authors (Horsch and Rietschel, 1977; Lie and Carbotte, 1978; Pickett, 1980) and, particularly in the context of tunneling data inversion, by Pickett and Klein (1981), Mitrović and Carbotte (1981a), and Kieselmann and Rietschel (1982). Following Mitrović and Carbotte, the required equations ($T=0$) for a general $N(E)$ variation, expressed as $\rho(E)=N(E)/N(0)$ are

$$\Delta(\omega)=[Z(\omega)]^{-1}\int_0^\infty n_1(\omega')[K_+(\omega,\omega')-\mu^*(\omega_c)\theta(\omega_c-\omega')]\,d\omega' \tag{4.41}$$

$$[1-Z(\omega)]\omega=\int_0^\infty n_2(\omega')K_-(\omega,\omega')\,d\omega' \tag{4.42}$$

Fig. 4.12. (a) The dotted spectrum referenced to the right-hand scale is the $\alpha^2F(\omega)$ function for Nb_3Sn after Wolf, Zasadzinski, Arnold, Moore, Rowell, and Beasley (1980), while the solid curve referenced to the center scale is the electronic density of states normalized to the density of states at the Fermi surface of Nb_3Sn due to Klein, Boyer, Papaconstantopoulos, and Mattheiss (1978). (b) Plot, on a broader scale, of the electronic density of states for Nb_3Sn of Klein, Boyer, Papaconstantopoulos, and Mattheiss (1978). (After Pickett and Klein, 1981.)

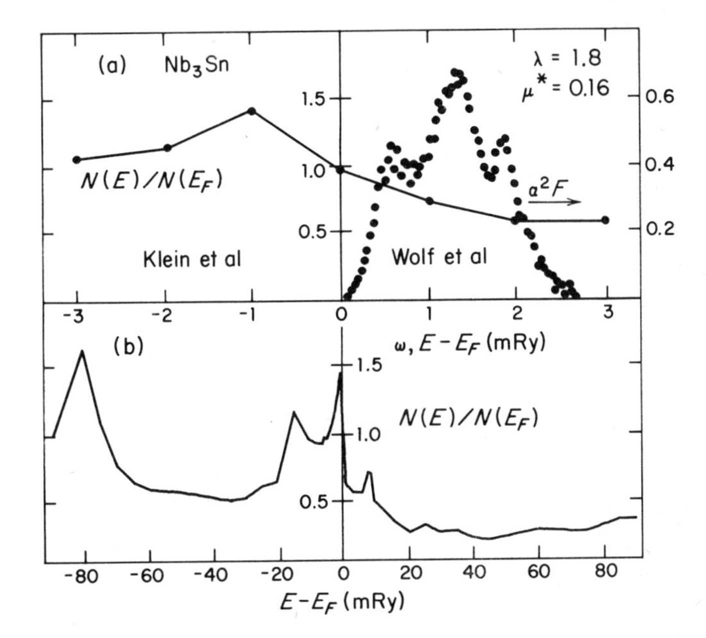

$$\chi(\omega)=\int_0^{\infty} n_3(\omega')K_+(\omega,\omega')\,d\omega' \tag{4.43}$$

where $K_{\pm}(\omega, \omega')$ are given by (4.13) and $\theta(x)$ is the unit step function. The $\rho(E)$ variation enters through functions defined as

$$n_1(\omega)=\left(-\frac{1}{\pi}\right)\text{Im}\int_{-\infty}^{\infty}\rho(\omega')\frac{\Delta(\omega)Z(\omega)}{D(\omega,\omega')}\,d\omega' \tag{4.44}$$

$$n_2(\omega)=\left(-\frac{1}{\pi}\right)\text{Im}\int_{-\infty}^{\infty}\rho(\omega')\frac{\omega Z(\omega)}{D(\omega,\omega')}\,d\omega' \tag{4.45}$$

$$n_3(\omega)=\left(-\frac{1}{\pi}\right)\text{Im}\int_{-\infty}^{\infty}\rho(\omega')\frac{\omega'+\chi(\omega)}{D(\omega,\omega')}\,d\omega' \tag{4.46}$$

where

$$D(\omega,\omega')=Z^2(\omega)[\omega^2-\Delta^2(\omega)]-[\omega'+\chi(\omega)]^2 \tag{4.47}$$

The energies are measured from the chemical potential, and χ is a function whose real part represents a shift of the chemical potential due to the electron-phonon interaction. The main effects of the energy dependence are that the pair density of states $P(\omega)\rightarrow N(0)n_1(\omega)$ and the quasiparticle density of states $N_T(\omega)\rightarrow N(0)[n_2(\omega)-n_3(\omega)]$ are modulated by the electronic density of states $\rho(E)=N(E)/N(0)$, in compari-

Fig. 4.13. Calculated effect on the tunneling conductance of Nb_3Sn, taking $\alpha^2F(\omega)$ from Shen (1972b), assuming variation in the electronic density of states $N(E)$ near the Fermi energy (dashed curve), and taking $N(E)$ as constant (dot-dashed curve). (After Mitrovic and Carbotte, 1981.)

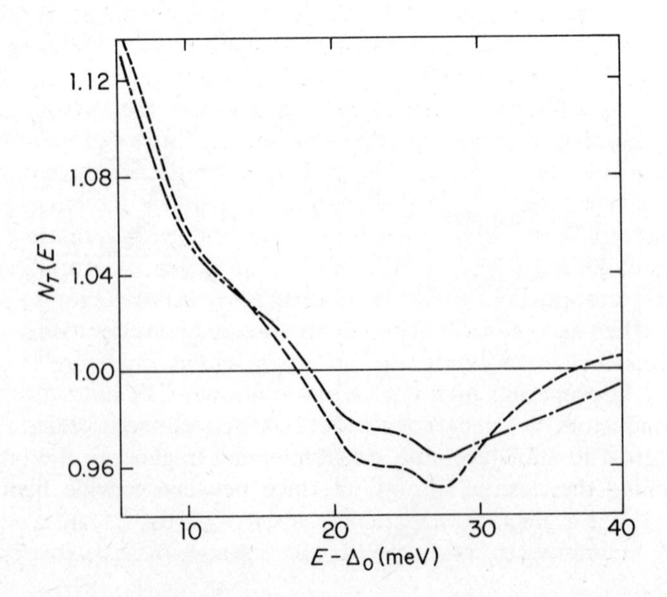

son with (4.11) and (4.12). A numerical simulation of the difference predicted in the tunneling density of states for Nb_3Sn for constant vs. varying $N(E)$, based on the $N(E)$ of Klein et al. (1978) and the $\alpha^2F(\omega)$ of Shen (1972*b*) is shown in Fig. 4.13. It is clear from this result that large changes in the $\alpha^2F(\omega)$ inferred from tunneling data would result, depending upon the assumption of constant or varying $N(E)$. According to Kieselmann and Rietschel (1982) in application to Nb_3Sn, a large increase in the inferred value of λ (from 1.8 to 2.3) should result from inclusion of the theoretical $N(E)$ variation. A severe difficulty in this area is to judge how directly applicable to typical thin-film samples will be a theoretically calculated $N(E)$ variation based upon a perfect single crystal. It seems beyond realistic expectation to invert tunneling data simultaneously for the $\alpha^2F(\omega)$ and $N(E)$ variations, although one would be pleased to be shown to be wrong on such a point.

4.6 Limitations of the Conventional Method

As a method for determining the Eliashberg function $\alpha^2F(\omega)$ for a metal, the conventional spectroscopy is limited to those metals which are strong-coupling superconductors. The phonon structure in the density of states varies as $N_T(E) \simeq \text{Re}[1+1/2(\Delta/E)^2]$; for a characteristic phonon energy $E_{ph} = k\theta_D$, and with the BCS relation $\Delta \simeq 1.76kT_c$, the magnitude of the structure should scale roughly as $\sim 1.5(T_c/\theta_D)^2$. This value is, e.g., for Nb with $T_c = 9.2$ K and $\theta_D = 276$ K, about 0.17%. After we adjust this scale to match the observed deviation in Nb of 1.2%, the percent deviations predicted for various metals are as collected in Table 4.1, which is similar to one given by McMillan and Rowell (1969).

It is clear from the table that the fraction of metals whose percent deviation is large enough (say >0.1%) to be reasonably measurable is rather small. This result is an obvious limitation of the McMillan-Rowell spectroscopy for measuring the Eliashberg function. At the end of the table, a few additional entries have been made: metals which are not superconducting at presently attainable temperatures but whose phonon structure has been observed nevertheless by the strategem of proximity tunneling. In Mg, in fact, an α^2F function has been published (Burnell and Wolf, 1982) based on superconducting tunneling spectra. This result demonstrates the possibility, in principle, of extending tunneling α^2F measurements to *all* metals excepting strong ferromagnets, with some further exceptions based on experimental considerations.

A second class of limitations to conventional tunneling arises from difficulty in preparing high-quality conventional CIS junctions on some superconductors S, even though the expected phonon strength is sufficiently great to allow accurate measurements. In general, the problem is in attaining the desired abrupt interface between a wide bandgap insulator I and a clean homogeneous superconductor S, and a variety of specific circumstances, often poorly understood, may be involved. One

Table 4.1. Expected Magnitudes of Phonon Deviations in $\sigma/\sigma_{BCS}-1$

Metal	T_c (K)	θ_D (K)	Deviation (%)
Be	0.026	1390	4.0×10^{-7}
Al	1.18	420	8.8×10^{-3}
Cd	0.52	209	6.9×10^{-3}
Zn	0.85	310	0.008
Ga	1.08	325	0.012
Sn	3.72	195	0.41
In	3.41	109	1.09
Tl	2.38	79	1.01
Pb	7.2	105	5.23
Hg	4.15	72	3.7
W	0.015	390	1.6×10^{-6}
Ir	0.1125	420	7.9×10^{-5}
Ti	0.40	425	9.8×10^{-4}
Zr	0.61	290	4.9×10^{-3}
Ru	0.49	550	8.8×10^{-4}
Os	0.66	500	1.9×10^{-3}
Mo	0.92	460	4.4×10^{-3}
Re	1.70	415	0.019
Ta	4.47	258	0.33
V	5.40	383	0.22
Tc	7.77	—	—
Nb	9.25	276	1.25
V_3Ga	15.9	310	2.9
V_3Si	17.0	530	1.14
Nb_3Sn	18.3	290	4.4
Nb_3Ge	23	378	4.12
Th	1.38	165	0.08
La (α)	4.88	151	1.16
La (β)	6.0	139	2.1
Mg	0	400	—
Cu	0	343	—
Ag	0	225	—
Au	0	165	—

fairly clear case is that in which attempts to grow or otherwise deposit an oxide barrier I on the superconductor S lead to degraded superconducting properties at its surface. Reactive metals such as La, V, and Nb are of this type, and the surface degradation may result from dissolved oxygen lowering the T_c or by the formation of undesired metallic phases such as VO or NbO as thin layers between the insulating barrier and the region of bulk superconductor properties.

It has gradually been realized that such thin interposed layers can have

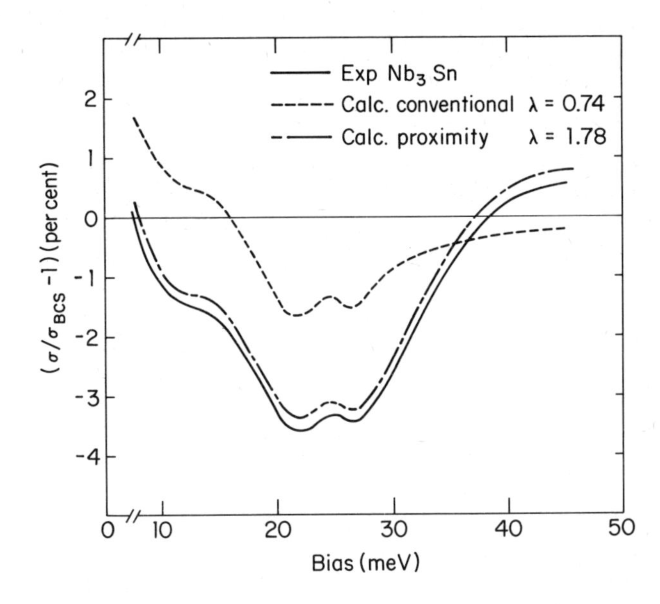

Fig. 4.14. The solid curve is the experimental reduced conductance for Nb_3Sn (after Moore, Beasley, and Rowell, 1978), while the upper dashed curve is obtained by using a conventional McMillan–Rowell inversion of the tunneling data. The dot-dashed curve incorporates the effect of a surface layer of weak superconductivity on the Nb_3Sn electrode. The major discrepancy between the solid and upper dashed curves is typical of a class of surface contamination problems which limit the applicability of the conventional tunneling spectroscopy. (After Wolf, Zasadzinski, Arnold, Moore, Rowell, and Beasley, 1980.)

rather subtle effects, scarcely detectable in the gap spectra, yet leading to apparently anomalous parameters when the McMillan-Rowell inversion analysis is performed. One characteristic indication of an interposed "proximity layer," it has been found (Arnold, Zasadzinski, and Wolf, 1978), is residual offset between the calculated and experimental reduced-conductance curves, as illustrated in Fig. 4.14. An additional limitation of conventional McMillan-Rowell spectroscopy, then, is the absence of any correction for such unintended proximity effects. In order to understand how such corrections can in fact be carried out, we turn to discussion of tunneling phenomena in inhomogeneous electrodes, principally NS bilayers.

5

Inhomogeneous Superconductors: Proximity Electron Tunneling Spectroscopy

5.1 Introduction

The first indication of the superconducting proximity effect was reported by Holm and W. Meissner (1932), who had observed zero resistance between pressed contacts providing an SNS sequence of metals. Later measurements (H. Meissner, 1960; Smith, Shapiro, Miles, and Nicol, 1961) verified the pioneering work, making clear that a supercurrent can, under suitable conditions, cross a "normal" N layer even micrometers in thickness. This result is illustrated in convincing detail in the more recent measurements of Clarke (1969), shown in Fig. 5.1, of a Pb–Cu–Pb sandwich, with a 0.55 μm Cu layer which behaves like a single superconductor at 2.98 K, while there is no evidence for superconductivity at any temperature in isolated copper. In the experimental situation shown, the Cu is alloyed slightly with Al to ensure the dirty limit. If the N layer is in the clean limit, e.g., by use of an annealed foil, the thickness through which a supercurrent can persist may be as great as 50 μm. The strength of superconductivity in the N layer can often be expressed as a penetration length K^{-1}, corresponding to an exponential decay $\exp(-Kx)$ of the order parameter $F(x)$ in N, with x measured into N from the NS boundary. The expression

$$K^{-1} = \frac{\hbar v_{FN}}{2\pi kT} \tag{5.1}$$

is found to apply at sufficiently high T. In the clean limit at low temperatures, however, the decay of order with x becomes nonexponential and much slower (Falk, 1963):

$$F(x) \propto \frac{1}{x_1 + x} \tag{5.2}$$

In the dirty limit, in contrast, the exponential decay length K^{-1} is governed by diffusion processes expressed by a mean free path ℓ_N and related diffusivity $D_N = \frac{1}{3} v_{FN} \ell_N$:

$$K^{-1} = \left(\frac{\hbar D_N}{2\pi kT}\right)^{1/2} \tag{5.3}$$

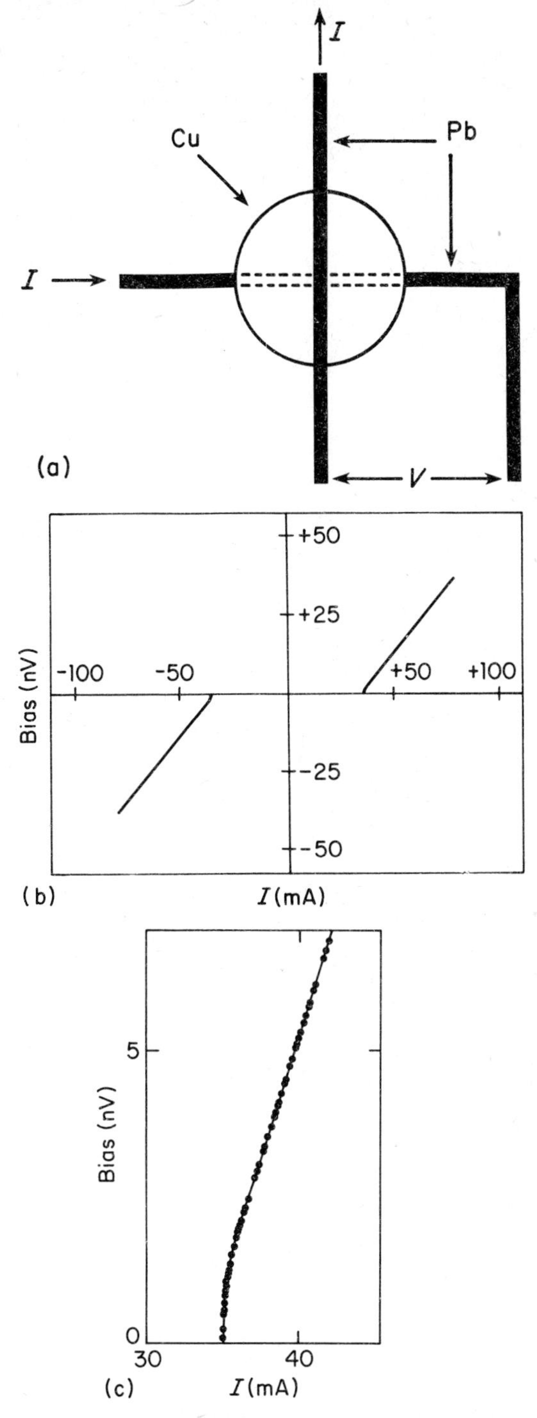

Fig. 5.1. (a) Configuration of Pb–Cu–Pb sandwich. (b) The I–V characteristic of the Pb–Cu–Pb sandwich measured at 2.98 K reveals a zero-voltage current of approximately 35 mA through copper of thickness 5520 Å. (c) An enlarged portion of the I–V characteristic of (b) at the onset of supercurrent. (After Clarke, 1969.)

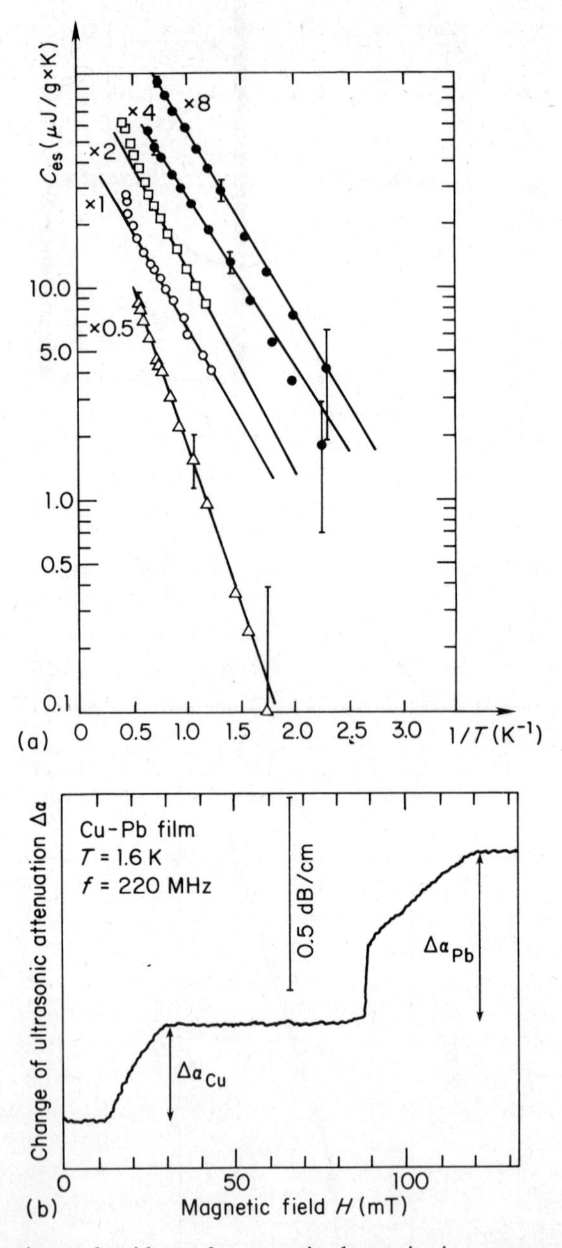

Fig. 5.2. Experimental evidence for a gap in the excitation spectrum of a normal metal in proximity with a superconductor. (a) Exponential dependence upon $1/T$ in the electronic contribution to the specific heat of a set of Ag–Pb–Ag sandwiches indicates an energy gap in the excitation spectrum of the sandwiches. The curves have been successively displaced by factors of two for clarity. The lowest curve (steepest slope) corresponds to the thinnest silver layers (1270 Å), the thickest silver layer being 2150 Å. In all cases, the lead films are 2500 Å in thickness. (After Manuel and Veyssie, 1973.) (b) Sharp changes in the ultrasonic

If the N layer has a nonzero transition temperature T_{CN} in isolation, then the difference $(T - T_{CN})$ enters the denominator of (5.1) or (5.3), leading to a divergence of K^{-1} as $T \to T_{CN}$. The systematics of these effects are described in detail by Deutscher and de Gennes (1969). Recent extensions of this work to explore dependences upon magnetic field and additions of specific magnetic and nonmagnetic impurities to N have been described by Hsiang and Finnemore (1980), Niemeyer and von Minnigerode (1979), and Paterson (1979).

From a microscopic point of view, the electronic specific heat data shown in Fig. 5.2a, plotted logarithmically vs. $1/T$, indicate a proximity-induced gap in the excitation spectrum of the N region of an SNS sandwich. The sample (S = Pb, N = Ag) showing the steepest slope (largest apparent gap Δ) is of 0.127 μm Ag thickness, compared with 0.16 μm and 0.21 μm for the curves of smaller slope. These results indicate that the N layer becomes superconducting in the usual microscopic sense, developing an excitation spectrum with a gap whose magnitude increases with the fraction of S metal in the sandwich. As we will see, observations of the superconductivity of N by electron tunneling support this picture, including a pair potential with strong-coupling structure near the phonon energies of N. Figure 5.2b reveals another aspect of induced superconductivity in the N layer, here 0.21 μm of Cu on 0.25 μm of Pb. In this case, separate critical parallel magnetic fields for flux penetration of the Cu and Pb layers (~30 mT and ~130 mT, respectively) are revealed by step increases in the attenuation of longitudinal ultrasonic waves of energy $\hbar\omega \ll 2\Delta$. Since the measured attenuation is proportional to the number of thermally excited quasiparticles, $n \propto [1 + \exp(+\Delta/kT)]^{-1}$, the two step increases reveal the distinct collapses of Δ in the Cu and Pb portions, respectively, of the Cu–Pb sandwich with increasing parallel magnetic field.

More detailed information about the induced superconductivity in N layers of NS sandwiches is contained in N-side tunneling data, as shown in Fig. 5.3. The original curves showing strong N-phonon features induced in Al by Pb (Chaikin and Hansma, 1976) are given in Fig. 5.3a, while in Fig. 5.3b similar behavior is revealed in Mg backed by thick Ta. It is notable that Mg itself has never been observed in the superconducting state, even though measurements have extended below 1 mK. The geometry of such tunneling structures is shown in Fig. 5.4a, with a simplified plot of the variation of $\Delta(x)$ (Fig. 5.4b) when the N layer is sufficiently thin and the S layer is thick.

---

attenuation for a Cu–Pb sandwich as a function of parallel magnetic field, at 1.6 K. The thickness of the copper layer is 2100 Å, that of the lead layer is 2500 Å. The sharp rise in attenuation at about 20.0 mT represents the penetration of the magnetic field into the copper film. Similarly, the sharp rise at about 90.0 mT signals the onset of magnetic penetration into the lead film. (After Krätzig, 1971.)

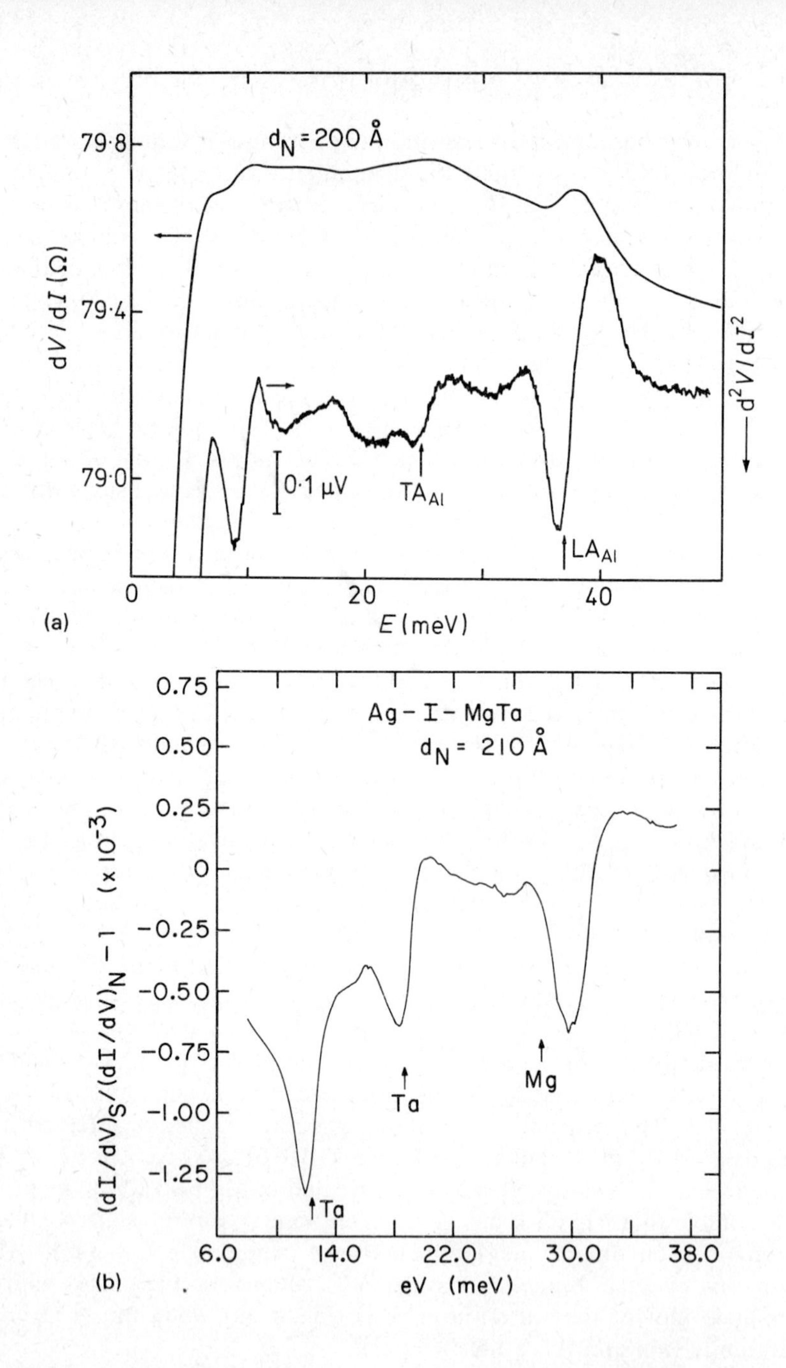

Fig. 5.3. Tunneling observation of induced pair potential in the N member of an NS proximity sandwich. (a) Greatly enhanced aluminum longitudinal phonon at 37 mV is observed by tunneling into a 200-Å aluminum film backed by a thick lead film, from an Al counterelectrode. (After Chaikin and Hansma, 1976.) (b) Observation of a strong magnesium pair potential in a 210-Å film deposited on a tantalum foil in an Ag–MgO–MgTa junction. Note that magnesium by itself is not a superconductor at attainable temperatures. (After Burnell and Wolf, 1982.)

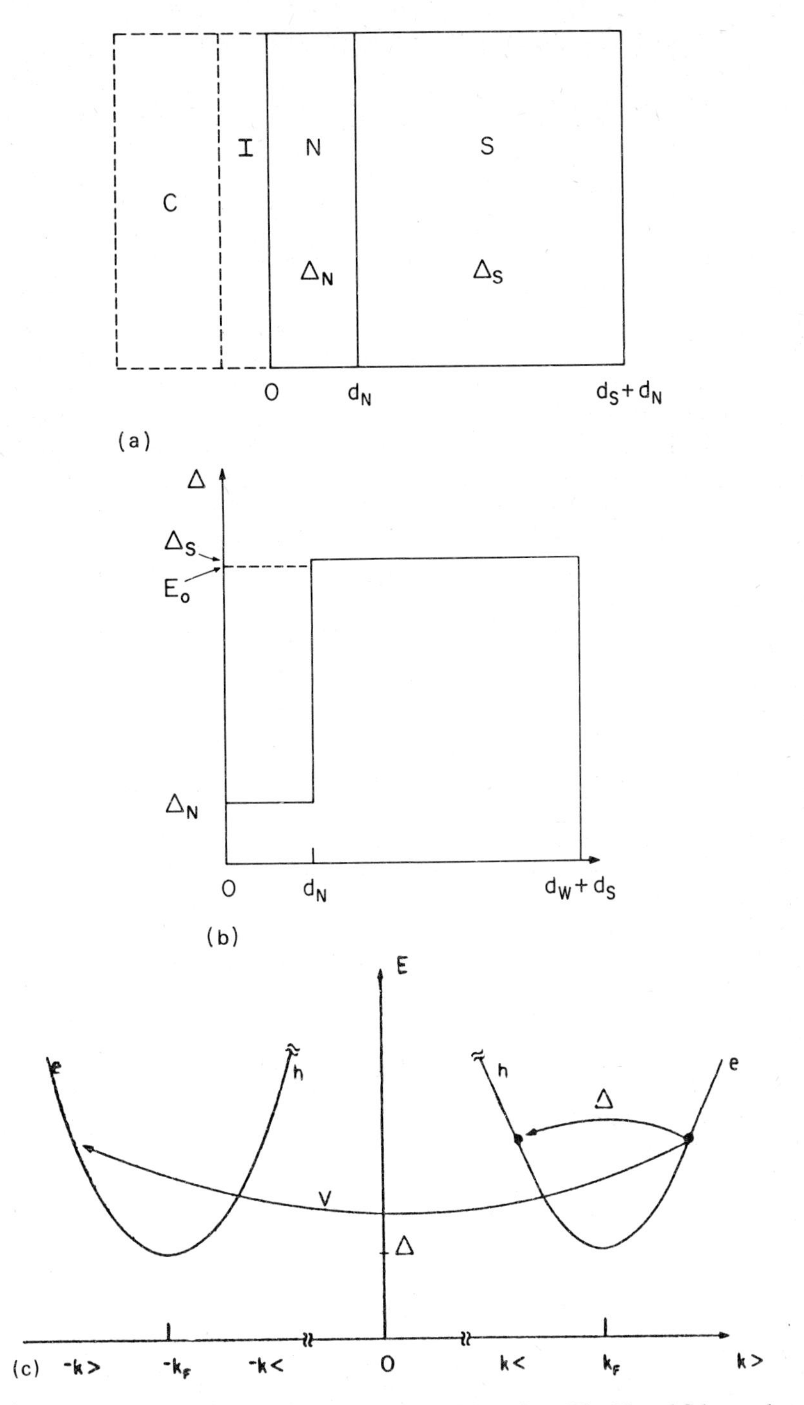

Fig. 5.4. (a) Schematic diagram of CINS tunnel junction. The N and S layers have thicknesses d_N and d_S, respectively. (b) Idealized variation of the pair potential Δ as a function of position in the CINS junction under the assumption $d_N \ll \xi$. (c) Energy vs. wavevector dispersion in a superconductor, with illustration of scatterings involved in the Andreev (Δ) and ordinary reflection (V) process.

5.2 Andreev Reflection

A quasiparticle scattering process which occurs at NS interfaces and is central to an understanding of tunneling phenomena in NS structures is known as Andreev, or electron-hole, reflection (Andreev, 1964). The simplest situation to visualize is one in which an electronlike quasiparticle in the N region, having energy $\Delta_N < E < \Delta_S$, propagates to the NS interface, as in Fig. 5.4b. A freely propagating quasiparticle state at $E < \Delta_S$ in S (i.e., for $x > d_N$ in Fig. 5.4b) is energetically impossible, corresponding to exponential decay, but reflection of the electronlike excitation $k > k_F$ to the degenerate holelike excitation $k < k_F$ can occur (Fig. 5.4c), the group velocity $v_g = 1/\hbar\, \partial E/\partial k$ of the hole being the negative of the original electronlike v_g. This process occurs as the original electron, under the influence of the anomalous or pairing potential step $\Delta_S - \Delta_N$ at $x = d_N$, joins a second electron (leaving a hole in N), with the resultant pair propagating to the right in S. The change in wavevector, $\Delta K \simeq 2\,(k_> - k_F)$, when $\mu_F \gg E$, is

$$\Delta K \simeq \frac{2(E^2 - \Delta^2)^{1/2}}{\hbar v_F} \tag{5.4}$$

This quantity is small, $\sim \xi_0^{-1}$ in the material in which the quasiparticles propagate, and becomes proportional to E for $E \gg \Delta$. For these reasons, Andreev reflection leads to interference effects as a function of E (bias voltage) in tunneling structures whose thickness is on the scale of the coherence length, typically 100–1000 Å. The same scattering process as described also occurs, to a certain extent, for incident quasiparticle energies $E > \Delta_S$. This feature of nonzero above-barrier reflection is familiar from elementary quantum mechanics.

The process described by Andreev (1964) is a consequence of the Bogoliubov equations, (4.4) and (4.5), when applied to the situation of a spatially varying $\Delta(x)$. Using these equations, one finds for the probability of reflection at the abrupt NS interface,

$$B^2 = \frac{E - \Omega}{E + \Omega} \qquad \Omega = (E^2 - \Delta^2)^{1/2} \tag{5.5}$$

which is of unit magnitude for $E < \Delta$, drops smoothly, and finally falls as $\Delta^2/4E^2$ for $E \gg \Delta$.

An application of the Bogoliubov-de Gennes equations which is relevant to NS tunneling is the demonstration of the de Gennes and Saint-James (1963) quasiparticle bound state. These authors showed that the process of total reflection of a quasiparticle of $E < \Delta_S$ in the N region of the geometry of Fig. 5.4 leads to the existence of at least one quasiparticle bound state, of energy $E_0 \leq \Delta_S$. The description of Fig. 5.4 via (4.4) and (4.5), shifting the x origin to the NS boundary, requires

$$\Delta(x)=\begin{cases}0, & -d_N<x<0\\ \Delta, & 0<x\end{cases} \tag{5.6}$$

$$V(x)=\begin{cases}\infty, & x<-d_N\\ 0, & x\geqslant -d_N\end{cases} \tag{5.7}$$

Then the two-component wavefunction in N can be written

$$\psi_N(x)=a\begin{pmatrix}1\\0\end{pmatrix}[e^{-ik_1x}+e^{ik_1(x+2d_N)}]+b\begin{pmatrix}0\\-i\end{pmatrix}[e^{ik_2x}+e^{ik_2(x+2d_N)}] \tag{5.8}$$

where

$$k_1^2=k_{Fx}^2+\frac{2mE}{\hbar^2} \qquad k_2^2=k_{Fx}^2-\frac{2mE}{\hbar^2} \tag{5.9}$$

and

$$k_{Fx}^2=k_F^2-k_y^2-k_z^2=k_F^2\cos^2\theta \tag{5.10}$$

In S, the wavefunction is

$$\psi_S(x)=c\begin{pmatrix}1\\-iB\end{pmatrix}e^{ik_1'x}+d\begin{pmatrix}B\\-i\end{pmatrix}e^{-ik_2'x} \tag{5.11}$$

where B is given by (5.5) and

$$\begin{aligned}k_1'^2&=k_{Fx}^2+\frac{2m\Omega}{\hbar^2}\\ k_2'^2&=k_{Fx}^2-\frac{2m\Omega}{\hbar^2}\end{aligned} \tag{5.12}$$

Continuity of ψ and ψ' at $x=0$ [generalized boundary conditions are considered by Bar-Sagi and Entin-Wohlman (1977)] leads to the condition for nontrivial solution for the coefficients a, b, c, d:

$$\frac{2d_N\Delta}{\hbar v_F}\cos\phi=(n\pi+\phi)\cos\theta \tag{5.13}$$

where $\cos\phi=E/\Delta$ and v_F is assumed common to N and S. This equation gives the spectrum of bound-state excitations E_n:

$$E_n=\Delta\cos\phi_n \tag{5.14}$$

The number of bound states n increases with the coefficient in (5.13) of $\cos\phi$,

$$\frac{2d_N\Delta}{\hbar v_F}\simeq\frac{2}{\pi}\frac{d_N}{\xi_{prox}}$$

with the definition of the proximity coherence length given by

$$\xi_{prox}=\frac{\hbar v_{FN}}{\pi\Delta} \tag{5.15}$$

As $d_N \to 0$, one bound state, E_0, always remains and closely approaches Δ in energy. For each index n in (5.13), a spectrum falling to lower energy results from the $\cos\theta$ dependence. A similar situation persists in the case $\Delta_N \neq 0$, treated in detail by Saint-James (1964). A careful treatment of transmission and reflection in the generalized case of an NIS interface has also been given (Demers and Griffin, 1971; Griffin and Demers, 1971). The latter has recently been applied to NS microconstriction contacts (Blonder, Tinkham, and Klapwijk, 1982) which we will discuss in Sec. 5.9.

5.3 Survey of Phenomena in Proximity Tunneling Structures

The purpose of this section is to present the experimental background of tunneling into NS and SN (Tomasch geometry) structures, as a basis for judging theoretical models and also for developing an intuition for the behavior expected in the regimes of parameters for optimum quantitative spectroscopic measurements on the N and S members of C–I–NS proximity structures.

The original observations of geometry-controlled interference oscillations in superconductive tunneling were made by Tomasch (1965*a*) in what we will designate as the C–I–SN geometry, even though the added N layer which enhanced the observations was not always present. The original dV/dI and d^2V/dI^2 spectra of Tomasch (1965*a*) on thick, unbacked Pb films are shown in Fig. 5.5. The experimental arrangement resembles that of Fig. 5.6a, with the positions of N and S in the figure interchanged. Tomasch soon (1966*a*) reported that similar d^2V/dI^2 oscillations in thick In films were enhanced fivefold in strength if a thin overlay of Ag (N) was deposited on the free In surface. The indexed points corresponding to points of maximum conductance dI/dV are quite regular in voltage, the period being inversely proportional to the thickness of the S film in the range 2.9 to 9.7 μm for Pb and 5.8 to 32.2 μm for In. The interpretation of the effect was provided by McMillan and Anderson (1966), who demonstrated that eV_n of the nth conductance peak could be more accurately located by the modified expression

$$E_n \simeq \left[\Delta^2 + \left(\frac{n\hbar v_F'}{2d}\right)^2\right]^{1/2} \tag{5.16}$$

where $v_F' = v_F/Z$ is the renormalized Fermi velocity in S. This distinction is important for Pb, a strong-coupling material for which $Z(0) = 1 + \lambda = 2.55$. The conductance peaks can be regarded as arising from resonances, or virtual-state peaks in the local density of states at the insulator-S interface, the latter corresponding to a constructive interference condition

$$\Delta K d = \frac{2d}{\hbar v_F'}(E^2 - \Delta^2)^{1/2} = n2\pi \tag{5.17}$$

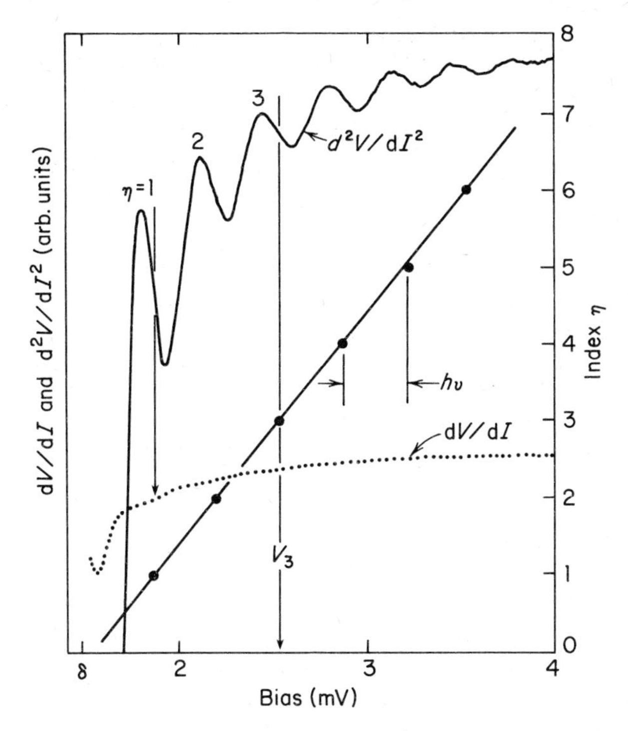

Fig. 5.5. Original report of oscillatory behavior in dV/dI and d^2V/dI^2 for an Al–Al_2O_3–Pb tunnel diode measured at 1.1 K. The thickness of the lead layer is 4.3 μm, and the energy spacing of the oscillations was found to vary inversely as the thickness of the lead layer. (After Tomasch, 1965*a*.)

using (5.4). This equation is the same as (5.16). Physically, the phase difference ΔKd accumulates between an initial wave and one which, after injection, crosses the S film, Andreev-reflects at the back and again at the insulator interface, and is a consequence of the wavevector difference ΔK between the electron- and holelike excitations of the superconductor. A typical spacing of the Tomasch oscillations is 2 meV for a 1 μm film (if $v_F' \simeq 10^8$ cm/sec); this spacing scales inversely with the thickness of the superconductor film. This type of process is shown as it occurs in an N layer in Fig. 5.6b in the sequence $1' \rightarrow 2' \rightarrow 1'$.

Returning to the NS geometry, Fig. 5.7 shows one of the first observations (Rowell, 1973) of the de Gennes–St. James bound-state structure (see also Bellanger et al., 1973), here oberved in thick Zn films ($d_N \simeq 2$, 3.5 μm) backed by Pb and observed with a Pb counterelectrode. The four distinct rises in $I(V)$ for $\Delta_{Pb} \leq eV \leq 2\Delta_{Pb}$ (which are located by peaks in dI/dV labeled A, C, E, G in the lower panel) can be identified as bound states, whose excitation energies in the Zn layer do not exceed Δ_{Pb}. The scattering sequence involved is 1–4 in Fig. 5.6b. The further oscillations to higher energy, first observed and interpreted by Rowell and McMillan

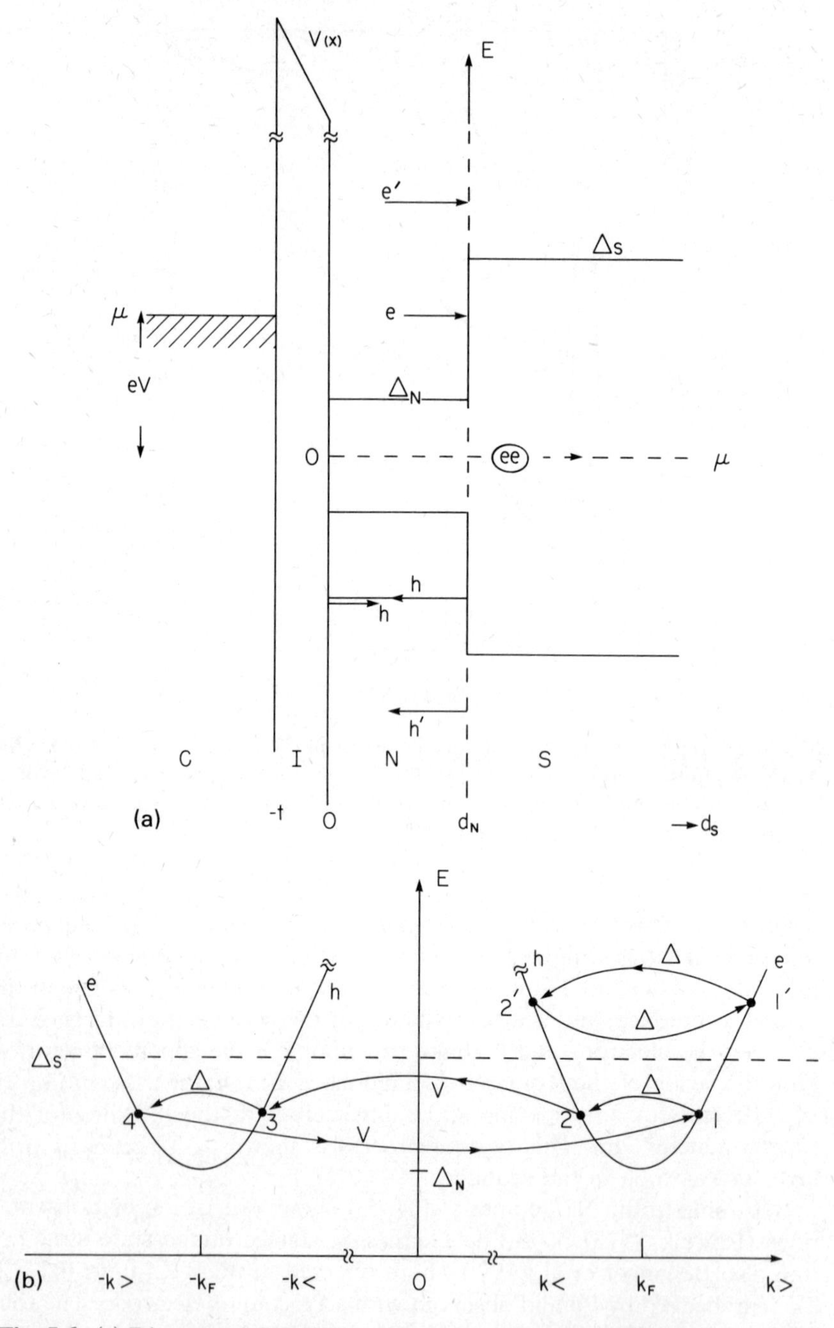

Fig. 5.6. (a) Diagram of C–I–NS junction under bias, illustrating Andreev reflection of quasielectron e at $E = eV$ from pair-potential jump $\Delta_S - \Delta_N$ at N–S boundary $x = d_N$. The result is hole h plus a pair in S. The reflection of the hole by the oxide barrier $x = 0$ is also shown. An above-barrier (Δ_S) reflection $e' \rightarrow h'$ is also illustrated. (b) Tomasch (2d) oscillations result from processes $1' \rightarrow 2'$

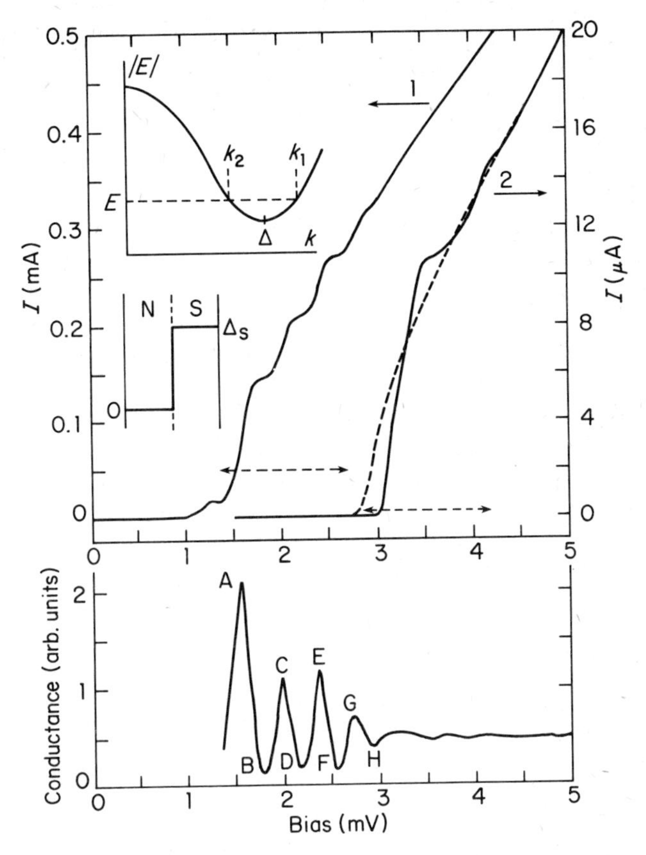

Fig. 5.7. Observation of bound-state structure in I–V (upper traces) and conductance (lower traces) spectra of two Pb–insulator–ZnPb junctions with Zn thicknesses of 3.5 and 2 μm, measured at 0.4 K; H is 0 for the solid curve and is 5 mT for the dashed line. The dashed arrows in the upper figure indicate the range in which the energy E of the quasiparticle injected into the zinc lies in the range $0 \leqslant E \leqslant \Delta(\text{Pb})$. (After Rowell, 1973.)

(1966), can be regarded as virtual or resonance levels in N (the Zn layer) caused by partial reflection at the NS interface.

The Rowell–McMillan oscillations arise from interference after *four* traversals of the N layer. This result would correspond to the sequence 1, 2, 3, 4, 1 in Fig. 5.6b, with the injection energy large compared with Δ_S. An initial wave at the insulator surface $x = 0$ Andreev-reflects at $x = d_N$ and returns to $x = 0$, where it reflects normally, assuming $\Delta_N = 0$. A further transit to $x = d$ and Andreev reflection are necessary before

---

occurring in a thick S layer. McMillan (4d) oscillations arise from scattering sequence shown as $1 \rightarrow 2 \rightarrow 3 \rightarrow 4 \rightarrow 1$ if E exceeds Δ_S; otherwise, the process contributes to bound-state formation.

interference with the initial wave occurs, as may be seen in Fig. 5.6b. The condition $2\Delta K d_N = n2\pi$ corresponds to the 4d (Rowell–McMillan) levels:

$$E_n = \left[\Delta_N^2 + \left(\frac{n\hbar v_F'}{4d_N}\right)^2\right]^{1/2} \qquad \Delta_N \simeq 0 \tag{5.18}$$

To the extent that the pair potential in N is nonzero, some Andreev reflection events may also occur at the N-insulator interface ($x=0$), giving added contributions at the 2d peaks governed by (5.16). This process is shown as the scattering sequence $1' \to 2' \to 1'$ in Fig. 5.6b.

In certain circumstances (Wong, Shih, and Tomasch, 1981), compound (Gallagher, 1980) oscillations may be observed, which can be regarded as beats between oscillations of the Tomasch and Rowell–McMillan types in structures where coherent reflections occur at both the insulator and vacuum interfaces and also if mean free paths against scattering are sufficiently great. Figure 5.8 illustrates these details and their theoretical fit in the encircled regions and also in the upper right where the dashed line (Gallagher theory) more faithfully reproduces the spectra than does the Wolfram theory (dash-dot curve), in which reflections from the vacuum surface of S are neglected. The spectra also show three prominent bound states in the 2.1 μm Zn film backed by 1.22 μm of Pb.

An additional feature of tunneling into the N side of the NS structure is shown in Fig. 5.9, after Rowell and McMillan (1966). In Al–I–Ag/Pb junctions with $d_{Ag} = 470$ Å, 1000-Å, phonon structures in σ, clearly due to Pb but attenuated and evidently modulated by an energy-dependent phase factor, are observed. The phase modulation is particularly noticeable in the sign change, between $d_N = 470$ and 1000 Å, of the Pb phonon feature at 10-mV bias, observed through the Ag films.

5.4 Specular Theory of Tunneling into Proximity Structures

In order to incorporate the phonon (strong-coupling) effects of interest into the theory of proximity structures such as indicated in Figs. 5.4 and 5.6, one must go beyond the Bogoliubov equations as originally given, (4.4) and (4.5). This task was first done in the important paper of Wolfram (1968), based on the equations of Gor'kov (1958, 1959). As Wolfram's work also forms the basis for later work on closely related problems (Arnold, 1978; Gallagher, 1980), it deserves mention. In this approach, the density of states at the tunneling surface is evaluated from the local density of states at x with energy ω and transverse momentum $k_\perp$:

$$\rho(k_\perp; x) = -\frac{1}{\pi} \operatorname{Im} G(k_\perp; x, x')|_{x=x'} \tag{5.19}$$

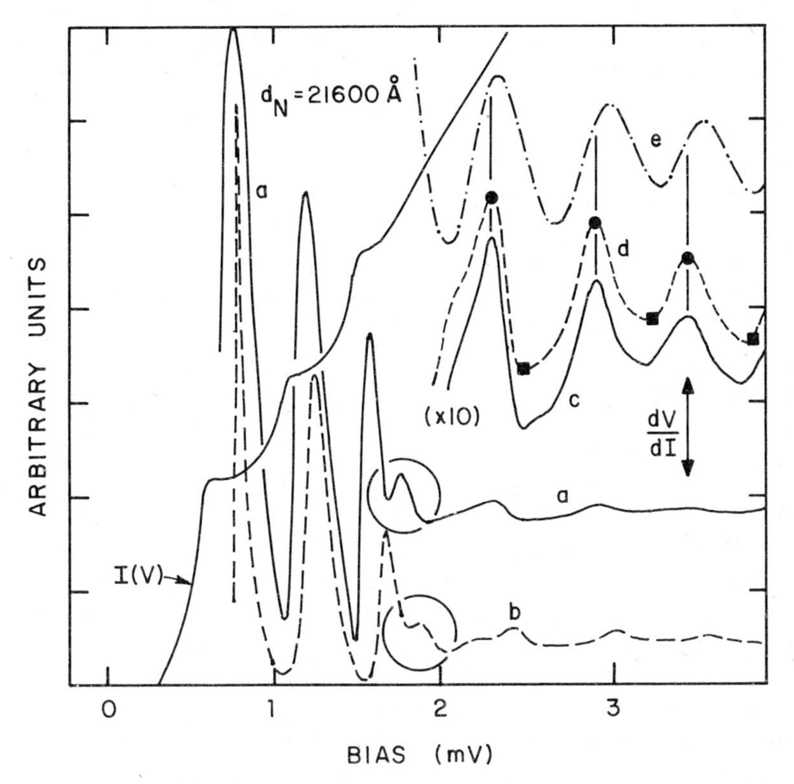

Fig. 5.8. Oscillatory phenomena in the I–V and dV/dI–V curves of a thick (2.16 μm) zinc layer on a Pb film 1.2 μm thick. De Gennes quasiparticle bound states occur below 1.6-mV bias, followed by oscillations to beyond 6 mV. The latter resonances, whose spacings ΔE are inversely proportional to the zinc layer thickness, are influenced in detail by both d_S and d_N and are accurately modeled by the theory of Gallagher (1980), which is indicated by the dashed line. The dash-dot line indicates modeling by the original theory of Wolfram (1968). (After Wong, Shih, and Tomasch, 1981.)

This equation requires solution of Gor'kov's equations (1958)

$$\left[\omega - \frac{p_x^2}{2m} + \mu_{Fx}\right] G_\omega(k_\perp; x, x') - \Delta F_\omega^+(k_\perp; x, x') = \delta(x - x')$$
$$\left[\omega + \frac{p_x^2}{2m} - \mu_{Fx}\right] F_\omega^+(k_\perp; x, x') - \Delta G_\omega(k_\perp; x, x') = 0 \qquad (5.20)$$

where $p_x = -i\hbar\,\partial/\partial x$ and the G (single-particle) and F^+ (pair-correlation) Green's functions are defined as Fourier transforms with respect to $k_\perp$ and ω of the correlation functions

$$G(\vec{r}, \vec{r}'; t) = -i\langle 0|\, \psi(\vec{r}, t)\psi^+(\vec{r}', 0)\,|0\rangle \qquad t > 0 \qquad (5.21)$$

$$F^+(\vec{r}, \vec{r}'; t) = -i\langle 0|\, \psi^\dagger(\vec{r}, t)\psi^\dagger(\vec{r}', 0)R\,|0\rangle \qquad t > 0 \qquad (5.22)$$

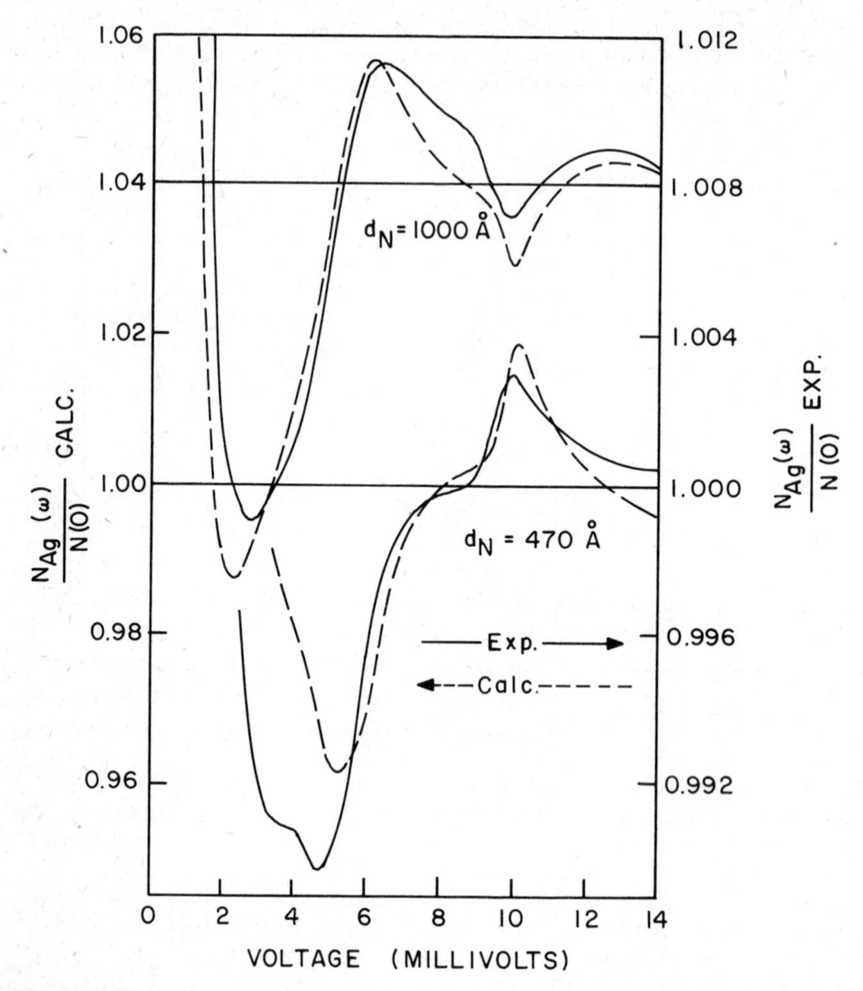

Fig. 5.9. Original observation of phonon structure due to the S metal in measurements made into a superimposed N layer. The conductance spectra shown are from junctions in which 1000-Å and 470-Å layers of silver are overlayed on a thick lead film. The structure visible in the range from 2 to 10 mV is produced by the lead phonons. Note that the magnitude of the experimental structure (right-hand scale) is about 5 times smaller than the predicted magnitude (left-hand scale). This loss of amplitude was attributed to scattering at the Pb–Ag interface and to scattering of quasiparticles in the silver layer. The energy period of the oscillatory structure here is ~22 mV, compared with about 0.9 mV in the previous figure. (After Rowell and McMillan, 1966.)

Here, R decreases the number of pairs by one. The coupled Gor'kov equations must be solved in the two regions subject to boundary conditions. Continuity of F^+ and G and of their normal derivatives is required at the NS boundary. Wolfram (1968) showed that, at the free surfaces, equivalent results are obtained by setting F and G zero, or by setting $\partial F/\partial x$ and $\partial G/\partial x$ zero. The free boundaries reflect the separate Green's functions into themselves: $F^+ \to F^+$, $G \to G$. However, the action of the pair potential jump at the NS interface is to carry $F^+ \to G$, $G \to F^+$. This result is clear from the structure of the two-dimensional Nambu matrix, with G functions on the diagonal and F^+ in off-diagonal positions. The gap function enters as the Pauli matrix

$$\hat{\tau}_1 = \begin{pmatrix} 0 & 1 \\ 1 & 0 \end{pmatrix}$$

which thus carries $F^+ \to G$ and vice versa.

Solution of the NS bilayer problem has been achieved only in the case of spatially constant pair potentials Δ_N, Δ_S. This solution is a good approximation only if one or both of d_N and d_S are very small compared with a coherence length. In the former case the thinner film has negligible effect on the pair-potential variation in the other.

Following Wolfram's notation, $N_1(E)$ measured at the surface of film 1 backed by film 2, assuming film 2 has sufficient quasiparticle scattering that coherent quasiparticle reflections from its back surface may be neglected, is

$$N_1(E) = \mathrm{Re}\left\{\left(\frac{E}{\Omega_1}\right)\left[1 + 2\frac{(\Delta_1/E)\psi + \psi^2}{1-\psi^2}\right]\right\} \tag{2.53a}$$

$$= \mathrm{Re}\left\{\left(\frac{E}{\Omega_1}\right)\left[1 + 2\sum_{n=0}^{\infty}\left(\frac{\Delta_1}{E}\psi^{2n+1} + \psi^{2n+2}\right)\right]\right\} \tag{5.23b}$$

where

$$\Omega_{1,2} = (E^2 - \Delta_{1,2}^2)^{1/2} \tag{5.24}$$

$$\psi = r(E)e^{-\gamma} \exp i\phi_1 \tag{5.25}$$

$$\phi_1 = \frac{2}{\hbar v_{F1}} Z_1 \Omega_1 d_1 = K_1 d_1 \tag{5.26}$$

$$r(E) = \frac{R_1\Delta_2 - R_2\Delta_1}{R_1R_2 - \Delta_1\Delta_2}$$

$$R_{1,2} = E + \Omega_{1,2}$$

$$\gamma_1 = \frac{2d_1}{l_1} \tag{5.27}$$

with d_1 and l_1, respectively, the thickness and quasiparticle mean free path in film 1.

These equations are valid for Δ_1, Δ_2 spatially constant but unrestricted in magntiude and thus apply to both NS and SN geometries. The ψ and ψ^2 terms in (5.23a) generate the 2d and 4d series of oscillations, in agreement with experiment and earlier analysis.

The result of Arnold (1978) is equivalent to (5.23a), in case N is the first layer. The exact expression for $N_T(E)$, now written explicitly in terms of Δ_N (first layer assumed thin: $d_N \ll \xi$) and Δ_S (thick backing layer, $\Delta_S > \Delta_N$), is

$$N_T(E) = \mathrm{Im}\left\{\frac{E/\Omega_N[iF(E)\cos(\Delta K d) + \sin(\Delta K d)] + i(\Delta_N/\Omega_N)G(E)}{\cos(\Delta K d) - iF(E)\sin(\Delta K d)}\right\} \tag{5.28}$$

where

$$\begin{aligned} F(E) &= \frac{E^2 - \Delta_S\Delta_N}{\Omega_S\Omega_N} \\ \Omega_{S,N} &= (E^2 - \Delta_{S,N}^2)^{1/2} \\ G(E) &= \frac{E(\Delta_S - \Delta_N)}{\Omega_S\Omega_N} \\ \Delta K d &= \frac{2Z_N d_N}{\hbar v_{FN}}\Omega_N + i\frac{d_N}{l_N} = RZ_N\Omega_N + i\frac{d_N}{l_N} \end{aligned} \tag{5.29}$$

A different convention has been adopted for the scattering length l, as is seen by comparison of the last of the equations in (5.27) with the last of the equations in (5.29). A more direct $N_T(E)$ expansion, valid for $E > \Delta_N$, Δ_S, is also given by Arnold:

$$N_T(E) \simeq \mathrm{Re}\left\{1 + \frac{1}{2}\left(\frac{\Delta_N}{E}\right)^2 + \frac{1}{2}\left[\frac{(\Delta_S - \Delta_N)^2}{E^2}\right]I(2\phi_N) + \left[\frac{\Delta_N(\Delta_S - \Delta_N)}{E^2}\right]I(\phi_N)\right\} \tag{5.30}$$

where the phase angle ϕ is defined as in (5.26) and the oscillatory integral function $I(\phi)$ is (McMillan, 1968*a*)

$$I(\phi) = \int_1^\infty \frac{dt}{t^2} D\left(\frac{1}{t}\right)e^{i\phi t} \tag{5.31}$$

where $D(1/x)$, $1/x = \cos\theta$, with θ the tunneling angle, gives the angular transmission of the barrier. In (5.29) and (5.30), the effect of scattering is included by adding an imaginary part $i(d/l)$ to the phase ϕ (5.26) formally appearing as a contribution to Im Z. If these factors are shown explicitly, they appear as $e^{-2d/l}$ and $e^{-d/l}$, respectively, in the third and fourth terms of (5.30), in Arnold's form of the theory. We will return to two other important extensions of the specular theory made by Arnold (1978), discussion of the bound-state spectrum and of the self-consistency between the pair potentials Δ_N and Δ_S.

Before doing so, we consider, following Gallagher, the effect of coherent reflections from the rear surface of film 2, leading to "compound geometrical resonances" identified by Wong, Shih, and Tomasch (1981) and shown in Fig. 5.8.

Gallagher's $N_T(E)$ for the N side, introducing normalized quantities

$$\varepsilon_{N,S} = \frac{E}{\Omega_{N,S}} \qquad \delta_{N,S} = \frac{\Delta_{N,S}}{\Omega_{N,S}} \tag{5.32}$$

is

$$N_T(E) = \mathrm{Re}\{D[\varepsilon_N \sin\phi_N \cos\phi_S + \varepsilon_S \sin\phi_S \cos\phi_N + \delta_N(\varepsilon_S\delta_N - \varepsilon_N\delta_S)\sin\phi_S(\cos\phi_N - 1)]\} \tag{5.33a}$$

where

$$D^{-2} = 1 - [\cos\phi_N \cos\phi_S - (\varepsilon_N\varepsilon_S - \delta_N\delta_S)\sin\delta_N \sin\delta_S]^2 \tag{5.33b}$$

Gallagher has shown that the expansion of (5.33), valid for $E \gg \Delta_N$, Δ_S and restricted to $\theta = 0$, is obtained from (5.30) by replacing $I(2\phi_N)$ with $\sin^2(\phi_S)/\sin^2(\phi_N + \phi_S)$ and $I(\phi_N)$ with $\sin\phi_S/\sin(\phi_N + \phi_S)$. Note that in the limit $d_S \to \infty$ with any finite scattering in S, one recovers (5.30), since $\sin\phi_S/\sin(\phi_N + \phi_S) \to \exp i\phi_N$, which is the $\theta = 0$ (forward-transmission) limit of $I(\phi_N)$ in (5.30a).

Another interesting limit which is obtained (Gallagher, 1982*b*; Arnold, Gallagher, and Wolf, 1982) from (5.33) is the limit d_N, $d_S \ll \xi$ (Cooper limit). It has been shown by Gallagher (1982*b*) that (5.33) reduces to

$$N_T(E) = \mathrm{Re}\left\{\frac{|E|}{[E^2 - \Delta_c^2(E)]^{1/2}}\right\} \tag{5.34}$$

where

$$\Delta_c(E) = \frac{\Delta_S(E)Z_S\tau_S + \Delta_N(E)Z_N\tau_N}{\tau_S Z_S + \tau_N Z_N} \tag{5.35}$$

with $\tau_{S,N} = d_{S,N}/v_{FS,N}$, and that this equation is valid for observations on either side of the NS bilayer. Extensions of this result have been discussed by Arnold, Gallagher, and Wolf (1982), who make use of a simple relation between the α^2F function of the Cooper limit bilayer and those of its constituents:

$$\alpha^2F(\omega)_c = \frac{\alpha_S^2F(\omega)_S\tau_S + \alpha_N^2F(\omega)_N\tau_N}{\tau_S + \tau_N} \tag{5.36}$$

The Cooper limit density-of-states expression can easily be deduced from the expression (5.30) modified for the fully specular bilayer case (Gallagher, 1980) by replacing $I(\phi_N)$ by $\sin(\phi_N)/\sin(\phi_N + \phi_S)$. The Cooper limit corresponds to small d_N, d_S and hence to small ϕ_N, ϕ_S. Thus,

$$\frac{\sin\phi_N}{\sin(\phi_N + \phi_S)} \simeq \frac{\phi_N}{\phi_N + \phi_S} \simeq \frac{\tau_N}{\tau_N + \tau_S} \tag{5.37}$$

which, when inserted in (5.30), leads to

$$N_T(E) = \mathrm{Re}\left(1 + \frac{1}{2}\frac{\Delta_c^2}{E^2}\right) \tag{5.38}$$

with Δ_c given by (5.35). This equation is the expansion of (5.34), as expected.

It can be seen from (5.37) and from the fact that $\phi \propto \Omega \simeq E$, for $E \gg \Delta$, that the accuracy of a Cooper limit description decreases with increasing energy E. In particular, formula (5.35) may be accurate in describing the gap region $E \simeq \Delta$ while failing in the phonon range of energies $E \simeq \hbar\omega_{ph}$, where $\phi(\hbar\omega_{ph}) \simeq \phi(\Delta) \times (\hbar\omega_{ph}/\Delta)$ is larger, because (5.37) may then be invalid. This question has been investigated numerically by Arnold, Gallagher, and Wolf (1982).

Another interesting feature of the specular bilayer calculation of Gallagher (1980) is its applicability to an equivalent superlattice generated by the reflection, which has layer thicknesses $D_N = 2d_N$, $D_S = 2d_S$, as shown in Fig. 5.10. In this case, however, $N_T(E)$ is sampled at the *middle* of one of the D layers. This subject will be discussed further in Chapter 9.

All that has been said thus far applies only to the idealized step–pair-potential bilayer. A less restricted analysis, valid only for $E \gg \Delta_S$ but for arbitrary $\Delta(x)$, has been given by McMillan and Rowell (1969). This analysis has been extended recently by Arnold and Wolf (1982), whose expression, which applies to a general $\Delta(x)$, is

$$N_T(E) \simeq \mathrm{Re}\left(1 + \frac{1}{2}\frac{P(E)^2}{E^2}\right) \tag{5.39}$$

Fig. 5.10. Equivalence of fully specular bilayer and extended superlattice of layer thickness $D_N = 2d_N$, $D_S = 2d_S$, with sampling at the middle of the N layer (arrow) corresponding to N-side sampling of bilayer (Gallagher, 1982a).

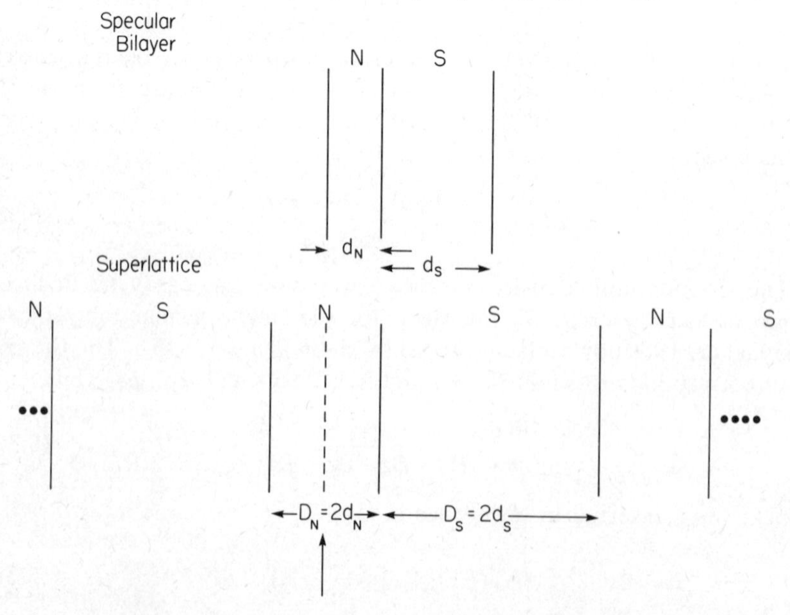

Here,

$$P(E) = -i\Delta K \int_0^{\infty} \exp(i\Delta Kx)\Delta(x)\,\mathrm{d}x \tag{5.40}$$

and

$$\Delta K = \frac{2Z\Omega}{\hbar v_F} \tag{5.41}$$

Applied to the stepwise $\Delta(x)$ of Fig. 5.4 and to (5.30), with only forward tunneling $[\theta = 0,\ I(\phi) = \exp(i\phi)]$, this equation gives

$$P(E) = \Delta_N(E) + [\Delta_S(E) - \Delta_N(E)]\exp(i\phi_N) \tag{5.42}$$

where ϕ_N is given by (5.26). It is easy to see from this result and (5.39) that, as $d_N \to 0$, $\phi_N \to 0$, one obtains

$$N_T(E) \simeq \mathrm{Re}\left\{1 + \frac{1}{2}\frac{[\Delta_S + i(\Delta_S - \Delta_N)\phi_N]^2}{E^2}\right\} \tag{5.43}$$

Thus, the influence of Δ_N cancels out to first order, remaining only as multiplied by the (small) phase angle $\phi(E) \simeq (d_N/\xi_{prox})(E/\Delta_S)$, where $\xi_{prox} = \hbar v_{FN}/2\Delta_S \simeq \pi/2(v_{FN}/v_{FS})\xi_{0S}$, ξ_{0S} being the usual BCS coherence length in S.

Another useful way of expressing the phase angle ϕ (Arnold, 1978) is as $\phi = RE$, where

$$R = \frac{2Zd}{\hbar v_F} \tag{5.44}$$

typically expressed in $(\mathrm{meV})^{-1}$. The period in energy of the $2d$ oscillation is $2\pi R^{-1}$, which works out to 1.03 meV for $d = 2\ \mu\mathrm{m}$, $Z = 1$, and $v_F = 10^8$ cm/sec, and to 1.03 eV for $d = 20$ Å. The R parameter in the two cases is 6.3 meV^{-1} ($d = 2\ \mu$m) and 0.0063 meV^{-1} ($d = 20$ Å).

In the equations for $N_T(E)$ above, Δ_N and Δ_S are assumed to be those functions resulting after the proximity effect occurs, effectively making N a superconductor by development of pair density in N and some reduction in pair density in S. Evaluation of these adjusted pair potentials is a separate problem. This question of self-consistency between N and S is discussed by Arnold (1978), who found that in the case of a thin N layer of weak coupling on a *thick* S layer, an easily interpreted approximate result could be obtained. The results for a *wholly induced* pair potential and renormalization function at $T = 0$ are

$$\Delta_N(E) = \frac{1}{Z_N(E)} \int_0^{\infty} \mathrm{d}E'\ \mathrm{Re}\left\{\frac{\Delta_S(E')}{[(E')^2 - \Delta_S(E')^2]^{1/2}}\right\} K_+(E, E')_N \tag{5.45}$$

$$Z_N(E) = 1 - \frac{1}{E}\int_0^{\infty} \mathrm{d}E'\ \mathrm{Re}\left\{\frac{E'}{[(E')^2 - \Delta_S(E')^2]^{1/2}}\right\} K_-(E, E')_N \tag{5.46}$$

where

$$K_{\pm}(E, E') = \int_0^{\infty} d\omega\, \alpha^2(\omega) F(\omega) \left\{ \frac{1}{E'+E+\omega} \pm \frac{1}{E'+E+\omega} \right\} - \mu^* \theta(E_c - E')\theta(\pm 1) \quad (5.47)$$

where E_c is an energy cutoff and $\theta(x)$ is the unit step function. These equations, written in N, are local Eliashberg equations with the pair density and density of quasiparticle states taken the same as in S. That is, the pair correlations across the Fermi surface in S are assumed to persist into N, in this thin-N limit: $d_N < \xi$. The first equation (5.45) is the strong-coupling counterpart to the familiar weak-coupling statement

$$\Delta(x) = \lambda(x) F \quad (5.48)$$

where the pair wavefunction F is constant over lengths less than ξ (as across N in the thin-N case considered) but the local coupling $\lambda(x)$, and hence $\Delta(x)$, can vary abruptly as the lattice changes, e.g., in crossing the NS interface. Note that while $\Delta_N(E)$ may be greatly enhanced by putting $\Delta_S(E)$ into the numerator of the bracketed factor in (5.45), as illustrated in Fig. 5.11, in the comparison equation for $Z_N(E)$, no such enhancement occurs; the peak of the bracketed function is shifted, but its weight at

Fig. 5.11. Calculation illustrating enhancement of pair-potential $\Delta_N(E)$ –induced in 20-Å Al film by proximity with a thick superconductor layer S. (After Wolf and Arnold, 1982.)

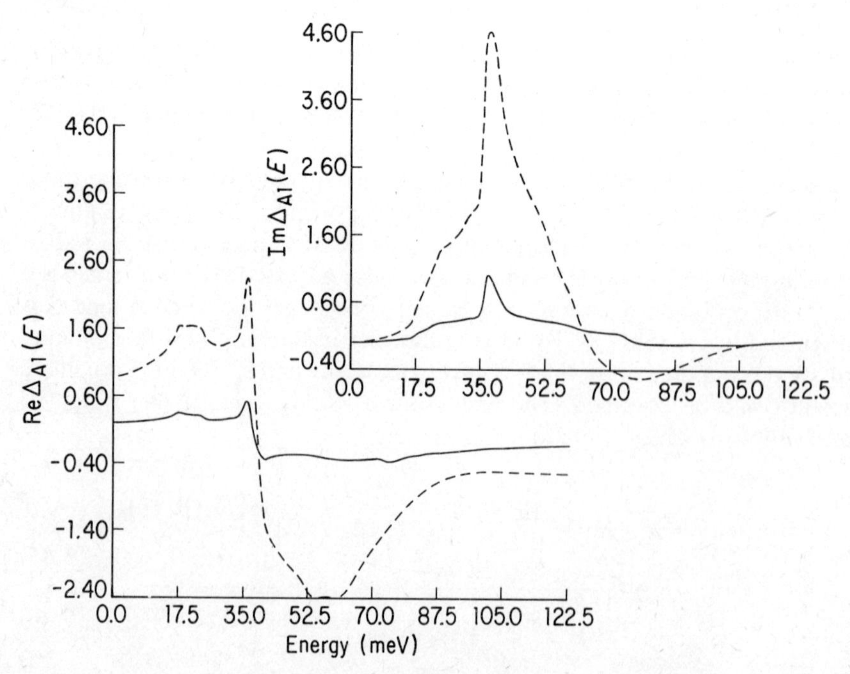

energies displaced from the peak is not much changed. There is essentially no "proximity effect" for the renormalization function, which is closely related to the inherent electron-phonon coupling $\lambda = N(0)\langle I^2\rangle/M\langle\omega^2\rangle$ (McMillan, 1968c), as indicated by the relation $Z(0) = 1+\lambda$. The constancy of the pair wavefunction F across the NS boundary may imply an enhanced degree of pair correlation across the Fermi surface in N, but certainly it does not change $N(0)$ or the other factors in λ.

Arnold (1978) also investigated the accuracy of the step-pair-potential approximation, using the results of his self-consistency calculations. The pair potential in S is expected to be slightly depressed at the interface with N, over a distance into S of order ξ_S. Writing the surface value of Δ_S as Δ_S^0, Arnold finds

$$\Delta_S^0 \simeq \Delta_S(1-\pi R\Delta_S) \tag{5.49}$$

with R from (5.44). This correction can be a small one; e.g., for $d_N = 20$ Å, $R \simeq 0.0063\ \text{meV}^{-1}$, and $\Delta_S = 1.5$ meV, one obtains $\pi R\Delta_S = 0.03$. A 3% reduction in Δ_S^0 at the interface is predicted.

A second topic which has received extended treatment is the form of $N_T(E)$ in the gap region, i.e., the bound-state spectrum, which is illustrated for a thin-N case of 20 Å of Al on Nb in Fig. 5.12, calculated from the $N_T(E)$ of Arnold (1978). The solid line shown corresponds to

Fig. 5.12. Calculated tunneling density of states in the gap region on assumptions of random tunneling and of specular tunneling (dashed curve). Note the presence of a gap between the bound-state spectrum ($E < \Delta_S$) and the continuum ($E \geqslant \Delta_S$). (After Wolf and Arnold, 1982.)

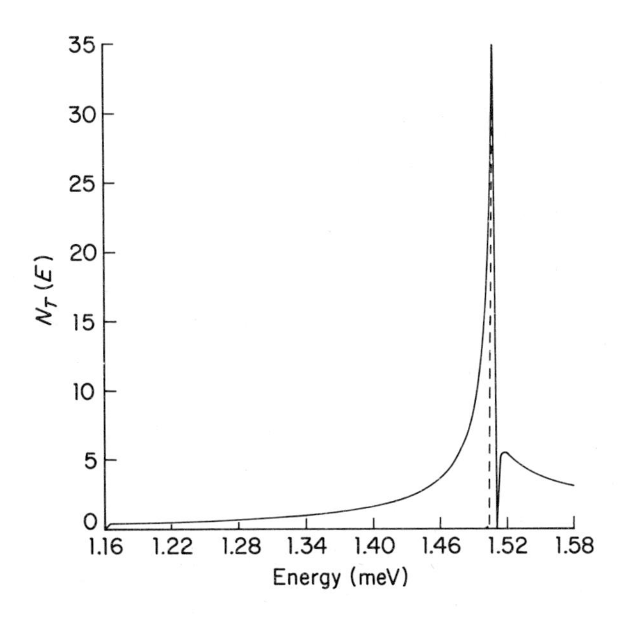

$D(1/x)=1$, i.e., no angular dependence of the tunneling matrix element, while the dashed line models the spectrum observed through a specular trapezoidal barrier. The tail extending to a low-energy cutoff at $(\Delta_N\Delta_S)^{1/2}$ corresponds to quasiparticle bound states propagating at nonzero angles θ from the normal, effectively experiencing a thickness greater than d_N. A feature of this spectrum not present in the original calculations of de Gennes and Saint-James (1963) and Saint-James (1964) is a small gap between E_0, the bound state at $\theta=0$, and the continuum of superconductor excitations above Δ_S. Discussion of the form of $N_T(E)$ above and below Δ_S is given by Arnold (1978). For small R, Arnold finds

$$E_0 \simeq \Delta_S[1-\tfrac{1}{2}(RZ_N\Delta_S)^2] \tag{5.50}$$

as the bound-state energy. Further insight into the magnitude of this gap can be obtained from the work of Bar-Sagi and Entin-Wohlman (1977) in Fig. 5.13, which plots E_0/Δ_S vs. d_N for several values of an assumed δ-function repulsive potential at the NS interface. The latter potential is intended to model the effects of an imperfect metallic connection between N and S, as might result, e.g., from a monolayer or so of adsorbed gas. The effect of the δ-function repulsive potential is parameterized by the resulting reflection coefficient $|r^2|$.

Our primary interest in the specular theory being discussed is as a means of treating phonon structures in $N_T(E)$ arising from both Δ_S and Δ_N. Figure 5.14 shows numerical calculations of $N_T(E)$, illustrated by the

Fig. 5.13. The energy of the bound state in the N side of the NS sandwich as a function of d_N and the reflection R produced at the NS interface by an assumed repulsive δ-function potential. The bound-state energy approaches the superconducting energy gap as d_N approaches zero. (After Bar-Sagi and Entin-Wohlman, 1977.)

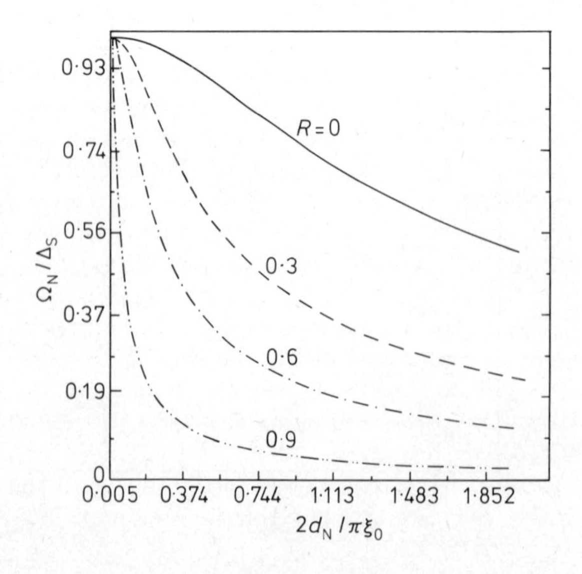

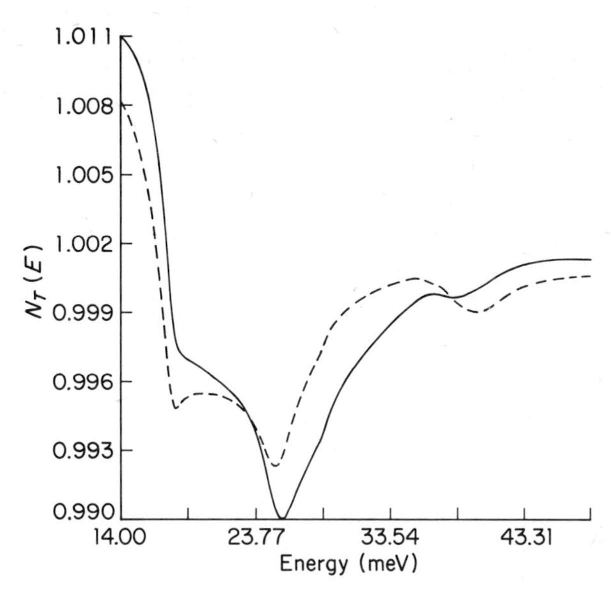

Fig. 5.14. Calculated Al phonon structure in $N_T(E)$, taking two thicknesses d_N (R-parameter values), illustrating growth of structure with increasing R. (After Wolf and Arnold, 1982.) Dashed curve, $d_N = 80$ Å; solid curve, $d_N = 20$ Å.

Al–Nb system, for $d_N = 20$ Å (solid) and $d_N = 80$ Å (dashes) assuming a perfect NS interface (reflection zero) and zero scattering ($d/l = 0$). The points to be noted here are that N-metal (Al) structure (recognized as the dip having maximum negative slope at 36 mV) increases in strength with increasing d_N [see (5.43)] and that the structure in $N_T(E)$ from S (Nb) phonons at 16 and 24 mV is characteristically distorted—"rotated counterclockwise" by the phase factors exp $i\phi_N(E)$ entering (5.23) and (5.30).

Analysis of such NS tunneling data requires the proximity $N_T(E)$ to take proper account of the effects mentioned above: strength of the N-phonon features and the modulation of the S-phonon structures by the phase factors. One assumption involved in applying these specular $N_T(E)$ functions to data (as has been successfully accomplished; see especially Fig. 5.8) is that the NS interface constitutes only a pair-potential barrier but does not scatter quasiparticles in the normal state of the coupled structure. That is, the calculation assumes no $V(x)$ (oxide) barrier at the NS interface and, in fact, neglects such small reflections as may arise from the discontinuity in Fermi velocities at the NS interface. As a test of the sensitivity of $N_T(E)$ to small amounts of real reflection, as must occur, model calculations are presented in the next three figures, after Wolf and Arnold (1982).

One must expect, in general, some reflection at the NS interface due to differences in the band structure of N and S. In a simple model, it will not be expected that the bandwidths E_{FS} and E_{FN} of the N and S metals agree exactly. While the Fermi energies must coincide after contact, the lower band edges will not match, and the free-electron Fermi wavevectors k_{FS} and k_{FN} will differ. Calculation of the reflection coefficient $|r^2|$ at such a potential jump $|E_{FS} - E_{FN}|$ gives

$$|r^2| = \frac{(k_{FS} - k_{FN})^2}{(k_{FS} + k_{FN})^2} \tag{5.51}$$

which is plotted as a function of E_{FS}/E_{FN} in Fig. 5.15a. Even for a factor of two for the difference in Fermi energies, the resulting reflection is seen to be only ~3%, again a small effect.

Fig. 5.15. (a) Calculated values of normal-state NS electron reflection coefficient $|r|^2$ on the free-electron model. (b) Electron transmission coefficient $|t|^2$ on a tight-binding model. (After Wolf and Arnold, 1982.)

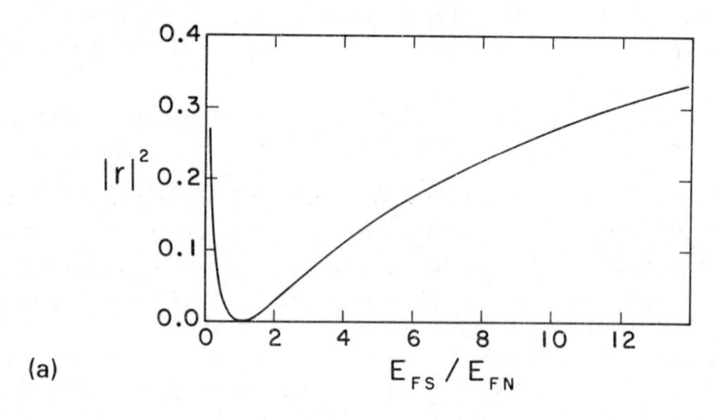

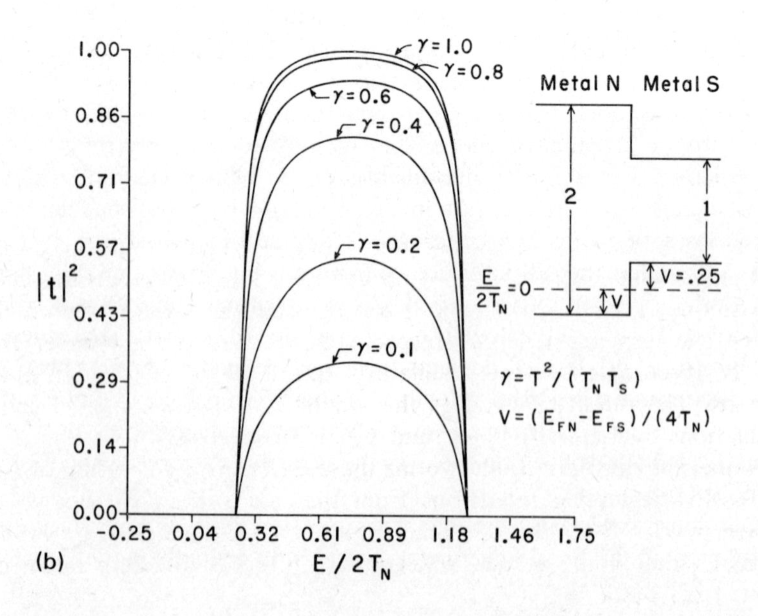

An alternative approach to the normal-state reflection question is via a tight-binding model calculation as sketched in the inset of Fig. 5.15b. A jump in bandwidth in passing from one tight-binding metal (N) to the other (S) implies a jump in nearest-neighbor overlap integrals from T_N to T_S. The symbol T represents the overlap at the interface between atoms of N and S, while γ is defined as

$$\gamma = \frac{T^2}{T_N T_S} \tag{5.52}$$

The resulting transmission $t^2 = 1 - r^2$, assuming two simple cubic metals with a (100) interface, is plotted versus electron energy $E/2T_N$. The transmission peaks for E at the midband position $E/2T_N = 0.75$ (note the offset zero in E) and falls reasonably slowly as γ decreases below the optimum value 1, corresponding to $T = (T_N T_S)^{1/2}$.

Turning to the effect of the reflection $|r^2|$ on the $N_T(E)$, consider first the bound states. As shown in Fig. 5.13, increasing $|r^2|$ reduces E_0. This effect can be described by an effective, increased R parameter:

$$R_{\text{eff}} = \frac{t^2}{(1-r)^2} R = \left(\frac{1+r}{1-r}\right) R \tag{5.53}$$

as if the N region were increased in width. This result, obtained by Arnold from the exact Green's function, confirms the result of Bar-Sagi and Entin-Wohlman (1977) on the position E_0 of the bound state.

In the phonon region, the expansion of the full $N_T(E)$ containing the normal-state reflection coefficient r^2 is

$$\begin{aligned} N_T(E) \simeq \operatorname{Re}\bigg\{ & 1 + \frac{1}{2}\frac{\Delta_N^2}{E^2} + \frac{1}{2}\frac{(\Delta_S - \Delta_N)^2}{E^2} \exp(i2\phi_N)(1-2r^2) \\ & + \frac{\Delta_N(\Delta_S - \Delta_N)}{E^2} \exp(i\phi_N)(1-r^2) + \frac{(\Delta_S-\Delta_N)^2}{E^2} r^2 \exp(i4\phi_N) \\ & + \frac{\Delta_N(\Delta_S - \Delta_N)}{E^2} r^2 \exp(i3\phi_N) \bigg\} \end{aligned} \tag{5.54}$$

The significance of the last two terms involving phase shifts $3\phi_N$, $4\phi_N$ is that normal-state reflections may occur before Andreev reflections events ("postponed Andreev reflection") corresponding to a larger path length. These two terms correspond, respectively, to six and eight traversals of the N layer and lead to oscillatory contributions of shorter energy periods. The reflection reduces the amplitudes of the usual 2d and 4d oscillations by factors of $1 - r^2$ and $1 - 2r^2$, respectively.

Numerical calculations illustrating these effects for a 50-Å film of Al on thick Nb (neglecting reflections from the rear surface of the Nb) are shown in Figs. 5.16a and 5.16b. It may be concluded that, in the thin, normal metal limit, normal reflections at the NS interface make only

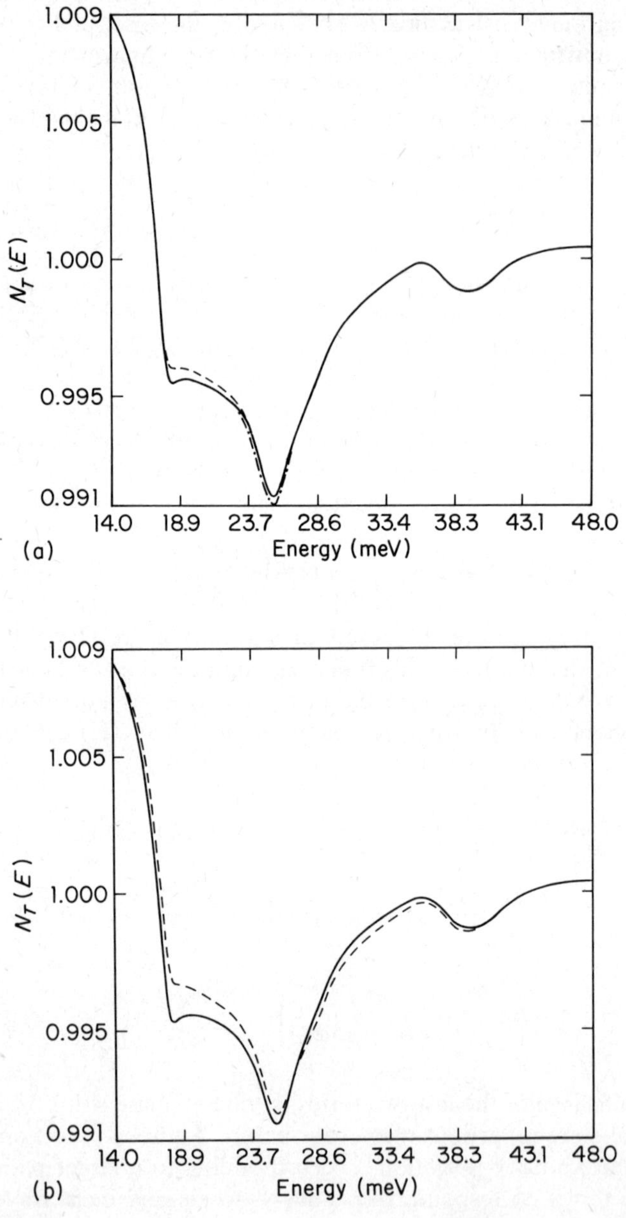

Fig. 5.16. Influence of assumed NS reflection coefficients r^2 on calculation of tunneling density of states $N_T(E)$, in NS bilayers whose N layers are 20 Å (a) and 80 Å (b) in thickness. In these figures, the solid curves are for $d/l = 0$. (After Wolf and Arnold, 1982.)

small changes in the $N_T(E)$ if the r^2 values are limited to a few percent, the expected range from Fermi velocity mismatch.

A further type of scattering, namely elastic scattering in the N region, will be discussed later, after we have presented the McMillan tunneling model of NS structures. This delay is appropriate because incorporation of elastic scattering into the specular model leads to expressions similar to those of McMillan (1968*b*), providing something of a conceptual bridge between the two approaches.

Finally, we return to the fully specular (Gallagher) case, with specular reflections assumed also at the rear surface of the thick S layer. In the case of a thick N region backed by an S layer of decreasing thickness, Fig. 5.17 shows $N_T(E)$ bound-state spectra calculated by Gallagher (1980). These curves show characteristic broadening (band formation) with decreasing d_S. This interesting effect is best understood by thinking of the superlattice analog $D_S = 2d_S$, $D_N = 2d_N$ of the fully specular bilayer, Fig. 5.10, and of overlap and band formation between bound states (in N) separated by barriers (S regions) of decreasing thickness.

Fig. 5.17. Local density of states at the surface of a normal metal of thickness $d_N = 5\pi\xi_S/2$ backed by superconductors of various thicknesses d_S. The mean free path for this calculation is chosen at $6.7 \times d_N$; the curves are displaced vertically by 112.6, 1.6, and 1.0 units. (After Gallagher, 1980.)

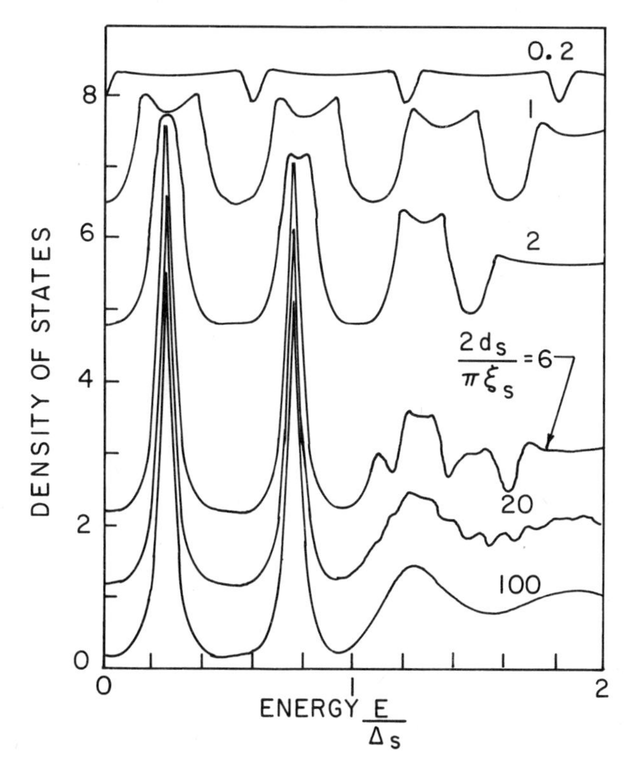

5.5 McMillan's Tunneling Model of Bilayers

In this model calculation, McMillan (1968*b*) assumed thin films of N and S having spatially constant pair potentials, separated by a tunnel barrier of *small* transmission σ. The coupling between the two is treated by the tunneling Hamiltonian, which is assumed to connect each state in N(S) with equal probability with each state in S(N). This model assumes fully random tunneling, quite the opposite of the specular transmission and reflection that we have been considering. In terms of film properties, one assumes enough scattering so that the films are dirty superconductors in the Anderson sense, and there are no momentum selection rules. Each state in N(S) has a tunneling lifetime width $\hbar/2\tau$ to disappear into the opposing film

$$\Gamma_N = \frac{\hbar}{2\tau_N} = \pi T^2 A d_S N_S(0)$$
$$\Gamma_S = \frac{\hbar}{2\tau_S} = \pi T^2 A d_N N_N(0) \tag{5.55}$$

The simultaneous self-energy equations for fixed pair potentials Δ_S^{ph} and Δ_S^{ph} (in films S and N, respectively) are

$$\Delta_N(E) = \left\{\Delta_N^{ph} + \frac{i\Gamma_N \Delta_S(E)}{[E^2 - \Delta_S^2(E)]^{1/2}}\right\} \Big/ \left\{1 + \frac{i\Gamma_N}{[E^2 - \Delta_S^2(E)]^{1/2}}\right\} \tag{5.56}$$

$$\Delta_S(E) = \left\{\Delta_S^{ph} + \frac{i\Gamma_S \Delta_N(E)}{[E^2 - \Delta_S^2(E)]^{1/2}}\right\} \Big/ \left\{1 + \frac{i\Gamma_S}{[E^2 - \Delta_N^2(E)]^{1/2}}\right\} \tag{5.57}$$

Here, $\Delta_{N,S}^{ph}$ represents the pair potential in N, S when isolated. The denominator bracketed terms in (5.56) and (5.57), respectively, are $Z_N(E)$ and $Z_S(E)$, which take on imaginary parts linear in the coupling σ, i.e., proportional to the matrix element T^2 or, equivalently, to the Γ values. The self-consistency conditions for the pair potentials at temperature T are

$$\Delta_N^{ph} = \lambda_N \int_0^{E_c} \mathrm{Re}\left\{\frac{\Delta_N(E)}{[E^2 - \Delta_N^2(E)]^{1/2}}\right\} \tanh\frac{E}{2kT}\, dE \tag{5.58}$$

$$\Delta_S^{ph} = \lambda_S \int_0^{E_c} \mathrm{Re}\left\{\frac{\Delta_S(E)}{[E^2 - \Delta_S^2(E)]^{1/2}}\right\} \tanh\frac{E}{2kT}\, dE \tag{5.59}$$

and the excitation spectra are

$$N_{S,N}(E) = \mathrm{Re}\left\{\frac{|E|}{[E^2 - \Delta_{S,N}^2(E)]^{1/2}}\right\}$$

In terms of the barrier transmission σ, the lifetimes may be expressed as

$$\tau_{\mathrm{N,S}} = \frac{l_{\mathrm{N,S}}}{v_{\mathrm{FN,S}}\sigma} \tag{5.60}$$

where $l_{\mathrm{N,S}}$ is the mean free path between collisions with the barrier in N, S. In general,

$$l_{\mathrm{N,S}} = 2B_{\mathrm{N,S}}d_{\mathrm{N,S}} \tag{5.61}$$

where $B_{\mathrm{N,S}}$ is a function of the ratio of mean free path to film thickness. Then, one can write

$$\Gamma_{\mathrm{N}} = \frac{\hbar v_{\mathrm{FN}}\sigma}{4B_{\mathrm{N}}d_{\mathrm{N}}} \tag{5.62}$$

and an analogous equation in S.

The transition temperature of the bilayer is determined by

$$\ln \frac{T_c^{\mathrm{bulk}}}{T_c} = \frac{\Gamma_{\mathrm{S}}}{\Gamma_{\mathrm{S}}+\Gamma_{\mathrm{N}}}\left[\psi\left(\frac{1}{2}+\frac{\Gamma_{\mathrm{S}}+\Gamma_{\mathrm{N}}}{2\pi kT_c}\right)-\psi\left(\frac{1}{2}\right)\right] \tag{5.63}$$

where ψ is the digamma function (McMillan, 1968*b*).

The McMillan tunneling model predicts that the same energy gap will be seen in $N_{\mathrm{T}}(E)$ measured from either side, of magnitude $\Omega_{\mathrm{N}} = \Delta_{\mathrm{N}}^{\mathrm{ph}} + \Gamma_{\mathrm{N}}$. The spectrum $N_{\mathrm{N}}(E)$ measured from the N side shows a BCS-like peak at Ω_{N} (or at Γ_{N} if the N metal has $\lambda = 0$, negligible inherent electron-phonon coupling). The relationship between this feature at Ω_{N} and the bound state of de Gennes and Saint-James (1963) is not close, for Γ_{N} depends linearly on σ/d_{N}, a widely varying ratio. For this reason, the Ω_{N} peak of the McMillan model is not obviously limited on the high-energy side by the value Δ_{S}, as is the de Gennes bound state in the limit $\sigma/d_{\mathrm{N}} \rightarrow \infty$. The $N_{\mathrm{T}}(E)$ spectrum observed for the S side has weak structure at Ω_{N} and a broadened BCS-like peak at Δ_{S}. Note that this model, which does not incorporate Andreev scattering and assumes small mean free paths in N and S, is not capable of predicting the oscillatory phenomena observed by Tomasch and others.

In spite of its deficiencies, the McMillan model has been widely applied to analyzing proximity effect experiments, especially relating to T_c, the gap Δ_0, and tunneling spectra in the gap region. An example is shown in Fig. 5.18 illustrating the T dependence of an effective gap parameter measured on several Sn films backed by 5000 Å of Pb in comparison with solid theory curves obtained from the tunneling model. Extensive experimental studies primarily devoted to gap spectra have been reported by Adkins and Kington (1966, 1969), Vrba and Woods (1971*a*, 1971*b*), Freake (1971), Wyatt, Barker, and Yelon (1972), Gray (1972), and Lykken and Soonpaa (1973). An additional literature relates to the addition of magnetic impurities to the N region of an NS sandwich (Dumoulin, Nédellec, and Guyon, 1972; Dumoulin, Guyon, and Nédellec, 1973, 1977), making use of extensions of the McMillan model to

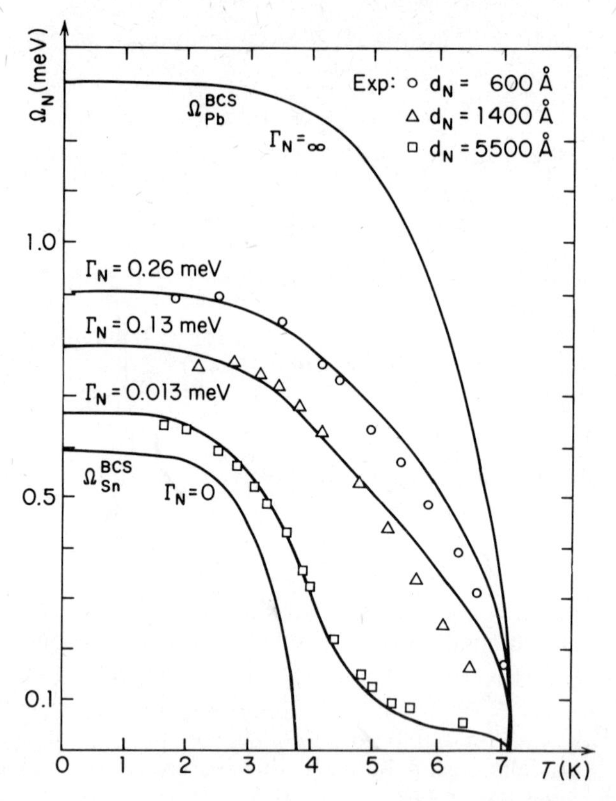

Fig. 5.18. Temperature variation of the energy gap of tin films backed by 5000 Å of lead. The experimental points are compared with theoretical curves from the McMillan model. (After Gilabert, Romagnon, and Guyon, 1971.)

include the $\vec{S} \cdot \vec{d}$ exchange interaction by Kaiser and Zuckerman (1970), Kaiser (1977), Machida (1977), and Machida and Dumoulin (1978).

A second area of extension of the McMillan tunneling model to treat phonon spectra, observed, e.g., by Toplicar and Finnemore (1977) and by Chaikin, Arnold, and Hansma (1977), has been made by the latter authors and by Mohabir and Nagi (1979*a*). Extension of McMillan's calculation to higher orders of perturbation theory is described by Mohabir and Nagi (1979*b*). In general, treatment of bilayer spectra containing both N and S phonon structure has not been successful. However, an excellent treatment of S phonon features in Pb films backed by Cd (Toplicar and Finnemore, 1977) was possible within the tunneling model.

We close this section by mentioning briefly an alternative approach to the proximity bilayer problem proposed by Ovadyahu and Entin-Wohlman (1979, 1980). In this approach, as in the McMillan tunneling model, Andreev scattering does not occur; hence, the well-known oscillatory effects are not treated. The "empirical" proposal of Ovadyahu and

Entin-Wohlman (1979) for the $N_T(E)$ at the N side of the NS bilayer is

$$N_T(E) = \text{Re}\left(\frac{Z_N(E)E}{\{[Z_N(E)E]^2 - \phi_N^2(E)\}^{1/2}}\right) \tag{5.64}$$

where the pairing self-energy in N is taken essentially as that in S:

$$\phi_N(E) = \beta(E)\phi_S(E) \tag{5.65}$$

Here $\beta(E)$, said to be a measure of the efficiency of coupling of N and S electrons, is expected to be weakly energy-dependent. These points are further discussed by Ovadyahu and Entin-Wohlman (1980) in the framework of the proximity effect theory of Silvert (1975). An additional discussion of the boundary condition at the NS interface has been given by Entin-Wohlman (1977). It has been shown (Wolf et al., 1981) that (5.64), when applied to foil-based junctions of minimized quasiparticle scattering, leads to distortion of the inferred $\alpha^2F(\omega)$.

5.6 Proximity Electron Tunneling Spectroscopy (PETS)

This name, PETS, has been given to the use of the specular theory, as described above, on C–I–NS junctions, fabricated to minimize quasiparticle scattering and with the objective of quantitatively determining the $\alpha^2F(\omega)$ functions of the constituent S and N superconductors. It should be emphasized that the specular theory can be applied quantitatively only to NS structures in which care has been exercised to produce long mean free paths and to enhance specular transmission at the NS interface. Thus far, it appears that the closest experimental approach to specular conditions has been achieved in junctions based on recrystallized foils of transition metals (Wolf, Zasadzinski, Osmun, and Arnold, 1980). There is good reason to expect, fortunately, that comparable or even improved NS-junction quality may be possible with S and N layers grown epitaxially, following demonstrations (Durbin, Cunningham, Mochel, and Flynn, 1981; Geerk, Gurvitch, McWhan, and Rowell, 1982) that coherent Nb/Ta superlattices can be grown on several single crystal faces of sapphire using ultrahigh-vacuum and sputtering methods. In these methods, the NS interface is formed by *in situ* deposition of N onto a clean and highly ordered surface of S.

In the foil technique, an atomically flat and clean S-metal surface is produced by heating the specimen to nearly its melting point in ultrahigh vacuum. This procedure releases most dissolved light-atom impurities and also allows recrystallization of the foil into flat grains whose lateral dimensions may be 0.1 cm or even greater. Various experimental evidence is now available indicating that deposition of soft metals (Al and Mg) onto hard (high melting point, high surface energy) transition metal foils under ultrahigh-vacuum conditions can lead to uniform coverage down to less than 10-Å deposit thickness. The first evidence comes from tunneling studies (Wolf et al., 1980; Rowell, Gurvitch, and Geerk, 1981) later confirmed by x-ray photoemission spectroscopy (XPS) studies (Kwo, Wertheim, Gurvitch, and Buchanan, 1982). Discussions of the thermodynamics which makes such uniform coverage favorable are given by Biberian and Somorjai (1979) and by Miedema and den Broeder (1979). The same coverage properties essential for PETS have been achieved in sputtered NS structures by Rowell, Gurvitch, and Geerk (1981).

We turn now to results obtained by using the foil tunneling PETS technique, applied to Nb, Ta, Mg, and Al and presented as a demonstration of the accuracy presently available from PETS under favorable conditions. Unless otherwise noted, the data that follows are obtained by using techniques outlined by Wolf et al. (1980) and analyzed in the Arnold (1978) specular theory using numerical techniques described by Arnold, Zasadzinski, Osmun, and Wolf (1980). Since extensive discussions of the PETS methods and results have been given recently (Wolf, 1982; Wolf and Arnold, 1982), the present section is rather brief.

In the application of PETS to study the superconductivity of a pair of metals N and S of weak and strong coupling, respectively, C–I–NS junctions of different N thickness d_N are required. A thin N junction having $d_N \leq 20$ Å is used to provide an $\alpha^2 F$ spectrum of S $[\alpha^2 F(\omega)_S]$ relatively free of N-phonon contributions. This sample is also used to obtain, from analysis of its bound-state spectrum, a value of the gap edge parameter of S at the NS interface, Δ_S^0.

A second sample of larger d_N, hence giving stronger phonon structure from N, is measured for $N_T(E)$. If the $\Delta_S(E)$ function is known, a numerical procedure (Arnold, Zasadzinski, Osmun, and Wolf, 1980) based on the exact $N_T(E)$, represented for $E > \Delta$ by (5.30), can be employed to obtain $\Delta_N(E)$ and $\alpha^2 F(E)_N$. Or if Δ_S has been taken approximately from an initial inversion of the thin-N sample, the more approximate Δ_N resulting from analysis of the second sample can be used to perform a second variational analysis of data from the first sample for $\alpha^2 F(E)_S$, including as input data in (5.30) the $\Delta_N(E)$ as approximated by the analysis of the second sample. An iterative analysis of this sort in Al/Nb samples was reported by Wolf, Zasadzinski, Osmun, and Arnold (1979) and in Al/V samples by Zasadzinski, Burnell, Wolf, and Arnold (1982).

The gap region spectra of foil-based junctions with $d_N \leq 100$ Å are typically BCS-like and are barely discernible from the spectra expected of conventional junctions on the same superconductor. This result is illustrated in Figs. 5.19 and 5.20 and is in quantitative accord with the Arnold exact $N_T(E)$ (Fig. 5.12, specular assumption). The minor difference between the thermally broadened dI/dV structure (taking a normal-state electrode at 1.38 K) and that obtained from BCS theory is displayed in Fig. 5.20. Clearly, these minor changes cause no problem in accurate determination of the underlying gap edge parameter Δ_S^0.

The first deviation from BCS dI/dV behavior is an increased peak

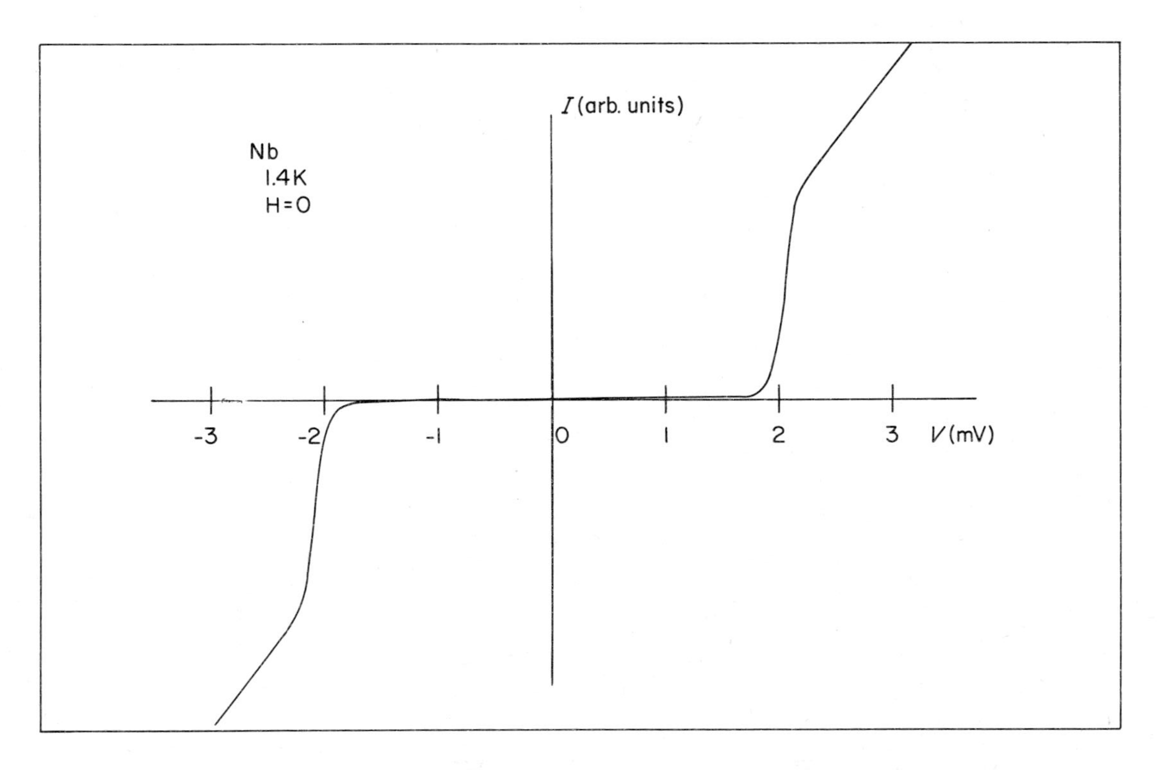

Fig. 5.19. Gap region characteristics of thin-N foil-based sandwiches are nearly indistinguishable from those of conventional C–I–S junctions, consistent with a single bound state at $E_0 \simeq \Delta_S$, when the NS interface is highly transmitting. The I–V characteristic is for 70 Å of Al on an Nb foil with an In counterelectrode. (After Wolf, Zasadzinski, Osmun, and Arnold, 1980.)

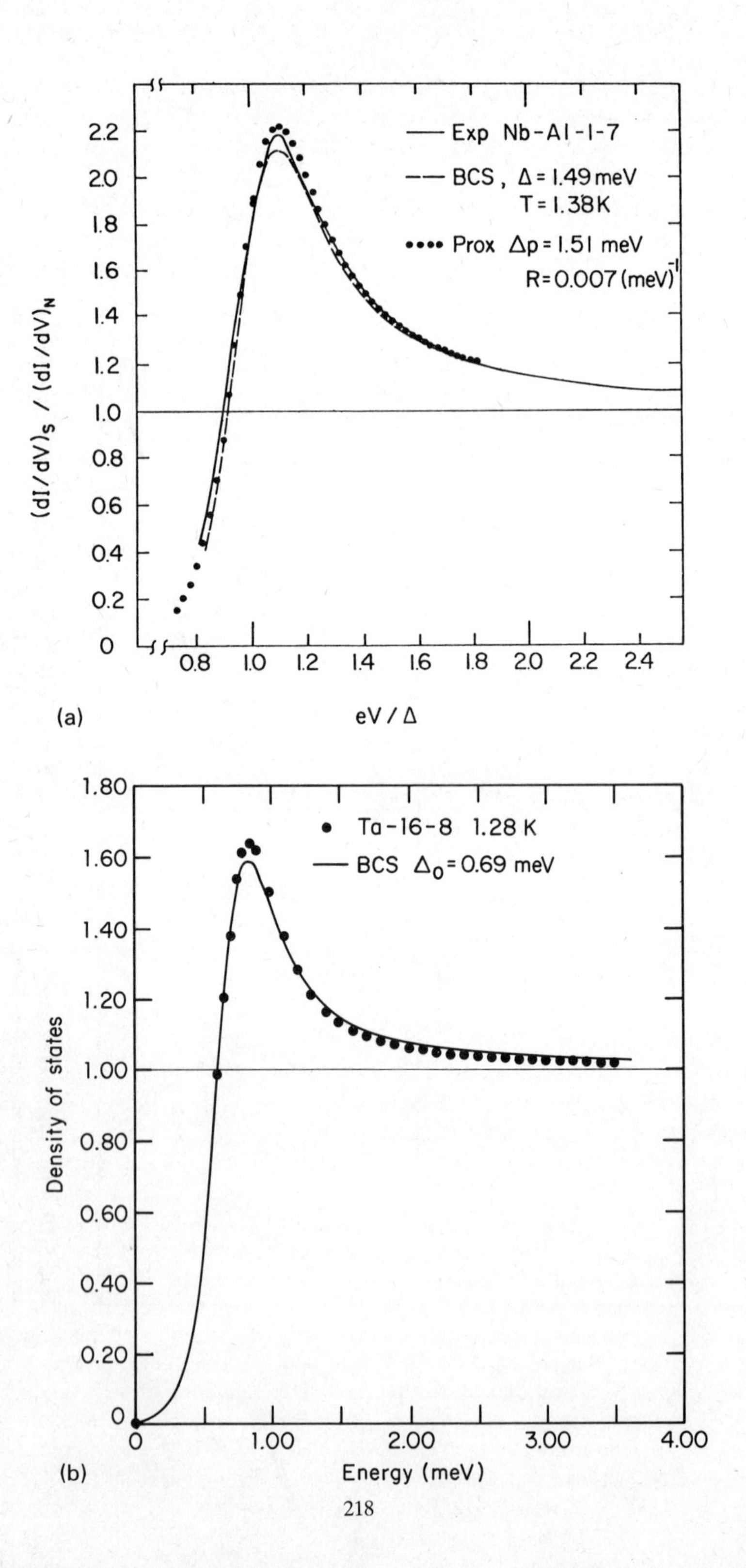

Exp Nb-Al-1-7
BCS, Δ = 1.49 meV
T = 1.38K
Prox Δp = 1.51 meV
R = 0.007 (meV)^-1
(dI/dV)s / (dI/dV)N
eV / Δ
(a)
Ta-16-8 1.28 K
BCS Δ0 = 0.69 meV
Density of states
Energy (meV)
(b)

height in the normalized experimental derivative σ; this feature is common to the McMillan tunneling model. Major differences between the two models, such as the bound-state line shape and the gap between E_0 and Δ_S^0, are not evident at $T \simeq 1$ K with a normal-state counterelectrode. These features have been observed by making use of the sharper energy resolution of a superconducting electrode and have been modeled semi-quantitatively by using the Arnold density of states in a study of the alloy $Nb_{0.75}Zr_{0.25}$ (Wolf, Noer, and Arnold, 1980). There is evidence that fully quantitative treatment of bound-state features for $d_N \geqslant 100$ Å (see, e.g., Freake, 1971) requires incorporation of a more correct treatment of elastic scattering of quasiparticles within the bulk of the N layer. This subject will be deferred to Sec. 5.7.

We turn now to the analysis of phonon features. Figures 5.21 and 5.22 are taken from a study of Al/Ta foil junctions (Wolf, Burnell, Khim, and Noer, 1981) which tests the degree to which junctions of different d_N, here 100 and 200 Å, can be consistently treated with the same superconducting functions $\Delta_N(E)$ and $\Delta_S(E)$. Figures 5.21a and 5.21b (solid curves) are experimental reduced conductances at $d_N = 100$ and 200 Å, respectively. In each case, the separate phonon features of Ta (~12 and 18 meV) and of Al (~37 meV) are clear; the Al feature becomes relatively stronger with the larger d_N value. The analysis of these two spectra (dashed lines) has assumed a bulk Al $\alpha^2F(\omega)$ (Leung, Carbotte, and Leavens, 1976) from which the induced Al pair potential $\Delta_{Al}(E)$ has been calculated using (5.45). The $\Delta_{Ta}(E)$ function required for this calculation was taken from analysis of an Al/Ta bilayer junction (not shown; see Fig. 7b of Wolf, Burnell, Khim, and Noer, 1981) with d_{Al} so small that the Al was totally oxidized in forming the tunnel barrier. The details of the α^2F_{Ta} used in (5.45) are not critical, but its overall strength (λ parameter) should be accurate. The energy of the prominent Al phonon structure in the calculated density of states (dashed lines), which is determined by the input bulk Al $\alpha^2F(E)$ function, does not precisely match that of the experiment; the line shape and strength are in reasonable agreement. With the $\Delta_N(E)$ contributions fixed, an extension of the McMillan-Rowell variational method (accommodating the proximity density of states) (5.30) was applied to determine the $\alpha^2F(E)$ for Ta (Fig. 5.22). The same $\alpha^2F(E)_{Ta}$ was then used in (5.30), having adjusted d_N to 200 Å and

---

Fig. 5.20. (a) Quantitative modeling of gap spectra (solid line) of a $d_N = 28$ Å Al–Nb proximity junction, using Arnold $N_T(E)$ (points) and BCS density of states (dashed line) and incorporating thermal smearing appropriate to 1.38 K. Bound-state features, unresolved at 1.38 K, are here nearly indistinguishable from conventional BCS peak structure. (After Wolf, Zasadzinski, Osmun, and Arnold, 1980.) (b) Density of states for 200 Å of Al on Ta foil, with Ag counterelectrode, is close to conventional thermally smeared BCS. This curve permits a good estimate of gap edge Δ_{S0}. (After Wolf, Burnell, Khim, and Noer, 1981.)

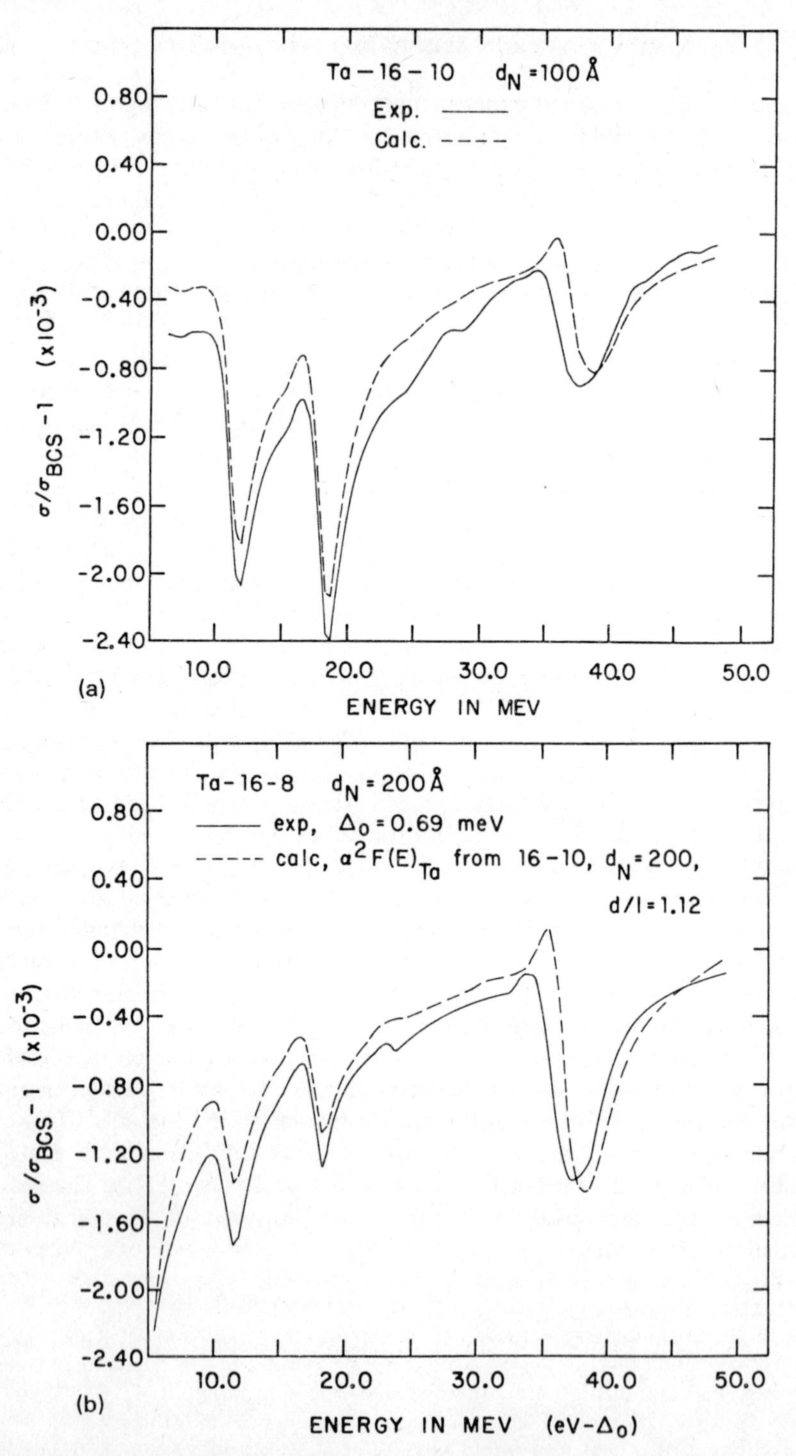

Fig. 5.21. Comparison of experimental (solid) and theoretical (dashed) $N_T(E)$ of thin-Al films on Ta foils (see text). (a) $d_{Al} = 100$ Å. (b) $d_{Al} = 200$ Å. (After Wolf, Burnell, Khim, and Noer, 1981.)

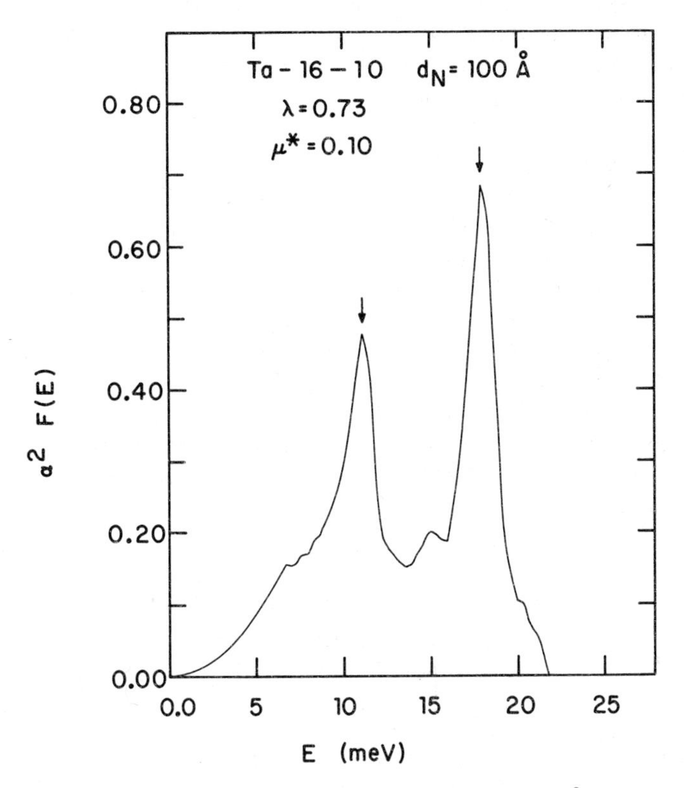

Fig. 5.22. Eliashberg function for Ta deduced from 100-Å AlTa proximity junction (Fig. 5.21a). Parameter values $\lambda = 0.73$ and $\mu^* = 0.10$ are close to those obtained conventionally, e.g., by Shen (1972). (After Wolf, Burnell, Khim, and Noer, 1981.)

choosing an optimum value of scattering d/l, to predict the dashed curve in Fig. 5.21b. The agreement at $d_N = 200$ Å is not precise but is roughly equivalent to that achieved in the 100-Å case. To this level of accuracy, which is limited by the inherent weakness of phonon effects in Ta, a consistent treatment of the two samples is achieved. The $\alpha^2F(E)$ for Ta shown in Fig. 5.22 is in good agreement with neutron scattering measurements as to peak positions and is also in good agreement with the earlier tunneling determination of Shen (1970), and the $d_N \simeq 0$ determination mentioned above with respect to the superconducting parameters λ and μ^*. The phonon peaks, in fact, in the $d_N = 100$ Å $\alpha^2F(E)$ function are better resolved than in the earlier tunneling determinations, with somewhat stronger weight at the higher-energy LO phonon position. In this regard, the PETS function is closer to the neutron scattering $F(\omega)$ (Woods, 1964). This improvement is attributed in part to protection of the prepared Ta surface from oxygen by the Al layer. A second possible contribution to improved resolution is the greater sampling depth of PETS (below).

An example of the determination of the $\alpha^2F(E)$ of the N layer is given in Fig. 5.23. The second-derivative spectrum of 70 Å of Al on a V foil is shown in Fig. 5.23a, leading to the α^2F for Al in Fig. 5.23b. To obtain this result, the experimenters employed analysis of a second Al/V bilayer of smaller d_N. The $\alpha^2F(E)$ for the 70-Å Al film is noticeably broader than the bulk Al $\alpha^2F(\omega)$ shown for comparison and also reveals a shift to lower energy of the *LO* peak more noticeable than that evident in Fig. 5.21. It is presumed that these differences arise from the conditions of epitaxy of Al on V. The epitaxy of Al on Nb is discussed by Hertel, McWhan, and Rowell (1982). Interestingly, the Mg *LO* peak observed in thin layers on Nb and Ta foils is shifted to higher energy (Burnell and Wolf, 1982). The broadening and shift of N-phonon features at small $d_N \leq 70$ Å, and thought to be consequences of coherent strains resulting from epitaxy, are complicating factors in the iterative analysis for N and S α^2F functions referred to above and first carried out by Wolf, Zasadzinski, Osmun, and Arnold (1979). Typically, the best correcting Δ_N function required in analysis of the small-d_N junction for Δ_S is broader in its phonon features than that obtained by analysis of the larger-d_N samples. One means of alleviating this problem is to numerically broaden the $\alpha^2F(E)$ deduced from the larger-d_N sample before generating from it the $\Delta_N(E)$ induced in the thinner-N sample (Zasadzinski, Burnell, Wolf, and Arnold 1982).

Clear confirmation of the accuracy which can be obtained through the PETS methods is given in Fig. 5.24, which directly compares PETS and conventionally obtained $\alpha^2F(E)$ functions for Nb. The PETS correction analysis applied to obtain the dashed curve (Arnold, Zasadzinski, Osmun, and Wolf, 1980) changed the λ value from 0.9 to 1.04; the latter value is very close to the value 1.03 obtained directly from the fully oxidized Al sample (solid curve) and very close also to values recently obtained by the Bell Laboratories group (see Hertel, McWhan, and Rowell, 1982).

The sampling depth in conventional electron tunneling can be expressed as an energy-dependent length

$$L_S(E) = \frac{\hbar v_{FS}}{2\,|Z_S(E)|\,E} \tag{5.66a}$$

$$= \left\{ \left[\frac{2E}{\hbar v_{FS}} \operatorname{Re} Z_S(E) \right]^{-2} + \left(\frac{l l_{ph}}{l + l_{ph}} \right)^2 \right\}^{1/2} \tag{5.66b}$$

Fig. 5.23. (a) Directly measured d^2V/dI^2 spectra for In–Al_2O_3–AlVa junction, $d_N = 70$ Å, with a 30.0-mT field to drive the In normal. The phonon features from V and Al are indicated. (b) Solid curves shows Eliashberg function $\alpha^2F(E)$ for a 70-Å Al film on V foil, from (a). The dashed line is the bulk Al $\alpha^2F(E)$ of Leung, Carbotte, and Leavens (1976). (After Zasadzinski, Burnell, Wolf, and Arnold, 1982.)

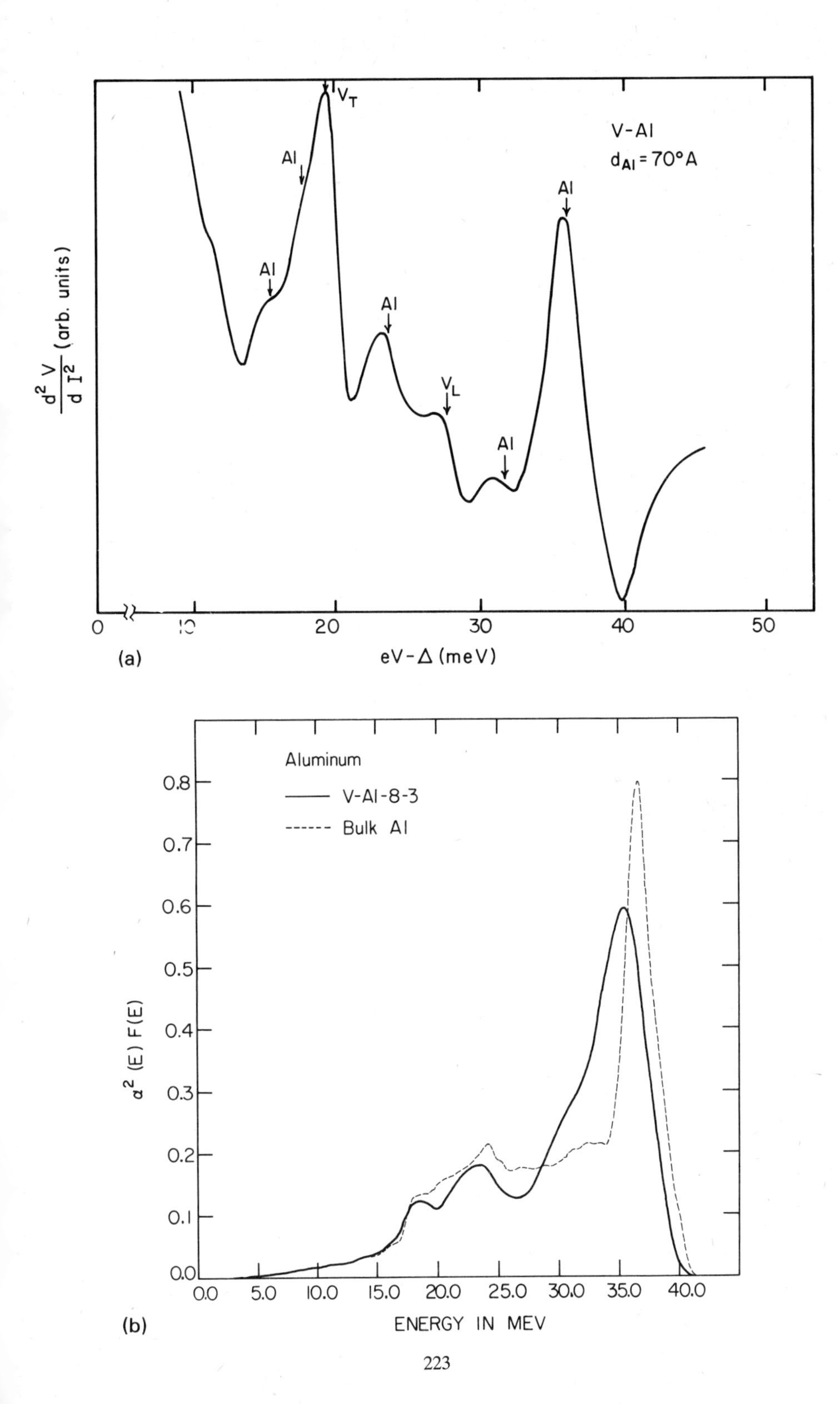

V-Al
d_{Al} = 70°A
V_T
V_L
Al
$\frac{d^2 V}{d I^2}$ (arb. units)
0
10
20
30
40
50
(a)
eV-Δ (meV)
Aluminum
V-Al-8-3
Bulk Al
0.8
0.7
0.6
0.5
0.4
0.3
0.2
0.1
0.0
α² (E) F(E)
0.0
5.0
10.0
15.0
20.0
25.0
30.0
35.0
40.0
(b)
ENERGY IN MEV

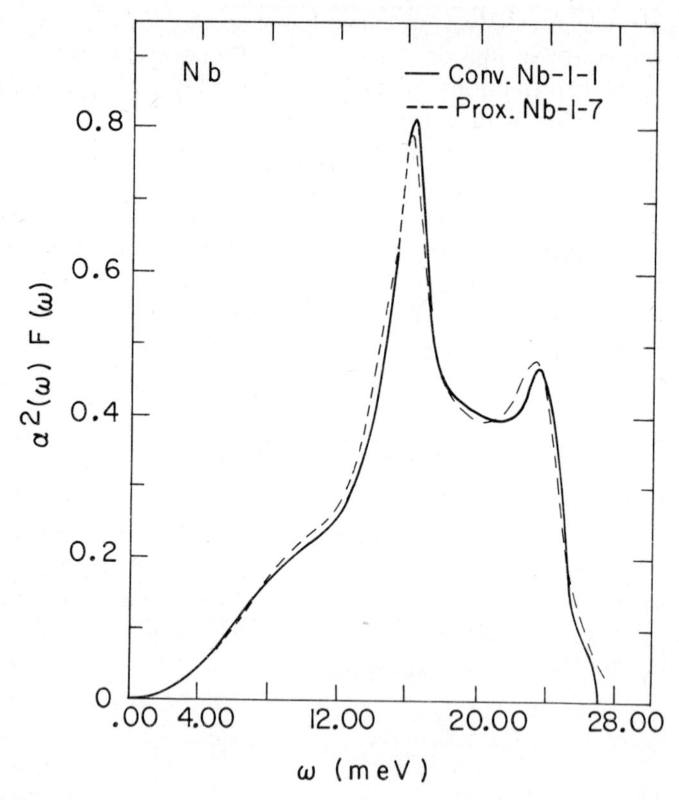

Fig. 5.24. Comparison of the Nb $\alpha^2F(\omega)$ function obtained from a $d_N = 0$, fully oxidized, AlNb "conventional" tunnel junction (solid curve) with that obtained from proximity junctions of the form In–AlO_x–AlNb, using the analysis of Arnold, Zasadzinski, Osmun, and Wolf (1980) (dashed line). Inferred values of λ are 1.03 (solid curve) and 1.04 (dashed). (After Khim, Burnell, and Wolf, 1981.)

where

$$l_{ph} = \frac{\hbar v_F}{2\,|\mathrm{Im}(ZE)|} \tag{5.67}$$

and l is the mean free path caused by impurities (McMillan and Rowell, 1969; Arnold and Wolf, 1982). This length is an energy-dependent coherence length into the electrode S of the C–I–S juntion and is further decreased by contributions to Im Z by scattering produced by phonons (l_{ph}) or by residual impurities etc. (l). The corresponding sampling depth into S for the C–I–NS proximity juntion has been recently considered. Arnold and Wolf (1982) find, by consideration of (5.39) and (5.40), that

$$\begin{aligned} L_{prox}(E) &= \frac{\hbar v_{FN}}{2\,|Z_N(E)E|} \\ &= (|\Delta K_N|)^{-1} \end{aligned} \tag{5.68}$$

is the result, which can also be expanded as in (5.66b). This expression follows from consideration of (5.40) as a Fourier transform of $\Delta(x)$, the range in x of contributions being limited by $|\Delta K|^{-1}$. The remarkable feature that emerges is that

$$\frac{L_{\text{prox}}}{L_{\text{S}}} \simeq \frac{|Z_{\text{S}}(E)|\, v_{\text{FN}}}{|Z_{\text{N}}(E)|\, v_{\text{FS}}} \tag{5.69}$$

typically exceeds unity, because of the characteristically smaller d-band Fermi velocities v_{FS} in transition metals. For Nb at the 24-mV phonon peak, this increased probing depth is found to be a factor ~6. For Nb_3Sn, an even greater advantage should be realized by use of an Al proximity layer.

5.7 Effects of Elastic Scattering in the N Layer

The treatment of quasiparticle scattering of mean free path l in the N layer has usually been limited to addition of a small imaginary part to the wavevector $k \to k \pm i/l$ (Wolfram, 1968), leading to a 2d decay factor $\exp(-2d/l)$. A more rigorous approach has been presented by Arnold (1981), who includes an elastic s-wave scattering lifetime τ in the thin-N layer in Z_{N} via the expression

$$Z_{\text{N}}(E) = Z_{\text{N}}^{\text{ph}}(E) - i\frac{\hbar}{2\tau}\frac{1}{[E^2 - \Delta_{\text{S}}^2(E)]^{1/2}} \tag{5.70}$$

Here, Z_{N}^{ph} is the phonon contribution, and the density of final states $N(E) = N_{\text{S}}(E)$ for scattering in N is taken as equal to that in S, appealing to experimental results, such as the BCS-like data of Fig. 5.20a and the expected similarity of the bound-state spectrum to the BCS spectrum for small d, as indicated in Fig. 5.12. Equation (5.70) is combined with a generalization of (5.45) for the pairing self-energy,

$$\Phi_{\text{N}}(E) = \Phi_{\text{N}}^{\text{ph}} + \int_0^\infty p_{\text{S}}(E')K_+(E, E')_{\text{N}}\, dE' \tag{5.71}$$

where $p_{\text{S}}(E)$ (4.11) is the pair density in S. Using a simplification for elastic scattering;

$$K_+(E, E')_{\text{N}} = \frac{i\hbar}{2\tau}\,\delta(E - E')$$

Arnold (1981) obtained a new result for $\Delta_{\text{N}}(E)$ including elastic scattering:

$$\Delta_{\text{N}}(E) = \frac{\Phi_{\text{N}}(E)}{Z_{\text{N}}(E)} = \frac{Z_{\text{N}}^{\text{ph}}\Delta_{\text{N}}^{\text{ph}} + i\Delta_{\text{S}}(\hbar/2\tau)N_{\text{S}}(E)}{Z_{\text{N}}^{\text{ph}}(E)E + i(\hbar/2\tau)N_{\text{S}}(E)} \tag{5.72}$$

In the extreme limit of short τ and large $N_S(E)$ (E in the gap region), (5.72) predicts $\Delta_N(E) \simeq \Delta_S(E)$: the pair potential in N can be strongly enhanced by elastic scattering. This expression, (5.72), is mathematically equivalent to the McMillan tunneling model expression (5.56), taking $\hbar/2\tau = \Gamma_N$, but is obtained from different physics and applied to a different case (strongly coupled N and S, semi-infinite S). Arnold (1981) emphasizes that the corresponding tunneling density of states seen from the N side differs substantially from that predicted by the McMillan model, because the specular expression indicated by (5.45) must be evaluated by using (5.72).

Some numerical results in this regime, with d/l now representing the elastic scattering strength incorporated via (5.72), are shown in Figs. 5.25 and 5.26. The enhancement of the low-energy region of $\Delta_N(E)$ via (5.72) is evident in the low-energy depression of the normalized conductance curves of Fig. 5.25. Effects similar to this one were reported by Bermon and So (1978b) in a study of NS structures with quench-condensed N layers, in which one would expect considerable elastic scattering. Gap region spectra similar to the calculated curves of Fig. 5.26 were reported by Freake (1971) and are shown in Fig. 5.27. The incorporation of Arnold's elastic scattering theory as outlined above has been shown by Shih, Khim, Arnold, and Tomasch (1981) to lead to an adequate treatment (in Fig. 5.28) of the Ag/Pb data of Khim and Tomasch (1979)

Fig. 5.25. Tunneling density of states calculated on the basis of elastic scattering theory, for $R = 0.02$, $\Delta_N^{ph} = 0$, and $Z_N^{ph} = 1$. The solid line for the curve, which is smallest in value from 10 to 16 mV, is for $d/\ell = 0$. The dashed line is for $d/\ell = 0.25$. The dash-dotted line is for $d/\ell = 1$. The second solid line is for $d/\ell = 5$. (After Arnold, 1981.)

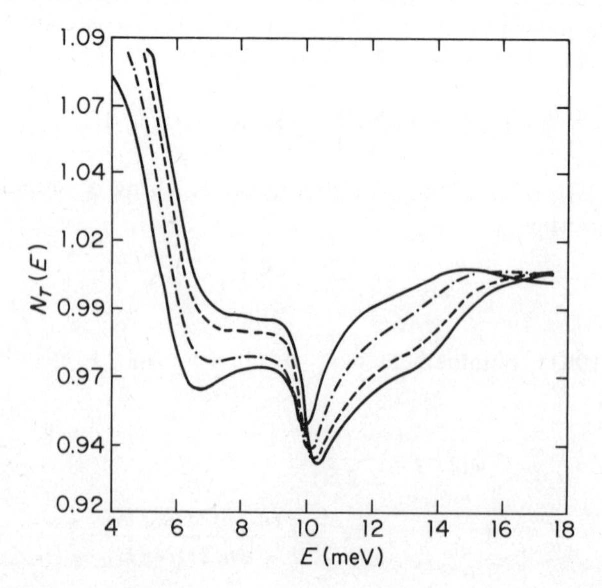

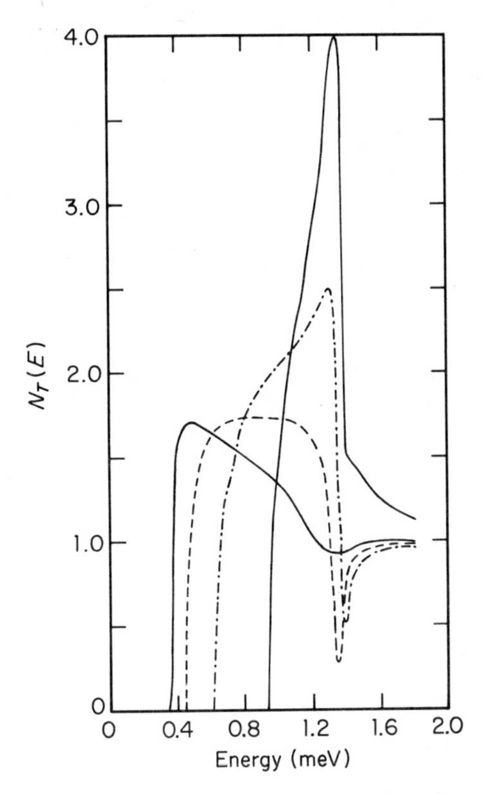

Fig. 5.26. Family of theoretical gap region spectra based on $N_T(E)$ of Arnold (1981), including elastic scattering in N region, assuming $d/\ell = 3$. Parameter values are $\Delta_S = 1.4$ meV, $\Delta_N^{ph} = 0.3$ meV, and $Z_N^{ph} = 1.2$. The R-parameter (thickness) values for the curves, in order of increasing energy of $N_T = 0$ intercepts, are 1, 0.5, 0.25, and 0.1 meV^{-1}. The general features of the spectra of Fig. 5.27 are reproduced by these curves. (After Arnold, 1981.)

without having to make unreasonable assumptions on the renormalization function in Ag backed by Pb.

5.8 Proximity Corrections to Conventional Results

One at the most perplexing characteristic results of the conventional inversion of tunneling data from transition metals (Shen, 1972*a*; Bostock et al. 1976) was a noticeable offset of the best calculated conductance above the experimental curve (see Fig. 5.29, where the effect is illustrated with recent Nb_3Al tunneling data) usually with an unreasonably small value of the coupling parameter λ. The Coulomb pseudopotential μ^* in such cases sometimes took unphysical negative values. While speculations

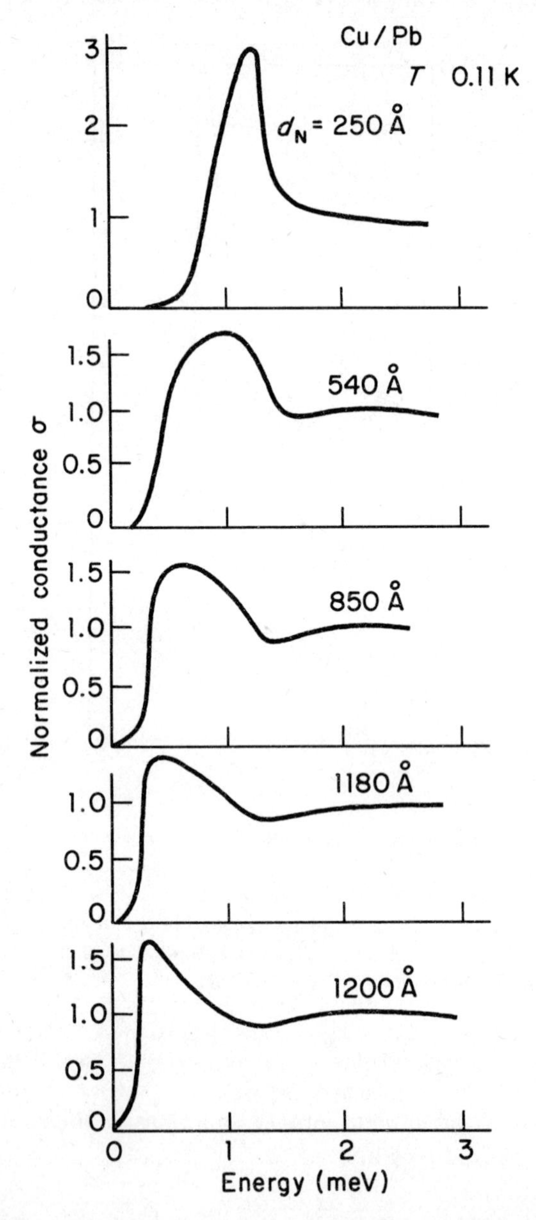

Fig. 5.27. Gap region spectra measured on Mg–MgO–Cu/Pb junctions of varying copper thickness d_N. The Pb film thicknesses are 7000 Å, and the measurements were made below 0.11 K. (After Freake, 1971.)

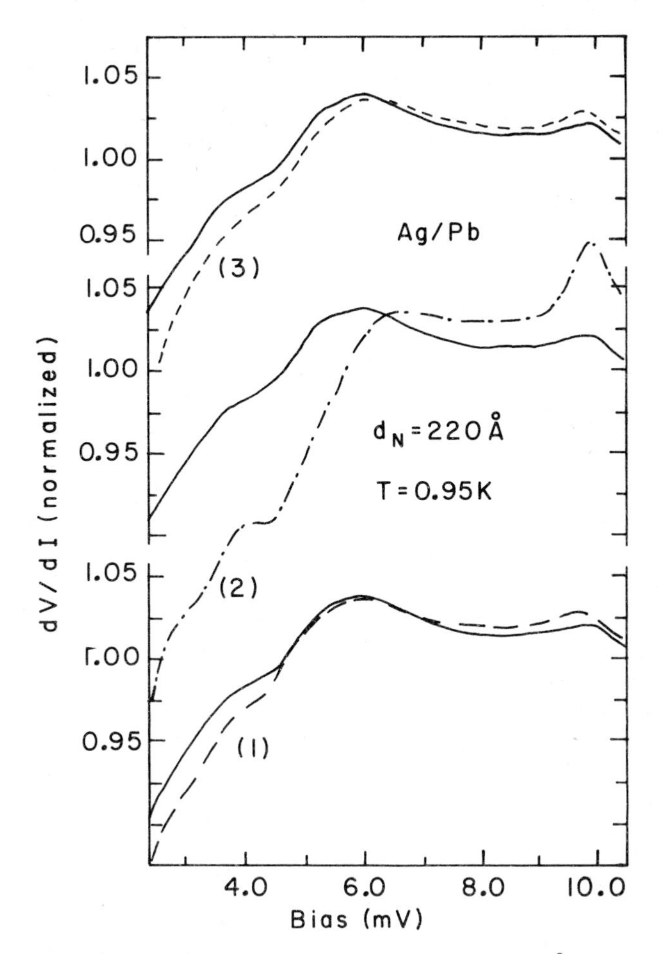

Fig. 5.28. The dV/dI spectrum (solid curve) of a $d_N = 220$ Å Al–Al_2O_3–AgPb junction (Khim and Tomasch, 1979) reanalyzed by Shih, Khim, Arnold, and Tomasch (1981), incorporating elastic scattering (Arnold, 1981) in curve 1, compound resonance theory (Gallagher, 1980) in curve 2, and both of these in curve 3. The calculations are carried out with a conventional renormalization function for Ag. (After Shih, Khim, Arnold, and Tomasch, 1981.)

of additional mechanisms of superconductivity in Nb, in particular, were advanced, a less profound origin of these effects was proposed by Arnold, Zasadzinski, and Wolf (1978) as an artifact of an unrecognized "normal metal" layer caused, e.g., by contamination of the layers of the tunneling electrode S immediately facing the insulator. It was shown by the latter authors that reanalysis of such data (Fig. 5.29) on a simplified proximity density of states (5.45), taking the expedient of setting $\Delta_N \simeq 0$ and adjusting the scattering parameter d/l, led easily to improvement in the fit and to conventional values of parameters. A careful discussion of the

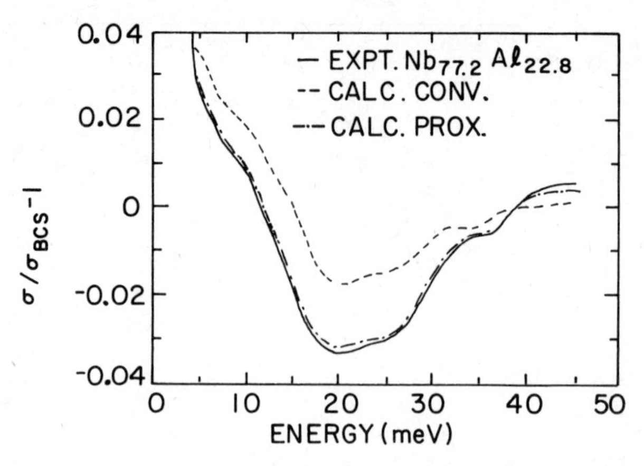

Fig. 5.29. The experimental and calculated tunneling densities of states from both conventional (dashes) and proximity inversion (dot-dashed) for an A–15 Nb_3Al junction of 22.8% Al, $T_c = 16.4$ K, and $\Delta = 3.15$ meV. The noticeable offset in conventional analysis (dashed) is reduced by incorporation of proximity corrections. (After Kwo and Geballe, 1981.)

offset caused by conventional analysis of proximity data containing appreciable scattering d/l is given by Arnold, Zasadzinski, Osmun, and Wolf (1980). Some typical physical origins of surface pair-potential depression and quasiparticle scattering at the surfaces of transition metals are discussed by Shen (1972*a*) and by Wolf, Zasadzinski, Osmun, and Arnold (1980), and in A–15 compounds by Wolf et al. (1980).

5.9 Further Applications of Proximity Effect Models

We have already mentioned several of the many applications of McMillan's tunneling model of the proximity bilayer to physically interesting situations. Here, we consider briefly two further situations.

The first is the analysis of data from Pb–Ir–Pb (SNS) junctions to obtain the BCS parameter $(NV)_N$ for Ir. The resistance-temperature plots of two such junctions (Clarke, Freake, Rappaport, and Thorp. 1972) are shown in Fig. 5.30, where the inset sketches the expected variation of the pair potential Δ at the NS interface and into N for the cases $V_N = 0$ and $V_N > 0$. Recall our discussion following Eq. (5.3) and note that the bulk T_c of Ir is 57 mK; then the variation $\Delta_N(x)$ (x measured into the N region, of width $2a$, from the Pb–Ir interface) should be given by

$$\Delta_N(x) = \Delta_S^0 \frac{(NV)_N}{(NV)_S} \exp(-k_N x) \tag{5.73}$$

$$K_N^{-1} = \left\{ \frac{\hbar v_{FN} l}{6[\pi k(T - T_{CN})]} \right\}^{1/2} \tag{5.74}$$

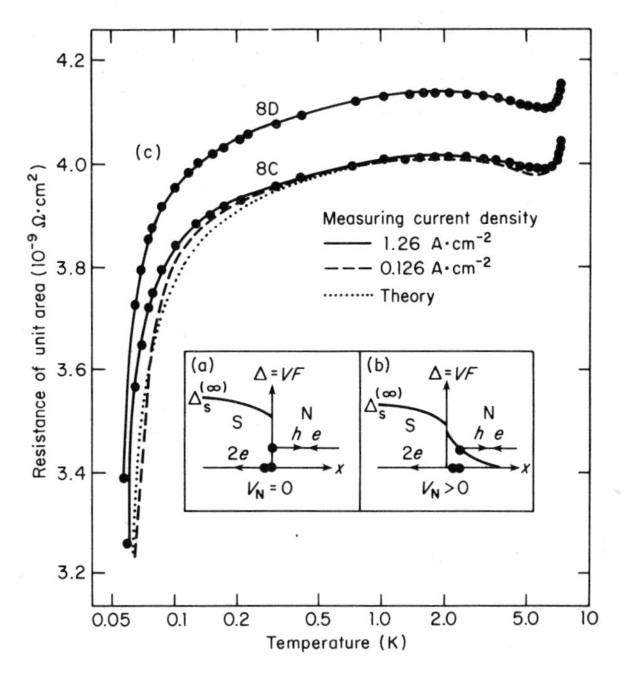

Fig. 5.30. (a) Variation of pair potential Δ across SN interface for $V_N = 0$. Incident electron is reflected at the SN interface. (b) Variation of pair potential Δ across SN interface for $V_N > 0$. Incident electron is reflected inside N. (c) Variation of resistance with temperature for two Pb–Ir–Pb samples. Thickness of Ir in sample 8C was 53 ± 1 μm, and in 8D was 54 ± 1 μm. The theoretical curve is for sample 8C. (After Clarke, Freake, Rappaport, and Thorp, 1972.)

where use has been made of the de Gennes (1964*a*) boundary condition for the interface values of the pair potentials $\Delta^0_{N,S}$,

$$\frac{\Delta^0_N}{(NV)_N} = \frac{\Delta^0_S}{(NV)_S} \tag{5.75}$$

In their analysis for $T \leq T_{CS}/2$, Clarke, Freake, Rappaport, and Thorp took $\Delta^0_S \simeq \Delta^{\text{bulk}}_S$ and $\Delta^{\text{bulk}}_S/kT_{CS} = 2.2$ for (strong-coupling) Pb. The Ir may be regarded as normal at $x > x_0$, where $\Delta_N(x_0) \simeq kT$; hence, the normal thickness is $2(a - x_0)$. As the temperature is lowered toward T_{cN}, the normal width, and hence the resistance, shrinks toward zero. The resistance of the SNS junction is calculated (dotted curve in Fig. 5.30) from the normal resistance R_0 of the slab of Ir as

$$R(T) = \frac{2(a - x_0)}{2a} R_0 \tag{5.76}$$

The choice $(NV)_N = 0.11$ for Ir and reasonable experimental estimates of other parameters yield the dotted curve, calculated for the specimen marked 8C, of thickness $2a = 53 \pm 1\ \mu\text{m}$. This result is equivalent to determination of the unknown parameter $(NV)_N$ via the relation

$$(NV)_N = 2\frac{(NV)_S}{\Delta_S^{\text{bulk}}}(kT)\exp[k_N a(1 - R(T)/R_0)] \tag{5.77}$$

The authors estimate that values of $(NV)_N$ as small as 10^{-3} can be estimated by this method. Extensions of this approach consider the effect of magnetic fields and magnetic impurities.

In the experiment of Clarke, Freake, Rappaport, and Thorp, the voltage drop across the N-metal region is essentially zero, with carrier energies of order kT and Andreev reflection occurring at x_0, where $\Delta(x_0) \simeq kT$.

A similar SNS model in a much different regime of bias for the microshort or microconstriction has been discussed by Klapwijk, Blonder, and Tinkham (1982) (see also Blonder, Tinkham, and Klapwijk, 1982), who find an explanation of the long-standing puzzle of subharmonic subgap structure and excess currents in imperfect S–I–S tunnel junctions. As sketched in Fig. 5.31, the superconducting electrodes S_1, S_2 are assumed to be in equilibrium, while the applied voltage appears across the microshort, typically less than 1000 Å in diameter, of spreading resistance R_0, which is treated in the perfectly clean limit so that carriers (electrons and holes), in crossing the short, gain the bias energy eV. The important role of Andreev scattering in the analysis is indicated by a modified "semiconductor model" plot for S_1 and S_2, in which the

Fig. 5.31. Sketch of a microshort model, incorporating barrier tunneling and Andreev scattering, to explain subharmonic structure. (After Klapwijk, Blonder, and Tinkham, 1982.)

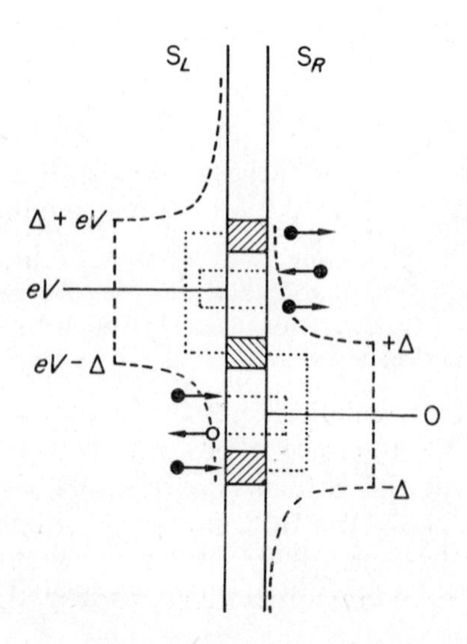

Andreev reflection coefficients are sketched (dashed lines). These coefficients are unity for $|E| \leq \Delta$ and fall off sharply for $|E| > \Delta$.

As emphasized by Klapwijk, Blonder, and Tinkham, there is a second effect closely related to the subharmonic structure (SHS), which is an excess current observed when at least *one* of the electrodes is superconducting (observation of SHS requires both electrodes to be superconducting). The excess current produces a nonzero current intercept at $V=0$ and is most easily understood in the microshort model with one electrode taken as normal. Consider the electron current flowing across a plane in the N region (short) just to the left of the NS_2 interface. This current consists of three components. First, electrons are flowing to the right, primarily in the energy range $0 < E < eV$ (no scattering in N) relative to the local chemical potential or that in S_2. This contribution is

$$\frac{1}{eR_0}\int_{-\infty}^{\infty} f_0(E-eV)\,\mathrm{d}E$$

Second, holes propagating to the right in N may be Andreev-reflected with probability $A(E)$ at the NS interface, leading to a reverse electron current

$$-\frac{1}{eR_0}\int_{-\infty}^{\infty} A(E)[1-f_0(E-eV)]\,\mathrm{d}E$$

Third, quasielectrons in S moving to the left may be transmitted by the SN interface with probability $1-A(E)$; this electron current is

$$I=\frac{V}{R_0}+\frac{1}{eR_0}\int_{-\infty}^{\infty} \mathrm{d}E[f_0(E-eV)-f_0(E)]A(E)$$

where the excess current represented by the integral term, resulting from the extra charge transmitted by the Andreev reflection process, is V/R_0 in the $T=0$ limit, just doubling the total current. For $eV/kT \ll 1$, the excess current is

$$\frac{4\Delta}{3eR_0}\tanh\frac{eV}{2kT}$$

while for large voltages $(eV-\Delta) \gg kT$, the excess current saturates to $4\Delta/3eR_0$ at all T. These results were obtained earlier by Artemenko, Volkov, and Zaitsev (1979) and by Zaitsev (1980), using Green's function methods, and relate to earlier observations such as those of Pankove (1966).

An extension of the calculation to cover a localized δ-function tunneling barrier at the NS interface allows interpolation between N–N–S microconstrictions and N–I–S tunnel junctions, as shown in a family of $\mathrm{d}I/\mathrm{d}V$ curves in Fig. 5.32. The BCS limit of the differential conductance is recovered, as one expects.

The subharmonic structure, as described in Chap. 3 (see Figs. 3.25 and

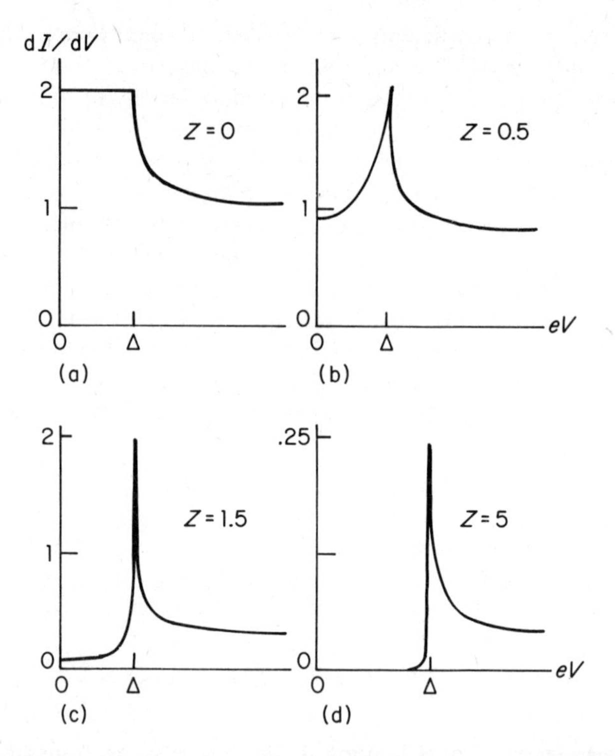

Fig. 5.32. Differential conductance versus voltage, at $T = 0$. Parameter H is the strength of the δ-function barrier, Z is a dimensionless barrier strength defined as $Z = H/(2\varepsilon_F/k_F)$, and dV/dI is in units of the normal-state resistance for no barrier, i.e., $Z = 0$. Characteristic dI/dV reaches a limit of 2 at $eV = \Delta$, independent of Z. (After Klapwijk, Blonder, and Tinkham, 1982.)

5.33) occurs only in S–I–S junctions, both electrodes superconducting, as a series of features in dV/dI at voltages subharmonically related to the energy gaps. By consideration of sequences of Andreev reflection events occurring at the S_1N and NS_2 interfaces of an S_1NS_2 microshort (see Fig. 5.31), Klapwijk, Blonder, and Tinkham (1982) have been able to generate complex subgap spectra (Figs. 3.25 and 5.33) resembling those observed. In this model, a subgap feature at V_n may be associated with the threshold for an n-fold Andreev reflection process, in each step of which a carrier crosses the constriction gaining energy eV, transporting one unit of electron charge, and finally being collected in one of the electrodes. An idea of the processes considered may be obtained by considering (Fig. 5.31) the upper electron (filled circle) leaving S_1 traversing N (dots) in the unshaded range of energies (for which a two-step process is likely), Andreev-reflecting with unit probability at the NS_2 interface, and exiting into S_1. A three-step process occurs for the lower

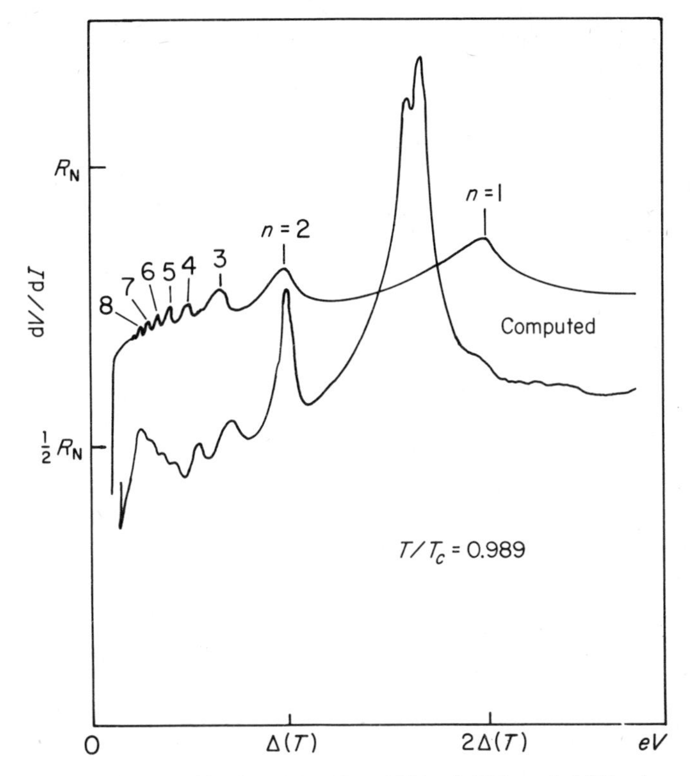

Fig. 5.33. Calculated dV/dI vs. eV, using $T/T_c = 0.989$ and $\Delta(T)$ given by the usual BCS expression, on the microshort model. The data shown below, from Gregers-Hansen, Hendricks, Levinsen, and Pickett (1973), $T/T_c = 0.989$, exhibit the shape exaggeration and voltage shifts characteristic of heating. An offset zero and arbitrary resistance units have been used for presenting the experimental data. (After Klapwijk, Blonder, and Tinkham, 1982.)

electron indicated as leaving S_1 and entering N in the dashed energy range, for which unit-probability reflection events occur in succession in S_2 and S_1, with the electron emerging (with high probability) in S_2 after crossing N the third time. Increasing the bias voltage may allow the particle to gain energy 2Δ (symmetric case) with fewer reflections: whenever eV passes through $2\Delta/n$, the $(n+1)$-fold reflection process disappears and the $(n-1)$-fold process opens up.

For an extensive review of other aspects of superconducting weak links, the reader is referred to the recent paper of Likharev (1979). The measurement and analysis of critical currents of SNS structures to probe the effects of magnetic field and magnetic impurities in N is discussed by Hsiang and Finnemore (1980) and references therein.

6

Superconducting Phonon Spectra and $\alpha^2F(\omega)$

6.1 Introduction

Methods for obtaining and analyzing tunneling data for the Eliashberg function $\alpha^2F(\omega)$ and the related quantities $\Delta(E)$, $Z_S(E)$, $Z_N(E)$, Δ_0, λ, and μ^* have been described, and we now turn to a summary of the results obtained for the major families of superconducting elements, alloys, and compounds. The systematics of superconductivity, particularly those aspects made apparent from tunneling studies, will also be considered briefly, as will the effects of several external parameters. In organizing the material, a distinction has been drawn between crystalline alloys and the truly amorphous metals, in which the pronounced softening of the spectra seems to require electronic $\alpha^2F(\omega)$ modifications, whereas in the crystalline alloys, the changes in $\alpha^2F(\omega)$ can be mainly accounted for in the phonon spectrum $F(\omega)$. We have also incorporated sections on extreme weak-coupling metals (e.g., Mg) and local-mode superconductors (e.g., Pd:H), even though there are relatively few examples at present. This material has been included since these groups fit the conceptual framework quite well and one expects further results in these areas to be forthcoming. Several interesting inhomogeneous systems, such as superconducting multilayers and lower-dimensional and organic metals, for which no complete tunneling phonon studies have yet appeared have been treated separately as unusual materials in Chap. 9. Tabulations of parameters are given in Appendix C.

6.2 s–p Band Elements

Metals in this class, having s and p valence electrons, are a good starting point, because they present fewer experimental problems in fabricating tunnel junctions to provide definitive data (see Fig. 6.1) and are also easier to handle from the point of view of calculations because of their closer approximation than transition metals to nearly free-electron character. A few examples of such metals are Hg, Sn, In, Tl, and Pb, for which tunneling $\alpha^2F(\omega)$ functions are plotted in Fig. 6.1 and Figs. 6.3–6.6. Figure 6.2 shows the fit to the experimental reduced conductance obtained from the variational calculation leading to the $\alpha^2F(\omega)$ function of Fig. 6.1. This function, obtained by Hubin and Ginsberg (1969) on a

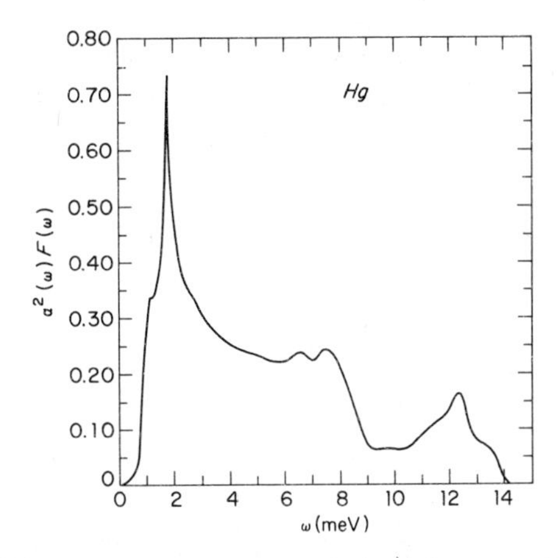

Fig. 6.1. The $\alpha^2F(\omega)$ spectrum of mercury derived from an Al–Al_2O_3–Hg junction measured at 0.35 K. Derived parameter values are $\lambda = 1.6$, $\mu^* = 0.11$, $\Delta_0 = 0.83$ meV, and $T_c = 4.19$ K. (After Hubin and Ginsberg, 1969).

junction for which the leakage current was less than 10^{-6} of the normal-state current, demonstrates the impressive overall energy resolution (experimental and numerical analysis) that can be achieved, with sufficient care, in determination of $\alpha^2F(\omega)$ by tunneling spectroscopy. Note that the cusp feature at 1.9 mV has a half-width Γ on the order of 0.1 mV, corresponding to a lifetime $\tau = \hbar/\Gamma$ on the order of 10^{-11} sec for the phonon modes involved near 1.9 meV. Such van Hove singularities arise from points in k space where $\vec{\nabla}_k(\omega)$ vanishes, since $F(\omega)$ is directly proportional to the k-space volume $\int d\vec{k}$ in a shell between constant-energy surfaces S_ω at ω and $\omega + d\omega$. Since the normal thickness dk_n of the

Fig. 6.2. Comparison of the measured and calculated reduced conductance for the Hg junction shown in Fig. 6.1. (After Hubin and Ginsberg, 1969.)

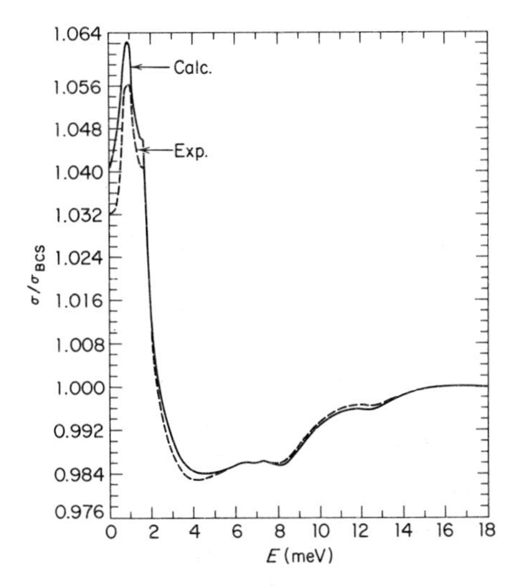

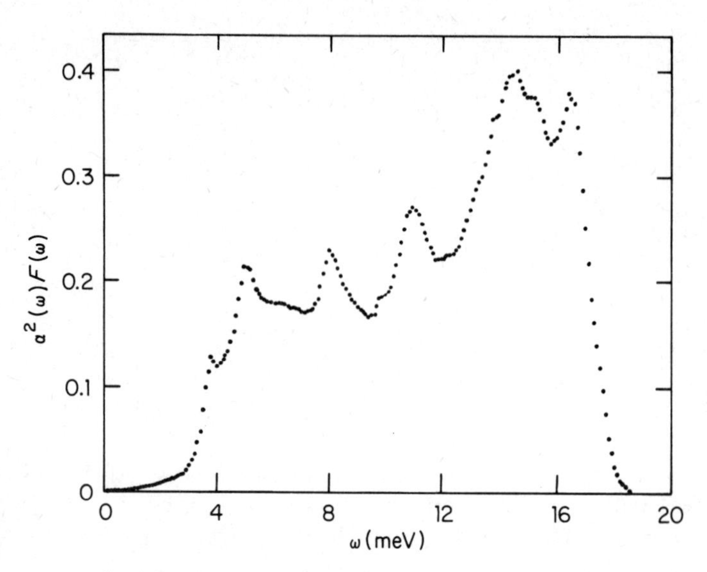

Fig. 6.3. The $\alpha^2F(\omega)$ for Sn measured from an Sn–SnO_2–Sn junction at 1 K. Parameters are $\lambda = 0.72$, $\mu^* = 0.11$, and $\Delta_0 = 0.606$ meV. (After McMillan and Rowell, 1969; Rowell, McMillan, and Feldmann, 1969.)

Fig. 6.4. The $\alpha^2F(\omega)$ of In determined from an Al–Al_2O_3–In junction at 0.35 K. Parameters are $\lambda = 0.805$, $\mu^* = 0.121$, and $\Delta_0 = 0.541$ meV. (After Dynes, 1970.)

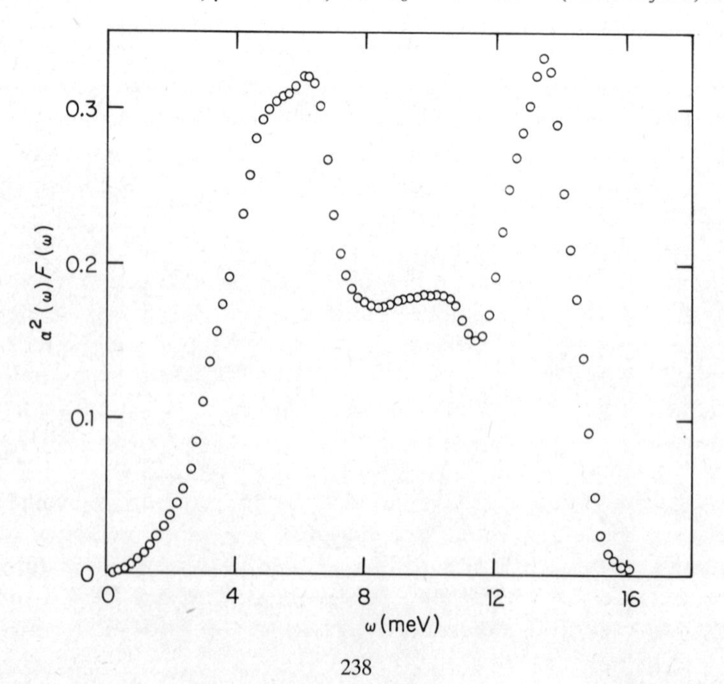

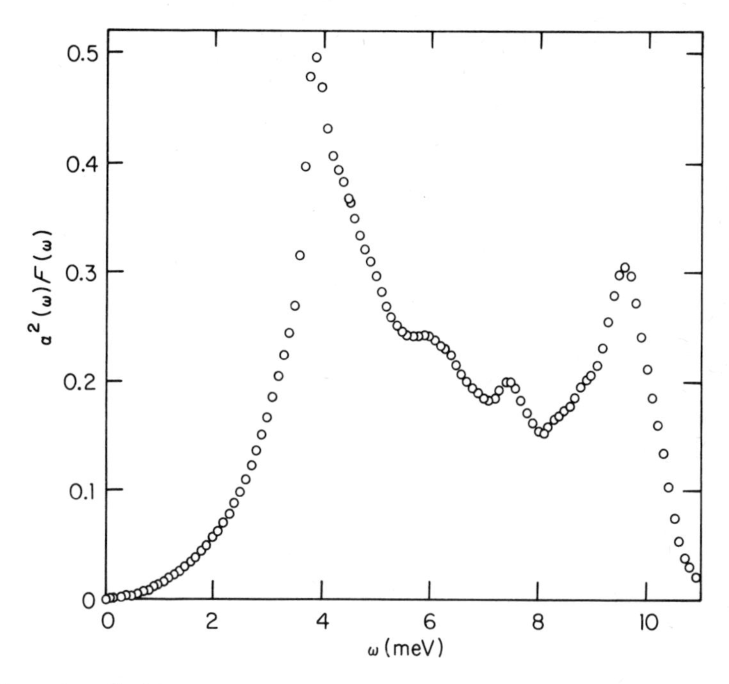

Fig. 6.5. The $\alpha^2F(\omega)$ of Tl measured from an Al–Al_2O_3–Tl junction at 0.35 K. Parameters are $\lambda = 0.795$, $\mu^* = 0.135$, and $\Delta_0 = 0.366$ meV. (After Dynes, 1970.)

shell is given by the relation $|\vec{\nabla}_k(\omega)|\,dk_n = d\omega$, one has, with $d^2k_t = dS$,

$$F(\omega) \propto \int_{S_\omega} \frac{dS}{|\vec{\nabla}_k(\omega)|}$$

which has van Hove singularities at critical points of $\omega(\vec{k})$, points where all the derivatives in the gradient vanish. These considerations have been carefully discussed in connection with tunneling $\alpha^2F(\omega)$ measurements by Scalapino and Anderson (1964).

The case of trigonal Hg also demonstrates the wealth of information that in principle is available in the tunneling $\alpha^2F(\omega)$ function, even when averaged over many crystalline directions in a polycrystalline film. In particular, the unusual low-energy singularity (Fig. 6.1) may be related to instability of the Hg lattice, suggested by its distortion from cubic into the trigonal structure. A great deal of detail is also shown in the $\alpha^2F(\omega)$ of tetragonal Sn in Fig. 6.3, while the succeeding cases of In and Tl are of the more familiar double-peaked form.

Having established that a great deal of information is available in tunneling $\alpha^2F(\omega)$ functions, one may ask about the accuracy of the information obtained by this technique. We will consider, in turn, the degree of agreement that may be expected with the closely related phonon density $F(\omega)$, determined by inelastic neutron scattering, and

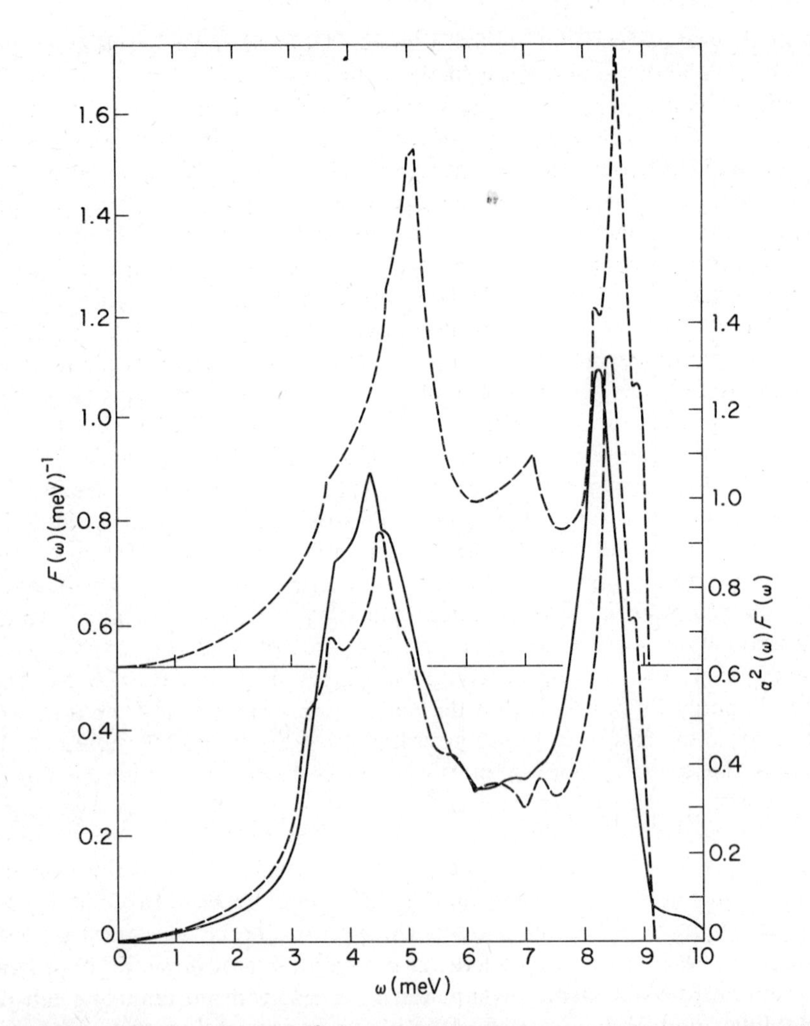

Fig. 6.6. Comparison of the $\alpha^2F(\omega)$ of Pb, shown as the solid curve, with $F(\omega)$ obtained from neutron scattering experiments. The upper dashed curve was calculated (Gilat, 1965) from dispersion curves measured along symmetry directions (Brockhouse et al. 1962), and does not agree well with either the tunneling result (lower panel, solid curve) (McMillan and Rowell, 1969) or with neutron scattering results involving a larger sampling of the Brillouin zone (Stedman, Almqvist, and Nilsson, 1967). The tunneling and improved neutron results (lower panel) are seen to be in rather good agreement.

with theoretical calculations of $\alpha^2F(\omega)$, obtained from pseudopotential band calculations making use of the neutron $F(\omega)$.

The comparison between the tunneling $\alpha^2F(\omega)$ function and the neutron scattering phonon density-of-states function $F(\omega)$ is typically quite close—e.g., as shown on the lower panel of Fig. 6.6. The accumulation of such data has led to a general expectation that the electron-phonon matrix element factor $\alpha^2(\omega)$ is a smooth and slowly varying function of energy. In occasional cases, a noticeable discrepancy in peak positions of $\alpha^2F(\omega)$ and $F(\omega)$ has initially been apparent (e.g., upper panel of Fig. 6.6). The origin of such large discrepancies has usually been traced to inaccuracy in measurement or analysis for $\alpha^2F(\omega)$ or $F(\omega)$, while residual energy differences on the order of a few tenths of a millivolt (or a few percent) presumably may come from genuine differences in the physical state of the separate specimens being measured. Neutron measurements are typically performed on large single crystals at 300 K, while tunneling usually involves a thin film on a massive substrate prepared at 300 K or higher and necessarily cooled to the vicinity of 4 K. Both the temperature difference itself and differential strains that the cooling may have caused are realistic sources of small discrepancy. A further possibility, which occasionally has been identified as the source of relatively large discrepancy (e.g., the initial, nearly 20% discrepancy in Fig. 6.6 for the lower-energy Pb peak), is that the peak in $F(\omega)$ may arise from a region in k space well away from the symmetry directions actually measured. If this is the case, inaccuracy may arise in the Born–von Karman fitting procedure used to interpolate $\omega(\vec{k})$ over the whole Brillouin zone from the restricted $\vec{k}$ directions of the measurements. In the cited case of Pb in Fig. 6.6, a further neutron scattering study which extended the original measurements (Brockhouse et al., 1962, fit to obtain $F(\omega)$ by Gilat, 1965) to off-symmetry directions, was performed by Stedman, Almqvist, and Nilsson (1967). The extended data provided substantially improved agreement (dashed curve, lower panel of Fig. 6.6) with the tunneling data of McMillan and Rowell (1965, 1969). Confirmation that this technique correctly resolved the original discrepancy in the case of Pb was provided by a further neutron scattering study (Roy and Brockhouse, 1970) in which polycrystals of Pb were measured directly and the Born–von Karman analysis was not required.

We have mentioned above some sample-related limitations of the typical thin-film tunneling determination of $\alpha^2F(\omega)$. On the other hand, an important advantage of the tunneling method is that the resulting $\alpha^2F(\omega)$ function inherently samples all parts of the Brillouin zone. Tunneling determination of phonon peak energies is also intrinsically direct and free of possible distortion by fitting procedures. The peak positions can, in fact, be read directly from the chart recorder plotting d^2I/dV^2 (negative peak) versus sample bias voltage. This must be measured with a high-impedance voltmeter across separate potential leads going directly to the two films in the cryostat. From this point of view, the tunneling method is extremely straightforward and reliable.

The question of agreement between inferred $\alpha^2F(\omega)$ functions and theory deserves some discussion. In calculations of $\alpha^2F(\omega)$, experimental neutron scattering data are usually used, and the peak positions in the resulting $\alpha^2F(\omega)$ will be determined by this data. More precisely, the measured $\omega_\lambda(\vec{k})$ for several phonon branches λ, interpolated over the whole Brillouin zone by application of a Born–von Karman analysis, is usually taken as the starting point. The $\alpha^2F(\omega)$ function in an isotropic case at $T=0$, given formally by Eq. (4.14), is conventionally expressed in a pseudopotential approximation appropriate to a nearly free-electron metal (Swihart, Scalapino, and Wada, 1965; Dynes, Carbotte, Taylor, and Campbell, 1969) as

$$\alpha^2(\omega)F(\omega) = \int_{q<2k_F} d\vec{q} \sum_\lambda L_\lambda(\vec{q}) \frac{\delta[\omega-\omega_\lambda(\vec{q})]}{(2\pi)^3 N} \tag{6.1}$$

where

$$L_\lambda(\vec{q}) = \frac{1}{4}\frac{m}{M}\frac{|\vec{q}\cdot\vec{\varepsilon}_\lambda(\vec{q})|^2}{k_F q \omega_\lambda(\vec{q})}|W(q)|^2 \tag{6.2}$$

and the pseudopotential form factor $W(q)$, given by

$$W(q) = \langle(\vec{k}+\vec{q})_F|\,W\,|\vec{k}_F\rangle \tag{6.23}$$

depends only on momentum transfer $\vec{q}$ from one point of the Fermi surface to another. Here, $\vec{\varepsilon}_\lambda(\vec{q})$ is the eigenvector of the phonon mode of branch λ and wavevector $\vec{q}$; m and M are, respectively, the electron and ion masses; and N is the number of atoms per unit volume. The evaluation of the matrix element $\vec{q}\cdot\vec{\varepsilon}_\lambda$ was done, e.g., by Carbotte and Dynes (1968) see also Dynes, Carbotte, Taylor, and Campbell (1969). This dot product in (6.2) implies $\alpha^2(\omega)$ for transverse phonons to be zero, apart from Umklapp contributions.

This procedure has been applied to the interesting case of trigonal Hg by Kamitakahara, Smith, and Wakabayashi (1977), as summarized in Fig. 6.7. The agreement, both in absolute values $[\alpha^2(\omega)]$ and in peak positions, is seen to be generally good. The calculated $\alpha^2F(\omega)$ gives $\lambda=2.0$ compared with $\lambda=1.6$ from experiment. The lower panel compares the empirical $\alpha^2(\omega)$ (solid curve), which is the ratio of the experimental curves in the upper panel, with the computed $\alpha^2(\omega)$ (dots). The enhancement of $\alpha^2(\omega)$ in the low-energy range (~2 meV) comes from the factor ω^{-1} in the expression for $L_\lambda(\vec{q})$.

6.3 Crystalline s–p Band Alloy Superconductors

The consequences of alloying can be crudely separated into effects on the phonon spectrum and on the Fermi surface, particularly $N(0)$, and on the

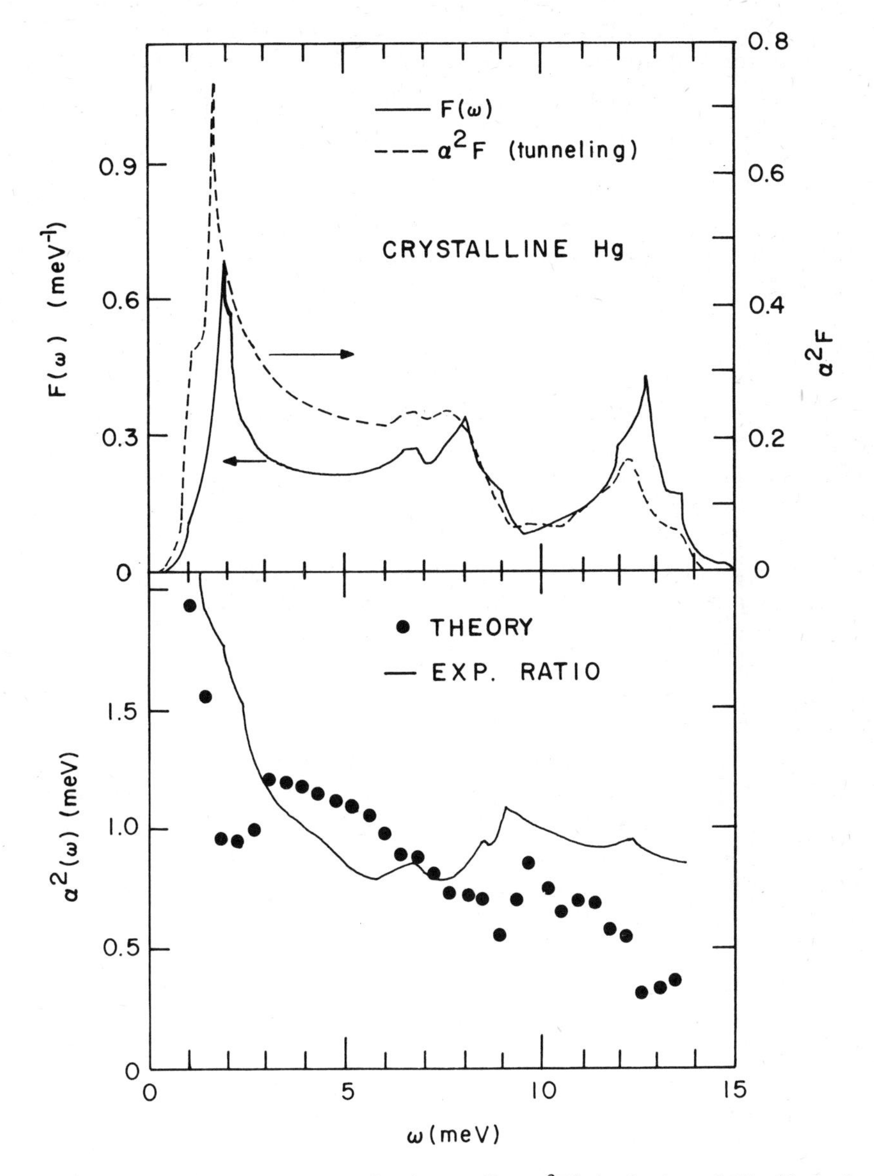

Fig. 6.7. Comparison, upper panel, of tunneling $\alpha^2F(\omega)$ of trigonal Hg (dashed curve, right scale) (Hubin and Ginsberg, 1969) and $F(\omega)$ from neutron scattering (solid curve, left scale) (Kamitakahara, Smith, and Wakabayashi, 1977). In the lower panel, the solid curve is the ratio of $\alpha^2F(\omega)$ to $F(\omega)$ from data shown above, while dots are calculations of $\alpha^2(\omega)$. A Born–von Karman model, assuming pseudopotential theory with a simple form for the electron-phonon matrix element, was used. (After Kamitakahara, Smith, and Wakabayashi, 1977.)

matrix elements entering $\alpha^2F(\omega)$. Typical effects on the phonon spectrum $F(\omega)$ are the appearance of new features, some of which we may refer to as local modes, and an overall broadening and possible shift of $F(\omega)$. As we will see later, in amorphous metals, as opposed to the crystalline alloys considered here, qualitatively different behavior occurs. Another pronounced shift to low energy (softening) typically occurs in $\alpha^2F(\omega)$, a large portion of which arises from changes in $\alpha^2(\omega)$ rather than through real changes in the vibrational spectrum $F(\omega)$.

The appearance of a sharp new feature in the $\alpha^2F(\omega)$ of a crystalline alloy of Pb with 3% addition of the lighter In atom is shown in Fig. 6.8. This separate peak in $\alpha^2F(\omega)$, centered at 9.7 meV, arises primarily from localized vibrations of lighter In atoms surrounded by complete nearest-neighbor sets of Pb atoms. Broadening into a band occurs as a consequence of In–In forces. A more detailed study of this feature in $\alpha^2F(\omega)$ of Pb:In alloys over a range of In concentrations (2, 4, 6, and 8%) is shown in Fig. 6.9. Analysis of these spectra reveals two discrete bands, each broadened, the second arising (Sood, 1972) from pairs of In impurities surrounded by Pb atoms. These spectra also reveal a progressive concentration broadening of sharp features associated with the van Hove singularities of the Pb host. This typical behavior can be regarded as a lifetime broadening associated with scattering of lattice waves from the impurities. The data for Fig. 6.9 have been discussed (Sood, 1972) in terms of the Green's function theory of Elliott and Taylor (1967).

Two types of systematic variation with alloying are of particular interest: varying the electron/atom ratio z with elements of nearly the same

Fig. 6.8. Measured $\alpha^2F(\omega)$ of PbIn alloy reveals phonon impurity band (solid curve, 9–10 meV); comparison curve (dashes) is $\alpha^2F(\omega)$ for pure lead. Concentration of indium is 0.03, and measurements are at 0.85 K in an Al–Al_2O_3–$Pb_{0.97}In_{0.03}$ junction. (After Rowell, McMillan, and Anderson, 1965.)

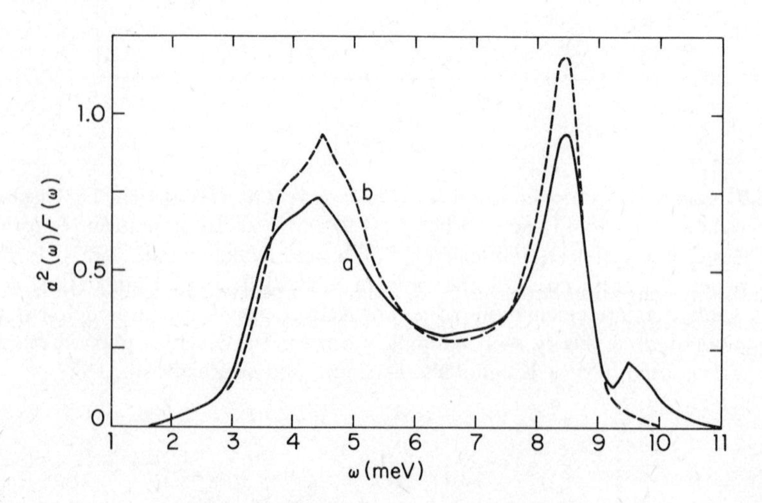

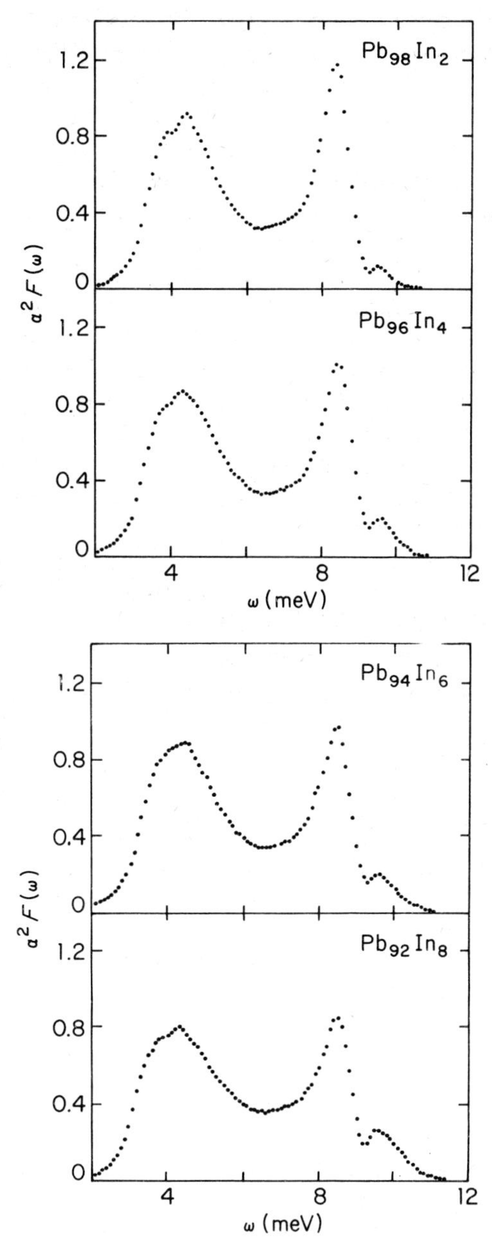

Fig. 6.9. Concentration dependence of the indium phonon impurity band in lead as determined from Al–Al_2O_3–PbIn junctions measured near 1 K. (After Sood, 1972.)

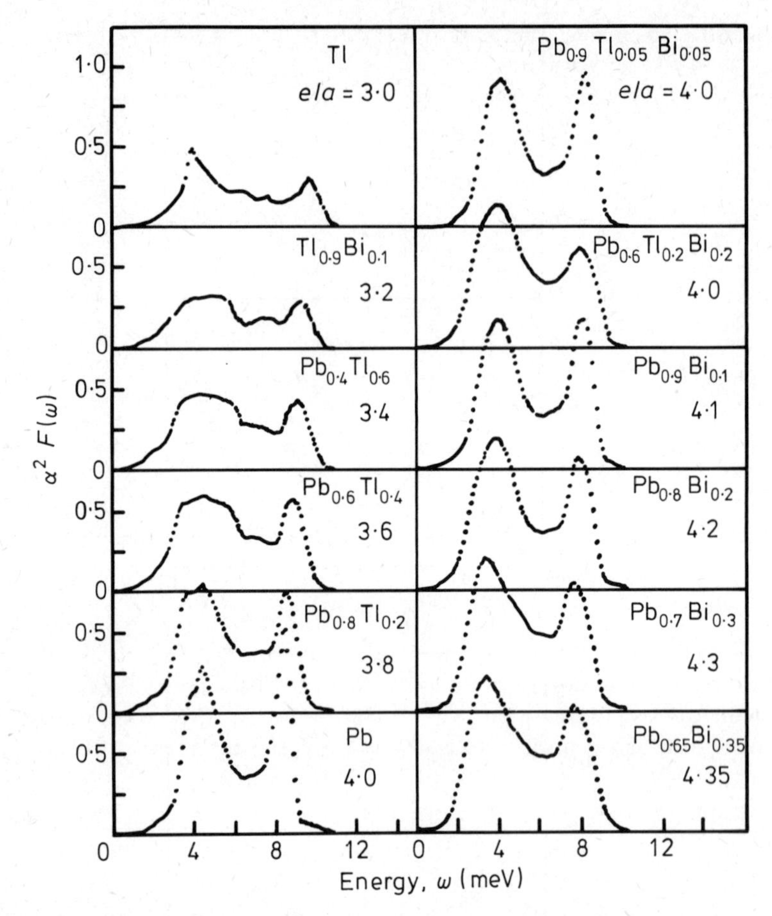

Fig. 6.10. Spectral functions $\alpha^2F(\omega)$ for the isoelectronic PbTlBi crystalline alloy system, in which the electron per atom *e*/*a* ratio increases, as indicated, in progressing from Tl (upper left) to $Pb_{0.65}Bi_{0.35}$ (lower right). (After Dynes and Rowell, 1975.)

mass, as in binary and ternary combinations of Tl, Pb, and Bi (Dynes and Rowell, 1975), and shifting the phonon spectrum by mass changes at a fixed electron/atom ratio z, as in the In–Tl system (Dynes, 1970). A sequence of $\alpha^2(\omega)F(\omega)$ functions resulting from a study of the former type is shown in Fig. 6.10, spanning the electron/atom range from 3.0 to 4.35 (see Table 6.1) and representing changes in T_C from 2.36 to 8.95 K and in λ from 0.795 to 2.13. The gross effect of increasing z is an increase in $\alpha^2(\omega)$ resulting from the increase in the rate of electron-phonon scattering possible with a larger Fermi surface. In addition, increasing the disorder in masses and in local force constants does lead to to a smearing effect on the α^2F function. This result is revealed in the comparison of the two isoelectronic members of the series, Pb and

Table 6.1. Parameters Obtained from Tunneling Measurements (Dynes and Rowell, 1975)

Alloy	z^a	$\langle\omega\rangle$ (meV)	$\langle\bar{\omega}^2\rangle$ (meV)	$\langle\omega^2\rangle$ (meV^2)	λ	Δ_0 (meV)	$2\Delta/kT_c$
Tl	3.0	5.03	6.08	30.6	0.795	0.366	3.6
$Tl_{0.9}Bi_{0.1}$	3.2	4.77	5.87	28.0	0.78	0.354	3.57
$Pb_{0.4}Tl_{0.6}$	3.4	4.79	5.89	28.2	1.15	0.805	4.06
$Pb_{0.6}Tl_{0.4}$	3.6	4.87	5.86	28.6	1.38	1.08	4.25
$Pb_{0.8}Tl_{0.2}$	3.8	4.84	5.75	27.8	1.53	1.28	4.37
Pb	4.0	5.20	5.94	30.8	1.55	1.40	4.51
$Pb_{0.9}Tl_{0.05}Bi_{0.05}$	4.0	5.00	5.74	28.7	1.56	1.42	4.58
$Pb_{0.6}Tl_{0.2}Bi_{0.2}$	4.0	4.56	5.40	24.6	1.81	1.50	4.80
$Pb_{0.9}Bi_{0.1}$	4.1	4.80	5.63	27.0	1.66	1.54	4.67
$Pb_{0.8}Bi_{0.2}$	4.2	4.44	5.36	23.8	1.88	1.61	4.70
$Pb_{0.7}Bi_{0.3}$	4.3	4.48	5.28	23.7	2.01	1.77	4.86
$Pb_{0.65}Bi_{0.35}$	4.35	4.31	5.24	22.6	2.13	1.84	4.77

[a] Here z is the electron/atom ratio.

$Pb_{0.9}Tl_{0.05}Bi_{0.05}$, both at four electrons per atom. Such smearing has also been observed directly in neutron scattering. In the case of $Pb_{0.4}Tl_{0.6}$, in particular, close agreement was found between $F(\omega)$ determined by neutron scattering and $\alpha^2(\omega)$ determined by tunneling (Rowell and Dynes, 1971).

For quantitative assessment of the changes in the $\alpha^2F(\omega)$ spectra resulting from alloying, moments of this function have been tabulated in Table 6.1. The nth moment $\langle\omega^n\rangle$ of $\alpha^2F(\omega)$ is defined as

$$\langle\omega^n\rangle \equiv \frac{2}{\lambda}\int_0^\infty \omega^n\alpha^2F(\omega)\omega^{-1}\,d\omega \tag{6.4}$$

while the average frequency is defined as $\bar{\omega} \equiv \langle\omega^2\rangle/\langle\omega\rangle$. The area under $\alpha^2F(\omega)$ is $(\lambda/2)\langle\omega\rangle$, sometimes referred to as A^2.

The In–Tl alloy data (Dynes, 1970), shown in Fig. 6.11, on the other hand, test the effect of phonon spectrum changes at a fixed electron/atom ratio z on the superconducting state. The influence of increased disorder at the center of the series on the breadth (smearing) of features in $\alpha^2F(\omega)$ is of particular interest. The relevant moments and related quantities are collected in Table 6.2 (Dynes, 1970). The evidently increasing smearing of features in the center of the series has been approximately quantified in the column labeled $\Delta\omega/\omega$ in Table 6.2.

These estimates were made by Dynes and Rowell (1969) by fitting the alloy $\alpha^2F(\omega)$ to functions generated from the pure constituent (In, Tl) $\alpha^2F(\omega)$ functions by convolution with a Lorentzian smearing function of width $\Delta\omega$. The width $\Delta\omega$ was assumed to increase linearly with ω in a given spectrum. The smearing inferred is quite severe; e.g., $\Delta\omega/\omega = 0.16$

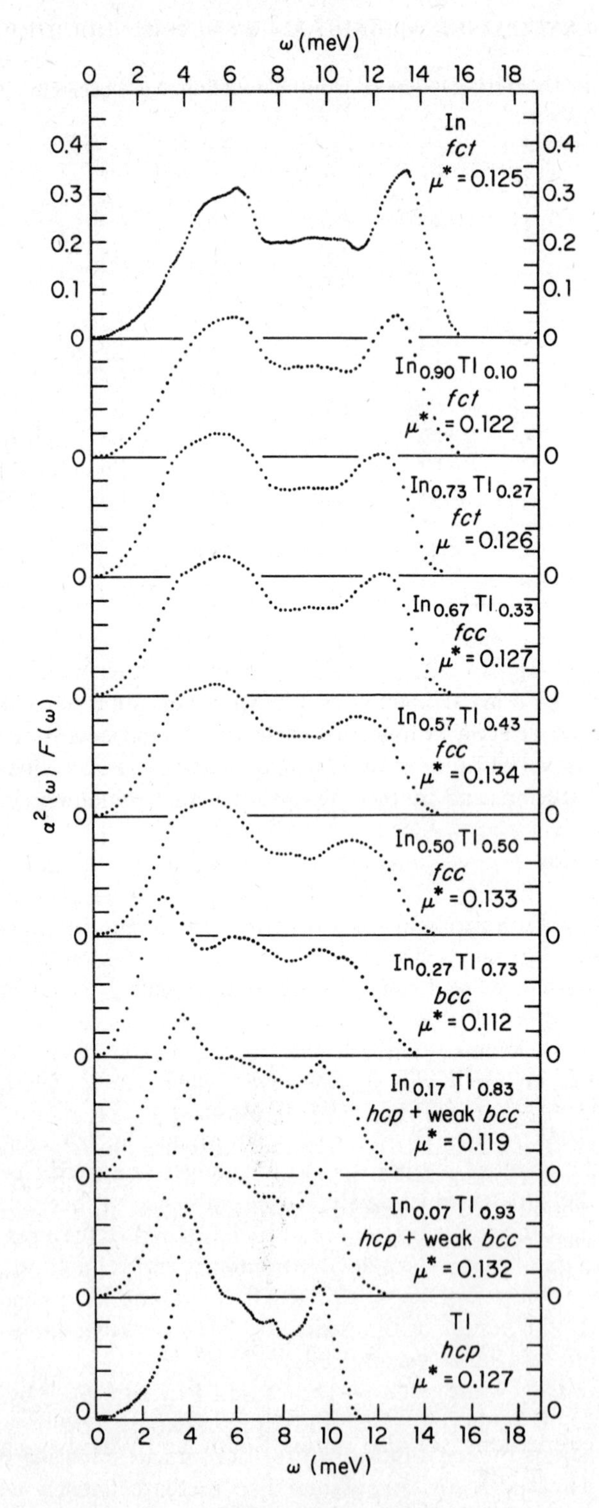

Fig. 6.11. Spectral functions $\alpha^2F(\omega)$ of InTl alloy system. (After Dynes, 1970.)

Table 6.2. Parameters Obtained from Tunneling Measurements (Dynes, 1970)

Alloy	$\langle\omega\rangle$ (meV)	$\langle\omega^2\rangle$ (meV^2)	$\Delta\omega/\omega^a$	λ	$T_{c\,\text{exp.}}$ (K)	Structure
In	8.86	61.17	0	0.834	3.40	*fct*
$In_{0.90}Tl_{0.10}$	8.41	54.30	0.026	0.850	3.28	*fct*
$In_{0.73}Tl_{0.27}$	7.67	44.19	0.077	0.933	3.36	*fct*
$In_{0.67}Tl_{0.33}$	7.81	46.00	0.08	0.899	3.26	*fcc*
$In_{0.57}Tl_{0.43}$	7.33	40.50	0.135	0.847	2.60	*fcc*
$In_{0.50}Tl_{0.50}$	7.20	39.32	0.15	0.835	2.52	*fcc*
$In_{0.27}Tl_{0.73}$	6.46	29.32	—	1.092	3.64	*bcc*
$In_{0.17}Tl_{0.83}$	6.30	29.45	0.055	0.980	3.19	*hcp* + weak *bcc*
$In_{0.07}Tl_{0.93}$	6.09	29.61	0.026	0.889	2.77	*hcp* + weak *bcc*
Tl	6.04	30.13	0	0.780	2.33	*hcp*

[a] See text.

implies coherent oscillation only over about six periods of ion motion. This magnitude of smearing had earlier been inferred (Dynes and Rowell, 1969) at the composition $Pb_{0.6}In_{0.2}Tl_{0.2}$, shown in Fig. 6.10.

Inspection of the systematics of Table 6.2 reveals that while $\Delta\omega/\omega$ rises linearly with alloyed concentration at each end of the concentration range, a more complex variation in λ and T_c occurs. The added complexity may arise in part from the changes in crystal structure which accompany the concentration variations, indicated in the table.

A further example of variation of spectra with alloying in the In–Sn system is given in Fig. 6.12. This alloy system unfortunately falls into four different crystal structures, producing discontinuities in the superconducting properties. Within each structure, however, increasing concentration of Sn (four-valent), replacing trivalent In, leads to an increase in electron/atom ratio. Within each structure, except in the final Sn phase, the T_c and λ are seen to increase with Sn concentration and e/a ratio.

The phonon energy width $\Delta\omega$ discussed above is a consequence of impurity scattering of phonons and disappears in a perfectly ordered crystal. An intrinsic source of phonon width has been shown (Allen, 1972; Allen and Cohen, 1972) to arise from electron-phonon coupling, by which a phonon may disappear ($T=0$) by generating quasiparticles. The rate of such processes and, hence, the intrinsic phonon width γ_q can be related (Allen, 1972) to the electron-phonon matrix element via the equation

$$\gamma_q = \pi \sum_k |M_{k,k+q}|^2\, \omega_q \delta(\varepsilon_k)\delta(\varepsilon_{k+q}) \tag{6.5}$$

which is valid at $T=0$. Comparison with earlier expressions for $\alpha^2F(\omega)$ leads to the expressions (see also Butler, Smith, and Wakabayashi 1977)

$$\alpha^2F(\omega) = \sum_{q,\lambda} \gamma_\lambda(\vec{q})\delta[\omega - \omega_\lambda(\vec{q})][2\pi N(0)\omega]^{-1} \tag{6.6}$$

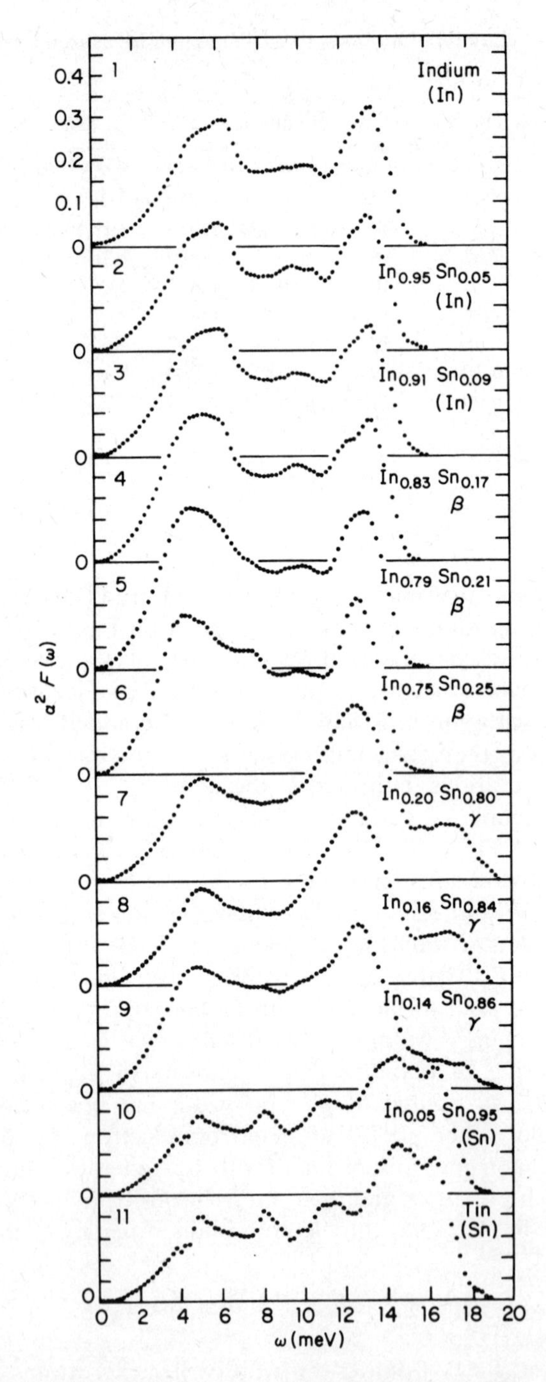

Fig. 6.12. Spectral function $\alpha^2F(\omega)$ of InSn alloy system. Note the changes in crystal structure in the panels numbered from the top; 1–3 are in the indium phase; 4–6 are δ phase; 7–9 are γ phase; and 10 and 11 are tin phase. (After Moore, Reed, and Brickwedde, 1977.)

and

$$\lambda = \sum_{q,\lambda} \gamma_\lambda(\vec{q})[\pi N(0)\omega_\lambda^2(\vec{q})]^{-1} \tag{6.7}$$

Inspection of (6.6) reveals that the factor $[\gamma_q/\pi N(0)\omega]$ inside the summation distinguishes $\alpha^2F(\omega)$ from the phonon spectrum $F(\omega)$; this factor is essentially the $L(\vec{q})$ of (6.1). Details of application of this method have been given by Allen and Dynes (1975*a*) and by Butler, Pinski, and Allen (1979).

6.4 Amorphous Metals

Qualitatively different $\alpha^2F(\omega)$ spectra are observed if the sample has been quench-condensed onto a cryogenic substrate, freezing it in a liquid-like amorphous state. The $\alpha^2F(\omega)$ curves of Fig. 6.13 (Knorr and Barth, 1970) illustrate the extreme broadening and pronounced enhancement of the low-frequency region, here produced in Pb films by condensing at 2 K (Fig. 6.13b) and by adding 10% Cu to the cryogenic deposit (Fig. 6.13c). A characteristic change brought about is the loss of the quadratic $\alpha^2F(\omega) \propto \omega^2$ behaviour at $\omega \simeq 0$ in favor of a linear or even possibly sublinear rise with ω. Because of the $1/\omega$ factor in the integrand of

$$\lambda = 2\int_0^\infty \alpha^2F(\omega)\omega^{-1}\,d\omega$$

these increases at the low-frequency end of $\alpha^2F(\omega)$ can produce a rather pronounced increase in λ. Corresponding to the curves of Fig. 6.13 (top to bottom), the λ values are 1.66, 1.91, and 2.01. In the case of amorphous Bi (Fig. 6.14), the softening of $\alpha^2F(\omega)$ leads to extremely strong coupling: $\lambda = 2.46$ and $2\Delta/kT_c = 4.59$. The T_c value, however, is a modest 6.1 K, mostly because of the small value of $\langle\omega\rangle$, which forms the prefactor to the exponential in the T_c equation: $\langle\omega\rangle_{Bi} = 2.86$ meV. Finally, a similar situation exists for amorphous $Pb_{0.45}Bi_{0.55}$ (Fig. 6.15, after Allen and Dynes, 1975*b*) for which $\lambda = 2.59$, $\mu^* = 0.12$, and $\langle\omega\rangle = 3.27$ meV.

In amorphous Sn, a similar extreme softening of $\alpha^2F(\omega)$, relative to the crystalline white tin, was observed in tunneling (Knorr and Barth, 1970). Evidence is available from the Mössbauer effect measurements of Bolz and Pobell (1975*a*, 1975*b*), however, that the true phonon spectrum $F(\omega)$ does *not* change enough in the amorphous state from its value in the crystalline state to account for the large change in $\alpha^2F(\omega)$. The Debye-Waller factor, relating to atomic motions, was found to be identical in its temperature dependence in the two phases. On the other hand, the observed isomeric shift in the Mössbauer line, as well as the Debye-Waller factor results, imply a change in the *electronic* properties of Sn, i.e., a change in $\alpha^2(\omega)$.

An alternative electronic explanation of extreme α^2F softening in the amorphous case (see the three previous figures) has been given by Poon

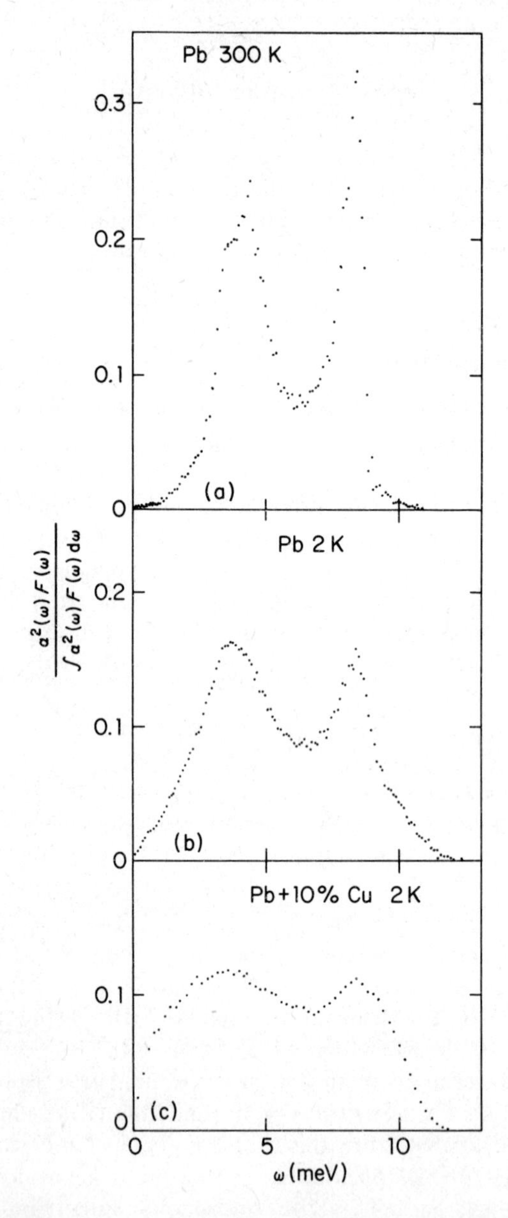

Fig. 6.13. Spectral function $\alpha^2F(\omega)$ for lead films with different degrees of disorder. (a) Lead deposited at 2 K and measured after anneal to 300 K. (b) The same film but measured before annealing. (c) Lead coevaporated with 10% Cu, deposited at 2 K, and measured before annealing. (After Knorr and Barth, 1970.)

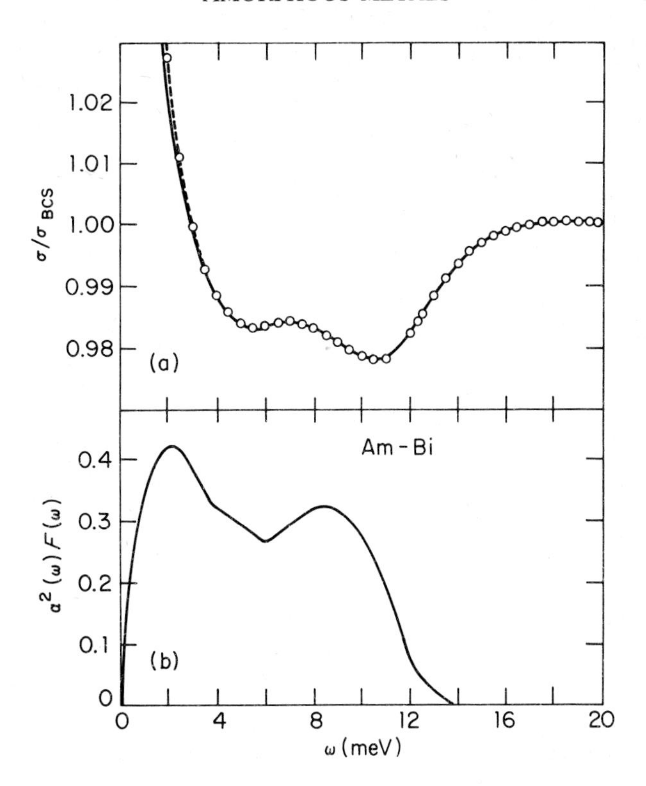

Fig. 6.14. Tunneling results for amorphous bismuth. (a) The solid line is the experimental reduced conductance; the dashed line is the calculated conductance. (b) The phonon spectrum $\alpha^2 F(\omega)$ for amorphous bismuth corresponds to $\lambda = 2.46$ and $\mu^* = 0.105$. The films were made by depositing bismuth onto a prepared Al–Al_2O_3 substrate held at 1.5 K. (After Chen, Chen, Leslie, and Smith, 1969.)

and Geballe (1978), based on the assumption that the samples have essentially no crystalline order. It is useful to expand discussion of the matrix element $g_{k,k'\lambda}$ of (4.14), closely related to the factor $L_\lambda(q)$ in (6.1) and (6.2), to see the consequence of a loss of lattice periodicity. In the rigid-ion approximation (see, e.g., Ziman, 1960, p. 182)

$$g_{kk'q} = \sum_\ell \left(\frac{\hbar}{2MN\omega_q}\right)^{1/2} \exp[i(\vec{k}' - \vec{k} + \vec{q}) \cdot \vec{r}_l](\vec{k}' - \vec{k}) \cdot \vec{\varepsilon}_q W_{k'k} \qquad (6.8)$$

where $\vec{R}_\ell$ is the ℓth atomic position and $W_{k',k}$ is the Fourier transform of the atomic pseudopotential (pseudopotential form factor), Eq. (6.3). In a perfect crystal, the sum over lattice sites ℓ in (6.8) vanishes unless $\vec{k}' - \vec{k} + \vec{q} = 0$ or $\vec{G}$, where $\vec{G}$ is a reciprocal lattice vector. In the former case, one has a normal (N) process and in the latter an Umklapp (U) process. The dependence of the Umklapp processes upon periodicity can

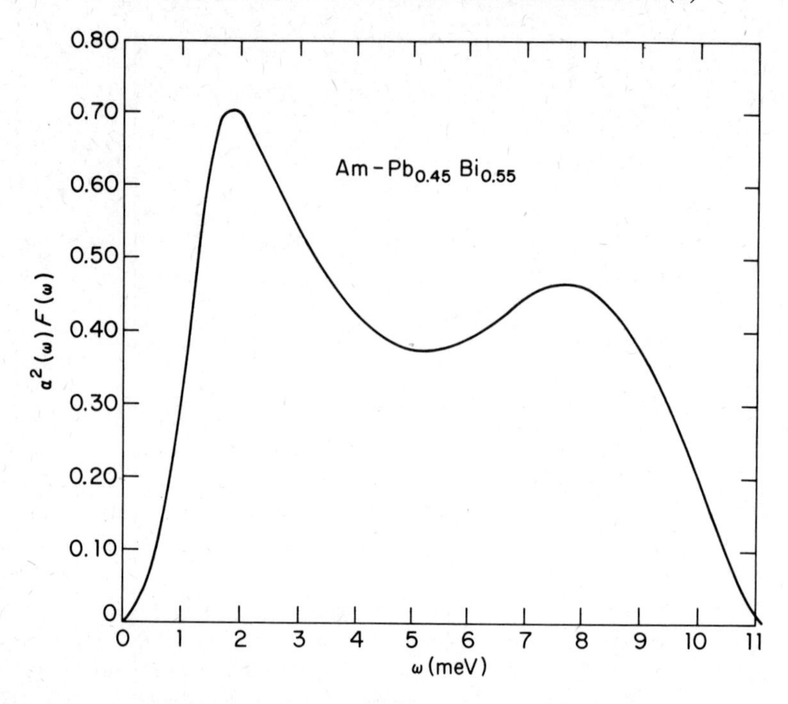

Fig. 6.15. The $\alpha^2F(\omega)$ for the amorphous alloy $Pb_{0.45}Bi_{0.55}$, quench-condensed onto a substrate held close to 4.2 K. The parameters that result are $\lambda = 2.59$ and $\mu^* = 0.116$. (After Allen and Dynes, 1975*b*.)

be quantitatively treated via the structure factor

$$A(\vec{k}) = \frac{1}{N}\sum_{\ell,\ell'} \exp[i\vec{k}\cdot(\vec{R}_\ell - \vec{R}_{\ell'})] - N\delta_{k,0} \tag{6.9}$$

This quantity can be determined from the x-ray intensity normalized by the appropriate atomic scattering factor. In a crystalline solid, $A(\vec{k})$ indeed has sharp peaks at $\vec{k} = \vec{G}$, recovering the selection rule for U contributions given above. In a disordered material, the peaks in $A(\vec{k})$ broaden. In an isotropic amorphous system, one finds

$$A(k) = 1 + \int_0^\infty 4\pi r^2[\rho(r) - \rho_0]\frac{\sin(kr)}{(kr)}\,dr \tag{6.10}$$

where ρ_0 is the mean density and $\rho(r)$ the actual density at radius r. In typical cases, the first maximum, k_0, in $A(k)$ occurs at $k_0a \simeq 7-8$, where a is the nearest-neighbor distance, with oscillations of decreasing amplitude at higher k, which reflect the loss of long-range order. Poon and Geballe (1978) find a generalized expression for $\alpha^2F(\omega)$:

$$\alpha^2F(\omega) = \sum_q \left[\frac{L(\vec{q}) + N(\vec{q})}{\omega}\right]\delta(\omega - \omega_q) \tag{6.11}$$

where

$$L(q)=\frac{\hbar}{2MN}\sum_{k,k'}A(|\vec{k}'-\vec{k}+\vec{q}|)\cdot|\vec{\varepsilon}_{\vec{q}}\cdot(\vec{k}'-\vec{k})|^2\cdot|W_{k',k}|^2 \times\delta(\varepsilon_k)\delta(\varepsilon_{k'})\left[\sum_k\delta(\varepsilon_k)\right]^{-1} \tag{6.12}$$

and

$$N(q)=\frac{\hbar}{2MN}\frac{\sum_k|\vec{\varepsilon}_{\vec{q}}\cdot\vec{q}|^2|W_q|^2\delta(\varepsilon_k)\delta(\varepsilon_{k+q})}{\sum_k\delta(\varepsilon_k)} \tag{6.13}$$

There are several comments to be made concerning this form of $\alpha^2F(\omega)$. First, clearly, only longitudinal modes contribute via normal processes in (6.13), for $\vec{\varepsilon}_q\cdot\vec{q}$ vanishes in transverse motion. Second, the Umklapp term $L(q)$ in (6.12) includes transverse-mode contributions. Third, the presence of $A(\vec{k}'-\vec{k}+\vec{q})$ in $L(q)$ in the disordered case [broadened peaks in $A(\vec{k})$] enhances the contributions at low ω (small $|\vec{q}|$). Therefore, it can be stated that nonconservation of lattice momentum, allowed by disorder, enlarges the phase space for $k\rightarrow k'$ scattering. Similar conclusions have been stated by Bergmann (1971) and by Keck and Schmid (1976).

By making use of results from the theory of liquid metals, Poon and Geballe are able to express $L(q)$ for small q in terms of the resistivity and to obtain an approximate expression for $\alpha^2F(\omega)$ valid for $\omega\simeq0$. This expression exhibits a linear, rather than quadratic, ω dependence:

$$\alpha^2F(\omega)=\left[\frac{e^2E_Fk_F^2N(0)\rho}{3\pi^3\langle v^3\rangle mM}\right]\omega\equiv A\omega. \tag{6.14}$$

Here, $N(0)$ is expressed in states/eV spin, v is the sound velocity, E_F is the Fermi energy, and ρ is the electrical resistivity. The assumption of small q such that $\omega=vq$ for the phonons has been made, and

$$\frac{1}{\langle v^3\rangle}=\frac{1}{3}\left(\frac{1}{v_l^3}+\frac{2}{v_t^3}\right) \tag{6.15}$$

where v_l and v_t, respectively, represent the longitudinal and transverse sound velocities.

Quite reasonable numerical agreement is obtained (Poon and Geballe, 1978) between theoretical and observed values of A in the linear, low-ω portion of $\alpha^2F(\omega)$. Since $F(\omega)\propto\omega^2$ at small ω, the quoted results, collected in Table 6.3, may also be described as demonstrating a region of small ω for which $\alpha^2(\omega)\propto\omega^{-1}$. Poon and Geballe (1978) (see also Poon, 1980) have found that this analysis still predicts resolved peaks in the longitudinal and transverse portions of $\alpha^2F(\omega)$. This feature, however, is apparently lost in the case of d-band amorphous transition metals, as shown in Fig. 6.16.

Table 6.3. Comparison of Theory and Experiment for the Coefficient A of Linear Region at Small ω in $\alpha^2F(\omega)$ for Amorphous Alloys (Poon and Geballe, 1978)

Amorphous Alloys	$A_{calc.}$ (eV^{-1})	$A_{exp.}$ (eV^{-1})	Reference
$Pb_{90}Cu_{10}$	320	300	*a*
$Pb_{95}Bi_5$	250	400	*b*
$Sn_{90}Cu_{10}$	200	150	*a*
Bi	1300	~1000	*c*
Tl(+Te)	115	170	*d*
$In_{80}Sb_{20}$	120	175	*e*
Ga	190	~180	*f*

[a] Knorr and Barth (1970).
[b] Ewert (1974).
[c] Krauss and Buckel (1975).
[d] Ewert and Comberg (1976).
[e] Comberg and Ewert (1976).
[f] Chen, Chen, Leslie, and Smith (1969).

These data from amorphous Nb and Mo stabilized by addition of N_2 to the quench-condensed deposits exhibit a complete *absence* of separate phonon peaks in the $\alpha^2F(\omega)$ curves (Fig. 6.16a). At the same time, there is no significant enhancement of the low-energy region of the electron-phonon coupling function $\alpha^2(\omega) = \alpha^2F(\omega)/F(\omega)$. These relatively structureless shapes of $\alpha^2F(\omega)$ are shown by Kimhi and Geballe (1980) to closely resemble $F(\omega)$ curves calculated numerically (Rehr and Alben, 1977) for a 500-atom amorphous model (Rahman, Mandell, and McTague, 1976). The calculations, which are summarized in Fig. 6.17a, assume a restricted nearest-neighbor harmonic form of the elastic energy:

$$V = \frac{1}{2} \sum_{i,j} \alpha_{ij} \, |(\vec{u}_i - \vec{u}_j) \cdot \vec{r}_{ij}|^2 \tag{6.16}$$

Here α_{ij} is the force constant between atoms i and j, $\vec{u}_i$ is the displacement of the atom from equilibrium, $\vec{r}_{ij}$ is a unit vector between atoms i and j, and the sum is restricted to nearest neighbor i and j. That such a simplified interaction energy is still quite accurate is suggested in the comparison of the two upper panels of Fig. 6.17a, comparing $F(\omega)$ calculated for Cu [from the neutron $\omega(\vec{k})$ measurements of Nicklow et al. (1967)], using (6.16) with a single force constant $\alpha = 27.9$ N/m in the middle panel, and using an accurate 18-parameter force constant model (top). When the same single force constant model (6.16) is applied to a 480-atom *fcc* sample, and the results smoothed ($\Delta\omega/\omega = 0.15$, $\Delta\omega$ is the full width at half-maximum), the lower panel reveals a recognizable approximation to $F(\omega)$ for Cu. The effects of disorder are introduced in Fig. 6.17b. Disorder in position, but a single force constant, leads to the smearing indicated in the top panel, which, however, preserves the two

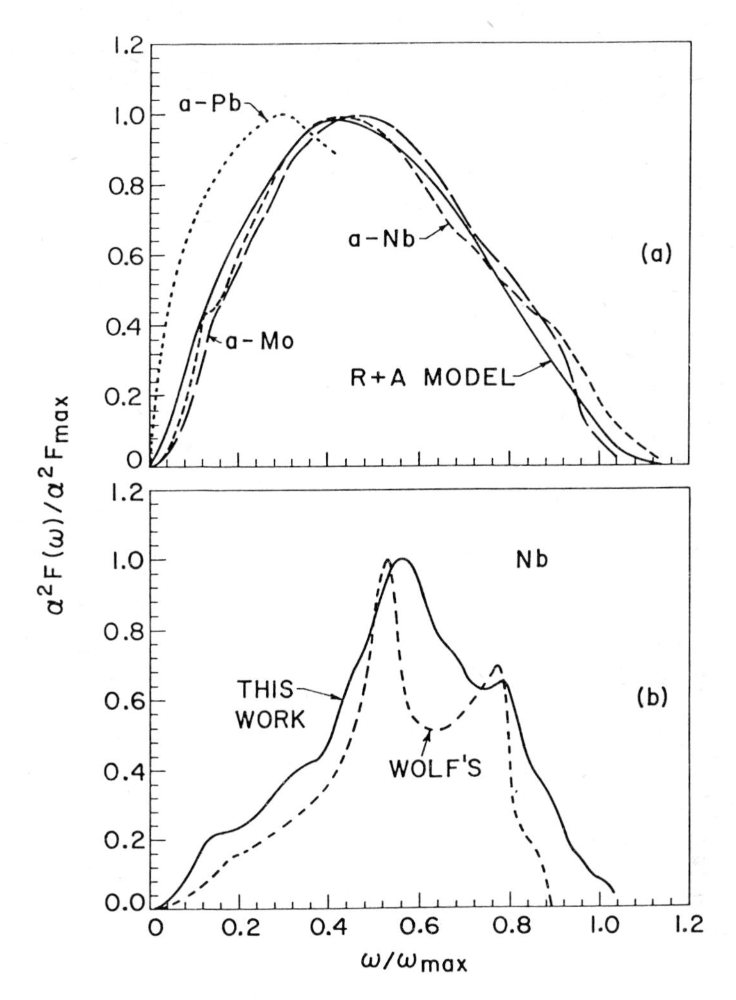

Fig. 6.16. (a) Eliashberg functions for amorphous (quench-condensed) Nb and Mo stabilized with N_2. Solid line is based on theory of Rehr and Alben (1977). (b) Result (solid line) of low-temperature anneal of the Nb film. Dashed line (Wolf, Zasadzinski, Osmun and Arnold, 1980) represents crystalline Nb. (After Kimhi and Geballe, 1980.)

peaks characteristic of crystalline structures. The middle and bottom panels show the additional effect of force constant disorder. A Morse potential, using a value 9.8 for the dimensionless anharmonicity parameter A for the force constant α

$$A = \frac{r}{\alpha}\frac{d\alpha}{dr}$$

was used in the center case, while a stiffer Lennard-Jones potential having $A = 19.5$ was used in the lower case. The latter, as shown

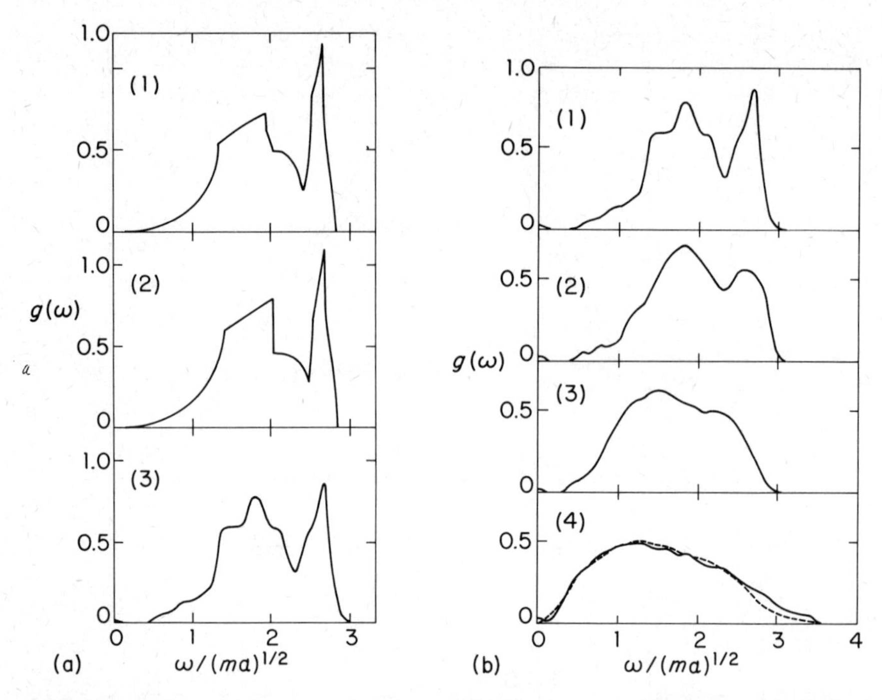

Fig. 6.17. Effects of finite crystal size and disorder on the vibrational density of states. (a) Vibrational density of states for (1) Cu, 18-parameter fit, (2) Cu, first-nearest-neighbor interaction only, (3) 480-atom $4 \times 5 \times 6$ *fcc* model. (b) Effect of disorder on the vibrational density of states (1) 480-atom *fcc* model, (2)–(4) 500-atom amorphous model with (2) constant-force constants ($A = 0$), (3) a Morse potential ($A = 9.8$), and (4) a Lennard-Jones potential ($A = 19.5$). (After Rehr and Alben, 1977.)

earlier, closely fits the $\alpha^2F(\omega)$ results observed for amorphous Mo and Nb stabilized with N_2. It is suggested (Kimhi and Geballe, 1980) that the qualitative difference in the spectra of amorphous s–p and d–band metals is attributable in the latter case to an interatomic potential which is more sensitive to changes in separation than that of simple metals. This result might come from strong overlap of core electrons in the d-band metals.

6.5 Transition Metals, Alloys, and Compounds

The first definitive tunneling study of any transition metal was the study of Ta by Shen (1970), leading to an $\alpha^2F(\omega)$ function in quite reasonable agreement with the neutron $F(\omega)$ function (Woods, 1964) and providing reasonable parameter values: $\lambda = 0.69$ and $\mu^* = 0.11$. The improvement in results was achieved with a carefully outgassed and recrystallized foil specimen (Shen, 1972*a*), on which thermal oxide was grown and a silver electrode evaporated. This procedure has been adopted as the basis for proximity effect studies of transition metals, leading to the $\alpha^2F(\omega)$ for Ta shown in Fig. 5.22, which differs only slightly from Shen's result, with $\lambda = 0.73$, $\mu^* = 0.10$, and a slightly stronger *LO* phonon peak.

A method essentially similar to Shen's sample preparation (Bostock, Agyeman, Frommer, and MacVicar 1973), however, when applied to Nb (Bostock et al., 1976) led to anomalous results: small values of λ and negative values of μ^*. While these anomalous results were initially discussed in terms of nonphonon attractive mechanisms of superconductivity (Bostock et al., 1976), this result has by now been ruled out at an appreciable level, and the source of the anomalous results [which were also obtained by Gärtner and Hahn (1976) and by Schoneich, Elefant, Otschik, and Schumann (1979)] traced to depression of the superconducting pair potential at the surface of the Nb. The surface pair-potential depression is a consequence either of dissolved oxygen in the Nb or of metallic (NbO) or semiconducting (NbO_2) interfacial layers interposed between the insulator Nb_2O_5 and bulk Nb (Arnold, Zasadzinski, and Wolf, 1978; Wolf, Zasadzinski, Osmun, and Arnold, 1980). We have already shown in Fig. 5.24 close agreement between $\alpha^2F(\omega)$ functions for Nb obtained via two schemes expressly designed to avoid surface contamination of Nb.

The curves of Fig. 6.18a (Geerk, Gurvitch, McWhan, and Rowell, 1982) nicely confirm the latter proximity PETS results. These are obtained, in fact, from sputtered Nb films, protected from oxygen by an extremely thin Al deposition immediately following the Nb deposition (Rowell, Gurvitch, and Geerk, 1981). The upper and lower spectra in Fig. 6.18a correspond to substrate temperatures of 800°C and <75°C, respectively, and to λ values 0.97 and 0.96. The μ^* value in each case is 0.15.

The solid circles in Fig. 6.18b are $F(\omega)$ determined from polycrystalline Nb by Gompf (1981), while the solid curve reproduces the $\alpha^2F(\omega)$ of Khim, Burnell, and Wolf (1981). Inasmuch as the tunneling results from three different methods agree quite well, and the neutron results shown here cannot be questioned on the grounds of the Born–von Karman interpolation [these are directly measured $F(\omega)$ points], this figure can be regarded as a definitive experimental comparison of $F(\omega)$ and $\alpha^2F(\omega)$, the most noticeable feature being the apparent drop of $\alpha^2(\omega)$ from about 3 meV at $\omega = 16$ meV to about 1.9 meV at 24 meV. This result remains at some variance with theory, as shown in Fig. 6.18c. Here, the solid curve, which represents $\alpha^2F(\omega)$ calculated in a rigid-muffin tin approximation (RMTA) method which incorporates experimentally determined phonon line widths (Butler, Smith, and Wakabayashi, 1977; Butler, Pinski, and Allen, 1979), would appear to imply $\alpha^2(24\,\text{meV}) \simeq 4.7$ meV, more than twice the experimental value from the previous figure. On an absolute scale, the experimental values of $\alpha^2(\omega)$ (above) agree at $\omega = 16$ meV and below with those calculated by Butler, Pinski, and Allen (1979); the latter are on the order of 3 meV for the bulk of the spectrum, with possibly a peak value of 4.5 meV in a narrow region near 22 meV. This discrepancy

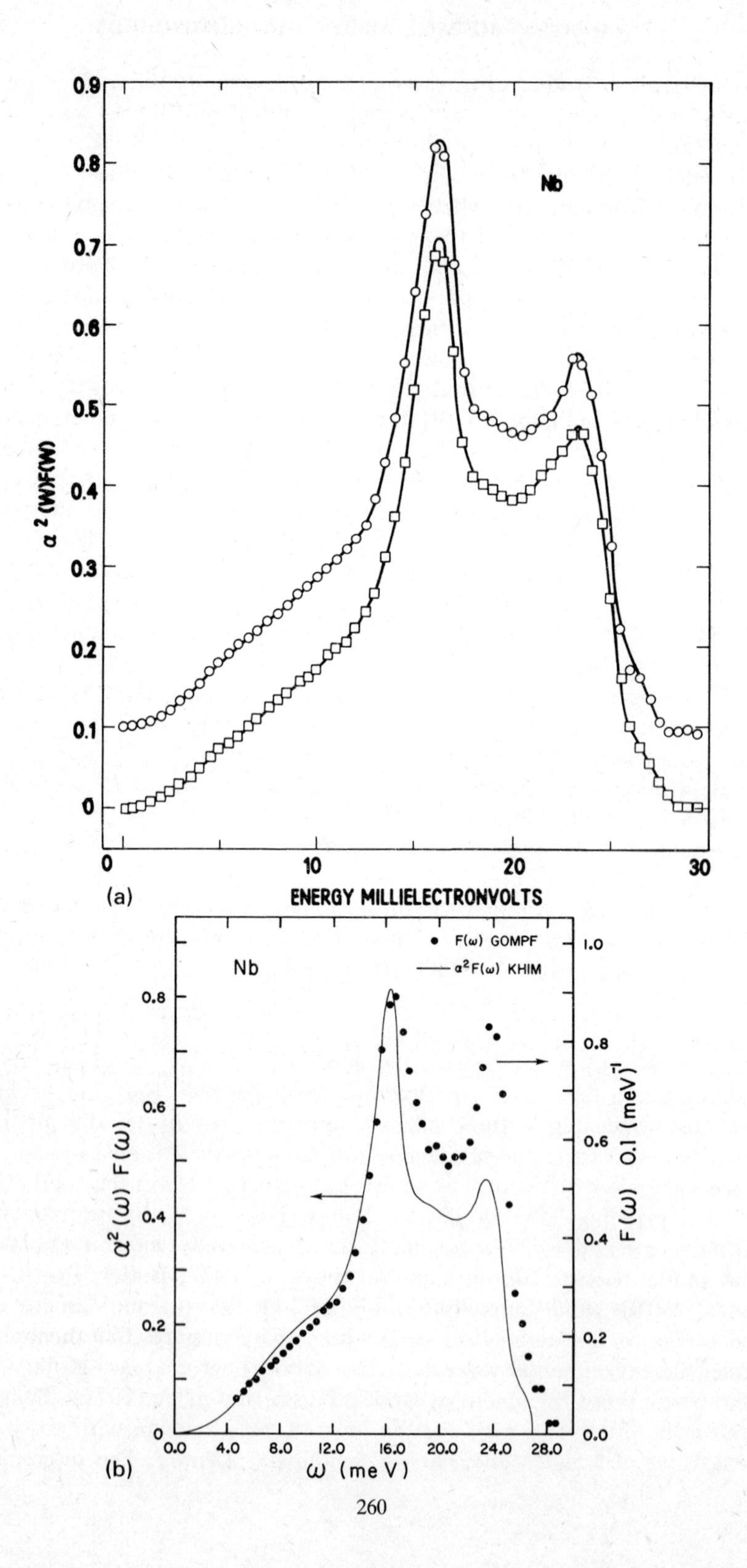

Nb
α^2(W)F(W)
ENERGY MILLIELECTRONVOLTS
(a)
F(ω) GOMPF
α^2F(ω) KHIM
$\alpha^2(\omega)$ F(ω)
F(ω) 0.1 (meV)$^{-1}$
ω (me V)
(b)

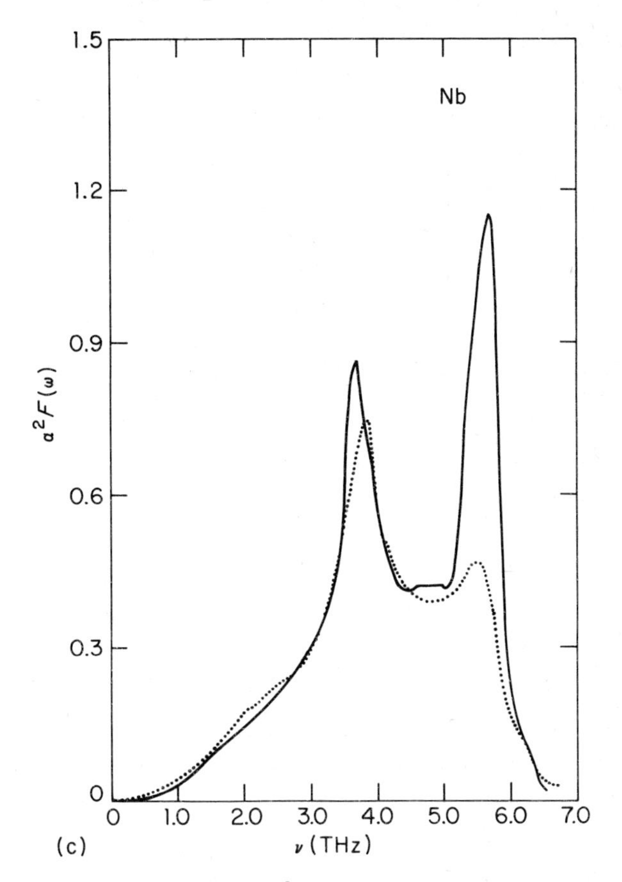

Fig. 6.18. Comparison of tunneling $\alpha^2F(\omega)$ and neutron-derived $F(\omega)$ for Nb. (a) Tunneling $\alpha^2F(\omega)$ of sputtered Nb films: upper curve, 600°C substrate temperature; lower curve, ambient substrate temperature. These results are similar to those shown in Fig. 5.26 obtained in two different ways from niobium foils. (After Geerk, Gurvitch, McWhan, and Rowell, 1982.) (b) The Nb $F(\omega)$ obtained from incoherent inelastic neutron scattering on polycrystalline material. (After Gompf, 1981.) (c) Calculated $\alpha^2F(\omega)$ of Nb from rigid-muffin tin calculations based upon neutron scattering measurements. (After Butler, 1980*b*; Butler, Pinski, and Allen, 1979.) The theoretical curve is shown solid in comparison with the tunneling $\alpha^2F(\omega)$ of Arnold, Zasadzinski, Osmun, and Wolf (1980).

in the strength of the upper phonon peak in $\alpha^2F(\omega)$ for Nb is characteristic of all the tunneling determinations, which, as we have seen, now agree closely. Its origin is not definitely known, but two possibilities are the rigid-muffin tin approximation used in the theory in calculating the matrix elements and a sampling depth in the experiment decreasing with the mean free path as the electron-phonon coupling increases in the 22 meV region.

While such a large relative reduction in the high-energy *LA* phonon

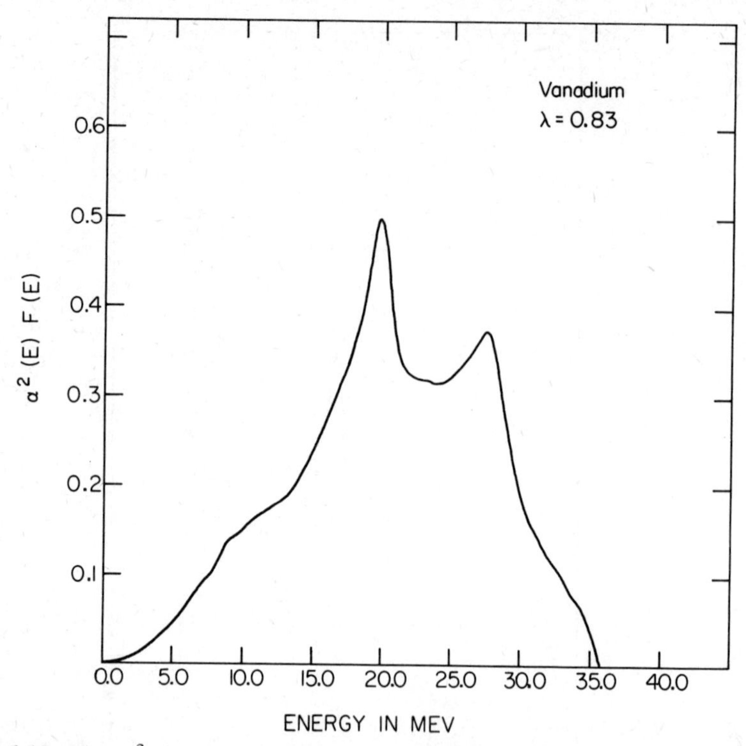

Fig. 6.19. The $\alpha^2F(\omega)$ of vanadium corresponding to $\lambda = 0.83$. (After Zasadzinski, Burnell, Wolf, and Arnold, 1982.)

region does not occur in the $\alpha^2F(\omega)$ of Ta, it does occur in V, as shown in Fig. 6.19. This spectrum was obtained from an Al/V proximity junction, as V is one of the most difficult metals to study using conventional tunneling methods. Since V and Nb are more difficult to clean than Ta, and since their superconducting properties are more rapidly degraded with oxygen and other gaseous impurities, the possibility remains that the more serious reduction of $\alpha^2F(\omega)$ in the *LA* phonon region of these two, relative to that of Ta and to the rigid muffin tin calculation, is an artifact of imperfectly cleaned Nb and V surfaces.

The $\alpha^2F(\omega)_\mathrm{p}$ (point-contact spectroscopy) function of crystalline Mo at 1.5 K is shown in Fig. 6.20 in comparison with calculations. There are no tunneling data in the crystalline state for Mo because of its low T_c, 0.915 K, and the weakness of phonon effects that this low T_c would imply. The point-contact function $\alpha^2F_\mathrm{p}(\omega)$ is expected to differ from the tunneling $\alpha^2F(\omega)$ by the presence of a geometrical structure factor $K(\vec{v}, \vec{v}')$, where $\vec{v}$ and $\vec{v}'$ are electron velocities (see, e.g., Yanson, Kulik, and Batrak, 1981; Ashraf and Swihart, 1982). The rather noticeable broadening present in the experimental spectra may arise from impurity scattering. Broadening the theoretical spectrum, on the assumption of a

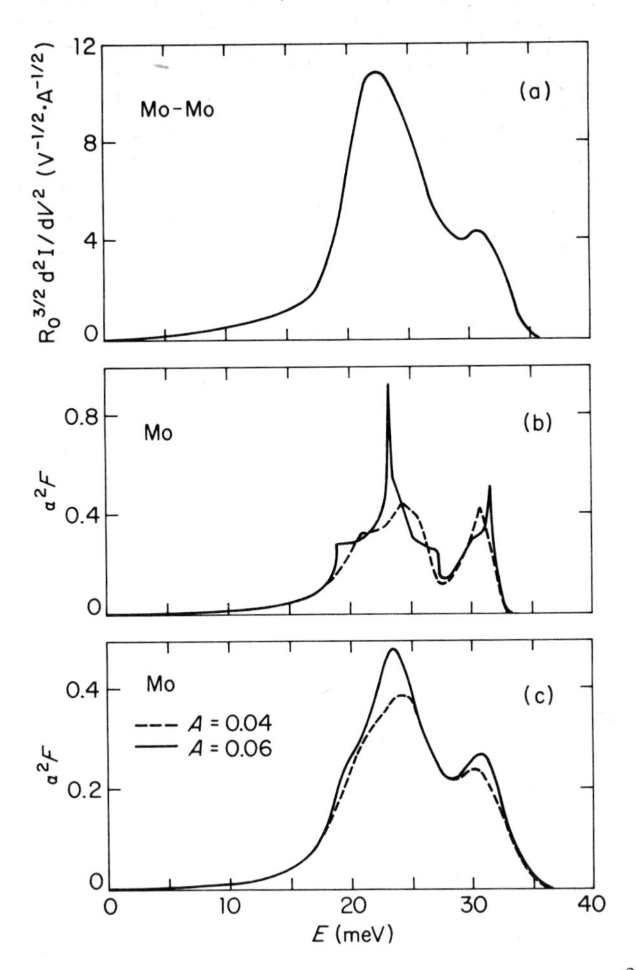

Fig. 6.20. Phonon spectra of Mo. (a) Measured point contact $\alpha^2F(\omega)$ from Mo–Mo contact immersed at 1.5 K. (b) Calculated $\alpha^2F(\omega)$ for Mo based on neutron spectra and rigid-muffin tin theory (solid curve) (Pinski, Allen, and Butler, 1978) and rigid–atomic sphere approximation (dashed curve) (Glotzl, Rainer, and Schober, unpublished). (c) The $\alpha^2F(\omega)$ curves of (b) subjected to thermal broadening, using an equivalent temperature $T=(T_0^2+A\varepsilon^2)^{1/2}$. (After Caro, Coehoorn, and DeGroot, 1981.)

width increasing with energy within the spectrum, leads to reasonable agreement with experiment, as shown in Fig. 6.20c.

An experimental $\alpha^2F(\omega)$ for the d–*hcp* form of the reactive element La (Lou and Tomasch, 1972) is shown in Fig. 6.21. This element has a $4f^05s^25p^65d^16s^2$ outer electronic shell, making it really a transition element. However, the proximity of the empty $4f^0$ levels to the Fermi energy—consistent with the fact that the next 14 electrons, filling out the lanthanide series, go into the 4f levels—might be expected to show up in

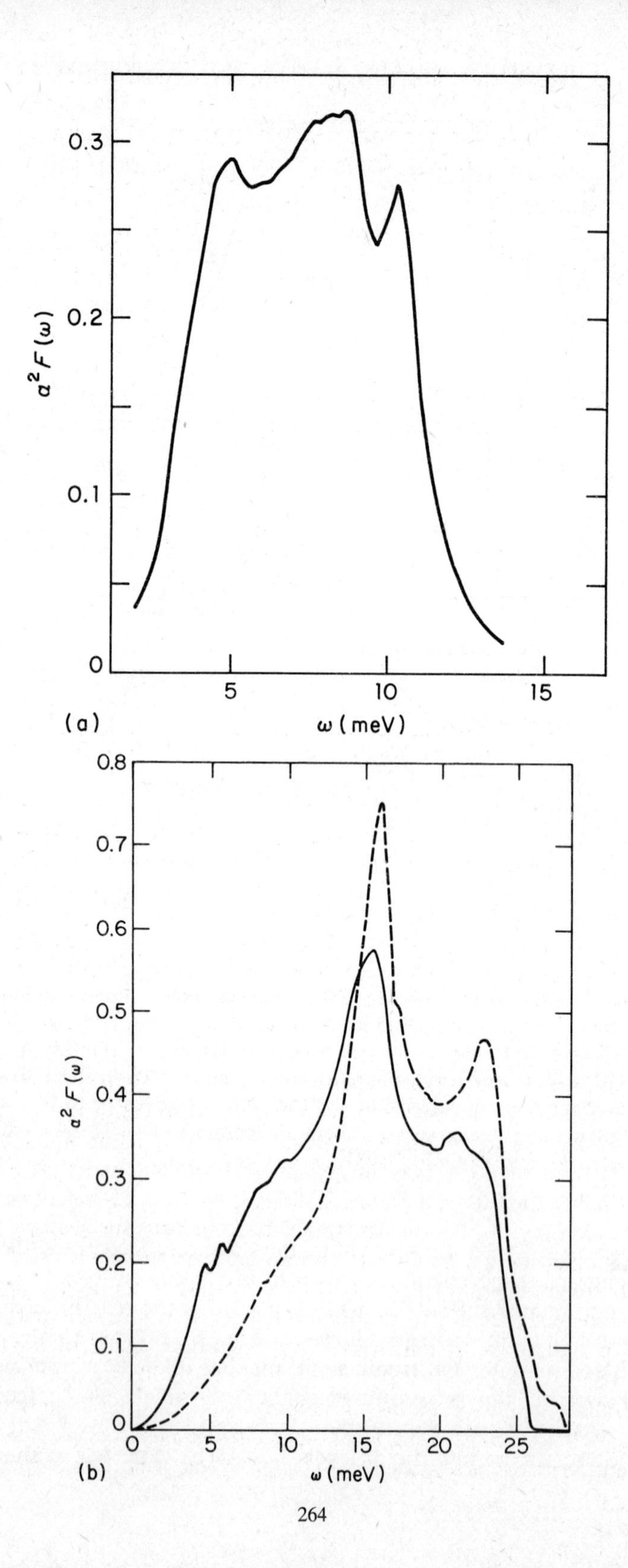

0.3
0.2
0.1
0
$\alpha^2F(\omega)$
5
10
15
(a)
ω (meV)
0.8
0.7
0.6
0.5
0.4
0.3
0.2
0.1
0
$\alpha^2F(\omega)$
0
5
10
15
20
25
(b)
ω (meV)

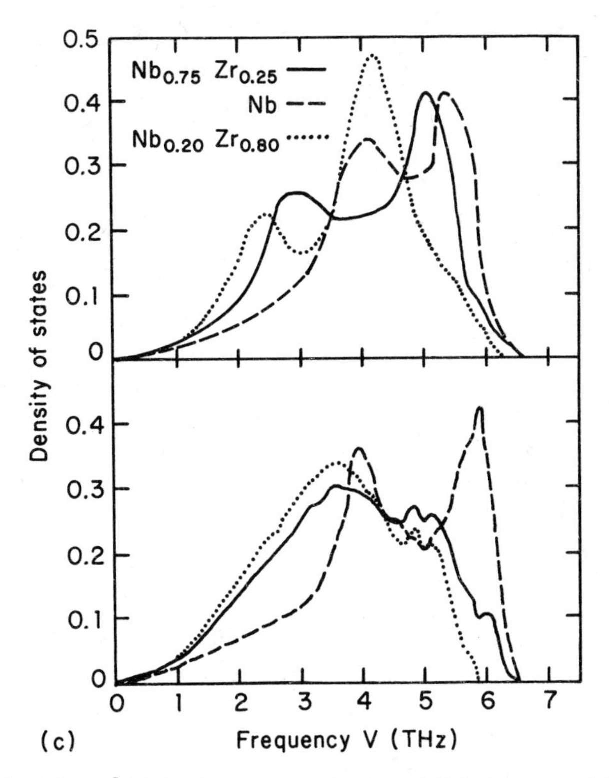

Fig. 6.21. (a) The $\alpha^2F(\omega)$ of La determined at 0.9 K from an Al–Al_2O_3–La junction corresponds to $\lambda = 0.77$. (After Lou and Tomasch, 1972). (b) Effect of addition of 25% Zr to Nb upon the $\alpha^2F(\omega)$ function (shown as solid) in comparison with pure Nb (dashed), as measured by Wolf and Noer (1979). (After Mitrovic and Carbotte, 1981.) (c) Phonon spectra $F(\omega)$ of NbZr alloys as determined by coherent, inelastic neutron scattering experiments, using time-of-flight methods. Solid curve is $Nb_{0.75}Zr_{0.25}$; dashed curve is pure Nb; dotted curve is $Nb_{0.2}Zr_{0.8}$. (After Gompf, Richter, Scheerer, and Weber, 1981.)

some fashion in the superconducting properties. There is, however, no strong evidence for unusual behavior in these spectra or in the results of the tunneling study of Wühl, Eichler, and Wittig (1973) in which the effect of hydrostatic pressure was investigated. From an experimental viewpoint, the reactivity of La makes it difficult to produce junctions without some weakening of phonon structure due to impure surface layers of the La. This result may account for the unusually small values of $\mu^* = 0.02$–0.04 obtained even after Lou and Tomasch applied an *ad hoc* amplifying factor $f = a(\omega/10.9)^2 + 1$ to the phonon spectra, where a is taken as 0.3 and ω is expressed in millivolts. This procedure leads to $\lambda = 0.77$ for d–*hcp* La.

Crystalline transition metal alloys such as NbTi, used in superconducting magnets, are important technologically because of the high critical fields and critical current densities which result from the reduced mean free paths. Other aspects of the strengthened superconductivity of these alloys have been probed by tunneling spectroscopy recently, using NbZr as a model system (Wolf, Noer, and Arnold, 1980). A comprehensive discussion of the systematic variation of the factors of $\lambda = N(0)\langle I^2\rangle/M\langle\omega^2\rangle$ along the 4d transition series has been given by Butler (1977) and is summarized in Table 6.4. The elements Z = 40–43 (Zr, Nb, Mo, Tc), which provide a strong variation in critical temperature (0.55 K, 9.22 K, 0.92 K, 7.86 K, respectively), are of primary interest.

In an overall sense, the variation of λ over these four elements (Table 6.4) follows the variation of the band density of states $N(0)$: the ratio $\langle I^2\rangle/M\langle\omega^2\rangle$ varies less rapidly than $N(0)$ and λ. One should note, however, that the numerical values in this survey (compiled by Butler, 1977) are necessarily of limited accuracy. For example, since tunneling values of λ are not available over the whole series, the quoted λ values have been obtained from the Allen-Dynes T_c formula with $\mu^* = 0.13$. The prefactor $\omega_{\log}$ required in the Allen-Dynes formula was obtained from neutron scattering $F(\omega)$ functions on the assumption $\alpha^2 = \text{constant}$. Finally, the noted inaccuracy in λ must be present to a smaller extent in $N(0)$, the band structure or bare density of states, because this quantity is dependent on λ. The density of states $N(0)$ is related to the measured specific heat coefficient γ by the relation

$$N(0) = \gamma[\tfrac{2}{3}\pi^2k^2(1+\lambda)]^{-1}$$

which makes $N(0)$ sensitive to errors in λ at large λ.

The comparative tunneling study of Nb and $Nb_{0.75}Zr_{0.25}$ is summarized in Fig. 6.21b and in Table 6.5, while Fig. 6.21c shows recent neutron scattering $F(\omega)$ curves for NbZr. The clear indication from the curves in Figs. 6.21b and 6.21c is that addition of 25% Zr alters the superconducting λ and T_c and coupling $2\Delta/kT_c$ primarily through phonon softening. The quantity $\langle\omega^2\rangle$ decreases upon alloying to only 0.70 of its value for Nb, accounting entirely for the increase in λ. Although the γ (specific heat) value for the alloy is indeed 20% larger than the γ value for Nb, this result can be accounted for entirely by the increase in λ, leaving $N(0)$ essentially unchanged. This latter behavior can be understood—even though the Fermi level in the alloy is shifted, with Zr additions, toward a region of higher density of states—by increased broadening of electronic levels and density-of-states peaks by disorder, as discussed by Testardi and Mattheiss (1978). The same kind of broadening effect is thought to *increase* $N(0)$ in *amorphous* Mo and thus to explain its high T_c of 8 K, in comparison with $T_c = 0.92$ K for crystalline Mo. Corroboration of these conclusions for $Nb_{0.75}Zr_{0.25}$ has been provided by Mitrović and Carbotte (1981), who have successfully calculated the known thermodynamic properties of the alloy, as well as the coupling parameter $2\Delta/kT_c$, from the tunneling $\alpha^2F(\omega)$ shown in Fig. 6.21b. The further calculation of the (phonon) softening of $F(\omega)$ (Fig. 6.21c, Gompf, Richter, Scheerer, and Weber, 1981), following the method of Varma and Weber (1979), is in

Table 6.4. Survey of Empirical Quantities Related to 4d Transition Metal Superconductivity (Butler, 1977)

	Y	Zr	Nb	Mo	Tc	Ru	Rh	Pd
T_c (K)	(0.0006)	0.55	9.22	0.92	7.86	0.48	(0.002)	
θ_D (K)[a]	256	290	277	460	454	550	500	270
θ_c (K)[b]	210	255	288	392	(355)	392	346	290
$\langle\omega^2\rangle^{1/2}$ (K)[c]	149	173	195	268	245	(264)	(239)	200
$\omega_{\log}$ (K)	132	142	161	241	223	(229)	(208)	(174)
$\langle\omega\rangle^{1/2}/\theta_c$	0.71	0.68	0.68	0.70	(0.69)	(0.69)	(0.69)	0.69
$\omega_{\log}/\theta_c$	0.63	0.56	0.56	0.63	(0.63)	(0.60)	(0.60)	(0.60)
θ_c/θ_D	0.62	0.88	1.04	0.83	0.78	0.69	0.69	1.07
γ (mJ/mole · K^2)	10.2	2.78	7.8	1.83	4.30	3.1	4.70	9.4
λ[d]	0.26	0.45	0.97	0.44	0.78	0.41	0.27	(0.7)
$N(0)$ (states/Ry)	23.3	5.53	11.41	3.66	6.96	6.33	10.66	15.9
$M\langle\omega^2\rangle$ (a.u.)	0.0722	0.100	0.129	0.252	0.218	0.258	0.215	0.156
$\langle I^2\rangle$ (a.u.)	0.00081	0.00814	0.011	0.0303	0.0244	0.0167	0.00544	
η (a.u.)[e]	0.019	0.045	0.125	0.111	0.170	0.106	0.058	

[a] Heiniger, Bucher, and Muller (1966).
[b] Values obtained from setting Debye specific heat function equal to experimental lattice specific heat.
[c] Obtained from $F(\omega)$ deduced from neutron scattering.
[d] Estimates based on McMillan's T_c equation with $\mu^* = 0.13$.
[e] $\eta = N(0)\langle I^2\rangle$.

Table 6.5. Tunneling Values of Superconducting Parameters of Nb and $Nb_{0.75}Zr_{0.25}$ (After Wolf, Noer, and Arnold, 1980)

	$Nb_{0.75}Zr_{0.25}$	Nb^a	Ratio
T_c (K)	10.67	9.25	1.15
$2\Delta/kT_c$	4.13	3.79	1.09
$\lambda = N(0)\langle I^2\rangle/M\langle\omega^2\rangle$	1.31 ± 0.1	0.93 ± 0.1	1.41
$\omega_{\log}$ (meV)	9.4	12.2	0.77
$\langle\omega\rangle$ (meV)	11.3	14.0	0.81
$\langle\omega^2\rangle^{1/2}$ (meV)	12.8	15.3	0.84
$M\langle\omega^2\rangle$ (eV/Å^2)	3.7	5.2	0.7
$N(0)\langle I^2\rangle$ (eV/Å^2)^b	4.84 ± 0.5	4.84 ± 0.5	1.0
γ (mJ/mole · K^2)	9.55^c	7.8	1.22
$N(0)$ (states/eV spin)^d	0.88	0.86	1.02
$\langle I^2\rangle$ (eV/Å)^2	5.5	5.6	1.0

[a] For a valid comparison, these values are from Nb proximity data corrected in an equivalent fashion (the $\Delta_N = 0$ approximation) as used for the NbZr data. Preferred values for Nb are given by Burnell et al. (1982).

[b] Inferred from the experimental values of λ and $M\langle\omega^2\rangle$, using the relation $\lambda = N(0)\langle I^2\rangle/M\langle\omega^2\rangle$.

[c] This value is obtained by averaging published values of 8.9 mJ/mole · K due to Heiniger, Bucher, and Muller (1966), and 10.2 mJ/mole · K due to Masuda, Nishioka, and Watanabe (1967).

[d] The equation $\gamma = \frac{2}{3}\pi^2 k_B^2 N(0)(1+\lambda)$ has been used.

qualitative agreement with the observations. Phonon dispersion curves in the alloy have been calculated by Simons and Varma (1980), while effects of force constant disorder specifically on $\alpha^2F(\omega)$ have been considered by Grunewald and Scharnberg (1979).

No other studies of comparable detail in transition metal alloys have been reported, although new work on these materials is being carried out using rapid-sputtering methods followed by Al metal overlayer deposition (Rowell, Gurvitch, and Geerk, 1981).

The study of transition metal compounds has been rather limited, although considerable progress has been made in recent years on members of the A–15 class including Nb_3Sn (Moore, Zubeck, Rowell, and Beasley, 1979; Wolf et al., 1980), Nb_3Al (Kwo and Geballe, 1981), and Nb_3Ge (Kihlstrom and Geballe, 1981). Apart from studies of A–15 compounds, phonon spectra have been reported for TaC (Zeller, 1972) and NbC (Geerk et al., 1975).

The first complete tunneling study of an A–15 compound was that to Shen (1972*b*), who formed Nb_3Sn as a surface reaction layer on a vacuum-outgassed foil of Nb. The $\alpha^2F(\omega)$ spectra showed three peaks, two resembling those of Nb at 16 and 25 meV, the third lying at about 9 meV. This structure has been confirmed, at least in its main features, by more recent tunneling work, shown in Fig. 6.22a, and by time-of-flight

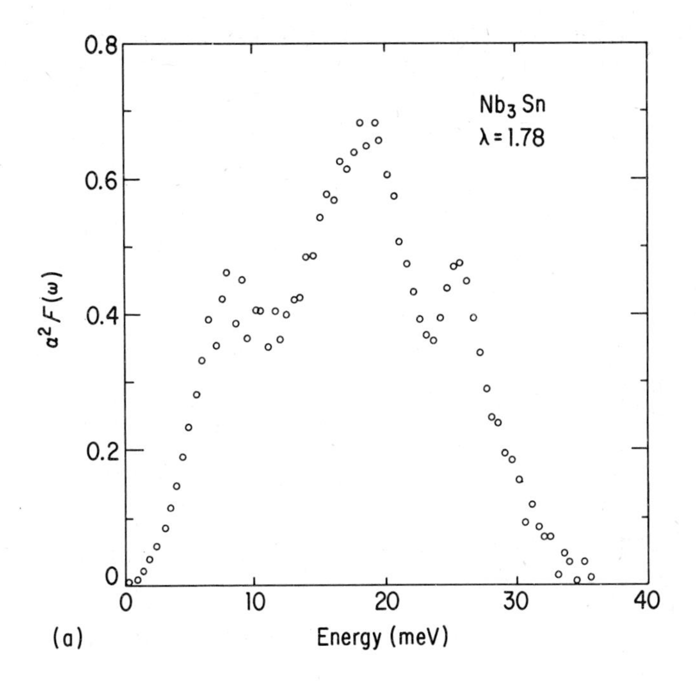

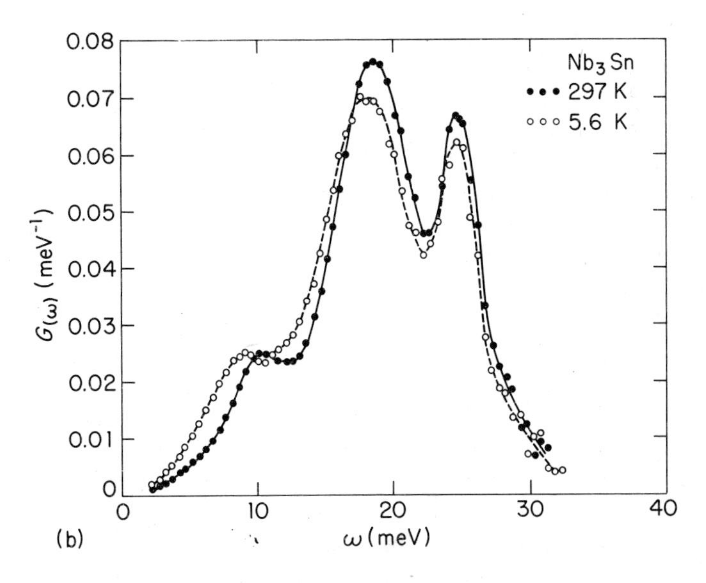

Fig. 6.22. Phonon spectra of Nb_3Sn. (a) The $\alpha^2F(\omega)$ determined from tunneling measurements on coevaporated films by Moore (1978). (After Wolf et al., 1980.) (b) Phonon spectra $G(\omega)$ of Nb_3Sn determined at two temperatures; solid points are 297 K; open points are 5.6 K. (After Schweiss, Renker, Schneider, and Reichardt, 1976.)

neutron determination of a generalized phonon density-of-states function $G(\omega)$ by Schweiss, Renker, Schneider, and Reichardt (1976), shown in Fig. 6.22b. The neutron scattering cross sections for Nb and Sn differ somewhat, so that $G(\omega)$ must differ, but not qualitatively, from the true spectrum $F(\omega)$. Overlooking this distinction, comparison of $\alpha^2F(\omega)$ and $G(\omega)$ in Fig. 6.22 suggests that $\alpha^2(\omega)$ for Nb_3Sn is a decreasing function of energy.

The neutron $G(\omega)$ spectra of Nb_3Sn at 5.6 K show a large, ~10%, softening of the 9 meV mode from its room temperature value. An even more pronounced emphasis on the lowest-frequency modes is observed in the tunneling $\alpha^2F(\omega)$ for Nb_3Al, as shown in Fig. 6.23. In this case, the degree of softening increases with approach to perfect stoichiometry, a feature also observed in Nb_3Ge and accompanied by an increase of the ratio $2\Delta/kT_c$ above the BCS ratio 3.53.

The recent tunneling study of Nb_3Ge of Kihlstrom and Geballe (1981) is summarized in Fig. 6.24. In this case, as in the cited tunneling work on Nb_3Al and Nb_3Sn, the A–15 films were coevaporated onto substrates held at high temperature, the tunnel barrier was formed by subsequent oxidation of a thin ~25–30 Å layer of Si deposited immediately after cooling of the film to 100°C, and the analysis of the phonon spectra to obtain

Fig. 6.23. Eliashberg functions for Nb_3Al films, characterized by $2\Delta/kT_c$ of 3.6 (open symbols) and 4.4 (closed symbols), compared with neutron-phonon density $G(\omega)$ (Schweiss, Renker, Schneider, and Reichardt, 1976). (After Kwo and Geballe, 1981.)

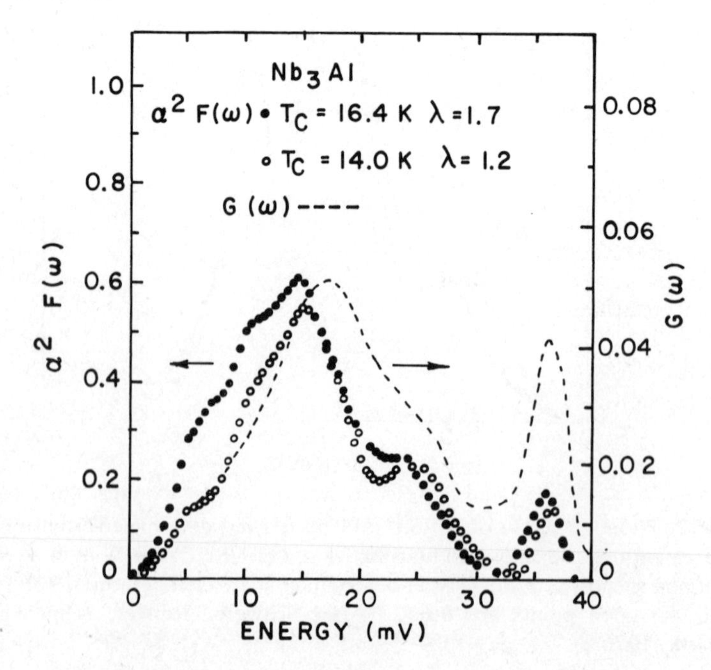

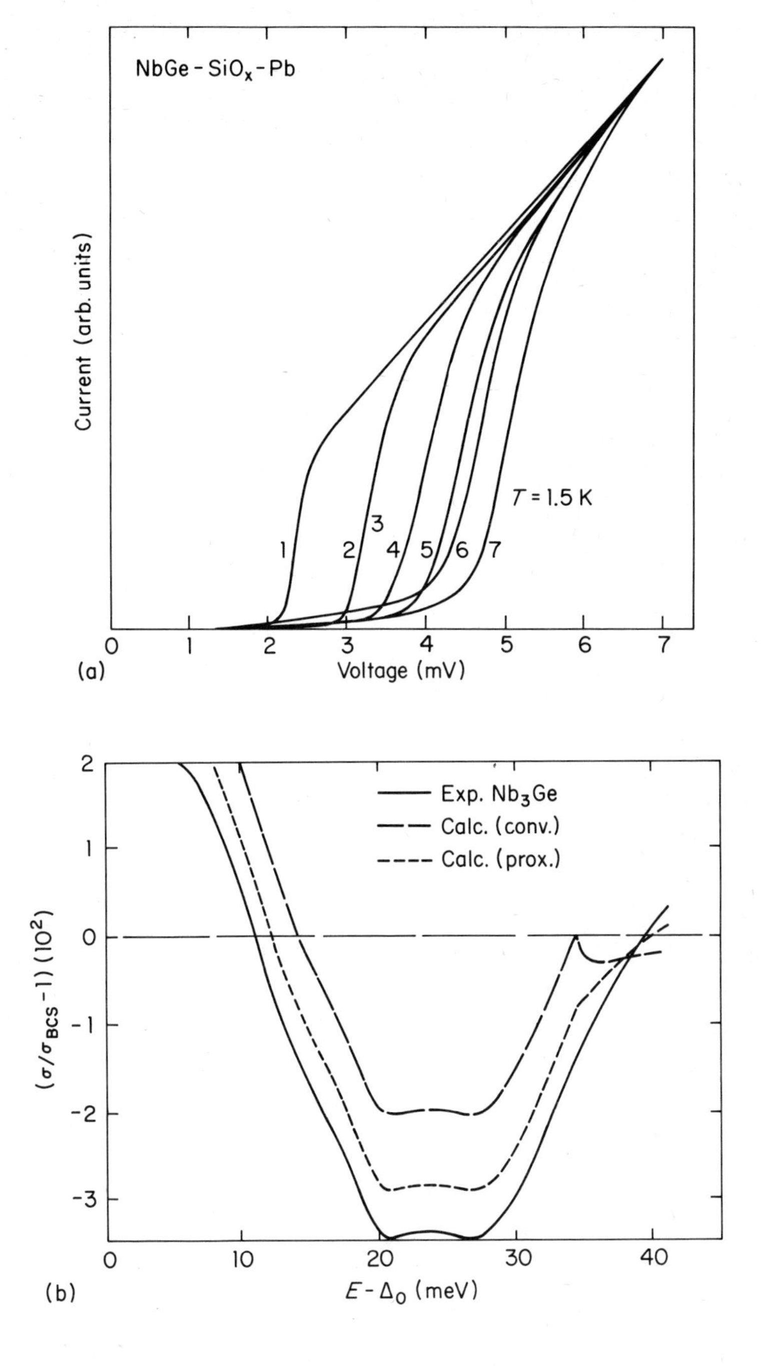

Fig. 6.24. Properties of A–15 $NbGe_x$ inferred from tunneling. (a) The I–V characteristics of $NbGe_x$–SiO_2–Pb junctions measured at 1.5 K. The Ge concentrations in samples 1–7 are, respectively, 16.7, 20.3, 20.4, 20.9, 21.8, 22.1, and 22.9%. (b) Reduced conductance in the phonon region of a typical $NbGe_x$ sample. Solid line is the experiment; dashed line is calculated on a conventional inversion; dot-dash line is calculated on a proximity inversion.

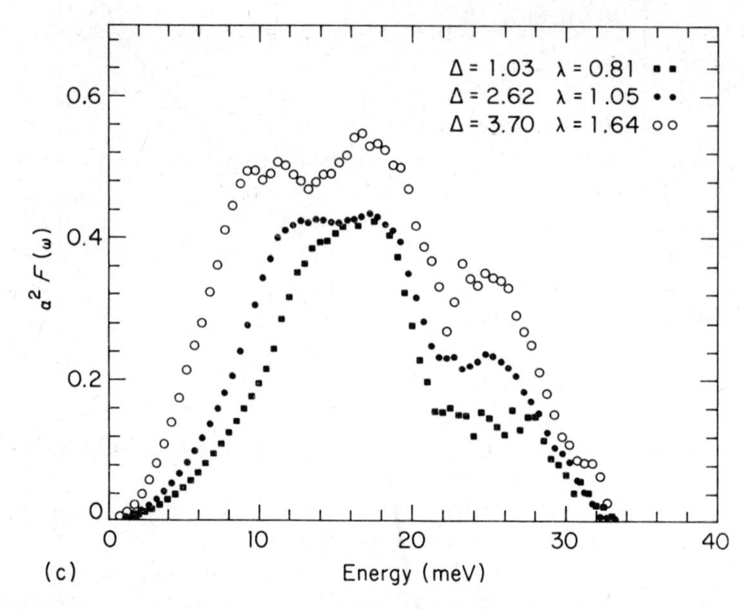

Fig. 6.24. (cont.) (c) Electron-phonon functions $\alpha^2F(\omega)$ for $NbGe_x$ samples with gaps of 1.03, 2.62, and 3.70 meV. The maximum value of λ in this set is 1.64. (After Kihlstrom and Geballe, 1981).

$\alpha^2F(\omega)$ was accomplished with proximity corrections applied in the $\Delta_N = 0$ approximation.

In the work on Nb_3Ge, which follows an earlier study by Rowell and Schmidt (1976), a strong increase in sum gap (Fig. 6.24a) is observed as the Ge fraction x in $Nb_{1-x}Ge_x$ increases from 0.167 (curve 1) to 0.229 (curve 7). The observed $Nb_{1-x}Ge_x$ gap at $x = 0.229$ is 3.62 meV (taking $\Delta_{Pb} = 1.4$ meV), corresponding to $T_c = 19.8$ K and $2\Delta/kT_c = 4.24$. Conductance curves are shown in Fig. 6.24b. Pronounced softening of $\alpha^2F(\omega)$ is evident in Fig. 6.24c as the λ parameter increases from 0.81, in a sample of $\Delta_0 = 1.03$ meV, to 1.64, in a sample with $\Delta_0 = 3.70$. The $\langle\omega^2\rangle$ values obtained decrease as λ increases, so that one identifies the increase in λ with $\alpha^2F(\omega)$ softening, as in the comparison of $Nb_{0.75}Zr_{0.25}$ with Nb. The degree to which the softening also occurs in the phonon modes themselves is apparently appreciable.

6.6 Extreme Weak-Coupling Metals

The proximity effect occurring in C–I–NS junctions, as we have seen in Fig. 5.3 and discussed in connection with (5.45)–(5.47), allows the possibility of strong enhancement of the pair potential $\Delta_N(E)$ in weak-coupling metals. Under certain circumstances, the tunneling spectra as described by (5.23) or (5.30) make possible determination of $\Delta_N(E)$ and of the underlying phonon spectrum $\alpha^2F(E)$ for the weak-coupling metal. Such a result for Al induced into a state of strong-coupling superconductivity has been shown in Fig. 5.23b.

In the context of (5.30), there are two cases which facilitate extracting $\Delta_N(E)$. In the first case, one observes phonons from both metals but has independent knowledge of $\Delta_S(E)$, permitting a variational determination of $\Delta_N(E)$. This procedure was applied in connection with the Al phonon data shown in Fig. 5.23.

A second simplified case occurs if the N layer provides sufficient quasiparticle scattering so that the $\Delta_S(E)$ contributions in (5.30) become negligibly small, leaving only

$$N_T(E) \simeq \mathrm{Re}\left[1+\frac{1}{2}\left(\frac{\Delta_N}{E}\right)^2\right]$$

which is the expansion for $E > \Delta_N$ of the usual strong-coupling expression for $N_T(E)$, $\mathrm{Re}\{|E|/(E^2-\Delta_N^2)^{1/2}\}$.

Recall that the third and fourth terms in Eq. (5.30), representing the 4d and 2d series of oscillations, are attenuated by factors $e^{-2d/\ell}$ and $e^{-d/\ell}$, respectively, which can effectively remove their contributions. The latter approach has been applied to obtain the $\alpha^2F(\omega)$ for Mg shown in Fig. 6.25a. A sequence of Mg/Nb foil-based bilayers was studied, yielding strong 27-meV Mg *LA* phonon responses similar to that shown for an Mg/Ta foil in Fig. 5.3b. At $d_N = 880$ Å, as shown in Fig. 2 of Burnell and Wolf (1982), the Nb phonon peaks become unobservably weak, allowing interpretation of the remaining features in terms of Mg phonons alone. The analysis provides the $\alpha^2F(\omega)$ shown, corresponding to $\lambda = 0.29$ and $\omega_{\log} = 18.9$ meV. By comparison with the neutron scattering function $F(\omega)$ of Pynn and Squires (1972), an $\alpha^2F(\omega)$ function is obtained which rises by about a factor 2 over the range 14 to 27 meV. This result is consistent with the expectation that transverse modes in simple metals should be relatively weakly coupled, as indicated by (6.2). From the tunneling results for Mg and an independent estimate $\mu^* = 0.16$ (Allen and Cohen, 1969), a prediction of the T_c of pure Mg of 0.3 mK, within rather wide bounds (0.02 to 7 mK), is obtained (Burnell and Wolf, 1982).

An additional study of Al, in proximity with Pb, is shown in Fig. 6.25b (Khim, 1979). In the calculation of this $\alpha^2F(\omega)$ function for Al, the Pb phonon-related structure in dV/dI and d^2V/dI^2 occurring below about 12 meV was deleted, and the simplified spectra were inverted using the conventional McMillan Stage III program. The results are in good agreement with the measured phonon density $F(\omega)$ for Al (Stedman, Almqvist, and Nilsson, 1967) and the calculated $\alpha^2F(\omega)$ (Leavens and Carbotte, 1972).

The induced superconductivity of Ag and Cu is revealed in proximity to Pb in measurements of Chaikin and Hansma (1976). Extension of this work has led to an estimate of $\lambda = 0.13$ for Cu (Wilson, Simon, McGinnis, and Chaikin, 1980), and an $\alpha^2F(\omega)$ function for Cu is given by Wilson (1980).

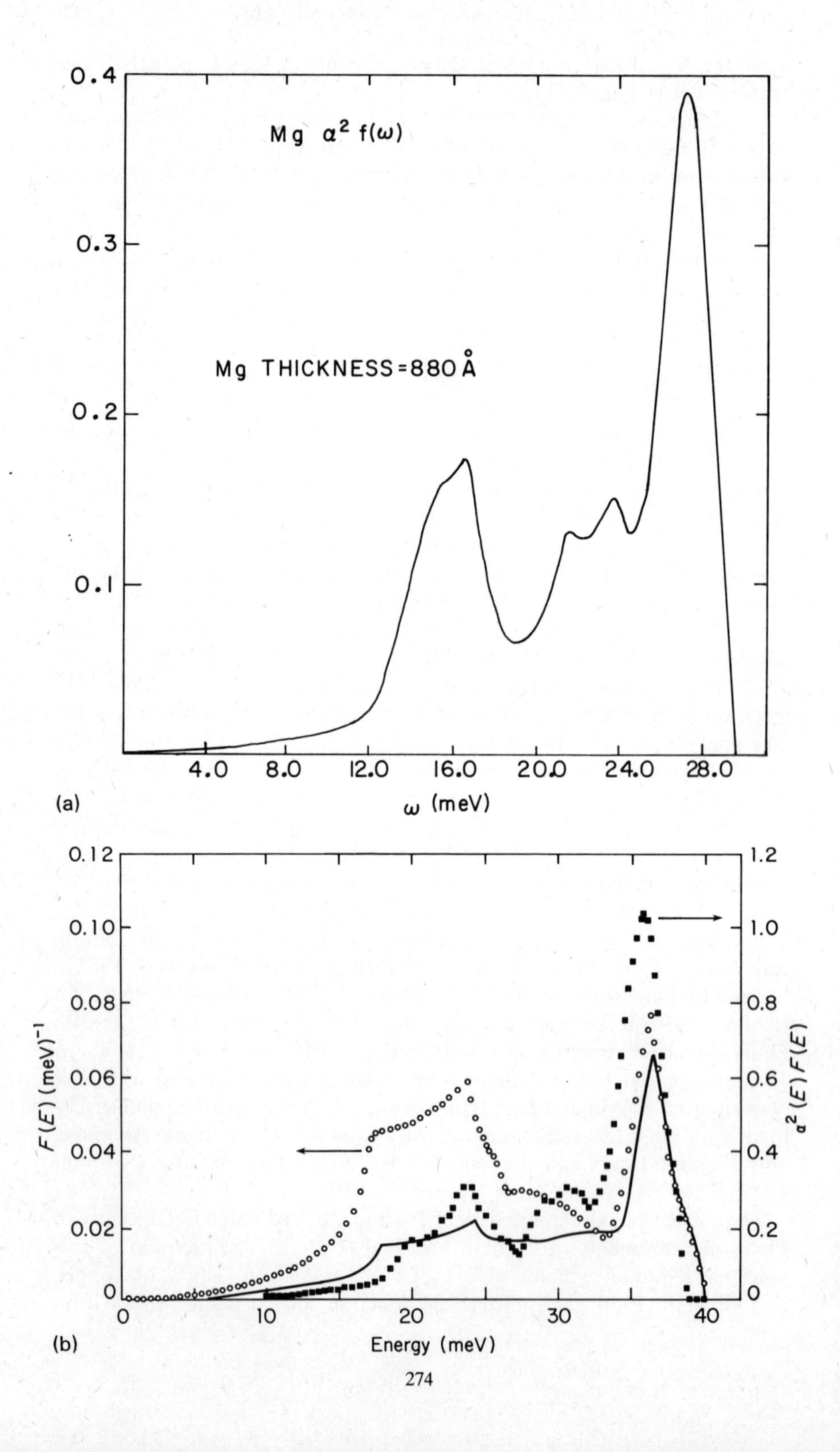
Mg α² f(ω)
Mg THICKNESS=880 Å
0.4
0.3
0.2
0.1
4.0
8.0
12.0
16.0
20.0
24.0
28.0
(a)
ω (meV)
0.12
0.10
0.08
0.06
0.04
0.02
0
1.2
1.0
0.8
0.6
0.4
0.2
0
F(E) (meV)⁻¹
α² (E) F(E)
0
10
20
30
40
(b)
Energy (meV)

6.7 Local-Mode Superconductors

The original observation of a local vibrational mode contributing to $\alpha^2F(\omega)$ of a superconductor was that of In in Pb, shown in Figs. 6.8 and 6.9. The light mass impurity introduces vibrational modes above the cutoff energy for the host lattice phonons, and these localized vibrations, analogous to shallow donors in the case of electronic states, provide additional electron-phonon coupling. Somewhat surprisingly, only a few cases of local-mode contributions in this sense have been reported in tunneling spectra. These are Pb:In (Rowell, McMillan, and Anderson, 1965); Tl:In (Granqvist and Claeson, 1974*a*, 1974*b*); PbTlIn and PbTl-BiIn (Jacobsen, 1979); Pd:H and Pd:D (Dynes and Garno, 1975; Silverman and Briscoe, 1975; Eichler, Wühl, and Stritzker, 1975*a*, 1975*b*); Al:H and Al:D (Dumoulin et al., 1976); and Pb:H and Pb:D (Nedrud and Ginsberg, 1981). One might possibly consider compounds such as NbC, NbN, or even Nb_3Al, with relatively large mass ratios and isolated high-frequency peaks in $\alpha^2F(\omega)$ (see Fig. 6.23) as belonging to this class. In many of these cases, the electron-localized phonon coupling appears to be quite strong.

This result appears to be the case, in particular, for the Pd:D system illustrated in Fig. 6.26. These results are the more striking because metallic Pd is not superconducting, presumably as a result of magnetic spin fluctuations, indicated by the high paramagnetic susceptibility. It is believed that the role of the H(D), in part, is to donate a portion of its electron to the Pd, shifting the Fermi energy to a position of lower $N(E)$, thus reducing the tendency to spin fluctuations. A mass dependence of the spatial extent of the vibrational wavefunction of H(D) may be involved in the observed anomalous isotope effect (Miller and Satterthwaite, 1975): the maximum T_c values for Pd:D and Pd:H, respectively, are 11 and 9 K. It is clear from Fig. 6.26b that a substantial contribution to the total electron-phonon coupling parameter λ must come from the localized mode, although a complete analysis appears not to be available.

A framework in which to treat the contributions to λ in cases of this sort has been proposed by Gaspari and Gyorffy (1972). The resulting expression for λ is

$$\lambda = \sum_j \frac{N(E_F)\langle I_j^2\rangle}{M_j\langle\omega^2\rangle} \tag{6.17}$$

where j indexes the different atoms in the unit cell. Application of this

---

Fig. 6.25. Some phonon results for weak-coupling metals. (a) The $\alpha^2F(\omega)$ for Mg obtained from an Ag–MgO–MgNb proximity junction with $d_N = 880$ Å and corresponding to $\lambda = 0.29$. (After Burnell, 1982.) (b) The $\alpha^2F(\omega)$ of Al obtained (squares) from an AlPb proximity junction. (After Khim, 1979.)

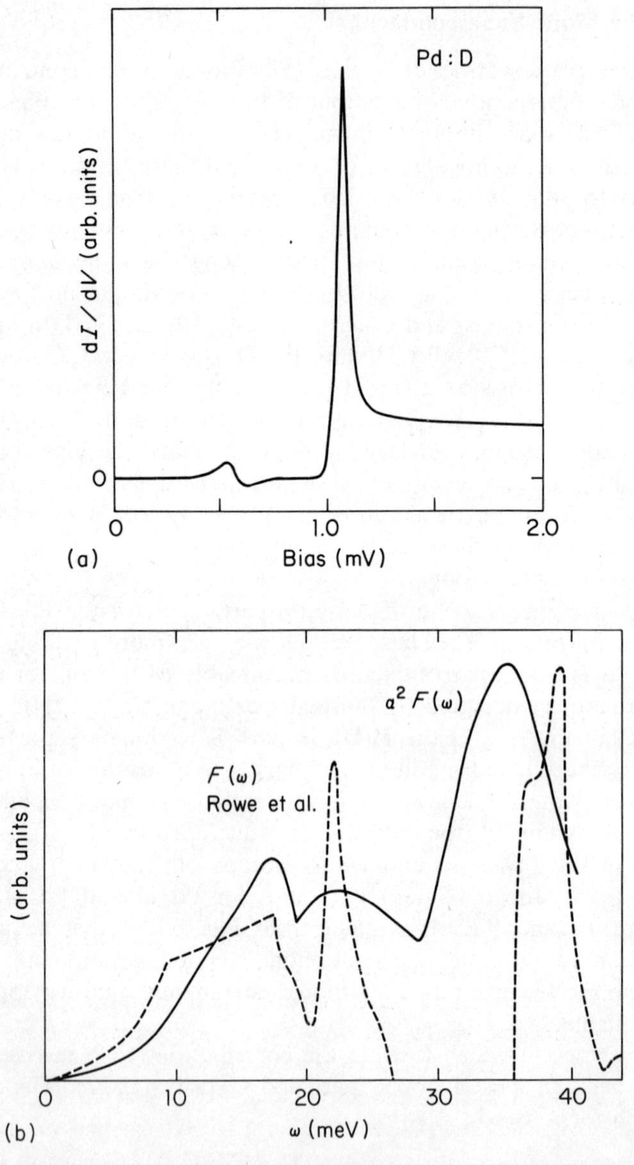

Fig. 6.26. Properties of Pd:D. (a) Conductance dI/dV of an Al–Al_2O_3–Pd:D junction measured at 1.1 K. These data at D/Pd = 0.9 provide $\Delta_0 = 0.82$ meV and ratio $2\Delta/kT_c = 3.6$. (b) Partial $\alpha^2F(\omega)$ of Pd:D shown as solid line in comparison with neutron $F(\omega)$ of D:Pd at D/Pd = 0.63. (Rowe et al., 1974). (After Eichler, Wühl, and Stritzker, 1975*b*.)

formalism, e.g., to NbN and NbC has been given by Gomersall and Gyorffy (1974).

6.8 Systematics of Superconductivity

The results of the full tunneling study of a superconductor—$\alpha^2F(\omega)$, μ^*, and the derivable functions $Z_{S,N}(E)$ and $\Delta(E)$—predict the behavior of several other bulk thermodynamic properties of the metal, including T_c and the isotope effect (Leavens, 1974), the critical field $H_c(T)$, conveniently expressed as the critical field deviation function

$$D(t)=\frac{H_c(T)}{H_c(0)}-\left[1-\left(\frac{T}{T_c}\right)^2\right] \tag{6.18}$$

(Fig. 6.27), and the specific heat ratio $C_{es}(T)/\gamma T$, conventionally plotted versus $t^2=(T/T_c)^2$. In addition, the systematic relations among the derived parameters λ, μ^*, $\omega_{\log}$, etc. and the measured Δ_0 and T_c can be examined.

A recent compilation of the thermodynamic properties of elemental superconductors Hg, In, Nb, Pb, Sn, Ta, Tl, and Al implied by the corresponding tunneling $\alpha^2F(\omega)$ results has been provided by Daams and Carbotte (1981), who have also included effects of gap anisotropy in their calculations.

An example of the change in a thermodynamic property that may be implied by a shift of spectral weight in $\alpha^2F(\omega)$ is given in Fig. 6.27a, after Mitrović and Carbotte (1981), which compares the function $D(t)$, eq. (6.18), calculated for $Nb_{0.75}Zr_{0.25}$ with that calculated for Nb. Even though the latter curve is in rough agreement with theoretical calculations (Butler, Pinski, and Allen 1980*b*) and with experimental results obtained recently for pure Nb by Kerchner, Christen, and Sekula (1981) (Fig. 6.27b), refinement of the experiments and of the analysis to include paramagnon effects may be needed; see also Daams and Carbotte (1980) and Baquero, Daams, and Carbotte (1981). Unfortunately, the corresponding experimental results for the alloy are apparently not available. In other cases as well, e.g., V_3Si (Mitrović and Carbotte, 1982) where the $\alpha^2F(\omega)$ is not well known but the critical field function $D(t)$ has been measured, such calculations can be useful to put bounds on acceptable $\alpha^2F(\omega)$ functions. It will be important to exploit this method as a working tool as better experimental tunneling and critical field data become available.

Similarly, comparison of experimental and calculated $D(t)$ functions for Nb has been used as a check on the strength of the paramagnon suppression of superconductivity in this case (Baquero, Daams, and Carbotte, 1981).

A useful representation of the systematic relations between the superconductor parameters T_c, $\omega_{\log}$, and λ is the plot (Allen and Dynes, 1975*b*) of $T_c/\omega_{\log}$ vs. λ, shown in Fig. 6.28a (omitting recent data for

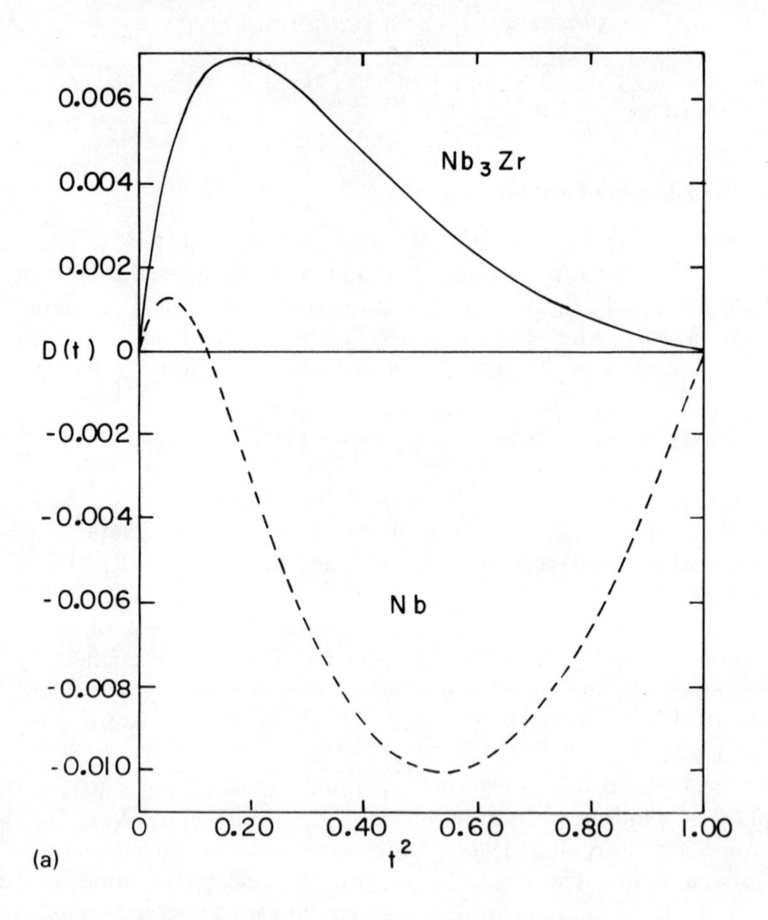

Fig. 6.27. (a) Magnetic field deviation function $D(t)$ calculated from tunneling data for Nb and $Nb_{0.75}Zr_{0.25}$. (After Mitrovic and Carbotte, 1981.) (b) The $D(t)$ for Nb, including effect of anisotropy. (After Kerchner, Christen, and Sekula, 1981.)

transition metals). It should be emphasized that plotting these data points required the λ and $\omega_{\log}$ parameters deduced from the tunneling $\alpha^2F(\omega)$ function but did not involve that most elusive and uncertain of quantities deduced from tunneling, namely μ^*. As a purely empirical representation of a large set of superconductors, a surprising degree of regularity is found. This regularity is summarized rather well by the Allen–Dynes T_c expression, including the prefactors, (4.19 and 4.21), with the choice $\mu^* = 0.1$. It is found that variation of μ^* in the Allen-Dynes expression creates a family of roughly parallel curves with a reduction in $T_c/\omega_{\log}$ on the order of 0.002 from an increase of 0.01 in μ^*.

The addition of points representing V and Nb and new results for Ta to the Allen-Dynes plot is shown in Fig. 6.28b, after Burnell et al. (1982).

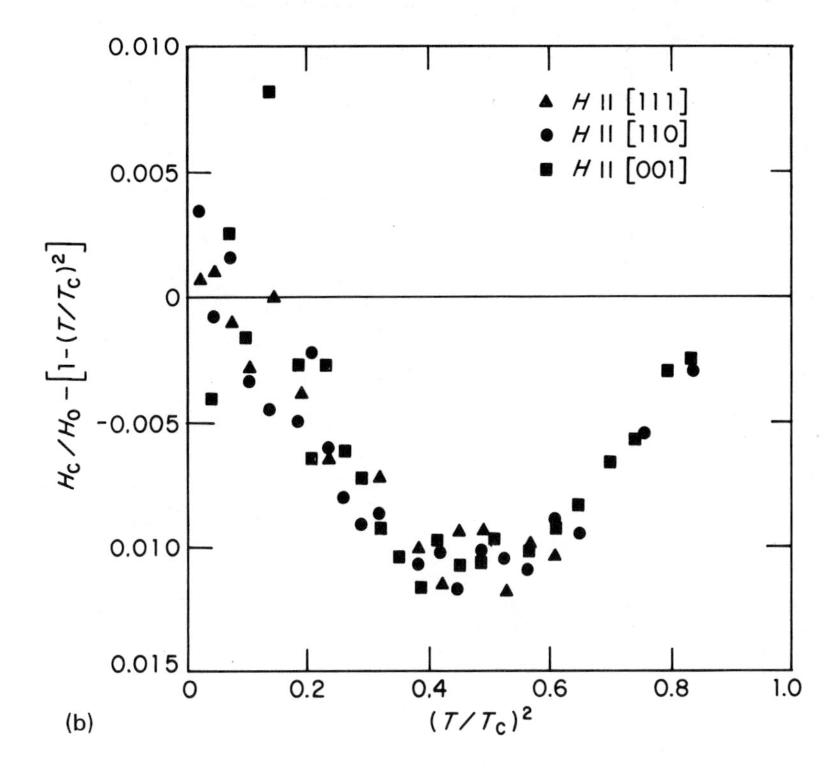

While the points for Ta fall close to the trajectory formed by the tabulated superconductors, the points for Nb (solid square) and V (solid triangle) are noticeably depressed. On the possibly naive assumptions (a) that the depression arises solely from spin fluctuations in V and Nb, (b) that the corresponding points would otherwise fall among the tabulated collection of points, and (c) that the rescaling analysis of Daams, Mitrović, and Carbotte (1981) is valid, the λ_S parameters for V and Nb have been estimated. This estimation is done simply by restoring the V and Nb points to the line (along trajectories shown dashed, generated by use of the Allen-Dynes formula for interpolation). The resulting parameters for Ta, Nb, and V are summarized in Table 6.6.

The resulting values for λ_S for Nb and V, 0.05 and 0.07, respectively, are much smaller than theoretical estimates, 0.21 and 0.34, respectively, given by Rietschel and Winter (1979). Intermediate values are given by Orlando and Beasley (1981). A possible resolution of the discrepancy in values of λ_S has recently been provided by Leavens and MacDonald (1983), who have incorporated in their reanalysis of the data shown in Fig. 6.28b an additional parameter describing the energy of the peak in the paramagnon spectrum. [To establish their rescaling analysis, Daams, Mitrović, and Carbotte (1981) took the paramagnon peak at 0.4 eV, a rather low energy.] The work of Leavens and McDonald suggests that the

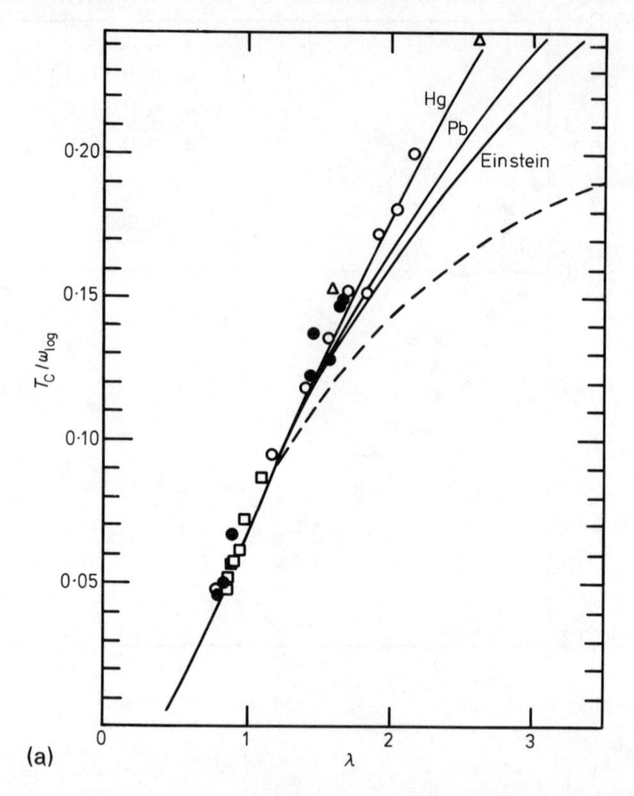

Fig. 6.28. Systematics of superconductivity as inferred from tunneling experiments. (a) Plot of $T_c/\omega_{\log}$ vs. λ reveals empirical regularity in the behavior of s–p superconductors. (After Allen and Dynes, 1975*b*.) (b) Some transition metal superconductors fall below the trajectory, as would be expected on the suggestion of spin fluctuations. Here, the solid circle represents V; the solid square represents Nb; the solid triangle represents Ta. Dashed lines connecting to open symbols correspond to paramagnon parameters $\lambda_S = 0.07$ for V and 0.05 for Nb. (After Burnell et al., 1982.)

estimated λ_S values for Nb and V in Table 6.6 should be increased by about a factor of 2 to the order of 0.1 and 0.15, respectively.

6.9 Effects of External Conditions and Parameters on Strong-Coupling Features

The properties of strong-coupling superconductors with various boundary conditions and under the influence of externally applied hydrostatic pressure, uniaxial strain, etc. are accessible in principle to tunneling study. The effect of temperature on the tunneling density of states in the phonon range has already been mentioned in Chap. 4.

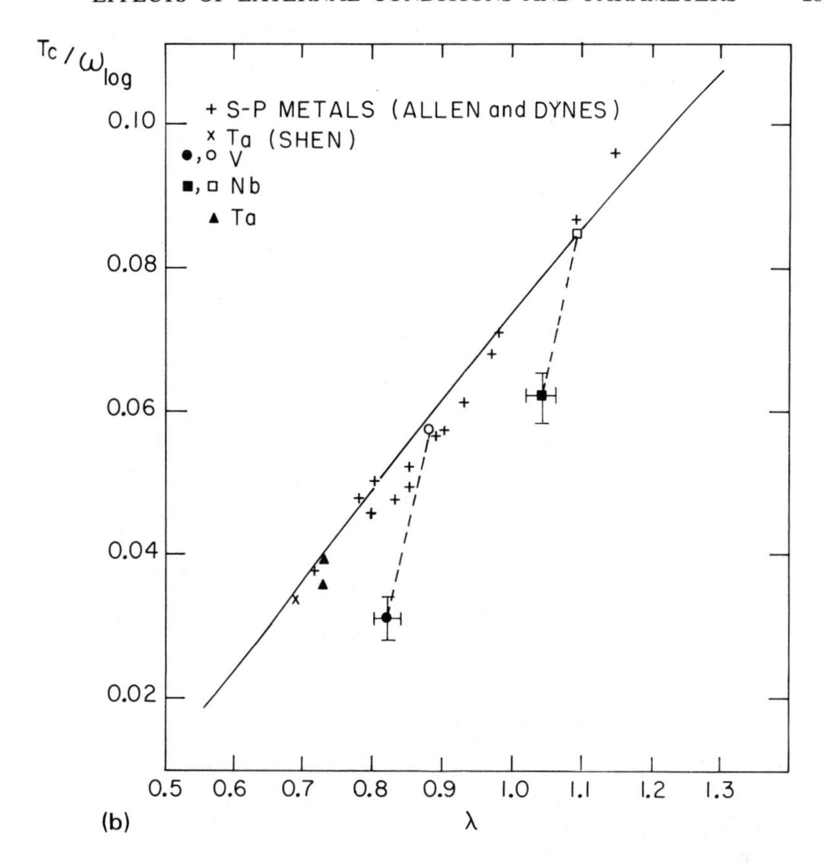

A somewhat novel example of such an external effect, resulting from backing of the Pb film under study by layers of Cd (Toplicar and Finnemore, 1977) is shown in Fig. 6.29. The motivation of this study of effects of an externally applied boundary condition was to look for possible changes in the phonon spectrum of the Pb due to its mechanical coupling with the Cd. The question of the possible importance of such effects in modifying the properties of lamellar PbCd (NS) structures, or in multifilamentary Nb–Cu wires, led to a practical interest to this question.

Table 6.6. Superconductive Parameters of Ta, Nb, and V (Burnell et al., 1982)

Element	T_c (K)	λ_{exp}	ω_{log} (K)	λ_{ph}	λ_S	$m^*/m = 1 + \lambda_{ph} + \lambda_S$
Ta	4.47	0.73	120	0.73	—	1.73
Nb	9.25	1.04	147	1.09	0.05[a]	2.14
V	5.35	0.82	172	0.88	0.07[a]	1.95

[a] See text.

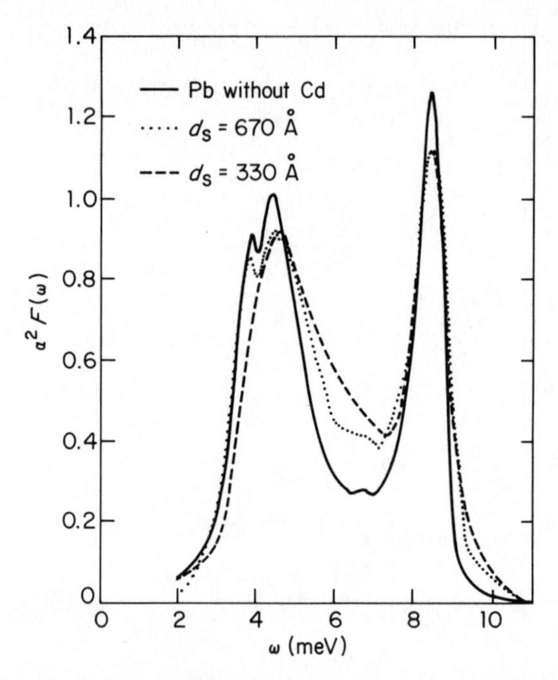

Fig. 6.29. Spectral functions $\alpha^2F(\omega)$ of Pb films backed by Cd. Solid curve is no Cd; dotted curve is for Pb thickness of 670 Å; dashed curve is for Pb thickness of 330 Å. The data have been analyzed by assuming a modified renormalization factor $Z(\omega)$ incorporating lifetime broadening representing escape of quasiparticles into the Cd layer. (After Toplicar and Finnemore, 1977.)

The central result of the work, suggested by the $\alpha^2F(\omega)$ curves of Fig. 6.29, is that the Pb–phonon-induced structure observed from the Pb side of the Pb–Cd proximity samples retains the bulk Pb features down to Pb thickness 300 Å. The deviations that are seen are mainly in the form of broadening which was treated by incorporating a lifetime-broadening term into the renormalization function in the Pb. Physically, this term is related to the lifetime against loss of an electron from the Pb into the Cd and is not related to mechanical coupling. This result seems to imply a diffusive electronic transmission at the NS boundary.

The application of hydrostatic pressure to a superconductor provides a variation in the lattice constant and, hence, of the electron-phonon coupling. A natural system in which to study the effects of cell volume on the $\alpha^2F(\omega)$ function, gap parameter, and T_c is Pb, and, indeed, a number of experimental (Franck, Keeler, and Wu, 1969; Hansen, Pompi, and Wu, 1973; Wright and Franck, 1977) and theoretical (Trofimenkoff and Carbotte, 1970) studies have been published for Pb.

The experimental difficulties with high-pressure tunneling measurements are considerable, stemming in large part from the fragile nature of

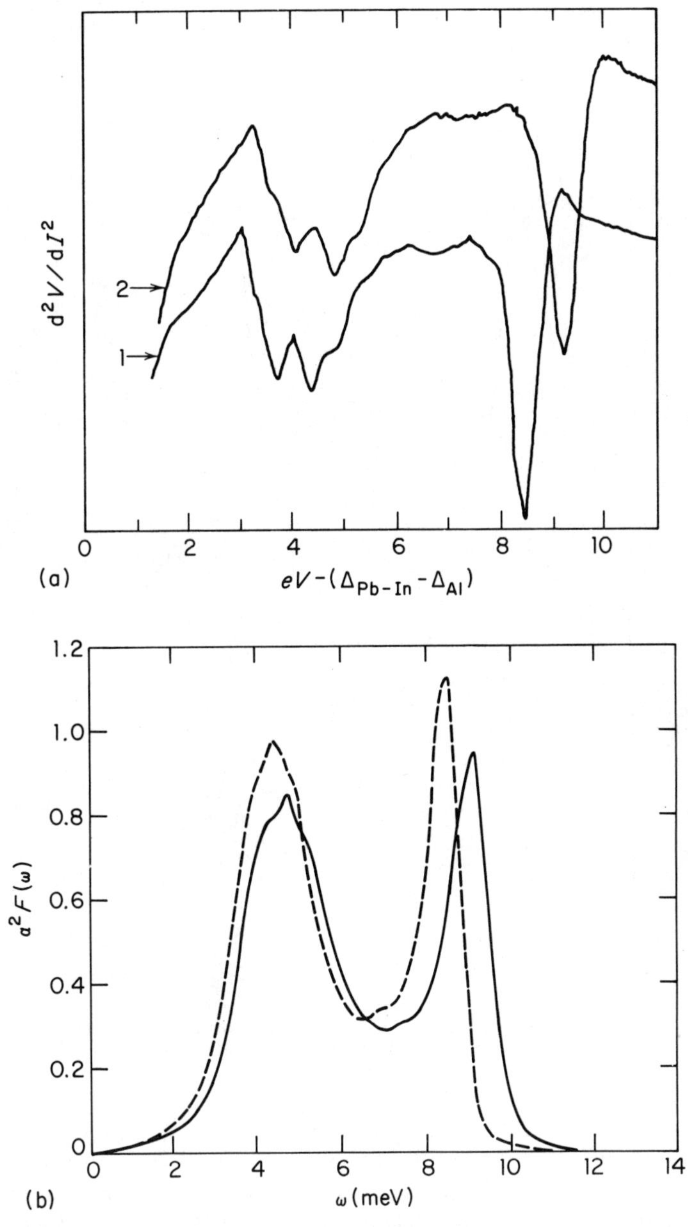

Fig. 6.30. Effects of pressure on the phonon spectra of Pb. (a) Directly measured d^2V/dI^2 curves of Pb: curve 1 is at zero applied pressure; curve 2 at a pressure of 12.2 kbar. (b) The $\alpha^2F(\omega)$ of Pb at zero pressure (dashed curve) and at 12.2 kbar (solid curve). (After Svistunov, Chernyak, Belogolovskii, and D'yachenko, 1981.)

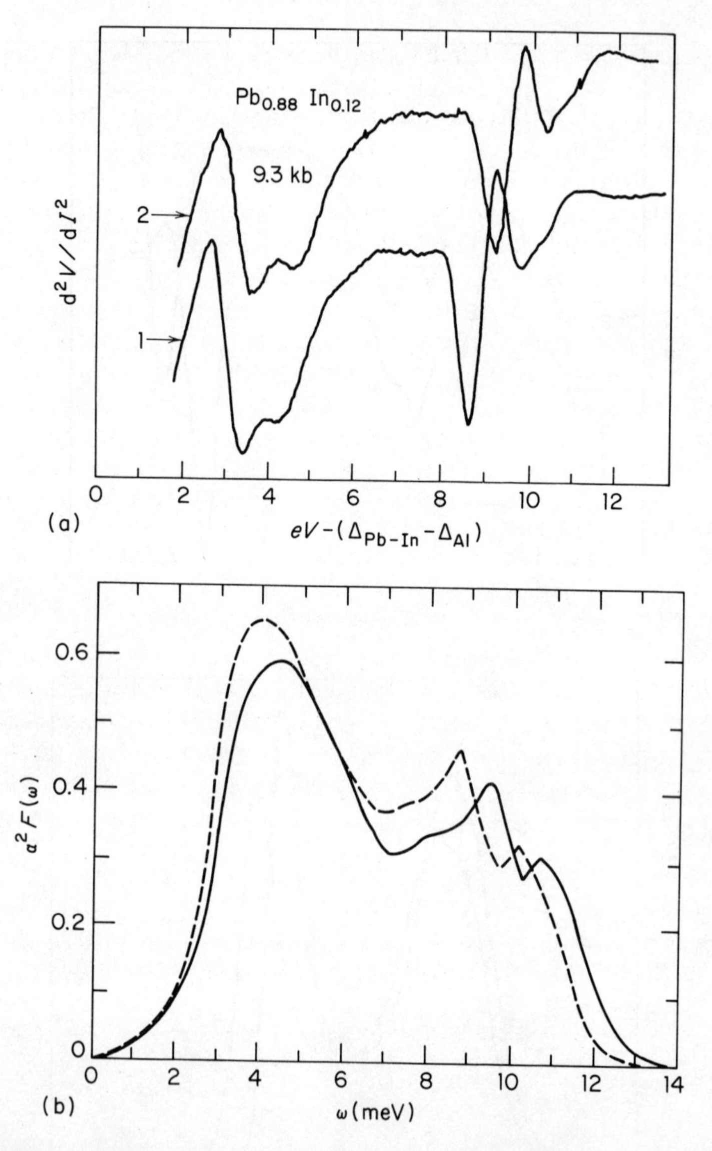

Fig. 6.31. Properties of $Pb_{0.88}In_{0.12}$ alloy under pressure. (a) Directly measured d^2V/dI^2 curves: curve 1 is at zero pressure; curve 2 is at 9.3 kbar. (b) The $\alpha^2F(\omega)$ of $Pb_{0.88}In_{0.12}$ determined from spectra in (a) at zero pressure (dashed curve) and at 9.3 kbar (solid curve). (After Svistunov, Chernyak, Belogolovskii, and D'yachenko, 1981.)

the tunneling barrier. A definitive study of Pb and Pb–In alloys up to 12 kbar has recently been reported by Svistunov, Chernyak, Belogolovskii, and D'yachenko (1981), whose data are shown in Figs. 6.30 and 6.31. The reliability of the Al–Al_2O_3–Pb tunnel junctions was found by Svistunov and colleagues to be improved by a fabrication procedure involving deposition of Al onto a cold (between 80 and 120 K) substrate, warming slowly in vacuum and oxidation in 0.1 atm of dry oxygen at 320 to 340 K for several minutes. Following deposition of the Pb or PbIn alloy (this was done rapidly, at 200–500 Å/sec), an anneal of the completed junction at ~320 K for hours or even days was found useful in reducing stress in the junctions. The data that result are in any case of the finest quality.

The directly traced d^2V/dI^2 spectra of Pb at 0 and 12.1 kbar applied pressure are shown in Fig. 6.30a, from which the hardening of the longitudinal phonon from 8.4 to about 9.2 meV is clearly seen. The corresponding 12.2 kbar $\alpha^2F(\omega)$ function (Fig. 6.30b) is seen also to be diminished in magnitude: the combination of hardening and reduction of magnitude in $\alpha^2F(\omega)$ causes a change in λ from 1.55 at zero pressure to 1.38 at 12.2 kbar. The parameters for Pb and for three In concentrations are collected in Table 6.7.

The corresponding behavior of $Pb_{0.7}In_{0.3}$ at 0 and 11 kbar is shown in Fig. 6.31. The phonon frequency shifts $\Delta\omega_i$ are discussed in terms of parameters $\gamma_i^* = \Delta\omega_i/\omega_i P$, where P is the pressure, and microscopic Grüneisen parameters $\gamma_i = \gamma_i^*/\kappa$, where κ is the compressibility. As the In concentration is increased, the fractional change in frequency of the transverse (low-energy) modes is always greatest, and the least change is

Table 6.7. Electron-Phonon Interaction Parameter Variations in Pb and Pb–In Alloys Under Pressure (Svistunov, Chernyak, Belogolovskii, and D'yachenko, 1981)

Interaction Parameter[a]	Pb P (kbar) 0	Pb P (kbar) 12.2	$Pb_{95}In_5$ P (kbar) 0	$Pb_{95}In_5$ P (kbar) 10	$Pb_{88}In_{12}$ P (kbar) 0	$Pb_{88}In_{12}$ P (kbar) 9.4	$Pb_{70}In_{30}$ P (kbar) 0	$Pb_{70}In_{30}$ P (kbar) 11
μ^*	0.123	0.126	0.132	0.129	0.143	0.147	0.143	0.143
λ	1.55	1.38	1.56	1.44	1.57	1.45	1.57	1.43
$\bar{E}$ (meV)^2[b]	23.2	23.8	23.9	24.2	24.5	25.1	25.0	25.8
d ln λ/dP (10^{-6} bar^{-1})	−9.2		−7.8		−8.2		−8.14	
d ln $\bar{E}$/dP (10^{-6} bar^{-1})	+2.2		+1.3		+2.7		+3.2	
d ln Δ_0/dP (10^{-6} bar^{-1})	−8.8±0.3		−7.9±0.3		−9.2±0.3		−8.3±0.3	

[a] Measurements at 1.17 K.
[b] $\bar{E} = \int_0^\infty \alpha^2F(\omega)\,d\omega$.

observed in the In local mode (~11 meV):

$$\gamma_t^* > \gamma_\ell^* > \gamma_L^*$$

The significance of the tabulated quantity $\mathrm{d}\ln\bar{E}/\mathrm{d}P$ is discussed in terms of the definition

$$\bar{E} = \int_0^\infty \alpha^2 F(\omega)\,\mathrm{d}\omega$$

and the McMillan (1968c) relation $\bar{E} = N(0)\langle I^2\rangle/2M$, which implies that $\bar{E}$ is independent of lattice frequencies, following an expression for $\alpha^2F(\omega)$ given by Carbotte and Dynes (1968). Svistunov and colleagues infer the relation

$$\frac{\mathrm{d}\ln\bar{E}}{\mathrm{d}P} = \frac{\mathrm{d}\ln\left[\int_0^1 |V(t)|^2\, t^3\,\mathrm{d}t\right]}{\mathrm{d}P}$$

where $t = q/2p_F$, with p_F the Fermi momentum, and $V(t)$ is the form factor of the screened local pseudopotential. Using an Animalu-Heine pseudopotential, a theoretical value $\mathrm{d}\ln\bar{E}/\mathrm{d}P = 1.5\times10^{-6}\,\mathrm{bar}^{-1}$ for Pb, comparing with $2.2\times10^{-6}\,\mathrm{bar}^{-1}$ experimentally, was obtained. While refinement of the theory for polyvalent metals may be required, the basic theoretical picture seems to be correct.

7

Nonequilibrium Effects in Tunneling

We now consider relaxations of a general assumption that heretofore has been made in this book, namely, that the physical systems of interest, and notably the tunneling electrodes, are essentially in thermal equilibrium. Strictly, this assumption cannot be a good one in any tunnel junction at finite bias V because the tunneling process itself typically provides injection of quasiparticles over a range, $0 \leqslant E \leqslant eV$, at a rate approximately varying with energy as $\exp[(E-eV)/E_0]$ [see Eq. (2.43)]. This injected distribution is distinctly nonthermal in nature. The degree to which the steady-state occupation $f_k(E)$ of electron states $\vec{k}$, E_k in the electrode differ from the thermal Fermi distribution $f_k^0(E, T)$ appropriate to the lattice temperature T depends on balance between the injection and the rates of relaxation processes, principally involving phonons, which act to restore thermal equilibrium. In this context, the question of possibly nonthermal distributions of phonons is also raised.

In examining such questions in this chapter, we will find that, although the assumption of equilibrium is generally the best starting point, in an interesting variety of cases, in both normal state and especially in superconducting systems, qualitatively new effects can be linked with the occurrence of nonequilibrium conditions. The superconducting cases are particularly fascinating because of the additional processes: pair breaking and quasiparticle recombination, made possible by the pair condensate. Further, the notion of branch imbalance, or unequal occupation of quasiparticle states $\vec{k}>$, $\vec{k}<$, is found to be especially useful in the superconducting case. The nonequilibrium state of a superconductor, as we will see, can actually be one of effectively *lower* temperature, implying enhancement of T_c and gap parameter. Finally, the superconducting state also allows, as first demonstrated by Gray (1978*a*), the construction of three-electrode transistorlike devices, now including the QUITERON (Faris, Raider, Gallagher, and Drake, 1983), whose function of amplification arises from controlled nonequilibrium conditions.

We turn first to discussion of nonequilibrium states in normal-state systems, sometimes generated or detected, however, by superconducting tunnel junctions.

7.1 Nonequilibrium in the Normal State

A basic form of nonequilibrium state occupation is generated by the tunneling process itself in a normal-state tunnel junction at bias eV. This

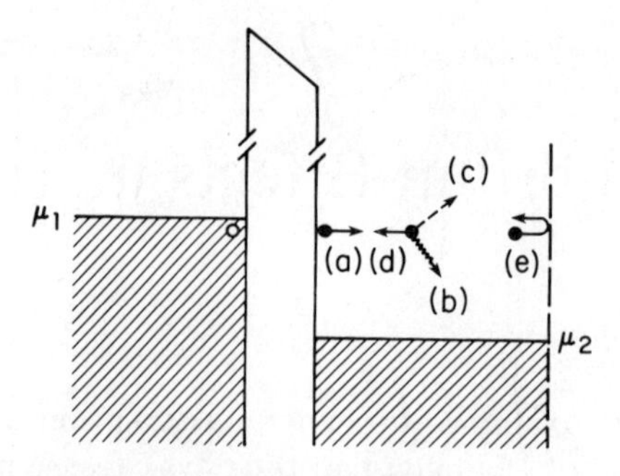

Fig. 7.1. Processes affecting (nonequilibrium) electrons (a) injected into the right-hand electrode in normal metal tunneling may include emission of a phonon (b), elastic scattering (c) and (d), and boundary scattering (e). (After Trofimenkoff, Kreuzer, Wattamaniuk, and Adler, 1972.) If the scattering rates associated with these processes are too small, then the occupation function f at the injection energy may exceed the equilibrium Fermi function value, blocking the particular states to further tunnel transitions.

effect has been studied by Trofimenkoff, Kreuzer, Wattamaniuk, and Adler (1972) and Adler, Kreuzer, and Straus (1975), as sketched in Figs. 7.1–7.3. Namely, the states $E(\vec{k})$ in the normal electrode in the range $0 \leq E \leq eV$ above the Fermi surface and specifically those with $\vec{k}$ lying in the "tunneling cone" of directions near normal incidence, receive injected electrons (Fig. 7.1). This tends to increase their occupation probability $f_k(E)$ and thus, via the Pauli principle, to block these states to further tunnel transitions. In the same fashion, the states at and below E_F in the other electrode may be depleted. This effect is described by Trofimenkoff and colleagues and Adler and colleagues by a blocking factor $g(E, T)$ representing the decrease, relative to the equilibrium case, in the occupation factor difference on the (left) between electrode 1 and 2 (right):

$$f_1 - f_2 = (f_1^0 - f_2^0)[1 - g(E, T)] \tag{7.1}$$

The blocking factor is evaluated as

$$g(E, T) = \frac{1/\tau_B}{1/\tau_B + [\tau_1(eV - E, T) + \tau_2(E, T)]^{-1}} \tag{7.2}$$

by simply requiring balance between injection and relaxation rates. Here, τ_1 and τ_2 are times for inelastic relaxation on left and right, and the inverse of the barrier tunneling lifetime τ_B is

$$\tau_B^{-1} = \frac{v_z P(v_z)}{2L} \tag{7.3}$$

This is the rate of tunneling of a given electron in the left electrode, of thickness L, through the barrier to the right. In each electrode the

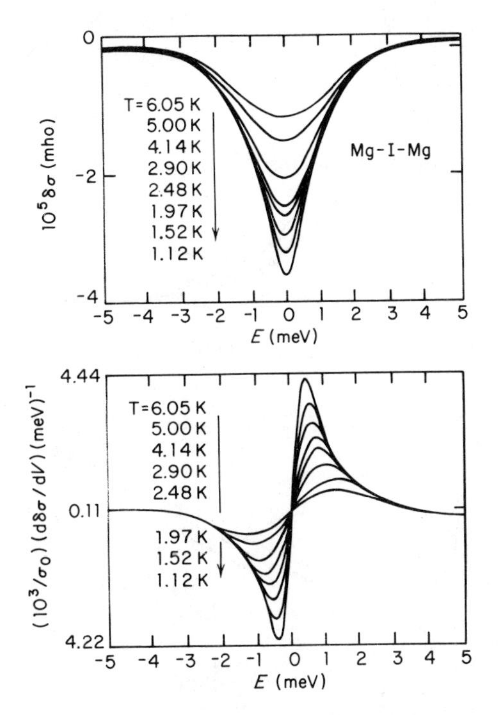

Fig. 7.2. Weak conductance minimum observed in Mg–MgO–Mg tunnel junctions, after subtraction of a background. (After Adler, Kreuzer, and Straus, 1975.) The conductance at zero bias, σ_0, is 6.56×10^{-3} mho for this junction, so that $\Delta\sigma/\sigma \simeq 10^{-2}$. Lower curves display second derivative $d\sigma/dV$.

expression

$$\tau^{-1} = \tau_{ep}^{-1} + \tau_i^{-1} \tag{7.4}$$

describes the total rate of relaxation due to electron-phonon processes (τ_{ep}^{-1}) and impurity scattering (τ_i^{-1}).

In (7.3), v_z is the component of the electron velocity normal to the barrier, and $P(v_z)$ is the corresponding barrier transmission probability. The possibility of structure near $V = 0$ in the tunneling conductance arises from the fact that τ_{ep} becomes very long at the Fermi surface ($E = 0$) because the phonon density of states $F(E)$ becomes small as $E \to 0$. In the approximation

$$\tau_{ep}^{-1} = \frac{\pi a}{\hbar}[(eV)^2 + (\pi kT)^2] \tag{7.5}$$

the conductance dip is described by

$$\frac{\sigma(0) - \sigma(V)}{\sigma(0)} = \frac{\tau_B^{-1}}{\tau_B^{-1} + \tau_i^{-1} + (\pi/\hbar)a[(eV)^2 + (\pi kT)^2]} \tag{7.6}$$

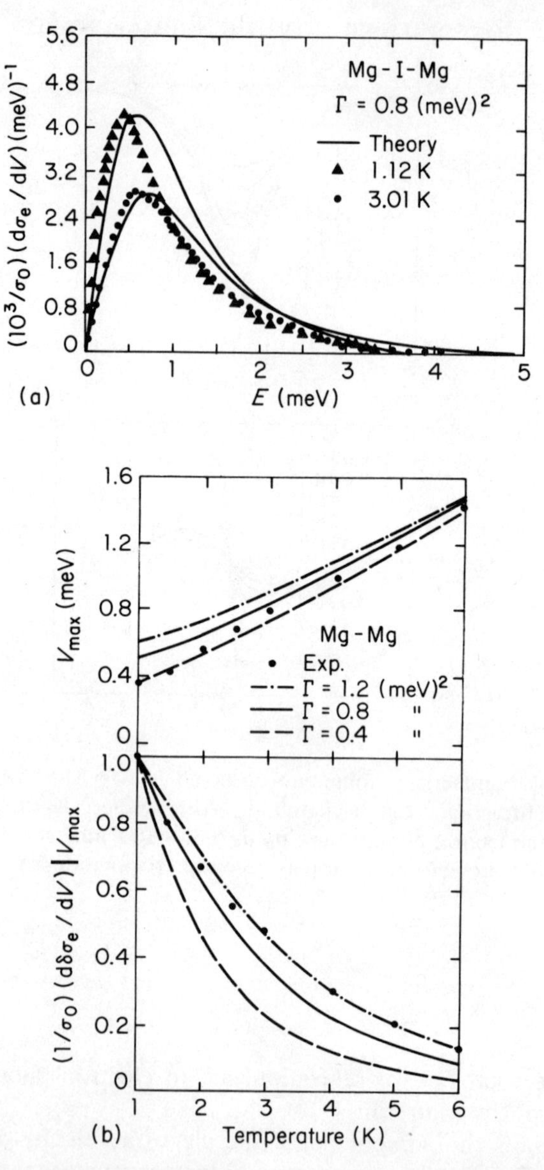

Fig. 7.3. Fit of conductance minimum of Mg–MgO–Mg junction to blocking model. (a) Comparison of derivative $d\sigma/dV$ data (triangles measured at 1.12 K, circles measured at 3.01 K) with theory (solid line) choosing $\Gamma = 0.8\,(\text{meV})^2$. (b) Plots of the experimental temperature dependence of the second-derivative peak position (V_{max}) (upper panel) and second-derivative peak magnitude (lower panel) shown as solid points in comparison with curves based on the blocking model with parameter values indicated. (After Adler, Kreuzer, and Straus, 1975.)

Observations of Mg–MgO–Mg junctions, shown in Fig. 7.2, are interpreted along the lines indicated. The T dependence of the experimental result comes partly from sharpening of the Fermi edge with falling T, neglected in (7.6), and partly from the T dependence of the relaxation times. For comparison of results with theory, an adjustable parameter $\Gamma \equiv \hbar/\pi a \tau_i$ is determined. The typically small magnitude, here $\delta\sigma/\sigma \simeq 10^{-1}$, and other features of the effect appear to be adequately described by the model, as illustrated in Fig. 7.3. Reasonable parameter values $\Gamma = 0.3\,(\text{meV})^2$, $a = 0.0015\,\text{meV}^{-1}$, $\tau_i = 2\times 10^{-10}$ sec, and $\tau_B = 10^{-7}$ sec are obtained with this fit to the conductance minimum.

Nonequilibrium distributions of phonons are also possible and of particular interest because of their role in quasiparticle recombination in superconductors. An ingenious experiment is illustrated in Fig. 7.4, after Dynes, Narayanamurti, and Chin (1971), in which Sn–SnO_2–Sn superconducting tunnel junctions, labeled (phonon) generator and detector, are cemented to opposite [1$\bar{1}$0] faces of a Ge:Sb single crystal. The paper shows how the sublevels of the Sb hydrogenic donor ground state are split by an applied [111] compressive stress, allowing the experimenters to vary the energy of maximum absorption of phonons. The trace in this figure, taken with the generator junction biased to $eV = 2\Delta_{Sn}$, shows a minimum detection rate of phonons at 1.09×10^8 dyn/cm^2, corresponding

Fig. 7.4. Observation of nonequilibrium phonon distribution generated by Sn–SnO_2–Sn tunnel junction biased above 2Δ (1.2 meV) and detected by a similar junction on the opposite (1$\bar{1}$0) face of the single crystal of germanium. Spectroscopy is achieved by tuning the phonon absorption of Sb impurities in the crystal with a (111) stress. A pronounced minimum in phonon transmission occurs at 1.1×10^8 dynes/cm^2 when the absorption energy equals 2Δ. (After Dynes, Narayanamurti, and Chin, 1971.)

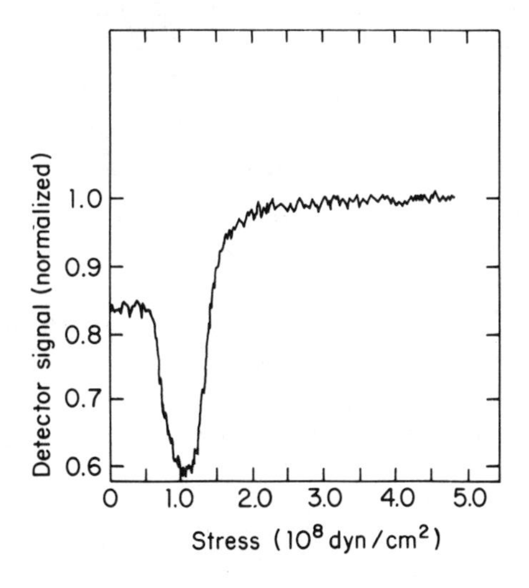

to absorption energy near to 1.2 meV, which corresponds to $2\Delta_{Sn}$. This demonstration of the tunable donor absorption as a phonon spectrometer also reveals that the Sn–SnO_2–Sn tunnel junction is an effective phonon generator at $\hbar\omega_{ph} = 2\Delta$.

Detection is accomplished with the Sn–SnO_2–Sn detector junction biased *below* the sum gap 2Δ, so that no current flows until quasiparticle pairs are generated by phonons, which must thus have energy $\hbar\omega_{ph} \geqslant 2\Delta$ to be detected. It was found, further, that increasing the bias of the generator junction beyond 2Δ, up to $eV = 10\Delta$, produced no measurable change in the spectrum from that of Fig. 7.4. The conclusion is that the higher-energy excitations (phonons and/or injected electrons) rapidly degenerate into superconducting pairs and multiple phonons of energy $\hbar\omega_{ph} \leqslant 2\Delta$. This result makes the tunnel junction biased to $eV = 2\Delta$, effectively a generator of monochromatic phonons at $\hbar\omega = 2\Delta$. By use of low normal-state resistances ($\sim$10 mΩ, with junction area $\sim$1 mm^2) phonon power levels of the order a few milliwatts are possible. Measurement of the ballistic propagation times of phonon pulses across the $\sim$0.4-cm width of the Ge bar confirmed the picture described and also allowed resolution of longitudinal (L) and fast transverse (F.T.) phonon contributions.

As we will see later, the situation inferred in the Sn–SnO_2–Sn generator illustrates an aspect of the problem of measurement of the quasiparticle relaxation time in a superconductor, such as one of the Sn films in the generator. At high bias, e.g., $eV = 10\Delta$, the observed output occurs in phonons of energy 2Δ. Hence, the primary phonon of energy 10Δ was trapped, presumably regenerating pairs of quasiparticles, which subsequently decayed, producing up to five phonons of energy 2Δ. The lifetime of the excess quasiparticle population can thus extend well beyond the lifetime for primary phonon generation.

A second point clearly illustrated by this experiment is that a distinctly nonthermal phonon distribution can persist, as in the insulating Ge bar, at distances the order of a centimeter or more from points of phonon generation.

7.2 Nonequilibrium in the Superconducting State

The phonon mean free path in a metal, in contrast to the insulating Ge, is likely to be much reduced by scattering of free electrons. In the superconducting state, however, for phonon energies $\hbar\omega_{ph} < 2\Delta$, this possibility for scattering again disappears, and ballistic propagation of phonons over millimeter distances becomes possible, as shown, e.g., in Fig. 7.5.

In this figure, ballistic propagation of phonons, generated by laser pulse heating of a (110) surface of a tetragonal β–Sn crystal and detected by Al–Al_2O_3–Al junctions after 1 and 4 mm traversals, is shown as a function of temperature. This particular detector is sensitive only to phonons of $\nu \geqslant 160$ GHz (0.66 meV). The rapid loss of detected phonon

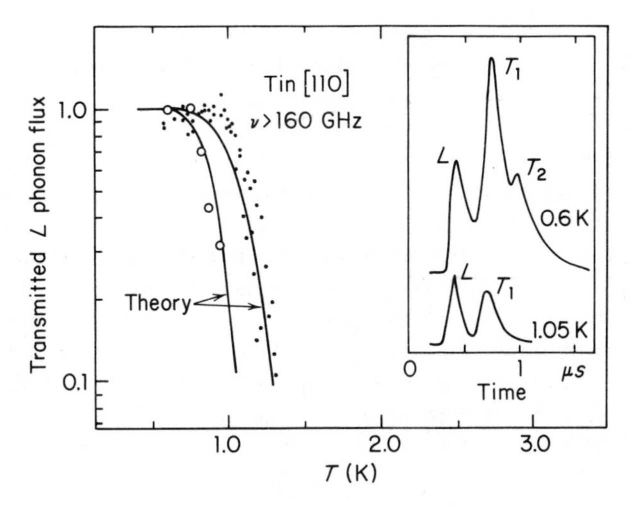

Fig. 7.5. Ballistic propagation of longitudinal (L) phonons, generated by a heat pulse, in the [110] direction in single crystal of Sn. Granular aluminum quasiparticle tunnel junctions are used as detectors. In the inset, the propagation times of the longitudinal and transverse phonons are clearly resolved at the lowest temperatures. As shown in the left, the intensity of the transmitted phonons drops sharply at higher temperatures. This is a consequence of increased scattering from the exponentially rising number of thermal quasiparticles. (After Pannetier, Huet, Buechner, and Maneval, 1977.)

flux between 0.6 and 1.3 K corresponds to absorption by the exponentially rising thermal quasiparticle population in the superconducting Sn: $n \propto [1+\exp(\Delta/kT)]^{-1}$.

More precisely, the ratio r of phonon mean free paths $\Lambda_{S,N}$ in the superconducting and normal states is given by

$$r = \frac{\Lambda_N}{\Lambda_S} = \frac{1}{\hbar\omega} \int \left(1 - \frac{\Delta^2}{EE'}\right) [f(E) - f(E')] N_T(E) N_T(E')\, dE$$

(Bardeen, Cooper, and Schrieffer, 1957; Bobetic, 1964), where $E' = E + \hbar\omega$, $f(E)$ is the Fermi function, and $N_T(E)$ is the BCS density of states of the superconductor. Pannetier, Huet, Buechner, and Maneval (1977) have used this expression and an assumed exponential attenuation of phonon flux, flux $\propto \exp(-d/\Lambda_S)$, to generate the theoretical curves shown in Fig. 7.5. The resulting fit corresponds to $\Lambda_N = 1.2 \pm 0.3\ \mu$m for L [110] 160-GHz phonons in Sn at 1 K.

As shown in the inset, measured propagation times allow resolution of the L, T_1, and T_2 acoustic branches ($q \ll \pi/a$) of the phonon dispersion in Sn.

Pannetier, Huet, Buechner, and Maneval (1977) have also demonstrated that the phonon mean free path in the superconducting Sn falls precipitously for $h\nu_{\rm ph} \geqslant 2\Delta_{\rm Sn}$, the onset of Sn pair breaking by the phonons. This task is accomplished with an Sn–SnO_2–Sn detector tunnel junction with a variable, parallel magnetic field H applied to reduce its energy gap $2\Delta_{\rm det}(H)$ and thus to adjust or tune the *minimum* detectable phonon energy $\hbar\omega_{\rm ph} \geqslant 2\Delta_{\rm det}(H)$. An applied field of less than the bulk critical field is ineffective in the Sn crystal because of the Meissner effect. The signal disappears abruptly when $2\Delta_{\rm det}$ exceeds the bulk value $2\Delta_{\rm Sn}$ (corresponding to [100] propagation) by the strong attenuation of all phonons of $\hbar\omega_{\rm ph} > 2\Delta_{\rm Sn}^{(110)}$ by pair breaking. The combination of tuned detector with a broadband phonon source, in fact, offers a precise means of measuring the bulk energy gap $2\Delta_{\rm Sn}$. It is interesting to note that clear and accurate measurements of the gap anisotropy of Sn are obtained in this experiment, in agreement with earlier work of Morse, Olsen, and Gavenda (1959). The present results are

$$2\Delta_{\rm Sn}^{(001)} = (3.2 \pm 0.1)kT_c$$

and

$$2\Delta_{\rm Sn}^{(110)} = (3.7 \pm 0.1)kT_c$$

The previous two experiments have detected pulses of nonequilibrium phonons coupled mechanically into and out of a medium through insulating layers. By fabricating a detector tunnel junction directly on a single-crystal electrode of Pb, Hu et al. (1977) were able to detect a pulse of quasiparticles. These were generated by a laser pulse on the opposite (111) face of the Pb crystal electrode, 0.87 mm in thickness. The detector signal (the Pb–I–Pb detector junction was biased in the region of thermal quasiparticle tunneling) shown in Fig. 7.6 changes its nature from a diffusive and relatively slow heat pulse at 2.95 K to an apparently diffusive but much faster quasiparticle pulse below 2 K. The diffusive, rather than ballistic, quasiparticle transport is consistent with the shape of the pulse and with the long transit time, ~ 1.2 μsec at 1.6 K for 0.87-mm thickness. This is about ten times longer than the expected ballistic transit time. It is also consistent with analysis based on estimates of the impurity scattering time, $\tau_i = 1.5 \times 10^{-10}$ sec, consistent with the measured resistance ratio, 20,000, of the 0.87-mm Pb crystal.

An additional point is that the measured quasiparticle transit time in Pb, ~ 120 nsec, which represents the lifetime of a propagating region of excess quasiparticle number, considerably exceeds the quasiparticle recombination time τ_R, which is about 3 nsec at 1.6 K (Kaplan et al., 1976; Dynes, Narayanamurti, and Garno, 1978). The point is that the phonons generated by the recombination events have an extremely short lifetime for regeneration of quasiparticles, and the coupled phonon-quasiparticle system decays more slowly. This characteristic situation was already mentioned at the end of Sec. 7.1. A full analysis of this interesting experiment has been given by Narayanamurti et al. (1978).

In a simple and elegant tunneling experiment on films of the extreme strong-coupling alloy $Pb_{0.9}Bi_{0.1}$, Dynes, Narayanamurti, and Garno

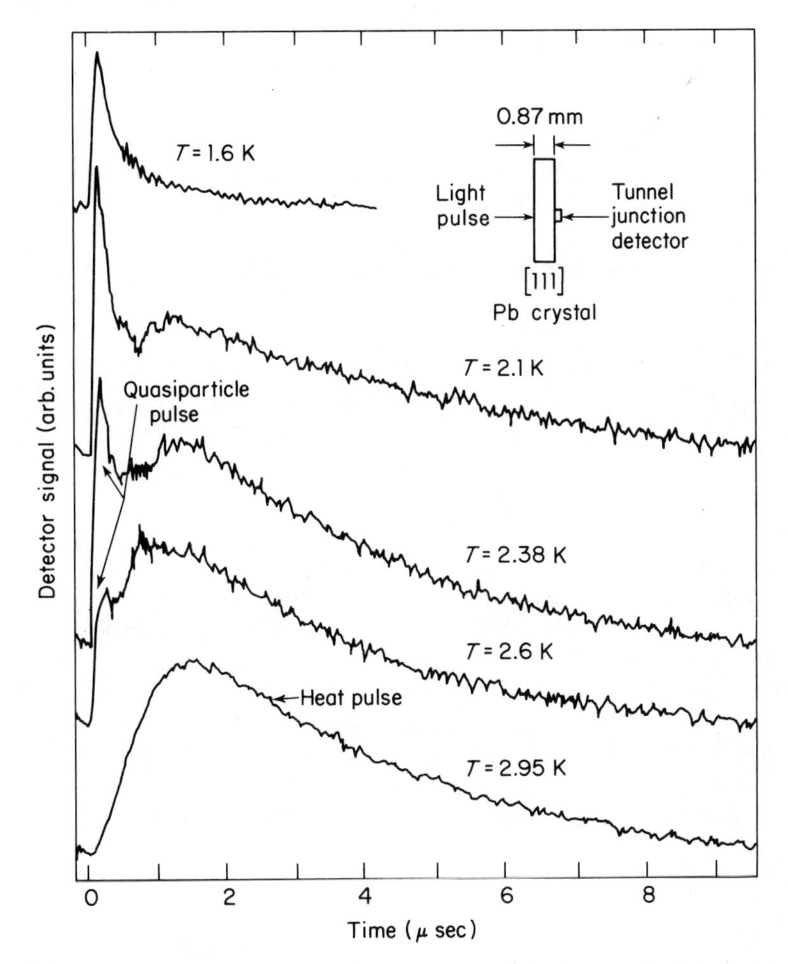

Fig. 7.6. A Pb–PbO–Pb tunnel junction, grown on a single crystal of Pb, 0.87 mm in thickness, used to detect nonequilibrium quasiparticle and phonon pulses. These pulses are generated by laser pulses on the opposite (left) side of the lead crystal. The separate and more rapidly propagating quasiparticle pulse is observed at temperatures below 2.6 K. (After Hu et al. 1977.)

(1978) have directly obtained the quasiparticle recombination time τ_R by measuring the corresponding temperature-dependent lifetime broadening of the energy gap edge. This effect can be seen directly in the I–V curves of the symmetric strong-coupling $Pb_{0.9}Bi_{0.1}$–I–$Pb_{0.9}Bi_{0.1}$ junction, shown as a function of temperature in Fig. 7.7a. The broadening shown here, in contrast to the finite discontinuities retained at higher temperatures in the Al–I–Al I–V curves of Fig. 3.10, arises from the large inherent width $\Gamma = \hbar/\tau_R$ of the quasiparticle states, arising, in turn, from strong electron-phonon coupling in this system. The strong temperature dependence is expected from the relation

$$\frac{1}{\tau_R} = \left(\frac{k_B T}{\Delta}\right)^{1/2} \frac{1}{\tau_0} e^{-\Delta/kT} \tag{7.7}$$

which corresponds to the two-particle nature of the recombination process; $1/\tau_R$ for a given particle depends on the number of other particles, $n \propto e^{-\Delta/kT}$.

These data have been analyzed using the S–I–S current expression

$$I = C_N \int_{-\infty}^{\infty} N_T(E) N_T(E+eV)[f(E) - f(E+eV)]\, \mathrm{d}E \tag{7.8}$$

The lifetime broadening is treated simply by adding an imaginary part Γ to the energy variable:

$$N_T(E, \Gamma) = \mathrm{Re}\left\{\frac{(E - i\Gamma)}{[(E - i\Gamma)^2 - \Delta^2]^{1/2}}\right\} \tag{7.9}$$

Differential I–V curves (data points in Fig. 7.7b) have been fit by variation of Γ, using (7.8) and (7.9). The excellent fits correspond to the values of Γ shown in the figure. The recombination times $\tau_R = \hbar/\Gamma$ are plotted (data points) semilogarithmically vs. $1/T$, as appropriate to (7.7), in Fig. 7.7c and compared with the curve calculated by Kaplan et al. (1976), based on the $\alpha^2 F(\omega)$ function measured for $Pb_{0.9}Bi_{0.1}$ by superconductive tunneling (Dynes and Rowell, 1975). The agreement is clearly excellent and does not involve parameters other than the measured Γ values.

7.3 Tunneling Measurements on Superconductors Under Nonequilibrium Conditions

A class of experiments in which a nonequilibrium state of a superconducting film can be generated in a controlled fashion and simultaneously monitored by tunneling is possible in the three-film, double-junction geometry indicated in Fig. 7.8. This type of experiment was first reported by Ginsberg (1962), followed by the work of Taylor (1963), of Miller and Dayem (1967), of Rothwarf and Taylor (1967), and of Gray, Long, and Adkins (1969). The objective was to study a thin central film S_2 driven into nonequilibrium by a high current density injector or generator junction S_1–I_1–S_2, biased beyond its sum gap $\Delta_1 + \Delta_2$, and sampled by a low current density detector junction S_2–I_2–S_e, typically biased below its sum gap $\Delta_2 + \Delta_3$, to collect the excess quasiparticles generated in S_2. Tunneling processes in such a double junction are sketched in Fig. 7.8b, while one possible experimental realization (Chi and Clarke, 1979) is illustrated in Fig. 7.8c.

Basic quasiparticle nonequilibrium states which may thus be created in the central film S_2 are indicated in Fig. 7.9. Equilibrium, of course, is characterized by occupation of quasiparticle states E_k according to the Fermi function $f_k(E, T) \equiv f_k^0$ evaluated at the lattice temperature T. Quasiparticle distributions f_k departing from equilibrium as $f_k - f_k^0 = \delta f_k$ can usefully be classified as departures δf_k that are even or odd under exchange of branch indices $k>$, $k<$. A special case of interest, indicated in Fig. 7.9b, is that in which an even nonequilibrium distribution is thermal in nature, i.e., can be quantitatively described by the Fermi

Fig. 7.7. Observation by tunneling of inherent recombination lifetime broadening of quasiparticle states in a strong-coupling superconductor. (a) The I–V characteristic of $Pb_{0.9}Bi_{0.1}$–I–$Pb_{0.9}Bi_{0.1}$ tunnel junction at several different temperatures. Note the gradual severe temperature smearing of the characteristic corner discontinuity at bias 2Δ, in contrast to its sharpness in weak-coupling Al–Al_2O_3–Al junctions (Fig. 3.10). A small magnetic field was applied to suppress the dc Josephson current.

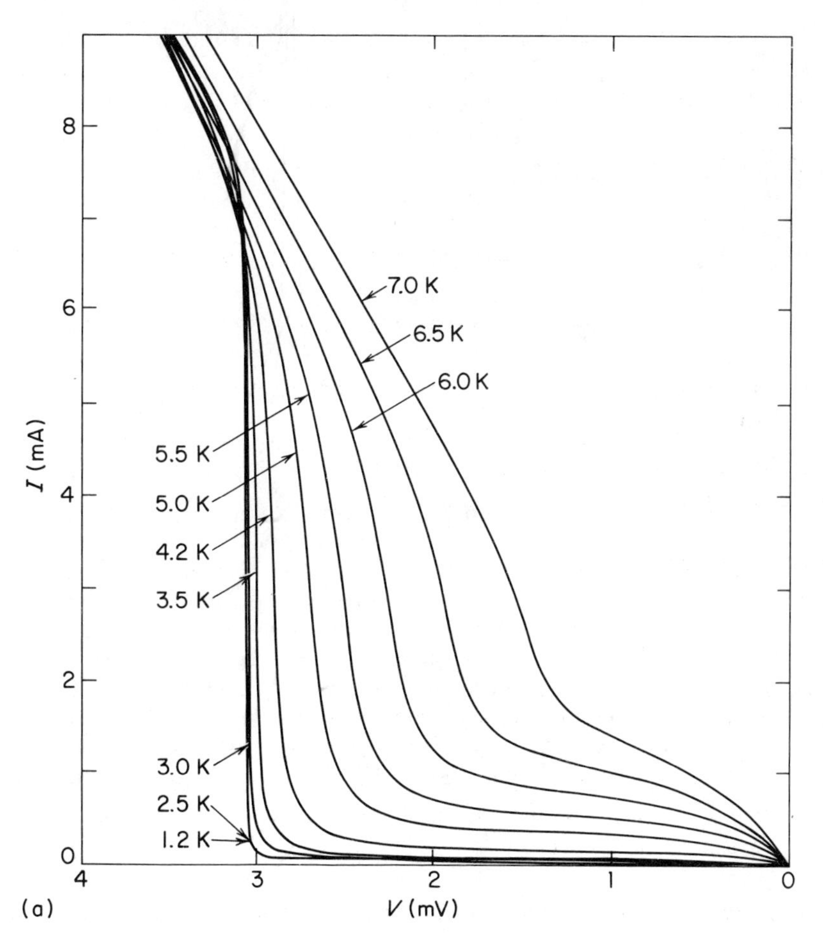

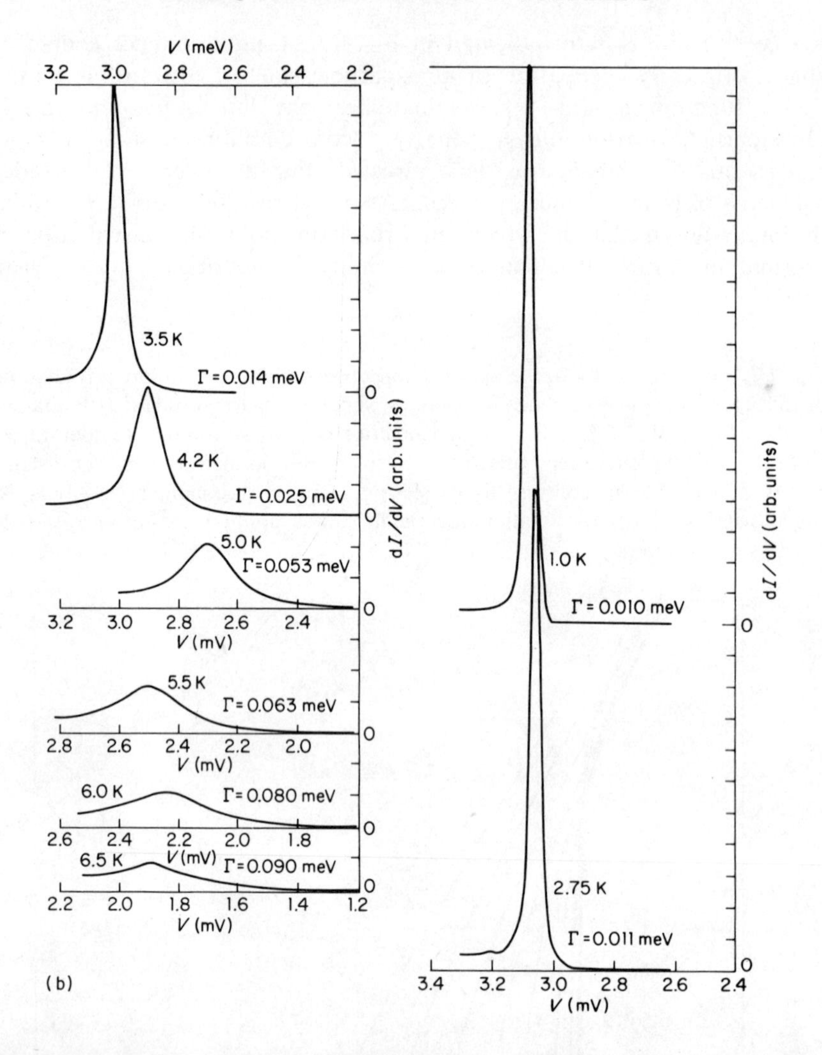

Fig. 7.7. (cont.) (b) The conductance dI/dV vs. V determined from the data shown in (a). The curves drawn through the data points are theoretical fits, assuming simply Lorentzian lifetime broadening. Note that the residual energy width at 1 K is only 10 μV; this result has been achieved, in part, by choice of alloys in which anisotropy effects are averaged out. (c) Plot of corresponding recombination time $\tau = \hbar/\Gamma$ on log scale vs. $1/T$ produces a straight line whose slope corresponds to an activation energy $\Delta = 1.48$ meV for PbBi. The theoretical curve is calculated, with no adjustable parameters, from the theory of Kaplan et al. (1976), in which the $\alpha^2F(\omega)$ function for the PbBi alloy was incorporated. (After Dynes, Narayanamurti, and Garno, 1978.)

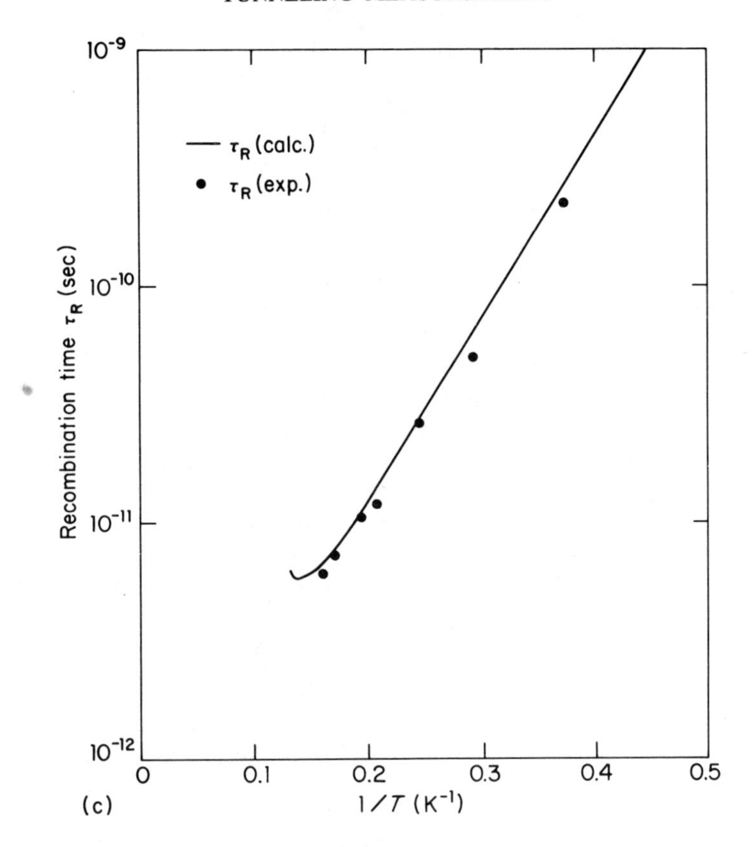

function using an effective temperature $T^* > T$. Since the electron and hole branches are equally occupied, no net quasiparticle charge density occurs. In contrast, the sketch in Fig. 7.9c illustrates the odd or branch imbalance mode of nonequilibrium. The single parameter by which this odd mode is characterized is the net quasiparticle charge Q^* (Tinkham and Clarke, 1972; Tinkham, 1972). To treat this situation quantitatively, we note that the effective charge, in units of e, associated with a quasiparticle excitation k is $q_k = \varepsilon_k / E_k$ (conventionally positive q for electrons), which changes continuously with k, is zero at k_F, and is -1 on the extreme of the holelike branch. The net charge Q^* is then

$$Q^* = \sum_k \delta f_k \frac{\varepsilon_k}{E_k} = 2N(E_F) \int_0^\infty [f_{k>}(E) - f_{k<}(E)]\, dE \qquad (7.10)$$

where the factor 2 accounts for both spin projections, understood in the k sum. As indicated in Fig. 7.9c, the branch imbalance mode requires a shift of the chemical potential for quasiparticles relative to that describing the pairs. This effect was first observed experimentally by Clarke (1972). A theoretical basis for discussion of these effects is given by Schmid and Schön (1975).

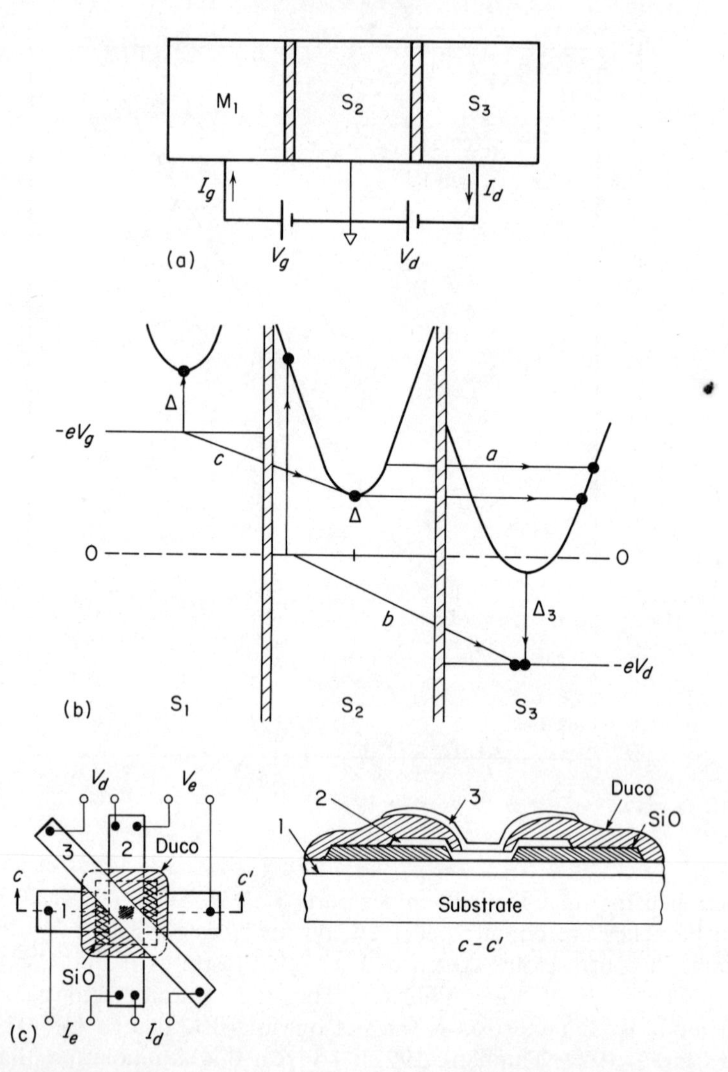

Fig. 7.8. Configuration of double-tunnel junction with a nonequilibrium state, set up in the middle film S_2 by injection from tightly coupled film 1 (the generator junction), is detected by weakly coupled junction S_2–I–S_3. (a) Biasing arrangement for double-tunnel junction configuration. (b) Schematic indication of tunnel transitions in double-tunnel junction with biases suitable for determination of τ_r. (After Paterson, 1978; Rothwarf and Taylor, 1967.) (c) Plan view of one form of the double junction at left; on right, section c–c' of the device, with film thicknesses exaggerated. Width of films 1 and 3 is 3 mm. (After Chi and Clarke, 1979.)

The first body of data to be accumulated from double-junction experiments such as illustrated in Fig. 7.8 concerns the quasiparticle pair recombination time τ_R, of which a summary for Sn is given in Fig. 7.10. Such results are obtained by using the steady-state analysis first outlined

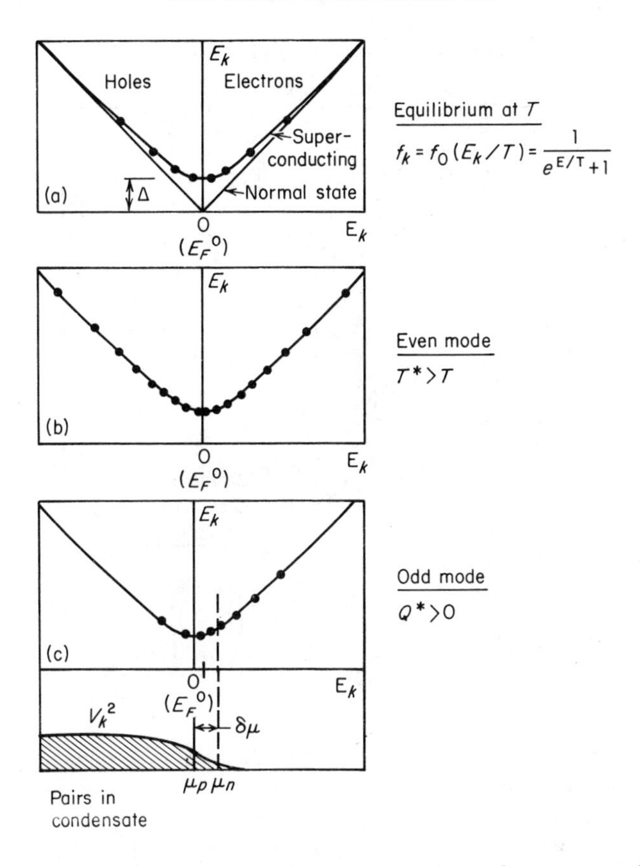

Fig. 7.9. Basic forms of quasiparticle nonequilibrium of a superconductor. (a) Dispersion curve of excitations E_k, dots suggesting thermal equilibrium occupation. (b) Schematic indication of excess quasiparticle population described by effective temperature $T^* > T$. (c) Schematic indication of quasiparticle distribution exhibiting branch imbalance corresponding to $Q^* > 0$. Note shift of μ_n and μ_p relative to the equilibrium value E_f^0. (After Tinkham, 1979.)

by Rothwarf and Taylor (1967) for the total quasiparticle density n and the density n_ω of pair-breaking phonons of energy $\hbar\omega \geqslant 2\Delta$ in the center film S_2. The experimentally measured quantity is the excess quasiparticle density Δn (assumed small compared with n) due to injection at known volume rate I_0 from the generator film. That is, $I_0 = I_i/eAd$, where I_i and A, respectively, are the injector junction current and cross-sectional area, and d is the thickness of the injected film S_2. Under the condition $eV_d < \Delta_2 + \Delta_3$, the excess quasiparticles Δn produce a directly measurable excess current ΔI_d in the detector junction according to the proportion

$$\frac{\Delta n}{n} = \frac{\Delta I_d}{I_d} \qquad (\Delta_3 \gg \Delta_2) \tag{7.11}$$

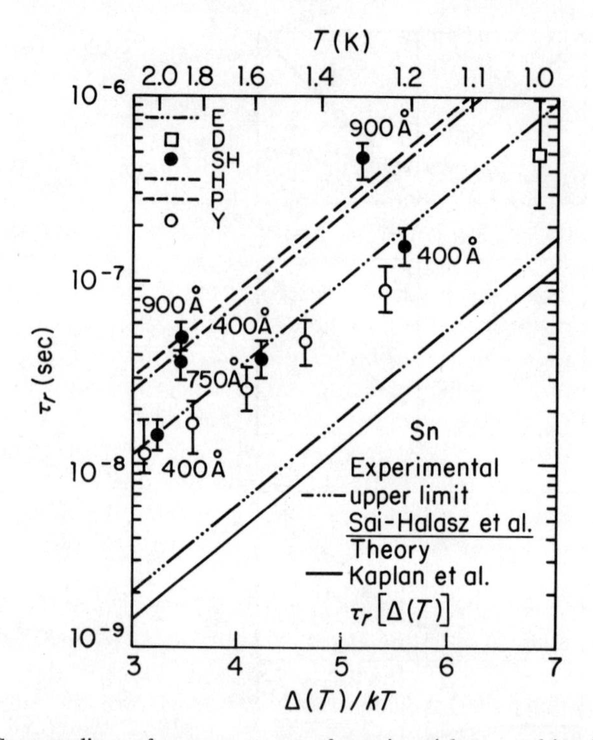

Fig. 7.10. Compendium of measurements of quasiparticle recombination lifetime τ_r for Sn compared with theory. Experimental conditions in the various experiments are of importance in relation to the phonon-trapping effect. Sources: E, Eisenmenger, 1969 (helium–4000 Å–sapphire); D, Dayem, 1972 (helium–3000 Å–saphire); SH, Sai-Halasz, Chi, Denenstein, and Langenberg, 1974 (vacuum–indicated thickness–quartz); H, Hu et al., 1977 (helium–3200 Å–glass); P, Parker, 1974 (helium–3500 Å–sapphire); open circles (helium–1600 Å–glass). (After Kaplan et al., 1976.)

The differential equation describing the excess population Δn in the injected film S_3 (Rothwarf and Taylor, 1967) is

$$\frac{\mathrm{d}\Delta n}{\mathrm{d}t} = I_0 + 2\Gamma_B \Delta n_\omega - 2\Delta n \Gamma_R \qquad (\Delta n \ll n) \tag{7.12}$$

where Γ_B is the pair-breaking rate produced by the excess phonon density Δn_ω and $\Gamma_R = 1/\tau_R$ is the recombination rate of interest. If the change in phonon population Δn_ω is negligible, the steady-state solution of (7.12) is simply

$$\Gamma_R = \frac{I_0}{2\Delta n}$$

In the more realistic case of appreciable phonon trapping, the true recombination rate is given by

$$\Gamma_R = \frac{1}{\tau_R} = \frac{I_0}{2\Delta n}\left(1 + \frac{\Gamma_B}{\Gamma_{es}}\right) \tag{7.13}$$

where Γ_B and Γ_{es}, respectively, are the pair-breaking rate by phonons and the escape rate of the phonons from the film. The phonon-trapping factor in parentheses, unfortunately, is typically neither near unity nor easy to measure or estimate, and it is believed to account, e.g., for most of the scatter evident in the τ_R data collected in Fig. 7.10.

It is important to realize that the distributions $f_{k>}, f_{k<}$ can be dramatically nonthermal in nature. This feature is illustrated in Fig. 7.11 in direct tunneling measurements, revealing quasiparticle injection into Al, after Kaplan, Kirtley, and Langenberg (1977). The lower d^2I/dV^2 traces show sharp structures in $f_{k<}$ or $f_{k>}$ produced by the quasiparticle injection (and depending upon its polarity) in the Al film 390 μV above its gap edge, observed at junction biases near $\pm$1.1 meV. Earlier indications of this sort of sharply nonthermal structure had been reported by Smith and Mochel (1975).

Direct determination of the nonequilibrium distribution is actually possible by inversion of the detector junction I–V curve, such as the upper curve in Fig. 7.10. Such an analysis (Willemson and Gray, 1978; Smith, Skocpol, and Tinkham, 1980) proceeds from the measured I–V

Fig. 7.11. Direct observation of branch imbalance under injection in an 850-Å aluminum film detected in a 2950-Å PbBi detector film. Detector junction I–V trace represented by upper curves and d^2I/dV^2 lower curves; note changes in lower curves depending upon electronlike or holelike nature of the injection. (After Kaplan, Kirtley, and Langenberg, 1977.)

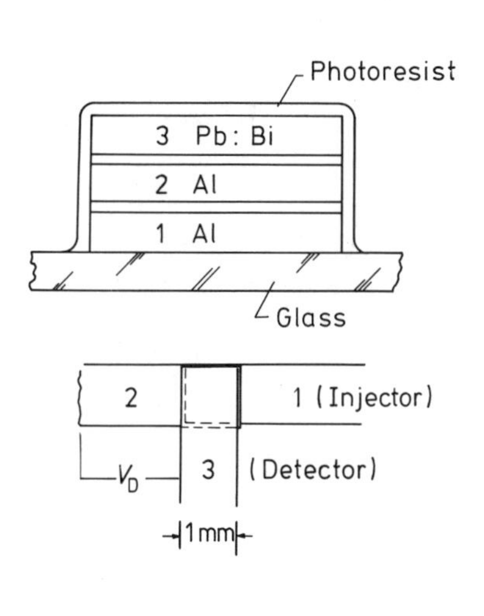

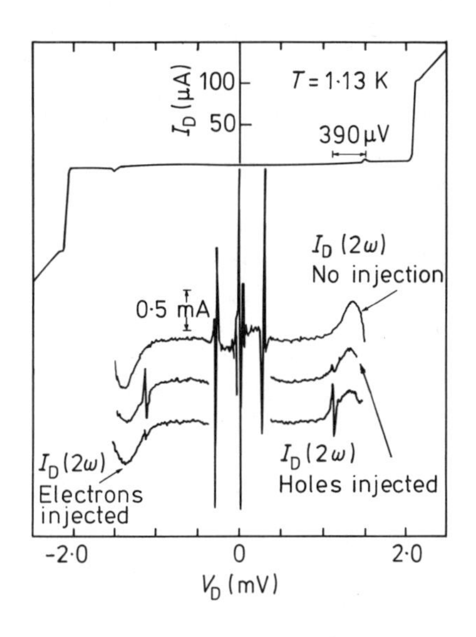

curve, using the relation

$$I(V)=\frac{G_{NN}}{e}\int_0^\infty \rho_{det}(E+eV)\rho_{Al}(E)f_{Al}(E)\,dE \tag{7.14}$$

which is valid when the detector film (S_3), typically Pb or PbBi, has a sufficiently large gap that its occupation function may be taken as zero. (A further advantage of a large gap Δ_3 is that recombination phonons from S_2 are energetically unable to generate pairs in S_3, which would complicate the results.) Here, G_{NN} is the conductance when both films are normal, and ρ_{Al} and ρ_{det} are the densities of states in the films S_2 and S_3, which can be determined from the tunneling spectra at zero injection, which ensures that $f(E)$ is simply the equilibrium Fermi function $f^0(E)$.

An example of the injection experiment and complete inversion analysis (Gray, 1980) is given in Fig. 7.12. In this case, the level of injection is greater than that in Fig. 7.11, such that in Fig. 7.12a both the top of the injected distribution (due to pair breaking, visible at 0.8 mV) and the disturbance due to quasiparticle injection (at 0.5 mV) are directly visible in the I–V curves. In both of these nonthermal features, there is evidence of branch imbalance (asymmetry of the response upon reversal of detector voltage V_d). On the other hand, the relative importance of branch imbalance is greatly reduced in the thermalized portion of the distribution nearer the Al gap edge $\Delta_0=0.199$ meV ($\times 10$ and $\times 1$ curves). These features are revealed precisely in the results of the inversion process in Fig. 7.12b, showing the even and odd quasiparticle distributions for a 300-Å central Al film injected at a rate $I_0=5\times10^{24}$ quasiparticles $cm^{-3}\cdot s^{-1}$. Note that the lower curve, when integrated over energy and multiplied by $2N(0)$, yields, from (7.10), the net charge Q^*. The gross behavior of the even distribution is an exponential decay $\rho(E)[f_>(E)+f_<(E)]\simeq\exp(-E/kT^*)$, with $T^*\simeq 0.89$ K, which may be compared with the measured substrate temperature 0.33 K, and critical temperature $T_c=1.33$ K.

There is, in fact, a literature devoted to theoretical discussion of what one must now regard, in view of the detail revealed in the past several figures, to be the gross features of nonequilibrium distribution functions. This work dates from the "μ^* model" proposed by Owen and Scalapino (1972), supported by work of Parker and Williams (1972) and of Fuchs, Epperlein, Welte, and Eisenmenger (1977), and includes the "T^* model" proposed by Parker (1975). Generalization to a "μ^*-T^* model,"

$$f=\left[1+\exp\left(\frac{E-\mu^*}{kT^*}\right)\right]^{-1} \tag{7.15}$$

has been found necessary by Willemson and Gray (1978) to approximate the measured distributions, as have been described above, with reasonable overall accuracy. The details are, however, beyond the reach of these models.

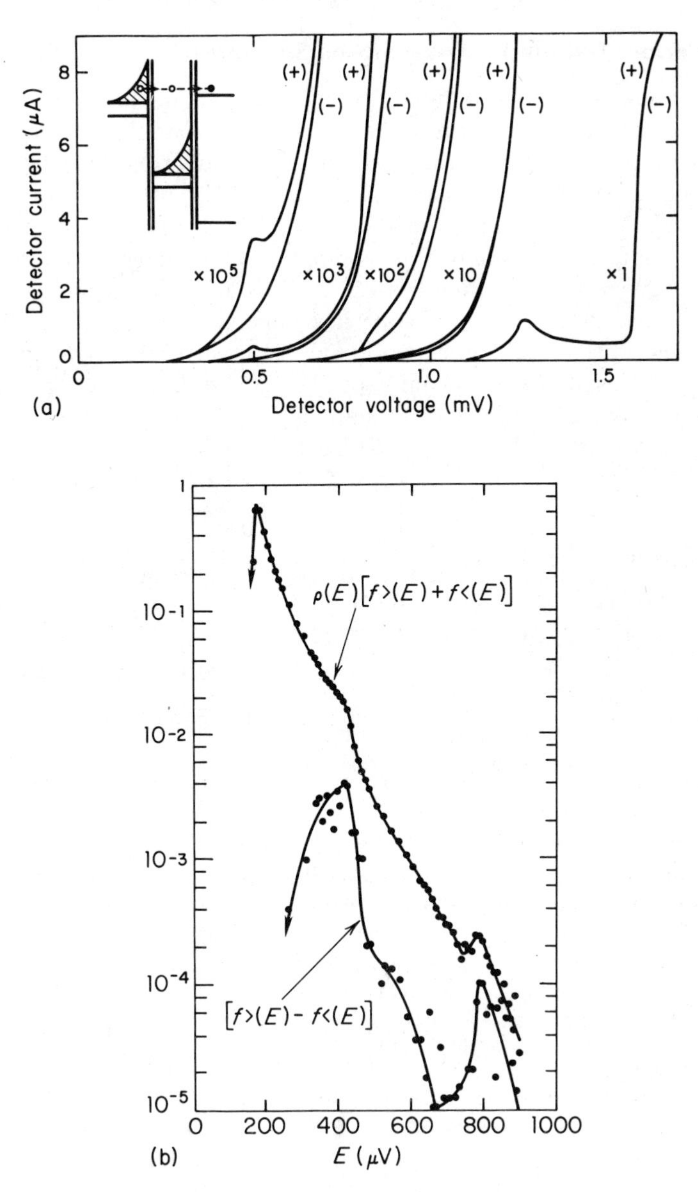

Fig. 7.12. (a) The *I*–*V* measurements of a detector junction in an Al–Al–PbBi double-junction structure under conditions of strong quasiparticle injection, $V_i = 0.76$ mV. (b) Quasiparticle distribution functions for Al at 0.33 K and $I_0 = 5.4 \times 10^{24}\ \mathrm{s^{-1} \cdot cm^{-3}}$ obtained by inverting detection junction *I*–*V* data similar to that shown in (a), using Eq. (7.11). (After Gray, 1980.)

7.4 Current Gain in Double-Junction Structures

In the double-junction configuration as discussed in the preceding sections, the detector junction has been regarded as a weakly coupled probe of S_2, with the expectation that the tunneling of quasiparticles from S_2 to S_3 can be neglected in the rate analysis, Eq. (7.12), for excess quasiparticles in S_2. An interesting case (Gray, 1978*a*) arises for low "detector junction" impedance, where this assumption is no longer valid. Suppose that the time τ_t for tunneling from S_2 to S_3 is actually much shorter than the recombination time τ_R existing in S_2 and S_3. This condition has the interesting consequence, as shown in Fig. 7.13 [where the S_2–S_3 (detector) junction is biased below its sum gap, $eV_d<\Delta_2+\Delta_3$], that *multiple* tunneling events, approximately τ_R/τ_t in number, are stimulated by single, initial quasiparticles injected from S_1. Current multiplication thus occurs. As shown, the quasiparticle of energy E in S_3 can next participate in a pair-exchange process in which a pair is formed in S_3, absorbing the quasiparticle of energy E plus a second quasiparticle resulting from decay of a pair in S_2, the result being a new quasiparticle in S_2 of energy $E'=E+eV$. The net result has been transfer of a second quasiparticle to S_3. The newly generated quasiparticle in S_2 at E' can again tunnel to S_3 at energy $E''=E'+eV=E+2eV$. This process continues until recombination occurs. The transfer of an initial quasiparticle from S_1 to S_2 then has stimulated transfer of τ_R/τ_t quasiparticles from S_2 to S_3.

The result is somewhat analogous to transistor action. This fact has

Fig. 7.13. Mechanism for current gain in SIS tunnel junction biased at eV_g less than 2Δ. Thermally excited electron in film *a* tunnels elastically into film *b*, increasing its energy by eV. This same particle can participate in the second process shown, in which it and a second electron, obtained by destroying a pair in film *a*, form a pair in film *b*; the quasiparticle left in film *a* has further increased its energy by eV and is now at energy 2 eV above its initial energy in film *a*. Since this process can continue until recombination occurs in either film *a* or *b*, current gain is possible. (After Gray, 1978*b*.)

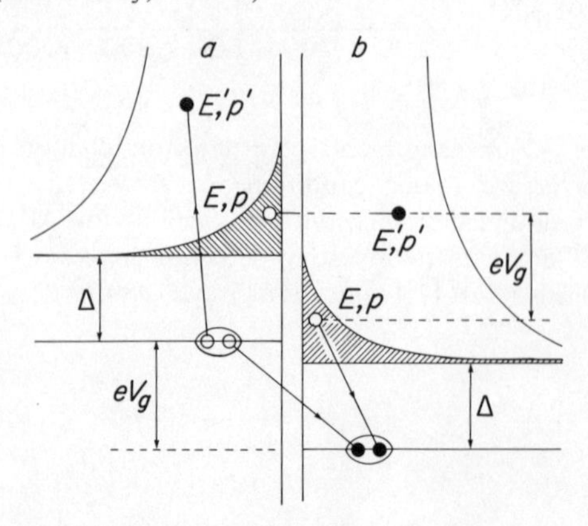

been demonstrated by Gray (1978*a*), as illustrated in Fig. 7.14. The tightly coupled (but not strongly biased) detector junction is now labeled the collector, while the injector junction is analogous to the emitter-base junction in a transistor. Depending upon the magnitude of the injector (pair-breaking) current (Fig. 7.14b), the subgap current in the collector junction is clearly seen to be incremented (Fig. 7.14c), and the ratio of these increments ΔI_c in the corresponding injector currents I_i in Fig. 7.14d amounts to a (temperature-dependent) gain of up to four. The temperature dependence arises from the relevant recombination-related times in S_2, as may be seen from the correct expresssion for the current gain,

$$G = \frac{\Delta I_c}{I_i} = \frac{\tau_R^*}{2\tau_t} \tag{7.16}$$

Here, the factor 2 has appeared from the previously neglected presence of thermal quasiparticles in S_2, and τ_R^* is the effective recombination time lengthened by the phonon-trapping effects: $\tau_R^* = \tau_R(1+\Gamma_B/\Gamma_{es})$, Eq. (7.13).

7.5 Gap Changes Under Nonequilibrium Conditions

As first pointed out by Eliashberg (1970), one should expect a close relation between a change in the quasiparticle distribution, $f_k(E)$, and the energy gap parameter, simply from the form of the BCS gap equation at finite temperature T, which contains the factor $[1-2f^0(E)]$ in the integrand:

$$\Delta(T) = N(0)V \int_\Delta^{\hbar\omega_c} \frac{\Delta}{\sqrt{E^2-\Delta^2}}[1-2f^0(E)]\,\mathrm{d}E \tag{7.17}$$

Eliashberg (1970) went further to make the conjecture that the *same* equation should apply for a nonequilibrium gap parameter Δ^* in the case of a nonequilibrium quasiparticle distribution function $f^*(E)$. Inspection of (7.17) makes obvious that the most influential changes in $f^*(E)$ are those near $E=\Delta$, for which the "pair density of states" $\Delta/\sqrt{E^2-\Delta^2}=uv$ is large. According to (7.17), at a given lattice temperature $T>0$, any nonequilibrium process which reduces $f(E\simeq\Delta)$ below its equilibrium value $f^0(E\simeq\Delta)$ would tend to increase Δ. In physical terms, this reduction will occur because fewer of the important gap edge pair states will be blocked by excitations. Two schemes for accomplishing this "unblocking" have been suggested. The first is extraction of quasiparticles into an adjoining film (Parmenter, 1961); in the second scheme quasiparticles are excited from the gap edge to regions of lower $N_T(E)$ by irradiation with quanta (photons or phonons) of $\hbar\omega \leqslant 2\Delta$. The latter restriction is necessary to avoid pair breaking, since $\hbar\omega \geqslant 2\Delta$ would also *create* quasiparticles and lead to gap reduction. Elaborations of the Eliashberg model have

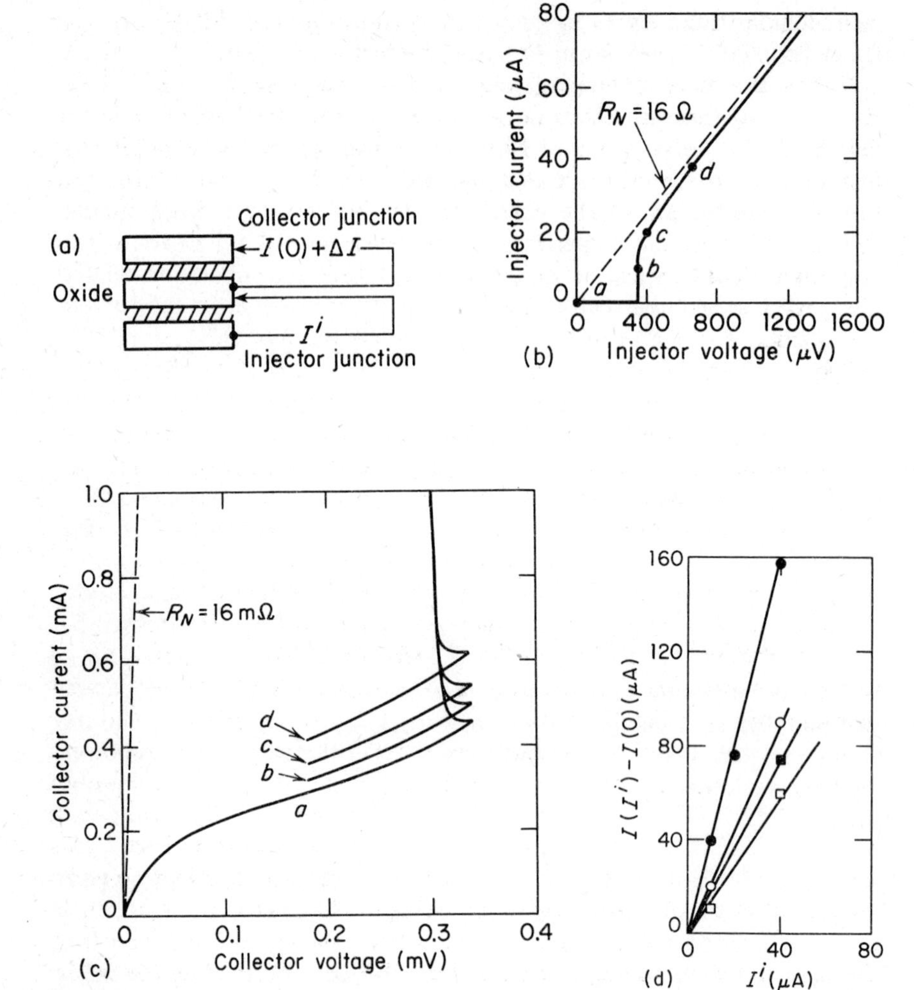

Fig. 7.14. (a) Stack of three 300-Å-thick aluminum films comprising the double-junction structure. The biases for the collector and injector junctions are independent. (b) The I–V characteristic of the injector junction showing the onset of pair breaking at $V = 2\Delta/e \simeq 360\ \mu\text{V}$. The labeled points refer to the injection current for the collector curves shown in (c). (c) The I–V characteristics at increased gain reveal the current due to thermally excited quasiparticles and injected quasiparticles. The various curves correspond to the injection currents as indicated in (b). (d) The increase ΔI in collector current vs. injector current I^i showing a temperature-dependent slope $\Delta I/I^i$ greater than unity. The temperature is determined from the quasiparticle current $I(0)$ in the collector, with $I^i = 0$: ●, 0.625 K; ○, 0.650 K; ■, 0.695 K; □, 0.703 K. (After Gray, 1978*a*.)

been given by Ivlev, Lisitsyn, and Eliashberg (1973), Chang and Scalapino (1977, 1978), and Chang (1978).

Enhancement of an energy gap observed by tunneling in a single-junction geometry was first reported by Kommers and Clarke (1977) (see also Hall, Holdeman, and Soulen, 1980) studying Al under microwave irradiation; in the double-junction case by Gray (1978*b*), using the tunneling mechanism of Fig. 7.13, again to enhance Δ_{Al}; and by Chi and Clarke (1979), also in the double-junction case, using an initial film S_1 of Al with a slightly larger gap than that of S_2 (also Al) to extract quasiparticles and thus enhance Δ_{Al} in S_2.

Consider, first, the tunneling enhancement experiment of Gray (1978*b*) in Fig. 7.15. The fractional enhancement of the gap of Al film S_2 at 1.2 K, on the order of 0.005, is shown in the upper curve (Fig. 7.15a) as a function of the generator junction bias, while the lower curve shows the generator current. The upward step in generator current identifies the onset of pair breaking in film S_2 and corresponds to the expected change from gap enhancement to gap reduction in film S_2. The physical process leading to the gap enhancement is that illustrated in Fig. 7.13, identifying the two films as S_1 and S_2. A calculation based on this process (Chang, 1978) predicts an effect in qualitative agreement with the experiment but about four times larger than observed. Additional results as a function of reduced temperature and a scaled curve from the Chang (1978) calculation are shown in Fig. 7.15b.

The original report of gap enhancement by microwave irradiation of a single tunnel junction composed of Al films on a BaF_2 substrate was that of Kommers and Clarke (1977), whose measured Al gap values with increasing microwave power at $\hbar\omega = 41.4\ \mu\text{eV}$ are shown in Fig. 7.16a. The sum-gap feature at $T/T_c = 0.99$ under irradiation corresponding to $\hbar\omega/2\Delta = 0.83$ clearly shifts to higher voltage. The remaining structure in the dV/dI curves was interpreted in terms of photon-assisted tunneling. These results were subsequently questioned (Falco, Werner, and Schuller, 1980), stimulating experimental confirmation of the original observations by Hall, Holdeman, and Soulen (1980), whose dV/dI curves are shown in Fig. 7.16b. The corresponding T_c enhancement in the Al film is shown in Fig. 7.16c, definitely confirming the original report.

Finally, the quasiparticle extraction experiment (Parmenter, 1961) was performed and carefully analyzed by Chi and Clarke (1979). The quasiparticles in S_2 are extracted into a tightly coupled film S_1, whose gap is slightly larger than that of S_2, with S_1 biased positively at $eV = \Delta_1 - \Delta_2$ to line up the "empty" BCS singularity in S_1 with that of S_2. The detector junction is used in the usual way to probe the properties of S_2. The measurements revealed up to an 8-μeV increase in Δ_2, or over 40%, at $T/T_{c2} = 0.998$. It is suggested by Chi and Clarke (1979) that phonons generated in S_1 by recombination may diffuse to S_2 and cause significant pair breaking, to limit the size of the observed enhancement of Δ_2.

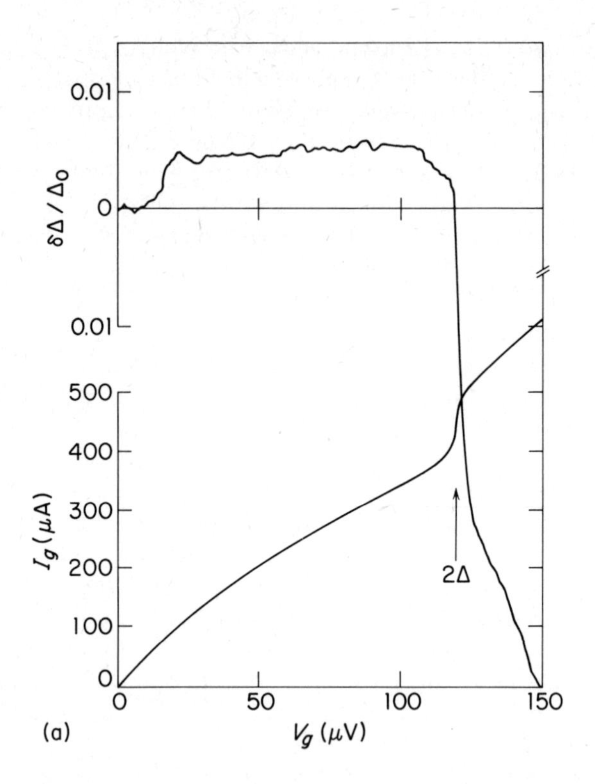

0.01
0
δΔ / Δ₀
0.01
500
400
300
200
100
0
I_g (μA)
2Δ
0
50
100
150
(a)
V_g (μV)

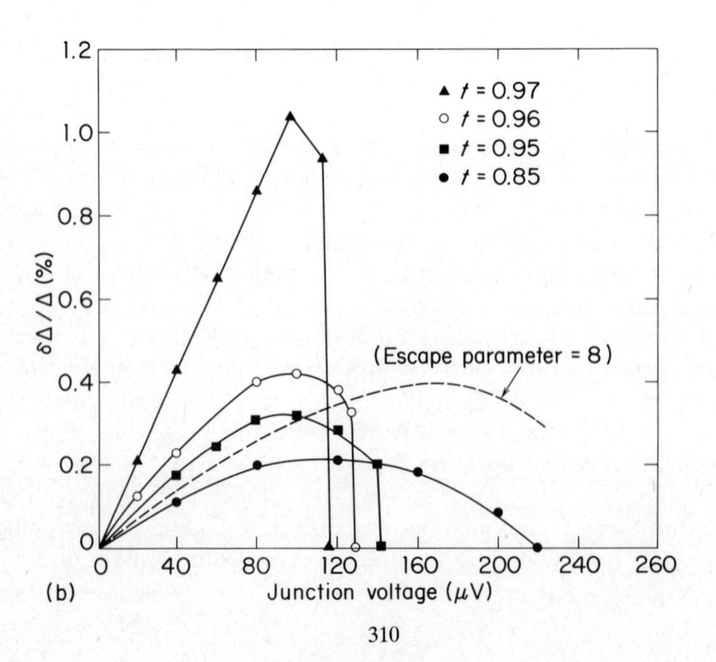

1.2
1.0
0.8
0.6
0.4
0.2
0
δΔ / Δ (%)
▲ t = 0.97
○ t = 0.96
■ t = 0.95
● t = 0.85
(Escape parameter = 8)
0
40
80
120
160
200
240
260
(b)
Junction voltage (μV)

Fig. 7.15. (a) Observation of enhancement $\delta\Delta$ of superconducting gap Δ by quasiparticle tunneling between identical superconductors biased at $eV<2\Delta$, as shown in Fig. 7.13. The fractional change in the gap as observed by the second detector junction is plotted on the upper scale as a function of the generator junction bias on the lower scale. Also shown is the generator junction current which increases at the sum-gap voltage 2Δ. (After Gray, 1978*b*.) (b) Enhancement of gap parameter Δ in a similar experiment in which the gaps of the aluminum films were made precisely equal. The dashed curve in this figure is based upon the calculation of Chang (1978), who assumed a phonon-trapping factor of 8.0. (After Gray, 1980.)

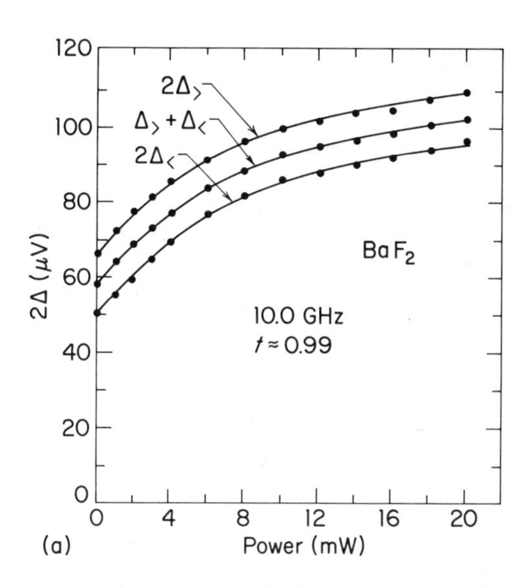

Fig. 7.16. Microwave enhancement of the energy gap in superconducting aluminum as observed by tunneling. (a) Variation of energy gaps of an Al–Al_2O_3–Al junction as a function of microwave power (in milliwatts) of frequency 10.0 GHz, measured at reduced temperature $t=0.99$. (After Kommers and Clarke, 1977.)

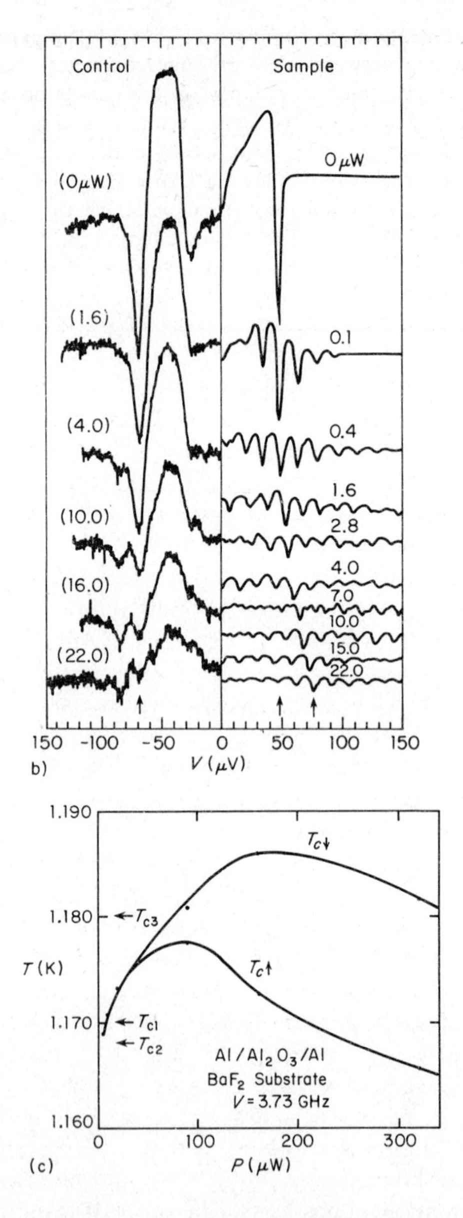

Fig. 7.16. (cont.) (b) Confirmation of the original measurements of Kommers and Clarke (1977) in dV/dI vs. V spectra of an Al–Al_2O_3–Al tunnel junction at 1.161 K shown at increasing levels of microwave power at frequency 3.72 GHz. In the uppermost curve with zero microwave power, the dV/dI peak locates the sum gap at 48 μV. In the next curve (labeled 0.1 μW), the structure which is seen is primarily splitting of the gap edge by microwave photon-assisted tunneling (the microwave photon has energy 15.4 μeV). With progressing microwave power, however, the basic sum-gap structure is seen to shift to 76 μV. A control junction in the same cryostat but shielded from the microwaves (left hand curves) does not

7.6 Instabilities of Superconductors Under Injection

The Eliashberg argument, Eq. (7.17), which we have earlier discussed in terms of gap enhancement, also implies that quasiparticle *injection* into the energy region $E \simeq \Delta$ of large $N(E)$ will be particularly effective in *reducing* the gap; this result has already been indicated in Fig. 7.15a. An earlier clear observation of large gap reduction under injection, in fact, leading to the onset of an instability, was reported by Fuchs, Epperlein, Welte, and Eisenmenger (1977). The gap of Sn in the double-junction geometry was observed to fall smoothly to 0.6 of its initial value, at which the instability occurred. The behavior was found to agree rather well with the μ^* model of Owen and Scalapino (1972). While it was assumed at the time that the transition observed at 0.6Δ probably was the first-order transition to the normal state implied by the Owen–Scalapino model, in retrospect this assumption seems unlikely. The state reached under injection was more likely an inhomogeneous state in which regions of a second, and substantially lower, gap appeared.

The appearance of a double gap under injection at $eV \simeq 2\Delta_{\mathrm{Al}}$ was first reported by Dynes, Narayanamurti, and Garno (1977) in double-Al junctions. The results were interpreted as the result of a fundamental instability under homogeneous injection. Some confusion arose about the interpretation of the result because the generator junction bias $eV \simeq 2\Delta_{\mathrm{Al}}$ was shown by Gray and Willemsen to be particularly susceptible, in the (unavoidable) presence of any small inhomogeneity in the films and hence of Δ_{local}, to break up into an inhomogeneous state. This result occurs because the injection level, sensitively dependent through $N_T(E, \Delta_{\mathrm{local}})$ on the local gap, would itself become inhomogeneous, driving toward amplification of originally present fluctuations in Δ. While such effects likely occur, the fundamental question now seems to be settled in favor of the original interpretation and the existence of a fundamental instability, independent of initial inhomogeneities, following the work of Iguchi and Langenberg (1980). These authors answered the objection of Gray and Willemsen by using higher-resistance generator junctions under conditions of more homogeneous injection easily achieved with the generator bias well *above* the region of sensitivity to Δ_{local} near the sum gap. The data of Iguchi and Langenberg (1980) also revealed multiple states; up to four were seen, as shown in Fig. 7.17a. One gap value was always close to the value of the homogeneous gap at the onset of instability. The work of Iguchi and colleagues was important in outlining a model which could reasonably explain the multiple-gap formation as the result of a diffusive

---

show this shift. (c) Plot of T_c of the Al films as a function of the microwave power. The upper curve locates the $T = T_{c\downarrow}$ at which superconductivity (structure in dV/dI) abruptly disappears with increasing T; the lower curve locates the $T = T_{c\uparrow}$ at which superconductivity abruptly reappears as T was decreased. (After Hall, Holdeman, and Soulen, 1980.)

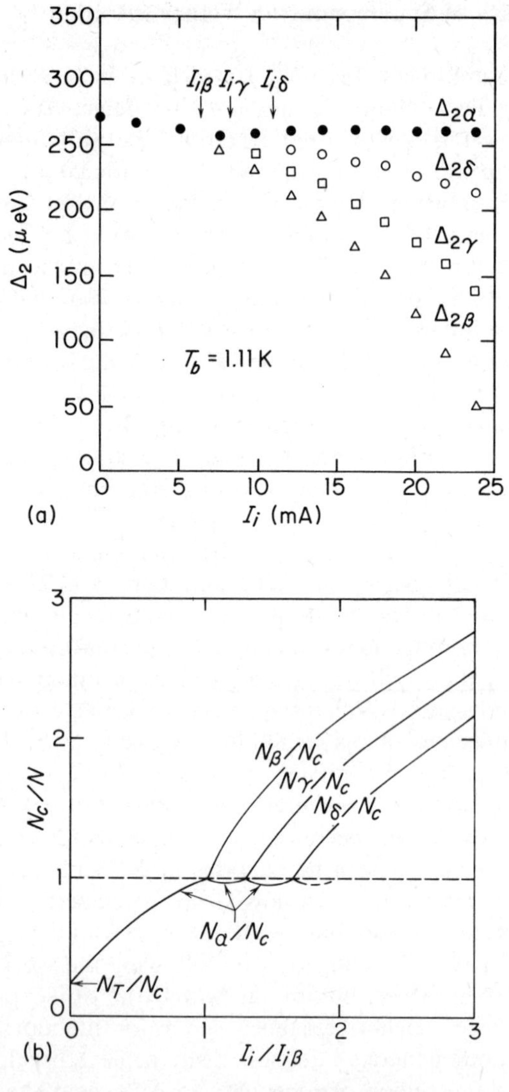

Fig. 7.17. Breakup of homogeneous superconductor under strong injection into multiple domains of lower gap values. (a) Dependence of Al gaps on injection current in PbBi–I–Al–I–Al double junction. Note that the sum-gap injector current is ~2 mA, less than currents at which the multiple gaps occur. (b) Schematic dependence of quasiparticle concentration on injection. (After Iguchi and Langenberg, 1980.)

quasiparticle instability. The same inhomogeneous behavior inferred from I–V measurements has since been observed in Sn films (Iguchi et al., 1981) and in Pb–In films (Akoh and Kajimura, 1981). Further, a spatially inhomogeneous distribution of recombination *phonons* has been resolved (Kotani, Suzuki, and Iguchi, 1982), if rather weakly in the initial report. This result seems to provide direct and only slightly tentative evidence for spatially inhomogeneous multiple-gap states under injection.

The analysis of Iguchi and Langenberg (1980) (and of Iguchi, Kent, Gilmartin, and Langenberg, 1981), leading to multiple-gap regions, is based on the Rothwarf–Taylor (1967) equations extended to the inhomogeneous case by addition of a quasiparticle diffusion term $\vec{\nabla} \cdot \vec{J}$, where $\vec{J}$ is the quasiparticle current density. The equations become

$$\frac{\partial N}{\partial t} = I_{qp} - 2RN^2 + \frac{2N_{ph}}{\tau_B} - \vec{\nabla} \cdot \vec{J} \tag{7.18}$$

$$\frac{\partial N_{ph}}{\partial t} = I_{ph} + RN^2 - \frac{N_{ph}}{\tau_B} - \frac{N_{ph} - N_{ph}^T}{\tau_{es}} \tag{7.19}$$

Here, N and N_{ph}, respectively, are quasiparticle and phonon concentrations; I_{qp} and I_{ph} are the corresponding injection rates; N_T and N_{ph}^T are thermal concentrations; R is the recombination coefficient; τ_B is the pair-breaking time of a phonon; and τ_{es} is the phonon time to escape.

Following Iguchi and Langenberg (1980), the steady-state solutions of the two equations are

$$N^2 = N_T^2 + \frac{(I_0 - \vec{\nabla} \cdot \vec{J})P}{2R} \tag{7.20}$$

$$I_0 = I_{qp} + \frac{2(P-1)}{P} I_{ph} \tag{7.21}$$

where $P \equiv 1 + \tau_{es}/\tau_B$ is the phonon-trapping factor, and $R = N_{ph}^T \tau_B / N_T^2$. The form for J is taken from the work of Scalapino and Huberman (1977):

$$\vec{J} = -\frac{2D}{\rho(E_F)\Delta_0}(N_c - N)\vec{\nabla} N + \frac{D\xi^2}{2}\vec{\nabla}(\nabla^2 N) \tag{7.22}$$

Here, D is the quasiparticle diffusion constant, Δ_0 is the zero-temperature equilibrium gap parameter, and ξ is the coherence length. For N exceeding the critical quasiparticle concentration N_c, the first term causes quasiparticle diffusion from regions of low N to high N, while for $N < N_c$, the flow reverses. This first term, which leads directly to diffusive instability in the model, was obtained by Scalapino and Huberman from the relation $\vec{J} = -\rho(E_F) D \vec{\nabla} \mu^*$, μ^* being discussed in somewhat qualitative terms to obtain the result given above. Combination of the three preceding equations leads strictly to a fourth-order nonlinear differential equation for N. Direct solution of this combined equation is not possible. It is, nevertheless, expected that the set of equations must lead to separate regions in each of which N varies only slightly. Idealizing this case to two regions of constant N, Iguchi and Langenberg (1980) proceeded to consider diffusion at the interfaces between the two regions. The normal component of J was approximated by

$$J_n = -\frac{2D}{\rho(E_F)\Delta_0}[N_c - N(s)]\left(\frac{\partial N}{\partial n}\right)_S \tag{7.23}$$

where $(\partial N/\partial n)_S$ is the spatial gradient of N along the direction n in the plane of the film and normal to the boundary S. The analysis proceeds by considering two distinct gap regions α and β sharing the boundary S at which the relation

$$2R(N^2(s) - N_T^2) = PI_0 \equiv \hat{I}_0 \tag{7.24}$$

is to be satisfied. It is shown by Iguchi and Langenberg (1980) that the solution of (7.20)–(7.24) has the behavior shown schematically in Fig. 7.17b. The homogeneous value of N rises with increasing injection to the critical value N_c at $\hat{I}_0 = I_\beta \equiv 2R(N_c^2 - N_T^2)$, at which the two branches appear. Further bifurcations occur with increase in $\hat{I}_0$, as shown in Fig. 7.17b and as observed in the experimental results, Fig. 7.17a.

8

Tunneling in Normal-State Structures. II

8.1 Introduction

The variety of normal-state tunneling situations which have been encountered experimentally is so great that the task of providing an overall framework within which to treat all cases is quite difficult. In extending the discussion of Chap. 2 to the many cases reported, we have attempted to classify these according to increasing complexity of the electronic interactions involved. Thus, cases of one-electron interactions are considered first, in the context of final-state electronic effects as well as in simple inelastic excitation effects. In the category of one-electron final-state cases, primary examples are quantum-size-effect electronic splittings in metal films and in accumulation layers at degenerate semiconductor surfaces; while excitation of zone boundary phonons in *pn* junctions and of molecular vibrations in doped M–I–M junctions are examples of simple inelastic assisted tunneling.

Beyond this relatively clear class of cases, we find further examples in which truly many-body final states are involved. Fundamental origins of inherently many-body final states include electron-electron interactions, which can lead in the extreme case to electron localization, electron-phonon self-energy effects; and many-body magnetic interactions associated with the Kondo effect.

Finally, as a testing ground for this body of theory, we summarize the various experimentally observed zero-bias tunneling anomalies and attempt to identify their physical origins.

8.2 Final-State Effects. I

8.2.1 Two-Dimensional Final States

It is useful to consider the case of M–I–M normal-state tunneling in which the width W of the second (free-electron) electrode is extremely small, with specular boundary conditions. This case is properly considered *two-dimensional* if, in the range of bias voltage of interest, a single quantum state describes the motion perpendicular to the film. To be specific, consider the electrode as a semi-infinite box of width W (in the x direction) with infinite potential barriers at the boundaries, requiring the electron wavefunction to vanish at $x=0$ and $x=W$. Then

$$E=\frac{\hbar^2}{2m}(k_t^2+k_x^2)=E_t+E_{nx} \tag{8.1}$$

where

$$k_y^2+k_z^2=k_t^2$$

and the boundary conditions require

$$\psi=W^{-1/2}\sin\left(\frac{n\pi x}{W}\right)\exp[i(k_y y+k_z z)], \qquad n=1,2,3,\ldots \tag{8.2}$$

Hence, one must have

$$k_x=\frac{n\pi}{W}$$

and

$$E_{nx}=\frac{\hbar^2\pi^2 n^2}{2mW^2} \tag{8.3}$$

The integer n specifies a two-dimensional band, for which the minimum total energy corresponding to zero transverse kinetic energy E_t is E_{nx}. The spacing between adjacent band edges can be estimated as

$$E_{n+1,x}-E_{n,x}\simeq\frac{\partial E}{\partial k_x}\Delta k_x=\hbar v_g(E)\frac{\pi}{W} \tag{8.4}$$

Evaluated at $W=1000$ Å and $v_g(E_F)=v_F=10^8$ cm/sec (corresponding to $n=275$), this equation gives $\Delta E_{nx}\simeq 21$ meV, which is an easily observable splitting. Since (in the free electron model) $\partial E/\partial k_x$ varies linearly with n, however, much smaller values of W would be required to similarly split, e.g., the $n=1$ and $n=2$ band edges.

The density of states for the two-dimensional motion at fixed n, $\rho_n(E-E_{nx})$, plays a more prominent role in the tunneling conductance than it does in the case of three dimensions. This feature can be seen in the expression for the tunneling current density:

$$J=\frac{2e}{(2\pi)^3\hbar}\int dk_x\, d^2k_t\frac{\partial E}{\partial k_x}[f(E)-f(E+eV)]D(E,k_t) \tag{8.5}$$

$$=\frac{2e}{h}\int_{-\infty}^{\infty}dE\,[f(E)-f(E+eV)]\int_0^E\rho_n(E_t)D(E,E_t)\,dE_t \tag{8.6}$$

where $\rho_n(E_t)$ is defined by the relation

$$\rho_n(E_t)\,dE_t=\frac{d^2k_t}{(2\pi)^2}$$

and reduces to

$$\rho_n(E_t)=\begin{cases}\dfrac{m^*}{2\pi\hbar^2}, & E_t>0\\ 0, & E_t=0\end{cases} \tag{8.7}$$

which is constant in the simplest case of a parabolic dependence of E_t on k_t. For a given value of n, the E_t integration in (8.6) collapses to a single value $E_t = E - E_{nx}$, by Eq. (8.1). Further, if we specify $T=0$, the effect of the Fermi functions is to limit the integration on E from 0 to eV:

$$J_n = \frac{2e}{h} \int_0^{eV} \rho_n(E - E_{nx}) D(E, E - E_{nx})\, dE \tag{8.8}$$

Differentiation of this expression, neglecting any variations in the D factor arising from voltage distortion of the barrier, gives the conductance contribution (BenDaniel and Duke, 1967)

$$\frac{dJ_n}{dV} = \frac{2e}{h} \rho_n(eV - E_{nx}) D_n \tag{8.9}$$

Here dJ_n/dV is a direct measure of the two-dimensional density of states of the nth band, and D_n similarly signifies the barrier factor at constant $E_x = E_{nx}$. In this case, the band edges E_{nx} would provide steps in dJ/dV, while peaks or oscillations in dJ/dV would arise if the voltage dependence of D_n is, in fact, important.

8.2.2 Quantum Size Effects in Metal Films

The first report of two-dimensional final states in tunneling was given by Jaklevic, Lambe, Mikkor, and Vassell (1971), who observed oscillations in the dI/dV of Al–Al_2O_3–Pb junctions at both 4.2 and 77 K in a (Pb-positive) bias range near 0.8 V. The oscillations corresponded to standing wave states across thin polycrystalline Pb films, evaporated at 77 K, and in the thickness range 200 Å $< W <$ 1000 Å. The effects were subsequently observed in Mg (Jaklevic, Lambe, Mikkor, and Vassell, 1972) and also in Au and Ag (Jaklevic and Lambe, 1975). In all cases, the range of splittings is ~10 to 100 mV.

One's first intuition might well be that such effects should be unobservable in evaporated films because of fluctuations in film thickness. Returning to the example of a 1000–Å film for which $\Delta E_n \simeq 20$ mV for $v_g \simeq v_F \simeq 10^8$ cm/sec, the free electron model implies $E_{nx} \simeq E_F \simeq 3$ eV. Then, from (8.3), one finds

$$\frac{\Delta E}{E_F} = -\frac{2\Delta W}{W}$$

so that $\Delta E = 10$ meV corresponds to a thickness fluctuation of only $\Delta W \simeq 2$ Å. Such a variation in thickness seems unavoidable, and might be expected to obscure the effect of interest.

The observability of the splittings in real polycrystalline films, as pointed out by Jaklevic, Lambe, Mikkor, and Vassell (1971), depends upon the facts that (a) the films are partially oriented (textured), with microcrystal [111] axes perpendicular to the substrate (Fig. 8.16), (b) the widths of the individual crystallites are integer multiples of the [111] cell dimension d: $W = Nd = N\sqrt{3}\, a_0$, and (c) in real materials, certain states, denoted commensurate states, are such that their energies E_{nx} are independent of $W = Nd$. Such a case occurs for k_x at the zone boundary, $k_x = \pi/d$, since $\partial E/\partial k_x = 0$. This is true, more fundamentally, because the electron wavelength $\lambda = 2\pi/k$ in this case is exactly $2d$. Hence, addition of one more atomic layer to the film ($N \rightarrow N+1$) is accommodated by adding precisely one more half-wave of ψ, which still matches the boundary condition $\psi(W) = \psi[(N+1)d] = 0$. Since the curvature of the wavefunction is the same, the kinetic energy $E_x = (-\hbar^2/2m)(\partial^2\psi/\partial x^2)$ is unchanged. A set of such commensurate states is defined (Jaklevic and Lambe, 1975) by

$$k_x = \frac{S}{Q}\frac{\pi}{d} \tag{8.10}$$

where S/Q is an irreducible fraction. In the case $S/Q = \frac{1}{2}$, such that $k_x = \pi/2d$, for example, one has $\lambda = 4d$, so that addition of *two* atomic layers accommodates one additional half-wavelength and leaves E_{nx} unchanged. Thus, for $S/Q = \frac{1}{2}$, the subset of crystallites of *even*-integer N will all have a level E_{nx}. However, in grains of *odd* N, the boundary condition $\psi(W) = 0$ requires a change of k_x. It turns out that the odd-N grains provide a set of levels exactly interleaving the even-N levels, so that the level spacing becomes

$$\Delta E = \frac{1}{2}\hbar v_g(\pi/d) \tag{8.11}$$

in this case. The data shown in Figs. 8.1 and 8.2 correspond to $S/Q = \frac{1}{2}$: the plot of ΔE vs. $1/W$ (W = film thickness) leads to a correct value for v_g for Pb, evaluated in the [111] direction at 0.8 eV above E_F.

The detailed line shape of the oscillations has been analyzed by Davis, Jaklevic, and Lambe (1975) and has been shown to be dominated by the change of the transmission factor D with bias voltage and with E_{nx}. The possibility that the finite lateral size of the grains (these are roughly cubes, and any interruption of the parallel electron motion at these boundaries is ignored in the analysis) may affect the line shape (Wolf, 1974) is apparently obscured by the dominance of the barrier factor. We will see shortly that in the case of a two-dimensional electron accumulation layer, on the other hand, the barrier factor plays a minor role and the measurements obey (8.9) rather well.

Extensions of the experiments on metal films to apply tensile stress—and, thus, discriminate between the band structure effects of interest and, e.g., inelastic excitations of organic molecules—have been reported by Jaklevic and Lambe (1973). This strain modulation technique has allowed identification, in addition, of band edge effects in Au, Ag, Sn, and Mg. An example of this technique is shown in Fig. 8.3.

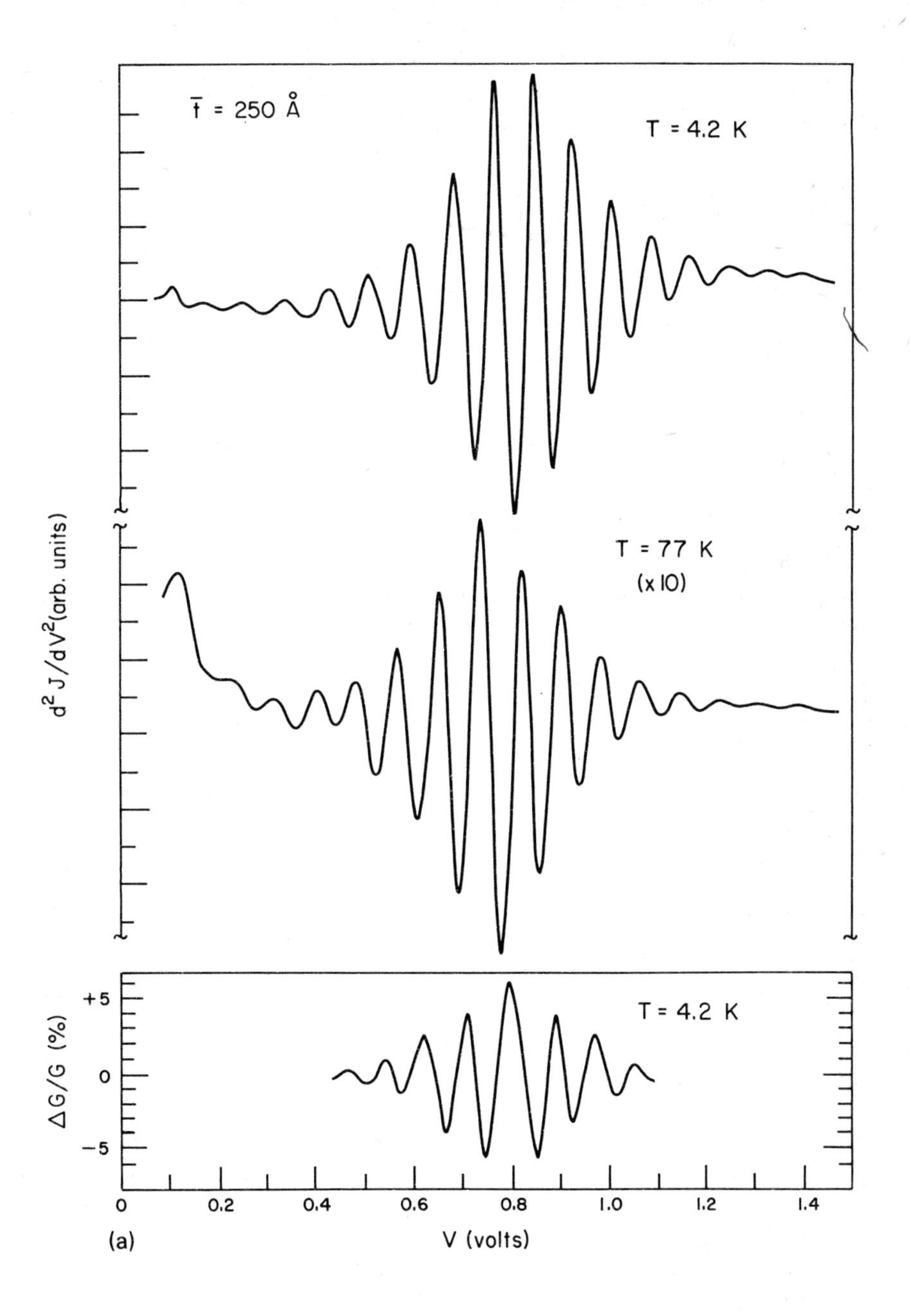

Fig. 8.1. (a) Electron standing wave splittings observed in d^2I/dV^2 measurements on Al/Al–oxide/Pb junction. The Pb layer is of 250 Å average thickness, regarded as a mosaic of oriented single crystals with closely similar individual thicknesses. Electron states lie 0.78 eV above the Pb Fermi level, at one-half the $L(111)$ zone boundary wavevector, corresponding to approximately 130 nodes across the crystalline "box." (After Jaklevic, Lambe, Mikkor, and Vassell, 1971).

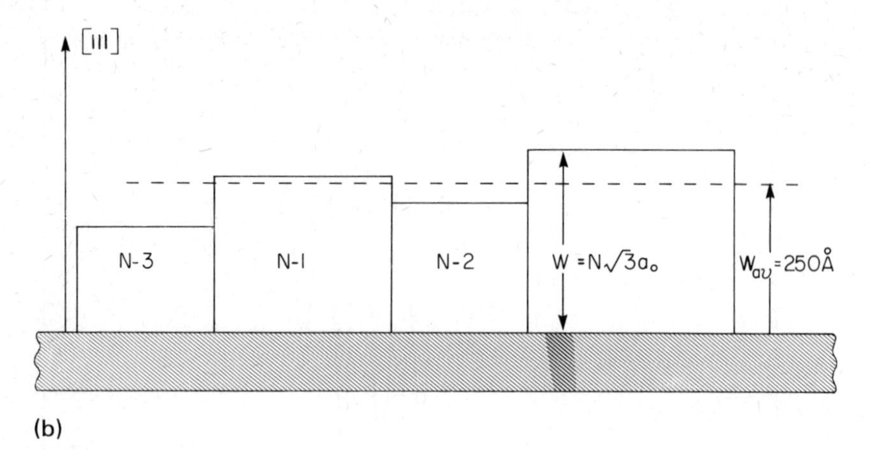

Fig. 8.1. (cont.) (b) Sketch of electrode, regarded as a mosaic of Pb crystallites, found to be oriented with [111] axis normal to substrate. (Vertical scale is exaggerated.) Observability of the standing wave effect depends upon the fact that thicknesses of individual crystallites differ only by integer multiples of the [111] cell dimension $d = \sqrt{3}\,a$, which must be commensurate with the electron wavelength $2\pi/k$.

8.2.3 Accumulation Layers at Semiconductor Surfaces

The band configuration shown in Fig. 8.4 illustrates the formation of a single ($n = 1$) two-dimensional band, labeled $E_b(V)$ at a semiconductor surface. The band bending results primarily from a distributed positive charge in the oxide barrier, used to form a tunnel junction onto an n-type degenerate (metallic) semiconductor, and is weakly affected by the bias V. The potential well $U(x, V)$ in which the one-dimensional wavefunction $\psi_1(x, V)$ is bound is the result not only of the external (bias-dependent) charge in the oxide but also of the bound-state charge distribution $e\,|\psi_1(x)|^2$ itself. Thus, the determination of $U(x, V)$ and $\psi_1(x, V)$ requires simultaneous solution of Schrödinger and Poisson (electrostatic) equations. The dependence of both functions upon bias V, through the distortion of the barrier shape, is indicated explicitly. An example of the spectacular tunneling spectra obtained by Tsui (1970*a*, 1970*b*, 1791*a*, 1971*b*, 1972, 1973, 1975*a*) is shown in Fig. 8.5, in which the semiconductor substrate is n-type InAs with $n \simeq 5.5 \times 10^{17}\ \mathrm{cm}^{-3}$. The inset in this figure shows, in d^2I/dV^2, the breakup of the conduction band (and also of the *surface* band) into Landau levels with an applied magnetic field (parallel to $\vec{J}$) of 35 T. Such Landau levels occur at

$$E_{\rm tn} = \left(\frac{e\hbar H}{m^* c}\right)\left(n + \frac{1}{2}\right), \qquad n = 0, 1, \ldots \tag{8.12}$$

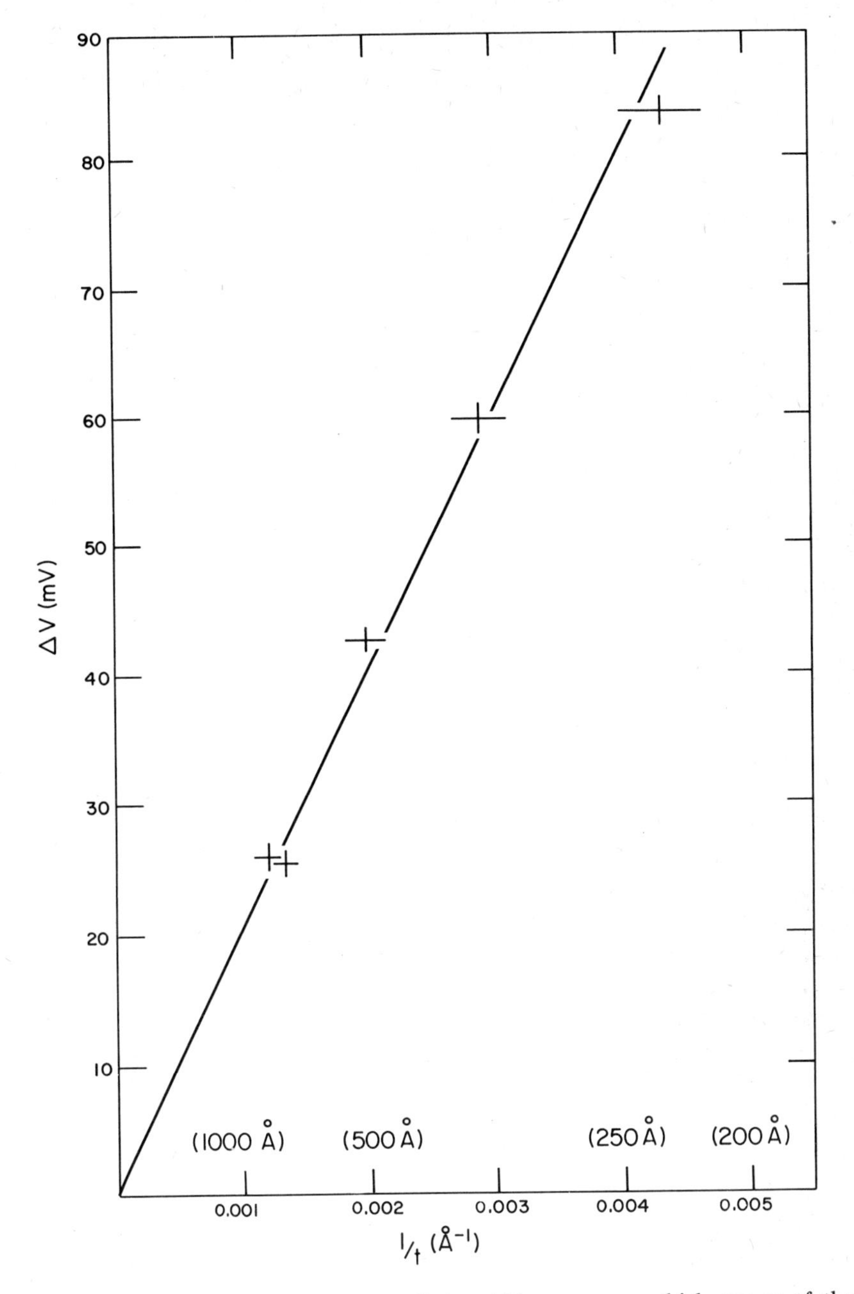

Fig. 8.2. Dependence of observed splitting ΔV on average thickness w of the metal film. The slope of the solid line corresponds to the electron group velocity in Pb of 2×10^8 cm/sec. (After Jaklevic, Lambe, Mikkor, and Vassel, 1971.)

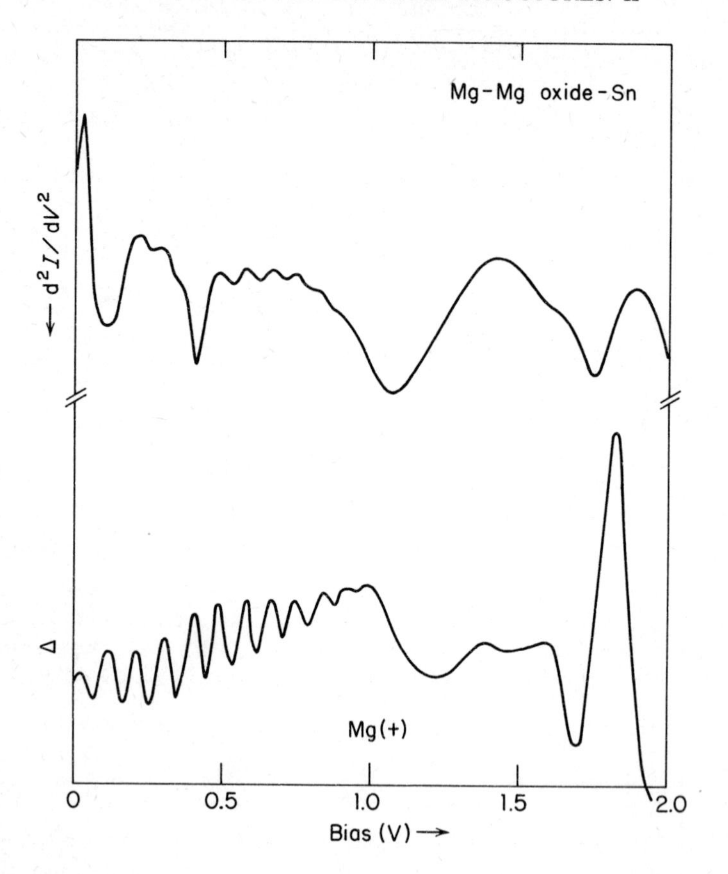

Fig. 8.3. Experimental d^2I/dV^2 spectra of Mg–MgO–Sn junction measured at 77 K (upper trace) reveals weak standing wave structure as well as molecular vibration effects. By subjecting the film to strain and taking the difference Δ of two curves, as shown in the lower trace, the molecular vibration spectra can be eliminated. The difference curve Δ reveals more clearly standing wave states in the range form 0 to 1.1 V, a broad band from 1.1 to 1.4 V (identified as a Fermi surface energy gap in the [001] direction of Mg), and a band edge of Sn observed near 1.8 V. (After Jaklevic and Lambe, 1973.)

measured from E_b. Negative bias in Fig. 8.5 corresponds to raising the counterelectrode Fermi level, thus probing higher kinetic energy $E_t = E - E_b$ in the surface band. Thus, the negative peak in d^2I/dV^2 at +168 mV, corresponding to a negative step in dJ_1/dV, marks the lower limit of the surface band, in agreement with (8.9). If one plots the positions of the oscillation features vs. magnetic field ($\vec{H}_{\|}\vec{J}$), as shown in Fig. 8.6, *two* sets of Landau levels are revealed, one set converging to 168 mV at $H = 0$, the other set converging to 95 mV. The former set is

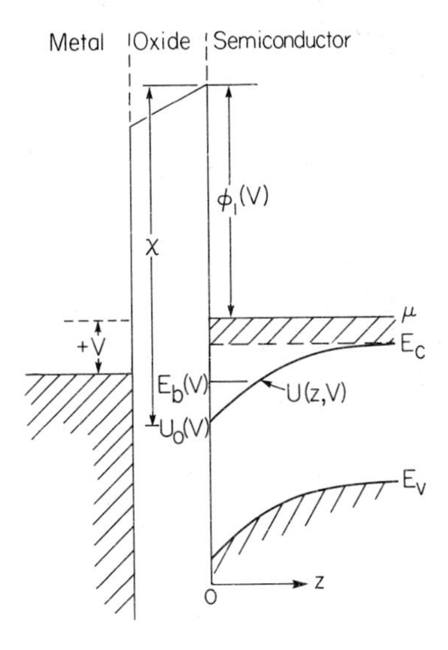

Fig. 8.4. Energy band diagram for a metal-oxide-semiconductor junction in which the degenerate n-type semiconductor has formed an accumulation layer. The potential $U(V, z)$ associated with positive charge in the oxide producing the accumulation layer has one bound-state level, donoted $E_b(V)$. (After Tsui, 1972.)

absent in the $\vec{H} \perp \vec{J}$ configuration and is therefore attributed to orbits in the surface band. The set converging to 95 mV is assigned to orbits in the bulk conduction band, and the 95-mV positive bias corresponds to the band edge E_c opposite the counterelectrode Fermi level. By use of (8.12), the slope of the plot of oscillation positions vs. H can be used to infer m^*, for both surface and bulk conduction electrons. The wealth of information obtained on n-type InAs from this line of research is summarized by Tsui (1972, 1975a). Similar but less extensive information in n-type PbTe has been obtained by Tsui, Kaminsky, and Schmidt (1974). In this case, three surface bound states E_{1x}, E_{2x}, E_{3x} are observed, each generating a surface band of two-dimensional character.

The application of the tunneling technique of Tsui apparently has been limited to accumulation layers on degenerate InAs and PbTe. This choice of materials is presumably suitable in having generally high electron mobilities $\mu = e\tau/m^*$ and large electron wavelengths $\lambda = h/p$ associated with small values of m^*/m (0.02 in InAs and 0.07 in PbTe). The severe problem of energy broadening due to scattering of the surface electrons by charge fluctuations arising in the oxide barrier and by other surface imperfections is thereby minimized. Also, the Landau-level splittings are larger and, hence, more easily observed as a consequence of the inverse dependence on m^* in (8.12).

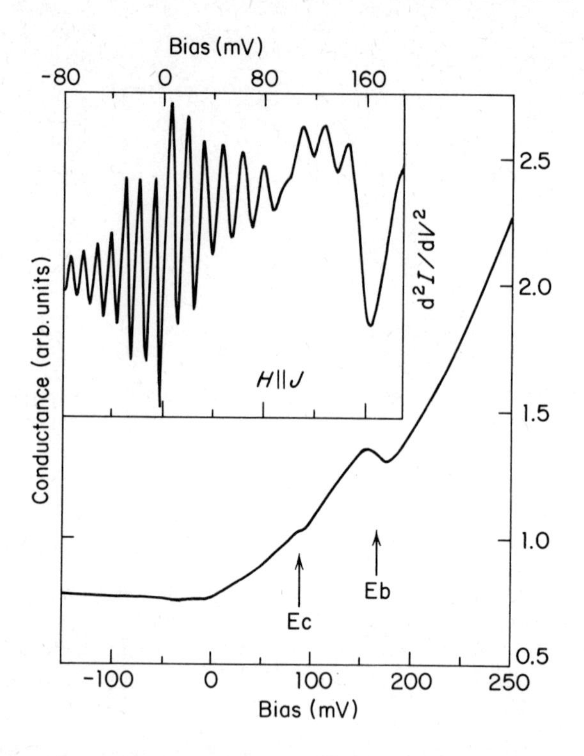

Fig. 8.5. Conductance dI/dV and (inset) d^2I/dV^2 spectrum of InAs ($n = 5.5 \times 10^{17}$ cm^{-3})/oxide/Pb tunnel junction at 4.2 K and with a 3.5-T magnetic field perpendicular to the junction. Landau levels extend in two sets from the prominent two-dimensional band edge at E_b and from the conduction band edge E_c. An increase in damping is discernible at -30 mV, corresponding to threshold for emission of phonons. (After Tsui, 1970*a*.)

A set of extremely interesting related experiments in (n-type) inversion layers on p-type nondegenerate silicon (Kunze and Lautz, 1982) has been briefly reported. The experimental structures are specially designed metal-oxide-semiconductor (MOS) tunnel junctions with thinned SiO_2 tunnel barriers. Heavily doped n-type surface channels surround the p–Si surface layer in which inversion is induced and connect this to the surface drain contact. This step is done in order to avoid series voltage drops in portions of the structure which would otherwise become highly resistive at 4.2 K. The choice of an Mg metal electrode in the Mg–SiO_2–p–Si junction, motivated by the unusually small work function of Mg, is reported to inherently generate a surface electron density in the Si on the order of 10^{13} cm^{-2}. Provision is also made to vary the Si-surface electric field with a third electrode covering the rear surface of the Si wafer. The

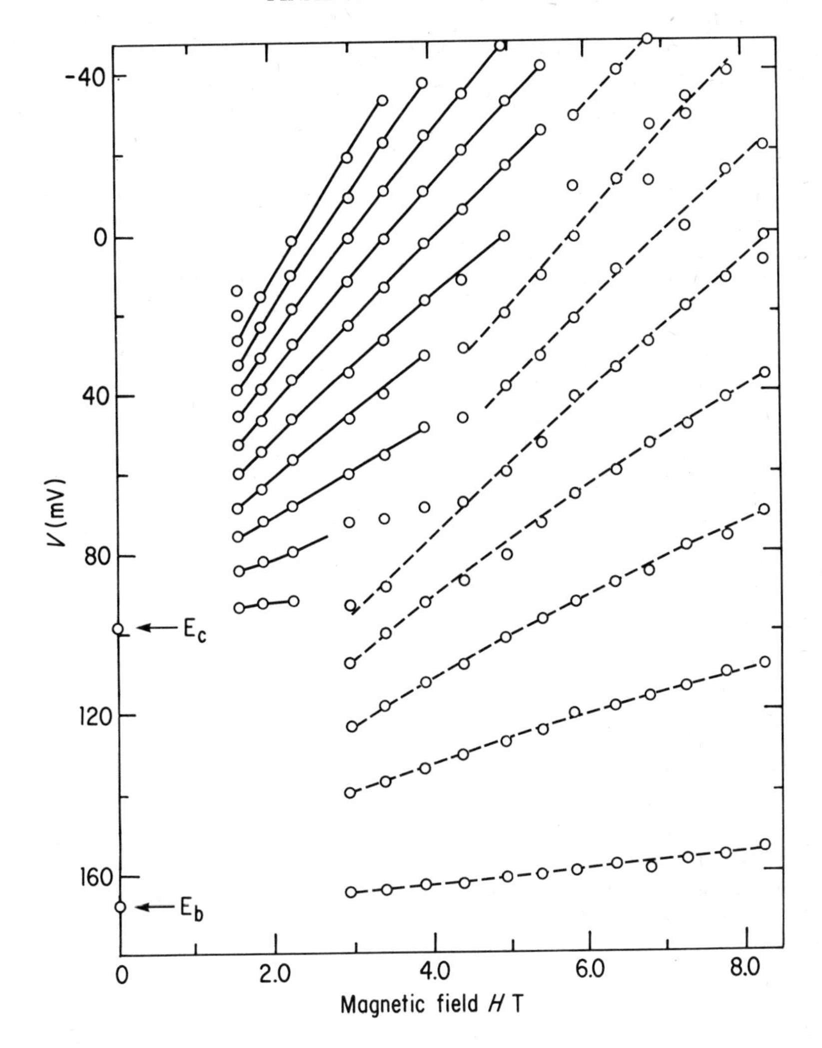

Fig. 8.6. Demonstration that two sets of Landau levels, arising from both the two-dimensional accumulation layer and the conduction band, are observed in the perpendicular field case. Only the latter are observed for a magnetic field parallel to the InAs surface. (After Tsui, 1971*a*.)

experiments have been carried out on both (100) and (111) faces of Si and reveal negative peak features locating various subband minima (bound states E_{nx}), again in accord with (8.9). In the (111) case, the equivalence of the six (100) conduction band minima in Si simplifies the observed structure to a single ladder E_{nx}, $n = 1, 2, \ldots$, of subbands, of which eight are observed as a function of substrate bias above the lowest subband. In the (100) case, two series of levels are observed. Further experiments of this sort under conditions of applied magnetic field or stress could be very useful.

8.2.4 Spin-Polarized Tunneling as a Probe of Ferromagnets

We have already discussed (in Chap. 3, in connection with Figs. 3.26–3.29) the use of extremely thin (high $H_{c\parallel}$) Al-film counterelectrodes (Meservey and Tedrow, 1971) in large, parallel magnetic fields $\vec{H}$ to produce shifted (by $g\mu_B H$) BCS superconducting densities of states for electrons of parallel and antiparallel spin projections. In an interesting series of experiments (Tedrow and Meservey, 1971, 1973; Paraskevopoulos, Meservey, and Tedrow, 1977), the spin polarization of electrons tunneling from ferromagnetic films has been inferred from the asymmetry of spin-split conductance spectra observed in these cases and shown in Figs. 8.7 and 8.8. The comparison of spectra in a nonmagnetic case (*Ni*Cr 12.5 at%) in Fig. 8.7 and a magnetic case (*Ni*Fe 76 at%) in Fig. 8.8 gives convincing evidence that the normal-state density-of-states difference in the ferromagnet,

$$\frac{\rho(E,\uparrow)-\rho(E,\downarrow)}{\rho(E)} \tag{8.13}$$

Fig. 8.7. Nonmagnetic alloy NiCr shows symmetric spin splitting consistent with equal populations of spin-up and spin-down electrons. Junctions are Al–Al_2O_3–alloy, with the Al film only 40 Å in thickness to provide an extremely high parallel critical field. (After Paraskevopoulos, Meservey, and Tedrow, 1977.)

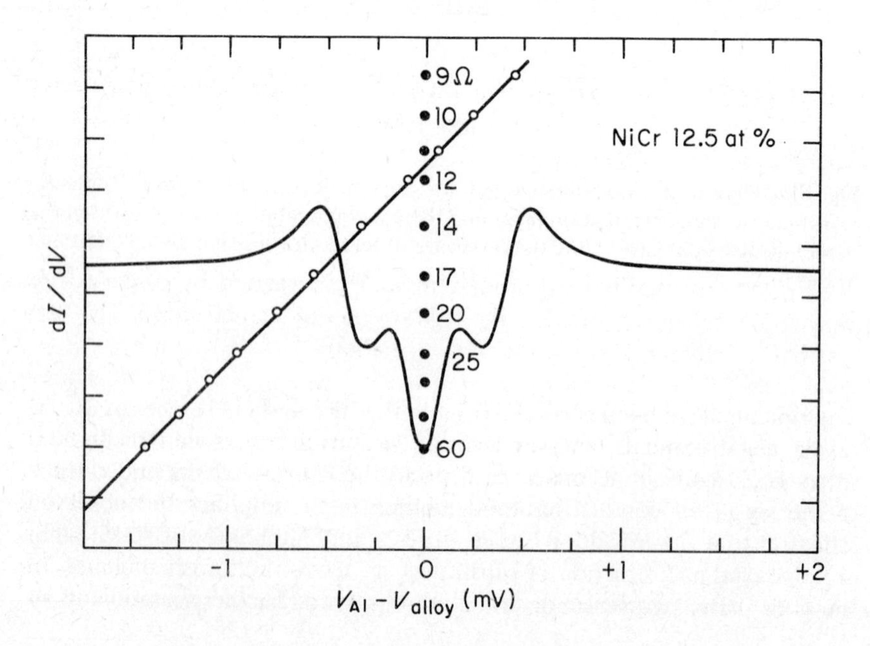

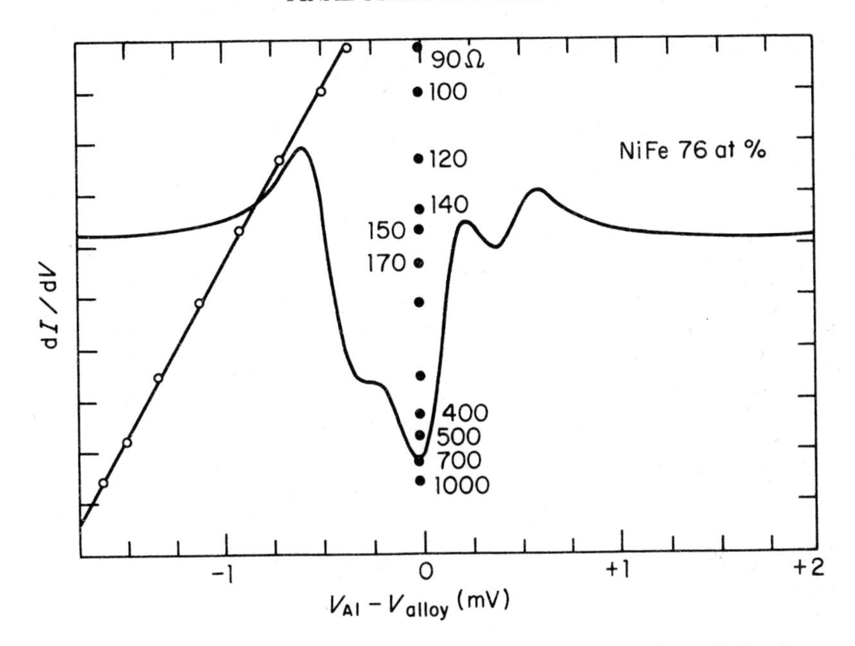

Fig. 8.8. Demonstration of the effect of ferromagnetic spin polarization on tunneling dI/dV spectrum in the energy gap range. Ferromagnetic NiFe alloy provides asymmetric dI/dV spectrum, resulting from unequal populations of spin-up and spin-down electrons in the alloy. This result represents observation of the normal metal density of states in a tunneling configuration. (After Paraskevopoulos, Meservey, and Tedrow, 1977.)

must be reflected as spin polarization of the tunneling electrons detected in the superconducting Al film. Such data are easily analyzed for the apparent spin polarization

$$P \equiv \frac{j\uparrow - j\downarrow}{j\uparrow + j\downarrow} = 2\alpha - 1 \tag{8.14}$$

where $j\uparrow$ corresponds to the tunneling current carried by electron spin magnetic moments parallel to the ferromagnetic magnetization. This task has been accomplished via the decomposition

$$G(V) = \alpha G_{\uparrow}(eV) + (1-\alpha)G_{\downarrow}(eV) \tag{8.15}$$

Here, $G_{\uparrow}(eV)$ and $G_{\downarrow}(eV)$ are the shifted BCS conductance structures, as described in Chap. 3, which incorporate spin-orbit effects and thermal smearing. A summary of tunneling polarization results from elements and alloys of the 3d transition series is given in Fig. 8.9 (Meservey, Paraskevopoulos, and Tedrow, 1978; Meservey, 1978). The dashed curves are plots of the saturation magnetic moment along the 3d series.

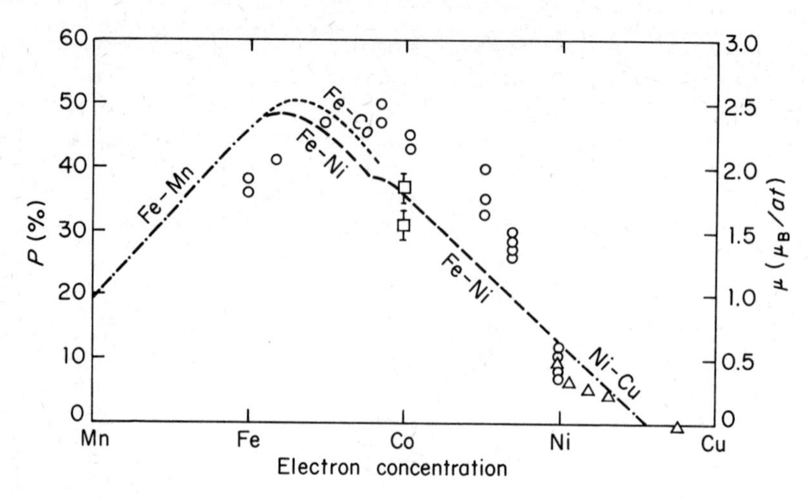

Fig. 8.9. Survey of spin polarization determinations by electron tunneling among elements and alloys of the 3d transition series. (After Meservey, 1978.)

The typically positive values of P have posed a severe problem, because elementary band theory explanations lead to *negative* values of P, as is evident from Fig. 2.19. In fact, for both Ni and Co, for which the tunneling value of P is positive, the Stoner-Wohlfarth theory shows the density of majority d-band states at the Fermi surface to be zero, the majority d band being entirely filled. Thus, apart from the s electrons, whose density is much smaller, one would expect, at a small bias, a total negative polarization of the tunneling current: $P = -1$.

While earlier photoemission measurements of P (Banninger, Busch, Campagna, and Siegmann, 1970) gave positive values in rough agreement with the tunneling values, more recent photoemission data on Ni, the most carefully studied metal, extending to the low-energy threshold (Eib and Alvarado, 1976; Kisker et al., 1979), now give negative P, in agreement with field emission experiments (Landolt and Campagna, 1977). Further, a calculation based on the Stoner-Wohlfarth theory (Moore and Pendry, 1978) now predicts quite well the observed photoemission spin polarization results, including the change of sign of P from positive at $\hbar\omega - \phi > \sim 0.1$ eV (on the Ni (100) surface) to negative P at lower energies. This picture contradicts that inferred by Meservey, Paraskevopoulos, and Tedrow (1978) from their tunneling results.

The origin of the now clearly anamalous barrier tunneling results seems to be in a larger relative contribution from the s electrons, which, by their hybridization with the d electrons, *do* acquire positive polarization P (Hertz and Aoi, 1973; Chazalviel and Yafet, 1977; Stearns, 1977). In a qualitative argument, it seems that the relative preference for the (positively polarized) s contributions would increase with reduced barrier transmission. Hence, the narrow vacuum barrier of the field emission tip may allow more of the (negatively polarized) d electrons to transfer. An even greater relative preference for d electrons would occur in the photoemission case, as the barrier effectively disappears for $\hbar\omega \geqslant \phi$. The relatively large area tunnel contacts with the thickest barriers and smallest D would show the least relative contribution from d electrons, i.e., are expected to be most selective toward s-electron contributions. Other factors in this complex set of comparisons are considered at length (but prior to some key research results) by Feuchtwang, Cutler, and Schmit (1978). This qualitative picture, as pointed out by Meservey (1978), is consistent with the scaling of P in the tunneling results of Fig. 8.9 with the saturation magnetization, because the s-electron polarization will be driven by that of the d electrons.

Information about (a) the depth of the tunneling probe and (b) the critical thickness of a film of Fe or Co to become ferromagnetic has been obtained (Tedrow and Meservey, 1975; Meservey, Tedrow, and Kalvey, 1980) in the "proximity effect" geometry shown in Fig. 8.10a. Here, ultrathin films of Fe, Co, and Ni are deposited at 77 K onto sapphire substrates and immediately overcoated with ~50-Å films of Al. These conditions have been argued by the above authors to lead to continuous Al films; this layer is then oxidized to form the tunneling barrier. The polarization detected by such a spin-split, thin-Al superconducting counterelectrode as a function of ferromagnetic film thickness is shown in Fig. 8.10b. The analysis (dashed lines) employs a simple expression, $P = P_\infty\{1-\exp[(d-d_0)/t]\}$, yielding best values $d_0 = -1.2$ Å, $t = 3.0$ Å for Fe; $d_0 = 1.4$ Å, $t = 3$ Å for Co; and (not shown) $d_0 = 7.3$ Å, $t = 10.2$ Å for Ni. These results seem to say that ferromagnetism occurs in Fe and Co films less than one atomic layer in thickness. The curves in Fig. 8.10b saturate at about $d = 10$ Å to values (see Fig. 8.9) characteristic of much thicker films. This result implies that the tunneling probe does not sample regions deeper than 10 Å from the insulator interface in this normal-state structure.

Spin-polarized tunneling measurements on rare earth metals have been reported by Meservey, Paraskevopoulos, and Tedrow (1978). Progress toward the desirable goal of making tunnel junctions on single-crystal faces of Ni has been reported by Meservey, Tedrow, Kalvey, and Paraskevopoulos (1979), with results similar to those obtained on films.

A survey of thin-film results in the heavy rare earth metals was given by Meservey, Paraskevopoulos, and Tedrow (1978). The values obtained for P (in percent) are Gd, 14; Tb, 6; Dy, 7; Ho, 8; Er, 5; Tm, 3; Yb, 0. These values, within rather large scatter, scale linearly with the effective conduction electron magnetic moment in these metals.

8.2.5 Other Bulk Band Structure Effects

The previous two sections have provided examples of observation of bulk normal-state band structure effects: the difference in density of states at E_F of spin-up and spin-down electrons in ferromagnets, and the observation

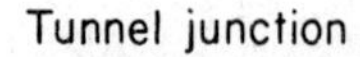

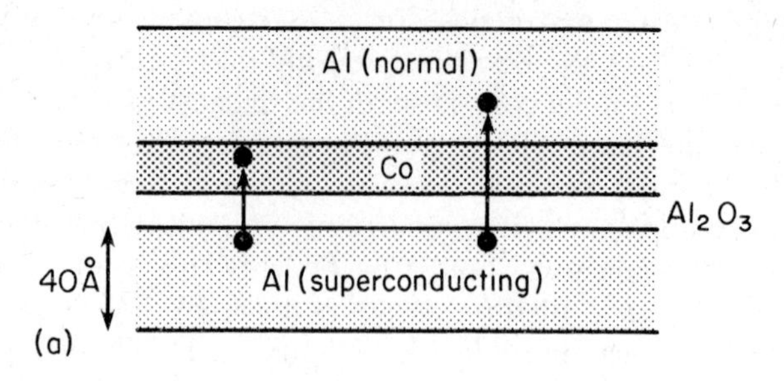

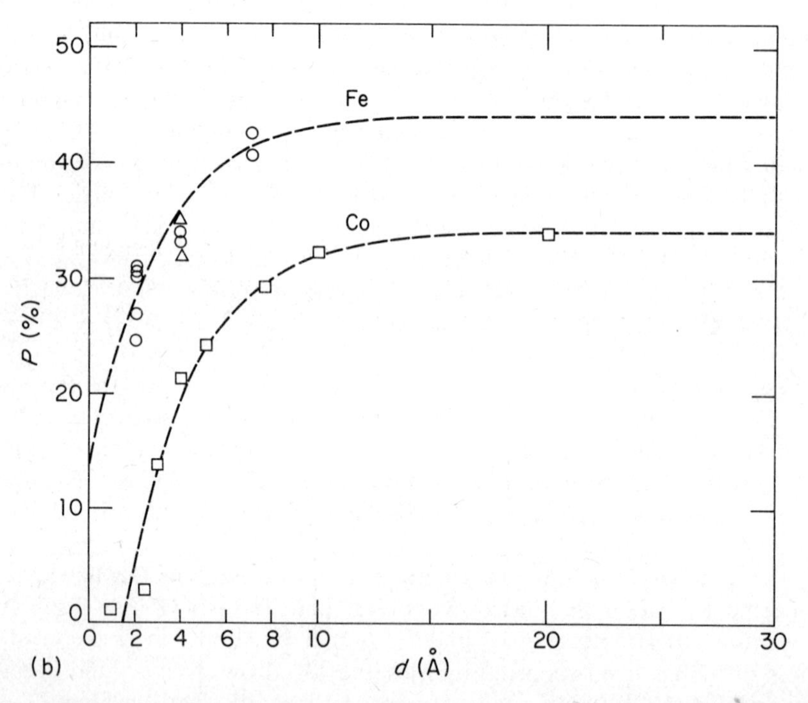

Fig. 8.10. Determination of a proximity effect in ferromagnetic metals. (a) Schematic diagram of the tunnel junctions. The ferromagnetic layer deposited at 77 K has an average thickness varied between 1 and 20 Å. The top layer of Al is 50 Å thick and remains in the normal state because of the proximity effect with the ferromagnetic film. The lower layer is a 40-Å film of aluminum deposited upon a cold substrate and oxidized to form the Al_2O_3 tunneling barrier. (b) The measured polarization of electrons tunneling into Fe (upper curve) and Co (lower curve) films of thickness shown on the abscissa, in angstroms. (After Meservey, Tedrow, and Kalvey, 1980; Tedrow and Meservey, 1975.)

in *parallel* magnetic field of *bulk* Landau levels in InAs. In the first case, an important caveat seems to be that the tunneling observation heavily weights the s-electron contributions over the d-electron contributions, because of their different exponential decay lengths in the tunneling barrier or, effectively, because of their different tunneling matrix elements. In the case of the Landau levels, observed even when the orbital motion of surface accumulation layer electrons is not possible, the effect seems to arise through direct measurement of $\rho(E)$, in the spirit of (8.9), rather than through variation of the barrier transmission factor. Tsui has suggested, in fact, that the bulk Landau levels observed (in orientation $\vec{H} \perp \vec{J}$) are those of electrons moving tangentially to the surface in "skimming orbits," for which (8.9) is still appropriate.

The largest effects of band structure observed in tunneling are certainly those which are associated with a change in the tunneling barrier transmission factor. A particularly clear example of this type of effect is in observations of Guétin and Schreder (1971*b*, 1972*b*) on Schottky barrier contacts to *n*-GaSb under hydrostatic pressure. As shown in Fig. 8.11, GaSb is, at zero pressure, a direct band gap semiconductor, but (dashed curves) the gap becomes indirect [with conduction band minima in

Fig. 8.11. Schematic view of the energy bands in a particular direction of k space for GaAs and GaSb. Solid curves and dashed curves refer to atmospheric pressure and high pressure, respectively. Note that in GaSb a pressure of about 10 kbar inverts the ordering of the conduction bands, turning the material from a direct band gap semiconductor to an indirect band gap semiconductor. (After Guétin and Schreder, 1971*b*.)

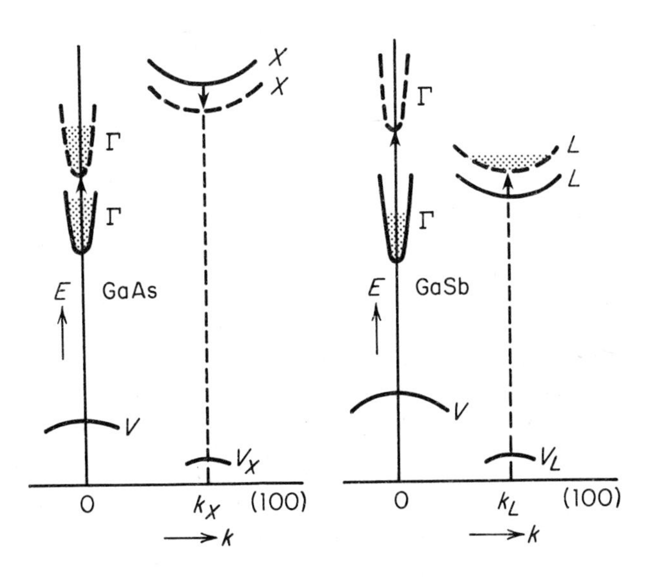

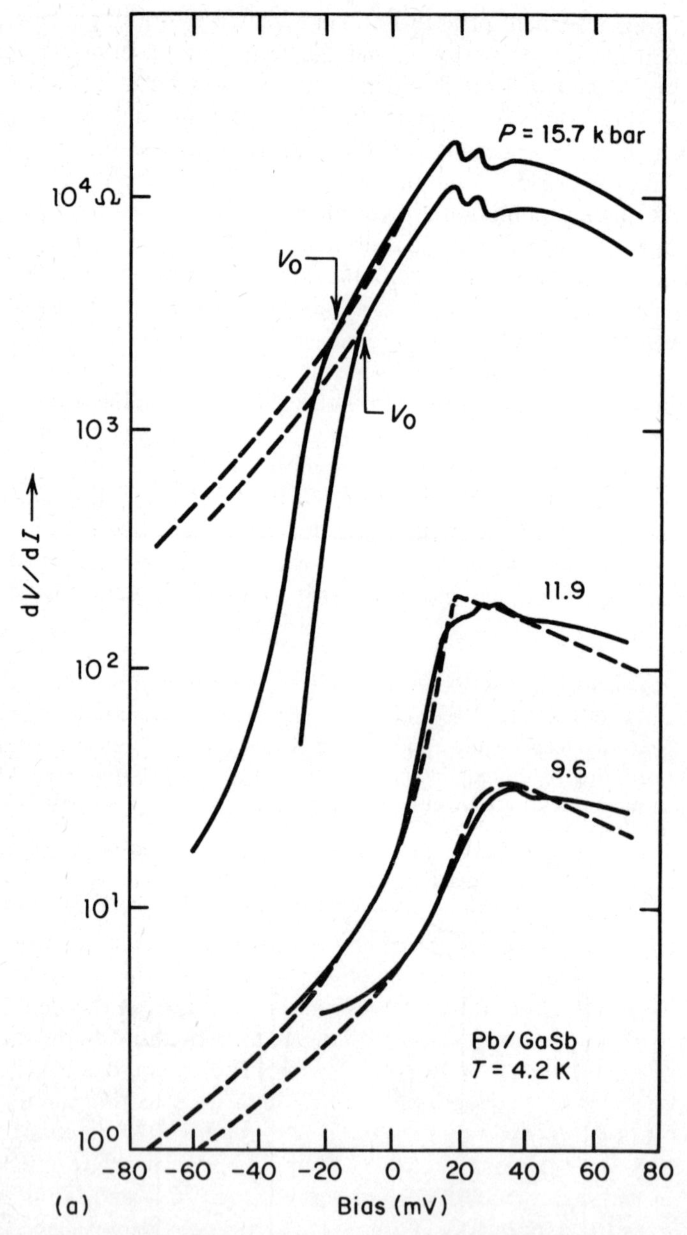

Fig. 8.12. (a) The dV/dI vs. V spectra of Pb/GaSb Schottky barrier contacts as a function of pressure. The great change in resistance which occurs between the 11.9- and 14.7-kbar curves is attributed to the reordering of the conduction bands, as shown in Fig. 8.11, and a smaller tunneling probability from the point L in the Brillouin zone. (After Guétin and Schreder, 1971*b*.) (b) Detail of the changes in the slope of the conductance produced by the pressure. The d^2I/dV^2 vs. V at 14.7 kbar produces two prominent peaks which correspond to emission of *LA* and *TO* zone boundary phonons. (After Guétin and Schreder, 1971*b*.)

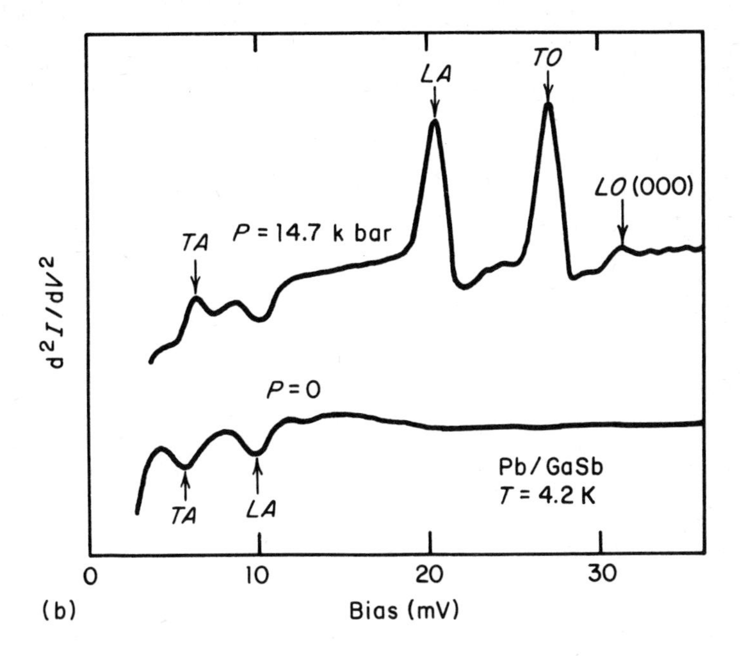

equivalent (111) directions] under hydrostatic pressure of order 10 kbar. The dramatic changes in the dV/dI spectra which accompany the band crossing are shown in Fig. 8.12a (note the logarithmic scale of dV/dI and that more than a factor of 10^3 increase in resistance occurs). Figure 8.12b shows, in d^2I/dV^2, the *LA* and *TO* zone boundary phonon structures that appear at high pressure, related to the indirect gap. It is clear that the major part of the profound change in the dV/dI spectra comes from the change in the transmission process, involving the change in the position in the GaSb Brillouin zone, rather than from any changes in the density of states. The phonon structures observed in the high-pressure (indirect gap) case are associated with $\vec{k}$ conservation: with phonon emission, the [111] electron transfers to a virtual state at Γ and tunnels to the metal. This process has been described by Davis and Steinrisser (1970) in connection with Ge Schottky barriers. The absence of the large pressure effects in similarly prepared GaAs (Guétin and Schreder, 1972*a*) and Ge (Guétin and Schreder, 1973) Schottky contacts under pressure is consistent with the lack of band crossings under pressure in these cases.

A related observation of two aspects of semiconductor band structure in Ge is shown in Fig. 8.13 (Conley and Tiemann, 1967). In this figure, dV/dI for an Au/*n*–Ge contact is shown, with positive *V* corresponding to raising the Ge bands. The interpretation of the dV/dI peak at +30 mV is that this peak corresponds to raising the degenerately doped Ge conduction bands so that their minima face the Au Fermi level. Within conditions discussed in Chap. 2, and believed applicable to Ge, this condition determines 30 meV as the Fermi degeneracy in the Ge electrode. The distinct change of slope at $V=-0.124$ eV or 0.154 eV above the (L) conduction band edge is interpreted as the onset of direct tunneling of electrons from the Au into the unoccupied $k=0$ (Γ_2') conduction band.

A much weaker band edge effect in Au is made evident in Fig. 8.14 (Jaklevic and Lambe, 1973) in the difference of strained and unstrained d^2I/dV^2 spectra. The feature 3.5 eV above the Au Fermi level is interpreted as a band edge at L_1, while an L_2' band edge is identified 0.7 eV below the Fermi level.

While all of the band structure effects in single-crystal semiconductor Schottky barriers are quantitatively reliable and reproducible, this has not been the case for band edges inferred for semimetals, notably Bi, as studied in succession, but with somewhat different results, in Al–Al_2O_3–Bi thin-film junctions by Esaki and Stiles (1965, 1966) and by Hauser and Testardi (1968). Extension of similar measurements to Bi and to BiSb alloys, using Sn–SnO_2 counterelectrode-barrier combinations, has been more recently reported by Chu, Eib, and Henriksen (1975), who find qualitative agreement in major features with the work of Hauser and Testardi (1968). Chu, Eib, and Henriksen (1975) find also rough agreement with a band structure calculation for Bi (Mase, 1958, 1959) and have reviewed the work in this area. The much better reliability achieved in the degenerate semiconductor cases may be attributable to the completely reproducible, oriented, single-crystal Schottky barrier contacts which are studied. By comparison, the semimetal films are polycrystalline, and the oxide barriers grown in these films may quite possibly grow differently on the (possibly) several different metal faces present. Quite plausibly, the distribution of crystallite orientation may differ from film to film. In terms of *superconducting* tunneling, such small effects are not noticeable because of the relatively large energy (bias) scale over which they occur, and because of their cancellation in obtaining the ratio G_S/G_N. It is a desirable possibility that, in further work of this type, epitaxial metal films may be used. This work still leaves open some questions along the lines raised by Dowman, MacVicar, and Waldram (1969) concerning the selection of particular states which are transmitted by a given tunneling barrier, which can be definitely answered in the case of the single-crystal Schottky barriers. In the possible extension of such work to epitaxial metal films, a check that the barrier oxides are amorphous, rather than also epitaxial, is desirable, because the selection of transmitted states by the amorphous oxide is more easily predictable.

In Schottky barrier contacts of In to vacuum-cleaved degenerate EuS—$1.4\times10^{19}\ \mathrm{cm}^{-3}$ Eu near and below the ferromagnetic transition, $T_c=2.15$ K—Thompson, Holtzberg, McGuire, and Petrich (1972) and Thompson, Penney, Holtzberg, and Kirkpatrick (1972) observed increases in $G(0)$ by factors of 10^2 to 10^3, which vary exponentially with the

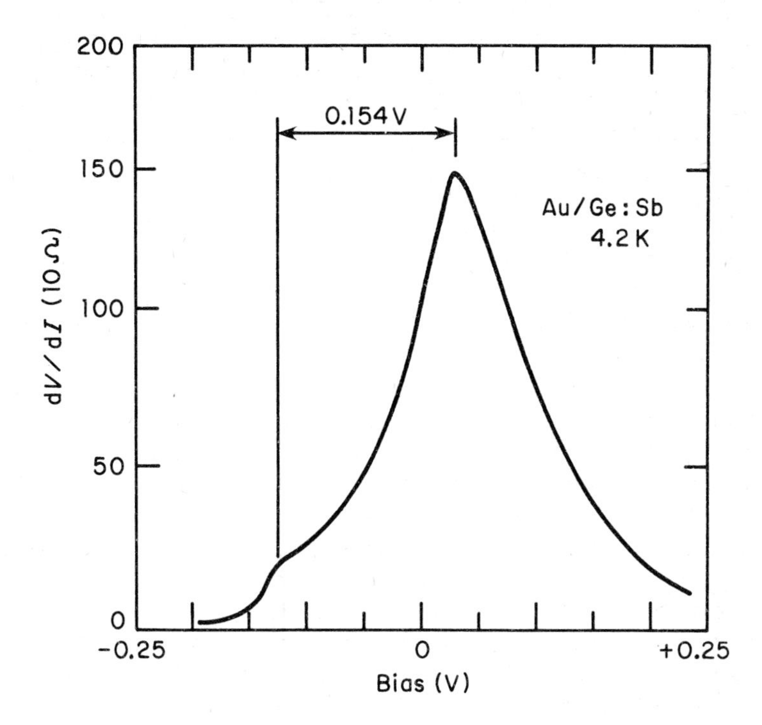

Fig. 8.13. The dV/dI vs. V for Schottky barrier contacts of gold to Ge:Sb reveals, at reverse bias of 0.124 V, the onset of tunneling transitions into the upper $k=0$ conduction band, thus locating this edge at 0.154 eV above the indirect band edge. The band structure of germanium is rather similar to that of GaSb under pressure as shown in Fig. 8.11. (After Conley and Tiemann, 1967.)

bulk magnetization, in turn due to alignment of electron moments in Eu 4f states. This effect is interpreted as a reduction in the Schottky barrier height by $\frac{1}{2}g\mu_B H_w$, where H_w is an internal Weiss molecular field. The conduction band is split by $g\mu_B H_w$, assumed larger than the Fermi degeneracy μ_F, into spin-up and spin-down bands; subsequent realignment of the lower band to the Fermi level occurs by transfer of electrons from the metal, which reduces the width of the barrier as well as its height.

This interpretation, which implies completely polarized conduction electrons, is confirmed in lower-concentration samples by capacitance measurements which indicate $\Delta V_B = -0.24 \pm 0.03$ eV at $T=0$. Observation of the onset of tunneling into the upper, reverse-spin conduction band, which should occur at $eV = -2\Delta V_B$, was not reported, however.

In the same Schottky barrier system, electron critical scattering by fluctuations in the ferromagnetic barrier has been observed by Thompson and colleagues (1972). The zero-bias conductance follows a Curie-Weiss

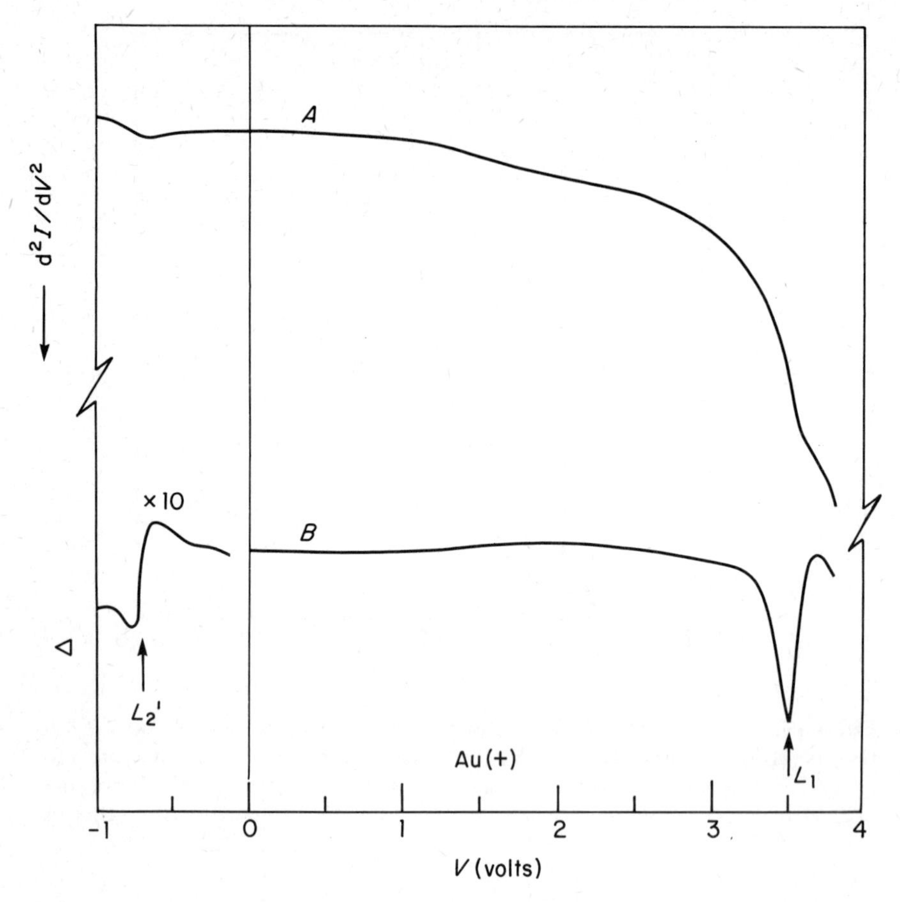

Fig. 8.14. Band edge structure in Au revealed in tunneling spectra of $Al–Al_2O_3–Au$ junction at 77 K. Curve B represents the difference between the d^2I/dV^2 curves with and without strain to remove band structure independent features. Assigned band edges are -0.7 eV (L_2') and $+3.5$ eV (L_1) relative to the Au Fermi level. (After Jaklevic and Lambe, 1973.)

temperature dependence, $G(0) = G_0(T - T_c)$, where the ordering temperature T_c is 25 K, in agreement with measurements of the bulk magnetization. The EuS in this case is doped to a concentration of $8 \times 10^{19}\,\mathrm{cm}^{-3}$ carriers, higher than the $3 \times 10^{18}\,\mathrm{cm}^{-3}$ in the In/EuS junctions discussed above. As suggested, this effect would seem to require an extension of tunneling theory; a basis for such an extension is proposed.

The absence of any structure in tunneling into magnetic metal films that could be attributed to d bands has been confirmed by careful studies of Ni, Pd, and Ni–Pd, and Ni–Cu alloys by Rowell (1969*a*) and by Hanscom (1970). These experiments were designed to shift the Fermi level through the d-band peaks, with no detected effect. While a brief report of a small effect (~3% change in dV/dI, with a small voltage shift in the background) upon hydrogenation of Pd has been given in terms of changes in d-band occupancy (Grant, Barker, and Yelon, 1969), such weak effects might alternatively arise from changes in the barrier asymmetry, introduced, in some way, by the hydrogenation. In a careful study of evaporated Cr–Cr_2O_3–metal junctions, Rochlin and Hansma (1970) observed no structure related to an antiferromagnetic gap and, in fact, no anomaly in $G(V)$ at $V=0$.

It is likely that the absence of easily observed d-band edge features is again related to the more rapid decay of these states in the barrier, as in the case of spin polarization measurements on Ni mentioned earlier. Again, in a careful study of tunneling into normal Cu:Cr and Cu:Fe alloys films (Dumoulin, Guyon, and Rochlin, 1970), no concentration-dependent structure, to 10^{-3} of the background conductance, was observed.

Other aspects of normal-state properties, somewhat arbitrarily judged to be of "many-body" origin, are taken up later, in Sec. 8.4. Some further band structure effects in unusual materials such as TTF–TCNQ have been reported and will be considered in Chap. 9.

8.3 Assisted Tunneling: Threshold Spectroscopies

The earliest observations of assisted tunneling of any sort would appear to be those of Holonyak et al. (1959) and of Esaki and Miyahara (1960) (see also Esaki, 1958, 1974) in which phonon energies of Si were observed (Fig. 1.4) in the I–V characteristic of *pn* junctions at low temperatures. The possibility of a theoretical treatment of such effects within the tunneling Hamiltonian model was mentioned by Bardeen (1961), and a general discussion of this theory has been given in Sec. 2.6. Since the work of Holonyak and of Esaki and Miyahara, a proliferation of observations of excitation thresholds $eV=\hbar\omega_{ph}$ where ω_{ph} may represent a variety of modes including plasmons, magnons, electronic excitations, vibrations of organic molecules, etc. have accumulated, and they represent probably the most extensive set of data from nonsuperconductive tunneling spectroscopy. One subset of this area of threshold spectroscopies, relating to vibrational modes of organic molecules in barriers (Jaklevic and Lambe, 1966), has become known as IETS (inelastic electron tunneling spectroscopy) and is reviewed separately in Chap. 10. The purpose of the following discussion is to survey the broader range of threshold spectroscopies.

8.3.1 Phonons

An excellent example of the use of threshold spectroscopy to locate semiconductor phonons is shown in Fig. 8.15a, after Payne (1965), in d^2I/dV^2 measurements on a Ge *pn* junction (Esaki diode). In fact, the

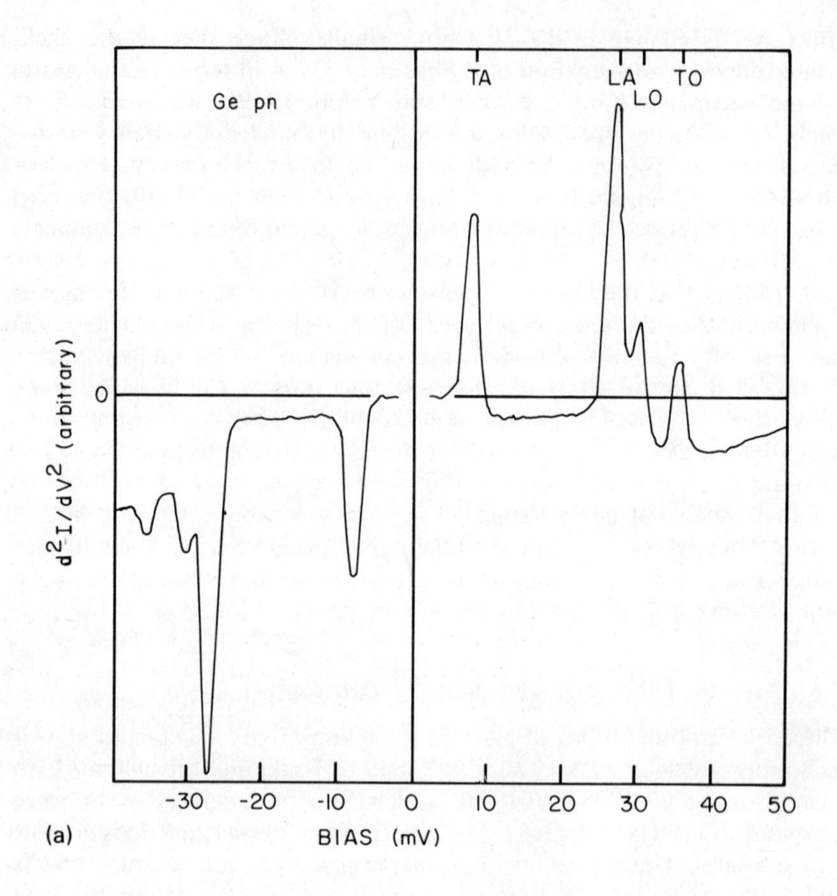
Ge pn
TA
LA
LO
TO
d²I/dV² (arbitrary)
0
-30
-20
-10
0
10
20
30
40
50
(a)
BIAS (mV)

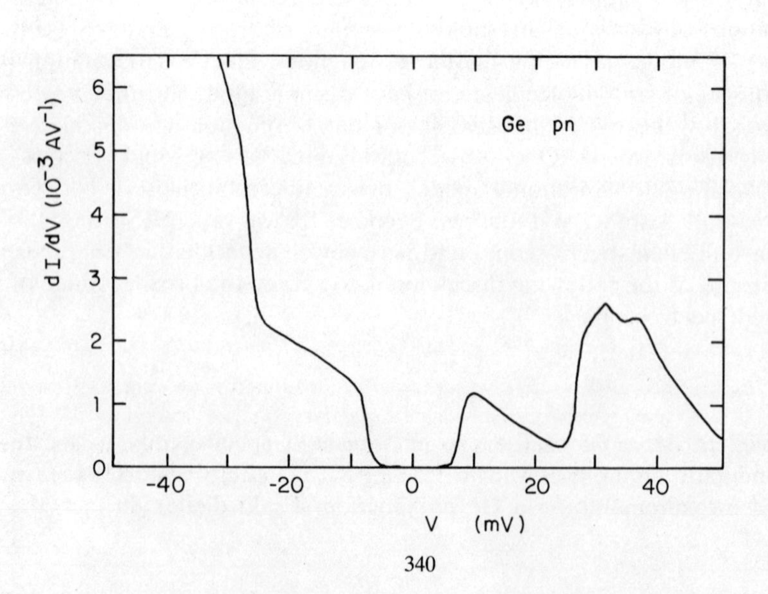
Ge pn
dI/dV (10⁻³ AV⁻¹)
6
5
4
3
2
1
0
-40
-20
0
20
40
V (mV)

strongest known examples of phonon thresholds in tunneling occur in cases similar to the Esaki diode (see Figs. 8.15 and 8.16), where *phonon assistance* is *required* for a tunneling transition to occur. In Fig. 8.15b, note that the (low-temperature) conductance below 8-mV bias is negligibly small. In this case, at very low bias, no elastic process can occur because it would violate the conservation of crystal momentum $\vec{k}$. Electrons which tunnel from the n-type Ge, as shown in Fig. 8.16, have large k values lying near the boundaries of the Brillouin zone in the six equivalent [111] conduction band "valleys." The final states, however, lie at the zone center, the location of the highest energy levels in the valence band of Ge. (The usually filled states near $k=0$ are made available for extra electron occupation in the degenerate p-type material by a heavy doping of an acceptor impurity such as boron.) Since the whole tunnel junction occurs within a perfect Ge crystal lattice, with only the dopant concentration changing with position, k is conserved, and the transition of the electron from $k=\pi/a$ (111) to $k=0$ must be accompanied by creation of a phonon of wavevector $k=\pi/a$ (111) (Fig. 8.16c) to preserve the initial wavevector. [Of course, at high enough temperature, a $k=-\pi/a$ (111) thermal phonon might be absorbed, permitting the electron transition at $eV=0$. Such absorption would most closely illustrate the concept of a phonon-assisted transition, but the same term is applied when the phonon is generated rather than absorbed.] Essentially, the entire low-temperature current in the Esaki diode (Fig. 1.4) is thus phonon-assisted. It turns out that there are four different phonons (the four energies occur in Fig. 8.15a), of wavevector $k=\pi/a$ (111), able to provide the assistance. These phonon energies, clearly revealed in the d^2I/dV^2 spectrum of Fig. 8.15a, are known also from inelastic neutron scattering measurements. The ω–k dispersion relations provided by the neutron scattering measurements of Brockhouse and Iyengar (1958) in Ge are shown in Fig. 8.16c.

A second illustration of threshold determinations in Ge is given in Fig. 8.17 in the case of a metal-semiconductor or Schottky barrier of a metal contact on degenerate n-type material. In this case, the conductance changes at threshold are relatively small, and a much larger elastic or nonassisted current flows even at $T=0$ and $V\simeq 0$. In this case, direct electronic transitions can occur from the large Fermi surface of the metal to the various conduction band valleys of the Ge, conserving, as required, the component of k parallel to the interface. The large observed nonas-

---

Fig. 8.15. (a) Second-derivative spectrum of a Ge tunneling *pn* junction (tunnel diode) at low temperature reveals zone boundary phonon energies. (b) First derivative dI/dV for Ge tunneling *pn* junction, showing phonon thresholds. The conductance is almost entirely phonon-assisted, as indicated by the extremely small conductance values for $|eV|<8$ meV. (After Payne, 1965.)

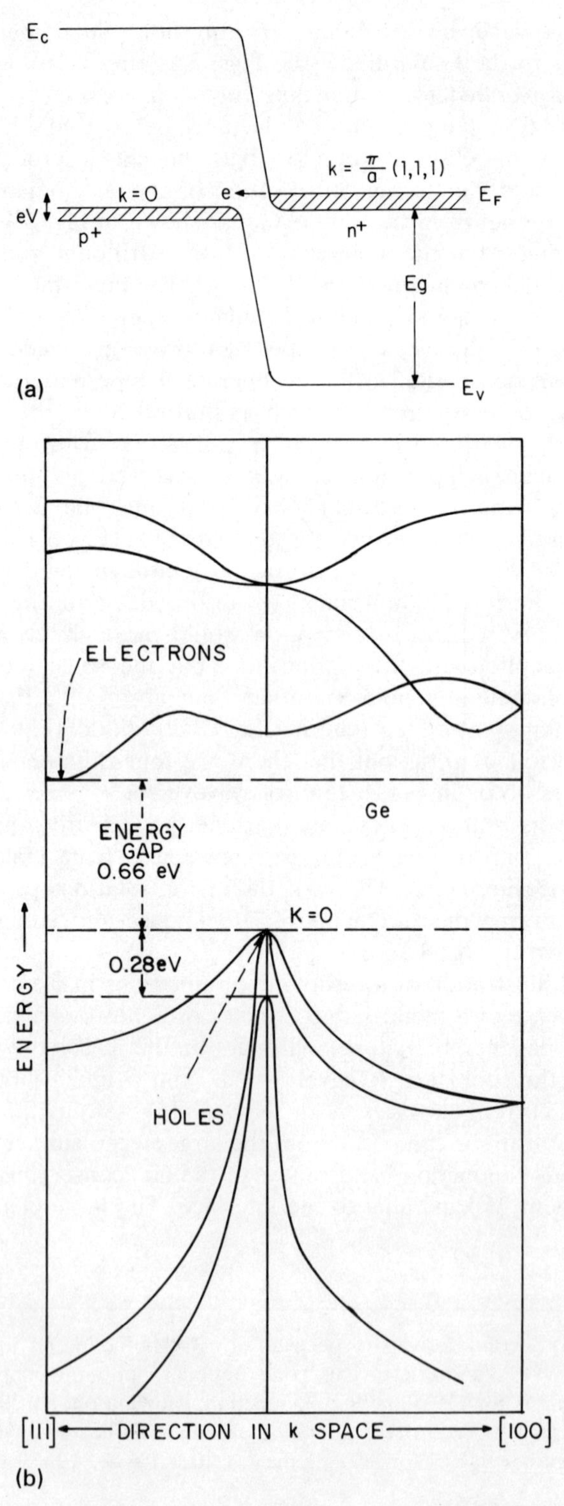

E_c
$k = \frac{\pi}{a}(1,1,1)$
k=O
e
E_F
eV
p^+
n^+
Eg
(a)
E_V
ELECTRONS
Ge
ENERGY
GAP
0.66 eV
K=O
0.28eV
ENERGY
HOLES
[111]
DIRECTION IN k SPACE
[100]
(b)

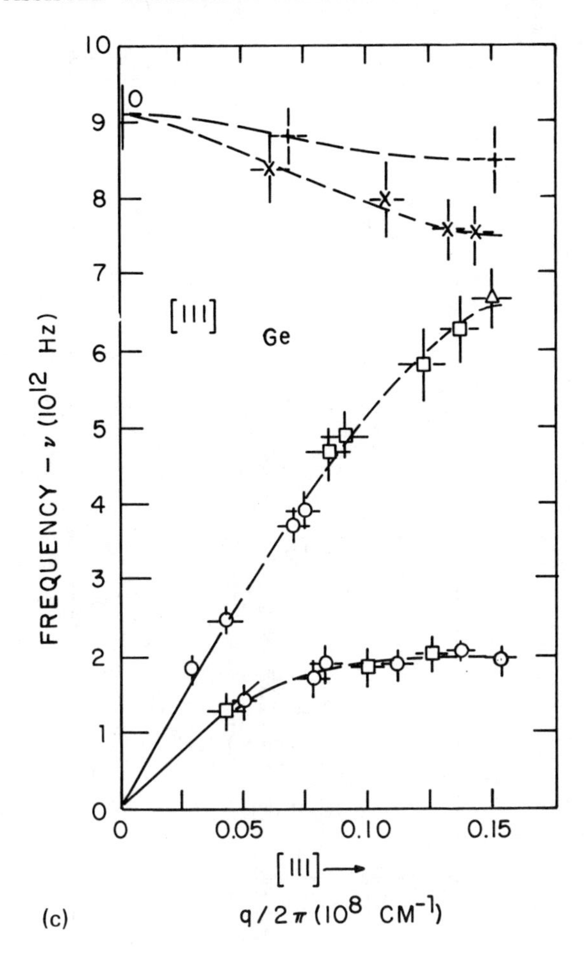

Fig. 8.16. (a) Schematic diagram of a Ge tunneling *pn* junction (Esaki diode). (b) Band structure of Ge, showing indirect gap with [111] valleys at the zone boundary. [After H. Brooks, "Advances in Electronics and Electron Physics," Vol. VII, L. Marton, Ed. (New York, Academic Press, 1955).] (c) Phonon dispersion curves in the [111] directions for Ge, determined by neutron scattering measurements. (After Brockhouse and Iyengar, 1958.)

sisted current may indicate that the wavevector values corresponding to the conduction band valleys are available in the metal, or that the selection rule is weakened by the fact that the two electrodes have separate crystal structures and may well be joined in a rather disordered interface region which may change the wavevector of the traversing electrons.

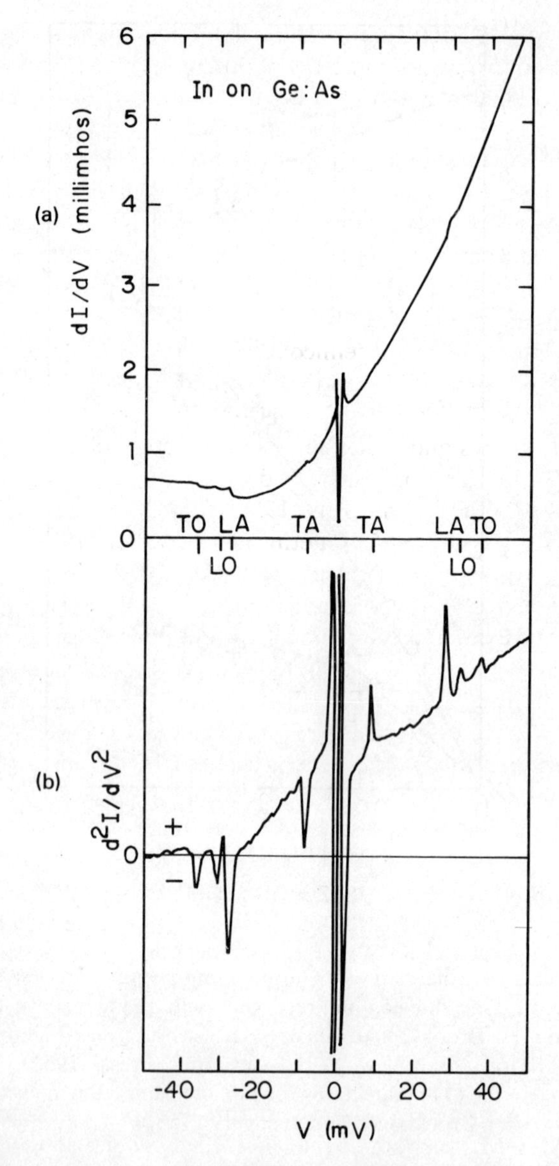

Fig. 8.17. (a) Conductance dI/dV for Schottky barrier contact on heavily doped metallic n-type Ge. (b) Second derivative $\mathrm{d}^2I/\mathrm{d}V^2$ of the same data reveals phonon features. (After Steinrisser, Davis, and Duke, 1968.)

As an inelastic excitation can occur for either sign of bias voltage, sharp increases in conductance, symmetric about $V=0$, and hence peaks in d^2I/dV^2, antisymmetric about $V=0$, are characteristic of an inelastic threshold process. The antisymmetry in d^2I/dV^2 (or d^2V/dI^2) and symmetry about $V=0$ of the step increases in the $G(V)$ characteristic of assisted tunneling are evident in Figs. 8.15 and 8.17. Conversely, structure of different symmetry about $V=0$ at a characteristic excitation energy may be presumed to arise by modification of the E–k relation of electron states, through interaction with the phonon or other mode or through a more complex self-energy effect.

A partial listing of the many semiconductors for which phonon energies have been measured includes CdS (Mikkor and Vassell, 1970; Losee and Wolf, 1969), GaAs (Mikkor and Vassell, 1970; Tsui, 1968; Conley and Mahan, 1967; Thomas and Quiesser, 1968; Guétin and Schreder, 1970), $Ga_{1-x}Al_xAs$, $GaAs_{1-x}P_x$, and GaP (Andrews et al., 1972), GaSb (Mikkor and Vassell, 1970; Guétin and Schreder, 1971*a*), Ge (Steinrisser, Davis, and Duke, 1968; Payne, 1965; Guétin and Schreder, 1972*a*), InP (Andrews et al., 1972), $In_{1-x}Ga_xP$ (Andrews et al., 1972; Korb et al., 1971), $KTaO_3$ (Sroubek, 1969; Johnson and Olson, 1971), Si (Holonyak et al., 1959; Esaki and Miyahara, 1960; Chynoweth, Logan, and Thomas, 1962; Wolf, 1968; Wolf and Losee, 1970; Cullen, Wolf, and Compton, 1970; Schein and Compton, 1971), SiC (Schein and Compton, 1970), SnO_2 (Mikkor and Vassell, 1970), and $SrTiO_3$ (Sroubek, 1969).

A related observation is of phonons excited in the insulating barrier layer of a tunnel junction. Examples of this type include AlN (Shklyarevskii, Yanson, and Zaporozhskii, 1974), Al_2O_3 (Rowell, McMillan, and Feldmann, 1969; Geiger, Chandrasekhar, and Adler, 1969), CdO (Giaever and Zeller, 1968*b*), CdS (Giaever and Zeller, 1968*b*; Lubberts, 1971), CdSe and CdS_xSe_{1-x} (Lubberts, 1971), CoO (Tsui and Kaminsky, 1973), Cr_2O_3 (Jaklevic and Lambe, 1970), Ge (Giaever and Zeller, 1968*b*; Ladan and Zylbersztejn, 1972), MgO (Klein et al., 1973; Adler, 1969), PbS (Giaever and Zeller, 1969), Y_2O_3 (Jaklevic and Lambe, 1970), ZnO (Giaever and Zeller, 1968*b*; Tsui and Kaminsky, 1973), and ZnS (Giaever and Zeller, 1968*b*).

Very weak normal metal phonon structure ($\Delta\sigma/\sigma \simeq 10^{-3}$) is seen from *electrodes* of M–I–M junctions, again identified by the symmetry of the conductance under reversal of bias sign. Specifically, the inelastic phonon contributions are extracted from the *even* conductance, defined as

$$\sigma_e = \tfrac{1}{2}[\sigma(V)+\sigma(-V)] \tag{8.16}$$

where $\sigma = dI/dV$, while self-energy effects are obtained from the *odd* conductance,

$$\sigma_0 = \tfrac{1}{2}[\sigma(V)-\sigma(-V)] \tag{8.17}$$

In the equivalent decomposition of $d^2I/dV^2 \equiv \sigma'$,

$$\sigma_0' = \tfrac{1}{2}[\sigma'(V) - \sigma'(-V)] \tag{8.18}$$

is identified with purely inelastic assisted tunneling, while

$$\sigma_e' = \tfrac{1}{2}[\sigma'(V) + \sigma'(-V)] \tag{8.19}$$

is related to self-energy effects. The normal-state phonon structure of metallic Al has been reported by several workers, including Léger and Klein (1969), Chen and Adler (1970), and Dayan (1978). Other metals studied in this fashion are Ag and Au (Chen and Adler, 1970), Bi (Esaki et al., 1968; Straus and Adler, 1974), In (Chen and Adler, 1970; Yanson, 1971), Mg (Adler, 1969*b*), Pb (Rowell, McMillan, and Feldmann, 1969; Wattamaniuk, Kreuzer, and Adler, 1971; Yanson, 1971; Klein et al., 1973), and Sn (Chen and Adler, 1970).

A discussion of quantitative aspects of analysis of inelastic metal phonon data, as applied to Pb–PbO–Pb junctions, has been given by Adler, Kreuzer, and Wattamaniuk (1971). This analysis is based on a two-channel stationary-state approach to inelastic tunneling, similar to a standard theory of inelastic atomic collisions, as described by Brailsford and Davis (1970). It is of interest because it gives an indication, within a one-electron picture, of the effects on the observed inelastic phonon spectra which depend specifically on the location x of the electron-phonon interaction, described mathematically by a position-dependent matrix element $M(x)$,

$$M(x) = \int \phi_1^*(\xi) H_{ev}(x, \xi) \phi_0(\xi)\, d\xi \tag{8.20}$$

In this fashion, a uniform treatment of phonon generation at the edge of the metal film or in the barrier itself becomes possible. The Hamiltonian for the system is taken as

$$H(x, \xi) = H_e(x) + H_v(\xi) + H_{ev}(x, \xi) \tag{8.21}$$

where the free-particle term $H_e(x)$ contains the barrier $V(x)$, and $H_v(\xi)$, with vibrational wavefunction $\phi(\xi)$, represents the lattice vibration. The total wavefunction $\Psi(x, \xi)$ is approximated as

$$\Psi(x, \xi) = \psi_0(x)\phi_0(\xi) + \psi_1(x)\phi_1(\xi) \tag{8.22}$$

where $\phi_0(\xi)$ and $\phi_1(\xi)$ are the ground and first excited-state wavefunctions of the vibration corresponding to energies E_{v0}, E_{v1}, respectively, such that $E_{v1} - E_{v0} = \hbar\omega$. The two stationary electron waves ψ_0 and ψ_1 are determined by substituting (8.22) into (8.21), which yields, setting $E_{v0} = 0$,

$$H_e(x)\psi_0(x)+M(x)\psi_1(x)=E\psi_0(x)$$
$$H_e(x)\psi_1(x)+M(x)\psi_0(x)=(E-\hbar\omega)\psi_1(x)$$

with $M(x)$ given by (8.20). These equations were solved by Brailsford and Davis (1970).

Taking the interaction $M(x)$ to be nonzero in only a narrow region of the electrode at the edge of the barrier, Adler, Kreuzer, and Wattamaniuk (1971) find a transmission probability

$$|T|^2=|T_0|^2+\left(1-\frac{\hbar\omega}{E}\right)^{1/2}|T_1|^2\,\theta(E-\hbar\omega) \tag{8.23}$$

The second term, when integrated over $F(\omega)\,\mathrm{d}\omega$ up to $\omega=E/\hbar$, with $F(\omega)$ the phonon density of states, gives the inelastic assisted conductance. The energy-dependent prefactor in (8.23) plays a noticeable role in matching the experimental observations from normal-state Pb–PbO–Pb and Al–I–Pb junctions with predictions based on knowledge of $F(\omega)$ for Pb from superconducting tunneling and neutron scattering (Wattamaniuk, Kreuzer, and Adler, 1971; Adler, Kreuzer, and Wattamaniuk, 1971).

Specifically, the latter authors obtain, from (8.23), the expression

$$\frac{1}{\sigma_0}\frac{\mathrm{d}\sigma}{\mathrm{d}V}\propto\frac{1}{2V^2}\int_0^{eV}\frac{|M(\omega)|^2\,\omega F(\omega)\,\mathrm{d}\omega}{(1-\omega/eV)^{1/2}} \tag{8.24}$$

in which $\sigma=\mathrm{d}I/\mathrm{d}V$; σ_0 is the elastic conductance evaluated at $V=0$, with V the bias voltage; and the matrix element $|M(\omega)|^2$ is identified with $\alpha^2(\omega)$. The factor $(1-\omega/eV)^{-1/2}$ accounts reasonably for small shifts in phonon energies observed in the inelastic Al–I–Pb $\mathrm{d}\sigma/\mathrm{d}V$ data of Adler and colleagues (1971) (4.4 meV $\rightarrow$ 4.8 meV for Pb transverse modes; 8.4 meV $\rightarrow$ 8.8 meV for Pb longitudinal modes). In a case like this, where the electrodes are dissimilar and only one has strong electron-phonon coupling, inversion of (8.24) yields the formula (Wattamaniuk, Kreuzer, and Adler, 1971)

$$F_{\mathrm{eff}}(\omega)\propto\frac{2}{\pi}\frac{\mathrm{d}}{\mathrm{d}\omega}\int_0^{\omega}\frac{V^{3/2}\,\mathrm{d}\sigma/\mathrm{d}V\,\mathrm{d}V}{(\omega-eV)^{1/2}} \tag{8.25}$$

through which an effective phonon spectrum can properly be extracted from normal-state $\mathrm{d}^2I/\mathrm{d}V^2$ data. A second success of the analysis represented by (8.24) is the correct prediction of a rather strong two-phonon harmonic at ~17 meV in the symmetrical Pb–I–Pb case. This effect is interpreted as tunneling accompanied by emission of one phonon at the edge of each of the two electrodes.

Turning away from inelastic (real phonon) effects to self-energy (virtual phonon) effects, an example of the information contained in the normal-state odd conductance from a strong-coupling metal (Pb) is shown in Fig. 8.18 (Rowell, McMillan, and Feldmann, 1969). Here, the difference

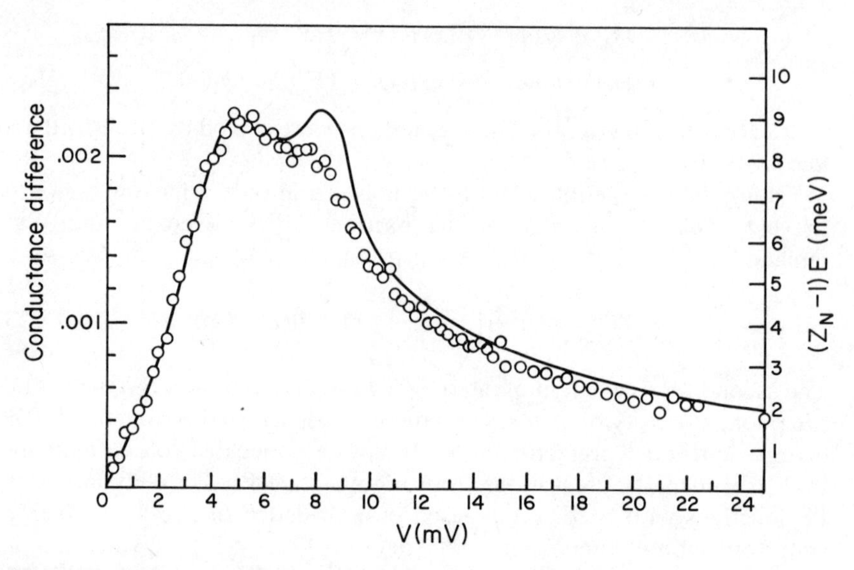

Fig. 8.18. Self-energy effect in Pb in normal metal tunneling (circles, left scale) compared with the self-energy function $(Z_N - 1)\ E$ inferred from superconducting tunneling (curve, right scale). The circles measure the difference between the odd conductance of a Pb–I–Pb junction at 1 K and the background conductance; the solid line is the energy dependence of the self-energy in Pb, $(Z_N - 1)E$. (After Rowell, McMillan, and Feldmann, 1969.)

between the odd-conductance and a background conductance curve (shown as data points) is compared with the function $[Z_N(E) - 1]E$, where the energy-dependent renormalization function $Z_N(E)$ has been obtained from analysis of superconducting tunneling. An effect similar in nature, identified by symmetry, rather than asymmetry in d^2I/dV^2, in degenerate *p*–Si was reported by Wolf (1968). Further discussion of such self-energy effects is deferred to Sec. 8.4.3.

8.3.2 *Inelastic Electron Tunneling Spectroscopy of Molecular Vibrations*

In an important pair of papers, Jaklevic and Lambe (1966) and Lambe and Jaklevic (1968) reported that evaporated Al–Al_2O_3–Pb junctions examined in d^2I/dV^2, with large modulation amplitudes and high gain in the bias range 0 to 0.5 V, displayed rather broad peaks at 120 and 450 meV, as shown in Fig. 8.19. These features coincide with the energies of OH "bending" and "stretching" modes observed in the infrared spectra of Al_2O_3. The individual peaks correspond to increments in $G(V)$ of order $\Delta G/G \simeq 10^{-2}$.

More important, if molecules such as water, acetic acid, methyl alcohol,

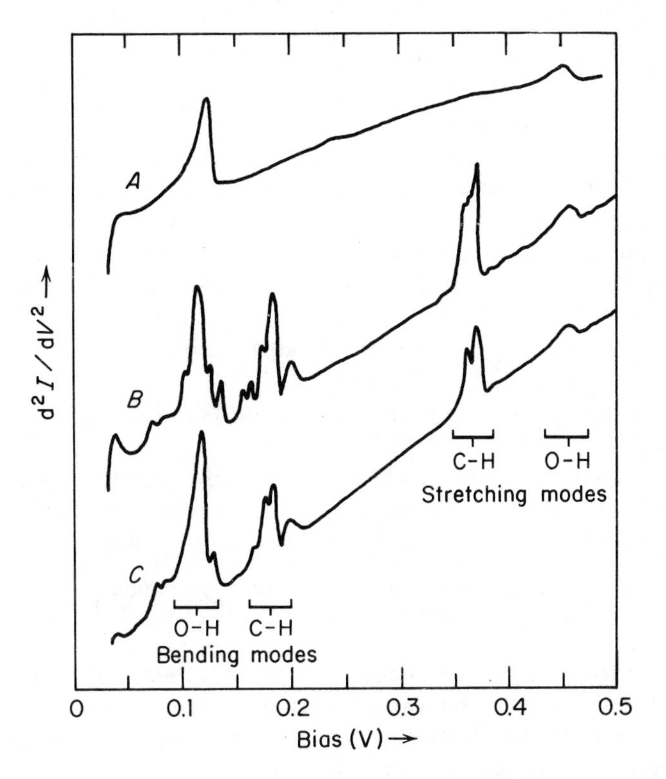

Fig. 8.19. Original observation of molecular vibrational spectra in tunneling (IETS). Recorder traces of d^2I/dV^2 for three Al–Al_2O_3–Pb junctions at 4.2 K. The zero of the vertical scale is shifted for each curve, and all three are normalized to the same arbitrary units. The largest peaks represent increases of 1% in dI/dV. Also indicated are intervals associated with the energy of IR-active molecular vibrational modes. Curve A is obtained from a "clean" junction. Curves B and C are obtained from junctions exposed to propionic acid [$CH_3(CH_2)COOH$] and acetic acid (CH_3COOH), respectively. (After Jaklevic and Lambe, 1966.)

or hydrocarbons (e.g., pump oil) were introduced in approximate monolayer amounts onto the Al oxide barrier (by gentle vapor deposition), *additional* peaks appeared in d^2I/dV^2, as shown in Fig. 8.19, which were closely related to the vibrational spectrum of the molecule in question. This interpretation was confirmed with observation of the expected shift in frequency by $1/\sqrt{2}$ upon deuteration: e.g., replacing a CH by a CD group. The original theoretical investigation (Scalapino and Marcus, 1967) treated the molecule as an assembly of oscillating electric dipoles (x component p_x) of characteristic resonant frequencies $\hbar\omega$. These dipoles are coupled to the tunneling electron via a dipolar interaction

$$\delta U(x) = \frac{2ep_x x}{(x^2 + r_\perp^2)^{3/2}} \tag{8.26}$$

and can thus be excited if a resonance condition $eV = \hbar\omega$ is satisfied. Here, x is the coordinate across the barrier of thickness t, and $r_\perp$ is $(y^2+z^2)^{1/2}$. In the calculation of Scalapino and Marcus, the term $\delta U(x)$ is added to the barrier potential $U(x)$ in the WKB tunneling exponent. Assuming $\delta U(x)$ to be small, and expanding the exponential, one obtains, for the matrix element T in the transfer Hamiltonian,

$$T \simeq \left[1 + \left(\frac{2m}{V_B}\right)^{1/2} \frac{ep_x}{\hbar t}\, g\left(\frac{r_\perp}{t}\right)\right] \exp\left[-(2mV_B/\hbar^2)^{1/2} t\right] \tag{8.27}$$

where V_B is the barrier height relative to the Fermi energy, and $g(x) = (1/x) - 1/(1+x^2)^{1/2}$.

This equation leads to an expression for the assisted conductance $\Delta G(V)$ (normalized to the elastic conductance) expressed as a function of the perpendicular distance $r_\perp$ of the electron to the molecular vibrator:

$$\frac{\Delta G}{G} = \frac{2m}{V_B}\left(\frac{e}{\hbar t}\right)^2 |\langle 1|\, p_x\, |0\rangle|^2\, g^2\left(\frac{r_\perp}{t}\right) \theta\left(V - \frac{\hbar\omega_1}{e}\right) \tag{8.28}$$

Here, the θ function [$\theta(x)$ is 1 for $x \geqslant 1$ and 0 otherwise] expresses the (threshold) condition that the excitation cannot occur unless energy $\hbar\omega_1$ is available, in which case the electron appears precisely at the Fermi energy of the opposite electrode after tunneling. This result yields increments in conductance $\Delta G/G \simeq 10^{-2}$, as observed, and points to an important feature of this effect, which is its extreme sensitivity in detecting small numbers of molecules. Quantitative measurements on adsorbed benzoic acid (Langan and Hansma, 1975) show that about $\frac{1}{30}$ monolayer can be detected; with a junction area 4×10^{-4} cm^2, this result corresponds to about 10^{10} molecules, or 2×10^{-12} g.

It has been found, both experimentally and theoretically (Lambe and Jaklevic, 1968), that Raman active modes can also be excited. The tunneling electron, which interacts more strongly with the molecule than does a photon, can also possibly interact with an induced electric dipole in the molecule. Calculation of the expected intensities in these two cases has been generalized by Kirtley, Scalapino, and Hansma (1976). The possible (but surprisingly minor) effects of the metal electrode and oxide layer in shifting the vibrational frequencies of molecules have been examined by Klein et al. (1973) and by Kirtley and Hansma (1976). Further details about this important and active area of research are deferred to Chap. 10.

8.3.3 *Inelastic Excitations of Spin Waves (Magnons)*

An interesting threshold observation is that (see Fig. 8.20) of inelastic spin-wave excitation in an (antiferromagnetic) NiO tunneling barrier,

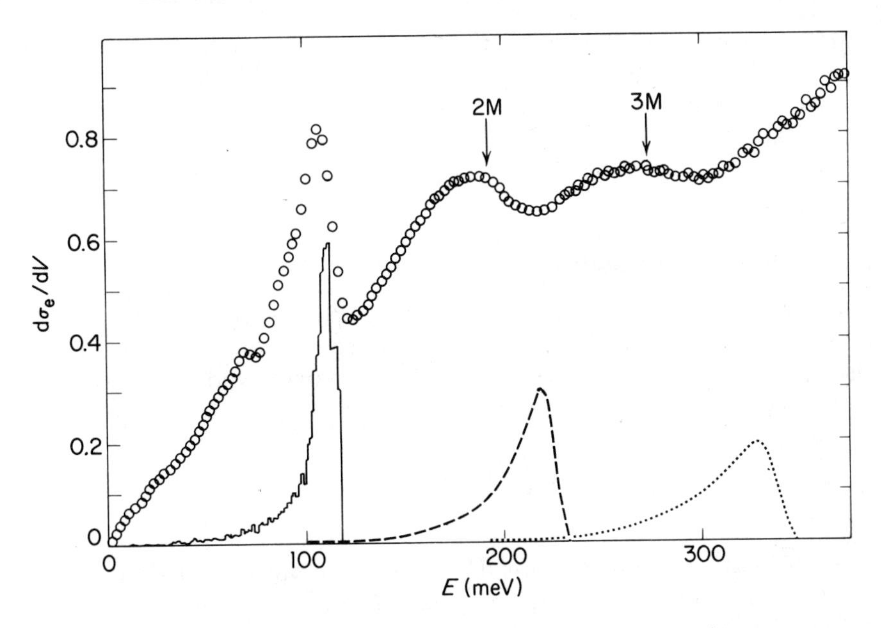

Fig. 8.20. Tunneling observation of magnon (spin-wave) excitations in ferromagnetic NiO grown as a thermal oxide barrier on an Ni single crystal. The circles represent the derivative of the even conductance; the peaks are interpreted as inelastic excitations of one, two,and three magnons, respectively. Modeling of this result is shown in the histogram and dashed and dotted curves at the bottom part of the figure. (After Tsui, Dietz, and Walker, 1971.)

which was grown thermally on a single crystal of Ni (Tsui, Dietz, and Walker, 1971). The NiO specimen is thus perhaps 50 Å thick. Nevertheless, the well-resolved peak at 107 meV and cutoff at 122 meV (in measurements at 1 K) in the derivative of the even conductance are in good agreement with the solid lower curve, which is the one-magnon density of states computed from the magnon dispersion relation, determined independently from inelastic neutron scattering (Hutchings and Samuelson, 1971) and from Raman scattering (Dietz et al., 1971). Further, the broadened higher-energy peaks at 190 and 270 meV compare reasonably (see arrows) with the two-magnon and (estimated) three-magnon spectra determined from Raman scattering. The shifts of the observed two- and three-magnon peaks to lower energy relative to the dashed and dotted curves (representing, respectively, the double and triple convolutions of the single-magnon density and hence corresponding to multiple emission of *non*interacting magnons) indicate that magnon-magnon interactions are strong. The roughly equal strengths of the one-

and two-magnon peaks indicate that the electron-magnon interaction is also strong. Using these results, Tsui suggests that the electron in NiO should be thought of as a *magnetic polaron* associated with perhaps two virtual magnons, in analogy with the more familiar case of a lattice polaron.

8.3.4 *Inelastic Excitation of Surface and Bulk Plasmons*

An additional example of inelastic threshold spectroscopy is provided by excitation of surface and bulk plasmons, which are collective motions of the electrons relative to the positive ionic charges, with characteristic frequency (in the surface case)

$$\frac{\omega_p}{\sqrt{2}} = \left(\frac{2\pi N_e e^2}{m^* \varepsilon}\right)^{1/2} \tag{8.29}$$

Here, N_e and ε are, respectively, the electron concentration and dielectric constant. The plasmon energy $\hbar\omega_p$ is ~ 10 eV in metals, but it may be less than 0.1 eV in degenerate semiconductors. The surface plasmon emission threshold was detected in Schottky barrier tunneling into the degenerate semiconductor *n*–GaAs (Tsui, 1969; Duke, 1969) at energies between 55 and 80 meV and compared with the theoretical electron concentration dependence between 4.2×10^{18} and 9.5×10^{18} cm^{-3}; these data were later analyzed by Ngai, Economou, and Cohen (1969) on the basis of the theory of Bennett, Duke, and Silverstein (1968). The data of Tsui (1969) (see also Tsui and Barker, 1969) are shown in Fig. 8.21. Other reports of plasmon excitation in semiconductor tunneling are given by Duke, Rice, and Steinrisser (1969), Duke (1969), Mikkor and Vassell (1970), and Guétin and Schreder (1972*b*). In metals, surface plasmon properties are well known from optical measurements, but as they generally lie in the 5–10 eV range, they have not been directly observed in tunneling. Somewhat indirect evidence of their excitation, however, is provided by measurements of light emission.

8.3.5 *Light Emission by Inelastic Tunneling*

Light emitted from a tunnel junction, presumably by radiative decay of surface plasmons excited by tunneling electrons, was first observed by Lambe and McCarthy (1976)—see Fig. 8.22a—and clearly identified by its high-frequency cutoff $\hbar\omega_{co} = |eV|$. The discoverers of the effect pointed to its possible utility as a method for the generation of light, especially if means of increasing the radiation efficiency beyond the range $\sim 10^{-4}$ at present can be found.

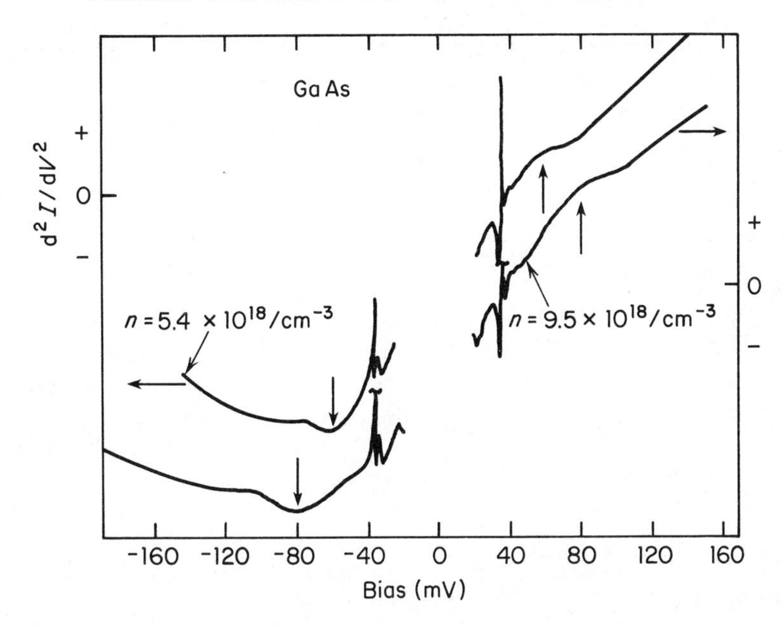

Fig. 8.21. Direct traces of second derivative d^2I/dV^2 for Pb/GaAs Schottky barrier junction measured at 1 K. The broad structures indicated by the vertical arrows are interpreted as excitation of surface plasmons in the degenerate semiconductor; the energies at which these occur are found to scale approximately as the $\frac{1}{2}$ power of the electron concentration which is indicated in the two cases. (After Tsui, 1969.)

The physical mechanisms responsible for the broadband radiation with sharply defined upper cutoff remain the subject of active research. It is generally believed that the tunneling electrons excite electromagnetic resonances of the junction structure through intermediate processes involving *surface plasmons and surface polaritons.* Coupling to the external radiation field is enhanced by deliberately roughening the surface. The possible intermediate steps that have been considered include the following:

1. Direct radiation from tunnel current fluctuations in the barrier region (Hone, Mühlschlegel, and Scalapino, 1978; Laks and Mills, 1979).
2. Excitation of slow-wave plasmon modes (Swihart, 1961) peaking in the barrier between the two metal films, which are coupled to the external radiation field by surface roughness (Lambe and McCarthy, 1976; Davis, 1977).
3. Excitation of fast surface plasmons at the metal film–substrate and

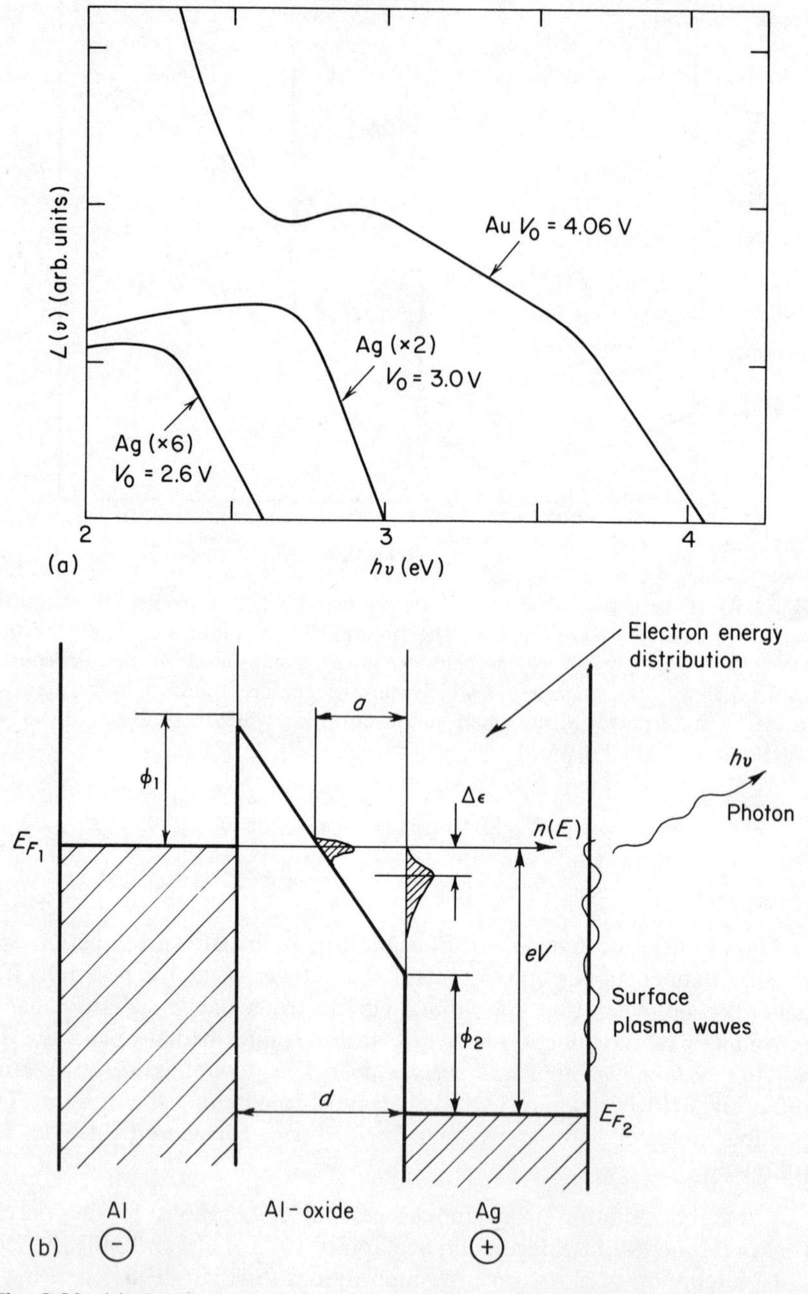

Fig. 8.22. (a) Original demonstration of light emission by inelastic tunneling in Al–Al_2O_3–Ag and Al–Al_2O_3–Au junctions at 77 K. The photon flux per unit frequency $L(\nu)$ as a function of photon energy $h\nu$ cuts off sharply at the quantum condition $eV = h\nu$. This effect was interpreted as a secondary consequence of plasmon excitation in the metal films. (After Lambe and McCarthy, 1976.) (b) Diagram of metal-oxide-metal tunnel junction under bias showing excitation of

metal film–vacuum interfaces, again followed by radiation if the proper spatial Fourier components of surface roughness are available (Kroo, Szentirmay, and Félszerfalvi, 1980, 1981, 1982; Kirtley, Theis, and Tsang, 1980, 1981; Laks and Mills, 1980; Walmsley, Quinn, and Dawson, 1982).

4. In specially prepared structures—e.g., an electrode approximated as an array of metallic spheres (Hansma and Broida, 1978; Hone, Mühlschlegel, and Scalapino, 1978)—the excitation of localized plasmon modes (Adams, Wyss, and Hansma, 1979) of these structures can lead to radiation.

We will limit our discussion to recent key experiments indicating the primary role of mechanism 3, as sketched in Fig. 8.22b, in many cases apparently leaving unclear to what extent mechanisms 1 and 2 contribute. First, the work of Kroo, Szentirmay, and Félszerfalvi (1980) convincingly correlated the emitted intensity of radiation with the traversal without scattering of the injected hot electron flux $I = I_0 \exp(-d/\lambda_B)$ across the thickness d of the tunnel electrode to its vacuum interface. Here, λ_B is the ballistic mean free path for the hot electrons, and I_0 is the tunnel-injected current at the film-barrier interface, monochromatized by the barrier factor D (see Fig. 8.22b). Secondly, Kirtley, Theis, and Tsang (1980, 1981) showed that the emitted light became sharply selected in frequency ω and angle θ if the junction is formed on a sinusoidally etched (diffraction grating) substrate whose periodicity λ_G corresponds to a visible light wavelength. Hence, the outer (vacuum) surface of the electrode is clearly generating the radiation, and analysis of the variation of the ω, θ dependence with grating spacing λ_G implies that the dispersion law of the plasmon is close to that of light: this is a fast surface plasmon rather than a slow mode. Finally, the recent work of Walmsley, Quinn, and Dawson, (1982) on junctions with randomly rough Au, Cu, and Ag electrodes provides evidence that the light intensity falls off rather sharply in Au and Cu electrodes at an energy $\hbar\omega$ corresponding to the onset of interband absorption, whereas this effect is unimportant in Ag because no interband absorption occurs until much higher energy. The argument involves the observed close similarity of the energy dependence of the electron energy loss function $\text{Im}(-1/\varepsilon)$ (ε the dielectric function of the metal), as calculated from optical measurements, to the voltage dependence of the radiation intensity. This view fits conceptually into the hot electron picture of Kroo, Szentirmay, and Félserfalvi either as an energy dependence of the surviving hot electron flux $I_0 \exp(-d/\lambda_B)$ through

---

fast (radiating) surface plasmon modes at outer surface by hot electrons after traversal of metal film. Angle tunability of $\hbar\omega$ radiating vs. emission angle can be achieved by building the junction on a holographically ruled optical diffraction grating. (After Kroo, Szentirmay, and Felszerfalvi, 1980.)

energy dependence of the ballistic mean free path λ_B, or as a nonradiative decay for the excited surface plasmon mode.

A possible further consequence of this process would be radiation from the single electron decays from the hot electron–excited higher bands of the metal, in films sufficiently smooth that the surface plasmon cannot radiate. Such a mechanism is simply a tunnel injection–excited inverse photoemission process, through which emission from direct optical interband transitions has recently been observed in ultrahigh-vacuum reflection geometries (Denninger, Dose, and Bonzel, 1982; Woodruff and Smith, 1982).

8.3.6. Spin Flip and Kondo Scattering

The first evidence for tunneling assisted by interaction of the electron spin $\vec{\sigma}$ with a localized magnetic state $\vec{S}$, e.g., via an exchange interaction (J)

$$H^I = -2J\vec{S} \cdot \vec{\sigma} \tag{8.30}$$

was observation of small temperature-dependent conductance peak anomalies at $V=0$, simultaneously reported by Wyatt (1964), in tunnel junctions involving oxidized Nb and Ta, and by Logan and Rowell (1964), in narrow Si *pn* tunnel junctions. The magnetic field splitting of the conductance peak was first demonstrated by Shen and Rowell (1967, 1968), making clear the involvement of magnetic moments. This result is shown in Fig. 8.23. The voltage and temperature dependences were approximately given by the logarithmic form

$$F(eV) \propto -\ln\left(\frac{|eV| + nkT}{D}\right)$$

with $n \simeq 1.3$ and D typically a few millielectronvolts for the semiconductors and ~ 10 meV for the Nb and Ta junctions.

In a detailed application of the transfer Hamiltonian theory, Appelbaum (1967) calculated the conductance assisted by interaction with magnetic moments $\vec{S}$, localized in the barrier near one electrode, assumed to be magnetically coupled to the conduction electron spins $\vec{\sigma}$ (in the electrode) via the exchange interaction, Eq. (8.30). This situation is sketched in Fig. 8.24a. The moments are assumed not to interact with each other.

Much of the preliminary formulation of Appelbaum (1967) has already been presented in Chap. 2 (Sec. 2.6), leading to (2.120) and (2.121) in which we will take H^I corresponding to the s–d interaction, (8.30).

It is important to emphasize that the conductance peak feature evident in Fig. 8.23 arises physically from forward scattering across the barrier by a resonant process formally equivalent to the elastic magnetic scattering process leading to a resistivity minimum at low temperatures in dilute magnetic alloys.

Kondo (1964) used perturbation theory applied to (8.30) to calculate

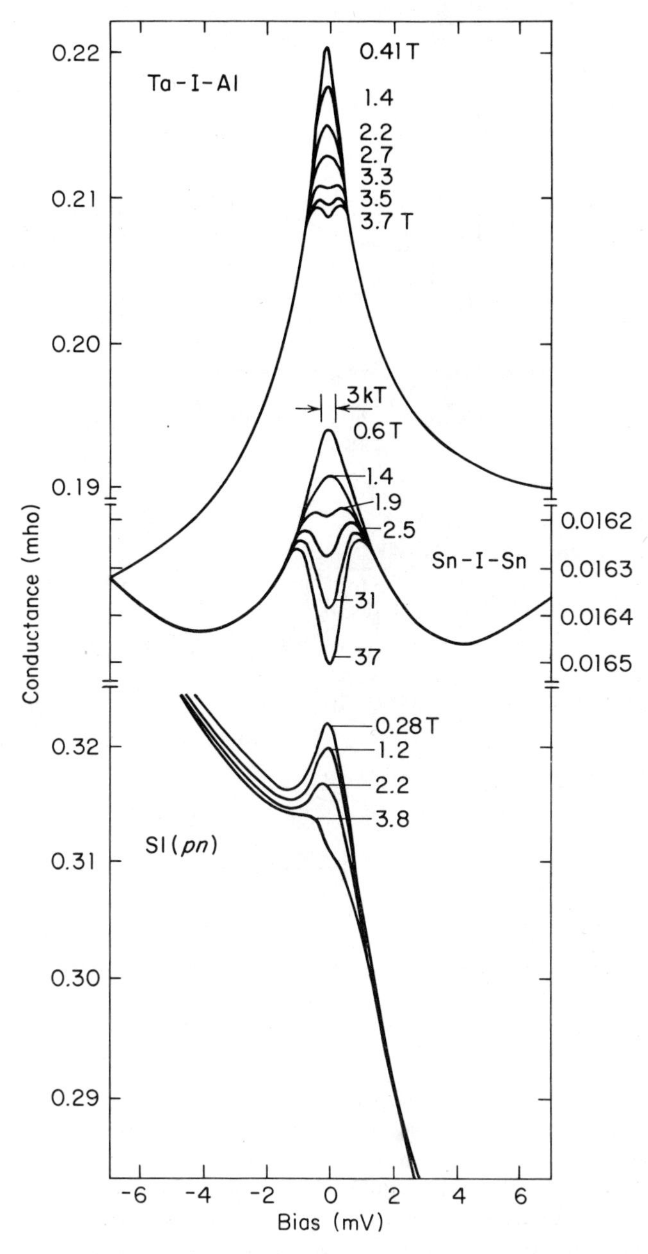

Fig. 8.23. Conductance vs. V plots for several junctions showing magnetic field–sensitive conductance peak anomaly at zero bias. Upper curve: Ta-oxide-Al junction; center: Sn–I–Sn junction; lower curve: silicon *pn* junction. (After Shen and Rowell, 1967.)

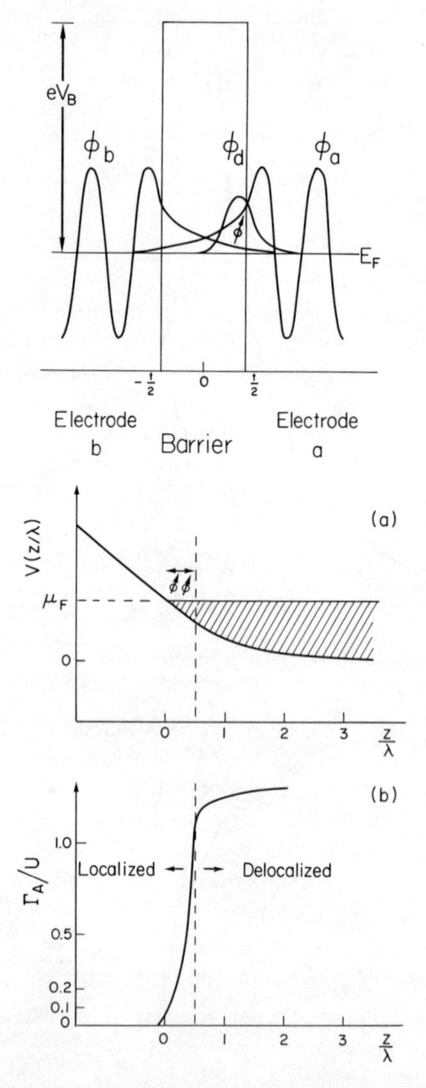

Fig. 8.24. Origin of the conductance peak anomaly by magnetic moment in orbital ϕ_d. (a) Overlap of conduction electrons in the opposing electrodes with localized orbital carrying a spin moment. (b) Suggested origin of magnetic moments found experimentally to be an intrinsic feature of Schottky barrier junctions on degenerate semiconductors. Moments (localized states) occur at the inner edge of the depletion region (Panel 1, $z/\lambda \simeq 0.5$) where the local carrier density has fallen below the Mott critical value. Here λ is the Thomas–Fermi screening length. Panel 2: The condition for localization can be described in terms of the ratio Γ_A/U which arises in the Anderson model for a localized magnetic state. (After Wolf and Losee, 1970.)

the contribution of the second term of Eq. (2.121) (second Born approximation):

$$w_{i,j} = \frac{2\pi}{\hbar}\left(\sum_{k\neq i} \frac{H^I_{ik}H^I_{kj}H^I_{ij}}{E_i - E_k} + \text{c.c.}\right)\cdot \delta(E_i - E_j) \tag{8.31}$$

An important feature of the calculation is the appearance in the final result of a strong temperature dependence. This feature comes from the Fermi functions which specify the occupation of electronlike and holelike intermediate states E_k and, because of the noncommuting property, $S^+S^- - S^-S^+ \neq 0$, of the raising and lowering operators for the localized spin. Kondo found anomalous temperature and energy dependences at the Fermi surface ($\varepsilon = 0$) described by a scattering rate $w \propto F(\varepsilon)$, with

$$F(\varepsilon) = \int_{-D}^{D} \frac{\tanh(\beta\varepsilon'/2)\, d\varepsilon'}{\varepsilon - \varepsilon'} \simeq -\ln\left[\frac{\varepsilon^2 + (k_B T)^2}{D^2}\right] \tag{8.32}$$

where $\beta = 1/k_B T$ and D is an energy cutoff. This equation is similar to the empirical form of Wyatt (1964).

The resulting perturbation series diverges below a characteristic temperature

$$T_K = \left(\frac{D}{k_B}\right)\exp\left[\frac{1}{2} JN(0)\right] \tag{8.33}$$

where J (which is negative) is the interaction constant and $N(0)$ is the density of states of one spin projection at the Fermi surface. The formal breakdown of perturbation theory is associated with a low-temperature bound state. Indeed, Appelbaum's (1967) tunneling Hamiltonian calculation obtains precisely the function $F(\varepsilon)$ of Kondo, Eq. (8.32), which gives a peak in tunneling conductance if the interaction constant J is negative (antiferromagnetic coupling).

Summarizing the predicted conductance $G(V)$ by using the notation of Appelbaum and Shen (1972), one has

$$G(V) = G_1(V) + G_2(V) + G_3(V) \tag{8.34}$$

where

$$G_n(V) = \int_{-\infty}^{\infty} g_n(\varepsilon) \frac{\partial f(\varepsilon - eV)}{\partial \varepsilon}\, d\varepsilon \tag{8.35}$$

and

$$g_1 = a_1 \tag{8.36}$$

$$g_2 = a_2\left\{S(S+1) + \frac{1}{2}\langle M\rangle\left[\tanh\left(\frac{\varepsilon+\Delta}{2k_BT}\right) + \tanh\left(\frac{\Delta-\varepsilon}{2k_BT}\right)\right]\right\} \tag{8.37}$$

$$g_3 = a_3(g_{31} + g_{32} + g_{33}) \tag{8.38}$$

$$g_{31} = \left\{S(S+1) - \langle M^2\rangle + \frac{1}{2}\langle M\rangle\left[\tanh\left(\frac{\Delta+\varepsilon}{2k_BT}\right) + \tanh\left(\frac{\Delta-\varepsilon}{2k_BT}\right)\right]\right\}\cdot F(\varepsilon) \tag{8.39}$$

$$g_{32} = \frac{1}{2}\left[S(S+1) + \langle M^2\rangle + \langle M\rangle \tanh\left(\frac{\varepsilon+\Delta}{2k_BT}\right)\right] \cdot F(\varepsilon+\Delta) \tag{8.40}$$

$$g_{33} = \frac{1}{2}\left[S(S+1) + \langle M^2\rangle + \langle M\rangle \tanh\left(\frac{\Delta-\varepsilon}{2k_BT}\right)\right] \cdot F(\varepsilon-\Delta) \tag{8.41}$$

The inelastic excitation energy, analogous, e.g., to $\hbar\omega_{ph}$ in the phonon case, is here the Zeeman splitting,

$$\Delta = g\mu_B H \tag{8.42}$$

where μ_B is the Bohr magneton and $\langle M\rangle$, which occurs in the g_2 and g_3 terms, is the expectation magnetization of the moment S in a magnetic field H.

The coefficients a_n are regarded as adjustable parameters. To leading order in the barrier factor, one expects a_1, $a_2 \propto e^{-2\kappa t}$ and $a_3 \propto e^{-2\kappa t} \cdot e^{-\kappa d}$, where d is the spacing between the local moment and the right-hand electrode. The leading terms in a_2 and a_3 are, respectively, $T_J^2 J$ in the earlier notation of Appelbaum (1967), to which terms exponentially smaller in κt and κd, but of the above energy dependence, must be added, corresponding to modifications of the electrode by the impurity (Appelbaum and Brinkman, 1970).

The G_2 term, Eq. (8.37), corresponds to excitation of a Zeeman transition of the local moment by the tunneling electron and thus to a threshold at $|eV| = \Delta$. Evidently, Δ is measured in the presence of coupling of the moment, via $H^I = -2JS \cdot \sigma$, to the right-hand electrode. This interaction is well known to give a g shift (Yosida, 1957)

$$g - g_0 = 2JN(0) \tag{8.43}$$

It is useful to recall the well-known Korringa (1950) relation of magnetic resonance:

$$\frac{\hbar}{T_1} = \text{constant} \times \left(\frac{\Delta H}{H}\right)^2 T \tag{8.44}$$

where T_1 is the spin-lattice relaxation time, $\Delta H/H$ is the Knight shift, T is the temperature, and the constant is as discussed, e.g., by Slichter (1963). By analogy, one expects a corresponding lifetime broadening of the Zeeman transition (Wolf and Losee, 1970) given by

$$\Gamma = \pi(JN(0))^2\Delta, \qquad \Delta \gg k_BT \tag{8.45}$$

where $\Delta = g\mu_B H$. These effects result from spin polarization of the electrons in the electrode by the magnetic field, assuming a Pauli susceptibility $2\mu_B^2 N(0)$; and from decay of the excited Zeeman level of the local moment by spin flip of a conduction electron of energy $-\Delta < \varepsilon < 0$.

At least semiquantitative agreement with the transfer Hamiltonian theory of the conductance peak has been reported in several types of tunnel junction, including the Ta–I–Al junctions (Wyatt, 1964; Appelbaum and Shen, 1972), magnetically doped M–I–M junctions (Lythall and Wyatt, 1968), and Schottky barrier junctions (Wolf and Losee, 1970), whose behavior is rather similar to that in the *pn* junctions shown in Fig. 8.23. It should be emphasized that in the theory (see also Ivezić, 1975) and in the physical systems citėd, the magnetic moments are presumed to lie *in the barrier*, at the edge of one of the electrodes. [The distinct case of moments lying *in the electrode* has been studied theoretically by Mezei and Zawadowski (1971*a*, 1971*b*) and by Appelbaum and Brinkman (1970) and experimentally by Bermon and So (1978*a*); discussion of this case is deferred to Sec. 8.4.4]

The localized moments in MIM junctions (Fig. 8.24a) are thought to originate typically from the unpaired d electrons of an isolated transition metal atom or ion. The moments in the Schottky barrier case are believed to occur inherently by electron localization (Mott transition) at the inner edge of the depletion region, as sketched in Fig. 8.24b. The Schottky-barrier conductance peak has been studied extensively in junctions formed by cleaving single crystals of Si in vacuum in the presence of rapidly evaporating metal, so that an atomically clean metal-semiconductor contact is formed. The semiconductor is metallic with a Fermi degeneracy of about 20 meV in its bulk, but it forms a depletion barrier layer at its surface about 100 Å in thickness. The Schottky barrier height at the metal interface is about 0.85 eV. The silicon crystals are typically free of transition metal impurities above 1 ppm. The donor atoms As or P, which have five sp bonding electrons, give up four of these to covalently bond (substitutionally) into the Si lattice. The remaining electron (for concentrations below the Mott critical concentration) is weakly bound in a hydrogenic fashion by the additional positive charge on the donor nucleus; because of the large dielectric constant and small effective mass, the Bohr radius scales to ~20 Å, and the binding energy becomes 45 mV for P and somewhat more for As. The electron spin is unpaired, producing a paramagnetic moment; the g value of the isolated donor has been measured as 2.00, and electron scattering experiments have shown that the cross section for spin-exchange collisions with a conduction electron at low temperature is $\sigma_x = 2 \times 10^{12}\ \mathrm{cm}^2$ (Feher and Gere, 1959). This value is a large one, close to that obtained by scaling the corresponding result for the hydrogen atom. Above the Mott critical concentration for delocalization of the electrons, about $4 \times 10^{18}\ \mathrm{cm}^{-3}$ in Si, corresponding to a screening length less than the effective Bohr radius, the material becomes effectively a semimetal. The depletion layer barrier, ~100 Å wide, is formed as donor electrons transfer to lower-energy surface states, leaving a positive space charge of ionized donors, as shown in Fig. 8.24b. The width of the transition from unoccupied to fully occupied donors, i.e., from barrier to electrode, is determined by the bulk screening length λ (10 Å in the case of $N_D = 1.6 \times 10^{19}\ \mathrm{cm}^{-3}$); from this result, one can estimate the number of localized moments as the number of occupied (neutral) donors whose local density is less than the Mott concentration. This density is $\sim 6\times10^{10}\,\text{cm}^{-2}$, close to an estimate of the required number of moments derived from the tunneling data (Wolf and Losee, 1970) using the value $\sigma_x = 2\times10^{-12}\,\text{cm}^2$. Further, one expects that these local moments interact via an antiferromagnetic exchange coupling

$$J = -\frac{8}{\pi N(0)}\left(\frac{\Gamma_A}{U}\right) \tag{8.46}$$

with the delocalized states, from application of Anderson's (1966) model for a localized magnetic state, and using the transformation of Schrieffer and Wolff (1966). From Anderson's (1966) estimate of $\Gamma_A/U \simeq 0.1$ for a localized moment, one deduces $JN(0) = -0.25$, i.e., a g shift of -0.5: the observed g shifts of ~ -1.0 would require $\Gamma_A/U = 0.2$. A semiquantitative self-consistent picture is thus obtained which links the origin of the moments with their surprisingly small measured g values, through (8.43) and (8.46).

Critical assessments of the level of agreement of the tunneling spectra with theory have been given by Appelbaum and Shen (1972) and, more recently, by Bermon, Paraskevopoulos, and Tedrow (1978) (Fig. 8.25), whose measurements on Fe moments in Al–I–Al junctions were extended to lower T (0.4 K) and to very high magnetic fields (18.1 T). As shown in Fig. 8.25, the zero-field conductance peak (Fig. 8.25a) varies accurately at $V = 0$ as $-\log T$ (Fig. 8.25b) and also fits the energy dependence of (8.32) in detail, as shown in Fig. 8.25c. However, the full range of measured $G(V)$ behavior up to 18.1 T, shown in Fig. 8.25d (solid lines), is not perfectly fit by the theory (dashed), even though the magnetic field-induced broadening Γ (Wolf and Losee, 1969) has been adjusted to optimize the fit at each field, rather than as constrained by (8.45). Inspection of the "well" produced by the field makes clear that broadening of the Zeeman transition ($|eV| = g\mu H$, the edges of the "well") increases monotonically with H, qualitatively as indicated by the simple theory. The empirical variation of Γ with H is indeed found to be linear but to display a significant $H = 0$ intercept, for which no explanation has been advanced. A second point of disagreement between experiment and theory (Bermon, Paraskevopoulos, and Tedrow, 1978) is in the degree of "overshoot" of the high-field $G(V)$ above the $H = 0$ curve, in the bias range $|eV| \geq g\mu H$. The latter authors do conclude that the $H = 0$, $G(V)$ data conform closely to the predictions of the Appelbaum-Kondo theory, providing a direct measure of the Kondo scattering amplitude in the perturbational limit. However, removal of the serious deficiencies in the high-field match of theory and experiment will evidently require either extension to the high-field case of more rigorous theories or a more realistic treatment of the moments. The theories of Appelbaum and Brinkman (1970) or Ivezić (1975) are possible candidates for this extension.

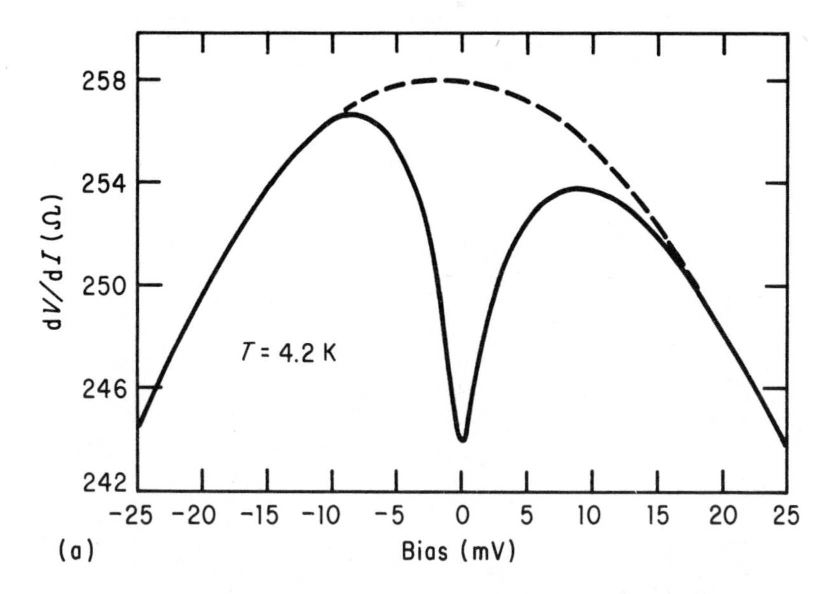

Fig. 8.25. Quantitative study of Kondo and spin-flip scattering in Fe-doped Al–Al_2O_3–Al tunnel junction. (a) The dV/dI spectrum at 4.2 K showing the conductance peak and extrapolated parabolic background resistance.

8.3.7 *Excitation of Electronic Transitions*

The obvious analog of tunneling excitation of a vibrational mode, namely, excitation of an *electronic transition* of the insulating barrier or of an impurity molecule adsorbed to the barrier, was first observed at $eV \simeq$ 1.5 eV in the copper phthalocyanine (CuPc) molecule in an Al–Al_2O_3–CuPc/Pb junction by Léger, Klein, Belin, and Defourneau (1972), as shown in Fig. 8.26. The observed feature is approximately a step increase in conductance of width ~0.25 eV, observed on a rapidly rising background conductance $G(V)$ because of the high bias voltage required, apparently a substantial fraction of the barrier height. So that the background effect was minimized, the logarithmic derivative $G^{-1}(\mathrm{d}G/\mathrm{d}V)$ $(G = \mathrm{d}I/\mathrm{d}V)$ was usually recorded, as discussed by Klein, Léger, Delmas, and de Cheveigné (1976), although the further derivative $\mathrm{d}[G^{-1}\,\mathrm{d}G/\mathrm{d}V]/\mathrm{d}t$ is recorded in Fig. 8.26. Studies of other molecular transitions have been reported by de Cheveigné et al. (1977) and Lüth, Ball, and Ewert (1978).

Observations of an electronic excitation of the tunneling barrier itself were first reported by Adane et al. (1975), whose curves revealing electronic transitions in Ho_2O_3 (0.64 eV) and in Er_2O_3 (0.81 eV) are shown in Fig. 8.27. These transitions have been identified as the $^5I_8 \rightarrow {}^5I_7$

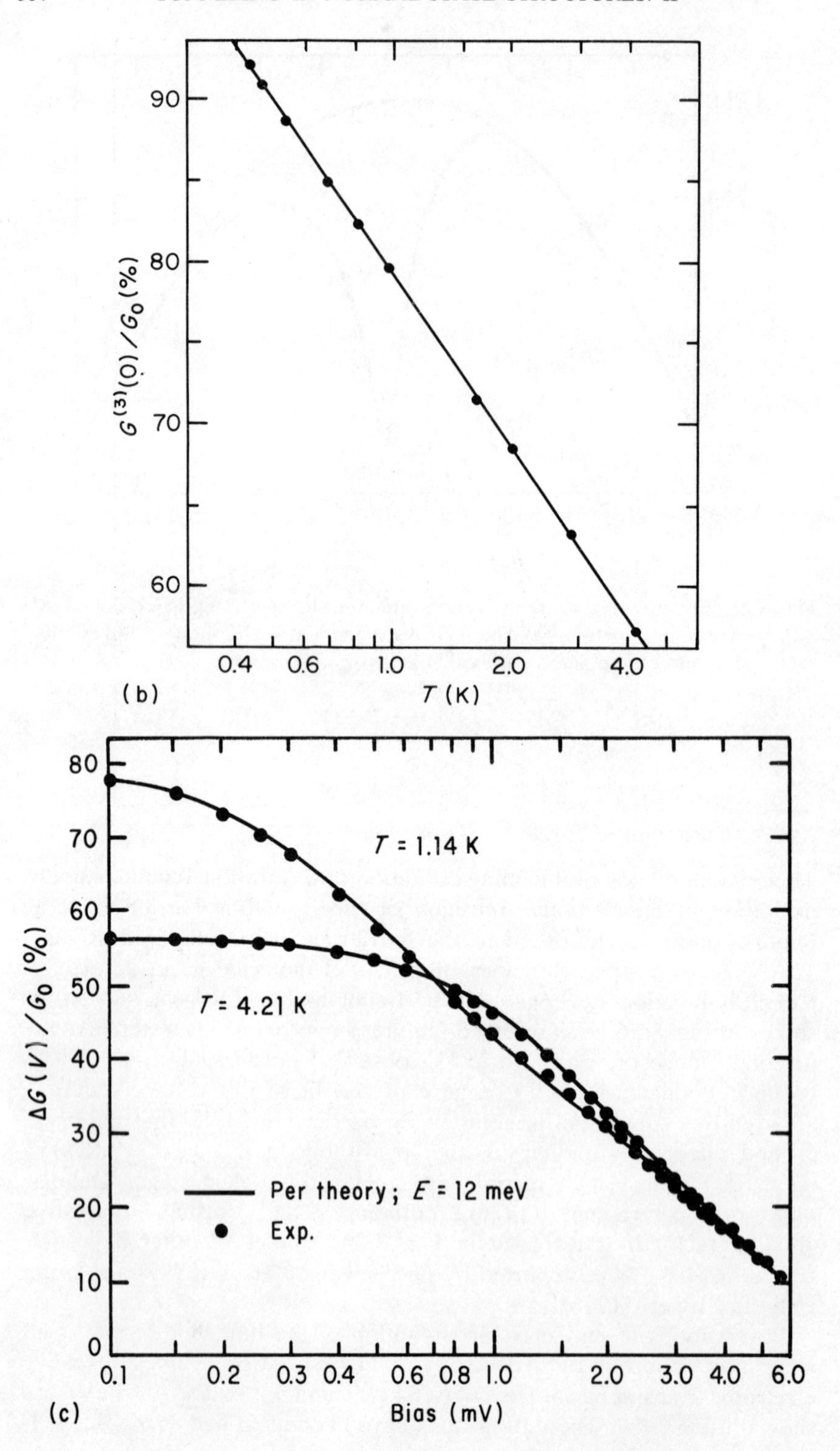

90
80
70
60
$G^{(3)}(0)/G_0$ (%)
0.4
0.6
1.0
2.0
4.0
(b)
T (K)
80
70
60
50
40
30
20
10
0
$\Delta G(V)/G_0$ (%)
T = 1.14 K
T = 4.21 K
Per theory; E = 12 meV
Exp.
0.1
0.2
0.3
0.4
0.6
0.8
1.0
2.0
3.0
4.0
6.0
(c)
Bias (mV)

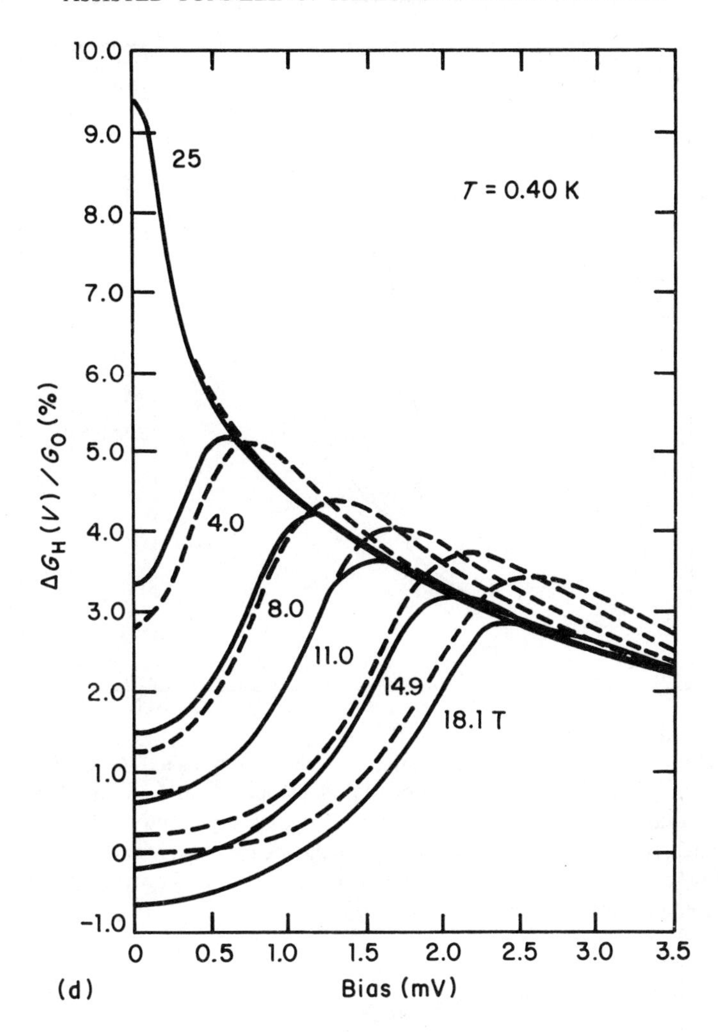

Fig. 8.25. (cont). (b) Demonstration of accurate $\log(T)$ dependence of the additional conductance at $V = 0$. (c) Demonstration of expected bias variation of conductance using perturbation theory (Kondo, 1964) as adapted to tunneling (Appelbaum, 1967). (d) Details of the magnetic field variation up to 18.1 T of the conductance associated with the magnetic scattering. Experimental curves are shown solid; dashed curves are based upon theory (Appelbaum, 1967); the high-field behavior can be approximately regarded as a broadened step-increase in conductance at $eV = g\mu H$, i.e., the condition for Zeeman transition of the moment in the applied magnetic field. (After Bermon, Paraskevopoulos, and Tedrow, 1978).

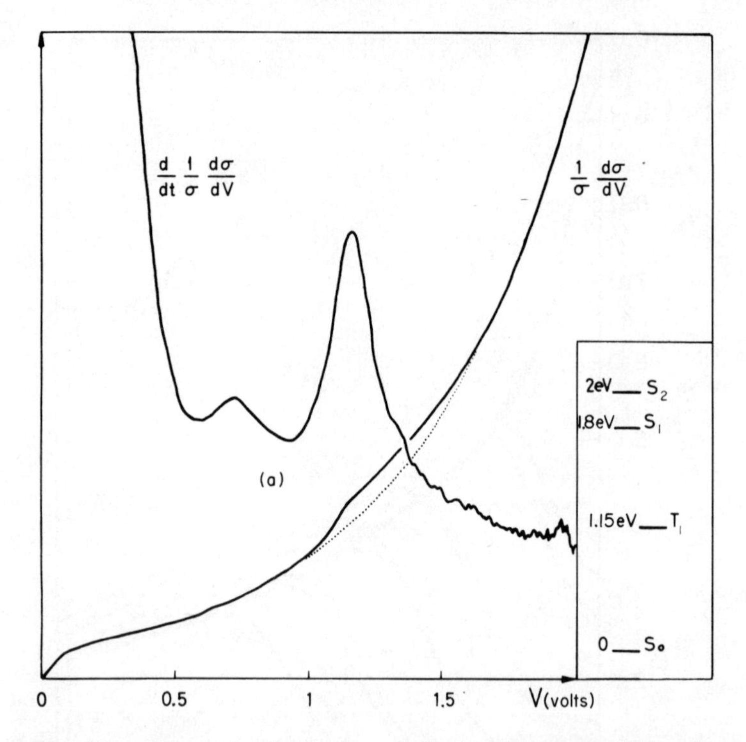

Fig. 8.26. Observations, in derivatives of the conductance $G(V)$, of singlet-triplet electronic transition of copper phthalocyanine (CuPc) in an Al-Al–oxide–CuPc-Pb junction at 4.2 K, at $V = 1.15$ V. The measurement of $d(G^{-1}\,dG/dV)/dt$ with linearly increasing current minimized the rising background in $G^{-1}\,dG/dV$. In subsequent measurements, the transition was also seen in reversed bias. (After Léger, Klein, Belin, and Defourneau, 1972.)

transition of the Ho^{3+} ion, and the $^4I_{15/2} \rightarrow {}^4I_{13/2}$ transition of the Er^{+3} ion.

Although the number of transitions observed in total is not very large, due in part to the limited range of bias voltage, typically $eV \leqslant 2$ eV, before breakdown or junction instability occurs—whereas most electronic transition energies exceed this value—several generalizations have been made.

First, the considerable broadening of all transitions is beyond that expected in free molecules or in single crystals of oxide. This result is generally attributable to local disorder in the thin films. Second, optical (dipole) selection rules are not obeyed. The optically allowed singlet-singlet and optically forbidden singlet-triplet transitions in molecules are generally of comparable strength. Further, the change of conductance

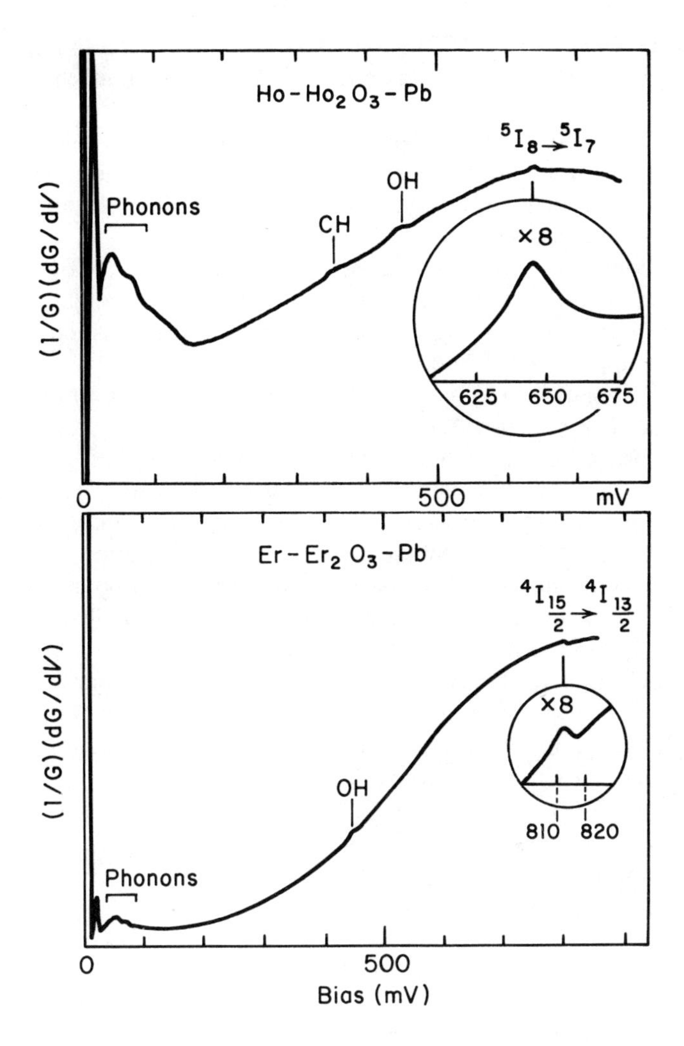

Fig. 8.27. The $(1/\sigma)(d\sigma/dV)$ vs. V for Ho_2O_3 and Er_2O_3. By amplification of the second derivative (×8) and expansion of the V axis, a peak can be seen at about 64 mV for HoO_3. According to optical results, this peak is due to the electronic transition $^5I_8 \rightarrow {}^5I_7$ for the Ho^{3+} ion. In the same way, a peak is observed in Er_2O_3, at about 810 mV, which is due to the electronic transitions $^4I_{15/2} \rightarrow {}^4I_{13/2}$ of the Er^{3+} ion. (After Adane et al., 1975.)

observed for these transitions is typically smaller, by more than an order of magnitude, than would be expected in the case of dipole allowed electronic transitions, on the basis of the known oscillator strengths and the assisted tunneling theory which has evolved from the work of Scalapino and Marcus (1967).

De Cheveigné et al. (1977) and de Cheveigné, Klein, and Léger (1982) have suggested the possible importance of an excitation mechanism based on exchange interactions between the tunneling electron and electrons in the molecule in order to explain the observation of (optically forbidden) transitions in which the molecular spin state is changed. The process envisioned has been established in the context of molecular electron-impact excitations (Matsuzawa, 1968).

The second point, namely that the allowed $S \rightarrow S$ transitions are not as strongly observed in the tunneling conductance as one might expect, has been discussed by the same authors as a barrier transmission effect. The argument applies to excitations of large energy (1 eV is typical of those observed, as shown in Table 8.1). The electron tunneling *after* excitation, having an energy E now reduced by $\hbar\omega \simeq 1$ eV, sees effectively a higher barrier, and consequently, its tunneling rate is substantially reduced. This reduction factor is estimated as ~ 30 for $\hbar\omega \simeq 1$ eV. This argument resembles that proposed earlier by Yanson, Bogatina, Verkin, and Shklyarevskii (1972) to explain asymmetry in excitation spectra from molecules near one side of the barrier. A comment on possible confusion of an electronic molecular transition and a size effect oscillation has been given by Jaklevic, Lambe, Kirtley, and Hansma (1977).

In conclusion, it should be said that the scope of barrier tunneling spectroscopy in study of electronic transitions is severely limited by the available range of energies, limited by the barrier heights of suitable barrier insulators to a few volts.

Table 8.1 Electronic Transitions of Molecules

Molecule	Transition	Type	$\Delta\sigma/\sigma$	References
Cu-phthalocyanine (CuPc)	1.15 eV	$S_0 \rightarrow T_1$		*a, c*
	1.8 eV	$S_0 \rightarrow S_1$		
Pentacene	0.9 eV	$S_0 \rightarrow T_1$	0.01	*b*
	1.9–2.1 eV	$S_0 \rightarrow S_1$	0.04	
β-Carotene	~1.3 eV	$S \rightarrow T$		*b*
Tetracyanin	~1.3 eV	$S \rightarrow S$	~0.01	*b*
Xenocyanin	~1.7 eV	$S \rightarrow S$		*b*
BDN[d]	~1.3 eV	$S \rightarrow S$	~0.01	*b*
H_2PC	~0.8 eV	$S \rightarrow T$		c
NiPC	~1.3 eV	$S \rightarrow T$		c

[a] Léger, Klein, Belin, and Defourneau (1972).
[b] de Cheveigné et al. (1977).
[c] Lüth, Ball, and Ewert (1978).
[d] bis(4-dimethylaminodithiobenzil)nickel.

8.4 Final-State Effects. II

There are a number of cases in which one-electron theory is not adequate to account for features of tunneling spectra related to final-state effects in one of the normal electrodes. We will refer to these loosely as many-body final-state effects.

In a fashion formally identical to that applied to the strong-coupling superconductor case, discussed in Sec. 4.3, these effects may be interpreted via a spectral function $A(\vec{k}, E)$. This function is necessary if interactions within the electrodes mix the one-electron states, so that a definite relation between the wavevector k and energy E is lost. The spectral function $A(\vec{k}, E)$ is the probability that an electron of wavevector $\vec{k}$ have energy E. The density of states is now given by (4.23)

$$N(E) = \int \frac{d^3k}{(2\pi)^3} A(\vec{k}, E)$$

which may differ, as in the superconducting case, from its values in the absence of interaction. Interactions typically result in both shift and broadening of one-electron states (see, e.g., Mahan, 1969; Hedin and Lundqvist, 1969). Term $A(\vec{k}, E)$ is calculated from the Green's function $G(\vec{k}, E)$ by the relation $A(\vec{k}, E) = (1/\pi)\,|\mathrm{Im}\, G(\vec{k}, E)|$. For example, when the energy shift and the lifetime, given respectively by the real part and -2 times the imaginary part of the self-energy, are both zero, the free-electron results are recovered from the limiting form

$$A(\vec{k}, E) = \left(\frac{1}{\pi}\right)\delta\left(E - \frac{\hbar^2 k^2}{2m^*}\right) \tag{8.47}$$

The tunnel current is now expressed via (4.26), which shows that tunneling measures the spectral function of the electrode. In a more general case, the $A(\vec{k}, E)$ can be expressed in terms of a relevant self-energy Σ_k, as

$$A(\vec{k}, E) = \frac{1}{\pi}\,\mathrm{Im}\,|G| = \frac{1}{\pi}\left|\mathrm{Im}\left(E - \varepsilon_k + \frac{i}{2\tau} - \Sigma_k\right)^{-1}\right| \tag{8.48}$$

where τ is the lifetime.

8.4.1 More General Many-Body Theories of Tunneling

The many-body theory of tunneling represented by (4.26) has been entirely successful in treatment of superconductors. This expression, (4.26), can be regarded as the most sophisticated version of the conventional tunneling Hamiltonian theory. Nevertheless, there are several deficiencies of this conventional theory, in which the tunneling current density J is expressed, in slightly different notation, as

$$J=\frac{4\pi e}{h}\sum_{k,q}|T_{k,q}|^2\int_{-\infty}^{\infty}\frac{dE_R}{2\pi}A_R(k,E_R)\int_{-\infty}^{\infty}\frac{dE_L}{2\pi}A_L(q,E_L)\cdot\delta(E_R-E_L-eV)[f(E_R)-f(E_L)] \tag{8.49}$$

Five deficiencies of this approach, which become apparent in cases, unlike superconductivity, where strong self-energy effects extend over a wide energy range, will be enumerated and discussed in turn. These deficiencies are as follows:

1. Possible "nonphysical divergences" of the matrix elements $|T_{kq}|^2$ when evaluated in the presence of strong self-energy corrections Σ_k, with the *unrenormalized* ("bare") values ε_k of particle energy in the tunneling exponentials, as the conventional theory requires (Appelbaum and Brinkman, 1969*a*, 1969*b*). That is, if $E_k=\varepsilon_k+\Sigma_k$, then ε_k occurs in the tunneling exponential barrier decay factor.

2. The spectral functions $A(\vec{k},E)$ as written in (8.48) are presumed to be *bulk* functions, while, in fact, the region being sampled is at the barrier interface, where the Σ_k and $A(\vec{k},E)$ will differ from the bulk. Such *local self-energy effects* were also described by Appelbaum and Brinkman (1969*a*, 1969*b*).

3. If the interface is *sharp*, Friedel-like oscillations of the self-energies Σ may result (*interface effects*, described by Zawadowski, 1967, and by Appelbaum and Brinkman, 1970). Equations for $G(\vec{k},E)$ in the presence of the interface must then be solved prior to evaluation of (8.48) and (8.49). This effect, like the previous effect, is not usually a problem in a superconducting case because of the large values, typically 0.1–1.0 μm, of the superconducting coherence length ξ.

4. A conceptual difficulty of the conventional theory is in distinguishing "assisted tunneling" and "final-state" contributions to the tunneling conductance. This question is particularly awkward if one imagines a molecular vibrator or magnetic moment, initially in the barrier, to be moved toward and past the electrode interface: at some ill-defined point, "assisted tunneling" must give way to a "final-state effect." This point has been treated by Brailsford and Davis (1970) and Davis (1970*a*) in a stationary-state formulation distinct from the work of Zawadowski (1967) and Appelbaum and Brinkman (1969*b*, 1970). The latter authors address the problem by absorbing all effects into an effective local density of states.

5. Finally, a fundamental shortcoming of the conventional theory is its unsuitability for treating the inherently nonequilibrium situation represented by the limit of a tunnel junction with vanishing barrier thickness. This point has been considered in the elaborate approaches initiated by Caroli, Combescot, Nozières, and Saint-James (1971*a*) and Feuchtwang (1974*a*), based on the nonequilibrium Green's function perturbation theory due to Keldysh (1964). In practice, for thick tunnel barriers, these

theories appear to confirm (e.g., see Ivezić, 1975) the results of the intermediate "ABZ" theory (Zawadowski, 1967; Appelbaum and Brinkman, 1969*b*, 1970) in addressing points 1–4 above.

The theories of Caroli et al. and Feuchtwang, in spite of their complexity—(which brings to mind a cautionary comment by none less than the theorist Zawadowski (1967), on an earlier "very suggestive method using Green's functions," that "the actual application of this method is not simple")—have nonetheless been successfully applied. The formulation of Caroli et al. has been applied with apparent success to the magnetic moment–Kondo problem (in the absence of a magnetic field) by Ivezić (1975), to (normal-state) electron-phonon effects in MIM junctions (Caroli, Combescot, Nozières, and Saint-James, 1971), and to the case of tunneling and phonon self-energy effects in metal-semiconductor contacts by Combescot and Schreder (1973, 1974).

The point of physics emphasized by Appelbaum and Brinkman (1969*a*, 1969*b*), mentioned in case 1 above is basically that the energy which determines the decay of a wavefunction across the barrier is the real (renormalized) energy

$$E = \varepsilon_k + \Sigma_k$$

rather than the formal unrenormalized ("bare") energy ε_k. In some cases, an appreciable (negative) electron self-energy $\Sigma(k, E)$ can arise from interaction, e.g., with phonons, to lower the real electron energy E from its "bare" value, $\varepsilon_k = \hbar^2 k^2/2m^*$. The point is that turning on such an interaction Σ must strongly affect the amplitude of a given electron state, at a point, say, $x = 2$ Å into the vacuum outside the metal (or in the barrier). If the state k_x has noninteracting x kinetic energy ε_{kx} and the interaction leads to a real self-energy Σ_x, then certainly the wavefunction tail will vary proportionally to $\exp(-\kappa x)$, with $\kappa = [(\hbar^2/2m^*)(V_B - \varepsilon_x - \Sigma_x)]^{1/2}$ and thus be exponentially reduced with a (negative) self-energy Σ_x. In the transfer Hamiltonian model, however, the exponent is constructed with the *noninteracting* or bare energy and can lead in extreme cases to unphysical results; i.e., Σ_x may become large (negative), forcing the bare energy, and hence the exponent and transmission factor, to unreasonably large positive values. Since the turning on of the self-energy Σ_x leads to an exponential change in ψ in the barrier, it is necessary to solve first for the correct wavefunction with all interactions and then to consider the decay of this function across the barrier. The tunneling Hamiltonian method overlooks this requirement by assuming that ψ in the barrier can, to sufficient accuracy, be taken at its noninteracting value.

In the view of Appelbaum and Brinkman (1969*b*), the tunneling object is a bare electron (not a quasiparticle) which is plucked from its dressed state at energy E on one side and inserted into the corresponding dressed state at $E \pm eV$ on the other. The probability of finding the bare electron at E accounts for the $A(\vec{k}, E)$ factors in the tunneling current formula. The derivation of Appelbaum and Brinkman (1969*b*) has been expanded upon by Feuchtwang, Cutler, and Nagy (1978).

The definite conclusion, as has been verified independently (Griffin and Demers, 1971), is that the tunneling Hamiltonian model must be reinterpreted with the real energy E inserted in the matrix element, with a consequence that purely frequency-dependent self-energy effects cannot enter the conductance via the transmission factor. The matrix element thus becomes energy-dependent, a feature which also appears in all extensions and alternatives to the transfer Hamiltonian model.

This problem was correctly avoided in the original Zawadowski (1967) calculation and its successors ("ABZ theory"). In these theories, the entire many-body (left- and right-hand) problems comprised by the two-electron, semi-infinite-barrier systems are first solved,and the current is then calculated based on these results. The current density J in the Zawadowski–Appelbaum–Brinkman (ABZ) theory is expressed as

$$J=\frac{e\hbar^3}{m^2\pi}\int d^3r\int d^3r'\delta(x-x_0)\delta(x'-x_0)\int_{-\infty}^{\infty} dE[f(E)-f(E+eV)] \times\left[\frac{\partial L}{\partial x}\frac{\partial R}{\partial x'}-\frac{\partial}{\partial x}\frac{\partial}{\partial x'}(LR)-L\frac{\partial}{\partial x}\frac{\partial}{\partial x'}R+\frac{\partial L}{\partial x'}\frac{\partial R}{\partial x}\right] \tag{8.50}$$

where $L \equiv \text{Im}\, G_L(\vec{r}, \vec{r}', E_L)$ and $R \equiv \text{Im}\, G_R(\vec{r}, \vec{r}', E_R)$ (with $E_L - E_R = eV$) are retarded Green's functions for the left and right electrodes, each evaluated assuming a thick barrier. The spatial derivatives arise from the matrix element of the current operator, in analogy to (1.4) and evaluated at x_0 in the barrier, between the two (L and R) Green's functions, which decay exponentially in the barrier and are oscillatory in the electrodes. Self-energy effects arising, e.g., from interaction with impurities are included in L and R by solution of Dyson equations. Slightly different versions of (8.50) are given by Appelbaum and Brinkman (1970) and by Davis (1970*a*). Examples of the solution of the equations for G_L and G_R at abrupt electrode-barrier interfaces are given by Appelbaum and Brinkman (1970). The charge density near the abrupt interface is known to exhibit Friedel oscillations, as shown in Fig. 8.28a. In an analogous fashion, the $G(\vec{r}, \vec{r}', E)$ functions decay exponentially with $|x-x'|$ when both x and x' lie in the barrier, but they oscillate with x as x varies within the electrode, reflecting at x' the oscillatory spectral function $A(\vec{k}, E, x)$. The oscillatory behavior is retained after solution of equations for G including interaction with, e.g., an impurity. In the expression for J, this variable x is integrated over the interface-induced oscillations in the spectral function. A direct consequence of the oscillation of G with x (in the electrode), with x' fixed in the barrier at the point of evaluating J, is shown in Fig. 8.28b for the case of a magnetic impurity. The second- and third-order contributions oscillate as a func-

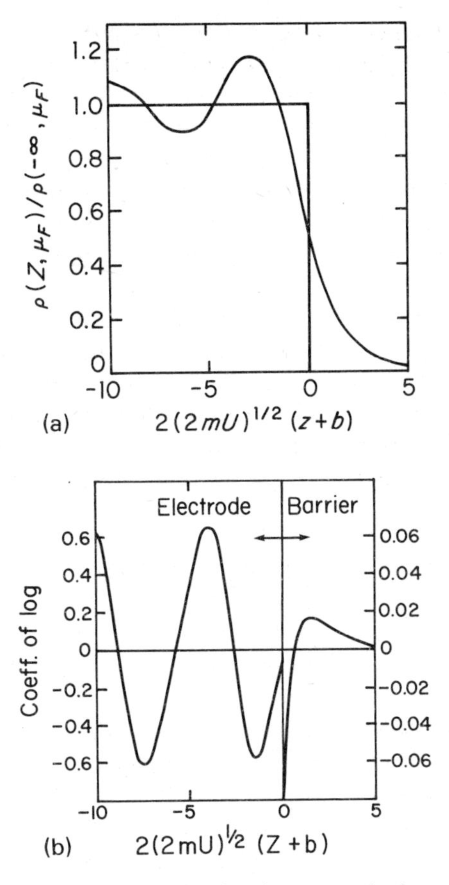

Fig. 8.28. (a) Calculated electron density in a metal electrode at the barrier interface shows a characteristic Friedel oscillation. This is evaluated assuming $\mu_F/U = 0.8$. (b) The coefficient of the logarithmic singularity, i.e., the Kondo scattering peak in the conductance as a function of the position of the spin with respect to the electrode-barrier interface. (After Appelbaum and Brinkman, 1970.)

tion of position *in the electrode* but, of course, assume positive values with the moment in the barrier. Any effect from randomly distributed moments (or other local interactions) in the electrode may thus be expected to average to zero.

The starting point of the calculations of Caroli, Combescot, Nozières, and Saint-James (1971), Caroli et al. (1971), and Caroli, Lederer, and Saint-James (1972) is quite different, namely, a localized or Wannier basis set, so that electron transport can be thought of as proceeding by hopping, and the metal-insulator interface is simply a plane at which the different localized states overlap. This technique removes, at least at the outset, any mystery about the boundary conditions at interfaces. Further, no effective Hamiltonian is employed, and the Keldysh theory is accurate in any degree of strong coupling including that of zero barrier width.

The results differ from the tunneling Hamiltonian results mainly in the form of the matrix element factor (transmission factor) $|T|^2$, which is weakly energy-dependent, confirming the discussion above in connection with the ABZ theory. The continuum limit of the Caroli theory corresponds to a $\psi=0$ boundary condition at the interface.

The Feuchtwang (1974*a*, 1974*b*, 1975, 1976) theory does not start from a localized basis and adopts a boundary condition, $d\psi/dx=0$, which differs from the continuum limit of the Caroli theory. In fact, Feuchtwang shows that all boundary conditions (at the interface between the *noninteracting* systems) which produce zero current are equivalent; this result effectively generalizes the Caroli theory. In both theories, densities of states are to be interpreted as local, position-dependent, state densities which must be averaged over. Further, the works of Caroli, Lederer, and Saint-James (1972), confirmed by the very thorough treatment (of the zero-width junction) by Feuchtwang (1976), provide the only serious attempts to take into account the actual three-dimensional nature of atoms comprising a real junction (or field emission tip). While the Green's functions $G(\vec{r}, \vec{r}', E)$ in the ABZ formulation, Eq. (8.50) [which is Eq. (2.22) of Appelbaum and Brinkman (1969*b*) and Eq. (2.1) of Davis (1970*a*)], may formally be written in vector variables $\vec{r}$, $\vec{r}'$ the treatment is not truly three-dimensional (Feuchtwang, 1976), in that lateral interactions [dependent upon $|\vec{\rho}|=|\vec{r}_\perp|=(y^2+z^2)^{1/2}$, if x is the barrier normal coordinate] are neglected. In a treatment of a field emission tip, where the atomic scale is obviously important, lateral interactions are clearly of some consequence. The expression [Eq. (4.2)] of Feuchtwang (1976) for a planar zero-width junction with translational symmetry parallel to the interface, including lateral dependences, is

$$
\begin{aligned}
J=\frac{2e}{\hbar B}\int_{-\infty}^{\infty} d\omega 2\pi[f_1(\omega)-f_2(\omega)]\iiiint_B ds\, ds_1\, ds_2\, ds_3 \\
\times[\Lambda^r(\vec{\rho}_1-\vec{\rho}_2,\omega)\Lambda^{r*}(\vec{\rho}-\vec{\rho}_3,\omega)]N_1(\vec{\rho}-\vec{\rho}_1,\omega)N_2(\vec{\rho}_2-\vec{\rho}_3,\omega)
\end{aligned}
\tag{8.51}
$$

where $B=\int_B ds$ is the area of the interface, with the area elements ds_i lying in four planes parallel to the interface, and $\vec{\rho}_i$ is a vector lying in the ith plane. The spectral densities N and "matrix elements" Λ are evaluated at the interface; i.e., the plane of the δ-function (zero-thickness) barrier. The latter quantity, Λ, is defined as

$$\Lambda^r(\vec{\rho},\omega)=-\left(\frac{\hbar^2}{2m}\right)^2\frac{\partial^2}{\partial x\,\partial x_1}G^r(x,x_1;\vec{\rho};\omega)\bigg|_{x=0=x_1} \tag{8.52}$$

while the spectral density functions are defined (in a case of no interactions, where ψ_i are eigenfunctions of energy ε_i) formally from

$$N(\vec{r},\vec{r}';\omega)\equiv\sum_i \psi_i(\vec{r})\psi_i^*(\vec{r}')\delta(\varepsilon_i-\omega) \tag{8.53}$$

Equation (8.53) is evaluated at $\vec{r} = 0\hat{\imath} + \vec{\rho}_j$ and, further, restricted to the case of translational symmetry, so that N depends only on $\vec{\rho} - \vec{\rho}'$. The relations of these spectral densities to the usual local density of states are discussed by Feuchtwang (1974*a*).

A complicating feature of (8.51) is that the product ("transmissivity") of Λ-transmission amplitudes in the integrand is dependent not only upon ω, as we now expect, but also upon the local densities of states (Feuchtwang, Cutler, and Schmit, 1978). The current is described as dependent upon a "transmissivity-weighted-average product" of the spectral densities of the uncoupled (infinite barrier) electrodes over the interface, rather than upon the product of their local energy densities of states at the interface. The generality of the various results appears not to be limited by the actual use in the calculation of the zero-width (δ-function) tunneling barrier, although this presumably makes it impossible to consider, e.g., assisted tunneling via magnetic moments in the barrier.

The work of Feuchtwang (1976) has apparently not been applied, as would appear to be useful, even to simple "three-dimensional" cases to compare its results with the ABZ theory in which three dimensionality is neglected. As it is stated to be limited to noninteracting systems, its application to many-body systems of primary physical interest is presently precluded.

8.4.2 *Tunneling Studies of Electron Correlation and Localization in Metallic Systems*

One of the most familiar cases of strong correlation and localization is the metal-to-insulator "Mott" transition in a degenerately doped semiconductor such as Si:P. The $T = 0$ extrapolated electrical conductivity $\sigma(0)$ in this case has been shown to drop precipitously (by a factor of 10^3 in a 1% concentration change) below a phosphorus concentration n, of $n_c = 3.4 \times 10^{18}\ \mathrm{cm}^3$, in the beautiful ultralow-temperature measurements of Rosenbaum, Andres, Thomas, and Bhatt (1980).

The results of this study are accurately described by

$$\sigma(0) = \sigma_c\left[\left(\frac{n}{n_c}\right) - 1\right]^{\zeta} \tag{8.54}$$

where, for Si:P, $\sigma_c = 260\ (\Omega \cdot \mathrm{cm})^{-1}$ and $\zeta = 0.55 \pm 0.1$. This expression is motivated by scaling theories; see, e.g., Abrahams, Anderson, Licciardello, and Ramakrishnan (1979). The detailed nature of this electronic localization transition is a topic of intense current research, which has recently been reviewed by Dynes (1982). Tunneling experiments to measure the density of states under comparable conditions of T and concentration have not been reported but would be of considerable interest.

In an earlier view (Mott, 1972), the Mott transition in a degenerate semiconductor occurs when the concentration of shallow impurities is reduced to the point where the Fermi-Thomas screening length λ equals the Bohr radius a^* of the shallow donor or acceptor impurity states and localization of the majority carriers into neutral, randomly distributed, hydrogenic impurity "atoms" takes place. The critical condition can also be stated as $\bar{r} \simeq 2.5a^*$, where $\bar{r}$ is the average nearest-neighbor spacing. In Schottky barrier tunneling into a semiconductor at the transition, one expects to observe a decrease in the conductance at the Fermi energy ($V=0$), corresponding to the disappearance of delocalized final states.

Such effects were reported in the Si:B system (Wolf et al., 1971) with the observation of a zero-bias conductance minimum with appreciable T and concentration dependence, similar to those shown in more recent results of Fig. 8.29 on Si:P. The conductance minimum was associated with the correlation "pseudogap" of the earlier theory (Mott and Davis, 1971).

This effect was thought to arise from the Coulomb energy U involved in placing a second majority carrier on the neutral impurity. In the pseudogap, a distribution of localized states, formed on statistically occurring pairs or larger clusters of the majority impurities, was thought to exist; transport between the localized states proceeds by thermally assisted hopping, leading to the $\exp(-T^{-1/4})$ law of Mott for the conductivity. Tunneling into bulk pseudogap states would give a zero-bias conductance minimum with $G(0) \propto \exp(-T^{-1/4})$. This behavior is approximately, but not well, obeyed in the data of Fig. 8.30 in panel (e), showing data of the concentration corresponding most closely to the critical value. The $T^{-1/4}$ law for the conductance at $V=0$ had earlier been reported in Schottky barrier contacts to n–Si compensated with Ga (Wolf, Wallis, and Adkins, 1975).

One of the original interpretations of such experiments was spectroscopic observation of a half-filled Hubbard band structure. If this interpretation were correct, removing some carriers from the system—e.g., by chemical compensation of donors by acceptors—should allow a shift of the Fermi level and a consequent shift of the structure from $V=0$. This shift was not observed in either the chemical compensation experiments of Wolf, Wallis, and Adkins (1975) or in irradiation compensation experiments such as those of Wimmers and Christopher (1981) shown in Figs. 8.29 and 8.30.

A more recent line of related theoretical and experimental research on disordered metallic alloys confirms and apparently explains the inherent Fermi surface location of these correlation effects and, thus, their $V=0$ location in tunneling spectra. In very general terms, Al'tshuler and Aronov (1979*a*, 1979*b*) have shown, in a disordered metal, that interference of the Coulomb interaction between conduction electrons and their elastic scattering from static disorder leads to a minimum in the density of states at the Fermi energy, with a characteristic $E^{1/2}$ dependence of $N(E)$:

$$N(E) = N(0)\left(1 + \sqrt{\frac{E}{\Delta}}\right) \tag{8.55}$$

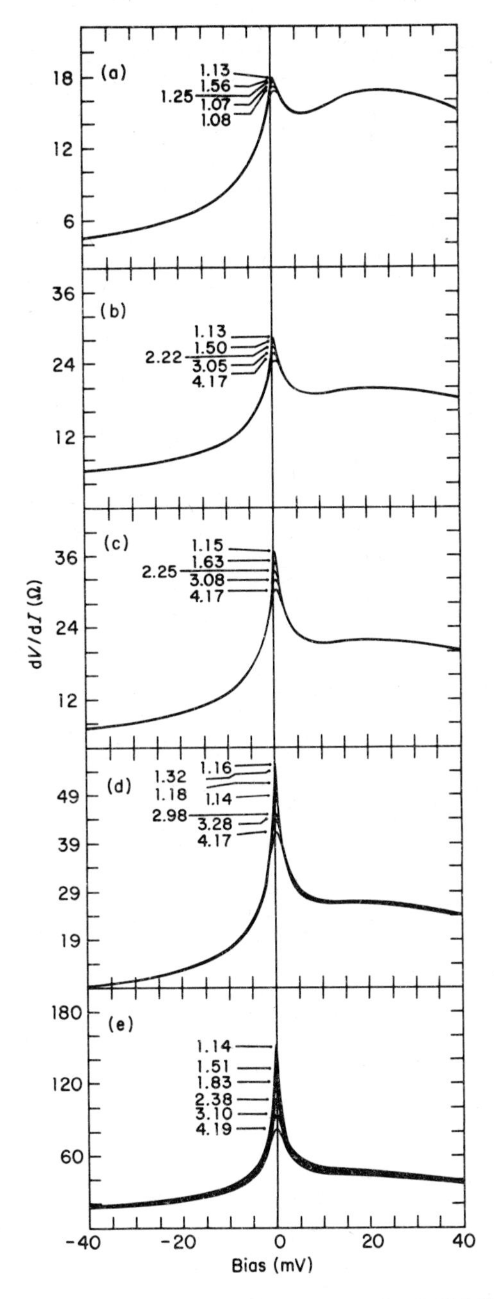

Fig. 8.29. Variation with temperature (as indicated) of dV/dI vs. bias V for a Schottky barrier junction of indium to Si:P ($N_D = 2.6 \times 10^{19}$ cm^{-3}) after radiation damage [proton fluence, $\times 10^{16}$ protons cm^{-2}, of 3.66, 4.12, 4.34, 4.69, and 5.26 in panels (a)–(e), respectively]. Increasing temperature dependence near zero bias is an indication of localized final states. (After Wimmers and Christopher, 1981.)

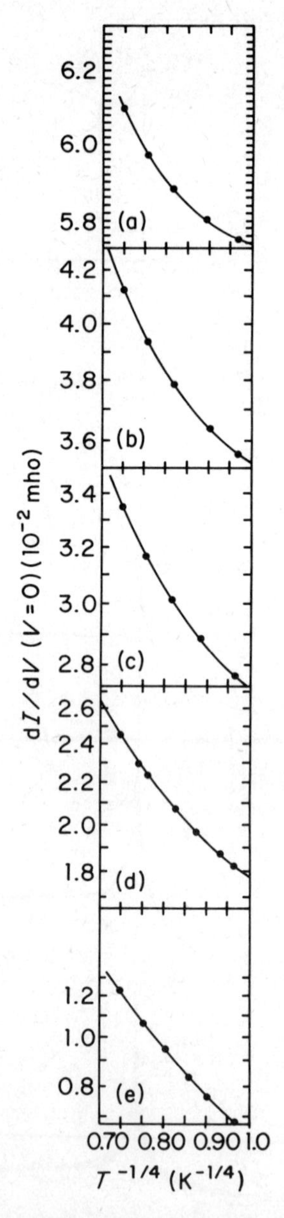

Fig. 8.30. The temperature dependence at $V=0$ for the set of curves, from a progressively radiation-compensated Si:P sample shown in the previous figure, plotted against $T^{-1/4}$. A straight line indicates the variable range hopping temperature dependence predicted by Mott. (After Wimmers and Christopher, 1981.)

Since E is measured from E_F, a characteristic cusp occurs at $E=0$. Here, $N(0)$ is the Fermi surface density of states and Δ has been termed (McMillan, 1981) a pseudo-correlation-gap parameter, which decreases as localization becomes stronger. Al'tshuler and Aronov emphasized the intimate relationship between this effect and the Anderson (1958) localization which should occur when the disorder-limited mean free path ℓ is reduced to the magnitude of the Fermi wavelength: $k_F\ell=1$.

Tunneling conductance measurements in rather good agreement with (8.55), even neglecting the presumably weak variation coming from the barrier voltage dependence, are now available in two experimental systems, following the pioneering work of Bermon and So (1978*c*), whose data were analyzed in the original paper (1979*a*) of Al'tshuler and Aronov. These more recent examples are the $Ge_{1-x}Au_x$ alloy system (McMillan and Mochel, 1981) and the granular Al system (Dynes and Garno, 1981). The $E^{1/2}$ dependence over a range of ~5 to ~30 meV in the granular Al data is clearly shown in Fig. 8.31. These data are rather similar to the original data of Bermon and So on Al–I(O_2)–Au junctions with O_2-doped, quench-condensed Au films. The $V=0$ region is excluded in Fig. 8.31 because of the superconducting transition of the Al films. This region is visible in Fig. 8.32, however, and indeed shows a cusp at $V=0$, whose sharpness is limited only by the resolution function ~3.5 kT. These data in the $Ge_{1-x}Au_x$ system for $0.08 \leqslant x \leqslant 0.20$, give insight into the metal-insulator transition which occurs between $x=0.12$ and $x=0.08$: note that by $x=0.08$, the value $N(0)$ has become zero on the scale of the measurements. The dI/dV data for metallic samples $x \simeq 0.12$, etc. are stated to be close in form to the dependence of (8.55), the exponent α for $(E/\Delta)^\alpha$ being best fit as $\alpha=0.6$. Since the effects occur precisely at $V=0$, the cusp feature and the gap (or pseudogap) in the insulating phase is a Mott correlation gap, rather than a semiconductor band gap. The accompanying electrical measurements on the same system (Dodson, McMillan, Mochel, and Dynes, 1981) show Anderson localization (in the sense described by Abrahams, Anderson, Licciardello, and Ramakrishnan, 1979) in the insulating region $x \leqslant 0.08$. McMillan and Mochel (1981) thus infer that in such amorphous materials, correlation (the tunneling gap) and localization (inferred from the film conductivity) are closely linked, which is a feature of the scaling theory of McMillan (1981). An explanation is not possible within any one-electron localization theory neglecting electron-electron interaction, for the latter is necessary to produce the observed density-of-states $E^{1/2}$ anomaly.

8.4.3 *Phonon Self-energy Effects in Degenerate Semiconductors*

Tunneling observations of optical phonon features in d^2I/dV^2 spectra from Schottky barrier contacts on several degenerate semiconductors have provided convincing evidence of self-energy interaction effects. These phonon features, as indicated in Fig. 8.33 and Fig. 8.34a and which

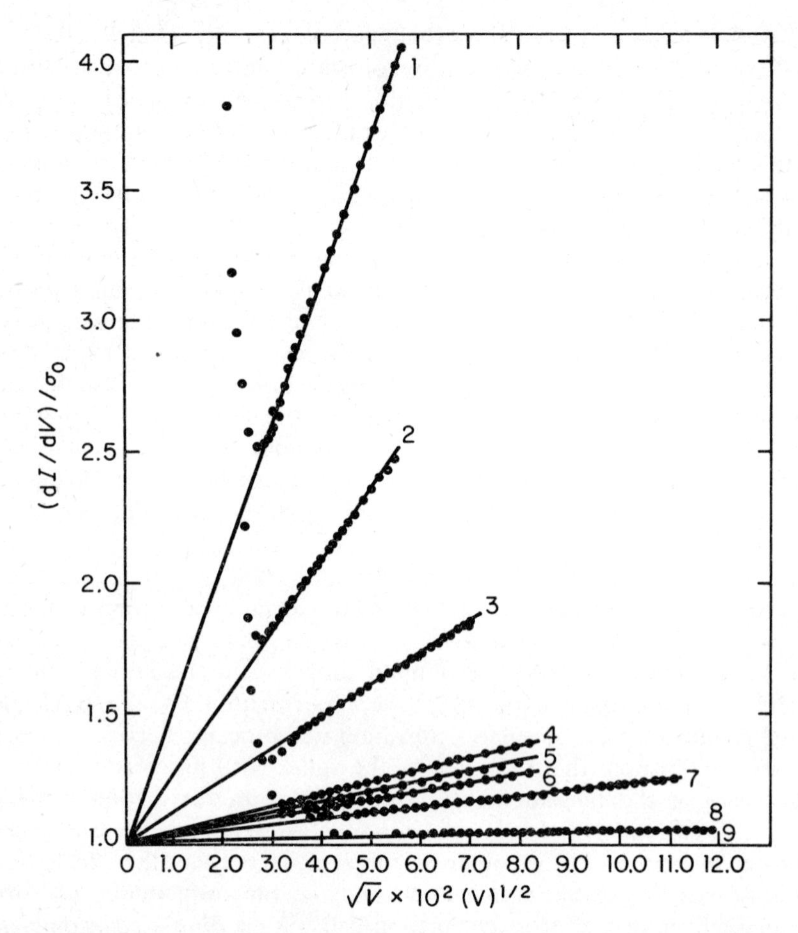

Fig. 8.31. The measured normalized conductance of selected Al–Al_2O_3–Al (granular) tunnel junctions as a function of $V^{1/2}$. (At lower bias, the superconducting energy gap structure dominates and has been omitted.) The resistivities of the various samples are 1, $4.26\times10^{-2}\,\Omega\cdot$cm; 2, $5.95\times10^{-2}\,\Omega\cdot$cm; 3, $1.89\times10^{-2}\,\Omega\cdot$cm; 4, $4.43\times10^{-3}\,\Omega\cdot$cm; 5, $9.84\times10^{-3}\,\Omega\cdot$cm; 6, $2.05\times10^{-2}\,\Omega\cdot$cm; 7, $3.20\times10^{-3}\,\Omega\cdot$cm; 8, $6.15\times10^{-4}\,\Omega\cdot$cm; 9, $3.28\times10^{-4}\,\Omega\cdot$cm. (After Dynes and Garno, 1981.)

are associated with Einstein-like $k=0$ optical vibrations, can be distinguished by their symmetry about $V=0$ from inelastic assisted tunneling features. The simple inelastic excitation of a phonon at $eV=\pm\hbar\omega_0$ would provide symmetric increases in $G(V)$ at $V=\pm\hbar\omega_0/e$ and, hence, antisymmetric peaks in d^2I/dV^2; whereas the observations in p–GaAs (Tsui, 1968) and p–Si (Wolf, 1968; Cullen, Wolf, and Compton, 1970) are of approximately *symmetric* structures in d^2I/dV^2. An additional notable feature common to these data is the large magnitude of the conductance

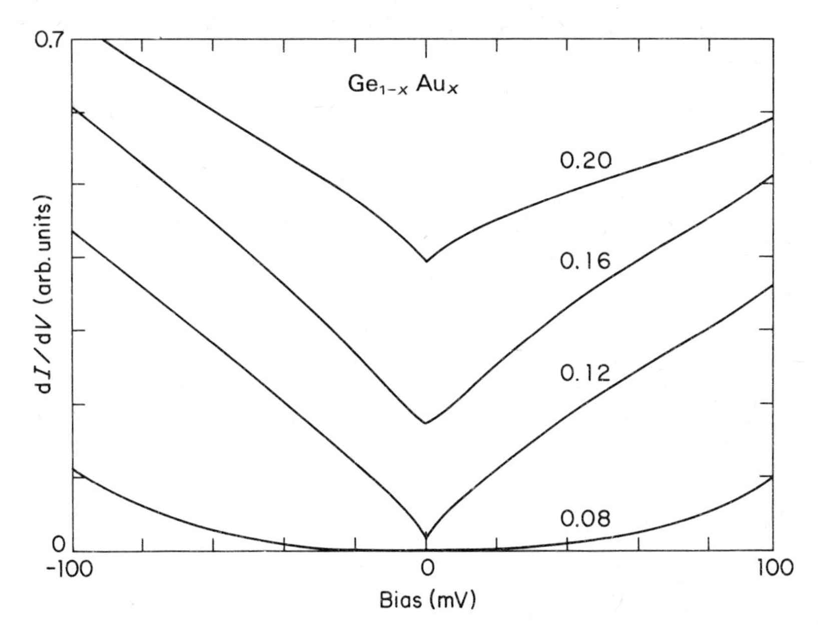

Fig. 8.32. The conductance dI/dV of $Al–Al_2O_3–Ge_{1-x}Au_x$ tunnel junctions vs. voltage on the $Ge_{1-x}Au_x$ electrode for four samples with $x = 0.08$, 0.12, 0.16, and 0.20. The conductance is normalized to unity at +0.3 V. (After McMillan and Mochel, 1981.)

effect, on the order of $\Delta G/G \simeq 0.15$ at 64 meV in the Si:B sample of Fig. 8.34a. This effect is much larger than any similar effect observed in the normal state spectra of junctions of strong-coupling metals like Pb. Note that the effect at reverse bias is, in the simplest approximation, a decrease in conductance. It is also pertinent that d^2I/dV^2, especially in forward bias, deviates from its background even at bias energies *below* the corresponding threshold energy. How can this result occur?

Consider an electrode in which electrons interact with a single Einstein phonon mode $\hbar\omega_0$. While injected electrons of energy $E < E_F + \hbar\omega_0$ cannot emit a real phonon, their behavior near this threshold can be modified by *virtual phonon emission.* The latter effect can be described by the real part of the self-energy, which enters the E–k relation of the electron, which can be approximated (Mahan, 1969) by

$$E(k) = \frac{\hbar^2 k^2}{2m^*} + g^2 \ln\left|\frac{E - E_F - \hbar\omega_0}{E - E_F + \hbar\omega_0}\right| \tag{8.56}$$

Here, g^2 is an electron-phonon coupling constant, and we assume $E_F > \hbar\omega_0$. This expression is also valid for $|E - E_F| > \hbar\omega_0$; the lifetime against emission of a real phonon above threshold is described by the imaginary part of the self-energy, which we have neglected here. In this expression, E represents the energy of the electron relative to the bottom of the band, rather than relative to the Fermi energy.

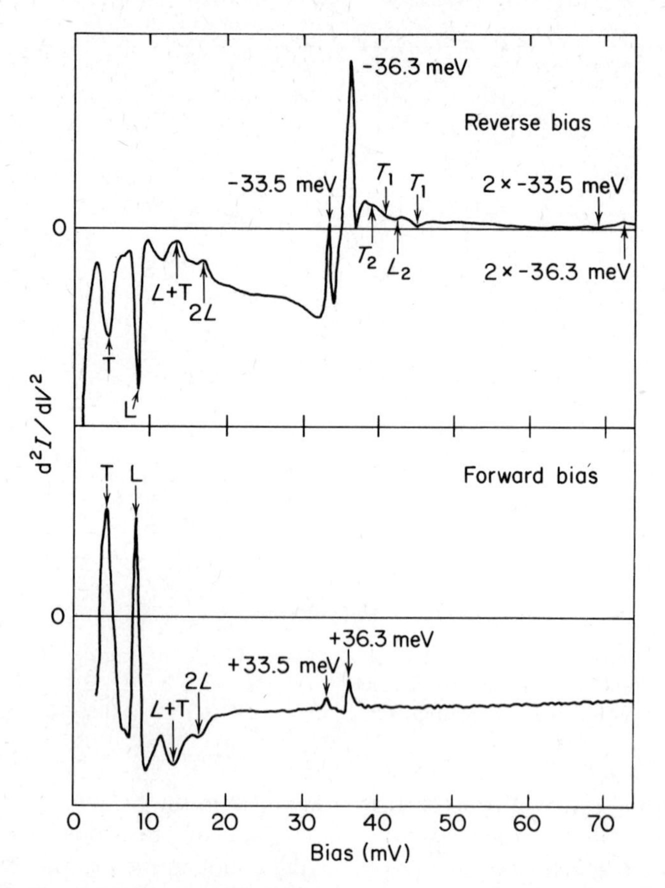

Fig. 8.33. Second-derivative d^2I/dV^2 curve from a p-type (Zn-doped) GaAs/Pb junction at 1 K. Bias is measured from the Pb gap. Here T and L indicate peaks due to the transverse and longitudinal structure in the superconducting density of states in Pb. (After Tsui, 1968.)

Physically, (8.56) reflects the tendency of the moving electron (or hole) to cause a co-moving polarization or distortion of the nearby ions from their usual lattice positions. (This polarization cloud can be expressed quantum mechanically as a cloud of virtual phonons.) As a result of this coupling to the ion motion, the $E(k)$ relation of the electron, and its consequences, notably the electron's group velocity, $\hbar^{-1}\, dE/dk$, may be strongly perturbed even below the crossing (at threshold) of the parabolic band $E = \hbar^2k^2/2m$ with the Einstein modes $E = E_F \pm \hbar\omega_0$. In order to deduce the consequences of such a perturbation, initial approaches were based on the conventional transfer Hamiltonian theory, Eq. (8.49), and were guided by an awareness of characteristic differences between the degenerate semiconductors with their Schottky barrier contacts and the more familiar metals in MIM junctions, discussed, e.g., by Hermann and

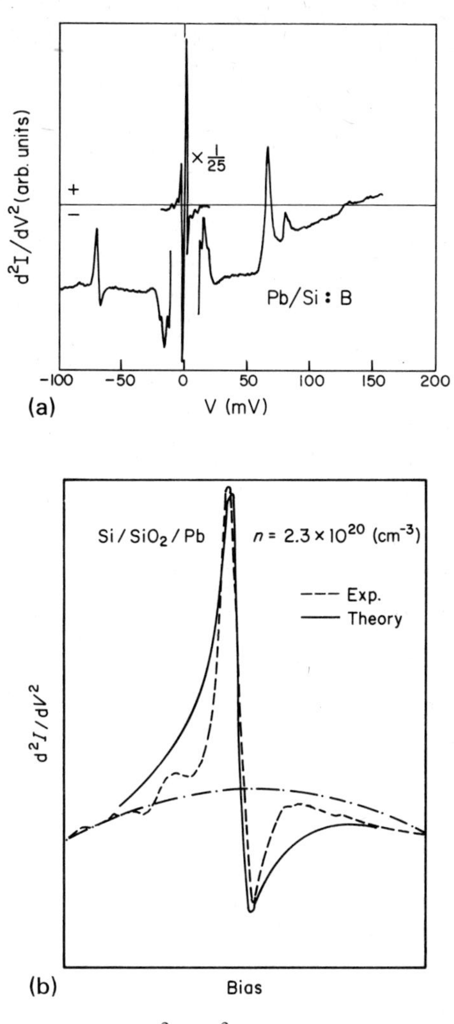

Fig. 8.34. (a) Second-derivative d^2I/dV^2 spectrum for In Schottky barrier contact to heavily doped Si:B. Note the approximate symmetry of phonon peaks at ± 64.9 mV, indicating a self-energy mechanism. Local-mode vibration of B impurity is present at 78-mV bias. (After Cullen, Wolf, and Compton, 1970.) (b) Calculated and observed effects of strong electron-phonon coupling in Si:B Schottky barrier tunneling d^2I/dV^2 spectrum, in forward bias.

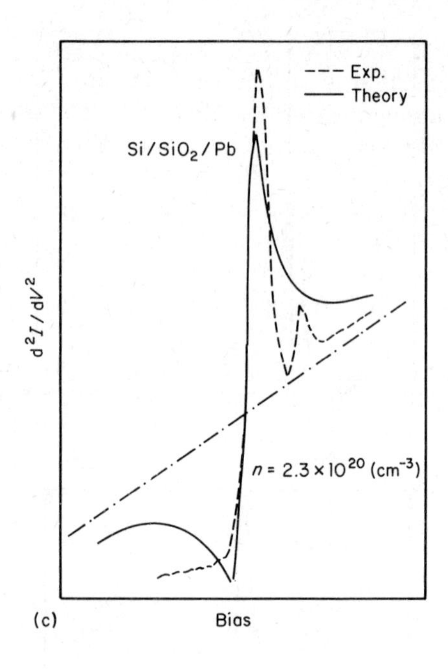

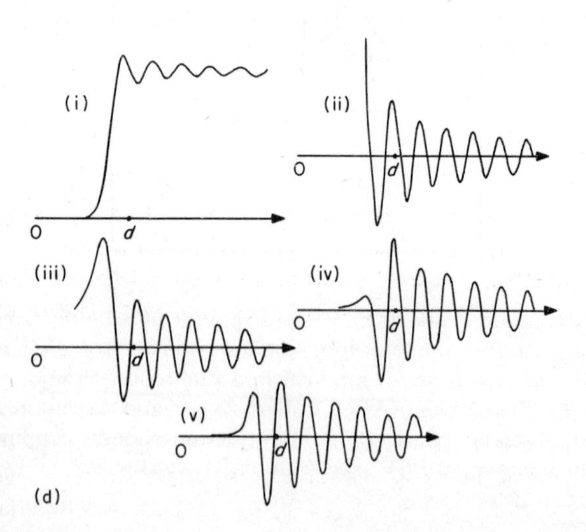

Fig. 8.34. (cont.) (c) Same as in (b), but reversed bias. (After Combescot and Schreder, 1974.) (d) General features of Green's functions—(i) $D(x)$, (ii) $G_1(x)$, (iii) $G_2(x)$, (iv) $D(x)G_1(x)$, and (v) $D(x)G_2(x)$—entering calculation [text Eq. (8.62)] from Combescot and Schreder (1974). Metal-semiconductor interface is at $x=0$; d is the edge of the Schottky barrier represented by $V(x)=A(x-d)^2$ for $0<x<d$.

Schmid (1968). First, the ratios $\hbar\omega_{ph}/E_F$ and Σ_{ph}/E_F, where Σ_{ph} is the relevant self-energy, can in semiconductors be large parameters, of order unity. A second unique feature is that the depletion barrier in the Schottky contact has very nearly the same phonon spectrum as the electrode. An added point is that the depletion barrier width is a strong function of both the applied bias voltage and the energy of the tunneling electron. If we define forward bias as that sign which reduces the barrier height, the (barrier thickness) distance seen by a Fermi surface electron at the metal surface to the classical turning point in the semiconductor decreases in reverse bias, while the barrier thickness seen by an electron at the semiconductor Fermi energy decreases in forward bias.

Application of (8.49) to this problem by Davis and Duke (1968, 1969) is summarized by Davis (1969) in the expression, valid at $T=0$,

$$G(V)=\frac{2e^2}{h}\rho_t\int_{-E_F}^{\infty} d\varepsilon_k \operatorname{Im} G(\varepsilon_k, -eV)\int_0^{E_t^0} dE_t D(E_t, \varepsilon_k) \tag{8.57}$$

where

$$\rho_t=\frac{m^*}{2\pi\hbar^2} \tag{8.58}$$

$$E_t=\frac{\hbar^2 k_t^2}{2m^*} \tag{8.59}$$

$$\varepsilon_k=\frac{\hbar^2 k^2}{2m^*}-E_F \tag{8.60}$$

In (8.57) the upper limit on E_t is given by

$$E_t^0=-eV-\operatorname{Re}\Sigma(-eV)+E_F \tag{8.61}$$

and k_t lies transverse to the tunneling direction. The main effect, through which good fits to the data on p–Si were primarily obtained by Davis and Duke (1968, 1969), was the pronounced sensitivity of $G(V)$ to the self-energy $\Sigma(eV)$ through the upper limit E_t^0 on the phase space (transverse k) integral. The reason for this sensitivity is that E_F is small, so that rapid changes in $\Sigma(-eV)$, which occur in the region $eV=\pm\hbar\omega_0$, noticeably affect the integral. In a metal, in contrast to a degenerate semiconductor, the E_t integration is not sensitive to Σ_{ph} through the upper limit. This is the case because E_F is typically $\sim 10^2$ larger, and the E_t dependence of D cuts off the contributions to the integral from the region $E_t \simeq E_F$. It is assumed by Davis (1970*b*) that the D factor is determined by the real energy E and not by the bare energy ε_k, avoiding the objection of Appelbaum and Brinkman (1969*a*) mentioned earlier. However, some dispute appears to remain on the form of the upper limit, Eq. (8.61), with Appelbaum and Brinkman (1969*a*) giving this limit as $E_t^0=E$, excluding the appearance of structure through Σ if this self-energy is strictly energy-dependent. However, weak momentum dependence of Σ_{ph} in this case cannot be excluded.

There are further reasons (see also Davis, 1970*b*) why a more general approach to this problem was needed. First, the possibility of contributions to $G(V)$ by emission of barrier phonons, which have the same energies, is not encompassed by (8.57), and the data of Fig. 8.34a at forward bias do suggest a possibly substantial inelastic contribution [positive step in $G(V)$]. Further, application of the wavefunction analysis of Davis (1970*a*) [confirmed by Appelbaum and Brinkman (1970) through application of the ABZ theory] to an Einstein vibrator whose position could be varied from the insulator into the electrode, indicated that symmetric positive steps in $G(V)$ became symmetric *negative* steps in $G(V)$ when the vibrator enters the electrode. The physical origin suggested for this result (Davis, 1970*a*) is interference between the direct elastic tunneling process and a two-step process involving virtual excitation of the vibrator, with the resulting sign of the $G(V)$ step depending upon relative magnitudes of various matrix elements. In view of the possible step contributions in $G(V)$ of differing sign from the barrier and electrode portions of the Si, and the above noted variation in tunneling path length (barrier thickness) with bias V, a treatment in which the whole system (barrier + electrode) is treated equally seems imperative.

Just such a treatment has been accomplished by Combescot and Schreder (1973, 1974) in a thorough application of the tunneling formulation of Caroli, Combescot, Nozières, and Saint-James (1971). In the work of Combescot and Schreder (1974), the phonon contribution in the d^2I/dV^2 spectrum is calculated from the expression

$$\frac{dG_{ph}}{dV} = \pm\frac{2e}{\pi\hbar} p_0 |g|^2 \int dx\, D_\pm(x)\left[\frac{\Gamma G_{1\pm}(x)}{(eV\mp\hbar\omega_0)^2+\Gamma^2} + \frac{(eV\mp\hbar\omega_0)G_{2\pm}(x)}{(eV\pm\hbar\omega_0)^2+\Gamma^2}\right] \tag{8.62}$$

where the constant p_0 is as discussed by Combescot and Schreder (1973), g is an optical deformation potential (Conwell, 1967), and Γ is a lifetime energy width. Here $D_\pm(x)$ and $G_{1\pm}(x)$, $G_{2\pm}(x)$ are, respectively, phonon and electron Green's functions, with $\pm$ indicating calculation of the relevant overlap integrals with the barrier under forward or reverse bias. The barrier is taken as a Schottky barrier: $V(x) = A(x-d)^2$, $0 < x < d$, including the dependence upon bias voltage; and calculations are done with WKB wavefunctions and also with the exact parabolic cylinder wavefunctions (Conley, Duke, Mahan, and Tiemann, 1966). The resulting d^2I/dV^2 curves calculated for Si:B with $2.3\times10^{20}\ \mathrm{cm}^{-3}$ B are compared with experimental curves (Cullen, Wolf, and Compton, 1970) in Figs. 8.34b and 8.34c. The curves in Fig. 8.34d show typical behavior of the relevant functions D, G_1, and G_2 entering the calculation. These curves make clear that contributions from all portions of the structure ($0 < x < d$ being the semiconductor depletion layer "barrier," $x > d$ the semiconductor "electrode") are incorporated on an equal footing. For further details, the reader is referred to Combescot and Schreder (1974).

Similar phonon self-energy effects have been observed in several other

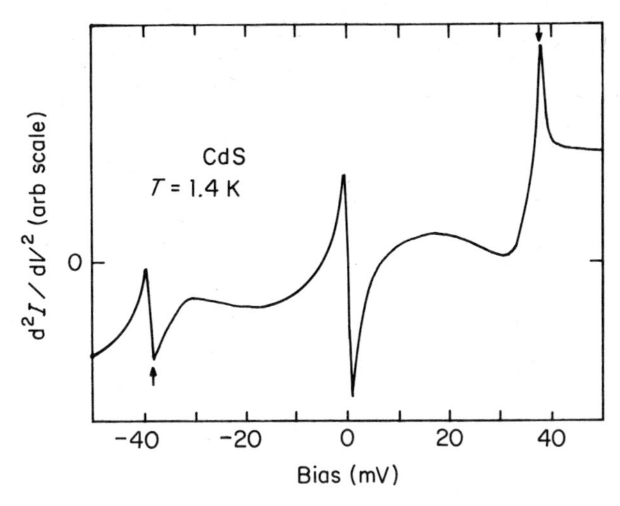

Fig. 8.35. Normal-state self-energy effects in the d^2I/dV^2 tunneling spectrum of an Au-degenerate CdS Schottky barrier contact. (After Losee and Wolf, 1969.)

materials, including GaAs (Conley and Mahan, 1967; Guétin and Schreder, 1971*b*; Mikkor and Vassell, 1970) and CdS (Losee and Wolf, 1969), from which the data of Fig. 8.35 is drawn. The relevant energies are summarized in Table 8.2.

A question still remains about the origin of normal metal phonon self-energy (Hermann and Schmid, 1968; Rowell, McMillan, and Feldmann, 1969) and phonon emission effects (Adler, 1969*a*). Upon application of the ABZ analysis to an MIM model with an Einstein vibrator, as mentioned earlier, Appelbaum and Brinkman (1970) found a negative step in $G(V)$, rather than the observed positive-step features, when the Einstein vibrator was located in the electrode. In the MIM case, the metal and insulator phonon spectra are distinct, so that positive contributions from insulator vibrations would be expected at phonon frequencies of the insulator but not at metal phonon frequencies. Suggestions for the origin of the observed, weak positive $G(V)$ steps at metal phonon frequencies in MIM junctions include (Appelbaum and Brinkman, 1970) the possibility of the vibration modes of the metal actually persisting into the oxide (of generally higher Debye energy), which has also been discussed by Klein et al. (1973); a further possibility is nonlocality of the electron-phonon interaction.

8.4.4 *Electron Scattering in the Kondo Ground State*

A more general expression for the s–d exchange Hamiltonian, Eq. (8.30), leading in perturbation theory (Kondo, 1964) to scattering $\sim -\ln T$, thus

Table 8.2. Semiconductor Self-energy Effects

Material	Mode	Energy (meV)	References
CdS (*n*)	*LO*		*a*
GaAs (*n, p*)	*LO*	36.3	*b–d*
(*p*)	*TO*	33.5	*c*
GaSb (*n*)	*LO*		*d, e*
Ge (*n*)	*LO*	38	*f*
(*p*)	*LO*	38	*g*
Si (*p*)	*LO*	64.2	*h*
(*p*)	B^{11}, B^{10} [j]	79.9, 77.4	*h*
(*n*)	*LO*	64	*i*

[a] Losee and Wolf (1969).
[b] Conley and Mahan (1967).
[c] Tsui (1968).
[d] Mikkor and Vassell (1970).
[e] Guétin and Schreder (1971*b*).
[f] Forest and Erlbach (1972).
[g] Steinrisser, Davis, and Duke, (1968).
[h] Wolf (1968); Cullen, Wolf, and Compton (1970).
[i] Schein and Compton (1970).
[j] Local-mode vibrations.

divergent at low T, is

$$H_{sd} = -\sum_{k,k'\alpha\beta} J_{k,k'} a^*_{k\alpha} \vec{\sigma}_{\alpha\beta} a_{k'\beta} \cdot \vec{S} \tag{8.63}$$

where $\vec{\sigma}$ is the Pauli spin operator, $\vec{S}$ represents the local moment, and $J_{kk'}$ is the k-dependent interaction. In order to make more explicit the possibility of k dependence in the coupling $J_{kk'}$, and to extend the discussion in Sec. 8.3.6 [see Eq. (8.46)] sketching the origin of the magnetic moment from a localized impurity, we follow Anderson (1966) in writing a Hamiltonian for a metal containing such an impurity:

$$\begin{aligned} H_{\text{Anderson}} = \sum_{k\sigma} \varepsilon_k a^*_{k\sigma} a_{k\sigma} + \sum_{\sigma} \varepsilon_d n_{d\sigma} + U n_{d\uparrow} n_{d\downarrow} \\ + \sum_{k_\sigma} V_{kd}(a^*_{k\sigma} d_\sigma + d_\sigma a^*_{k\sigma}) \end{aligned} \tag{8.64}$$

Here, a^* and d^*, respectively, are creation operators for band and "d" electrons, ε_d is the energy of the singly occupied localized d state, and U is a Coulomb correlation energy experienced by a second electron on the same d site, which thus occurs at energy $\varepsilon_d + U$. The interaction V_{kd} with the continuum of band states implies a width Γ for the ε_d, $\varepsilon_d + U$ states given by

$$\Gamma = \pi \langle |V_{kd}|^2 \rangle N(0) \tag{8.65}$$

The interesting point is that a transformation of (8.64) (Schrieffer and

Wolff, 1966) leads precisely to (8.63) and provides the desired identification

$$J_{kk'} = \frac{2V_{\mathrm{kd}}V_{\mathrm{dk'}}U}{\varepsilon_{\mathrm{d}}(\varepsilon_{\mathrm{d}} + U)} \tag{8.66}$$

The k dependence implied here will be strongly influenced by the radius a^* of the localized wavefunction ϕ_{d}. One line of experimental work of considerable interest is thus to measure the k dependence of $J_{kk'}$. We will return to this question after discussing the low-temperature limit, which, obviously, is incorrectly predicted by the perturbation theory result.

As indicated in Sec. 8.3.6, the physical origin of the tunneling conductance peak contribution is forward Kondo scattering across the barrier from a magnetic impurity in the barrier but close enough to one electrode that the V_{kd} term in the Anderson Hamiltonian (8.64) is appreciable. This result is approximated in perturbation theory by

$$G_3(eV) \simeq -\int_{-\infty}^{\infty} \frac{\partial}{\partial \varepsilon} f(\varepsilon + eV) \int_{-D}^{D} \frac{\tanh \beta\varepsilon'}{\varepsilon' - \varepsilon} \, d\varepsilon' \, d\varepsilon \simeq -\ln\left[\frac{(eV)^2 + (kT)^2}{D^2}\right] \tag{8.67}$$

where $\beta = (kT)^{-1}$ and the cutoff D is associated with the width of the conduction band. This expression is unphysically divergent at $V = T = 0$. This divergence has been removed in a variety of subsequent nonperturbative treatments (see, e.g., Hamann, 1967; Bloomfield and Hamann, 1967), culminating in the work of Wilson (1975) and Andrei (1980). This large body of work confirms the earlier expectations that at temperatures far below the characteristic Kondo temperature,

$$T_K = k_B^{-1} |2JN(0)|^{1/2} D \exp \frac{-1}{|2JN(0)|} \tag{8.68}$$

a spin-compensation cloud forms, leading to a singlet many-body ground state, with the consequence that all of the apparently divergent properties saturate at finite values. For example, the susceptibility χ is logarithmic at high T; $\chi \simeq \ln(T/T_K)$, behaves in a Curie-Weiss fashion near T_K,

$$\chi(T) \simeq \frac{1}{k_B}\left(\frac{0.17}{T + 2T_K}\right), \qquad 0.5T_K < T < 16T_K \tag{8.69}$$

(Goetze and Schlottmann, 1973), and has at $T = 0$ the limiting value $\chi(0) = 0.103/kT_K$ (Wilson, 1975; Andrei and Lowenstein, 1981).

The scattering rate measured by $G_3(eV)$ in the tunneling experiment (assuming the moment lies in the barrier) is generalized in the more advanced theories to the t matrix, which is given by Hamann (1967) as

$$t(\omega) = \frac{1}{2\pi i N(0)}\left\{1 - \frac{X}{[X^2 + S(S+1)\pi^2]^{1/2}}\right\} \tag{8.70}$$

where

$$X = \ln \frac{\omega + iT}{iT_K} \tag{8.71}$$

and where the specific T_K expression of Hamann (1967) is understood. The $\omega = 0$ limit of this scattering rate approaches a constant which has been identified with the unitarity limit for s-wave scattering, as $T \to 0$, in line with the behavior noted above. The high-energy expansion of this function contains powers of $\ln(|\omega + iT|/D)$, consistent with the perturbation result.

Evidence for just such a low-temperature saturation of the scattering rate, as is predicted by (8.70), was obtained in measurements of $G(0)$ in the Schottky barrier Si:P moment system, as shown in Fig. 8.36. Emphasis is placed on the observed difference between the temperature dependence of Au/Si:As and Au/Si:P junctions, which correlates with the observed difference in maximum g values and hence minimum

Fig. 8.36. Temperature dependence of vacuum-cleaved Au/Si: 1.6×10^{19} cm^{-3} As and Au/Si: 1.6×10^{19} cm^{-3} P junctions. The significant difference between the two cases is believed to be the stronger coupling $-JN(0)$ of the P local moment, determined from its g shift, which leads to deviation from log T behavior in the case of the P moment but not in the case of As moments. (After Losee and Wolf, 1971.)

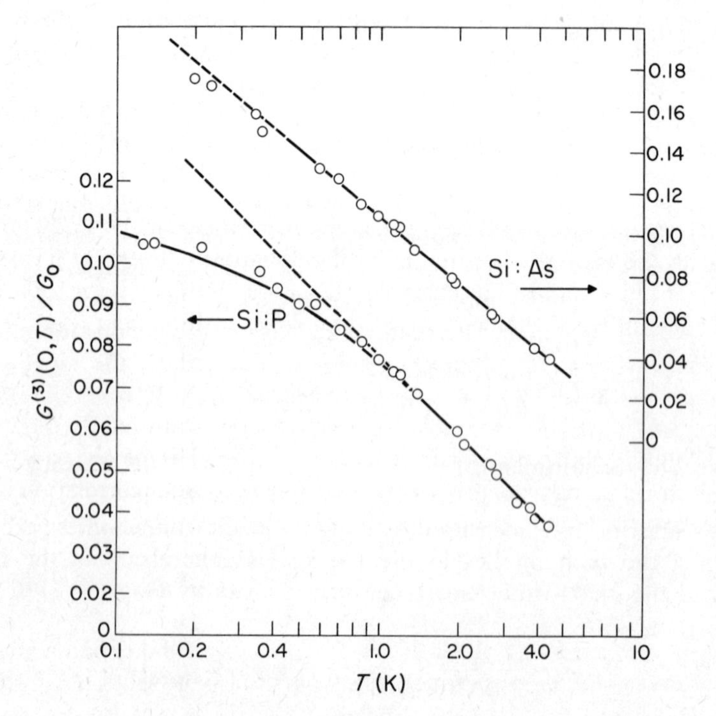

coupling strengths $-JN(0)$ as inferred via (8.43). The deviation from $-\ln T$ in the Si:P case is believed, therefore, to indicate the strong-coupling Kondo scattering case and is fit to the Hamann scattering rate, (8.70), with $T_K = 3.4$ K. More details on this research and additional data in the Schottky barrier system Si:Sb, which also shows deviation from the simple logarithmic behavior, well fit by the Bloomfield-Hamann (1967) scattering t matrix, have been given by Wolf (1975).

A limit to the analysis of these data, however, is imposed by the observation, especially in the Si:Sb case, of an apparently broad distribution of $JN(0)$ values, presumably indicating a variety of local environments. A second point of concern in pushing these analyses too far comes with the prediction of Solyom and Zawadowski (1968*a*, 1968*b*), subsequently confirmed by Appelbaum and Brinkman (1970), that a *change of sign* of the Kondo scattering conductance $G_3(eV)$ is to be expected if the moment actually lies within the electrode. The circumstances of the Anderson moments in the Schottky barrier junctions are not sufficiently well defined to rule out the complicating possibility, as suggested by the spread of g values, that some of the moments might lie in the electrode. This possible complication, and the theoretical work of Mezei and Zawadowski (1971*a*, 1971*b*) have stimulated more recent experiments (Bermon and So, 1978*a*) in which moments are implanted within the electrode at a fixed distance d from the insulator. The data shown in Fig. 8.37, after Bermon and So (1978*a*), are obtained in such a case—of a layer of Cr atoms imbedded at various depths d, $10\ \text{Å} \leq d \leq 50\ \text{Å}$, within the Au electrode of an Al–Al_2O_3–Au tunnel junction. In the inset, a plot of the estimated magnitude of the conductance *decrease* at $V = 0$, plotted against the depth d, is used to infer a characteristic coherence length ξ_Δ of the Kondo ground state in the Au:Co system. Let us consider how such an inference may be possible.

In general, the saturation of the susceptibility and other parameters, at temperatures much below T_K, has been taken to imply that a strong correlation between the impurity spin and conduction electrons with opposite spin direction is built up. This "spin compensation cloud" plausibly explains the apparent decrease of the local moment as the temperature is lowered. The spatial structure of the conduction electron disturbance around the impurity spin—and particularly the range of the correlation function—are of interest. Several types of coherence length may be envisioned. If the only electrons strongly scattered by the impurity are those whose energy from the Fermi level is of order kT_K, the characteristic Kondo energy, one would expect spatial correlation effects to exist over a length given approximately by

$$\xi_K = \frac{\hbar v_F}{kT_K}$$

where v_F is the velocity at the Fermi surface. A second coherence length ξ_0 is determined by the conduction electron bandwidth D from the

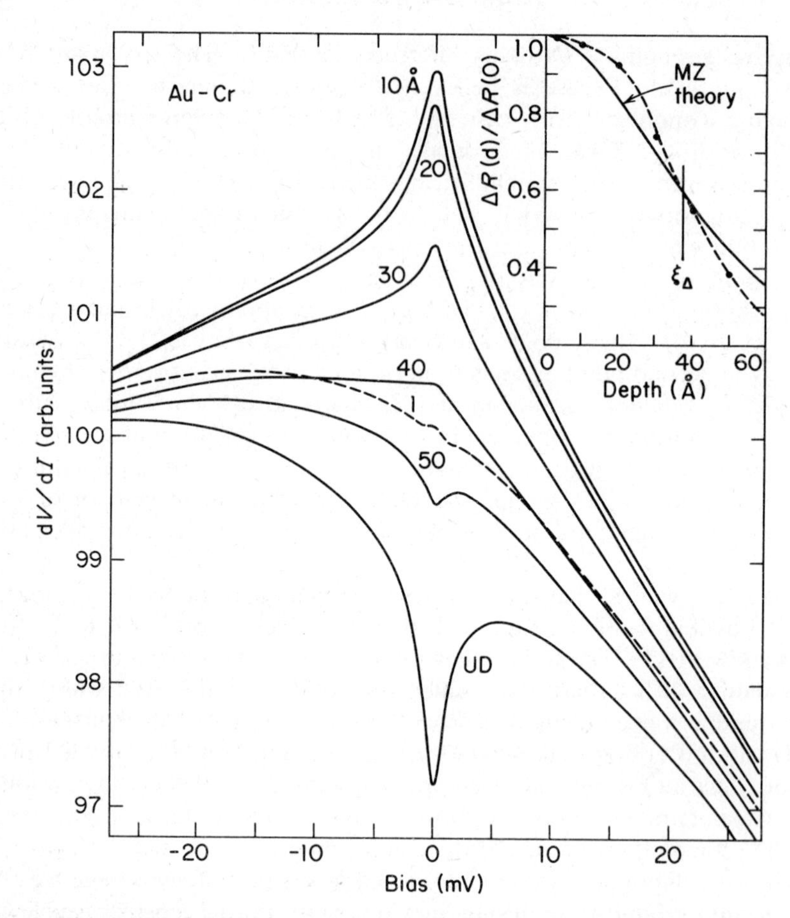

Fig. 8.37. Spectrum dV/dI measured at 2.1 K for an Al–I–Au tunnel junction into which a 0.6 Å layer of Cr has been implanted at depths (indicated) varying from 10 to 50 Å. Note that the curve marked UD, the junction without implanted chromium, shows a dV/dI minimum (conductance peak) which is believed to be extraneous, possibly due to molecular oxygen adsorbed in the junction fabrication. With the magnetic impurity chromium implanted at the optimum depth of 10 to 20 Å, this feature turns into a sharp conductance minimum (resistance peak). The inset shows the dependence of the resistance peak amplitude (dots connected by dashed curve) on the implantation depth. The solid line is obtained from the theory of Mezei and Zawadowski (1971*a*). (After Bermon and So, 1978*a*.)

relation

$$\xi_0 = \frac{\hbar v_F}{D}$$

This distance, in a metal, is on the order of one atomic distance. The energy D occurs here because the exchange coupling parameter $J_{\vec{k}\vec{k}'}$ in the s–d Hamiltonian, Eq. (8.63), is assumed constant with an energy cutoff determined by the conduction bandwidth D. If, however, the width of the energy region (denoted by Δ) over which $J_{\vec{k}\vec{k}'}$ changes appreciably is considerably smaller than D ($\Delta \ll D$), then the appropriate coherence distance will be

$$\xi_\Delta \simeq \frac{\hbar v_F}{\Delta}$$

where ξ_Δ is much greater than ξ_0. In the work of Mezei and Zawadowski (1971*a*, 1971*b*), the k dependence of $J_{kk'}$ entering (8.63) is given by

$$J^{\ell}_{\vec{k}\vec{k}'} = (2\ell + 1) J_\ell P_\ell(\cos\theta_{\vec{k}\vec{k}'}) F(k) F(k') \tag{8.72}$$

where ℓ is the angular momentum of the scattered states, P_ℓ is a Legendre polynomial, $\theta_{\vec{k}\vec{k}'}$ is the angle between the momenta of the incoming and outgoing electrons, and J_ℓ is a coupling constant. The precise form of $F(k)$ is unknown, but may plausibly be assumed to be Lorentzian:

$$F(k) = \frac{\Delta^2}{\Delta^2 + (\varepsilon_k - \varepsilon_0)^2}$$

Thus, Δ is a measure of the energy range over which $J_{\vec{k}\vec{k}'}$ is appreciable and

$$\xi_\Delta = \frac{\hbar v_F}{\Delta} \tag{8.73}$$

is the corresponding spatial coherence length, roughly the radius of the spin-compensation cloud around a given impurity. To proceed further, we must reconsider the tunneling experiment and, particularly, the role of the interface effects mentioned earlier in connection with the ABZ theory. We recall that, in the work of Zawadowski (1967), the M–I–M junction problem is divided into separate left- and right-hand problems. This separation is done by extending the insulator to positive infinity in the first case and negative infinity in the second, calculating the appropriate left- and right-hand Green's functions, and finally expressing the tunneling current in terms of these Green's functions and their spatial derivatives. The method has the important advantages of avoiding the phenomenological parameters of the tunneling Hamiltonian method (Appelbaum, 1967) and of providing explicit expressions for the dependence of the tunneling conductance on the *position* of the magnetic moment in the barrier or in the electrode. Solyom and Zawadowski (1968*a*, 1968*b*) employed this formalism to treat the magnetic impurity problem. However, by use initially of inappropriate mathematical approximations, they incorrectly obtained a conductance *dip* in the case of $J<0$ for the impurity in the barrier. This result is contrary to experiment as well as to Appelbaum's result. Appelbaum and Brinkman (1970), using the same (ABZ) Green's function theory (independently derived by them by using an alternative method), found that, for a rectangular barrier, the sign of both the second-order term $G^{(2)}$ and the third-order term $G^{(3)}$ was positive for moment locations in the barrier (except in the case of $G^{(3)}$ for the magnetic state very close to interface) and an oscillating function of moment position when this lies in the electrode. The period of the latter oscillation is on the order of π/k_F, where k_F is the Fermi momentum. Solyom and Zawadowski (1968*a*, 1968*b*) had obtained a conductance dip for the impurity in the barrier through neglect of the real part of a Green's function in their calculation, so that all points of disagreement were resolved. It is the nature of this formulation (splitting the problem into separate left- and right-hand problems) that the only effect of the impurity is to renormalize the density of states in the adjoining electrode. The enhancement in $G(V)$ obtained for the impurity in the barrier is a result of the interference of outgoing and incoming waves. Further confirmation has been provided by the theoretical work of Ivezić (1975), who applied the hopping model of tunneling proposed by Caroli, Combescot, Nozières, and Saint-James (1971*a*), which is based on the non-equilibrium perturbation formalism of Keldysh. This calculation is more general than any of those previously attempted.

Ivezić shows that his result for the square barrier reduces to that of Appelbaum and Brinkman (1970) only if the impurity is located sufficiently close to one electrode that the contribution to the propagator involving the spectral density of the other electrode may be neglected. For the impurity in the electrode, on the other hand, Mezei and Zawadowski (1971*a*, 1971*b*) have calculated the dependence of the tunneling current for arbitrary scattering amplitude $t(\omega)$. Taking explicitly into account the momentum dependence of the coupling $J_{\vec{k}\vec{k}'}$, they found rapid oscillatory spatial dependence (on the scale of π/k_F) of the tunneling conductance on the position d of the impurity relative to the electrode-barrier interface, in agreement with Appelbaum and Brinkman (1970). However, they also found additional damping factors associated with the momentum dependence of $J_{\vec{k}\vec{k}'}$ that lead to an expression for the renormalized density of states containing two parts: an oscillatory and a negative nonoscillatory part. For an impurity distribution in a metal (where k_F is large) extending over more than a few monolayers, the oscillatory term would average to zero (although this result would not necessarily be true in a degenerate semiconductor, where k_F is small). However, even in a metal, the nonoscillatory term should be observable. If the impurity layer is within a distance $\xi_\Delta = \hbar v_F/\Delta$, Eq. (8.74), of the interface, this term leads to a depression in the density of states and to a tunneling *resistance* dV/dI directly proportional to the imaginary part of the scattering amplitude, even in the strong-coupling limit. In the theory of Mezei and Zawadowski (1971*b*), believed to be relevant to the data of Fig. 8.37, the dependence of $G(d, V)$ on the position d of the impurity layer is given by (for $eV \ll \Delta$)

$$\frac{G(d, V) - G_0}{G(0, V) - G_0} = e^{-d/\xi_\Delta}(2 - e^{-d/\xi_\Delta}) \tag{8.74}$$

where G_0 is the elastic background conductance and ξ_Δ is the previously defined momentum coherence length. Thus, one expects the amplitude of the resistance peak to decay with a characteristic coherence length ξ_Δ as the embedded impurity layer moves further into the electrode.

The analysis applied to the Au:Cr data indicates $\xi_\Delta \simeq 37$ Å and $\Delta = 0.22$ eV, and the d dependence is in reasonable agreement with (8.74) inferred from the Mezei-Zawadowski (1971) theory. The narrow dV/dI peak spectrum at $d = 10$ Å is within one's expectation for a low-T_K system such as Au:Cr, although the presence of the spurious conductance peak in the undoped (UD) curve definitely is not. Bermon and So tentatively attributed this result to inclusion of (paramagnetic) molecular O_2 at the barrier–Au interface during the cryogenic junction fabrication steps. However, they show that the *difference* of the dV/dI spectra does behave closely as expected from the high-temperature (perturbation) expression. A further satisfactory aspect of the results is a correlation between the appearance of the dV/dI peak and the known appearance or nonappearance of a local moment in the corresponding bulk alloy system. Thus, implanted Co and Ni produce no appreciable effect in the tunneling spectra, consistent with alloy measurements indicating that Ni does not produce a moment in Au and with the expectation in the Au:Co system that T_K is so large that the $G^3(eV)$ feature would be too broad to observe. Other systems for which Bermon and So (1978*a*) reported Kondo parameters ξ_Δ and D are Au:Fe (33 Å, 0.25 eV), Ag:Fe (35 Å, 0.23 eV), and Au:Mn (36 Å, 0.23 eV). Two general conclusions are that the effective energy width Δ is much less than any true metal bandwidth D and that the Kondo coherence length appears to be typically in the range 20–40 Å.

An additional body of experimental results obtained by depositing generally submonolayer amounts of magnetic atoms at the metal-insulator interface will be referred to in connection with zero-bias anomalies. These results have also been more completely summarized by Wyatt (1974) and by Wolf (1975).

8.5 Zero-Bias Anomalies

The term *zero-bias anomaly* generally applies to any departure from the smooth conductance background arising from the basic barrier transmission process, including the distortion of the barrier potential with applied bias voltage. These features were considered at length in Chap. 2. A second connotation of the term *zero-bias anomaly* arises from the third word, implying a less than complete understanding of the origin of the effect. Thus, we have chosen to describe what is sometimes referred to as the conductance peak anomaly, in the previous section under Kondo scattering, with the conviction that the present degree of understanding has removed it from the class of anomalies. We begin with a survey of all the effects, starting with the most anomalous.

8.5.1 *Giant Resistance Peak*

The strongest and perhaps most puzzling zero-bias anomaly is the giant resistance peak (Rowell and Shen, 1966), which is shown in Fig. 8.38, as

Fig. 8.38. (a) The dV/dI–V spectrum for a Cr–I–Ag junction at 0.9 K. The bias voltage scales are $A = 0.2$ mV/division, $B = 1.0$ mV/division, $C = 5$ mV/division, and $D = 20$ mV/division. (b) The dV/dI vs. V plots for a Cr–I–Ag junction at various temperatures. Here, $E = 0.9$ K, $F = 20.4$ K, $G = 77$ K, and $H = 290$ K. The bias voltage scale is 10 mV/division. (After Rowell and Shen, 1966.)

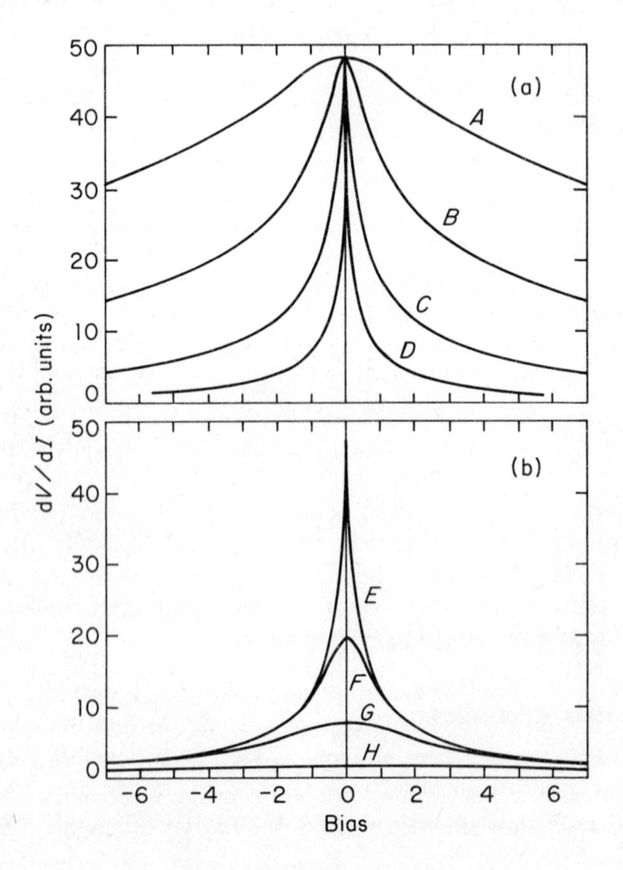

measured at 0.9 K in a Cr–I–Ag tunnel junction fabricated by oxidation of a Cr film. This effect is so strong as to dominate any background conductance, indicating that essentially all of the tunnel current is involved in the anomalous behavior. The value of dV/dI at $V = 0$ is as much as 100 times its value at 100 mV, depending on the measurement temperature. The dependence of dV/dI on V and T is stated to be homogeneous in these variables and essentially logarithmic. The effect is possibly related to the magnetic nature of Cr metal or its possible oxides CrO_2 (ferromagnetic) or Cr_2O_3 (antiferromagnetic). In fact, Mezei and Zawadowski (1971*b*) have suggested that the effect is due to Kondo scattering from magnetic moments embedded in the Cr electrode or at the electrode-insulator interface, as discussed in the previous section. The main difficulty with this interpretation seems to be the magnitude of the effect, in relation to the well-defined limit (unitarity limit) for the t-matrix scattering from a given impurity: max $|t(\omega)| = (1/\pi)N(0)$. The only way to reach the observed magnitude $\Delta R/R \simeq 100$ is to follow an argument of Mezei and Zawadowski (1971*b*) that locally $N(0)$ is depressed, making t larger. No strong evidence for the local depression of $N(0)$ has been offered.

A second explanation for the giant resistance peak is along the lines of the Giaever-Zeller two-step tunneling model to be considered below. In this interpretation, it is not obvious how the logarithmic dependence would arise, but the large magnitude of the effect then becomes relatively easy to understand.

Giant resistance peak anomalies of a possibly similar origin have been reported by Mezei (1967), by Wyatt and Lythall (1967), and, in connection with deliberately introduced impurities, by Lythall and Wyatt (1968), Mezei (1969), Kroo and Szentirmay (1970, 1971), El-Semary and Rogers (1972*a*, 1972*b*), and El-Semary, Kaahwa, and Rogers (1973). The latter authors performed an experiment similar to that of Bermon and So (1978*a*), finding a strong resistance peak with introduction of Ni into Al electrodes. However, El-Semary, Kaahwa, and Rogers (1973) reported that a strong resistance peak also occurred upon introduction of Sn into Al electrodes. The latter observation is inexplicable in terms of magnetic effects, as it is unlikely that a magnetic moment should accompany Sn in Al. This point raises questions of insertion, in fact, of partially oxidized particles of dopant metal, making more likely a Giaever-Zeller two-step tunneling mechanism.

8.5.2 *Semiconductor Conductance Minima*

Proceeding in a survey of zero-bias anomalies in an ordering that gives precedence to incompletely understood effects, we find that the type of semiconductor conductance minima shown in Fig. 8.39a, after Hall, Racette, and Ehrenreich (1960), ranks high on the list. The *pn* junction devices on typical (direct band gap) Group III–Group V compounds

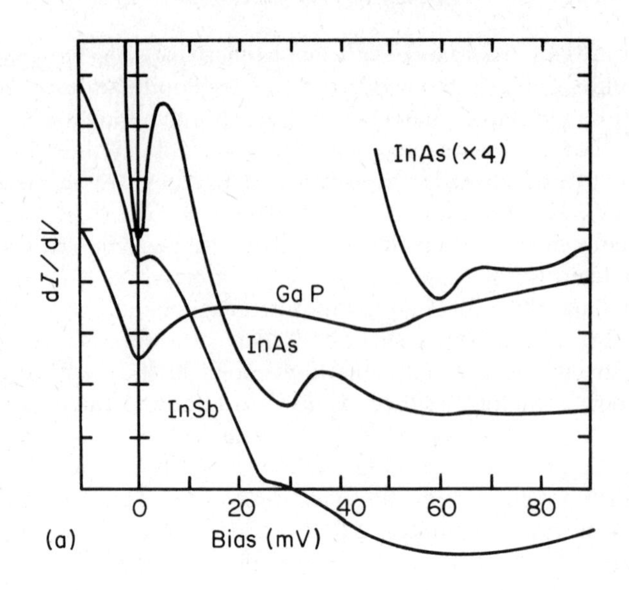
InAs (×4)
dI/dV
Ga P
InAs
InSb
0
20
40
60
80
(a)
Bias (mV)

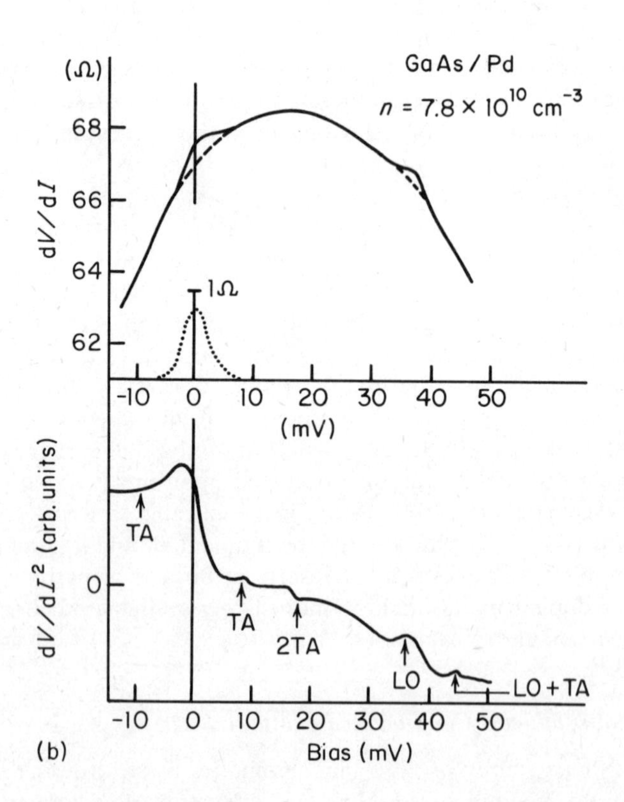
(Ω)
GaAs / Pd
n = 7.8 × 10^10 cm^-3
68
66
64
62
dV/dI
1Ω
-10
0
10
20
30
40
50
(mV)
dV/dI^2 (arb. units)
TA
0
TA
2TA
LO
LO + TA
(b)
Bias (mV)

GaSb, GaAs, InSb, InAs, and GaP show strong, long-wavelength, optical phonon features, while the width of the $G(V)$ minimum at zero bias suggests that this feature may be associated with emission of acoustic phonons. These features occur in junctions on heavily doped (metallic) material and are not strongly temperature-dependent. For this reason, the effect is not associated with the Mott transition. However, the possibility remains that the effect is partly related to the correlation gap in the metallic regime, as discussed in Sec. 8.4.2. A theoretical model of the $V=0$ minimum in terms of acoustic phonon emission was presented by Duke, Silverstein, and Bennett (1967). In this interpretation, the width of the conductance minimum should be related to the range of acoustic phonon energies available, which can be related to the velocity of sound, $(1/\hbar)(\partial E_{\mathrm{ph}}/\partial k)$; $G(V)$ might also, on this model, be expected to show some image of the cutoff of the phonon spectrum at its zone boundary value. [Such an effect was observed by Thomas and Queisser (1968); see Fig. 8.39b.] The coupling to the collective modes of the barrier (phonons) is assumed by Duke et al. to occur in the barrier by the interaction of the tunneling electron with ionized impurity states. Elaborations of this inelastic phonon emission mechanism, including rough fits to data, are given by Bennett, Duke, and Silverstein (1968) and by Duke (1969). The $T=0$ conductance contribution arising by single–acoustic phonon emission is given by these authors as

$$G(eV)=fG_0\left(\frac{E_0}{\hbar\omega_D}\right)^2\left\{\left[\cosh\left(\frac{eV}{E_0}\right)-1\right]\theta(\hbar\omega_D-|eV|)+\left[\cosh\left(\frac{\hbar\omega_D}{E_0}\right)-1\right]\theta(|eV|-\hbar\omega_D)\right\} \tag{8.75}$$

where $\hbar\omega_D$ is the acoustic phonon cutoff, G_0 is the background conductance, f depends on the concentration of ionized impurities and the resulting potential, and E_0 is estimated from the data as described by Duke, Silverstein, and Bennett (1967).

Related experimental results on GaAs–Pd Schottky junctions consistent with the phonon emission model are shown in Fig. 8.39b (Thomas

---

Fig. 8.39. (a) The dI/dV spectra of tunneling *pn* junctions on several III–V semiconductor compounds (as labeled) show characteristic dI/dV minima at zero bias which are rather broad and independent of temperature. These materials generally have strong electron-phonon coupling. (After Hall, Racette, and Ehrenreich, 1960.) (b) Demonstration of zero-bias conductance minimum in GaAs/Pd Schottky junction as arising from the acoustic phonon emission model (Duke, Silverstein, and Bennett, 1967). Upper solid curve measured at 2 K, dashed at 20 K. Dotted curve in the upper portion is the difference in resistance dV/dI between the two cases. Lower curve: second derivative d^2V/dI^2 vs. V, demonstrating that the width of the conductance minimum correlates precisely with the cutoff in the TA phonon spectrum of GaAs. (After Thomas and Queisser, 1968.)

and Queisser, 1968). Particularly convincing for this interpretation are the evidences of the acoustic phonon cutoffs indicated *TA* and 2*TA* in the figure. At the same time, it seems impossible to rule out contributions of a purely electronic nature in these complicated spectra.

8.5.3 *Assorted Maxima and Minima in Metals*

In earlier sections, we have discussed narrow, weak conductance minima interpreted as a consequence of electrode nonequilibrium under tunnel current injection (Sec. 7.1); a $V^{1/2}$ conductance minimum (or even a giant resistance peak) in dirty metals due to fundamental localization effects when the electron mean free path is short (Sec. 8.4.2); the occurrence of conductance minima or maxima from magnetic moments in the insulator or in the electrode metal (Secs. 8.3.6 and 8.4.4); and, in addition, the weak conductance minimum due to emission of quanta in a continuum such as the acoustic phonons discussed in Sec. 8.5.2. We will reserve for the next section (8.5.4) discussion of probably the best-understood resistance peak mechanism, due to Giaever and Zeller (1968*a*), and here finish a survey of the remaining anomalies which have been observed.

Returning to the results of depositing submonolayer amounts of transition metal impurities onto prepared Al–Al_2O_3 substrates, Figs. 8.40a and 8.40b, respectively, survey results obtained with Ag and Al host metal electrodes, after Cooper and Wyatt (1973). The choice of host metal plays a strong role in what is observed, for of the seven dopants Ti, V, Cr, Mn, Fe, Co, and Ni, only Ti and Ni produce a $G(V)$ peak in Ag (others producing dips, or in the case of Co, no feature at all), while in Al only Mn produces a dip (all others producing peaks). Some of these results overlap earlier work, agreeing with those of Nielsen (1970), Morris, Christopher, and Coleman (1969), Bermon and Ware (1971), and Kroo and Szentirmay (1970). If these data were interpreted strictly within the point of view of Mezei and Zawadowski—in which conductance peaks arise from magnetic moments in the insulator and dips from moments in the electrode—one would conclude that all cases, excepting Co in Ag, produce moments, those in Ag being more likely in the metal electrode, those in Al being more likely in the oxide. A different point of view is shown to be successful by Wyatt (1974), whose categorization (Table 8.3) of the results emphasizes the occurrence of the peaks.

In Table 8.3 (after Wyatt, 1974), the existence vs. nonexistence of moments on the 3d atoms in question, in junctions and in alloys, is revealed to be *symmetric* about the center of the series, strongly suggesting correlation with the spin S resulting from electrons of the 3d shell. The value of S is taken to be $n/2$ ($n<5$) and $(10-n)/2$ ($n>5$), where n is the number of electrons in the shell. Further, the S—d exchange constant J varies inversely with S in the generalization of (8.66) due to Schrieffer (1967),

$$J=-\frac{V_{kd}^2}{2S}\frac{U}{|\varepsilon_d|(U+\varepsilon_d)} \tag{8.76}$$

Table 8.3. Localized Moments in Junctions and in Alloys[a]

Dopant	n^b	Junctions		Alloys		Nominal Spin S
		Ag	Al	Au	Al	
Ti	2	√	√	X	X	1
V	3	X	√	?	X	$\frac{3}{2}$
Cr	4	X	√	√	X	2
Mn	5	X	X	√	?	$\frac{5}{2}$
Fe	6	X	√	√	X	2
Co	7	X	√	?	X	$\frac{3}{2}$
Ni	8	√	o	X	X	1

[a] Checks signify the existence of moments, and crosses signify nonexistence; the circle indicates probable existence of a magnetic moment. (After Wyatt, 1974.)
[b] Number of electrons in 3d shell.

Fig. 8.40. (a) The dI/dV for several doped Al–Al_2O_3–Ag tunnel junctions. Only Ti and Ni dopants produce a conductance peak at $V = 0$. On the left-hand side of the figure, the equivalent fraction of a monolayer is given. In the center, dI/dV ($V = 0$) is given in $m\Omega^{-1}$. The scale of the vertical axis is the same for each curve; however, the origins are shifted.

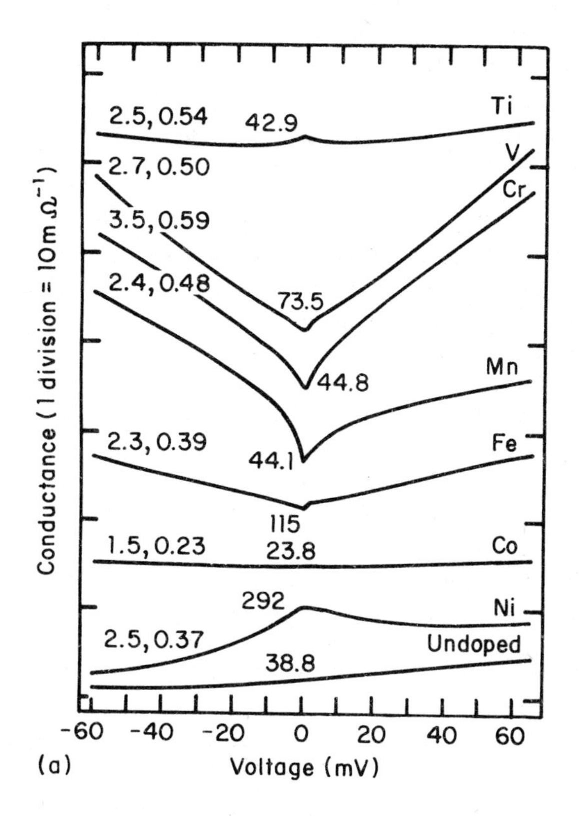

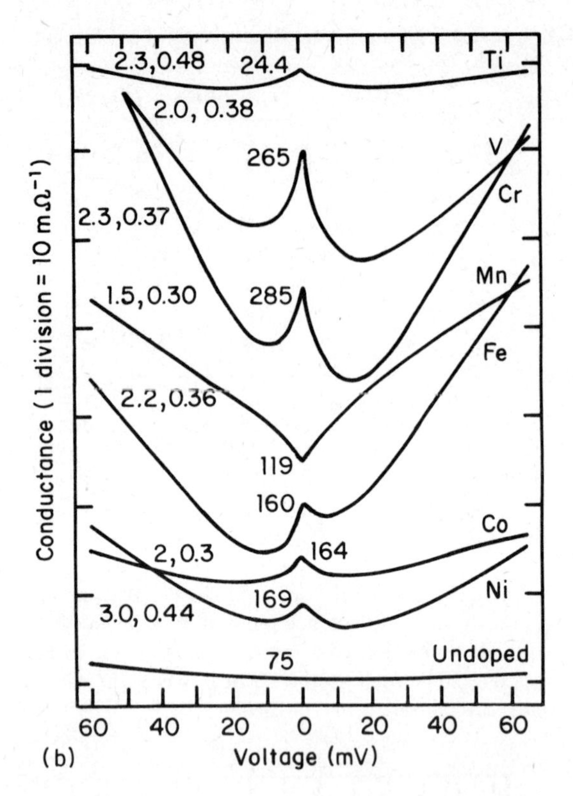

Fig. 8.40. (cont.) (b) The dI/dV is shown for doped $Al-Al_2O_3-Al$ tunnel junctions. Only Mn does not show a conductance peak at $V=0$. Labeling of dopant concentration is as in (a). (After Cooper and Wyatt, 1973.)

applicable to an atom with more than one 3d orbital. Here, $|\varepsilon_d|$ is the energy of the 3d orbital relative to the Fermi energy E_F.

It is argued by Wyatt (1974) that (8.76) will lead to large J and correspondingly strong $G(V)$ peak contributions at the *ends* of the 3d series, because S is large and $|\varepsilon_d|$ small in these cases; this observation indeed correlates correctly with the greater tendency to observe the peak anomaly toward the ends of the 3d series. The greater likelihood (Table 8.3, after Wyatt, 1974) of finding the peak in Al hosts rather than in Ag is attributed to the larger $N(0)$ of Al than of Ag, which is relevant because the basic coupling parameter is $JN(0)$. In the table, alloy results for an Au host are used instead of an Ag host, on the basis of available data and the expected similarity of Ag and Au. The basic comparison in the table between existence of moments in junctions and bulk alloys is interpreted as due to inherently stronger coupling in the alloy than in the junction, where the moment is presumably at the surface of the metal and where it interacts with fewer neighbors than in the bulk. Thus, moments are observable in Au alloys by the increased coupling relative to the junctions, while in Al alloys, the coupling is so strong that the moments become completely spin-compensated and unobservable. A careful discussion of experimental factors which may also influence the junction results is given by Wyatt (1974).

The optimum doping of the 3d element onto the previously oxidized Al electrode is found to be approximately 0.4 monolayer. The effect of interaction between moments has been observed (Wyatt, 1971; Wallis and Wyatt, 1972) and understood on a useful local-field model described by Wyatt (1973). It is found that the range of concentration in which $G(0)$ departs, at low T, from $-\ln T$ to a constant value, as expected in the Kondo strong-coupling case, is actually very narrow; above this range, one sees the signature of interaction between moments, which is a $G(0)$ which goes through a maximum and then falls with decreasing temperature. A close parallel exists between this behavior and related bulk behavior. The only case, in the class of doped junctions, where departure from $-\ln T$ has been linked to possible Kondo strong coupling is a Cr-doped Al/I/Al junction measured to 0.3 K by Nielsen (1970).

The weakest point in understanding the systematics in Fig. 8.40 and in Table 8.3 is the prediction of the conductance minima, which, by comparison with the undoped curves, do have a relation to the dopant which is not clarified by Wyatt's discussion. A possible interpretation of these effects may lie along the lines suggested by Al'tshuler and Aronov (1979*a*). Rather similar V-shaped conductance minima, also apparently lacking firm interpretation, were reported by Shen (1969) and Rochlin and Hansma (1970) in Cr–Cr_2O_3–M junctions. As noted by the latter authors, this later work on Cr did not reproduce the giant resistance peak obtained earlier (Shen and Rowell, 1968) in Cr–Cr_2O_3–Ag junctions. Clearly, the details of the interfacial layers, reflecting variations of the fabrication conditions, are involved in the different behavior.

Finally, departing from metals, a number of reports of tunneling into amorphous semiconductors have appeared, typically revealing a strong and temperature-dependent conductance minimum at $V=0$. Notable among these is the study of amorphous germanium by Osmun (1975). These effects are undoubtedly related to localization and hopping transport in the energy gap states of the semiconductor. Unfortunately, however, their study has not been particularly fruitful in clarifying properties of the amorphous state. A more extensive discussion of this area has been given earlier by Wolf (1975) and in references therein.

8.5.4 The Giaever-Zeller Resistance Peak Model

The sketches of Fig. 8.41a (Giaever and Zeller, 1968*a*; Zeller and Giaever, 1969) show a junction in which well-defined, oxidized Sn particles are embedded in a tunnel barrier between two Al film electrodes. The fabrication is deliberately such that tunneling from one Al electrode to the other most likely proceeds via the intermediate Sn

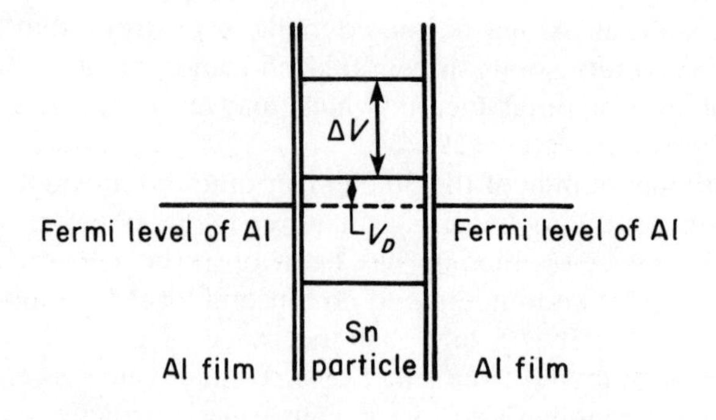

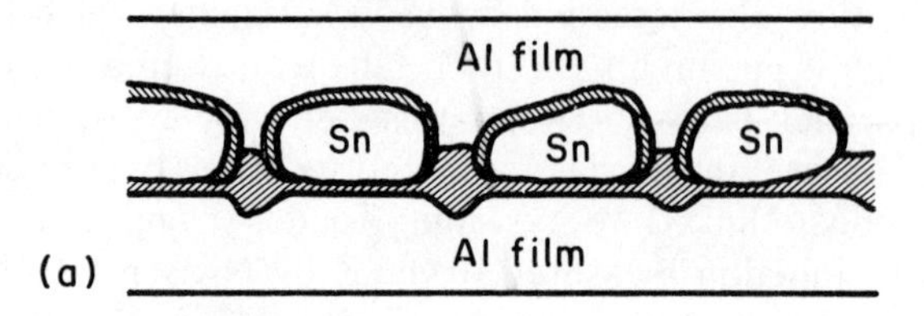

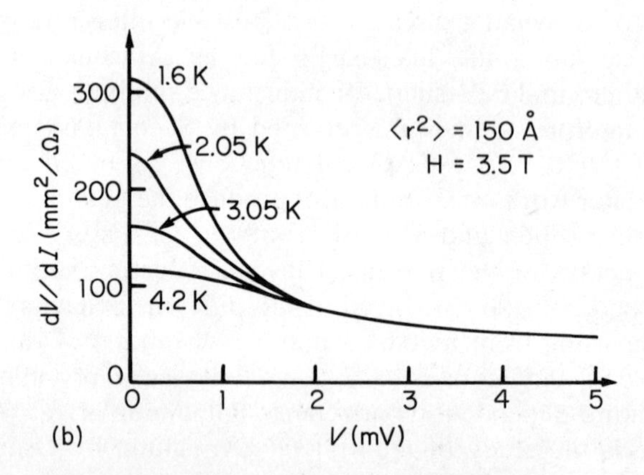

Fig. 8.41. Two-step tunneling; a case of real intermediate states in the barrier. (a) Model and level scheme of Sn particles in a tunnel junction. In the level scheme shown, $\Delta V = e/C$ is the voltage change of the particle caused by addition of one electron. The experimental structures are generated by depositing tin onto an oxidized aluminum film under conditions providing agglomeration, allowing the Sn particles to oxidize, and covering the assembly with a second aluminum film. (b) The $\mathrm{d}V/\mathrm{d}I$ spectra of such a junction at temperatures ranging from 1.6 to 4.2 K, with an applied magnetic field of 3.5 T to keep the Sn particles in the normal state. The particles have an rms radius of 150 Å. Note the general similarity of these spectra with those shown in Fig. 8.38. (c) Essential features of the model advanced by Giaever and Zeller (1968*a*) to explain the resistance peak. Topmost, dynamical conductivity as a function of voltage for a single particle; middle, for a set of particles with fixed capacitance C; lower, for a set of particles with a distribution of C values. (After Zeller and Giaever, 1969.)

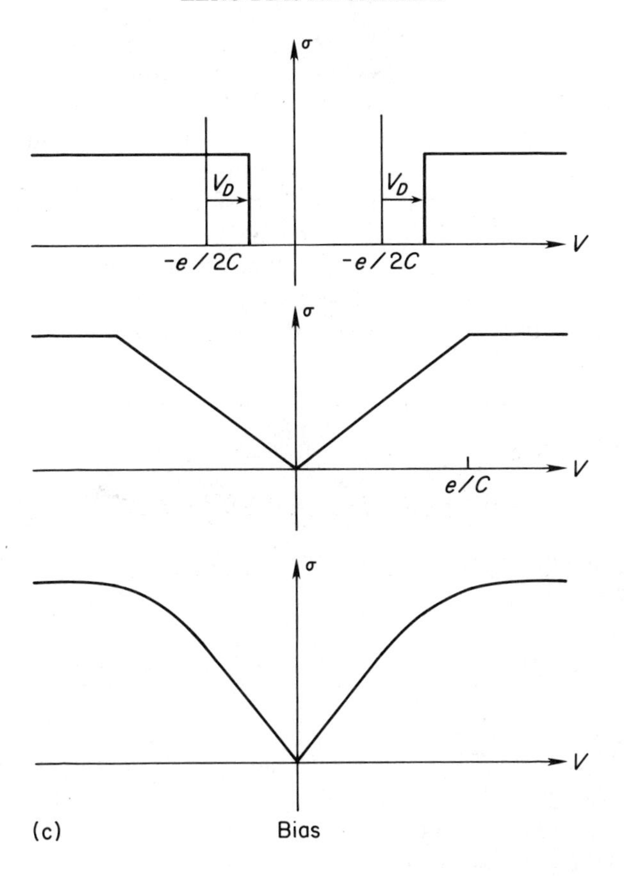

particles. The Sn particles themselves are sufficiently decoupled from the Al films to be regarded as disclike capacitors, of capacitance C, whose charging voltage for a single electron charge, therefore, is $\Delta V = e/C$. The magnitude of $\Delta V = e/C$ was set by Giaever and Zeller (1968*a*), by choice of C, as controlled by the Sn particle radii, to be of the order of millivolts.

The dV/dI spectra obtained from such junctions display a reproducible and controllable temperature-dependent resistance peak anomaly. This result is illustrated in Fig. 8.41b for Sn particles of typical radius $\sqrt{\langle r^2 \rangle} =$ 150 Å, held in the normal state by a 3.5-T magnetic field. The particle size distributions were determined by electron microscopy. Zeller and Giaever (1969) thus experimentally demonstrated and fully interpreted a "giant resistance peak anomaly" controllable by variation of the particle radii and the barrier oxide thicknesses; they went further to suggest that the same sort of mechanism on a more microscopic scale could account for the giant resistance peak reported by Rowell and Shen (1966) and Shen and Rowell (1968). This point has apparently never been settled.

Following Giaever and Zeller (1968*a*), the origin of the effect, regarded as a conductance minimum, comes from a distribution of threshold biases $V_c = (e/2C) + V_D$, as shown in Fig. 8.41c, for transfer of an electron across the barrier via an intermediate state on an Sn particle. Here, V_D is a random variable representing offset of the highest filled electron level of the Sn particle from the Fermi level of the electrode. The relevant energies here are pure classical Coulomb energies, and greatly exceed the energy spacing of electron quantum-states at the particle sizes used. For macroscopic particles, V_D is zero, but because of charge quantization, $|V_D|$ is not zero for small particles: rather, one expects V_D randomly distributed in the interval

$$-\frac{e}{2C} < V_D < \frac{e}{2C} \tag{8.77}$$

Now, given V_D, the energy required for the charge transfer step, i.e., placing an extra electron on a particle, is $\Delta E = \frac{1}{2}C(V_f^2 - V_i^2)$, V_f and V_i being the final and initial particle voltages. Therefore, one has to provide an energy

$$\Delta E = \frac{1}{2}C\left[\left(V_D + \frac{e}{C}\right)^2 - V_D^2\right] = e\left(V_D + \frac{e}{2C}\right) \tag{8.78}$$

The battery voltage V_C needed to cause this charge flow, doing work $\Delta E = eV_C$, is thus $V_C = V_D + (e/2C)$, which is, from the assumption of (8.77), randomly distributed from $0 < V_C < e/C$. This situation is indicated in Fig. 8.41c, center panel. The expression first given by Giaever and Zeller (1968*a*) for the resulting conductance $G(V)$ is

$$G(V) = \text{constant} \int_0^\infty dC\, Cf(C) \int_{-e/2C}^{e/2C} dV_D\, C\left\{\theta\left[V - \left(\frac{e}{C} - 2V_D\right)\right] + \theta\left[V - \left(\frac{e}{C} + 2V_D\right)\right]\right\} \tag{8.79}$$

where θ is the unit step function and $f(C)$ is the measured distribution of particle capacitances, which was determined by Giaever and Zeller from the electron micrographs of the particles, assuming $C \propto r^2$, r being the particle radius. The distribution typically produces a rounded $G(V)$ minimum, as shown in Fig. 8.41c (lower panel). Typical data plotted as conductances are shown in Fig. 8.42a; such curves can generally be nicely matched within the model.

A characteristic feature of this mechanism is that the conductance at $V = 0$ varies linearly with temperature, as shown experimentally in Fig. 8.42b. This feature follows from the model, since the conductivity at $V = 0$ is proportional to the number of particles in an excited state, which should vary as $\exp(-eV_C/kT)$, since charge is transported to or from an electrode as these particles relax or are excited. In the case of interest,

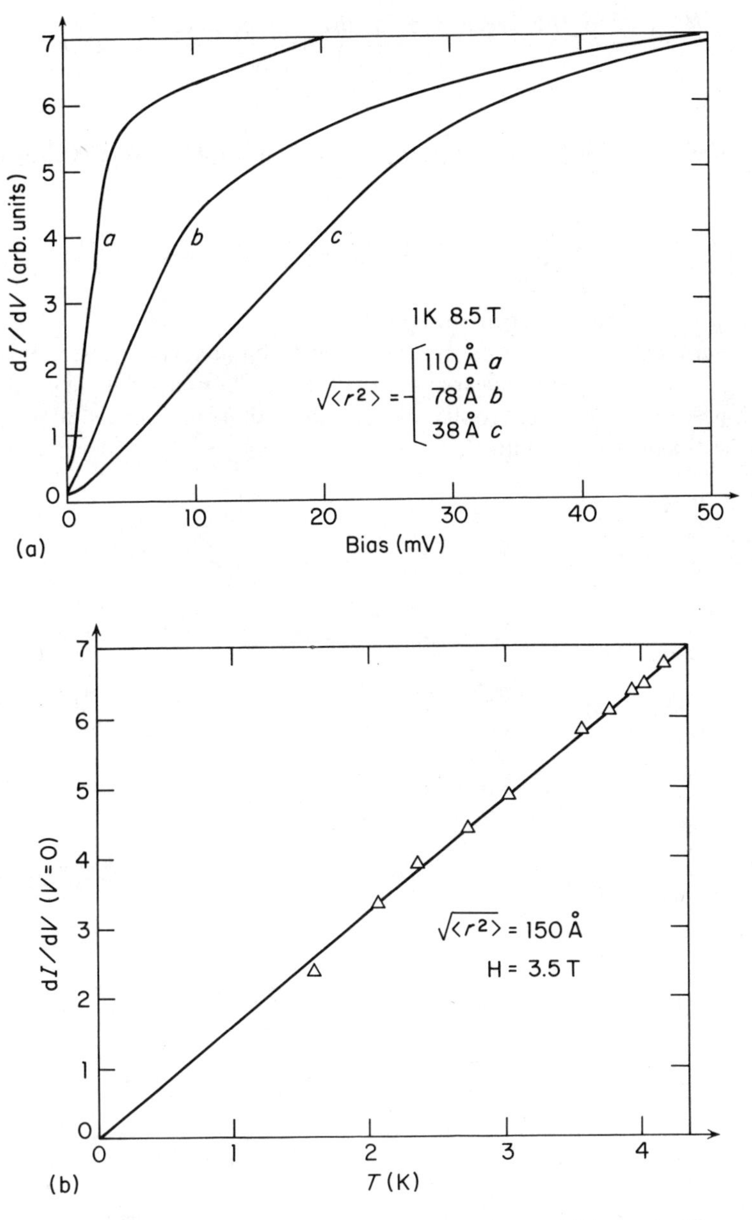

Fig. 8.42. (a) Plot of dI/dV vs. V for particles at 1 K and 8.5 T, of different rms radii 110, 78, and 38 Å, respectively, in cases a, b, and c, showing roughly linear dependence of dI/dV on voltage as predicted by the simple model. (b) Demonstration, with $B = 3.5$ T, of the linear temperature dependence of dI/dV at $V = 0$. (After Zeller and Giaever, 1969.)

$kT \ll e^2/C$. Hence, the expansion for the conductance

$$G(0, T) = \int_0^\infty n(V_C) \exp\left(\frac{-eV_C}{kT}\right) dV_C \tag{8.80}$$

where $n(V_C)$ is the area of all particles with threshold voltage V_C, reduces to

$$G(0, T) \simeq n(0) \int_0^\infty \exp\left(\frac{-eV_C}{kT}\right) dV_C \simeq T \tag{8.81}$$

which agrees with experiment.

In summary, all features of this cleverly engineered resistance peak system can be accounted for at least semiquantitatively. The further exploitation of this system to study superconductivity in small metal particles will be taken up in Chap. 9.

9

Unusual Materials and Effects

We have collected in this chapter several topics of an unusual nature. These are typically areas of considerable current research interest, to which tunneling methods have already been applied, at least in a preliminary fashion, with more important contributions judged to lie ahead. At the end, we also briefly mention a few rather speculative topics which, if confirmed, might also lead to new areas of research.

9.1 Artificial Superlattices

Introduction of a new periodicity into a crystalline structure by creation of repetitive layers ABAB. . . , with perfect epitaxial interfaces, would make possible, in principle, artificial crystalline materials with new electron and phonon band structures. If the imposed superlattice period (width of the AB bilayer) is λ, a new Brillouin zone boundary in the related direction in reciprocal space will appear at $k = \pi/\lambda$. This boundary will occur, e.g., at a submultiple $1/n$ of the original zone boundary wavevector, π/a, by choice of $\lambda = na$, supposing the lattice constant is a in both regions. Consequently, electron and phonon bands should be folded back into the new reduced Brillouin zone with boundary $k = \pi/na$. New interband transition energies can thus be generated.

The partial realization of these effects has come first in repeated layers of the lattice-matched semiconductor system GaAs–$Al_xGa_{1-x}As$, prepared by molecular beam epitaxy (Cho, 1971). Direct observation of new electron transitions is reported by Dingle, Gossard, and Wiegmann (1975), while the phonon folding is reported by Narayanamurti et al. (1979) and by Colvard, Merlin, Klein, and Gossard (1980).

In this semiconductor system, in which the interfaces can be made as narrow as one monolayer, the GaAs conduction band is lower than that in $Al_xGa_{1-x}As$ by about 0.2 eV at $x = 0.2$. Hence, the sytem may be regarded as one of GaAs quantum wells spaced by $Al_xGa_{1-x}As$ barriers. The barrier and well thicknesses are separately controllable.

The first tunneling experiment in the closely related GaAs–AlAs system is believed to be that of Chang, Esaki, and Tsu (1974) (see also Esaki and Chang, 1974), already shown in Fig. 2.23, in which transmission resonances occur in a barrier containing a single well. Metallic superlattices have been reported in Ni–Cu by Gyorgy et al. (1980) and in Nb–Ta by Durbin, Cunningham, Mochel, and Flynn (1981) and by Geerk, Gurvitch, McWhan, and Rowell (1982). In an earlier report by Schuller

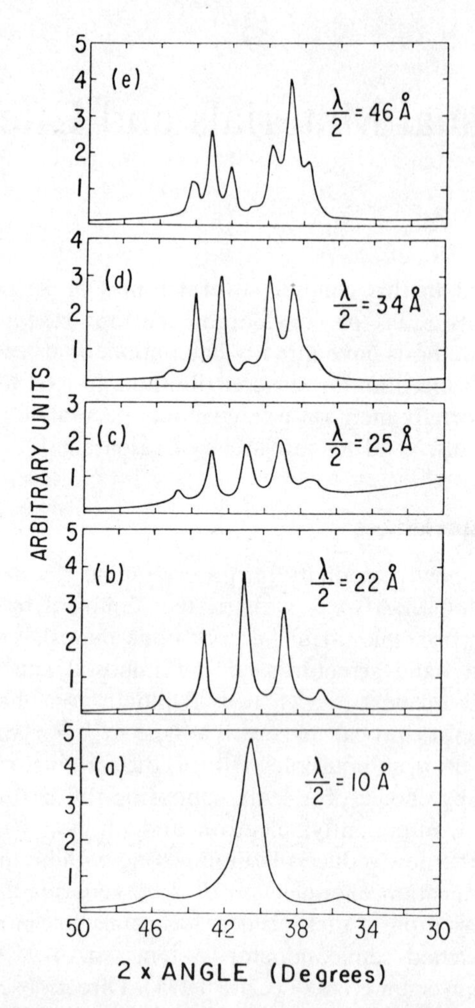

Fig. 9.1. X-ray spectra demonstrate formation of superlattice in sequentially sputtered Nb–Cu structure. While the coherence distance perpendicular to the layers is many times the supercell dimension λ, the coherence distance parallel to the layers has been determined to be rather short in this case. (After Schuller, 1980.)

(1980), order approaching that achieved in the above systems (having common structures and similar lattice parameters) was seen in repeating layers of Nb (*bcc*) and Cu (*fcc*) for which x-ray evidence of structural coherence perpendicular to the layering direction is shown in Fig. 9.1. The Nb (110) and Cu (111) directions line up in this case, termed a layered ultrathin coherent structure (LUCS). The degree of coherence in the transverse directions in the Nb–Cu system is severely limited, however. In the Nb–Ta system, on the other hand, single-crystal superlattices can be grown in three different crystalline directions (Durbin et al., 1982) by proper choice of substrate (several faces of Al_2O_3 and MgO are used).

The tunneling results of Hertel, McWhan, and Rowell (1982) on the NbTa superlattice system are the most extensive published at this time on any metallic superconducting superlattice. The tunneling data reveal a BCS-like gap structure, with $2\Delta/kT_c = 3.7 \pm 0.1$, and phonon structure recognizable as arising primarily from weighted contributions of the individual $\alpha^2F(\omega)$ functions of Nb and Ta. The results can be understood in more detail within a generalization of the superconducting Cooper limit, to be discussed below. The expected phonon structure for perfect superlattices of NbTa, under what appear to be the experimentally relevant assumptions (no band structure or phonon structure folding; each metal retaining its own properties) has been calculated from bulk Nb and Ta $\alpha^2F(\omega)$ functions by Arnold, Gallagher, and Wolf (1982). This study covered roughly the same range of superlattice periods λ as the ~18 to ~300 Å reported in the work of Hertel, McWhan, and Rowell (1982).

While there is no evidence in the present Nb–Ta tunneling spectra for phonon zone folding, or for electronic superlattice effects, these features may eventually be observable as the sharpness of the interfaces is improved. However, true superconducting superlattice effects, such as "zone folding" as a consequence of the spatial modulation of the superconducting pair potential $\Delta(x)$, are more easily observed. The ease of observation of specifically superconducting superlattice effects follows from the typically large values of the superconducting coherence length ξ, $\sim 10^3$ Å. On this scale, $\sim 10^2$ greater than that of the typical Fermi wavelength $2\pi/k_F$, the presently realized interfaces are sufficiently sharp for observation of superlattice effects.

The first consideration of the consequences of a periodically modulated $\Delta(x)$ was given in the work of van Gelder (1969), who extended the familiar Kronig-Penney treatment to superconductors. An approach to the same superlattice problem sufficiently general to treat the strong-coupling superconducting phonon effects of central interest in the work of Hertel, McWhan, and Rowell (1982) is possible through the equivalence (Gallagher, 1982*a*) of the superlattice to a related double-layer problem with fully specular boundary conditions. The strong-coupling tunneling density of states for the equivalent specular bilayer has been given exactly by Gallagher (1980), assuming constant pair potentials in each region, as occurs for layer thicknesses much less than ξ. The NS bilayer, of layer widths d_N and d_S, is expanded by the assumed specular (mirrorlike) bilayer boundary conditions, into an equivalent superlattice of period

$$\lambda = D_N + D_S = 2d_N + 2d_S \tag{9.1}$$

as shown in Fig. 5.10. In this equivalence, the N-vacuum interface is mapped to the center of the superlattice N layer. Hence, the tunneling density of states at the center of the N layer of the superlattice should be given by the expression for N-side tunneling into the bilayer (Gallagher, 1980), which, when expanded for $E \gg \Delta_N, \Delta_S$ is

$$N_T(E) \simeq \mathrm{Re}\left[1 + \frac{1}{2}\frac{\Delta_N^2}{E^2} + \frac{1}{2}\frac{(\Delta_S - \Delta_N)^2}{E^2}\frac{\sin^2(k_S d_S)}{\sin^2(k_N d_N + k_S d_S)} + \frac{\Delta_N(\Delta_S - \Delta_N)}{E^2}\frac{\sin(k_S d_S)}{\sin(k_N d_N + k_S d_S)}\right] \quad (9.2)$$

Here, Δ_N and Δ_S are functions of energy and should be calculated self-consistently, including the proximity effect, as was done by Arnold, Gallagher, and Wolf (1982). This effect tends to raise the magnitude of Δ_N and to lower that of Δ_S. Here,

$$k_{S,N} \simeq \frac{2EZ_{S,N}}{\hbar v_F}, \qquad E \gg \Delta_{S,N} \quad (9.3)$$

and (9.1) is assumed to apply.

The relevant conditions and formulas for the *strong-coupling Cooper limit* follow from this expression, in the limit of small arguments of the "phase shift" sine functions;

$$\frac{\sin(k_S d_S)}{\sin(k_N d_N + k_S d_S)} \simeq \frac{k_S d_S}{k_N d_N + k_S d_S} \equiv f_S, \qquad k_S d_S \ll 1, \qquad k_N d_N \ll 1 \quad (9.4)$$

with a similar definition for f_N. Here, $f_{S,N}$ are identified, using (9.3), as the fractions of time spent in (S, N) in the same sense as in the familiar weak-coupling Cooper limit. In this limit, the $N_T(E)$ function, (9.2), simplifies directly to

$$N_T(E) \simeq \mathrm{Re}\left[1 + \frac{1}{2}\left(\frac{f_S\Delta_S + f_N\Delta_N}{E}\right)^2\right] \quad (9.5)$$

from which one identifies the Cooper limit pair potential

$$\Delta_C(E) = f_S\Delta_S(E) + f_N\Delta_N(E) \quad (9.6)$$

The corresponding strong-coupling formula, in fact, turns out to be

$$N_T(E) = \mathrm{Re}\left\{\frac{|E|}{[E^2 - \Delta_C(E)^2]^{1/2}}\right\} \quad (9.7)$$

The formula, which we have merely guessed from the expansion of the equivalent specular bilayer $N_T(E)$ result, has actually been obtained exactly (Gallagher, 1982*b*) by consideration of the $N_T(E)$ in the gap region ($E \simeq \Delta_S^0, \Delta_N^0$) in weak coupling. The validity of the Cooper limit in the phonon range is, however, more restricted than it is in the gap region, where the condition d_N, $d_S \ll \xi$ is sufficient. In the phonon range, the validity of (9.4) requires $k_N d_N + k_S d_S \ll 1$. Combined with (9.3), and identifying $\xi_0 = \hbar v_F/\pi\Delta$, one obtains the more stringent condition for the superlattice period $\lambda = 2(d_N + d_S)$,

$$\lambda < \xi_0 \cdot \frac{\Delta}{E} \tag{9.8}$$

This condition for the use of the Cooper limit, (9.7), in the phonon range is more restrictive, requiring smaller λ, since $E \simeq \hbar\omega_{ph}$ is typically an order of magnitude larger than Δ.

This conclusion seems at least qualitatively in agreement with the results of Hertel, McWhan, and Rowell (1982) in that a BCS-like gap region consistent with (9.7) was observed, while deviations from the Cooper limit behavior in the phonon range were present. Specifically, the Cooper limit relation for the spectral function is (Arnold, Gallagher, and Wolf, 1982)

$$\alpha^2 F(\omega)_C = g_N \alpha^2 F(\omega)_N + g_S \alpha^2 F(\omega)_S \tag{9.9}$$

where

$$g_{N,S} \equiv \frac{(D/v_F)_{N,S}}{(D/v_F)_S + (D/v_F)_N} \tag{9.10}$$

Numerical calculations for the NbTa superlattice system comparing Cooper limit results with exact results are given by these authors and are found to agree well with the data of Hertel et al. in the case when (9.8) is satisfied. A second factor of importance in the data which is not treated in the calculations is quasiparticle scattering, which has the effect of weakening phonon contributions from deeper layers relative to those from the uppermost layer.

An additional point of numerical significance in use of (9.2) when λ exceeds the value (9.8) for the Cooper limit is that since the bilayer equivalent $N_T(E)$ probes the superlattice N layer at its center, the most appropriate superlattice to be used experimentally would be terminated by a layer (facing the tunnel barrier) of thickness half that of the remaining layers. The correction in the $N_T(E)$ for a superlattice of equal layer thickness, as appropriate to experiments reported, has been investigated numerically through use of (5.39) and (5.40).

9.2 Ultrathin-Film and Small-Particle Superconductors

The properties of very thin film and small-particle superconductors are of interest from both basic and practical points of view. We have already found several of the consequences on physical properties of one or more small dimensions of a metallic specimen. Evidence of splitting of the electronic quantum states by the effect of boundary conditions (size quantization) has been discussed in Secs. 8.2.1 and 8.2.2, while an important consequence of confining electrons to a small particle (Sec. 8.5.2) was found to be the classical Coulomb localization energy $Q^2/2C$, where the charge Q is quantized in units of e and C is the capacitance of the small particle.

The question of alteration of the phonon spectra of small particles on thin films is also of interest. A further consideration in dealing with a very thin tunneling electrode is that nonequilibrium of the electron or phonon system may be more likely, with a smaller volume in which relaxation can occur. This topic has been discussed in Sec. 7.1 in connection with junctions on thin normal films, in which a small zero-bias conductance minimum was associated with nonequilibrium resulting from electron injection. The same idea is part of the general strategy in the double-tunnel junction devices (Secs. 7.2–7.5) used to study nonequilibrium effects in a thin-film superconductor. Finally, scattering, as may occur at surfaces (more frequently encountered in thin films), and electron-electron interactions are both contributing factors in electron localization, discussed in Sec. 8.4.2 in connection with the $E^{1/2}$-dependent zero-bias anomaly and leading, in more extreme cases, to formation of a pseudogap at E_F, as observed by tunneling in p-type Si (Wolf, Losee, Cullen, and

Fig. 9.2. The inverse of dV/dI is plotted against bias for different thicknesses of a Tl film in an Al–I–Tl tunnel junction. The uppermost curve was taken just after the Tl film had become continuous (thickness 11.8 Å). Note the change of scale at 1–1.25 meV. Two extra peaks are evident at 1.05 and 2.10 meV above the peak at the superconducting gap ($2\Delta_{\text{thin}} = 0.95$ meV). For thick films, these features are completely absent, while the usual phonon-induced structure is seen at 4–5 meV. The derivative curves are arbitrarily normalized to unity at 4.5 meV and measured at 0.3 K. (After Granqvist and Claeson, 1973.)

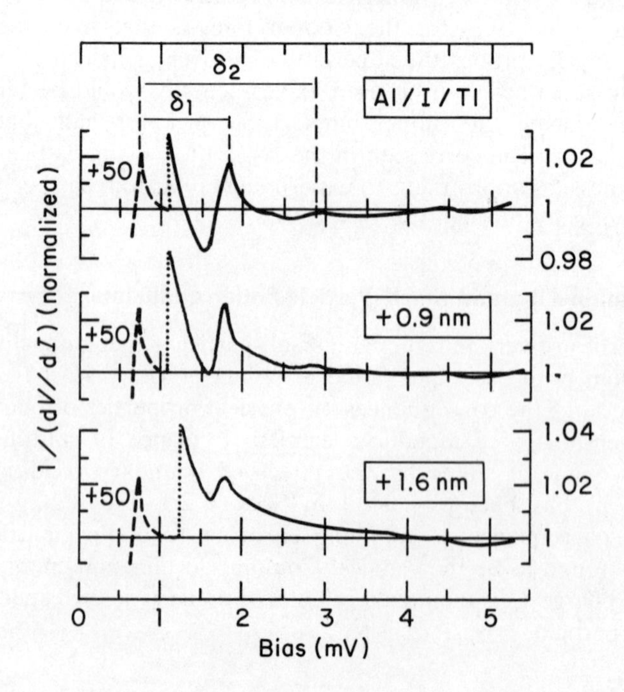

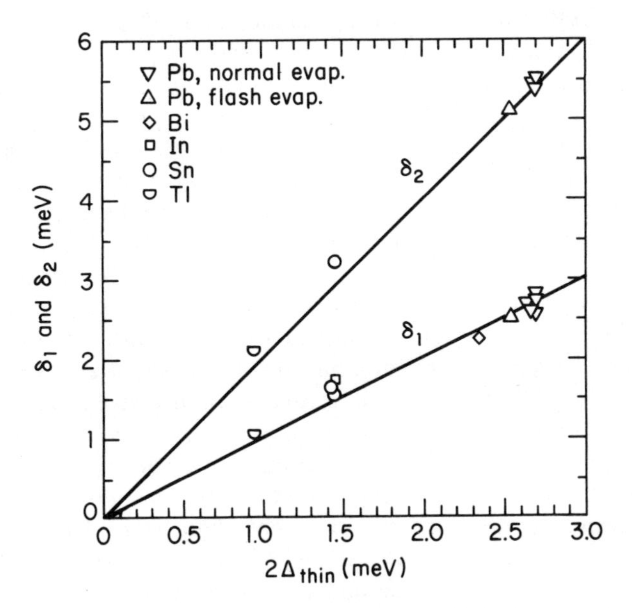

Fig. 9.3. The locations of the extra peak features of Fig. 9.2 are plotted against the gap energies of the investigated thin films. Terms δ_1 and δ_2, as defined in Fig. 9.2, are seen to be energies which scale with the gap energy. (After Granqvist and Claeson, 1973.)

Compton, 1971) and, more recently, in Ge_xAu_{1-x} alloys by McMillan and Mochel (1981).

Very interesting results from the study of ultrathin superconducting films by Granqvist and Claeson (1973) are shown in Figs. 9.2 and 9.3. The new features are extra peaks in the superconducting conductance at energies $\delta_1 = 2\Delta$ and $\delta_2 = 4\Delta$ measured from the gap edge, universally present in ultrathin, amorphous, quench-condensed superconducting films, including Pb, Bi, In, Sn, and Tl. These features are stronger (up to $\Delta G/G \simeq 0.1$) than superconducting phonon self-energy effects in the thinnest films (electrical continuity of the film is taken as a criterion to terminate evaporation), which vary in thickness from ~60 to ~120 Å. The effects were most clearly seen in samples made by first coating a single-crystal quartz substrate at room temperature with Ge to promote smoothness (Strongin, Thompson, Kammerer, and Crow, 1970), followed by room temperature deposition and oxidation of the Al. As shown in Fig. 9.2 for Tl films, additional film thickness increments (9 and 16 Å) decrease the size of these peaks, which are measured *in situ* at 0.3 K. Annealing the films to room temperature destroys the peaks.

The data of Fig. 9.3 demonstrate that these peak features correspond reliably in all cases to excess energies 2Δ and 4Δ measured from the gap edge of the ultrathin film. Granqvist and Claeson (1973, 1974*a*, 1974*b*) have suggested that these peak features may be related to electron decay by stimulated emission of recombination phonons, regarded as a "new channel for electron tunneling (which) opens up at $eV \simeq 3\Delta_{\text{thin}} + \Delta_{\text{Al}}$." Alternatively, the authors suggest that the extra peaks may be due to decay of the nonequilibrium quasiparticles by excitation of a Cooper pair.

The reason for loss of the effect after annealing is not known. Possibly, the annealing and grain growth reduce the effect by enlarging the volume into which the electrons tunnel, hence reducing the degree of nonequilibrium.

The question of the minimum *film* thickness at which superconductivity can occur is usually rather ill defined experimentally because of the tendency of thin films to be nonuniform in thickness and to occur as islands at low thicknesses. (We will return to an interesting counterexample to this generalization.) For metallic *particles*, on the other hand, the Zeller-Giaever (1969) experiment, discussed in Sec. 8.5.2, seems to provide well-defined experimental conditions. In this work, oxidized Sn particles, whose distributions of sizes were characterized by analysis of electron micrographs, were embedded in a tunnel barrier between Al electrode films. The experimental techniques allowed formation and

Fig. 9.4. The dV/dI characteristics for Sn particles in the normal and superconducting states at 1.6 K, measured in junction structures of the type shown in Fig. 8.41a. The increase in dV/dI at $H = 0$ occurs because the energy required for charge transfer, i.e., to localize an electron on the particle, is increased by Δ, half the superconducting energy gap. (After Zeller and Giaever, 1969.)

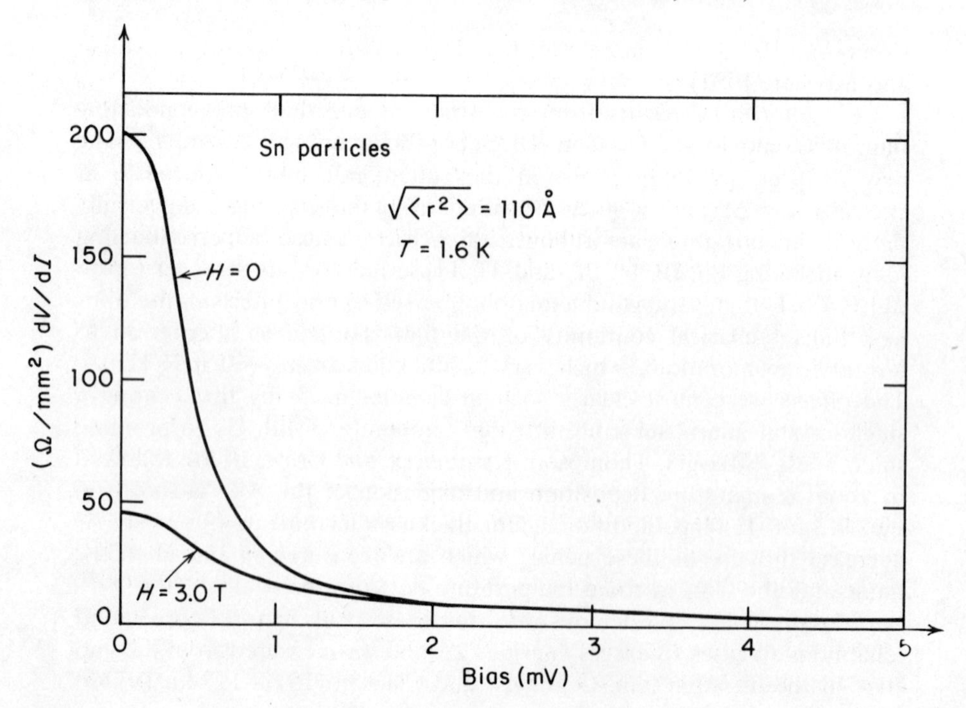

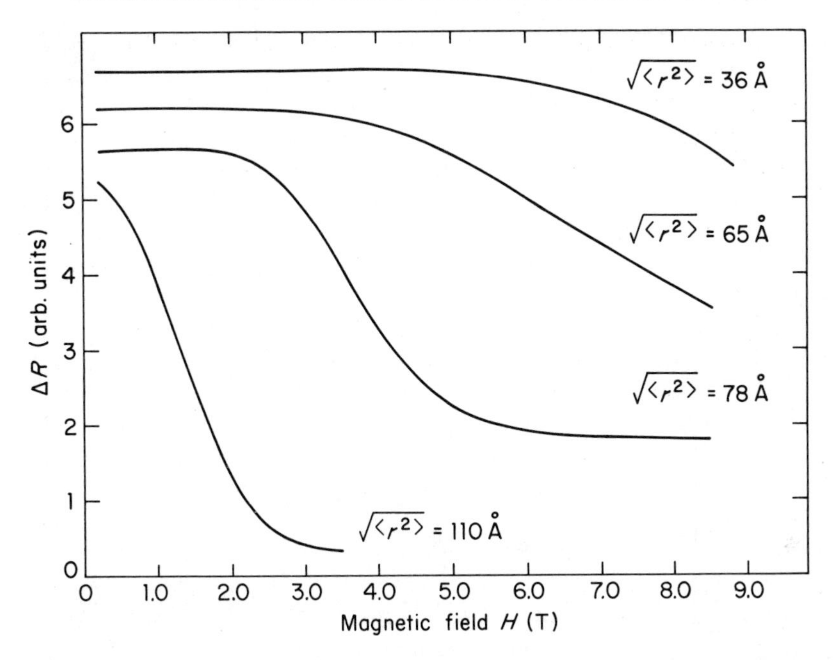

Fig. 9.5. Magnetic field dependence of the junction resistance measured at zero bias and 1 K. It is seen that, with very small particles, an extremely high magnetic field is required to reduce the junction resistance. Critical magnetic fields in excess of 9 T are implied. (After Zeller and Giaever, 1969.)

characterization of extremely small particles: Zeller and Giaever (1969) include data from a tunnel junction in which the rms Sn particle size was determined to be 18 Å, with an estimated maximum radius of about 25 Å. The next question is whether the experiment allows the superconductivity of these particles to be established. As shown in Fig. 9.4, the answer to this question is definitely yes; for the application of a magnetic field strongly reduces the magnitude of the characteristic $\mathrm{d}V/\mathrm{d}I$ peak structure. The occurrence of superconductivity in the Sn particle adds Δ_{Sn} to the critical bias V_c required to transfer one electron to the particle. In the limit of large particle size, Zeller and Giaever indeed found that the I–V characteristic of their junctions resembled that of two Al–I–Sn tunnel juntions in series. Thus, the large reduction of the $\mathrm{d}V/\mathrm{d}I$ peak in a 3T applied magnetic field (Fig. 9.4) is easily understood as the result of driving the Sn particles into the normal state. In any such junctions in which the characteristic reduction of the resistance peak with field can be observed, superconductivity must occur in the particle set, which is destroyed with increasing field. Note in Fig. 9.5 that such an effect begins to occur in particles of rms size 30 Å at about 8 T. Zeller and Giaever (1969) defined the effective critical magnetic field H_c^* to be that at which half the total field-induced change has occurred. Such a change was

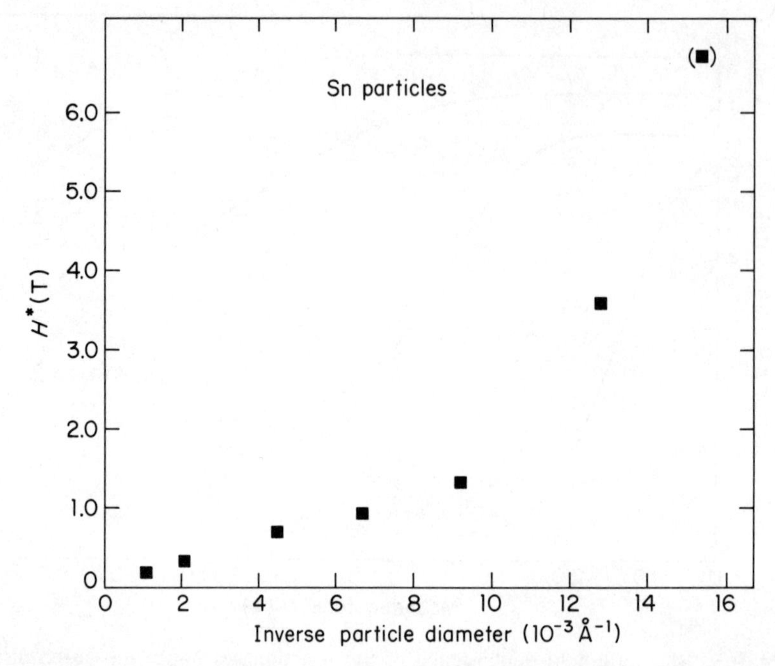

Fig. 9.6. The critical field H^*, defined as the midpoint of the decrease in resistance with increasing field (as shown in Fig. 9.5) is plotted against the inverse average particle radius. The conclusion drawn from this plot is that particles of rms radius of the order of 30 Å are still superconducting and have a critical field H^* of more than 10 T. (After Zeller and Giaever, 1969.)

observed even in the 18-Å rms Sn-particle junction, containing some particles of maximum radius 25 Å, from which superconductivity in 50-Å-diameter Sn particles (embedded in a dielectric oxide matrix) was inferred. The behavior of H_c^* vs. inverse average particle radius $[\langle r^2 \rangle]^{-1/2}$ is shown in Fig. 9.6.

Zeller and Giaever point out that a critical radius near 25 Å, as observed, for superconductivity in Sn is consistent with a criterion advanced by Anderson (1959) that superconductivity should disappear when the spacing of the electronic energy levels in the particle becomes comparable to the superconducting energy gap.

Further tunneling experiments on superconducting films even thinner than those mentioned earlier have been performed by Sixl, Gromer, and Wolf (1974). These results are summarized in Figs. 9.7–9.10. The remarkable success of Sixl et al. in carrying tunneling measurements down to film thicknesses <20 Å (see Fig. 9.7) of quench-condensed Al and In films (and to nearly the same thickness in Pb films) is attributed to an unusual composite barrier. This barrier consisted of a monolayer or so of SiO or naphthalene quench-condensed at 4.2 K upon a prepared partial

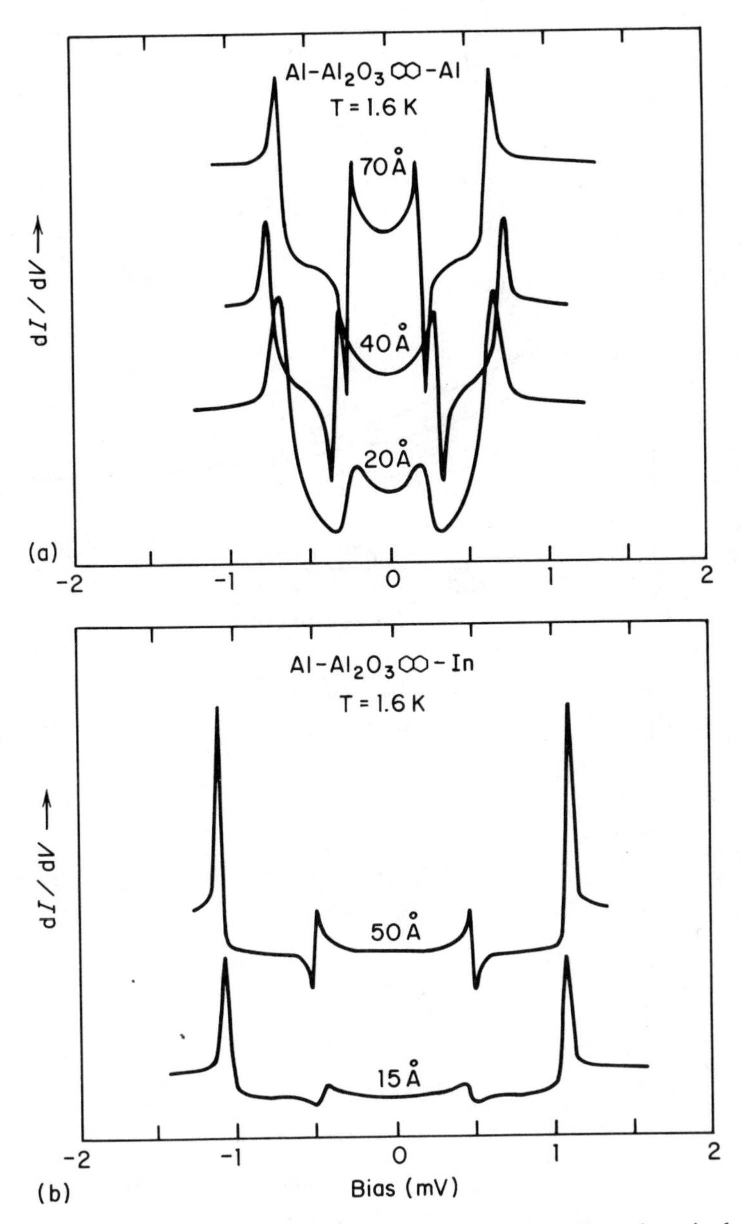

Fig. 9.7. (a) Gap spectra dI/dV of ultrathin Al and In films deposited upon prepared Al–Al_2O_3-naphthalene barrier insulators. The initial aluminum film is deposited at 300 K upon glass and oxidized only to approximately 15 Å. It is then coated with a monolayer of naphthalene to promote uniform coverage of the subsequently deposited ultrathin aluminum or indium film. Note that the gaps of the two aluminum-film electrodes differ substantially and that the gap of the ultrathin aluminum film achieves a maximum near 40 Å thickness. (b) Similar results for indium final films at thicknesses of 15 and 50 Å. (After Sixl, Gromer, and Wolf, 1974.)

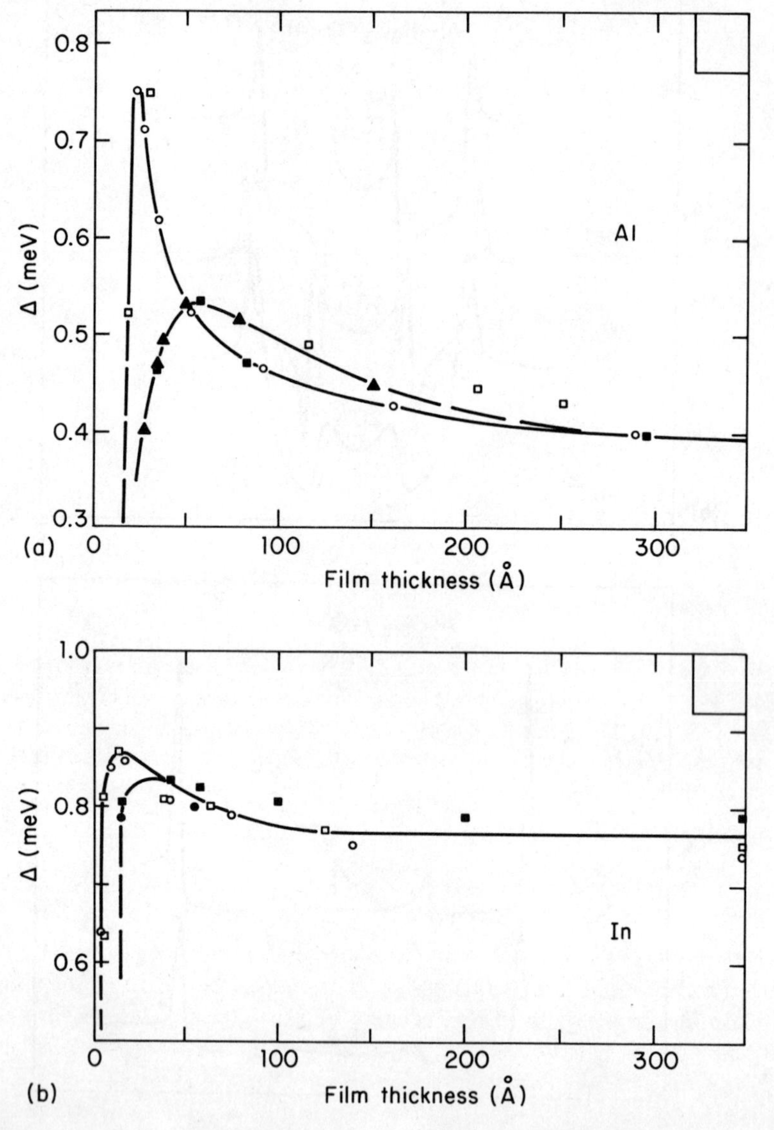

Fig. 9.8. (a) Plot of the energy gap of the ultrathin aluminum films vs. their thickness. The curve with the open symbols is obtained with SiO as the supplemental insulator; the closed symbols represent measurements with the naphthalene as the supplemental insulator. In measurements not shown, the T_c of ultrathin aluminum films deposited with the SiO supplemental insulator peaks at slightly more than 4.5 K. (b) Similar data for ultrathin indium films deposited in the same fashion upon SiO- or naphthalene-covered Al–Al_2O_3 barriers. (After Sixl, Gromer, and Wolf, 1974.)

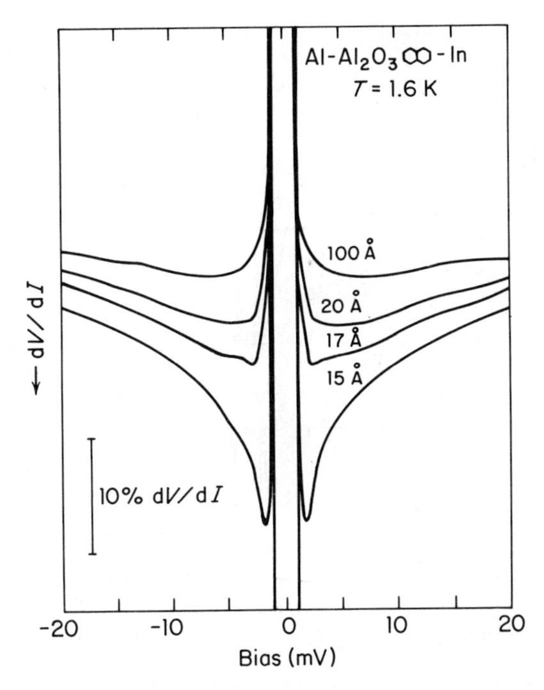

Fig. 9.9. Measurements of dV/dI (plotted downward) at higher sensitivity and with a wider bias range for ultrathin In films upon the Al–Al_2O_3-naphthalene barrier structure. The feature of interest is the broad increase in resistance dV/dI in the lower bias range for the thinnest film. Note that the thicker In films properly show a typical phonon structure at about 14 mV. (After Gromer and Sixl, 1974.)

barrier: Al_2O_3 of about 10-Å thickness, grown previously on the first Al electrode at room temperature. A particularly sensitive transmission optical-monitoring method was used to measure the ~20-Å thickness of the quench-condensed SiO or naphthalene, as shown in the original paper. Tunneling measurements were then carried out *in situ* at 1.6 K.

The tunneling derivative dI/dV spectra in Fig. 9.7 are quite impressive and convincing as to the correctness of basic interpretation. These data also lead to systematic dependences of T_c and inferred gap parameter Δ with film thickness. Note that the gap parameter in the Al and In films (Fig. 9.8) goes through a maximum with decreasing thickness, matching similar behavior observed in T_c (not shown), and that the detailed variation in the thinnest films depends on the choice SiO vs. naphthalene. The suggestion of "pair breaking" broadening of gap structure at the smallest thickness is present, especially in the 20-Å curve of Fig. 9.7a. The variations observed for Pb films differ qualitatively; here, monotonic decreases occur in both Δ and T_c with decreasing film thickness.

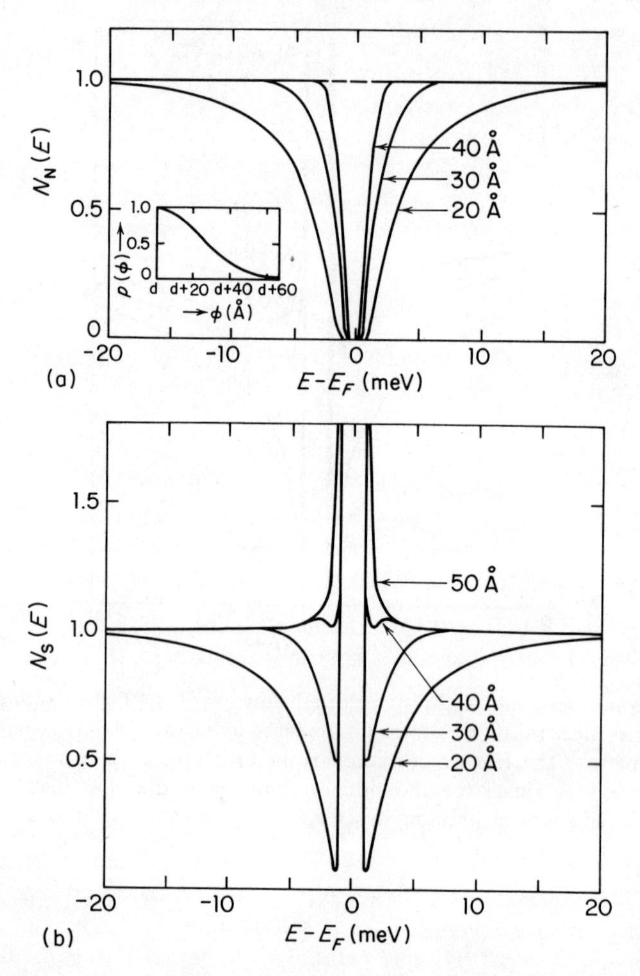

Fig. 9.10. Calculation of the normal and superconducting tunneling density of states of the ultrathin films on the assumption of size quantization of electronic states in small disc-shaped particles comprising these films. (a) The calculation of a normal density of states near the Fermi energy for films of thicknesses indicated, based on the assumption that the particle sizes have the distribution shown in the inset. (b) Calculated tunneling density of states in the superconducting film incorporating the results of (a). (After Sixl, Gromer, and Wolf, 1974.)

Measurements over a wider bias voltage range in dV/dI at higher sensitivity in Fig. 9.9 show a conductance minimum zero-bias anomaly which appears at ~20-Å metal-film thickness and thereafter increases rapidly with decreasing thickness. Such behavior was found, with minor variation in details, in all cases: In, Al, and Pb films, each with either SiO or naphthalene as the partial barrier. It is assumed that this effect arises in the normal-state conductance (of course, the normal-state conductance is difficult to measure because of the high sheet resistance of such ultrathin films), the measured quantity being $G(V)^{-1} = [G_N(V) \cdot N_T(eV)]^{-1}$. This effect is stated as being temperature-independent and not related to the Giaever-Zeller model. The authors, in their work published prior to the work of Al'tshuler and Aronov (1979*a*), have considered models for this behavior based on size quantization of the electronic states across the film, supposed to be composed of a distribution of disc-shaped particles. The results of this model calculation are shown in Fig. 9.10. The inset in Fig. 9.10a shows the assumed distribution function for the disc diameter; all discs are assumed to have the measured film thickness. In spite of the apparent success of this approach, one doubts its uniqueness and might, alternatively, consider the observed results as a consequence of a possibly two-dimensional localization transition of the sort discussed by Al'tshuler, Aronov, and Lee (1980) and McMillan (1981), resulting from scattering and electron-electron interactions, and possibly influenced in this case by the increase in surface scattering. In view of the current interest in localization phenomena, reconsideration of these unique tunneling results in more adequate theoretical terms might well be worthwhile.

9.3 Lower-Dimensional and Organic Conductors

Two-dimensional or layered metals and semimetals, especially those which are superconductors, have received considerable attention because of interest in the effects of their anisotropy upon the superconducting properties, such as the critical magnetic fields.

The tunneling spectra of the layer compound $NbSe_2$ have been reported by Morris and Coleman (1973). Evaporated carbon barriers were used to form junctions of $NbSe_2$ with In and Pb counterelectrodes. The $NbSe_2$ gap, measured as $2\Delta = 1.24 \pm 0.04$ meV at 1.1 K and $H = 0$, is observable up to 6.5 T, with H parallel to the layer. This gap, obtained by tunneling perpendicular to the layers, is smaller by a factor 1.7 than a value of 2.15 meV obtained optically for motion parallel to the layers. The latter value corresponds to $2\Delta/k_B T_c = 2.7 \pm 0.1$, where the measured T_c is 7 K. No measurement of the superconducting phonon structure of $NbSe_2$ has yet been reported.

Tunneling experiments involving the layered semimetal $Bi_8Te_7S_5$, which can be cleaved into thin, atomically smooth layers, have been reported by Lykken and Soonpaa (1973). This material, which resembles Bi_2Te_3, has fivefold layers of thickness 9.82 Å, themselves bound together only weakly, leading to the property of easy cleavage. Proximity sandwiches of the form $Pb/I/Bi_8Te_7S_5/I/Pb$ were studied by Lykken and Soonpaa (1973) in which the barriers are thought to arise from water vapor adsorbed on the semimetal surface. The thickness of the single-crystal semimetal layers was varied from 30 to 300 Å.

In spite of the great activity in the field of organic metals and superconductors and other nonconventional conductors, only fragmentary reports of tunneling measurements in these materials have appeared. In one of the earliest reports (Chaikin, Hansma, and Greene, 1978), crystals of the polymeric metal $(SN)_x$ were incorporated in several different types of tunnel junction and studied down to 30 mK, well below the superconducting transition temperature of $(SN)_x$, ~0.27 K. No evidence of the superconducting gap of $(SN)_x$ was reported. However, a broad resistance peak zero-bias anomaly was observed, consistent within the Zeller–Giaever picture discussed in Chap. 8, with charging energies on the order of 1.5 to 10 meV. The suggested presence of small particles of $(SN)_x$ at its surface is regarded as not implausible in view of what is known about the morphology of the $(SN)_x$ films.

Tunneling measurements on single crystals of the organic, one-dimensional conductor TTF–TCNQ have been reported by Leo (1981), who used a point-contact tunneling method (Thielemans, Leo, Deltour, and Mehbod 1979) and carried out measurements over the temperature range 30–250 K. A wide (~0.1-V) and roughly parabolic conductance minimum feature was observed. This feature shifted toward zero bias in the temperature range 230 to 48 K and became noticeably wider at 36 K. A sharp drop in the conductance $G(0)$ was observed in a narrow range at 53 K. This sharp change was correlated in temperature with the (Peierls) metal-insulator transition in the material, believed to produce a gap at the Fermi level below 53 K. These results are thus consistent with what is known about TTF–TCNQ but appear to provide little new information. One difficulty is in separating the effects in $G(V)$ due to the expected pseudogap structure from those arising simply from the barrier voltage dependence. The pressure contact barrier is apparently effective but a bit hard to characterize.

More detailed information has been obtained by tunneling in the case of superconducting organic salts which are members of the recently generated $(TMTSF)_2X$ family (Parkin, Ribault, Jérome, and Bechgaard, 1981; Bechgaard, et al., 1981). In this family, the salt with $X = PF_6$ becomes superconducting ($T_c = 1$ K) at 11 kbar pressure, while the ClO_4 salt has $T_c = 1.2$ K at atmospheric pressure. Schottky barrier tunnel contacts to $(TMTSF)_2PF_6$ have been studied by More et al. (1981) (see also Blanc, More, Roger, and Sorbier, 1981). Some of the results of More et al. are shown in Fig. 9.11.

These junctions have been prepared by evaporating (at room temperature) GaSb:Te ($N_D > 3 \times 10^{18}$ cm^{-3}) to ~500-Å thickness onto a natural face, parallel to the chain axis, of a $(TMTSF)_2PF_6$ single crystal. Ohmic contacts to the semiconductor (a dot ~0.3 mm in diameter) are subsequently made by evaporating an Sn overlay 500 Å in thickness. Finally, ohmic contacts to this Sn layer, and to the organic conductor, are made with drops of silver paint. The dV/dI peak structure shown, with qualitatively BCS-like dV/dI minima occurring (an expanded x-axis trace is shown dotted) at ~±1.8 mV, was measured at 0.05 K, 11 kbar, and a modulation voltage ≤0.1 mV. The inferred gap, $\Delta \simeq 1.8$ meV, is much

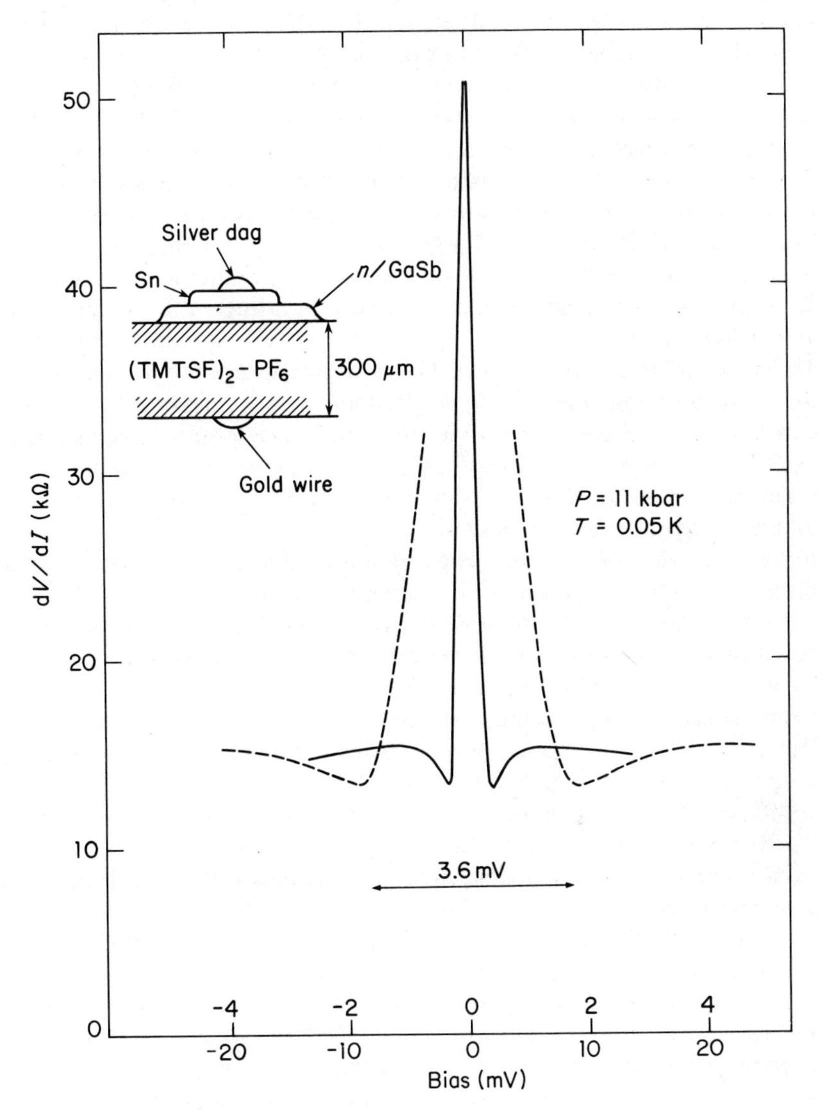

Fig. 9.11. The dV/dI spectrum for a $(TMTSF)_2PF_6$–GaSb tunnel contact at 0.05 K and 11 kbar gives an indication of a possible superconducting gap in the organic superconductor. The junction electrode structure is shown in the inset. (After More et al., 1981.)

larger than normally expected for $T_c \simeq 1$ K. The most puzzling feature of this dV/dI peak structure, however, is that it is scarcely affected at $T = T_c \simeq 1$ K and persists in recognizable if diminished form up to ~15 K. While the most obvious and conservative conclusion might be that this structure has nothing whatever to do with the superconducting state (admittedly, leaving a problem in the interpretation of the BCS-like peak features), the suggestion of More et al. is that the peak arises not from the full superconducting state, $T_c = 1$ K, but from fluctuating 1–d superconducting regions, persisting even to ~40 K, more in line with the observed $2\Delta \simeq 3.6$ meV. This suggestion and related experimental results (Ng et al., 1982) have been discussed at greater length in subsequent reports by Jérome (1982*a*, 1982*b*), adhering to a description of the results of Fig. 9.11 as a pseudogap related to superconducting precursor effects. To the extent that a 1–d system is required for such an interpretation, as implied by Schultz et al. (1981), some doubt is raised by the recent observation of (manifestly 3–d) de Haas-van Alphen oscillations in this material (Kwak and Schirber, 1983; see also Kwak, 1982). There is a great deal of promise for further tunneling experiments, including variations in experimental approach, to study these exciting new materials.

Another material of great scientific as well as practical interest is polyacetylene $(CH)_x$, which can be made highly conductive by exposure to I_2 or AsF_5 vapors (Chiang et al., 1977; Mele and Rice, 1981). Recent point-contact barrier tunneling measurements by Leo, Gusman, and Deltour (1982) are potentially of particular importance in probing the proposed soliton mechanism of conduction (Su, Schrieffer, and Heeger, 1979). Briefly, it is believed that electrons occupying π orbitals in the $(CH)_x$ carbon chain would lead to metallic behavior with a half-filled band if the C–C distances were all equal. However, the one-dimensional system is known to undergo a Peierls distortion (dimerization), the result of which is described as an alternation of single and double C–C bonds along the chain. This distortion is accompanied, in trans-$(CH)_x$, by an energy gap of order 1.4 eV. For a given chain, two degenerate (extended) ground states, A and B, must exist, because, from the translational symmetry, one could interchange all double and single bonds (or slide the bonds along the chain by one C–C spacing) with no cost in energy. The junction or one-dimensional wall between such A and B regions is described as a kink or soliton. This soliton is believed to provide a bound electronic state whose energy lies at the center of the Peierls gap, and which is neutral when occupied by one electron. When the electron is removed by acceptor doping (with an anion like AsF_5), the soliton becomes positively charged; conversely, donor doping leads to a negatively charged soliton. The point in this conceptual framework to which the tunneling experiment is relevant is that relatively heavy (~1 to 10%) AsF_5 doping leads, according to Mele and Rice (1981), to increasing and finally metallic conduction associated with a midgap band of states. These features, for which some rather direct experimental evidence has already

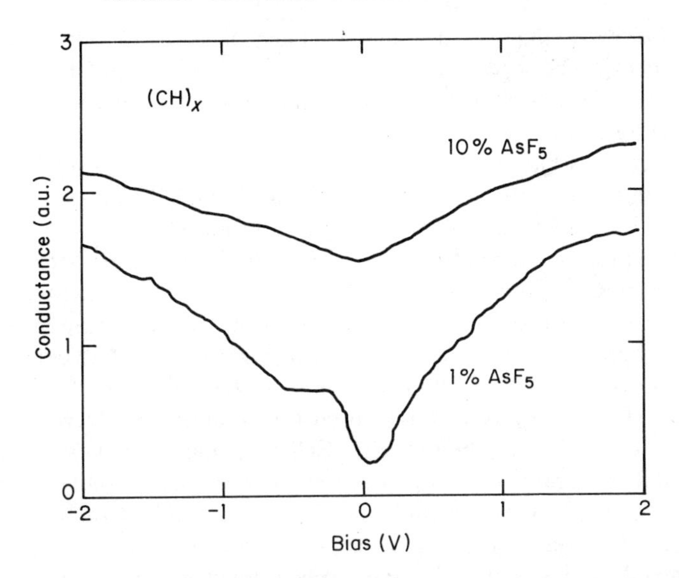

Fig. 9.12. Conductance measured between oxidized aluminum electrode and doped polyacetylene. Negative bias corresponds to higher-energy electrons injected into the polyacetylene. Note the large energy scale, and note that doping with AsF_5 is known to increase the conductivity of the polymer. (After Leo, Gusman, and Deltour, 1982.)

been given [see Heeger and MacDiarmid (1981) for a summary], would be directly observable by tunneling spectroscopy. The results of such measurements are shown in Fig. 9.12. The occurrence of a resolved second minimum in the lower $G(V)$ curve (1% AsF_5 dopant) at about -0.6 V bias (energies above E_F in the CH_x) is interpreted as delineating the postulated midgap band. Similar, but not quite as pronounced, effects are shown by the same authors in I_2 doped $(CH)_x$. Again, no attempt has been to made to correct the $G(V)$ spectra for voltage dependence of the barrier transmission associated with the point contact composed of an oxidized Al barrier-counterelectrode configuration. It seems likely that more information will become available from further tunneling measurements, perhaps under simultaneous optical irradiation.

9.4 Mostly Ternary Superconductors

One of the most interesting ternary superconductors is $ErRh_4B_4$ (Fertig et al., 1977), which exhibits superconductivity in the temperature range $0.92\,K < T < 8.5\,K$; apparently exhibits both ferromagnetism or a spiral magnetic state *and* superconductivity for $0.92\,K < T < 0.98\,K$, and becomes a normal-state ferromagnet below 0.92 K. A variety of tunneling techniques have been applied to study the gap structure of this compound. The $ErRh_4B_4$–Al–I–Pb proximity junction I–V characteristics reported by Rowell, Dynes, and Schmidt (1980) showed a gradual reduction in sum gap, consistent with loss of the $ErRh_4B_4$ gap, in the vicinity of

1.05 K; below 0.9 K, the results are typical of tunneling from Pb into a normal metal. The inferred gap for the sputtered film of $ErRh_4B_4$ was ~0.7 meV, which was surprisingly small, only ~0.5 of the BCS value. Improved values of the energy gap have been reported by Poppe (1981), who quotes $2\Delta/kT_c = 3.8$ using a point-contact method on single crystals; and by Umbach, Toth, Dahlberg, and Goldman (1981), who conclude that $2\Delta/kT_c$ achieves a maximum value of at least 4.2, clearly indicating strong-coupling superconductivity. The latter study was performed on sputtered $ErRh_4B_4$ films overcoated with Er (10–25 Å thick) which were subsequently oxidized to form the tunneling barrier. The Josephson effect in tunnel junctions comprising one $ErRh_4B_4$ electrode has been studied by Kuwasawa, Rinderer, and Matthias (1979) and by Umbach and Goldman (1982). The dc Josephson current is found to decrease sharply to zero at 1.1 K, i.e., before the $ErRh_4B_4$ gap disappears, which is interpreted in terms of rapid increases of the susceptibility χ and the penetration depth λ at the high-temperature boundary of the magnetic coexistence region. When the Josephson current, in the temperature range near its maximum, ~1.4 K, was quenched by application of a magnetic field, the zero-bias resistances were found to be only two to five times the normal resistance. This rather low ratio, smaller than would be expected for tunneling into a BCS superconductor with $T_c \simeq 8.5$ K, is tentatively attributed to partial gaplessness resulting from pair breaking by Er^{3+} magnetic ions in the barrier region. Further complications regarding the surface, including a possibility of magnetic domains, are discussed in the original paper. So far as is known, no phonon structure in tunneling into $ErRh_4B_4$ has been observed.

The Chevrel phase superconductors have also been studied by tunneling in only a very preliminary fashion, revealing, however, that both $PbMo_6S_3$ and $Cu_{1.8}Mo_6S_8$ have very strong coupling, in the range $4 \leq 2\Delta/kT_c \leq 5$ (Poppe and Wühl, 1981). Ingenious, bulk tunneling sample preparation methods and a variety of artificial barriers, as well as an Al-probe point-contact method, are described by Poppe and Wühl (1981). The d^2I/dV^2 spectra reveal phononlike structures in the same range as phonon energies inferred from neutron scattering; however, in neither case could the derivative spectra be quantitatively reduced to give $\alpha^2F(\omega)$. Further comments on this work have been offered by Pobell, Rainer, and Wühl (1982).

A report of successful tunneling spectroscopy into coevaporated Bi_3Ni films using Al counterelectrodes has been given by Dumoulin, Nédellec, Burger, and Creppy (1982). This study was motivated by interest in the possibility of local-mode behavior suggested by the large mass ratio $M_{Bi}/M_{Ni} \simeq 4$, making the system analogous, in this aspect, to the earlier studied PdH system (Arzoumanian, Dumoulin, Nédellec, and Burger, 1979). Indeed, contributions from the Bi vibrations in this orthorhombic system are seen to be clustered in the range 1–11 meV. This portion of the spectrum has been deconvoluted, leading to a partial $\alpha^2F(\omega)$ corresponding to $\lambda = 1.15$ and $\mu^* = 0.12$. Weak contributions at 17 and 23 meV are tentatively attributed to Ni vibrations.

Another interesting compound containing Bi is the superconducting oxide $BaPb_{1-x}Bi_xO_3$ (Cox and Sleight, 1976), which exhibits $T_c = 11.5$ K in the range $0.2 < x < 0.3$. Tunneling studies on single crystals of this material, at $x = 0.25$, have been reported (Batlogg et al., 1982). The observed gap $\Delta = 1.65$ meV corresponds to $2\Delta/kT_c = 3.5 \pm 0.1$. Again, very low energy phonon structures are observed, particularly in the range $\hbar\omega_{ph} = 2$ to 4 meV. The presence of the low-energy phonons is expected to explain the unusual negative thermal expansion (observed below 50 K) and the non-Debye behavior of the specific heat (Methfessel, Stewart, Matthias, and Patel, 1980), as well as the reported rise of resistivity with decreasing temperature, and other evidence of strong electron-phonon coupling. The earlier suggestions of a real space-pairing mechanism for this system (Rice and Sneddon, 1981) are now seen as less likely in view of an extensive band structure study by Mattheiss and Hamann (1982). This calculation places the Fermi energy of the system in the center of low-mass (and low-density-of-states) bands derived from Pb and Bi 6s and O 2p states. This result apparently explains the extremely small, but finite, specific heat anomaly (originally overlooked) which has now been seen in $BaPb_{1-x}Bi_xO_3$. While the new results make the material less likely to derive its superconductivity from an exotic mechanism, the origin of sufficiently strong electron-phonon coupling to provide a BCS superconductor at $T_c = 11.5$ K with so low a value of $N(0)$ remains an actively pursued question.

Other unusual superconductors being studied experimentally, as well as generating theoretical interest, include $SrTiO_3$ and $CeCu_2Si_2$. The former, which, on the basis of the tunneling measurements of Binnig, Baratoff, Hoenig, and Bednorz (1980), is believed to be a two-band superconductor, exhibits superconductivity down to extremely low carrier concentrations $N_D \simeq 10^{19}$–10^{20} cm^{-3}, produced by Nb doping. The Fermi degeneracy μ_F of the electronic system is variable through the choice of doping density N_D. Over most of the range, μ_F is less than the energy $\hbar\omega_{L_4} =$ 100 meV of the highest longitudinal optical phonon, which is believed (Baratoff et al., 1982) to provide strong electron-phonon coupling.

The highest $T_c \simeq 0.7$ K in Nb-doped $SrTiO_3$ occurs at $N_D = 1.5 \times 10^{20}$ cm^{-3}; T_c then falls rapidly with increasing N_D as μ_F exceeds $\hbar\omega_{L_4}$. It is known from infrared reflectivity measurements that the plasma frequency ω_p exceeds ω_{L_4} in the N_D range in which T_c drops, but below this N_D range, the coupling to the L_4 modes is believed to be essentially unscreened. According to Baratoff et al., (1982) this observation suggests the importance of virtual plasmon exchange in the pairing interaction, as developed in the theory of Takada (1980).

The exchange of acoustic plasmons in the superconductivity of Nb_3Sn has been suggested by Tütto and Ruvalds (1979); however, the experimental basis for this suggestion appears to have been largely removed through the reanalysis of Nb_3Sn tunneling data by Wolf, Zasadzinski, Arnold, Moore, Rowell, and Beasley (1980). See, however, Yu and Anderson (1984).

The superconducting pairs in $CeCu_2Si_2$ appear to be composed of heavy fermions (Rauschschwalbe et al., 1982; Steglich et al., 1979) with $m^* \simeq 10^3 m_e$, on the basis of extremely high values of the specific heat coefficient γ and the critical field slope $(dB_{c2}/dT)_{T_c}$. This material has superconducting $T_c \simeq 0.5$ K but, above $T \simeq 10$ K, shows localized magnetic moments associated with the Ce. Below 10 K, the magnetic moments disappear, leaving a "heavy mass Fermi liquid." The equivalent compound containing La shows neither the anomalously large electron masses nor superconductivity. Tunneling experiments would be of considerable interest in this system, especially in view of the possibility that $CeCu_2Si_2$ exhibits p-wave pairing.

The use of point-contact Josephson tunneling to distinguish p- vs. s-wave pairing has been discussed and demonstrated by Pals and van Haeringen (1977) and references therein. It would be useful to establish a similar definitive test also for the bipolaronic form of superconductivity proposed by Alexandrov and Ranninger (1981) and by Pedan and Kulik (1982); see also Anderson (1975) and Bishop and Overhauser (1981).

An extended experimental program revealing enhanced superconductivity in Al-formvar sandwiches has been reported by Mancini et al. (1979) (and references therein), which are discussed in terms of the excitonic mechanism of pairing (Ginzburg, 1970; Allender, Bray, and Bardeen, 1973; Vujicić, 1979).

10

Molecular Vibrational Spectroscopy by Tunneling: IETS

10.1 Introduction

Observations of inelastic electron tunneling spectra from organic molecules adsorbed to tunneling barriers were first reported by Jaklevic and Lambe (1966). Peaks in d^2I/dV^2, in the bias range $0.1 \leqslant V \leqslant 0.5$ V, as shown in Fig. 8.19, corresponded to vibrational modes of organic barrier contaminants, including acetic acid, propionic acid, and diffusion pump oil. This subject, now one of the most important applications of tunneling spectroscopy, has already been introduced in Sec. 8.3.2 as an example of an assisted tunneling threshold spectroscopy, and a sketch was given of the original theoretical treatment (Scalapino and Marcus, 1967; Lambe and Jaklevic, 1968). This early work correctly predicted inelastic threshold steps at $eV = \hbar\omega_{ph}$ of magnitude $\Delta\sigma/\sigma \simeq 10^{-2}$, as observed, on the basis of a straightforward tunneling electron–(molecular) vibrating dipole interaction, Eq. (8.26),

$$\delta U(x) = \frac{2ep_x x}{(x^2 + r_\perp^2)^{3/2}}$$

Here, p_x is the x component of the electric dipole moment of the molecule, as sketched in Fig. 10.1a. The molecule is situated at the edge of the barrier ($x = 0$, $r_\perp = 0$), where x, $r_\perp$ are, respectively, the perpendicular and transverse distances of the tunneling electron from the molecule. Figure 10.1b models a second overall effect of a monolayer of molecules as an additional dielectric layer, contributing to the barrier and reducing its transmission.

To understand the basic threshold effect, note that electrons at and above the threshold energy $eV \geqslant \hbar\omega_{ph}$ for the particular dipole vibration are still able to tunnel elastically, as below threshold. However, transfer can now occur also via the inelastic channel; by exciting the molecular dipole vibration, the electron may then tunnel to the Fermi energy ($E = 0$) of the opposite electrode. At first glance, the physics of this excitation process might appear straightforward enough that the spectra could be easily interpreted to infer the detailed situation of the molecule, whether in the barrier or at the metal surface, and also details of the 20 Å barrier itself. These microscopic details, of course, are predictable a priori only with considerable uncertainty. If the physics of the interaction is taken as known, a process of modeling these details and comparison

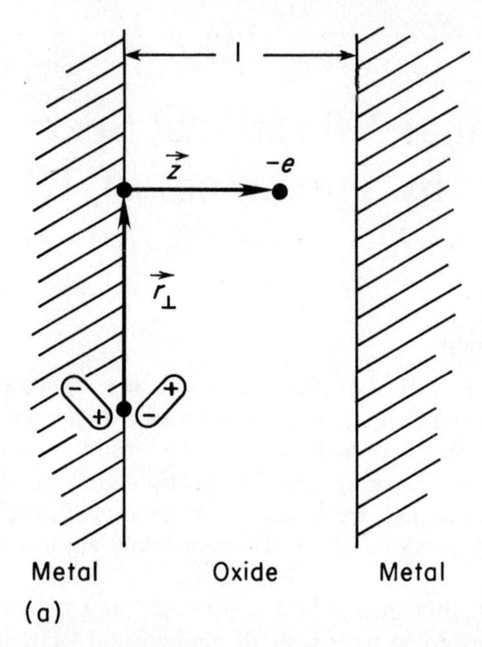

Fig. 10.1. Simple models for the effects upon tunneling of a monolayer of organic molecules adsorbed to the surface of the Al_2O_3 in an Al–Al_2O_3-organic-Pb tunnel junction. (a) Interaction of a tunneling electron with the molecule represented as a polar vibrator, inducing an image dipole in the adjacent electrode. (After Lambe and Jaklevic, 1968.) (b) Modeling the reduction in background conductance. Inset shows the usual trapezoidal barrier augmented by a narrow step potential associated with the molecules. The solid line in the figure is the measured normalized conductance of an Al–Al_2O_3-organic-Pb tunnel junction; measured $G(0)$ is 38.5 mmho · mm^{-2}. Dashed line is the mean of the two voltages required to produce a given experimental conductance. The points represent the calculated normalized conductance of a junction with assumed barrier parameters $\phi_1 = 0.7$ V, $\phi_2 = 4.2$ V, $d = 8.45$ Å, $\phi_3 = 8.8$ V, and $s = 2.53$ Å and a calculated zero-bias conductance of 38.6 mmho · mm^{-2}. (After Walmsley, Floyd, and Timms, 1977.)

with details of the tunneling data and other structural information might be expected to lead to improved understanding of the bonding, orientation, etc. of the molecule at the insulator-metal interface. Indeed, such a modeling approach has been central to the use of IETS to study molecular adsorption and reactions on oxide (and metal) surfaces. Understanding of this subject has developed to the point, however, that it has been again realized that the desired modeling to infer the molecular situation is reliable only to the extent that the basic interaction of the electron with the molecule is surely known. Thus, an important line of basic research in this area over the past few years has been to test and to refine the treatment of the electron-molecule interaction itself, which is only approximated by $\delta U(x)$ as given above and as Eq. (8.26).

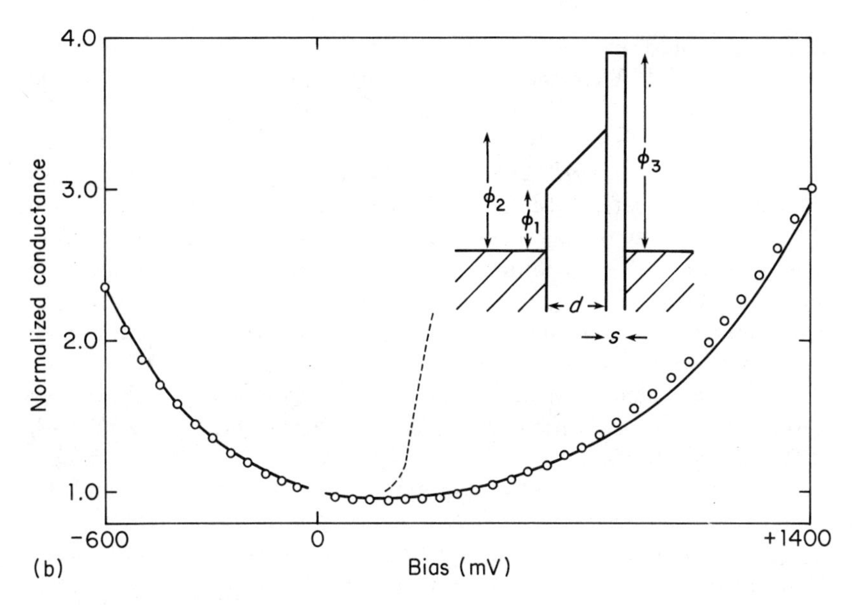

Fig. 10.1. *(continued)*

Several deficiencies inherent in this form of interaction immediately come to mind: (a) the electron is treated as a point charge, (b) the separation $(r_\perp^2 + x^2)^{1/2}$ is assumed so great that the molecule is adequately represented as a point dipole, and (c) the interaction is assumed sufficiently weak that the electron does not perturb the molecule (as it might by inducing a polarization). Section 10.4 is devoted to several improvements in the understanding of the basic interactions involved.

Important developments have also taken place in the techniques of preparing junctions with molecular impurities. This area is one of particular importance in IETS, for two objectives, difficult to achieve even separately, are often sought in the same devices. First, a workable and optimally characterized tunnel junction is necessary, whose oxide, in particular, will withstand bias up to 0.5 V, or, with $t \simeq 15$ Å, electric fields up to $\sim 3 \times 10^7$ V/cm. Secondly, the dopant molecules in their locations on the oxide may also be intended to model the situation of molecules in a technologically useful environment, as on the surface of an oxide catalyst. Further, so that the utility of IETS is broadened to include study of molecules adsorbed on metallic catalysts, an important advance in junction fabrication, pioneered by Hansma and co-workers (reviewed by Kroeker, 1982), now allows study of molecules adsorbed on small metal particles, these being suitably embedded in the tunneling barrier. Other important advances in junction fabrication have been made, the technique of external doping (infusion) of molecules into junctions (Jaklevic and Gaerttner, 1977) being just one further example.

Because of the importance of the details of junction fabrication to the central focus of IETS research, the situation and environment of the adsorbed molecule will be reviewed separately in Sec. 10.2. In the following section, basic properties of IETS spectra are reviewed, preparing a base of information before consideration, in Sec. 10.4, of refinements of the theoretical treatment of the electron-molecule interaction and its effects in tunneling spectra. Finally, a brief summary of current research topics will be given in Sec. 10.5. For a more complete coverage of current research topics, the reader is referred to specialized articles in recent volumes edited by Wolfram (1978) and by Hansma (1982) as well as a chapter written by Walmsley (1980). Other earlier reviews of the enthusiastic and characteristically interdisciplinary research in this area have been given by Hansma (1977), Keil, Graham, and Roenker (1976), and Weinberg (1978).

In view of the wide extent of experimental work in this area and the several excellent specialized reviews cited, the present, rather short survey will attempt particularly to relate topics in IETS to the broader framework of tunneling in normal-state structures.

10.2 Introduction of Molecules

As a preliminary to this topic, it may be well to review what is known about the typical Al_2O_3 surface upon which molecules are deposited. The oxide is grown upon evaporated Al, whose grain size, depending upon conditions, is typically on the order of micrometers (Kroeker, Kaska, and Hansma, 1979). The resulting oxide is known to be amorphous in the range of thicknesses $t \leq 20$ Å of the (partially hydroxylated) oxides used in IETS, although detailed *structural* studies of the oxides seem not to have been reported in the IETS literature. In lack of thorough studies of the structure of the thermally—or plasma discharge—grown films, we note that a brief report (Durbin, 1982) from reflection electron diffraction (REED) inspection of a thermally grown aluminum oxide indeed showed an amorphous deposit. Further, the absence of the many and distinct phonon excitations of crystalline Al_2O_3 in the IETS "clean" spectra supports the accepted picture (Lewis, Mosesman, and Weinberg, 1974; Bowser and Weinberg, 1977) of an amorphous alumina barrier, with typical OH^- surface contaminants. However, this and other properties of the barrier are no doubt dependent on the detailed conditions of oxide growth. Recently, Liehr and Ewert (1983) have reported that *in situ* oxide growth with high purity O_2 on Al films prepared in ultrahigh-vacuum produces "clean junction" d^2I/dV^2 spectra with phonons of *bulk crystalline* alumina Al_2O_3, different from the more typical traces such as that shown earlier as curve A in Fig. 8.19.

While the question of the dependence of film morphology on preparation deserves further work, rather extensive studies have been reported comparing background conductance spectra of thermal and plasma grown barriers to results from trapezoidal model barrier calculations; see, e.g. Fig. 2.12. The chemical and catalytic properties of the typical IETS oxides have been quite thoroughly studied as a function of preparation, with methods including XPS and low-energy electron energy loss spectroscopy (LEELS). Concerning the chemical composition, the ubiquitous peak at 455 meV certainly indicates at least surface OH^- groups. However, Bowser and Weinberg (1977) have questioned the identification of the prominent 117-meV peak in the clean junction d^2I/dV^2 as resulting from the AlO–H bending mode of an aluminum trihydroxide (Geiger, Chandrasekhar, and Adler, 1969), which was thought to imply that the insulator was really a bulk hydroxide, rather than an oxide, of aluminum. The finding of Bowser and Weinberg is that the 117-meV peak is an Al–O stretching mode of Al_2O_3. These workers conclude that plasma oxides grown at temperatures higher than 150°C are truly oxides, while there is some evidence for bulk hydroxide incorporation in films grown at lower temperatures. In both cases, however, there is strong evidence for OH^- groups attached at the surface of the oxide film, which may act as binding sites for adsorption or chemisorption of molecules. Such adsorbates may also provide electric dipole layers and rather strongly affect barrier heights as inferred from the background conductance (Dragoset, Phillips, and Coleman, 1982). For this reason, these authors suggest that OH-containing oxides produce higher barrier heights and thus allow improved observation of high-energy vibrations. In Fig. 10.1b, the WKB tunneling conductance calculation is extended beyond the trapezoidal barrier to model the effect of the addition of a molecular monolayer upon the elastic background conductance $G(V)$. Extensive use of such calculations has been made by the Virginia group; e.g., see Dragoset, Phillips, and Coleman (1982) and references therein.

The OH^- content also influences catalytic properties of alumina (Evans, Bowser, and Weinberg 1980). The correspondences between aluminum oxides used in IETS and those used commercially as catalysts have been investigated by Hansma, Hickson, and Schwarz (1977), by Evans, Bowser, and Weinberg (1980) using XPS methods, and by Dubois, Hansma, and Somorjai (1980) using high-resolution electron energy loss spectroscopy (LEELS); a summary of such investigations has recently been given by Dubois (1982).

Two basic methods of introducing molecules onto the surface of the grown oxide are from the vapor or in a liquid drop (Hansma and Coleman, 1974). In the latter case, the excess solvent is usually driven off centrifugally by placing the junction on a spinner. In a third technique, that of infusion (Jaklevic and Gaerttner, 1977), molecules are introduced in a combination of the first two methods into the completed junction through the vapor phase in a high-humidity environment. In most cases (excepting cases of cryogenic substrates) (Cederberg, 1981), the adsorption process is believed to occur on a set of active sites of the oxide surface, and the fractional coverage θ of these sites achieved depends upon the organic vapor pressure or solution concentration, binding energy, and temperature, very much along the lines originally set forth by Langmuir (1918). Evidence suggests that in most cases a preferred molecular orientation with respect to the oxide surface exists. In some cases, a chemical reaction is known to occur between the molecule and the oxide surface. An example is sulfonic acid, determined to adsorb on alumina as the methyl sulfonate ion $CH_3SO_3^-$ (Hall and Hansma, 1978; Kirtley and Hall, 1980). The C–S axis of this molecular ion is oriented perpendicular to the surface, to which it is bonded through the SO_3^- tripod. Similarly, acetic acid adsorbs as the acetate ion (Lewis, Mosesman, and Weinberg, 1974), and benzoylchloride adsorbs as benzoate ion (Walmsley, McMorris, and Brown 1975).

The great majority of all IETS work has been performed, as is indicated by the cited examples, by using Al_2O_3 barriers; the next most-frequently cited oxide is probably MgO (Lambe and Jaklevic, 1968; Klein et al., 1973; and Geiger, Chandrasekhar, and Adler, 1969; which more recently has been extensively and fruitfully studied as a catalytic substrate by Walmsley, Nelson, Brown, and Floyd (1980; and references therein). Brief reports of work with V and Cr oxides (McBride, Rochlin, and Hansma, 1974) on Cd oxide (Giaever and Zeller, 1968*b*), and on rare earth oxides (Adane et al., 1975) have been given, while the absence of detectable organic IETS spectra has been recorded by Walmsley, Wolf, and Osmun (1979) on Nb oxide and by Refai and Wolf (1982) on ZnO. In the latter two cases, the unobservability of molecular peaks appeared to be related to the low barrier heights achieved at the Nb- and Zn-oxide interfaces. The rather restricted range of substrates suitable in IETS remains a problem to be overcome, possibly by composite barriers such as those cited in connection with Fig. 9.7.

A more complex and remarkably successful Al_2O_3 barrier-doping scheme utilizing small (30-Å) metal particles has been carried out by Hansma, Kroeker, and co-workers following the demonstration of Hansma, Kaska, and Laine (1976) that Rh-on-alumina catalysts could be effectively incorporated into tunnel junctions and usefully studied. The barrier structures were generated on the usual aluminum oxide surface by evaporation of extremely small thicknesses (~4 Å) of Rh, which, under the influence of its own surface tension, formed dispersed Rh particles of diameter in the range 30 Å, as was nicely documented in electron microscope work. Kroeker, Kaska, and Hansma (1980, 1981) then demonstrated that exposure of the Al_2O_3:Rh structure to CO led to its adsorption on the Rh particles, and, finally, that the vibrations of the CO molecules on the Rh particles could be observed in the resulting IETS spectra in the usual fashion. As has been summarized by Dubois (1982), the Rh:CO/Al_2O_3 surfaces yield the same vibrational spectra of CO when probed by low-energy electron energy loss spectroscopy as when probed by IETS.

In all cases in IETS, the doped oxide surface is covered by evaporating metal to form the counterelectrode. The most common choice is Pb.

Three reasons supporting this choice given by Weinberg (1978) are (a) the high T_c of Pb, 7.2 K, allows high-resolution spectra to be taken at the convenient temperature of liquid helium, 4.2 K, (b) the ionic radius of Pb is sufficiently great that it does not have a tendency to diffuse through the oxide barrier, and (c) Pb is evidently sufficiently inert chemically that it does not react with the adsorbed molecules in any noticeable fashion. Studies of the effects of the overcoated counterelectrode on the vibrational frequencies of hydroxyl ions and benzoate ions, respectively, on alumina have been reported by Kirtley and Hansma (1975*b*, 1975*a*). In comparison with frequencies obtained by other methods (typically infrared spectra and Raman spectra measured in solutions), the IETS frequencies are systematically lower, by as much as 4% for OH-stretching vibrations but essentially not at all for ring deformation vibrations. The magnitudes of the shifts, studied using Pb, Ag, and Sn counterelectrodes, were found to vary systematically as R^{-3}, where R is the atomic radius of the counterelectrode metal. An approximate understanding of these effects (Kirtley and Hansma, 1976) was found in an image dipole model. Other metals which have been used for counterelectrodes include Al and Mg; detailed comparisons of spectra taken with counterelectrodes of several of these metals have been given recently by Lau and Coleman (1981).

10.3 Properties of IETS Spectra

The IETS spectra, reflected as symmetric step increases in conductance $\Delta\sigma/\sigma \simeq 10^{-2}$ at thresholds $eV = \pm\hbar\omega_{ph}$, would be optimally observed in the d^2I/dV^2 derivative spectrum, which would consist of peaks approximately antisymmetric in bias sign at $eV = \pm\hbar\omega_{ph}$. Experimentally, however, it is easier to measure $d^2V/dI^2 \equiv -\sigma^{-3}\, d^2I/dV^2$, and the results are quite similar since the conductance $\sigma = dI/dV$ is generally slowly varying (see, e.g., Fig. 10.1). For the purposes of this chapter, we will regard $-d^2V/dI^2$ as an equivalent measure of the IETS spectrum d^2I/dV^2.

Before considering the properties of the vibrational peak spectra, we note briefly that the effects of molecular incorporation on the bias dependence of the background conductance, as illustrated in Fig. 10.1, are quite straightforward; the first effect is simply a substantially increased scale of resistance. A typical increase of two to three orders of magnitude is reported by McBride and Hall (1979). Usually, experimenters reduce the oxidation time for junctions to be doped in order that the overall resistance level remain similar to that of clean junctions.

The $-d^2V/dI^2$ spectrum of a typical clean alumina junction (Fig. 8.19, curve A) is rather flat, with a linearly increasing term in V, reflecting the derivative of the roughly parabolic $G(V) = G_0 + \alpha(V - V_0)^2$ typical background conductance. Dominant sharp features of the clean $-d^2V/dI^2$ spectrum are the Al–O-stretching mode at 117 meV and the OH-stretching vibration at 447 meV. The former is intrinsic to all AlO_x barriers, while the latter is believed due to free OH^- ions at the surface of the alumina, resulting from adsorption of H_2O present at low levels in the oxidation plasma. In the case of the clean MgO barrier (Walmsley, Nelson, Brown, and Floyd, 1980), the 117-meV peak feature is replaced by two peaks at 53 and 80 meV, attributed to MgO vibrations; the 455.6-meV feature also typically present on MgO is again identified with OH^-, also adsorbed to the barrier surface from the oxidation ambient.

The width of such peaks with normal-state counterelectrodes (in the limit of small ac modulation) may be regarded as resulting from an inherent width convoluted with a thermal function arising from the Fermi level widths in the two electrodes. As shown by Lambe and Jaklevic (1968), the thermal resolution function appropriate to d^2I/dV^2 measurements is a bell-shaped peak with width at half-height of $5.44\,kT$. This "slit-width" function arises from differentiation of the Fermi factors $f(E)$ in the integral giving the current contribution I_i at the threshold, $eV_i = \hbar\omega_{ph}$, as

$$I_i = \int_{-\infty}^{\infty} dE\, f(E)[1 - f(E + eV - eV_i)] \tag{10.1}$$

$$= \frac{Ce(V - V_i)\exp(v)}{\exp(v) - 1} \tag{10.2}$$

where C is a constant and $v = e(V - V_i)/kT$. Taking derivatives with respect to voltage V, one obtains

$$\frac{d^2I}{dV^2} = C\frac{e^2}{kT}\left[e^v \frac{(v-2)e^v + (v+2)}{(e^v - 1)^3}\right] \tag{10.3}$$

At the usual measurement temperature 4.2 K ($kT = 0.362$ meV), the resultant width is 1.97 meV, or 15.9 cm^{-1}. The effect of finite modulation amplitude V_ω in the usual second-harmonic detection scheme is shown (Klein et al., 1973) to result in a similar bell-shaped broadening function

$$G(E) = \begin{cases} \dfrac{2eV_\omega}{3\pi}\left[1 - \left(\dfrac{E}{eV_\omega}\right)^2\right]^{3/2}, & |E| < eV \\ 0, & |E| \geq eV \end{cases} \tag{10.4}$$

which has a full width at half-height of $1.22\,eV_\omega$.

If one combines these two effects, the instrumental half-height width appropriate to IETS with normal-state electrodes can be taken as

$$\delta V = [(1.22eV_\omega)^2 + (5.44kT)^2]^{1/2} \tag{10.5}$$

The predicted temperature dependence was verified by Jennings and Merrill (1972), while the modulation effect was established by Klein et al. (1973). A comparison of IETS and IR spectra of cyclohexanol is given in Fig. 10.2, in which the inset shows resolution of a pair of CH-stretching modes at 353 meV by lowering the IETS measurement temperature from 4.2 to 1.2 K. Note that the same modes are resolved in the IR spectrum and agree nicely in position with the Raman spectra. However, the

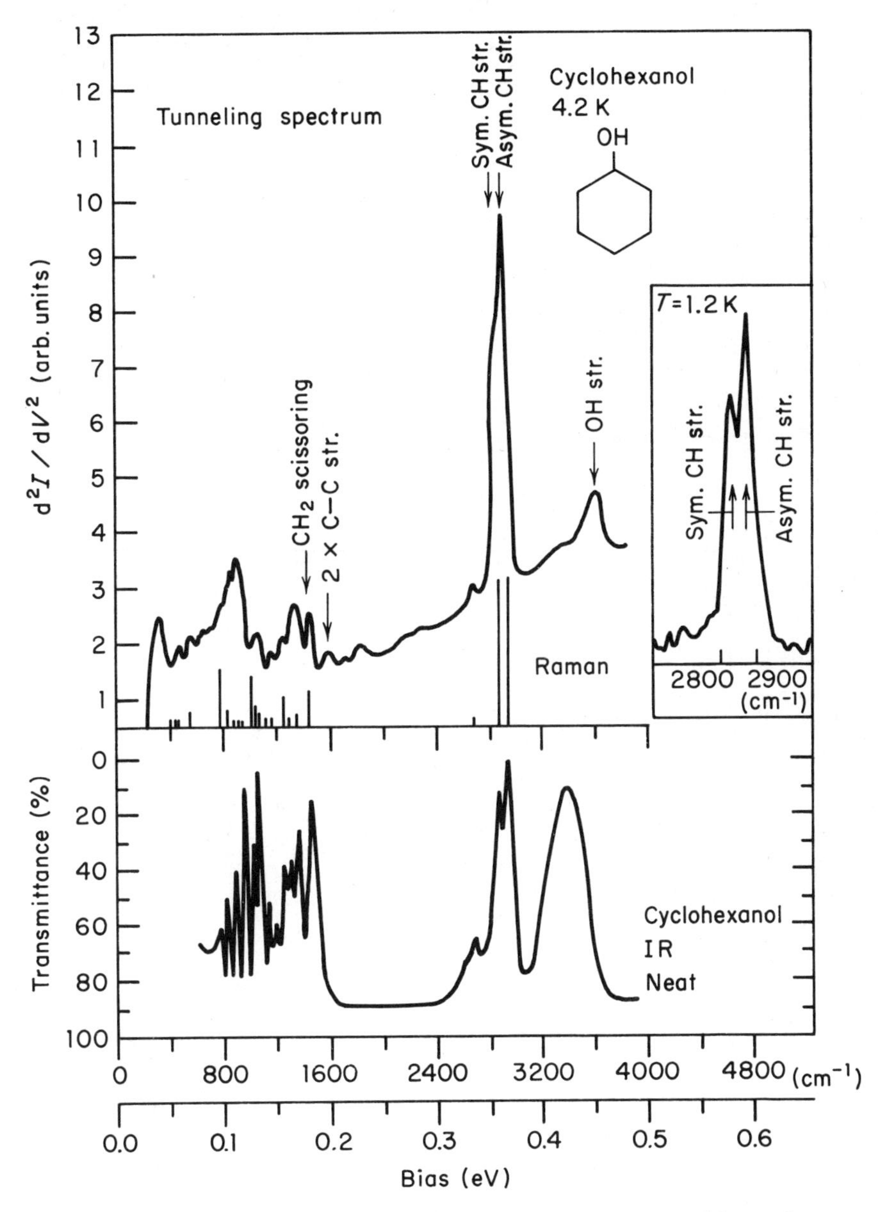

Fig. 10.2. Typical comparison of result of IETS spectroscopy (shown here as d^2I/dV^2) vs. conventional infrared absorption spectroscopy for cyclohexanol molecule adsorbed to alumina. The inset shows improved resolution in the IETS spectrum when measured at 1.2 K. (After Simonsen and Coleman, 1973.)

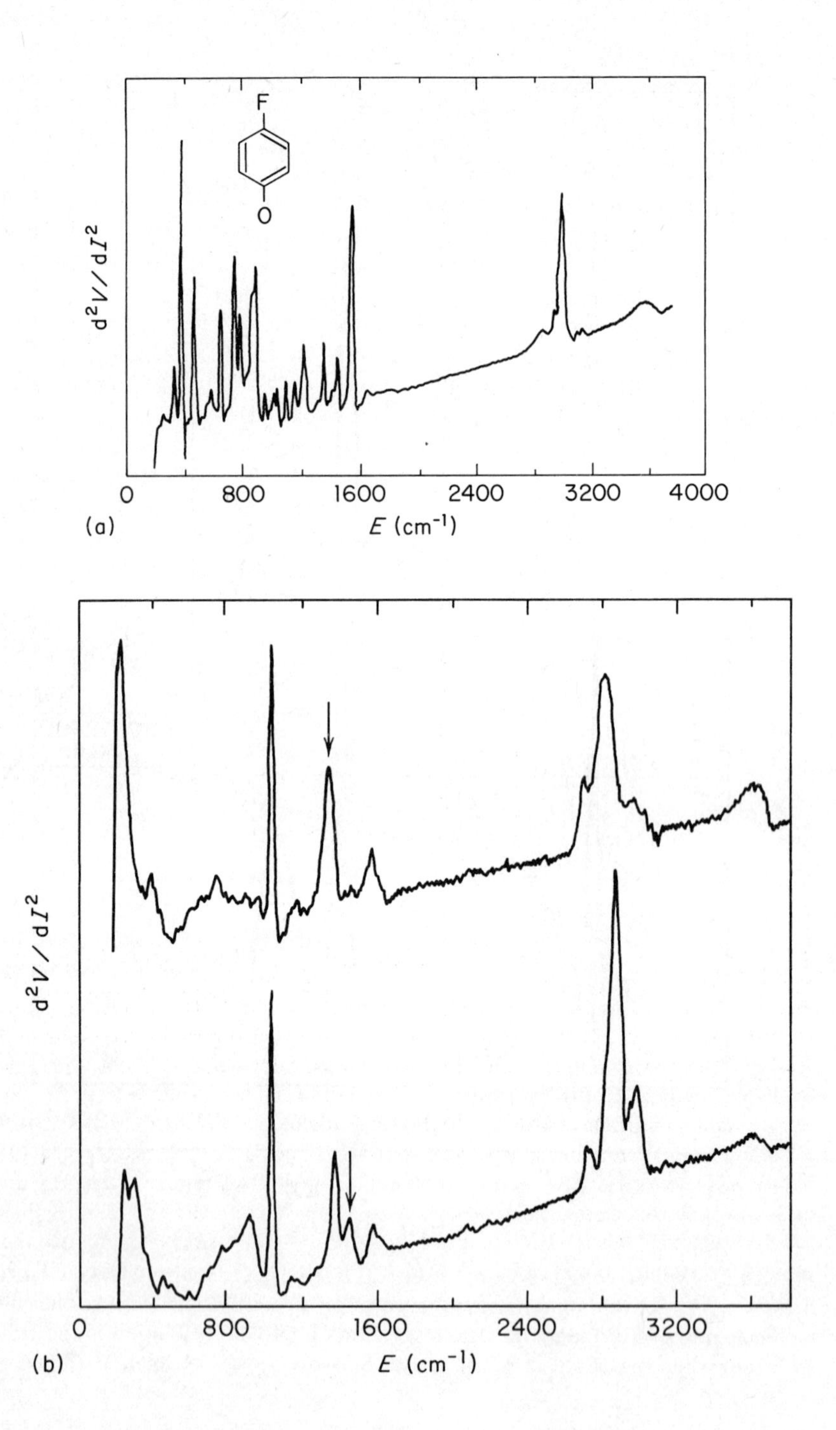
F
O
d^2V/dI^2
0
800
1600
2400
3200
4000
(a)
E (cm^{-1})
d^2V/dI^2
0
800
1600
2400
3200
(b)
E (cm^{-1})

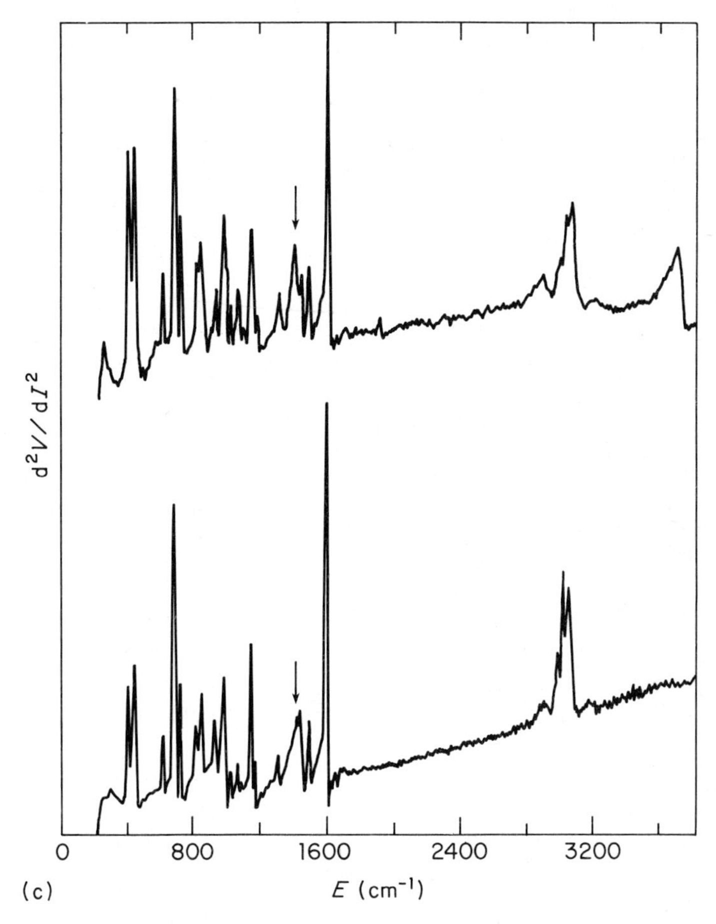

Fig. 10.3. Comparison of typical IETS spectra. (a) d^2V/dI^2 spectrum of *p*-fluophenylate ion at 2 K on plasma-grown Al oxide. This figure can be compared with Fig. 1.9b concerning the same ion without the fluorine attachment; the most noticeable change in the spectrum is near 3000 cm^{-1}, in the region of the CH-stretching vibrations. (After McMorris, Brown, and Walmsley, 1977.) (b) Inelastic electron tunneling spectra at 2 K of formic acid. Upper trace, Mg–MgO–Pb junction; lower trace, Al–Al_2O_3–Pb tunnel junction. The arrow denotes the line which is shifted down, when adsorbed on MgO, to coincide with another line. (After Walmsley, Nelson, Brown, and Floyd, 1980.) (c) A similar comparison to that in (b) revealing minor changes in the IETS spectra of benzaldehyde at 1.1 K on MgO (upper trace) and on Al_2O_3 (lower trace). (After Walmsley, Nelson, Brown, and Floyd, 1980.)

observed splitting, about 50 cm^{-1} or ~6 meV, is well above the instrumental resolution estimate, (10.5), leading one to suspect an inhomogeneous line width, possibly due to variation in adsorption sites in the sample. While this situation may be typical, it is not always the case, and sharper lines are evident, e.g., in Fig. 10.3a, which shows the extremely sharp spectrum resulting from attachment of a fluorine atom to the simple phenol ring (the spectrum of Fig. 1.9b) to produce the *p*-fluophenylate ion.

Minor changes in vibrational spectra from the corresponding free-molecule spectra are expected as the result of their adsorption to the oxide substrate; such changes may be used to deduce the type of bonding that occurs. For example, in the case of the simple acid HCOOH (formic acid), the spectra of which are shown on alumina and MgO tunnel barriers in Fig. 10.3b, the original work of Lewis, Mosesman, and Weinberg (1974) demonstrated the IETS spectrum to be identical with the IR vibrational spectrum earlier established for aluminum formate. This result indicates that a reaction to generate the formate ion ($HCOO^-$) has occurred in the adsorption process. Lewis, Mosesman, and Weinberg (1974) noted that this IETS spectrum contains all the vibronic transitions of the HCOOH molecule *except* those involving the OH group, implying bonding to the surface of the sort Al*OOCH, where the Al* resides in the oxide surface. Since this process breaks the OH bond in the original formic acid molecule, releasing H, which may form H_2O with surface OH^- ions, it represents an example of catalytic decomposition of a molecular species by the AlO_x surface. Similar conclusions were drawn by Lewis, Mosesman and Weinberg (1974) regarding acetic acid (CH_3COOH) adsorbing as acetate ion (CH_3COO^-) and water H_2O adsorbing as OH^-. All of these cases are, thus, properly termed chemisorption, or chemical bonding to the substrate, rather than adsorption. In an additional example, benzoic acid chemisorbs as benzoate ion (Kirtley, Scalapino, and Hansma, 1976), the same ion being produced also from benzaldehyde (Simonsen, Coleman, and Hansma, 1974).

Returning to Fig. 10.3b, Walmsley, Nelson, Brown, and Floyd (1980) have here sought out the effects of the substrate change from Al_2O_3 (lower trace) to MgO (upper trace) on the details of adsorption of formate ion from formic acid. The arrows in this figure locate the only major difference, a vibration, assigned as a symmetric CO_2^- stretching motion, whose distinct higher frequency (1456 cm^{-1}) on alumina has shifted down (to coincide with a second line) at 1370 cm^{-1} on MgO. The first conclusion (applying also to benzaldehyde, Fig. 10.3c, propionic acid, and phenol) is that chemisorption of the related ions occurs in the same fashion on MgO as on Al_2O_3. The small changes noted in the ionic vibrational spectra are attributed to stronger electrostatic binding of the ions to MgO, which has a more strongly ionic character than does Al_2O_3. These particular spectra also beautifully exhibit the resolution and detail that are available in IETS at its best.

An aspect of the d^2I/dV^2 peak spectra that has been puzzling since the earliest IETS work is the lack of accurate symmetry in strength of the modes at opposite signs of bias. In fact, the peaks are always stronger on Al_2O_3 when the Al electrode is biased negatively. This effect was first emphasized and given what is now believed to be a qualitatively correct interpretation by Yanson and Bogatina (1971) (see also Yanson, Bogatina, Verkin, and Shklyarevskii, 1972). The effect is illustrated in Fig. 10.4. The basic idea is that in Al negative polarity, electrons tunneling through the barrier from the Al to the counterelectrode encounter the molecules at the oxide-counterelectrode interface and lose energy only *after* traversing the oxide. Since the barrier transmission probability increases with increasing electron energy, the resulting inelastic current is greater than that in the opposite polarity, when the electron must traverse the barrier *after* losing energy to the molecule. This argument predicts a degree of intensity asymmetry increasing with the energy $\hbar\omega$ of the transition being observed. The data points in Fig. 10.4b indeed agree with this trend of increasing asymmetry with bias. This result suggests that careful observations of asymmetry (see, e.g., Muldoon, Dragoset, and Coleman, 1979) could be used to test the implicit assumption above, namely, that the electron-molecule interaction is essentially local in nature. In fact, the upper curve, based on an electron-dipole interaction only slightly refined over that first given by Scalapino and Marcus (1967) (this is the partial charge interaction of Kirtley, Scalapino, and Hansma, 1976), evidently reflects a potential of about the right range to explain the data shown, for benzoate ion on alumina (Kirtley, Scalapino, and Hansma, 1976). This subject will be discussed further in Sec. 10.4.

The dependence of the strengths of the IETS spectra on the surface density of molecules is typically slightly superlinear. This fact was first established by Langan and Hansma (1975), using radioactive tracer techniques to determine the surface concentration. In a solution-doping experiment, the surface concentration is proportional to solution concentration over a wide range and finally saturates as chemisorption sites are fully occupied. Cederberg and Kirtley (1979) have quantitatively compared the solution-doping results of Langan and Hansma (1975) with the Langmuir isotherm model. The role of available surface sites is important; e.g., Walmsley, Nelson, Brown, and Floyd (1980), in their comparison of Al_2O_3 and MgO substrates, found in vapor doping that MgO would typically bind approximately four times the surface concentration of molecules (as inferred from the standard quartz microbalance technique) as would Al_2O_3, both oxides being prepared in an O_2 plasma discharge in the same evaporation system. The origin of the slightly superlinear dependence of intensity $\Delta\sigma/\sigma$ upon fractional coverage θ [typically expressed as $\Delta\sigma/\sigma \simeq \theta^x$, with $x \simeq 1.3$, but see also Cederberg (1981)] has not been definitively established.

A plausible suggestion, however, is that in the dependence of $\Delta\sigma/\sigma$ on

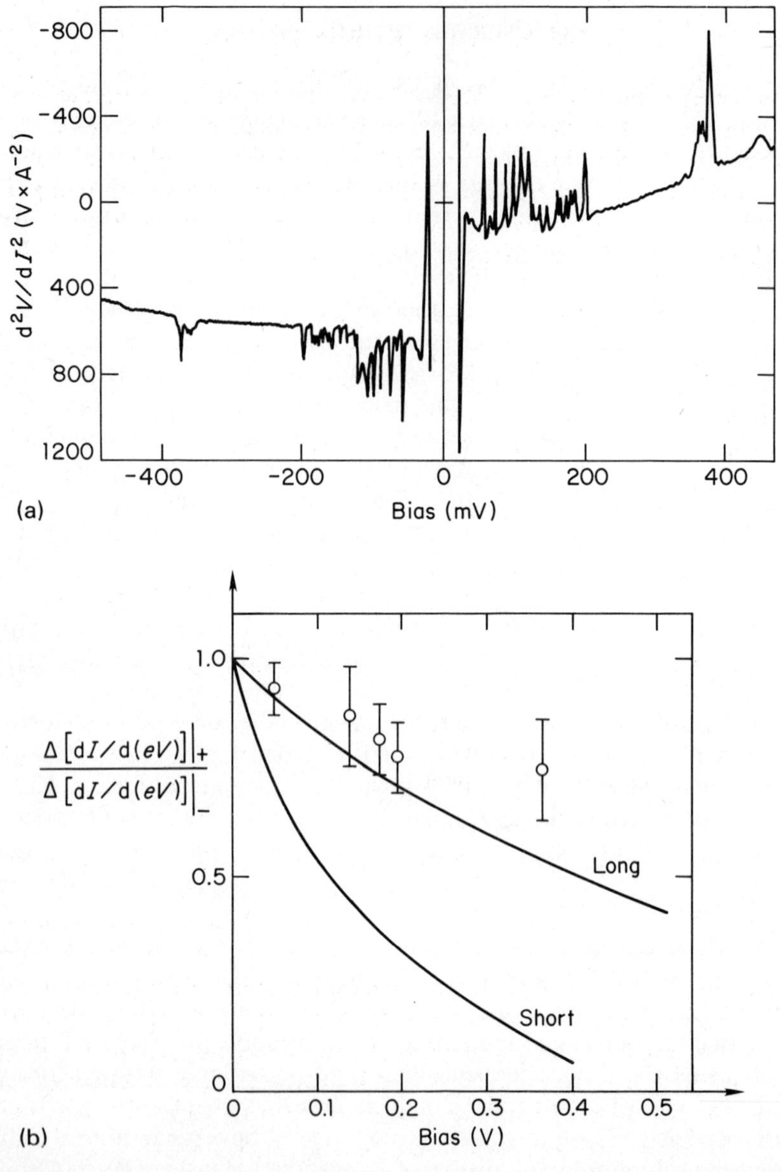

Fig. 10.4. (a) Illustration of the asymmetric strengths of IETS peaks under change of bias sign. An Al–Al_2O_3–Pb junction doped with *m*-cresol observed at temperature 2 K. (After Walmsley, Timms, and Brown, 1976.) (b) The ratio of the integrated peak intensities for opposite polarities vs. the magnitude of the transition energy (applied voltage). For positive polarity, the electrons tunnel from the electrode in direct contact with the dopant layer (Al positive in relation to Pb for standard junctions). The dots are experimental measurements for a benzoate dopant (Kirtley, Scalapino, and Hansma, 1976). The lower theoretical curve is based on an interaction of short range. The upper curve is the result of the long-range interaction (partial-charge) theory of Kirtley, Scalapino, and Hansma (1976). The remaining discrepancy between experiment and theory may be due to barrier asymmetry. (After Hansma and Kirtley, 1978.)

coverage, $\Delta\sigma$ linearly increases with the number of adsorbed molecules per unit area, while the denominator σ decreases, as molecules tend to increase the barrier thickness and reduce the elastic conductance. While this proposal of Cederberg (1981) seems to require test against more data than are presently available before being adopted with certainty, Cederberg's expression for the peak intensity,

$$I = \frac{(\Sigma_i/\Sigma_f)\times\theta}{\theta+(1-\theta)(\Sigma_e/\Sigma_f)} \tag{10.6}$$

has the a priori merit of being expressed in what most likely are the physically relevant parameters. Here, Σ_e, Σ_f, respectively, are the elastic conductances per unit area for undoped and doped portions of the junction, while Σ_i is the inelastic channel conductance, and θ is the fraction of the surface area covered by dopant. This expression, as one might expect, fits the available data equally well as does the conventional expression $I = I_0\theta^x$. The observed superlinearity arises explicitly in (10.6) because the term in the denominator decreases as θ increases, since (Σ_e/Σ_f) exceeds unity. The physics of the assisted tunneling process is contained in the prefactor ratio Σ_i/Σ_f of (10.6). This topic is considered next.

10.4 Calculations of IETS Spectra

The first objective of theory in IETS is to predict the overall strength of the peaks, including their asymmetry and other features. A more challenging task, and one important to further development of the IETS technique, is to predict the relative intensities of the various vibrational modes $\hbar\omega$ as a function of the position and orientation of the molecule in the barrier. This more difficult task is necessary if the observed intensities, taken together with a knowledge of the structure and vibrational modes of the free molecule, are to be used to infer the detailed situation of the adsorbed molecule, to make IETS really valuable as a probe of surface chemistry. Needless to say, the success of such an approach requires a correct treatment of the basic interaction of the tunneling electron with the molecule.

It now appears that the transfer Hamiltonian theory of Kirtley, Scalapino, and Hansma (1976), not too different in form from the original model of Scalapino and Marcus (1967), is capable of filling this important role. We will, therefore, sketch the steps that have led to this theory and summarize its application by Kirtley and Hall (1980) to sulfonic acid and by Godwin, White, and Ellialtioglu (1981) to formic acid—to demonstrate its use in inferring molcular orientation from the relative intensities of vibrational lines.

The theory of Kirtley, Scalapino, and Hansma (1976) (KSH) is a straightforward application of the Bardeen-Cohen-Phillips-Falicov transfer Hamiltonian method, in which the Coulomb interaction of the tunneling electron with partial charges $Z_j e$ on each atom of the molecule is accounted for separately. The interaction with each partial charge $Z_j e$ is taken to be long range and hence is screened with the oxide dielectric constant ε. An important feature of the calculation is a proper three-dimensional accounting of off-axis electron tunneling transitions, by which both IR and Raman active modes are found to be excited *without* need to assume a molecular polarizability α in the induced dipole interaction

$$\delta U(x) = -4e^2 \alpha x^2 (x^2 + r_\perp^2)^{-3} \tag{10.7}$$

as had been done in the early treatment of Lambe and Jaklevic (1968).

The basic interaction with partial atomic charges $Z_j e$ of KSH is simply

$$V(\vec{r}) = \sum_j \frac{-e^2 Z_j}{|\vec{r} - \vec{R}_j|} \tag{10.8}$$

where $Z_j e$ and R_j are the partial charge and position, respectively, of the jth atom in the molecule. This interaction is more accurate at small electron-molecule distance $\vec{r}$ than is a dipole approximation. In order to allow transitions of energy difference $\hbar\omega$, KSH isolate the component of $V(\vec{r})$ which oscillates at the frequency of the normal mode of interest, for which the atomic positions are given by

$$\vec{R}_j = \vec{R}_j(0) + \delta\vec{R}_j(t) \tag{10.9}$$

with

$$\delta\vec{R}_j(t) = \delta\vec{R}_j(0) e^{i\omega t} \tag{10.10}$$

Since the $|\delta\vec{R}_j|$ are small, KSH expand the potential $V(\vec{r})$ to obtain the interaction potential

$$V_I(\vec{r}) \simeq \sum_j -e^2 Z_j \, \delta\vec{R}_j(0) \cdot \vec{\nabla}_j \left(\frac{1}{|\vec{r}_j - \vec{R}_j(0)|} \right) e^{i\omega t} \tag{10.11}$$

This expression has the form of a sum of dipole potentials located on each atom. The *partial* charges, which can reflect uneven sharing of electrons involved in bonds, can be inferred from IR data.

When the image charges in each electrode are included, along with dielectric screening in the oxide, (10.11) gives

$$V_I(\vec{r}) = \sum_{n=-\infty}^{\infty} \sum_j \frac{-e^2 Z_j \, \delta\vec{R}_j \cdot \vec{\nabla}_j}{\varepsilon} \left(\frac{1}{|\vec{r} - \vec{R}_j - 2ntx|} - \frac{1}{|r - R_j - (2nt - 2a_j)x|} \right) e^{i\omega t} \tag{10.12}$$

Here, ε is the oxide dielectric constant, t is the barrier thickness, and $a_j x$ is the x component (into the barrier) of $\vec{R}_j$. Following the transfer Hamiltonian approach, the matrix elements of (10.12) between plane wave states of energy difference $\varepsilon_k - \varepsilon_{k'} = \hbar\omega$ on opposite sides, extended into the barrier by using the WKB approximation, are calculated. These elements, when squared, are written

$$|M|^2 = \left|\sum_j M_{kk'j}\right|^2 = G(\varepsilon, \varepsilon', \theta, \theta', \phi, \phi') \tag{10.13}$$

where ε, ε' are electron energies, and (θ, ϕ) and (θ', ϕ') are polar angles describing incoming and outgoing electrons. These angles are integrated over in the calculation of the current. As summarized in the original paper (Kirtley, Scalapino and Hansma, 1976) and by Kirtley (1980), adoption of reasonable numerical values for the relevant physical parameters produces semiquantitative agreement with spectral features, including the peak intensities, their dependence upon molecular orientation, and the bias asymmetry for benzoic acid dopants. Comparable success has since been reported in treatment of methylsulfonate dopant (Kirtley and Hall, 1980) and, except for C–H vibrations, in formic acid by Godwin, White, and Ellialtioglu (1981).

Before describing the use of the theory to infer the molecular orientation, we return to the question of the basic interaction. In their use of the Coulomb interaction, (10.8), KSH have neglected short-range exchange-correlation effects. These would be important if the tunneling electron had a tendency to form a negative ion, i.e., an electronic bound state on the molecule. This neglect is justified by the results of careful treatments of the consequences of important short-range interactions (Davenport, Ho, and Schrieffer, 1978; Kirtley and Soven, 1979). If such short-range interactions were important, strong enhancement of IETS peaks should occur when the energy of the injected electron approached that of a negative-ion molecular bound state (or scattering resonance) for an additional electron; however, such effects have never been observed. A second consequence of strong short-range interaction should be observation of more pronounced structure at *harmonics* of the basic molecular vibration frequencies than are observed. Finally, much greater asymmetry of intensities is predicted than is observed (see Fig. 10.4b). Thus, the partial-charge interaction approximation of KSH is supported not only by a (nearly) correct prediction of the intensity asymmetry (Fig. 10.4b) but also by fundamental calculations properly including multiple scattering and exchange-correlation effects. The weakest point in the treatment of KSH, according to the recent work of Godwin, White, and Ellialtioglu (1981), is in its prediction of C–H vibration intensities, and those other vibrations involving H. This discrepancy is attributed to charge distortion along the C–H bond. A more sophisticated interaction than the partial-charge model, possible along the lines recently proposed by Rath and Wolfram (1978) is indicated for bonds containing H. In all other cases, Godwin, White, and Ellialtioglu (1981) find the KSH theory adequate.

The first demonstration of the use of IETS intensities as interpreted in the KSH theory to deduce a molecular orientation is that of Kirtley and Hall (1980), confirming the perpendicular orientation of the $CH_3SO_3^-$ ion on alumina. This linear orientation, with SO_3^- bonding to Al^+ ions, had been proposed earlier (Hall and Hansma, 1978) on the basis of vibrational frequencies.

The principle that allows inferences on molecular orientation within the KSH theory is basically that the potential from a vibrating electric dipole, whose moment is perpendicular to the barrier, is reinforced by its image in the electrode, while cancellation tends to occur from the image of the dipole oriented parallel to the barrier. However, this rule applies strictly only if the dipole is at the edge of the barrier and coincident with the image plane [which is possibly shifted from the actual metal–barrier interface, see Kirtley, Scalapino, and Hansma (1976)]. At any rate, the rule fails and its reverse become true as the electric dipole moves to the center of the barrier; from this location, parallel electric dipole moments contribute more strongly than do perpendicular moments, but not so strongly as perpendicular moments at the edge of the barrier. The origin of the 'center of the barrier' rule is not as transparent but is attributed by Kirtley and Hall (1980) to two effects. First, in the center of the barrier, where image effects are weaker, the fact that the full dipole potential is an odd function of the coordinate along the dipole tends to cause cancellation, independent of orientation. Second, off-axis scattering and contributions from dipoles of differing positions both allow destructive interferences, and these evidently are more severe in the perpendicular orientation.

A further generalization noted by Kirtley, Scalapino, and Hansma (1976) is that intensities from Raman and IR active modes tend to be of the same order of magnitude; this difference from the Scalapino and Marcus (1967) theory, in which Raman modes did not contribute without invoking polarizability, seems due both to consideration of off-axis electron tunneling and scattering and to more accurate treatment of close electron-molecule encounters. In such close encounters, different atoms, in the partial-charge approach, can contribute at different strengths, contrary to the *molecular* dipole assumption of the earlier work.

The modeling of $CH_3SO_3^-$ IETS spectra (Kirtley and Hall, 1980) was based on the known C_{3v} symmetry and dimensions of the molecule, leading to 18 vibrational normal modes. Alternative assumptions were made for the molecular orientation: with the C–S bond parallel and perpendicular to the barrier and for spacings near 1.5 Å for the distance to the Pb electrode. In each case, the intensities of the various modes were calculated, making use of IR data to determine the relevant partial charges $Z_j e$, as explained in the original paper. The best agreement is obtained with the C–S bond perpendicular, the SO_3^- group bonded to the alumina, and a distance $d = 1.5$ Å between the H atoms (at the other end of the molecule) and the Pb-electrode image plane. While the results offer a clear choice of orientations, the agreement is not perfect, and a reversal of intensities of a pair of C–H stretch modes near 170 meV remains. However, neglect of molecular distortion due to the chemisorption and of polarizability interactions; the assumption of a mathematically planar Pb-electrode imaging surface rather than e.g. a local dimple into the Pb—all would indeed make perfect agreement appear suspect. In spite of these simplifications, the viability of the KSH model as a workable research tool seems to have been established.

In a subsequent similar study of the simpler $CHOO^-$ (formate) molecule ion, Godwin, White, and Ellialtioglu (1981) considered a set of five modes, three of which involve C–H vibrations. Quite reasonable agreement was obtained on predicted intensities for the two modes (C—O and C=O stretch) not involving C–H vibrations, but poor results are again found for the C–H vibration intensities. Thus, as is not uncommon in scattering methods, results for H obtained via the present IETS technique have to be discounted, but present indications are that other modes are reliably treated.

10.5 Applications and Current Research in IETS

A broad area of interest in IETS research is that of surface reactions among molecules, particularly those reactions which may be catalyzed by the oxide or by extra particles embedded in the oxide. Fundamental questions of surface chemistry which are in principle accessible to IETS methods include, as we have already seen, the nature of the resultant surface-adsorbed species A^* and also the nature of reaction intermediate species A^*B^* if a second reactant molecule (B) is adsorbed (B^*) and, by means of surface mobility, encounters A^*. The resultant species C (or possibly C+D) desorbed in such a catalytic reaction may be of commercial importance. Two examples of commercially important surface-catalyzed reactions are the cracking of heavy hydrocarbons over hot platinum, to produce octane and other constituents of gasoline, and the catalytic oxidation of sulfur dioxide to sulfurtrioxide on finely divided platinum or on vanadium pentoxide, V_2O_5. Addition of water to sulfurtrioxide, produces sulfuric acid, a chemical of economic importance.

A new technique which should prove helpful in studies of surface reactions is the *infusion* method by which molecules can be doped into a completed junction. As this method was originally presented (Jaklevic and Gaerttner, 1977), a Pb counterelectrode of thickness 600 to 1000 Å acted much like a membrane through which molecules could pass. The infused molecules are possibly carried into the barrier region on layers of adsorbed water a few molecules thick, under the recommended conditions of high (70–98%) humidity. This transport presumably occurs along the surfaces of cracks or grain boundaries through the metal film. The infusion doping of completed Al–AlO_x–Pb junctions with D_2O, DCOOD, and HCN is illustrated in Fig. 10.5, after Jaklevic and Gaerttner (1977). Exploitation of this technique to demonstrate a surface chemical reaction occurring in the completed tunnel junction after infusion was reported by Jaklevic (1980), whose results are shown in Fig.

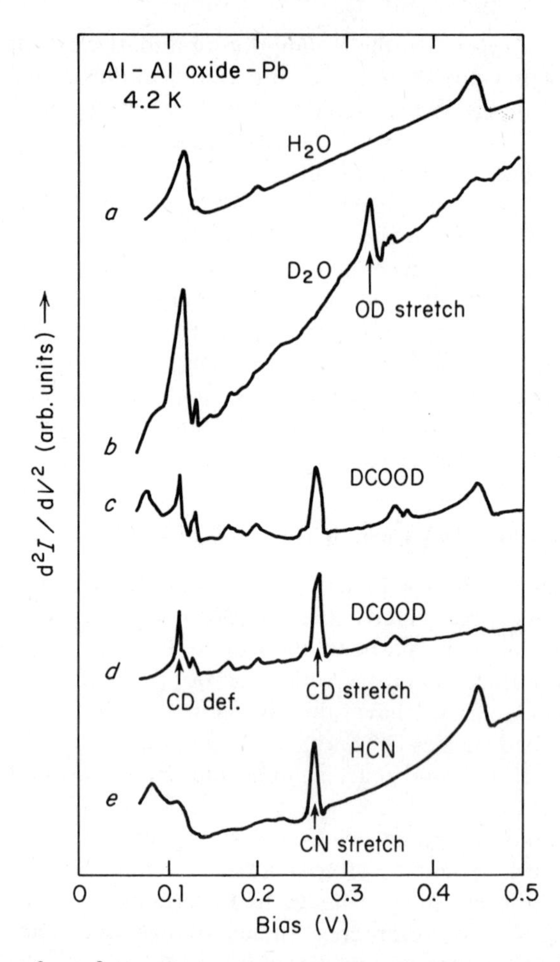

Fig. 10.5. The d^2I/dV^2 (IETS) spectra showing peaks due to molecular vibrations of species inside the barrier: *a*, clean junction; *b*, spectrum of OD-stretching mode induced by exposure of completed junction to D_2O vapor; *c*, completed junction exposed to deuterated formic acid vapor; *d*, formic acid spectrum for a junction doped conventionally; *e*, spectrum produced by exposure to HCN vapor. (After Jaklevic and Gaerttner, 1977.)

10.6. The original chemical is propiolic acid, HC≡CCOOH (infused from 3% solution in 95% relative humidity). The triple carbon-carbon bond (characteristic of acetylene) gives rise to three distinct features, which are pointed out in the original IETS spectrum (curve *A* in Fig. 10.6). These features disappear upon aging. The resultant (reactant) spectrum (*B*) is evidently identical to the spectrum *C*, which also results from direct infusion of acrylic acid (H_2C=CHOOCH). These spectra are also in good agreement with the summed IR and Raman lines of sodium acrylate (bottom). The surface reaction is considered to be hydrogenation from

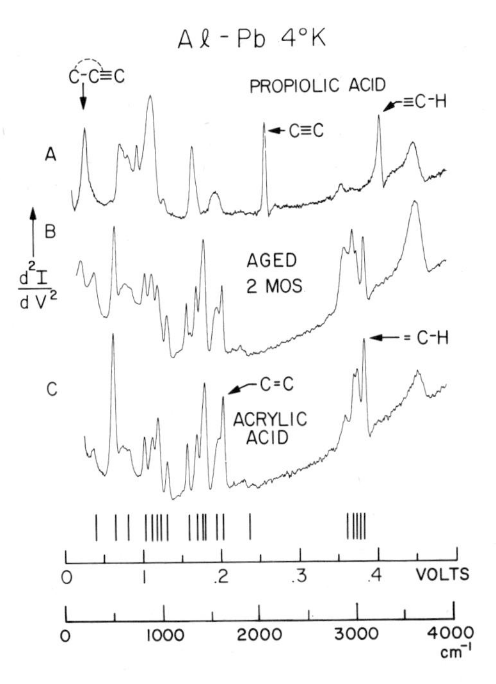

Fig. 10.6. Observation of a chemical reaction occurring in the barrier's dopant layer. Curve *A*, spectrum of propiolic acid; *B*, spectrum measured on the same junction after storage for two months; *C*, spectrum of junction doped deliberately with acrylic acid. Note the similarity between curves *B* and *C*. The line spectrum shown beneath curve *C* represents the infrared spectrum of acrylic acid and identifies this species in the junctions of curves *B* and *C*. (After Jaklevic, 1980.)

OH^- resident on the surface. It was found that the conversion also could be achieved in a few minutes by heating the junction containing propiolic acid to 170°. The IETS spectrum was found always to consist only of the sum of lines of the two species; providing no evidence for an intermediate species.

An earlier report of a similar surface reaction, the hydrogenation of muconic acid, was given by McBride and Hall (1979). These workers initiated the hydrogenation reaction on the acid-doped alumina surface with a controlled heating step in the presence of water vapor. The water vapor acted as a source of H by chemisorbing to OH^- binding sites on the alumina surface. The process is described by McBride and Hall as catalytically induced transfer hydrogenation from water vapor of surface muconate (the chemisorption product on alumina of muconic acid) to surface adipate. No evidence for intermediate species (other than the surface muconate and OH^- ions) was present in the measurements.

Interestingly, a reaction intermediate species in the catalytic formation of hydrocarbons from CO has been observed by IETS in the much more complex situation of the alumina-supported metal catalyst Rh, the latter embedded in a tunnel junction as described previously. The identification of an ethylidene ($CHCH_3$) intermediate species in formation of hydrocarbons from CO on alumina-supported Rh particles by Kroeker, Kaska, and Hansma (1980) followed their systematic development of relevant IETS techniques (Hansma, Kaska, and Laine, 1976; Kroeker, Kaska, and Hansma, 1979). The resulting proposed identification has stimulated parallel studies using electron energy loss spectroscopy, as has been reviewed by Kroeker (1982). In a similar study, the adsorption of ethanol on alumina-supported Ag particles has been reported by Evans, Bowser, and Weinberg (1974). It appears that the avenue to an important set of further IETS studies of catalysis by supported metal particles is now open.

Tunneling research has also been applied to the supported complex catalysts—e.g., $Zr(BH_4)_4$ supported on alumina. Studies of this particular system with IETS methods in connection with polymerization reactions of ethylene, propylene, and acetylene have been reported in a series of articles by Evans and Weinberg (1980*a*, 1980*b*, 1980*c*). In a related area, the structure of the adsorbed dimer $[RhCl(CO)_2]_2$ on alumina has been studied, and the results have been compared with those of previous work on CO:Rh by Bowser and Weinberg (1981). This important line of IETS research is reviewed by Weinberg (1982).

11

Applications of Barrier Tunneling Phenomena

11.1 Introduction

In the fundamental sense, a large fraction of all electronic processes on the microscopic level, where wavefunctions and quantum mechanics are essential, may be regarded as involving tunneling. For example, electronic conduction in a band is a process of repetitive tunneling from classically allowed regions (near the nuclei) through Coulomb barriers in the interstitial regions, where the wavefunctions decay exponentially. In this sense, applications of tunneling phenomena encompass much of chemistry and condensed matter physics. However, the domain of *barrier tunneling*—as implicitly associated in this book with geometries in which only one dimension is microscopically small, providing the barrier $V(x)$, while in the other two dimensions the structure is macroscopically planar—is greatly restricted. The difficulty in experimentally realizing the required regularity in the microscopic barrier $V(x)$ over the two macroscopic dimensions of solid-state structures is reflected, at least in part, in the award of the Nobel Prize in physics to Esaki, Giaever, and Josephson in 1973.

Restricting consideration to the barrier tunneling structures which have been considered in the preceding chapters, we may already point out many *diagnostic* applications of tunnel junctions. These applications have been largely in the category of deduced normal-state and superconducting properties of materials, principally metals and semiconductors. Phonon spectra and energy gaps are obvious examples. A more fundamental group of applications already discussed has been use of Josephson tunneling structures and the basic Josephson relation $\hbar\omega = 2eV$ to examine the exactness of pairing $e^*/e = 2$ in superconductivity; assuming this result, to measure with improved accuracy the ratio $2e/h$, as shown in Fig. 11.1; and assuming only constancy of e/h, to standardize voltage measurements against frequencies.

The use of *tunnel junction devices*, composed of single or multiple junctions, to perform specific functions has been mentioned occasionally but remains to be systematically discussed. Examples of tunnel junction devices as electronic circuit elements include the Esaki diode, as an oscillator or nonlinear element, and the Josephson junction, as a magnetically actuated bistable element (switch) from which families of computer logic circuits have been synthesized. The latter applications exploit

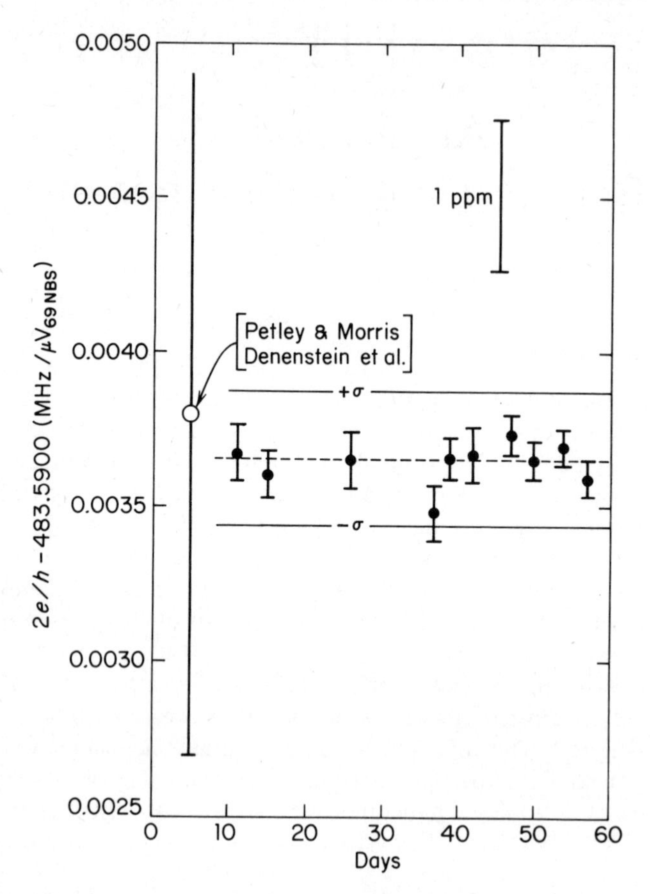

Fig. 11.1. Illustration of the use of Josephson effect in determination of the fundamental constant ratio $2e/h$. Here experimental values of $2e/h$ as a function of time of reported observation are displayed (after Finnegan, Denenstein, and Langenberg, 1970) in comparison with earlier results.

the magnetic field dependence of the maximum Josephson supercurrent shown in Fig. 11.2.

Examples of tunnel junction devices as *sensors* include the use of normal-state and superconducting junctions as light or microwave detectors, the use of superconducting junctions as phonon generators and detectors, and, especially, the use of pairs of Josephson junctions as magnetic interferometers or in the SQUID (superconducting quantum interference device). The latter class of devices has allowed measurements of magnetic field (and magnetic field gradient), current, and voltage at unprecedented sensitivity. Further superconducting junction devices which we have mentioned earlier (Sec. 7.4) are Gray's transistor (1978*a*) and the QUITERON (Faris, Raider, Gallagher, and Drake, 1983).

A different type of application of the tunneling process based on its

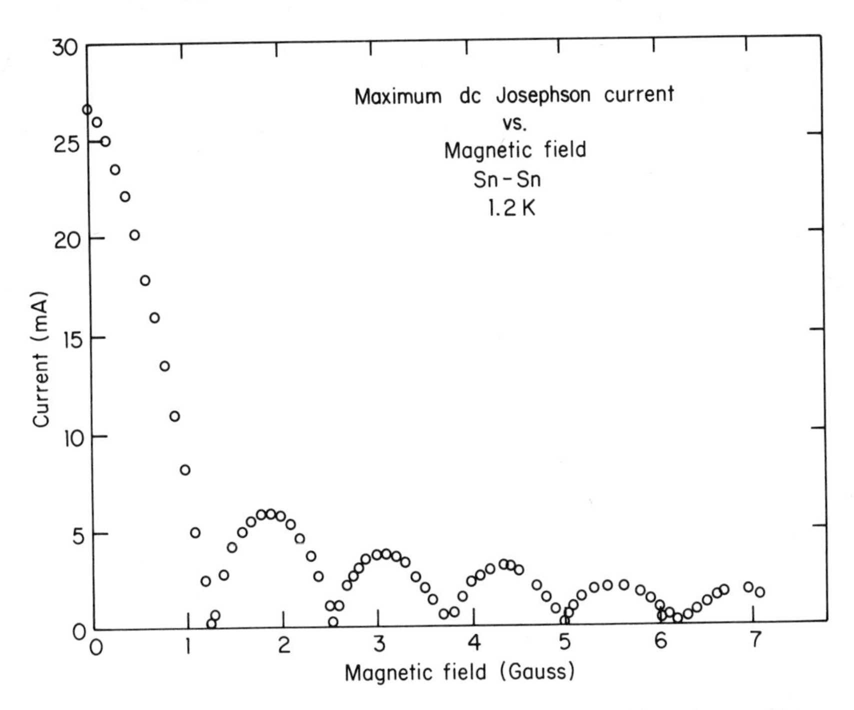

Fig. 11.2. The maximum supercurrent in a Josephson tunnel junction oscillates as shown with applied magnetic field, in the same Fraunhofer diffraction pattern behavior shown earlier, and can be used as the basis for switching elements for computers. In this case, a control current may provide the magnetic field which switches the junction. (After Langenberg, Scalapino, and Taylor, 1966.)

short range is the scanning tunneling microscope (Binnig, Rohrer, Gerber, and Weibel 1982*a*, 1982*b*, Binnig and Rohrer, 1982). Several other scanning or imaging devices based on the tunneling process will be mentioned.

The aim of this chapter is somewhat limited: to point out only the basic concepts for the digital and analog applications (principally based on the SQUID) of Josephson junctions, and for tunnel junctions as detectors, for these topics have already been extensively reviewed; and to describe only a relatively few novel and possibly promising applications from the broad remaining range of diagnostic and device applications of barrier tunneling phenomena.

11.2 Josephson Junction Interferometers

The concept of the Josephson junction interferometer is illustrated in Fig. 11.3, while its first experimental realization, by Jaklevic, Lambe, Silver, and Mercereau (1964, 1965), is illustrated in Fig. 11.4a.

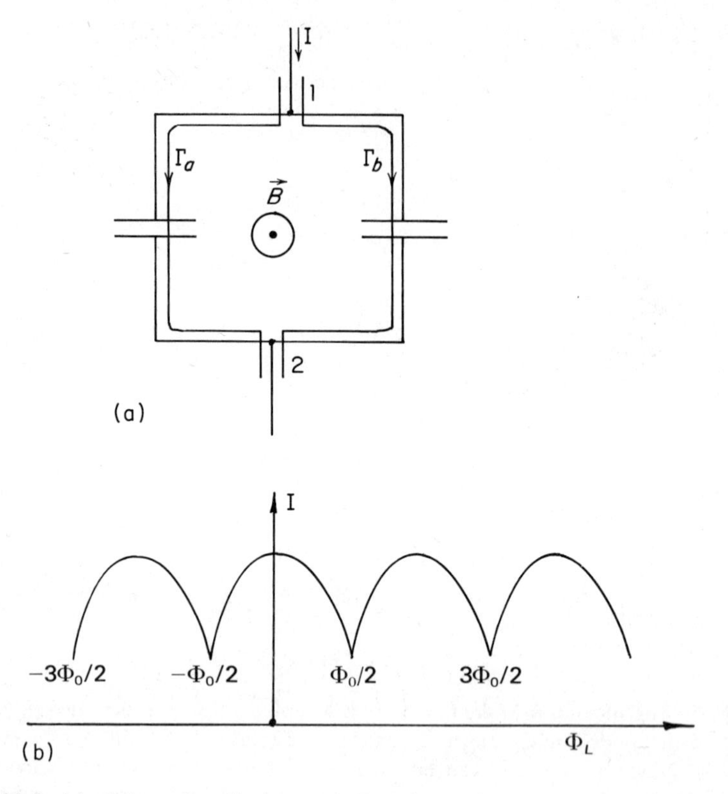

Fig. 11.3. (a) Schematic diagram of Josephson junction interferometer. (b) Response of interferometer current to increasing magnetic flux Φ_L threading the loop.

As first mentioned in Sec. 3.1, the supercurrent density J in a single Josephson junction is given by $J = J_0 \sin \phi$, where J_0 is the maximum supercurrent density and ϕ is the phase difference between the two superconductor electrodes. In the geometry of Fig. 11.3a, the total current $I = I_a + I_b$ is the sum of those through a and b junctions. The phase difference $\Delta\phi$ between points 1 and 2 (see Fig. 11.3a) must be the same over either path Γ_a or Γ_b; thus, using Eq. (3.4), one has

$$\Delta\phi = \phi_a + \frac{2\pi}{\Phi_0}\int_{\Gamma_a} \vec{A} \cdot \vec{d}r = \phi_b + \frac{2\pi}{\Phi_0}\int_{\Gamma_b} \vec{A} \cdot \vec{d}r \tag{11.1}$$

where $\Phi_0 = h/2e$ and $\vec{A}$ is the magnetic vector potential. But this equation can be rewritten, letting $\phi_b - \phi_a = \Delta\phi$, as

$$\Delta\phi = \frac{2\pi}{\Phi_0}\left[\int_{\Gamma_b} \vec{A} \cdot \vec{d}r - \int_{\Gamma_a} \vec{A} \cdot \vec{d}r\right] = \frac{2\pi}{\Phi_0}\int_{\Gamma} \vec{A} \cdot \vec{d}r \tag{11.2}$$

where Γ is the closed-loop path $\Gamma_b - \Gamma_b$. Hence, using Stokes's theorem,

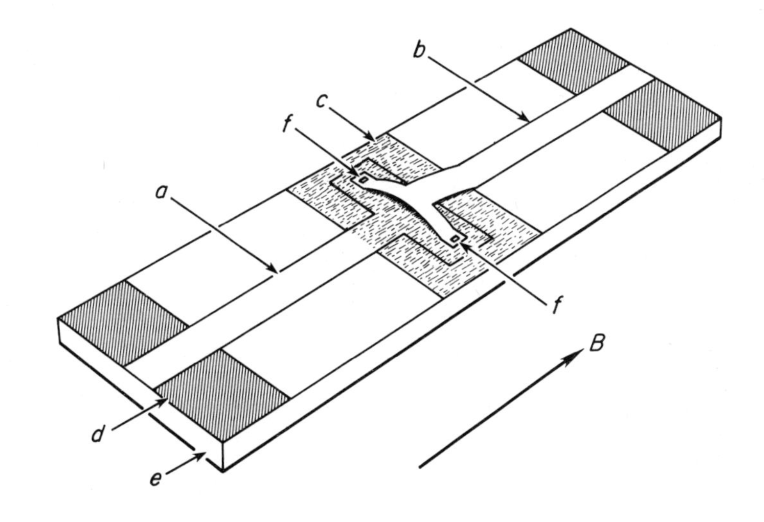

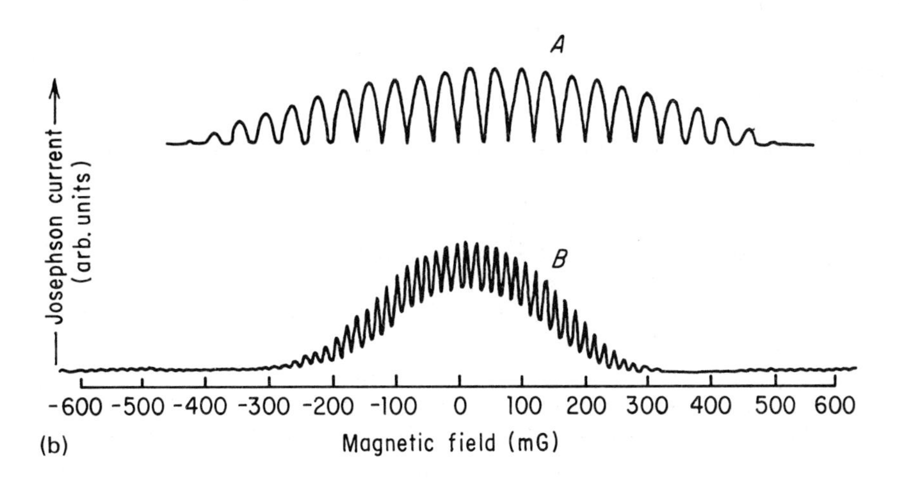

Fig. 11.4. (a) Schematic of junction interferometer employed by Jaklevic, Lambe, Mercereau, and Silver (1965). A magnetic field B is applied parallel to the long direction of the substrate. (b) Experimental traces of the maximum Josephson current vs. magnetic field showing interference and diffraction effects. The field periodicity is 39.5 mG for A and 16 mG for B. In both cases, the junction separation w is 3 mm and the junction width is approximately 0.5 mm. (After Jaklevic, Lambe, Mercereau, and Silver, 1965.)

one finds

$$\Delta\phi = \frac{2\pi}{\Phi_0}\int \mathrm{B}\cdot d\mathrm{A} = 2\pi\left(\frac{\Phi_L}{\Phi_0}\right) \tag{11.3}$$

where Φ_L is the magnetic flux threading the area A_L of the large loop Γ, neglecting flux from the current I itself. We arbitrarily set $\phi_a = \delta + (\Delta\phi/2)$, which requires $\phi_b = \delta - (\Delta\phi/2)$, and neglect the flux cutting the individual junctions compared with Φ. Then

$$I = I_a + I_b = I_0\left[\sin\left(\delta + \frac{\pi\Phi_L}{\Phi_0}\right) + \sin\left(\delta - \frac{\pi\Phi_L}{\Phi_0}\right)\right]$$

Hence, the maximum current $I_{\max}$, reached when $\delta = \pi/2$, is

$$I_{\max} = 2I_0\left|\cos\frac{\pi\Phi_L}{\Phi_0}\right| \tag{11.4}$$

where $I_0 = AJ_0$, A being the junction electrode area. In a more complete treatment, (11.4) is multiplied by

$$\left|\sin\left(\frac{\pi\Phi_j}{\Phi_0}\right)\left(\frac{\pi\Phi_j}{\Phi_0}\right)^{-1}\right| \tag{3.104}$$

where Φ_j is the smaller flux through each junction. The response of the interferometer current I to increasing flux $\Phi_L \gg \Phi_j$, (11.4) (see Fig. 11.3b), is oscillatory, with its first minimum at $\Phi_L = \Phi_0/2$. This function can be an extremely rapid function of magnetic field if the loop area A_L is large, as follows from the small size of the flux quantum, $\Phi_0 \simeq 2.07\times10^{-7}\ \mathrm{G\cdot cm^2}$. If the area A_L is 1 cm^2, e.g., one period of the current oscillation in (11.4) occurs for $A_L B/\Phi_0 = 1$, or $B \simeq 10^{-6}$ G. This effect was first observed in structures of smaller loop area, as shown in Fig. 11.4, by Mercereau and co-workers at the Ford Scientific Laboratory. The modulation of the interference pattern by the $(\sin x)/x$ single-slit diffraction envelope is also discernible in these data.

11.3 SQUID Detectors

The dc SQUID (Zimmerman and Silver, 1966; Clarke, 1966) is a two-junction interferometer optimized for the measurement of small magnetic fields. With use of a superconducting input coil coupled to the loop area A_L, measurements of current and voltage are possible. If the input coil is composed of two reversed pickup coils connected in series, magnetic field gradients can be measured.

Such measurements are usually arranged with feedback from the SQUID output to a second input coil, so that the SQUID loop sees zero or constant flux; i.e., the instrument is used as a null detector. In the SQUID application of the two-junction interferometer, it is desirable to modify the junction devices to avoid the inherently hysteretic I–V curves

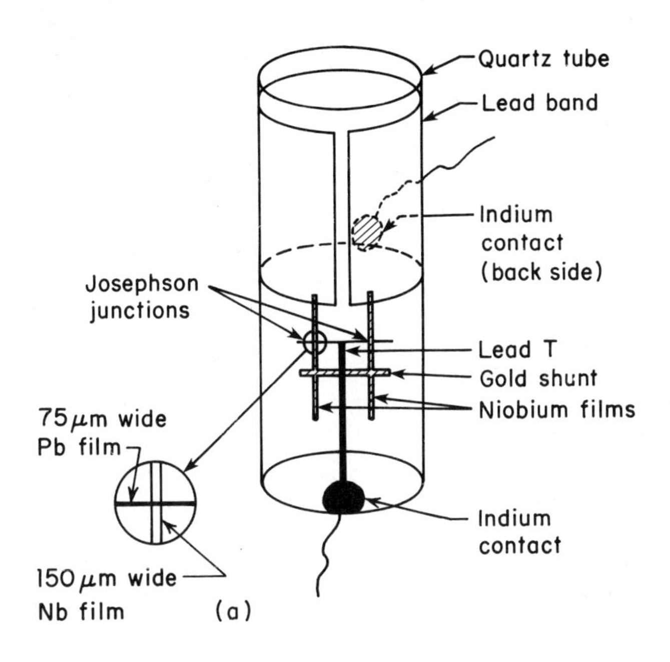

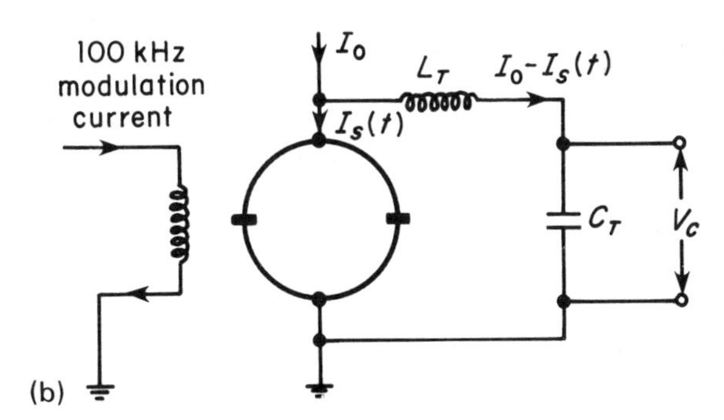

Fig. 11.5. (a) Configuration of a thin-film tunnel junction SQUID. The substrate is a fused quartz tube with an o.d. of 3 mm and with a Pb/In band evaporated around the circumference. This process is followed by deposition of the Au shunt film and by sputter deposition of two Nb films separated by 1.2 mm. After oxidation, a T-shaped film 3000 Å thick of Pb/In is deposited, forming the two Josephson tunnel junctions. (b) Schematic diagram showing incorporation of the SQUID into a basic measuring circuit utilizing an ac modulation coil and a resonant tank circuit operated at 100 kHz. (After Clarke, Goubau, and Ketchen, 1967.)

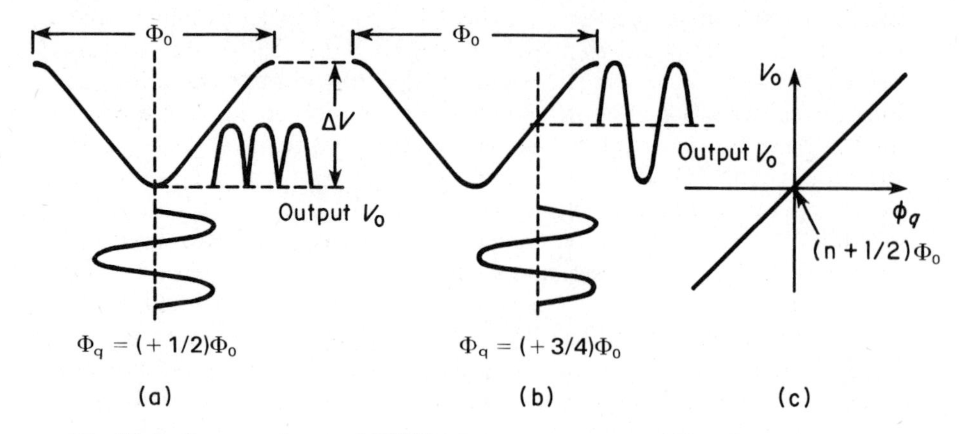

Fig. 11.6. Output voltage of SQUID in response to an applied flux at frequency ν_0. (a) The quasistatic flux is $\Phi_q = (n + \frac{1}{2})\Phi_0$, and the output voltage is predominantly at $2\nu_0$. (b) Here, $\Phi_q = (n + \frac{3}{4})\Phi_0$, and the output voltage is ν_0. (c) The output voltage V_0 at ν_0 as a function of Φ_q. (After Clarke, Goubau, and Ketchen, 1976.)

shown in Fig. 11.6, which also apply to two junctions in parallel, as in Fig. 11.3. In the work of Clarke, Goubau, and Ketchen (1976), described in Figs. 11.5–11.7, this modification was accomplished by shunting the individual junctions (Hansma, Rochlin, and Sweet, 1971) with resistors in the form of gold films (Fig. 11.5a). The nonhysteretic I–V characteristic makes feasible a detection circuit using a modulation field (Fig. 11.6) and lock-in detection, which in the work of Clarke, Goubau, and Ketchen (1976) was carried out at $\nu_0 = 100$ kHz. The device is dc-biased above the critical current as shown in Fig. 11.6a, where it is seen that the I–V dependence oscillates between the two limiting curves shown as Φ

Fig. 11.7. Output of the SQUID in a flux-locked condition at the temperature of the liquid helium, 4.2 K, with a measurement bandwidth of dc to 0.25 Hz. The sensitivity of the instrument is on the order of 10^{-9} G. (After Clarke, Goubau, and Ketchen, 1976.)

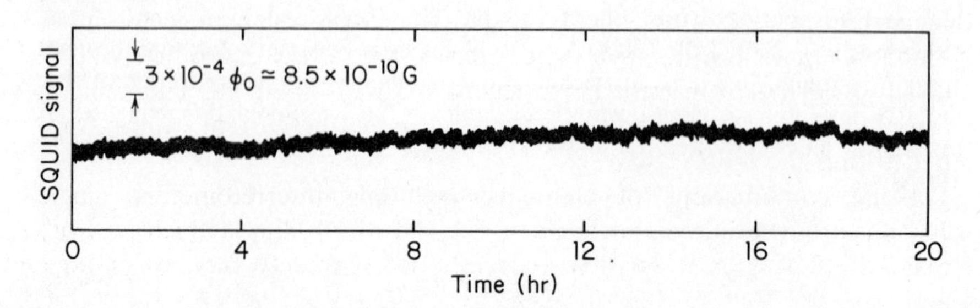

increases. Normally, the feedback field is controlled to maintain Φ at $(n+\frac{1}{2})\Phi_0$, as determined by a null of the ν_0 component of the lock-in-detected voltage V_c (Figs. 11.5b and 11.6b); this process is described as flux-locked operation. The magnetic field resolution of the device operated as described in detail by Clarke, Goubau, and Ketchen (1976) is on the order of 10^{-9} G, as shown in Fig. 11.7. The energy resolution of the device operated with a 24-turn input coil is stated as 7×10^{-30} J/Hz.

While the device of Clarke, Goubau, and Ketchen, (1976) is certainly practical, the intent in describing it has been primarily to illustrate the basic ideas. In fact, most commercial SQUID devices are rf SQUIDs (Zimmerman, Thiene, and Harding 1970) in which a single loop containing a junction is employed and rf detection is used. Discussion of this case is beyond the scope of the present chapter. For further details, see Solymar (1972) or van Duzer and Turner (1981) and references therein.

Application of these devices has been widespread, as may be seen from the above general references or from books such as those edited by Halbohm and Lubbig (1977), and by Deaver, Falco, Harris, and Wolf (1978).

11.4 Logic Applications of Josephson Devices

In logic applications, the inherently bistable nature of the Josephson junction I–V characteristic is exploited. In one basic configuration flux dependence of the maximum current of the single Josephson junction [Fig. 11.2, Eq. (3.104)] is used to switch the junction along a load line from the superconducting ($V=0$) state to a resistive state at $V<2\Delta/e$. This behavior is utilized, e.g., to switch currents in memory cells and also to sense the existence of flux in such cells.

The two-junction interferometer (Fig. 11.3a) can be used as an elementary memory cell. Even in the absence of net current I vertically through the two paths Γ_a and Γ_b of Fig. 11.3a, there can exist within the cell stable circulating current states corresponding to trapping of 0 or $\pm n$ units of flux Φ_0. The states—e.g., $n=0$, $n=1$—may be regarded as the basic states 0, 1 of a latching binary logic and be detected through the effect of the associated magnetic field on a Josephson junction sensor close to the cell (not shown). Reading in the circulating current can be accomplished by applying current I along the vertical path and simultaneously switching one of the junctions, say b, into the resistive state by means of current in a control line (not shown) whose magnetic field drives junction b into the resistive state. Hence, the vertical current I will be totally diverted from path Γ_b to follow path Γ_a. If, then, junction b is returned to the superconducting state, a circulating current can be trapped within the loop.

Using combinations of elements, including interferometers, single Josephson junction switches and sensors, and control lines (which generate magnetic fields), it is possible to synthesize complete families of logic circuits. Examples of such circuits are those to perform logical OR and AND functions with various numbers of inputs. An excellent source for further information is the March 1980 number of the IBM *Journal of Research and Development,* particularly articles by Anacker (1980), Matisoo (1980), Gheewala (1980), and Faris, Henkels, Valsamakis, and Zappe (1980).

In these applications, there are two inherent advantages of Josephson junctions over their counterparts in the silicon and GaAs semiconductor technologies. The first is lower dissipation level stemming basically from the ratio $\Delta/E_g \simeq 10^{-3}$ of the bias voltage scales, where the typical semiconductor band gap is $E_g = 1$ eV. This may ultimately allow higher device-packing density (permitting reduced transit and cycle times) within tolerable operating temperature limits. The importance of this factor will increase as the lithographic advances permit smaller device size. Secondly, one has the inherently short switching time $\tau = RC$ of a Josephson junction. These factors favor small devices of low resistance and low capacitance, factors which are optimized in edge junctions (Heiblum, Wang, Whinnery, and Gustafson, 1978; Havemann, 1978; Howard et al., 1979; Kleinsasser and Buhrman, 1980; Broom, Oosenbrug, and Walter, 1980). In these devices, the tunnel junction is formed on the edge or end of a deposited film which has been cut off obliquely, e.g., by vapor phase etching, leaving a tapered surface typically at a 45° angle to the substrate plane. If the film is 2000 Å thick, then the junction area might be a rectangle of width $w = \sqrt{2}\,2000$ Å $= 2830$ Å and length ℓ in the range 0.2–10 μm. Evidently, very small areas, in the range 6×10^{-10}–3×10^{-8} cm^2 can be fabricated in this fashion. Junction areas as low as 10^{-10} cm^2 have, in fact, been reported (Heiblum et al., 1978; Jackel et al., 1981), although precise reproducibility of this area over large numbers of junctions, necessary in the computer application, would seem a formidable problem. In any case, RC times, using edge junctions, as low as 1.6 psec have been reported (Vettiger, Moore, and Forster, 1981) in a device with a total Josephson current of 0.58 mA, corresponding to current density 3×10^4 A/cm^2. This switching time is shorter than times available in semiconductor devices.

Other important applications of Josephson devices exist in high-speed analog and analog-digital electronics. These applications have been summarized by Hamilton, Lloyd, and Kautz (1981).

11.5 Detection of Radiation

There are long histories and extensive literatures associated with photodetection by both normal-state tunnel junctions and by superconducting tunnel junctions. The normal junction prototype is probably the classical metal-semiconductor square law detector in the form of a metal "cat whisker" point contact on a galena (PbS) crystal used as an rf detector in a "crystal radio." Operation of this detector, like that of its primary present-day successor, the conventional tungsten-silicon microwave diode (which, however, is not a tunneling device), depends upon nonlinearity in the I–V relation, characterized conventionally by

$$S = \frac{\mathrm{d}^2 I/\mathrm{d}V^2}{\mathrm{d}I/\mathrm{d}V} \tag{11.5}$$

This nonlinearity parameter is closely related to the figure of merit of the direct detector, the classical current responsivity

$$R_i = \frac{\Delta I_{\mathrm{dc}}}{\Delta P_a} = \frac{S}{2} \tag{11.6}$$

where ΔI_{dc} is the change in dc output for an increment ΔP_a in absorbed power. Note that for a device obeying the ideal exponential law,

$$I = I_0\left[\exp\left(\frac{eV}{kT}\right) - 1\right]$$

the S parameter is just e/kT.

A second requirement for an rf detector is a low internal shunt capacitance C such that $\mathrm{d}I/\mathrm{d}V/\omega C \gg 1$ at frequencies of interest. This condition, favored by the small contact area, had—at least until the advent of the edge junction of Heiblum, Wang, Whinnery, and Gustafson (1978) and Havemann (1978)—best been satisfied in the mechanically unstable point-contact type of device.

A second, related mode of detection is heterodyne detection, mixing the signal at ω_S, with a local oscillator signal at ω_{LO}. The desired output is (usually) the lower difference frequency $\omega_S - \omega_{LO}$, termed the intermediate (IF) frequency, at which subsequent amplification at reduced bandwidth can be accomplished. Clearly, the mixing effect still depends on large nonlinearity S. The case of direct detection is formally obtained as mixing to zero frequency with $\omega_S = \omega_{LO}$. A classical figure of merit for heterodyne detection is the available conversion efficiency,

$$L^{-1} \equiv \frac{P_L}{P_S} = \left[\frac{\partial I(V, P_L)}{\partial (P_L^{1/2})}\right]^2 \frac{R_D}{8} \tag{11.7}$$

where $R_D = [\mathrm{d}I/\mathrm{d}V]^{-1}$ and P_L and P_S are, respectively, the available powers (assuming perfect matching) at the local (IF) and signal (S) frequencies. Here, $L^{-1} > 1$ (conversion gain) can occur classically only if a negative resistance appears in the I–V characteristic. Conversion gain can be important in attaining low-noise mixing, because the overall noise temperature of the heterodyne receiver can be shown to be

$$T_R = T_D + \frac{T_{\mathrm{IF}}}{L^{-1}} \tag{11.8}$$

Here, T_D and T_{IF}, are, respectively, the noise temperatures of the detector (mixer) and IF amplifier. Since it is common for the term in T_{IF}

to be dominant, a large value of L^{-1} may be the only way to obtain a low overall noise temperature T_R.

In tunneling devices, the nonlinearity S required in both direct and heterodyne detection can arise in several different ways. The typical weakly quadratic behavior of the background conductance $G(V) = G_0 + \alpha V^2$ of the MIM tunnel junction, as studied by Brinkman, Dynes, and Rowell (1970), offers a minimum level of S, which can be increased by choosing a tunnel barrier I of low height. Such low barriers form, as we have noted, rather easily with transition metals, with oxides such as NbO_x (Walmsley, Wolf, and Osmun, 1979; Brunner, Ekrut, and Hahn, 1982), which make them less than fully suitable for tunneling spectroscopy; this feature is of advantage in application to detection. An extremely useful, but still poorly characterized, tunneling device which presumably involves such a low barrier is the Ni–W freestanding, optical antenna, point-contact optical detector (Dees, 1966; Sanchez, Davis, Liu, and Javan, 1978). This device is now used as a heterodyne detector of laser light up to about 200 THz frequency ($\lambda = 1.5\ \mu m$) (Baird, 1983; Pollock et al., 1983).

A second tunneling mechanism for obtaining larger curvature S, and corresponding large detective responsivity R_i, is to use a Schottky tunnel diode of very small contact area, which may be akin to the historical cat whisker detector mentioned earlier. However, the frequency response of this device appears inherently to have a much lower cutoff, possibly due to the nature of conduction in the series-spreading resistance region (necessarily a region of relatively low carrier density $n < 10^{20}\ cm^{-3}$) connecting the small-area Schottky barrier tunnel contact to the doped semiconductor. A rather extensive body of information on this type of device had accumulated consequent to its use with a superconducting counterelectrode for near-millimeter wave detection ["super-Schottky detector"; see, e.g., Silver, Chase, McColl, and Millea (1978)], where it appears to be effective up to about 30 GHz.

If cryogenic operation is not an impediment, the use of one or two superconducting electrodes offers the possibility of vastly enhanced nonlinearity S, in NIS or SIS devices operated, respectively, near biases Δ/e and $(\Delta_1 + \Delta_2)/e$. Considerable advance, both experimental and theoretical, has occurred in the past few years, specifically for the SIS quasiparticle mixer biased at the sum gap edge (e.g., see Richards and Shen, 1980; Phillips and Dolan, 1982).

11.5.1 SIS Detectors

The subject of SIS quasiparticle photodetection has followed an interesting and surprisingly slow line of development (at least until recently). The original proposal of SIS photodetection (Burstein, Langenberg, and Taylor, 1961) was couched in the quantum terms of "photo injection" of carriers (quasiparticles) by optical excitation across the energy gap of the superconductor, the quasiparticles being subsequently collected by low-voltage bias, $eV < 2\Delta$. The proposed quantum mechanism of Burstein, Langenberg, and Taylor, (1961) imposed a long-wavelength detection limit $\lambda_{max} \leqslant hc/2\Delta$, estimated as $\lambda_{max} \leqslant 3.9$ mm for Al–I–Al junctions and $\lambda_{max} \leqslant 0.46$ mm for Pb.

The next relevant development, again understood in *quantum* terms, was the splitting of the SIS sum-gap edge by an imposed microwave field (Dayem and Martin, 1962; Tien and Gordon, 1963). In spite of this history, however, until fairly recently (Richards, 1978), experimental work had largely been limited to the NIS case in the form of the super-Schottky detector (McColl, Millea, and Silver, 1973; Silver, Chase, McColl, and Millea, 1978); these NIS results and projected work on SIS quasiparticle detectors (Richards, 1978) were both being considered in wholly *classical* terms. It appears possible that discovery of the Josephson effect and the original application of that effect to rf detection (Grimes, Richards, and Shapiro, 1966, 1968) diverted attention for about fifteen years from the SIS quasiparticle detector. Now, however, the research and development emphasis has returned to the SIS quasiparticle detector, following key theoretical work of Tucker and Millea (1978) and Tucker (1979, 1980). This work showed (unexpectedly, in the recent context) from a purely quantum mechanical approach, expanding upon the original work of Tien and Gordon (1963), that the S–I–S junction as quasiparticle heterodyne detector is, in fact, capable of conversion gain $L^{-1} > 1$. With the subsequent observations of this effect, the S–I–S gap edge-biased quasiparticle detector now appears to be the most promising superconducting rf detector. These devices are now being developed for heterodyne detection at 115 GHz in radio astronomy, where a detector noise temperature T_D approaching the quantum limit $T = h\nu/k$ can be expected (Shen, Richards, Harris, and Lloyd, 1980; see also Phillips and Dolan, 1982). This surpasses the performance of all other detectors including cooled Schottky diodes, super-Schottky diodes, InSb bolometer mixers (see Phillips et al., 1981), and Josephson mixers (Taur and Kerr, 1978; Richards and Shen, 1980; Taur, 1980).

It has also been established in general by the Tucker analysis (applied to direct detection, rather than heterodyne detection) that, in the case of sufficiently strong nonlinearity S, a quantum regime is entered in which unit quantum efficiency (one charge per absorbed quantum) can be attained. Thus, in the SIS case, the circle has finally all but returned, in 1983, to the original quantum picture of Burstein, Langenberg, and Taylor (1961), with most of the details properly filled in. The Josephson mixer, on the other hand, after intense and extensive study, has been established as being inherently noisy (Taur, Claasen, and Richards, 1974; Tucker, 1980) and probably incapable of attaining the quantum noise limit.

The quantum limit of detection is attained in general for nonlinearity S satisfying

$$\frac{S}{2} > \frac{e}{\hbar\omega} \tag{11.9}$$

(Tucker and Millea, 1978; Tucker, 1979, 1980). In this case, the responsivity is given by $R_{\hbar\omega}$:

$$R_{\hbar\omega} = \frac{e}{\hbar\omega}\left[\frac{I(V+\hbar\omega/e) - 2I(V) + I(V-\hbar\omega/e)}{I(V+\hbar\omega/e) - I(V-\hbar\omega/e)}\right] \tag{11.10}$$

This expression can be compared with (11.5) and (11.6) by noting that the limits (derivatives) have been replaced by finite differences. If the value of $\hbar\omega$ is effectively larger than the width of the corner in the SIS I–V characteristic at $eV = \Delta_1 + \Delta_2$, so that $I(V+\hbar\omega/e) \gg I(V)$ while $I(V-\hbar\omega/e) \simeq I(V)$, then R_i takes on the quantum limiting value $R_i = e/\hbar\omega$ (one charge per absorbed photon). A similar change is required in the value of detector resistance, $R_D = (\mathrm{d}I/\mathrm{d}V)^{-1}$, in (11.7) for the conversion efficiency, the new result being

$$R_D = \frac{2\hbar\omega}{I(V+\hbar\omega/e) - I(V-\hbar\omega/e)} \tag{11.11}$$

These quantum limit modifications apply generally, but they have first been fully observed, including the negative-resistance effect, in the SIS quasiparticle mixers (Kerr, Pan, Feldman, and Davidson, 1981; McGrath et al., 1981; Smith et al., 1981), as shown in Fig. 11.8. In this figure, the (Dayem and Martin) step structure results from the 36-GHz local oscillator photon, and the negative-resistance region for the $n = 1$ photon absorption is clearly evident. As pointed out in the 1981 references

Fig. 11.8. The I–V measurements on an Sn–SnO_2–Sn tunnel junction under irradiation with 36-GHz photons show a negative resistance region on the first step below the sum gap 2Δ, as theoretically predicted by Tucker (1980). This effect allows large gain in a heterodyne SIS tunnel junction mixer. (After Smith et al., 1981.)

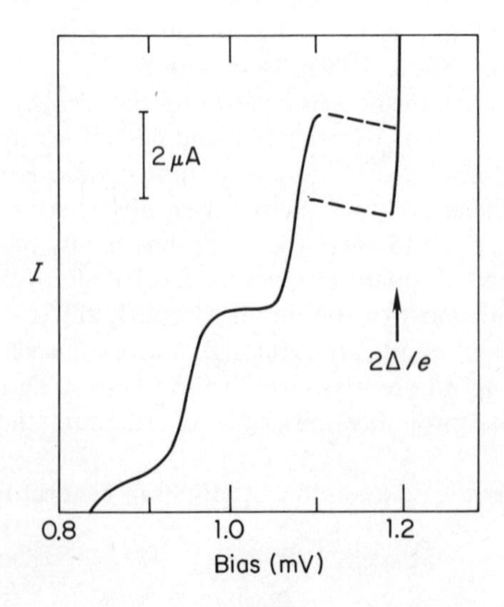

above, the important implication of the observed negative resistance is that a properly matched mixer should have unlimited conversion efficiency (gain) L^{-1}, allowing a low-noise temperature via (11.8).

11.5.2. Josephson Effect Detectors

In the original work of Grimes, Richards, and Shapiro (1966, 1968), analysis of rf detection by Josephson junctions was based on the known sensitivity of the Josephson current to magnetic flux, i.e. to electromagnetic radiation. In fact, the $V=0$ current is depressed, and current steps (Shapiro, 1963) occur at voltages $eV=n\hbar\omega$. For detection purposes, a nonhysteretic junction (such as the shunted junction shown in Fig. 11.6a) is used, with a current bias point such that the voltage state exists. The effect of rf radiation, referring to Fig. 11.6a, is to depress the I–V curve qualitatively as shown in connection with dc SQUID operation. This depression effect is utilized, with a current bias established, to extract a voltage signal dependent on the amplitude of the applied rf. A thorough review of the extensive work following this initial demonstration has been given by Richards (1977); a comparison of the resulting performance with that expected from SIS mixers is offered by Richards and Shen (1980). Other detector applications of the Josephson effect, in a parametric amplifier and a temperature sensor in a composite bolometer detector, are also discussed by Richards (1977).

11.5.3 Optical Point-Contact Antennas (High-Speed MIM Junctions)

The original device of Dees (1966) was a fine tungsten wire cat whisker in contact with a polished metal plate, used for detection of submillimeter waves. Although the detailed mechanism of detection was (and still is) unclear, the extremely low capacitance ($\sim 10^{-2}$ pF) and low spreading resistance were essential factors in an extremely low RC constant, on the order of 10^{-13} sec. The device has evolved (further references to earlier work are given by Heiblum, Wang, Whinnery, and Gustafson, 1978; Sanchez, Davis, Lin, and Javan, 1978) into a fairly standard configuration which is an ultrafine W wire with an etched point (comparable in dimensions to a conventional field emission tip), in contact with a flat Ni anode. Mechanical stress, or possibly electrical pulsing, is used to establish a contact of somewhat enhanced mechanical stability through the layers of oxide on both members of the structure. The resulting barrier, in some cases, is believed to be only 7–10 Å in thickness (Liu, Davis, and Javan, 1979). In use as an infrared optical mixer, the W wire whisker is freestanding, and angular measurements have revealed the expected antenna patterns at relatively large wavelengths. Sanchez, Davis, Liu, and Javan (1978), e.g., show results obtained with a whisker antenna of length 1.95 mm as a detctor of 311 and 337 μm, radiation, analyzed within conentional antenna theory. At wavelengths of the same order, support can be given for a detection mechanism involving tunneling. Several such mechanisms are discussed by Heiblum and colleagues (1978). At optical wavelengths, however, clear evidence has been given by Elchinger, Sanchez, Davis, and Javan (1976) that photoemission over the oxide barrier, and not a nonlinear I–V rectification mechanism, is dominant. The useful cutoff of the present devices appears to be around 1.5 μm; however, see also Pollock et al. (1983). In spite of serious attempts to analyze the operative mechanisms in these useful devices in the cited works of Heiblum et al. and Sanchez et al., a clear picture of all aspects of their structure and operation is apparently not yet available; see, e.g., Feuchtwang, Cutler, and Miskovsky (1981). Efforts to produce fast MIM antenna detectors based on the edge-junction geometry have been reported (Heiblum, Wang, Whinnery, and Gustafson, 1978; Yasuoka, Heiblum, and Gustafson, 1979), but the devices appear thus far not to have improved upon the performance available in the W-Ni whisker antennas. This whole area is of considerable interest from both practical and fundamental points of view, and it may eventually benefit from the edge-junction fabrication techniques further developed in connection with the recent Josephson computer programs.

11.6 Scanning Tunnel Microscope

Measurement of surface topography on an atomic scale has recently been achieved (Binnig, Rohrer, Gerber, and Weibel, 1982*a*, 1982*b*; Binnig and Rohrer, 1982) with an ingeniously scanned point probe from which the vacuum tunneling current to the metallic specimen is detected. A servomechanism is used to maintain the tunneling current constant by adjusting the vertical height of the probe from the surface, thus obtaining a measure of its topography with spatial resolution on the order of the tip radius. From the results, which evidently show a spatial resolution on the order of a few angstroms, much sharper than the overall radii of the ground points of the initial tips, the authors infer the presence on these larger tips of "mini tips" evidently less than the order of 10 Å in radius, formed by "cleaning" or "spot welding" operations against the surface of the specimen. The use of such poorly characterized composite scanning tips was preferred by Binnig, Bohrer, Gerber, and Weibel (1982) to the alternative use of etched field emission tips (whose radii can be reduced to ~100 Å, with a reproducible smooth tip shape) because the latter are rather long and narrow, which was assumed to make microphonic vibration a more severe problem.

The success of the recent work both in obtaining topography and in the first convincing demonstration of true vacuum tunneling as the current mechanism, depends upon the ingenious suspensions and scanning schemes devised by Binnig, some of which are shown in Fig. 11.9. Isolation from vibrations is accomplished by mounting the whole vacuum system on a stone slab isolated from building vibrations. The tunneling

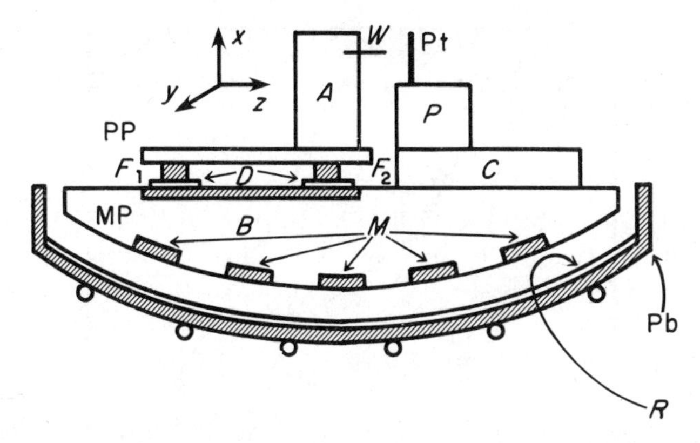

Fig. 11.9. Schematic diagram of a point-contact vacuum tunneling unit capable of scanning. Tunneling occurs between point *W* and platinum plate Pt. Support *A* rests on movable table PP which is a piezoelectric plate capable of contracting in the horizontal direction and supported by two feet F_1 and F_2, which can be clamped independently or released from metal plate MP from which they are insulated by dielectric *D*. By alternated clamping and unclamping of the feet coupled with contractions of the table, unit *A* (called the louse) can be walked across its horizontal resting surface in steps down to the order of 100 Å. Additional piezoelectric drive allows the point *W* to be moved continuously with a resolution of about 0.2 Å. By operation of the device in a constant-current mode while horizontally scanning and applying feedback control to the perpendicular distance, an ultrafine topography of the surface of Pt can be obtained. So that vibrations are reduced, the tunneling unit is mounted on permanent magnets *M* levitated on a superconducting bowl of lead Pb, which is cooled directly by liquid helium. A normal conducting sheet *R* between the lead and the magnets acts as an eddy current damper. (After Binnig, Rohrer, Gerber, and Weibel, 1982*a*.)

probe and sample are mounted on a magnetically (or mechanically) suspended and damped metal plate (MP in Fig. 11.9), which is mechanically isolated from the vacuum system by use of only very fine (0.06-mm-diameter) wires carrying the tunneling signals and actuating the scanning drives of the movable tip and the support table PP (called the "louse," for it walks across the plate MP). Careful mechanical sinking of these wires to the large plate MP before connection to the tip support is important.

Piezoelectric drives are used for the tip and in the body of the "louse." Three "feet" of this unit can either glide across a dielectric film on the metal table MP or be dielectrically clamped to it by application of a voltage difference. With an appropriate pulse sequence (Binnig and Gerber, 1979) actuating body contraction and foot clamping and unclamping, one can arrange a walking motion of the "louse" across the metal table. Fine position of the *x*, *y*, and *z* position of the tip is controlled to a sensitivity of about 2 Å/V by piezodrive *P* (made of commercial piezoceramics such as Phillips PXE–5 or Vibrit, available through Siemens).

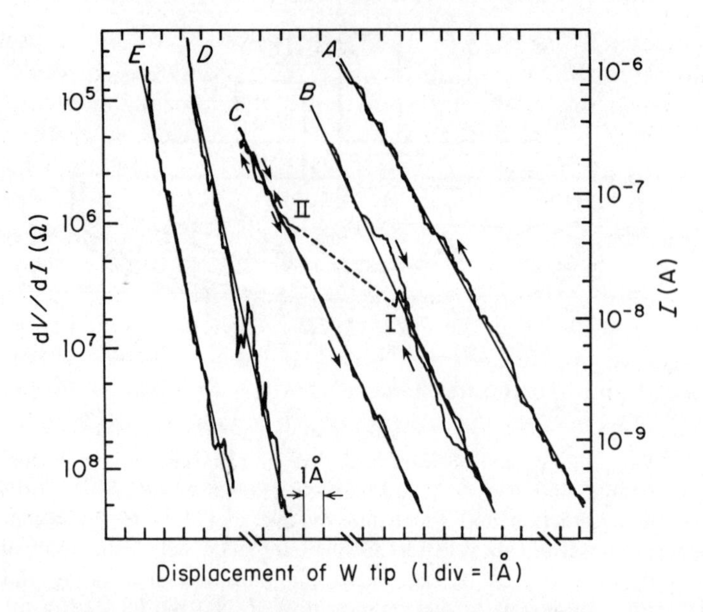

Fig. 11.10. Tunneling resistance and current (left and right scales, respectively) vs. displacement of Pt plate in measurements under vacuum at room temperature. The steepest curves (*E* and *D*) were obtained after cleaning the point and platinum plate and correspond to a work function ϕ of 3.2 eV, on the assumption of Fowler and Nordheim tunneling (Fowler and Nordheim, 1928). While the observed work function is lower than expected for cleaned platinum and tungsten, it is sufficiently high that, taken in addition to the observed exponential dependence of current upon spacing, a vacuum tunneling mechanism is convincingly demonstrated. (After Binnig, Rohrer, Gerber, and Weibel, 1982*a*.)

These drives have been used, as shown in Fig. 11.10, to verify an exponential dependence of current upon tip-specimen spacing, t. The slope of such a plot (of $\log I$ vs. spacing t) determines the work function ϕ; it is seen in Fig. 11.10 that surface-touching cleaning steps between curves C and D appreciably increased the ϕ value from about 0.6 eV (ABC) to about 3.2 eV (in DE), a value in line with the expected value $\phi \simeq \frac{1}{2}(\phi_{\mathrm{Pt}} + \phi_{\mathrm{W}}) \simeq 5$ eV for clean Pt (sample) and W (tip). The expected refinements of this instrument—to allow UHV clean surfaces and argon ion bombardment, and to define in a more satisfactory fashion the minitip geometry—will lead to an extremely useful tool.

11.7 Further Application of IETS

At present, IETS is one of the most active areas in tunneling spectroscopy, which has been applied, as we have earlier mentioned, to determinations of surface chemical species on AlO_x and MgO, with some promising initial results, in addition, in determining the molecular adsorption geometry. This type of information is expected to be useful in deducing intermediate steps in chemical reactions among molecules on a surface. While this work can be viewed as fundamental surface physics research, there is also interest from the point of view of application in several aspects of this work. As we have already mentioned in Chap. 10, the AlO_x surface generated in the tunnel junctions has been shown to closely resemble in its physical and chemical properties the so-called γ alumina used as a catalyst and also as a support for dispersed metal particle catalysts (Hansma, Hickson, and Schwarz, 1977).

Several recent IETS studies have been reported whch touch upon other areas of strong practical interest. As a first example, aspects of corrosion of Al metal and also the mechanism by which an inhibitor (formamide, $HCONH_2$) acts to block this corrosion have been addressed in two papers by Ellialtioglu, White, Godwin, and Wolfram (1980, 1981). The chemical causing corrosion in these studies is carbon tetrachloride, CCl_4, which has been reported to corrode Al (with production of $AlCl_3$) at the appreciable rate of 20 in per year (Stern and Uhlig, 1952*a*, 1952*b*). The point of view taken by Ellialtioglu et al. is that exposure of Al metal to boiling anhydrous CCl_4, the conditions employed by Stern and Uhlig, will still allow formation of thin protective oxide layer, not essentially different from that in the IETS modeling. Within this approach, many molecular species on the AlO_x surface were identified through their vibrational spectra: these included CCl_4, AlCl, $AlCl_2$, $AlCl_3$, C_2Cl_6, the free radical $\cdot CCl_3$, and the complex $CCl_3^+[AlCl_4]^-$. The results were interpreted to suggest that the corrosion process occurs by reaction of the solvent with defect sites which expose Al atoms. These atoms could occur at oxygen vacancies in the AlO_x or at other imperfections such as grain boundaries.

Ellialtioglu, White, Godwin, and Wolfram (1981) then studied the inhibition of CCl_4 corrosion of Al by doping AlO_x barriers with the inhibitor formamide, and also with dilute solutions of formamide in CCl_4. The spectra are interpreted to show that formamide preferentially chemisorbs to the active oxygen vacancy sites on the AlO_x which expose Al atoms, thus blocking the attack of the CCl_4. A model was presented by these authors describing the interaction of the formamide molecule with the surface to form an aluminum-nitrogen bonded surface species.

Another very practical surface-related phenomenon, lubrication, has been touched upon in IETS work as recently reported by Bayman and Hansma (1980). This work studied chemicals known as diols, characterized by two OH groups capable of bonding to a surface. These chemicals are used in emulsions to lubricate metallic aluminum during the manufacturing operations of rolling and cutting. The IETS work showed that both OH groups of the diol react with the alumina surface by giving up protons, just as had earlier been found for the single OH groups of alcohols (see, e.g., Evans, Bowser, and Weinberg, 1979) and of acids (Klein et al., 1973) on alumina. In comparison of IETS spectra from an alcohol, an acid, and a diol based on the same (hexadecane) molecule,

Bayman and Hansma found the diol to give a significantly larger signal intensity than either the acid or the alcohol. From this result, they inferred that the diol adsorbs to alumina more readily than do the alcohol or acid. With the additional information that each diol has *two* points of bonding, i.e., at the two OH groups, between which runs a rather long carbon chain [thirteen CH_2 groups: $(CH_2)_{13}$], Bayman and Hansma (1980) concluded that chemisorption of such molecules on thinly oxidized aluminum metal might indeed lead to lubrication properties consistent with the known application of these materials.

Another out-of-the-way application of IETS has been to probe properties of the molecular constituents of epoxies which react to form adhesive bonds (White, Godwin, and Wolfram, 1978). The work was aimed particularly at those adhesives used to bond metals. The authors point out that in technical applications most metal surfaces are coated with an oxide of thickness of 100 to 200 Å. The molecules studied in IETS by White, Godwin, and Wolfram (1978) were actually those employed in a commercial epoxy system (Hercules 3501, produced by Hercules, Inc., of Magna, Utah); these are tetraglycidycl 4,4′ diaminodiphenyl methane (DPM) and diaminodiphenylsulfone (DPS). Both DPM and DPS have side arms largely composed of CH_2 and NH_2 groups, respectively. In the reaction forming the rigid adhesive, extensive cross-linking occurs between the side arms of these rather large molecules. This process, and also an effect of deterioration of the bond ("hydrothermal aging"), was followed in the IETS spectra. The aging process, linked to the presence of water at the adhesive-epoxy interface, was correlated with a pronounced change in the observed low-frequency modes and was tentatively identified with the formation of hydrogen bonding of the NH_2 groups to the oxide surface.

The use of IETS has also been shown capable of characterizing biological molecules (Simonsen and Coleman, 1973), including nucleic acid derivatives, as reported by Clark and Coleman (1976). These authors pointed to the fact that IETS can, in fact, correctly identify nucleotides using only micrograms of starting material. Larger molecules, such as polynucleotides, are not as easily identified, however. If a molecule is so large that its interior is essentially out of range of the dipolar interaction, or that its modes are so numerous as to be beyond the resolution limit, difficulties in IETS can be anticipated. One might also expect, with molecules of diameter exceeding the typical oxide thickness, that most of the tunnel current would go through regions between molecules, cutting down on the strength of observed IETS peaks. With smaller molecules, however, the great sensitivity of IETS should be an important advantage.

The biological application to identification of extremely small samples has been made more attractive recently with the demonstration by Hansma and Hansma (1982) that IETS analysis can be obtained from subnanogram samples: the limit is estimated by these authors to be $\sim 10\,\mathrm{pg} = 10^{-11}\,\mathrm{g}$. The technique of applying, under a microscope, a tiny (nanoliter) drop of solution, using a 130-μm-diameter Pt wire loop, to a tunneling area ($100\,\mu m \times 0.08\,\mu m$) seems in fact to be reasonably straightforward. The drawbacks to the utility of the technique, which are several, are in the area of identifying the compound from the complex spectra, especially if other impurity compounds are present. The authors point out that large collections of IR and Raman spectra of biological molecules, analogous to those presently available for organic molecules and facilitating identification of the latter from IETS spectra, are not presently available.

APPENDIX A

Experimental Methods of Junction Fabrication and Characterization

The steps in obtaining a tunneling spectrum and full characterization of a superconductor, including the Eliashberg function $\alpha^2F(\omega)$, can be summarized as those of junction *fabrication* and characterization, *measurement* of tunneling spectra [e.g., $I(V)$, $\mathrm{d}V/\mathrm{d}I$, and $\mathrm{d}^2V/\mathrm{d}I^2$], and *analysis* to obtain $\sigma = \mathrm{d}I/\mathrm{d}V$, $\mathrm{d}^2I/\mathrm{d}V^2$, the reduced conductance $\sigma/\sigma_{\mathrm{BCS}} - 1$, and, finally, the desired $\alpha^2F(\omega)$, $\Delta_{\mathrm{S}}(E)$, $Z_{\mathrm{S}}(E)$, and $Z_{\mathrm{N}}(E)$ functions. The present survey of experimental methods concerns the fabrication and measurement steps, while analysis is treated separately in Appendix B. The discussion is oriented primarily toward the goal of obtaining superconductive and IETS phonon spectra. Emphasis on recent advances in experimental methods seems appropriate, since several good sources on conventional methods are available. The basic reference in many respects is the experimental section of the review of McMillan and Rowell (1969). An introductory article by Giaever (1969) in the Burstein-Lundqvist volume is also recommended. The book by Solymar (1972) contains sections on the fabrication of both quasiparticle tunnel junctions and Josephson devices. The experimental tunneling review of Coleman, Morris, and Christopher (1974) is a recommended source, and there is also useful, experimental information in the IETS reviews of Hansma (1977, 1982) and of Weinberg (1978).

Apart from the interesting point-contact tunneling schemes, to be considered separately, producing a tunnel junction involves three steps: obtaining the first electrode, producing a sufficiently thin but uniform and continuous barrier (without degrading the surface of the first electrode), and providing the second electrode without damaging the barrier. Typically, the electrodes are deposited as thin films, but several useful methods are available allowing one of the electrodes to be a single crystal. The tunneling barrier is typically produced by thermal or glow discharge oxidation of the first electrode: important exceptions include Schottky barriers; the oxidation of subsequently deposited ultrathin metal proximity layers such as Al, Mg, Eu, Lu; and deposit of artificial barriers such as Si. The counterelectrode again is usually a deposited thin film.

A.1 Thin-Film Electrodes

The deposit of thin films of metals (possibly intermetallic compounds) is usually accomplished by high-vacuum evaporation from thermal or elec-

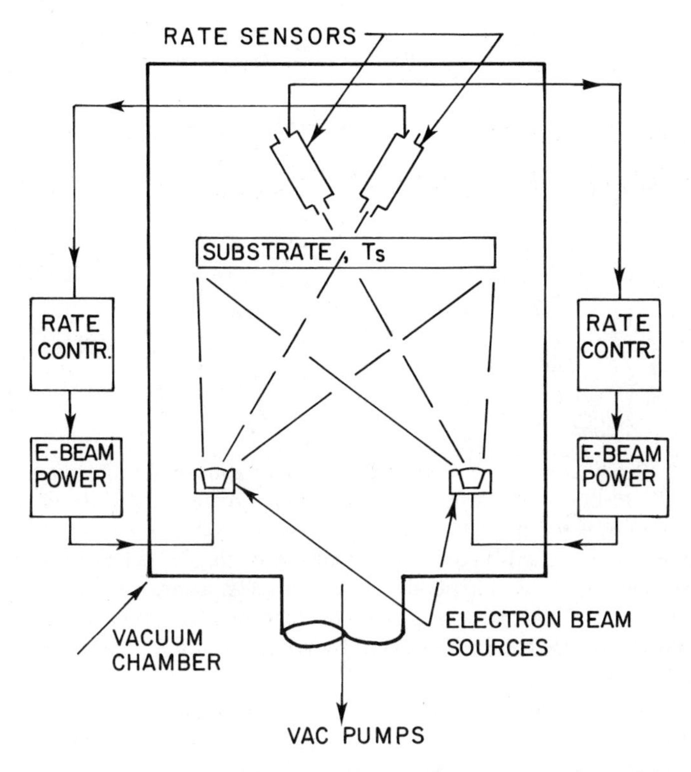

Fig. A.1. Essential features of an electron beam coevaporation system used in fabrication of A–15 superconductor films. An additional thermal source (not shown) is sometimes used finally to deposit ~10 Å of Si to improve the insulating properties of the subsequently grown oxide barrier. (After Hammond, 1975.)

tron beam sources or by sputtering in an inert atmosphere; reactive sputtering and chemical vapor deposition (CVD) are also viable methods. A general summary of such methods is given by Larson (1974); more extensive sources are Mayer (1955), Holland (1956) (vacuum evaporation), Chopra (1969), and Maissel and Glang (1970). Sputtering and CVD deposition methods and film-etching (removal) methods are emphasized in the recent edited volume of Vossen and Kern (1978).

Apparatus used in recent tunneling studies illustrates the state of the art in evaporation (in Fig. A.1) and in sputtering (in Fig. A.2).

A.1.1 Evaporated Films

The diffusion-pumped electron beam coevaporation system of Fig. A.1 (after Hammond, 1975) is used to prepare A–15 structure superconductor films (e.g., Nb_3Sn, Nb_3Al) by simultaneous evaporation of the constituents from separate electron beam sources onto heated substrates. In

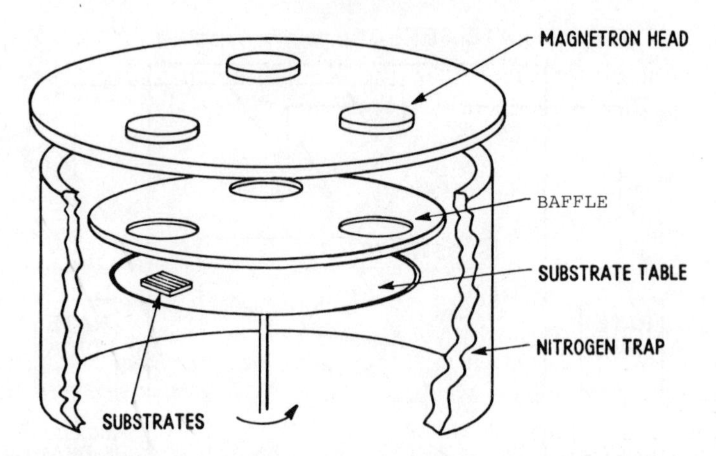

Fig. A.2. Schematic features of a three-magnetron sputtering apparatus used by Rowell. Rotation of substrate table permits growth of multilayer structures, which can be covered finally with a thin deposit of Al. Split magnetron sputtering sources or a less restrictive baffle permit alloy growth. The extended nature of the sputtering sources and the possibility of variable angles of incidence on the substrate with rotation may promote coverage of microscopically rough metal films with Al. (After Geerk, Gurvitch, McWhan, and Rowell, 1983.)

electron beam-heated sources, a focused high-energy (~1 keV) electron beam locally heats and melts the charge, held in a water-cooled hearth. Very simple electron beam–heating schemes have also been described for deposition of films of Nb and Ta (Neugebauer and Ekvall, 1964) and also for simply melting wires of such refractory materials to produce small single-crystal substrates (Gärtner, 1976). Since in these schemes only the charge itself becomes extremely hot, undesirable thermal release of gases (e.g., from the boat or basket, in thermal evaporation) is substantially reduced. Electron beam evaporation is therefore especially suitable for refractory metals, such as the transition metals. Tantalum, e.g. (melting point, 2996°C), requires a temperature of 2590°C to reach a (sublimation) vapor pressure of 10^{-4} torr. In such a case, evaporating the charge by heating it in a boat made of a different metal is virtually impossible, although self-heating a free Ta foil with a large current could be used to produce an appreciable Ta flux by sublimation. Typically, however, electron beam guns are operated with the charge in a locally molten state (e.g., $T > 2500$°C for Nb). The turbulent motion of the molten charge tends to produce fluctuations in the rate of evaporation, which must be accurately controlled if codeposition with a uniform and predictable stoichiometric ratio is required.

In this case, modified ion gauge tubes sampling the particle flux have been developed (Hammond, 1975) as evaporation rate monitors. The outputs from these rate monitors are compared with preset values and the differences used in a feedback circuit to correct the evaporation rate through the power supplied to the electron beam source. This control system has been described by Hammond (1975).

A.1.2 Film Thickness Measurement

The scheme of Hammond (1975) described above is a deposition rate monitor, the output of which could be integrated to obtain the film thickness. A more common method of determining deposit thickness is with a quartz crystal microbalance. This device is an exposed wafer of quartz with an evaporated metal electrode which is incorporated as the frequency-determining element (typically 5 MHz) of an oscillator circuit. The resonant frequency is electronically counted to the nearest hertz during deposit of metal in order to detect the small frequency change resulting from mass loading of the oscillating quartz crystal. The deposit thickness is expressed typically by an empirical law as $\Delta t \simeq 2\Delta f/\rho$, where Δt is the thickness of the film in angstroms, ρ is its mass density in grams per cubic centimeter, and Δf is the frequency change in hertz. Calibration of such devices can be achieved by accumulating a thick deposit: for $\Delta t \simeq 10^4$ Å or more, Δt is determined by weighing the substrate before and after deposit; in thicknesses $\Delta t \simeq 10^3$ Å, it is determined by measuring the deposit thickness by using an optical interferometer (a typical commercial instrument is the Angstrometer, marketed by Varian Associates) or by the mechanical deflection of a stylus drawn over the edge of the film. Typical of commercial instruments of the latter type is the Dektak, manufactured by Sloan Technology Corporation. Other methods suitable for thin films include measurement of the film resistance (typically $t \geq 100$ Å before electrical continuity occurs), optical reflection or absorption (Mayer, 1955), or shift in the polarization of reflected light in the method of ellipsometry (see, e.g., Knorr and Leslie, 1973).

A.1.3 Substrate Temperature

The morphology of deposited metal films is influenced by several parameters, including the temperature of the substrate. In general, a very low temperature inhibits mobility of the deposited atoms, limiting the growth of microcrystals, and, in some cases, creating an amorphous deposit. High temperatures aid growth of microcrystals. Thus, Pb deposited onto a 4.2-K substrate may result in an extremely smooth amorphous film [see Fig. 6.13 after Knorr and Barth (1970)], while Pb deposited on a substrate held at 77 K [Fig. 8.1, after Jaklevic, Lambe, Mikkor, and Vassell (1971*a*)] is known to be composed of crystallites of typical dimension comparable to the deposit thickness with the [111] direction normal to the substrate [but with unspecified azimuthal orientation; such a film is said to have [111] texture]. The variation in individual crystallite thickness in the [111] direction is about 10% in the average thickness range ~100 to 1000 Å. The intermediate substrate temperature 77 K thus produces a rather smooth but polycrystalline film, and higher substrate temperatures would be expected to produce larger crystallites and a rougher film, whose roughness would increase with thickness. The typical dull appearance of thick Pb films is an indication of such conditions. The scale of temperature required to achieve microcrystals of a given size depends on the metal being deposited (and also on other factors, including the evaporation rate). Thus, for the growing of good films of transition metals (Nb or Ta) or transition metal compounds (Nb_3Sn, Nb_3Al), substrate temperatures in the range 800–1000°C are frequently found necessary. A useful guide to the degree of order in a grown film is the residual resistance ratio (RRR), usually defined as the resistance at 300 K divided by the resistance at 4.2 K (or just above T_c, if the metal is a superconductor). The RRR is essentially the electron mean free path at 4.2 K expressed as a multiple of its 300 K value, which is usually phonon limited. Experimental values typically range from unity to a few hundred. Heating of substrates is conveniently accomplished with quartz-iodine tungsten lamps, whose radiation is focused by reflectors mounted inside the vacuum chamber. Higher temperatures can be achieved with radiation from bare W filaments and if care is taken to limit the thermal conduction from the substrate via its mounting. Cooling of substrates to near 77 K is possible by passing liquid nitrogen through the substrate holder using thermally insulated vacuum feedthroughs.

A.1.4 *Sputtered Films*

A convenient way to avoid the extreme temperatures needed to thermally evaporate refractory metals is to use dc or rf sputtering, usually in a purified argon atmosphere and typically at pressures in the range 10^{-3} to 1 torr. In sputtering, the source or target material is biased negatively at $-V_0$, and the desired flux of atoms to the substrate is generated by positive ion collisions onto the target at energy eV_0 from the ionized gas. Advantages of this method include the ability to produce films of compounds having nearly the same stoichiometry as the target, even though the constituent metals may individually possess very different vapor pressures (which would make thermal evaporation of the compound difficult to control), and convenience in controlling the rate of sputter deposition, by varying the target voltage or inert gas pressure. Conditions optimizing the production of highly ordered and pure films of refractory metals are roughly those maximizing the sputtering rate and minimizing the sputtering gas pressure (to reduce impurity incorporation from water and other impurities in the inert gas) consistent with kinetic energies of arriving atoms on the substrate not exceeding a few electron volts at optimum growth temperature (to avoid damaging the growing film). One solution of the experimental problem posed by the conflicting requirements is achieved in dc magnetron sputtering. Here, a magnetic field converging toward the source (target) traps electrons in the target region, greatly increasing the supply of ions and hence the ejected flux in this vicinity. This enhancement is still consistent with low target voltage V_0 (500–1000 V), which facilitates achieving a nearly thermalized particle flux, and relatively low sputtering gas pressure ($2–4 \times 10^{-3}$ torr). The magnetron sputtering apparatus shown schematically in Fig. A.2, capable of producing good sputtered films of refractory metals and intermetallic compounds, is of this type. An excellent source of information on modern sputtering configurations is the edited volume of Vossen and Kern (1978).

In special cases, *reactive sputtering* (see, e.g., Thornton and Penfold, 1978) may be used to advantage. For example, nitrides such as the cubic superconducting compounds NbN (Komenou, Tanaka, and Onadera, 1971; Keska, Yamashita, and Onadera, 1971) and VN can be sputtered by using N_2 (or N_2-argon mixtures) as a reactive sputtering atmosphere and Nb as the target.

A.1.5 *Chemical Vapor–Deposited Films*

In a typical procedure for chemical vapor deposition (CVD) of a metallic film, reactant gases flow into a quartz tube in a tubular furnace which is adjusted to provide an axial temperature gradient, and the substrates are located in the cooler region of the furnace. The reaction leading to deposit of metal or a metallic compound is frequently hydrogen reduction of a gaseous chloride: e.g., the reaction $5H_2 + 2NbCl_5 \rightarrow 2Nb + 10HCl$ could be used to deposit Nb. Many variations are possible (see, e.g., Kern and Ban, 1978, and references therein), and excellent films and crystals have been achieved. Notably, superior Nb_3Sn films can be made in CVD by hydrogen reduction of $NbCl_4$–$SnCl_4$ mixtures at 900–1200°C (Enstrom, Hanak, and Cullen, 1970). Films of Nb_3Ge have been similarly prepared by H_2 reduction of chlorides typically at 900°C (Newkirk, Valencia, and Wallace, 1976). As a further example, Oya, Onadera, and Muto (1974) have reported epitaxial deposit of NbN films on single-crystal MgO substrates at about 1000°C by reaction of gaseous $NbCl_5$ with NH_3 and H_2 in an open-flow system. Finally, carbides such as NbC (Caputo, 1977) and VC can be produced similarly, with the carbon introduced via CCl_4, CH_4 or other simple hydrocarbons.

A.1.6 *Epitaxial Single-Crystal Films*

Epitaxial films are grown upon single-crystal substrates of suitable lattice constant; growth, at a suitable temperature, occurs with long-range order stabilized by and commensurate with that of the substrate. An early review of this phenomenon is that of Pashley (1967); more recent sources are Arthur (1976), Mathews (1975), and Cullen, Kaldis, Parker, and Rooymans (1975). Some idea of the flexibility in achieving epitaxial relationships is indicated by the fact that Nb and Ta can be grown

(Durbin, Cunningham, and Flynn, 1982) in four different single-crystal film orientations on four different faces of single-crystal sapphire (Al_2O_3). The possibility of growing single-crystal Nb–Ta superlattices starting from individual epitaxial films on MgO and Al_2O_3 has also been demonstrated (Durbin, Cunningham, Mochel, and Flynn, 1981). Not surprisingly, many properties of epitaxially grown films, due to the higher degree of crystalline order, are vastly superior to those of their polycrystalline counterparts.

A.2 Foil and Single-Crystal Electrodes

An intermediate between the typical thin film and a true single-crystal electrode is the metal foil. Foils of pure (zone-refined) transition metals of thickness in the range 40–100 μm produced by cold rolling of vacuum-melted material are convenient as self-supporting tunnel junction electrodes. As obtained from the manufacturer, these foils typically have small grain sizes and substantial amounts of dissolved atmospheric gases. However, by the simple procedure of rigorous heating in ultrahigh vacuum, the gases can be driven off. *Recrystallization* of the initially microscopic grains constituting the flat foil into very large, flat crystallites (lateral dimensions of millimeters) can be readily achieved by simple heating for *bcc* metals and quite possibly in other cases. An easy method of heating a foil is to pass a high current ($I < 100$ A for 40 μm $\times$ 1 cm $\times$ 3 cm foil might allow 2000°C to be reached) through it; smaller samples can be heated by electron bombardment. Optical pyrometry is convenient in monitoring the temperature of such processes, a main practical consideration being to avoid melting the foil. These methods have been described by Shen (1970, 1972*a*), by Strongin (1972), and, more recently, by Wolf, Zasadzinski, Osmun, and Arnold (1980). A particularly convenient vacuum system, capable of achieving extremely high pumping rates for the copious condensable gases (N_2, O_2) released by a heated foil has been described by Shen (1972*c*): one version of this system is sketched in Fig. A.3.

This type of system, which has an ultimate pressure near 10^{-10} torr, has been found especially useful in preparing proximity junctions, where an atomically clean interface between the foil and a second metal (proximity) layer is required. The second (N) layer is evaporated *in situ* onto the recrystallized surface as soon as its temperature has fallen to a suitable value. A particularly useful characterization method, including those with such overlayers, for foils is with a scanning electron microscope (SEM) in a channelling mode, by which the crystalline orientation of the individual crystallites can be determined. Possible epitaxial relationships can be found by channelling inspection of the opposite side of the underlying foil (S) and comparison of the orientations of the N and S crystals. For a recent description of this method, see Joy, Newbury, and Davidson (1982).

The counterelectrode in a foil-based junction is evaporated through a

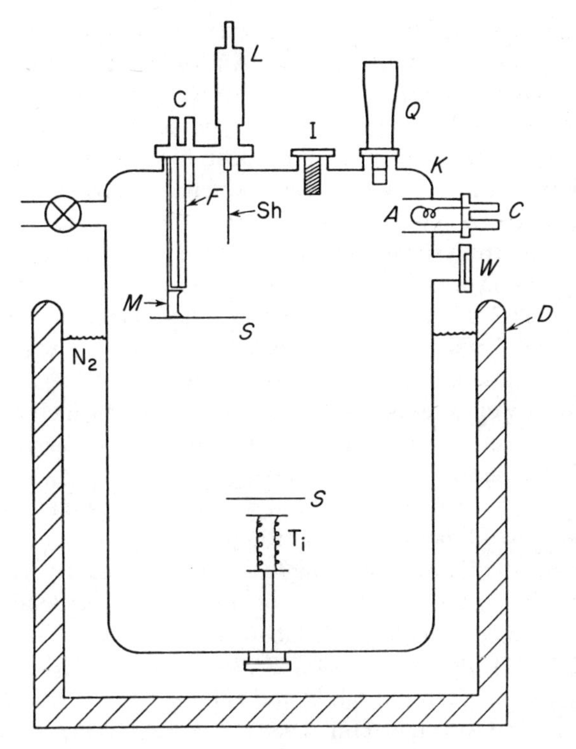

Fig. A.3. Schematic diagram of an ultrahigh-vacuum chamber (after Shen, 1972*c*) used for preparing Al and Mg proximity layers on transition metal foils. Here, *K* represents a Type 304 stainless steel keg (acid container), with several $1\frac{1}{2}$-in i.d. and one 4-in i.d. flanged tubulation. Filaments of Mo–Ti are used to deposit titanium metal onto the walls, maintained at 77 K by immersing *K* in an insulated container (*D*) filled with liquid N_2. The foil (*F*), shielded from Ti by shield *S*, is heated slowly to near its melting point by current from copper feedthroughs (*C*). The pressure during this process, monitored by the ion gauge *I* and quadrupole mass spectrometer (*Q*), is typically 10^{-9} torr. Aluminum is evaporated onto the cooled foil from the source (*A*), monitored by a quartz microbalance (*M*). Deposit on the foil is controlled by the shutter (Sh) positioned (not to scale) by the linear motion feedthrough (*L*). (After Wolf, Zasadzinski, Osmun, and Arnold, 1980.)

mask after the tunneling barrier is grown and after suitable insulation (masking) of the foil has been applied to define the tunnel junction area.

In many cases, the foil specimens mentioned above can be single crystals over the tunnel junction area. However, the orientation of the crystalline axis in recrystallized foils is usually not easily specified before the tunnel junction fabrication is completed.

Single-crystal electrodes of known orientation have been used in a variety of experiments, although seldom in full spectroscopic studies (see, however, Lykken, Geiger, and Mitchell, 1970). Extensive tunneling studies have been carried out by using degenerate (metallic) semiconductor single-crystal tunnel electrodes. An example is the Esaki diode junction, which can be generated by diffusing a high concentration of p-type impurities into a degenerate n-type semiconducting material, or vice versa. For purposes of tunneling spectroscopy, both thermally grown oxide barriers and direct metal-semiconductor (Schottky barrier) contacts have been used. In the Schottky barrier, the transfer of electrons (in n-type material) from donor impurities to the metal-semiconductor interface leaves behind a distributed, positive space charge barrier (Conley and Mahan, 1967; Conley, Duke, Mahan, and Tiemann, 1966; Steinrisser, Davis, and Duke, 1968). Schottky barriers of a variety of metals deposited onto vacuum-cleaved Si containing 5×10^{18} to 2×10^{19} donors cm^{-3} provide a well-understood tunneling system in which the Si acts as a single-crystal electrode and also as a single-crystal barrier (Wolf and Losee, 1970). Oxidized barriers on degenerate InAs and several other semiconductors have been employed in elegant tunneling studies (Tsui, 1971*a*) of the two-dimensional electron gas.

In the context of tunneling, single crystals of metals have been employed primarily in studies of the anisotropy of the superconducting energy gap. Metals studied include Al (Blackford, 1976), Ga (Gregory, Averill, and Straus, 1971), Pb (Blackford and March, 1969), Nb (Bostock et al., 1976; Schoneich, Elefant, Otschik, and Schumann, 1979), Sn (Zavaritskii, 1964, 1965; Blackford and Hill, 1981), and Re (Ochiai et al., 1971).

The procedure employed by Zavaritskii (1964, 1965) for Sn was to crystallize the molten metal between glass plates. Blackford (1976) (see also Blackford and March, 1969; Blackford and Hill, 1981) has had success in crystallizing vacuum-melted bulk samples of Al, Pb, and Sn, held in shape before crystallization simply by surface tension. In a subsequent annealing step, recrystallization occurs, and flat facets of several crystallographic orientations form, which may be sufficiently large to accommodate a tunnel junction. The orientation under individual tunnel junctions can be determined by Laue back-reflection x-ray analysis. This type of crystallization procedure becomes more difficult with metals of high melting point. For example, following earlier work by MacVicar and Rose (1968), Schoneich et al. (1979) report annealing an Nb single crystal (4.5 cm in length, 0.6 cm diameter) inductively to 2300° (measured on a pyrometer) in a UHV chamber at 1×10^{-10} torr. A relatively simple approach to vacuum melting and small crystals of Nb has been described by Gärtner (1976). An Nb wire is mounted in a UHV chamber, biased positively to accelerate electrons provided by a nearby tungsten filament. The tip of the Nb wire melts and recrystallizes in an approximately spherical shape. Such a small sample is likely to be of higher purity than the larger samples because outgassing from adjacent portions of the vacuum chamber is minimized and because a smaller sample cools more rapidly, decreasing the time during cooling in which it is hot enough to redissolve impurities from the residual gas.

Cleavage of metal crystals to form single-crystal tunneling electrodes has scarcely been reported, except in the context of layered materials which can be peeled apart. These include $NbSe_2$ (Frindt, 1972; Clayman, 1972; Morris and Coleman, 1973) and $Bi_8Te_7S_5$ (Lykken and Soonpaa, 1973).

Cleavage of a wider range of metals becomes possible when the metal is cold; while this fact has been exploited, e.g., in LEED and photoemission work (Palmberg, 1967; Baker and Blakely, 1972), it has apparently not been used in connection with tunneling experiments.

A.3 Characterization of Tunneling Electrodes

It is often desirable to characterize an electrode film which is the object of study in tunneling measurements by other techniques of surface chemical analysis, structural analysis, etc. In this section, we are primarily concerned with methods which can be used *in situ* to characterize a film (or single-crystal surface) before the tunneling barrier is formed.

We have seen, especially in connection with normal-state tunneling, that differing interface conditions, related to the properties of only a few monolayers near the interface, can have profound effects. The same sensitivity to conditions at an interface, again relating essentially to surface conditions in interfacial layers, occurs in proximity tunneling. In this case, specular transmission at the NS interface of a C–I–NS junction leads to results (de Gennes and Saint-James, 1963; Arnold, 1978) differing from those expected (McMillan, 1968*b*) if a weak barrier or simply diffuse scattering occurs at the NS interface. Also, in cases of high T_c, short–coherence length superconductors, serious reduction in the gap parameter can occur in even a thin disordered or contaminated electrode surface region just below the tunneling barrier, generally diminishing the superconducting $\Delta(E)$ and related propeties observed by tunneling. It has even been suggested (Rowell, 1980) in A–15 superconductors such as Nb_3Ge, where $\xi \simeq \hbar v_F/\pi\Delta \simeq 70$ Å (for $v_F \simeq 10^7$ cm/sec and $\Delta \simeq 3$ meV), that relaxation of even a perfect crystal near its surface could lead to measurable reduction in the surface pair potential. The surface properties of a tunneling electrode are thus often of interest as a guide to whether or not subsequently obtained tunneling spectra will truly reflect the bulk properties of the material under study. A typical experimental apparatus for analyzing surface properties is shown in Fig. A.4. A basic tool for surface chemical analysis, Auger spectroscopy (reviewed, e.g., by Chang, 1971), is available to identify atoms in the first few atomic layers of the surface. The same instruments, an electron gun and an electron energy analyzer of the cylindrical mirror (CMA) type described by Palmberg (1974), are used to perform reflection electron energy loss spectroscopy (ELS). A comprehensive review of ELS is given by Froitzheim (1977); for a recent application to transition metal surfaces, see Schubert and Wolf (1979). The information available from ELS includes plasmon energies,

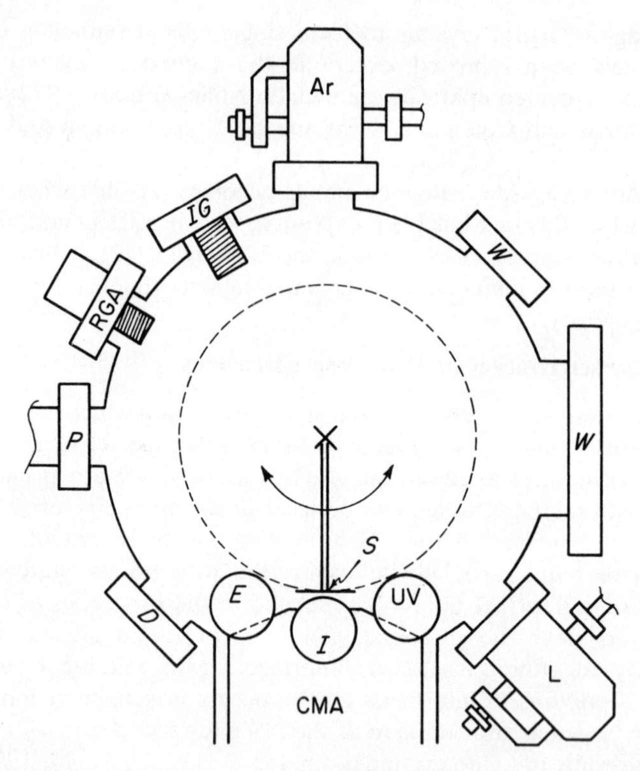

Fig. A.4. Schematic diagram of a vacuum chamber, operated with base pressure 10^{-10} torr, for cleaning, analyzing, and depositing N layers on foil samples *S*. Foil *S* can be rotated to large window *W* for pyrometer observations of outgassing (heating) by currents up to 80 A. As shown, *S* is at the focal point of the electron gun *E*, argon ion sputter gun *I*, windowless helium resonance lamp UV, and double-pass, cylindrical mirror, electron energy analyzer CMA. Rotation of *S* to position *D* allows deposition of Mg, Al, or other metal on *S*, monitored by a quartz microbalance (not shown). Residual gas analyzer RGA and ion gauge IG monitor pressure, while leak valve L permits introduction of gases. (After Schubert and Wolf, 1979.)

which are in the simplest case a measure of the "free" electron concentration N_e, through the relation $\hbar\omega_p = \hbar[4\pi N_e e^2/m]^{1/2}$. On a more sophisticated level, the quantities measured in ELS are the volume and surface loss functions, defined, respectively, as $\text{Im}(-1/\tilde{\varepsilon})$ and $\text{Im}[-1/(\tilde{\varepsilon}+1)]$, where $\tilde{\varepsilon} = \varepsilon_1 + i\varepsilon_2$ is the complex dielectric function. These quantities depend sensitively on the bonding and collective electronic properties, in contrast to the essentially atomic information provided by Auger spectroscopy. Thus, ELS can be especially useful if the electrode is a compound which may or may not be metallic, depending not only upon the stoichiometry actually achieved (which can be estimated from Auger spectra) but also on the crystalline phase. A useful addition to a system such as that shown in Fig. A.4, in fact, would be reflection high-energy electron diffraction (RHEED) (Pashley, 1956), from which the lattice parameter and crystal structure of a polycrystalline deposit can be determined *in situ.* An additional probe of electronic properties very close to the surface (~10 Å) available in the system of Fig. A.4 is ultraviolet photoemission spectroscopy (UPS), which, in simplest terms, probes the density of filled electron states below the Fermi level. This measurement also is extremely sensitive to surface contamination such as O or N. Useful reviews of UPS are given by Plummer (1975) and Feuerbacher and Fitton (1977). Other surface spectroscopies that might be applied in a system of the type shown in Fig. A.4 include x-ray appearance potential spectroscopy (APS) (Park, Houston, and Schreiner, 1970; Park, 1975) or inverse photoemission spectroscopy (IPES) (see, e.g., Dose, Fauster, and Scheidt, 1981; Woodruff and Smith, 1982). These methods probe empty states above the Fermi level and hence are complementary to the photoemission spectroscopy.

Other useful methods of analysis of films or foils generally require transfer of the sample to a different apparatus; these include the usual x-ray methods, transport measurements, etc. We will briefly describe two methods, SEM channelling and Rutherford backscattering, which are particularly useful in characterizing metal tunneling electrodes.

Electron channelling, recently reviewed by Joy, Newbury, and Davidson (1982), allows determination of the crystal structure of a film or other solid surface on a microscopic scale in a scanning electron microscope (SEM). The patterns observed in the SEM arise from modulation of the backscattered electron intensity when the angle θ of incidence (varying in the usual scanning of the beam) crosses the Bragg angle given by $2d \sin\theta = n\lambda$. For channelling measurements in an SEM at a 30-keV beam energy, corresponding to a typical Bragg angle of 40 mrad, the electron beam is usually reduced in its angle of convergence (defocused) to between 0.5 and 3.0 mrad. Characteristic channelling patterns are then visible in the usual SEM display of scattered intensity vs. position, revealing Bragg planes and, thus, the symmetry of the microcrystal being scanned. The usual SEM topographic image of precisely the same region can also be observed simply by increasing the beam convergence cone angle to more than the Bragg angle, ~40 mrad. A microscopic crystal structure identification is thus possible, which can usefully be applied to the region of a tunneling electrode specifically facing the tunneling barrier.

The technique of Rutherford backscattering is appropriate to diagnosing the composition depth profile of a layer or superimposed layers on the thickness scale of 0.1 to 1 μm, presumed homogeneous over a cross-sectional area of 1 cm^2 or so. In this technique, monoenergetic α particles ($m = 4u$), e.g., at $E_0 = 1.8$ MeV (usually from a van de Graaf generator), undergo elastic nuclear reflection against individual nuclei (mass M) in the film. The energy spectrum of the 180° back-reflected α particles is measured: if the film is composed entirely of mass M atoms, then the highest reflected α-particle energy will be $E_{r0} = E_0(1 - m/M)^2/(1 + m/M)^2$. This expression is derived directly from conservation of energy and momentum in the 180° elastic collision event. Measurement of E_{r0} thus determines M. The energies of α particles reflected from mass M nuclei at depth x into the sample are found to be systematically shifted to lower energies in amounts predictable from tabulated energy loss functions, $-dE/dx$. A homogeneous film of mass M and thickness t then produces a reflected energy spectrum extending from E_{r0} to $\sim E_{r0} - 2t(dE/dx)$, a characteristic step function shape from which t can be deduced. Computer analyses of such spectra permit accurate modeling of film composition vs. depth, as done, e.g., for Nb_3Ge by Testardi et al. (1975). Reviews of this method have been given by Nicolet, Mitchell, and Mayer (1972) and by Mayer and Ziegler (1974).

A further technique useful for characterizing the degree of local order in a material is EXAFS [extended x-ray absorption fine structure; see, e.g., Kincaid and Eisenberger (1975) and references therein], from which, in principle, the radial distribution function of neighbors around specific atoms in a structure can be obtained. For example, application of this method to sputtered Nb_3Ge films is reported by Brown et al. (1977).

A.4 Preparation of Oxide Tunneling Barriers

From an experimental point of view, the tunneling barrier is usually the most critical junction element, as well as the most difficult to characterize. Its small thickness, ~15 to 100 Å, requires high dielectric strength and tends to limit independent chemical analysis to the rigorous methods of surface physics. Large changes in electron barrier transmission may result from defects providing localized electron states, from fluctuations in thickness, as well as from defects such as pinholes in evaporated barrier films. While the thermally grown oxide is the most widely used barrier, single-crystal barriers, either a surface depletion layer of a semiconductor (Schottky barrier) or a wafer of layer compound, are alternatives which have been used in tunneling studies. Such barriers permit more complete characterization than do themal oxide barriers, whose beauty lies in their convenience.

A.4.1 Thermal Oxide Barriers

From the early work of Fisher, Giaever, and their colleagues, as described by Giaever (1969), it was established that the simplest of approaches to oxidation of Al, Cr, Ni, Mg, Nb, Ta, Sn, and Pb, "oxidation in ordinary air at room temperatures," can lead to insulating oxide barriers suitable for tunneling measurements of the superconducting gap. According to Giaever (1969), Cu, La, Co, V, and Bi are "difficult," while Ag, Au, and In were rated as "impossible" for the purpose of growing thermal oxide tunnel barriers. A useful collection of oxidation conditions found suitable in several cases is available in the review of Coleman, Morris, and Christopher (1974).

In approaching the conditions appropriate to oxidize a given metal, one should first obtain a highly resistive "capacitor junction," from which a capacitance estimate of the oxide thickness can be obtained and used to guide changes in the oxidation time or temperature. Inability to obtain a high-resistance junction with even a thick oxide is a pessimistic indication of the prospect of making a thin (~20-Å) barrier with good properties.

A different experimental approach to monitoring oxide growth to obtain a suitable barrier thickness is through measurement of the optical properties of the growing film. Figure A.5 (Sixl, Gromer, and Wolf, 1974) illustrates an *in situ* evaporator-cryostat equipped with a light source, monochromator, and detector. This apparatus is used for monitoring thicknesses of ultrathin Al, Pb, and In metal films through their optical absorption in a parallel geometry as described by Mayer (1955). The same apparatus is utilized to monitor deposit of ~20-Å insulator (tunnel barrier) films, using a method of interference colors (Heavens, 1965).

A sensitive optical instrument to monitor film thickness is the ellipsometer (see, e.g., Passaglia, Stromberg, and Kruger, 1964; Smith, 1969), which precisely measures rotation of the plane of polarization of linearly

Fig. A.5. Specialized apparatus for deposition of ultrathin and controllable amounts of three different materials on a cryogenic substrate, with means for determination of the thickness of the deposit. A substrate is indicated at point *C*, a liquid helium temperature shield is at *B*, and liquid nitrogen temperature shields are indicated as *A*. Bellows are employed to allow positioning of the selected evaporation source, minimizing the solid angle of room temperature radiation as seen by the cold substrate. (After Sixl, Gromer, and Wolf, 1974.)

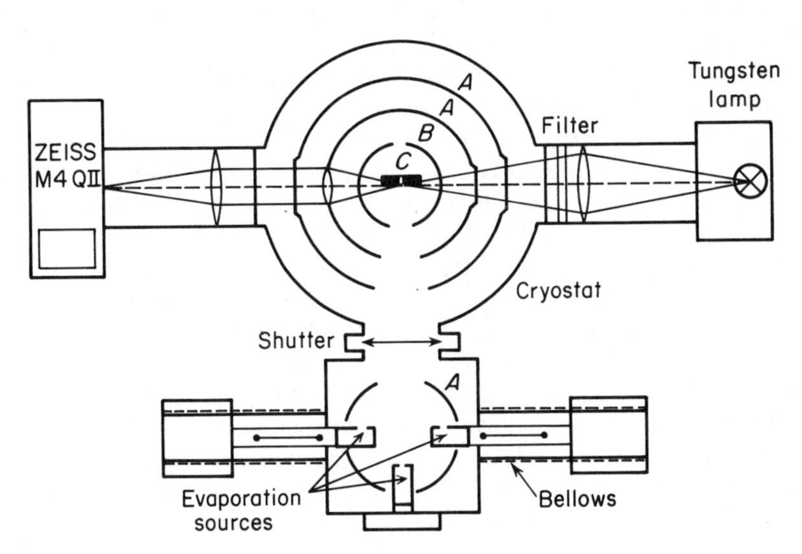

polarized light incident upon the growing film. This method can be made extremely sensitive to changes in film thickness. For example, Knorr and Leslie (1973), using an automated instrument (Ord, 1969) with an He–Ne laser source, state that the plane of polarization of the reflected light can be measured to ±0.01°, while a 10-Å film leads to a change in polarization angle of 0.53°. These numbers imply that a change in average film thickness of ~0.2 Å is detectable. Identification of precisely what film it is whose thickness has changed, however, is not always possible. For example, simple surface adsorption of gas or water vapor might easily give a large signal in such an instrument, and this signal could be confused with growth of the desired stoichiometric oxide film.

The physics of the growth of thin, self-limiting oxide layers on certain metals is understood in outline, although details of the oxidation conditions and particularly the defect structure of the initial metal can have a large influence in specific cases. An excellent and fairly recent source of information on mechanisms is the book by Fromhold (1976). A large body of experimental information and interpretation is also available in other books, including those of Kofstad (1966) and Samsonov (1973).

The most successful tunnel barriers, as listed by Giaever (1969), are examples of oxide films which grow at a rate decreasing with thickness (self-limiting). This property has the desirable consequence of producing uniformity of film thickness, for in its thinner regions, the oxide will grow faster, thus tending to correct their thickness deficiency. The self-limiting, low-temperature growth processes are empirically described by two logarithmic laws for the thickness d: $d = K_1 \log(t + t_0) + C_1$ and $1/d = C_2 - K_2 \log t$, where d is the thickness and t is the time. At high temperatures, a parabolic growth law, $d^2 = K_3 t + C_3$, is known, which is associated with rate limiting by a thermal diffusion process. Another well-known limiting case is linear growth, $d = K_4 t + C_4$, which is associated with situations in which some feature of the surface, independent of d, limits the rate.

The cases of self-limiting oxide growth, of primary interest in connection with barriers for tunneling, has been associated with rate limiting by the tunneling of electrons from the oxide-metal interface to the outer oxide surface, where incident oxygen molecules must be transformed into negative ions. This electron tunneling current associated with film growth is expected to vary with d as $\exp(-2\kappa d)$ and is the central feature of a model first advanced by Mott (1939, 1940) (see also Fehlner and Mott, 1970). A later proposal of Mott (1947), expanded upon by Cabrera and Mott (1949), described rate limitation by nonlinear, field-assisted diffusion. This proposal was based on the assumption of rapid electron equilibrium across the growing film (established either by tunneling or thermionic emission), with rate limitation by the necessary diffusion of the negative oxide ions through the oxide to the oxide-metal interface. This diffusion, moreover, was assumed strongly affected by an internal electric field resulting from the electron equilibrium mentioned above—hence the term *nonlinear diffusion.* Numerical investigations of more general models based on the earlier work cited above have been carried out by Fromhold and Cook (1967) and are discussed by Fromhold (1976). The results from the generalized theory show rate limiting, at first controlled by nonlinear diffusion (Mott and Cabrera theory) and later by electron tunneling (as first proposed by Mott) and leading again to a logarithmic growth law.

It is certainly to be expected that the growth rates will generally be increased by crystalline imperfections. The role of other parameters is not as easily guessed, but it is known experimentally that humidity is one parameter which can profoundly influence the oxidation rate. For example, oxidation of pure (freshly evaporated) aluminum *in situ* in an ultrahigh-vacuum environment with pure O_2 gas is generally reported to proceed extremely slowly at room temperature; the rate certainly increases with increasing humidity. Other factors, including the orientation, can be important. For example, the initial stages of oxidation of pure single-crystal Al have been reported as anisotropic, with an initially stable chemisorbed O_2 state Al (111), reported Flodström et al. (1978), and a transition from an initially amorphous oxide on Al (111) transforming to a crystalline oxide at elevated temperatures (Lynn, 1980).

A remarkable increase in the rate of oxidation of Pb with increasing humidity is shown in Fig. A.6a (after Garno, 1977). Garno's paper is recommended for a discussion of practical steps in growing reproducible oxide barriers—and also in applying evaporated counterelectrode films without damaging the grown oxide—to produce excellent SIS *I–V* characteristics at 1.2 K, as shown in Fig. A.6b. A summary of the recommendations of Garno for making good Pb–I–Pb junctions includes the use of heat shielding during the counterelectrode evaporation, scrupulous use of only filtered oxygen as input to the gas humidifier and furnace (anhydrous $CaSO_4$ filters were used), and use of the lowest possible oxidation temperature to avoid irreproducible results and defects.

Other "chemical" influences, besides humidity, on the rate and/or quality of oxidation have been reported and suggest further research. One of these is the reported beneficial effect of acetic acid vapor in air oxidation of In (Hebard and Arthur, 1977). A second material reported to oxidize better in the presence of acetic acid vapor is Nb_3Sn (Rudman et al., 1979).

In principle, similarly grown surface-insulating layers of nitrides or other compounds should provide workable tunnel barriers on some metals. Apart from brief reports of AlN grown on Al (Lewicki and Mead, 1966; Shklyarevskii, Yanson, and Zaporozhskii, 1974), there seems to be almost no literature on this approach. The possibility of using "reactive" excited-state N_2 (e.g., see D'Silva, Rice, and Fassel, 1980) to grow nitride layers might be worth investigation.

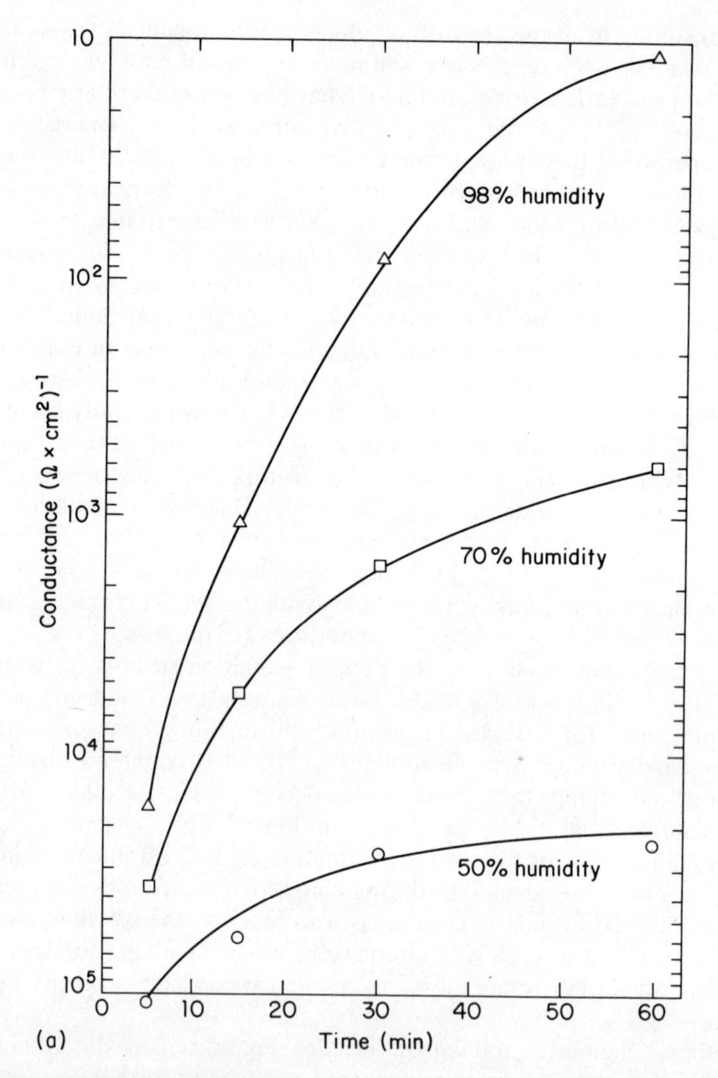

Fig. A.6. (a) Influence of humidity on the rate of growth of oxides on Pb film. (b) Illustration of I–V characteristics at 1.2 K for a good Pb–PbO–Pb junction. Shielding the substrate from the full heat of the evaporation source, filtering of the dry oxygen before introducing it into the oxidation chamber, and maintaining the temperature of the oxygen flow near room temperature were found important in achieving a high yield of Pb–PbO–Pb junctions of excellent quality. (After Garno, 1977.)

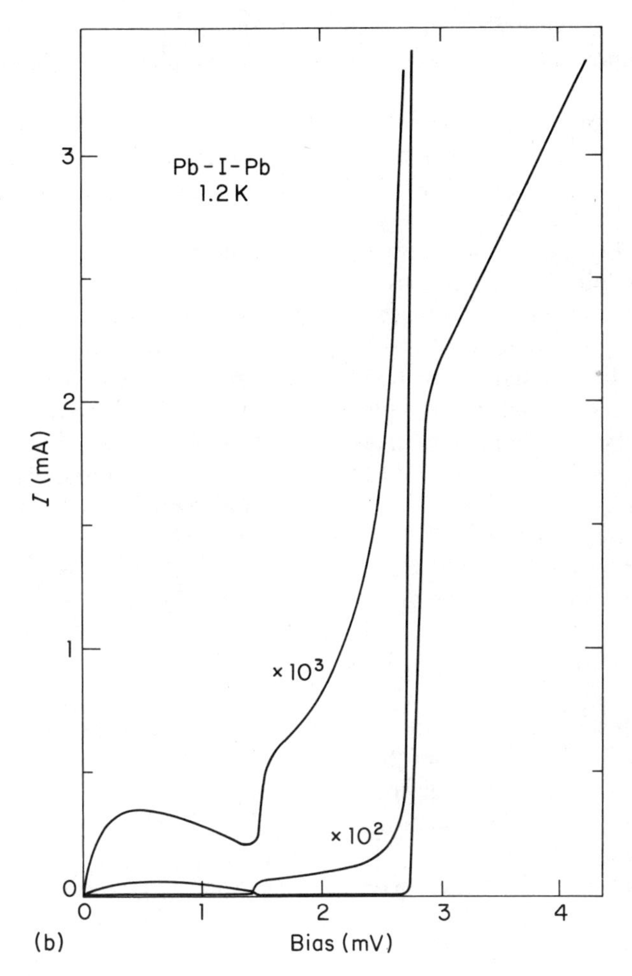

Fig. A.6. (*continued*)

A.4.2 Plasma Oxidation Processes

In plasma processes, oxygen is supplied to the surface of the oxidizing metal in the form of negative ions with additional kinetic energy deriving from the discharge, itself typically driven by voltages between a few hundred to 1000 V. There are two principal variants of the plasma oxidation process: using a dc voltage to sustain the discharge (Miles and Smith, 1963) and using an rf voltage (Greiner, 1971). These processes assist in oxidation in three ways: by providing ions, rather than molecules, which are needed in the growing oxide layer and thus accomplishing one possibly limiting step; by adding some control over the electric field, assisting diffusion across the oxide layer; and by providing a kinetic energy larger than kT, making the growth process less sensitively dependent upon temperature and possibly avoiding need for heating the sample. An important practical advantage is that the plasma oxidation, which usually does not require heating, can be done *in situ* with no need to expose the sample before or after oxidation to the laboratory atmosphere. This improved control over the atmosphere during oxidation is a great advantage, e.g., in IETS, to avoid molecular contamination (see, e.g., Walmsley, 1980), and in the oxidation of easily contaminated surfaces, such as Nb (e.g., see Hohenwarter and Nordman, 1982).

The conditions employed by Miles and Smith (1963) included an Al wire cathode, negatively biased relative to the baseplate (at ground potential) by between a few hundred and 1000 V, with a shield to block any direct line-of-sight path between this Al cathode and the oxidizing surface. It is essential that the film being oxidized lie in the negative-glow region of the discharge. The pressure of O_2 was ~50 μm, with oxidation times varying between a few seconds and tens of minutes. In most applications of this glow discharge oxidation technique, particularly in IETS, the metal film is left at ground potential. However, a discussion of the application of a specific bias potential to the Al film, as is applied in liquid-phase anodization, is given by Miles and Smith (1963). Variations on this dc method achieved by using an O_2-helium atmosphere and by use of water vapor instead of O_2 have been reported by Magno and Adler (1976).

An interesting and important use of an rf plasma to provide precisely repeatable oxide thicknesses for Josephson junctions was reported by Greiner (1971). In the rf plasma, consisting of Ar with an admixture of O_2 with again peak voltages in the range 200–1000 V, competing processes of oxidation (by O_2^- or O^- ions) and sputtering (by Ar^+ ions) occur in the portions of the rf cycle corresponding, respectively, to oxide positive and negative. Thus, following Greiner (1971), the overall rate dx/dt (where x is the oxide thickness) is assumed to be given by the difference of two terms:

$$\frac{dx}{dt} = \left(\frac{dx}{dt}\right)_{\text{oxidation}} - \left(\frac{dx}{dt}\right)_{\text{sputter}}$$

The point here is that $(dx/dt)_{\text{oxidation}}$ is a rapidly decreasing function of depth [corresponding to the same self-limiting behavior as that in thermal oxidation, which was observed experimentally in dc plasma growth by Miles and Smith (1963)], while $(dx/dt)_{\text{sputter}}$ is essentially independent of oxide thickness and is easily controlled by varying the Ar partial pressure, voltage, etc. Hence, the ratio of the two can be used to set the limiting oxide thickness.

In more detail, following Greiner (1971), suppose the oxidation rate decays with thickness as $K\exp(-x/x_0)$, where K and x_0 are process-dependent parameters, while $(dx/dt)_{\text{sputter}} = R$ is a constant. The balance of the two terms produces a limiting thickness $x(t) = x_L$, given by the condition $(dx/dt)_{\text{total}} = 0 = Ke^{x/x_0} - R$, or $x_L = x_0 \ln(K/R)$. The implication that the limiting oxide thickness in the rf sputter-oxidation process can be preset easily with process parameters was demonstrated by Greiner (1971). Increasing O_2 partial pressure increases tunnel junction resistance, while increasing rf power reduces the resistance. The latter dependence indicates a greater sensitivity of sputtering rate R than of oxidation rate to the incident ion energy. The time constant τ of the oxide thickness equilibration can be deduced from the growth law $x(t)$, obtained by integration of dx/dt (above):

$$x(t) = x_0 \ln\left[\frac{K}{R} - \left(\frac{K}{R} - e^{x_i/x_0}\right)e^{-Rt/x_0}\right], \qquad \text{for } x > 0$$

where x_i is the initial oxide thickness. Hence, the time constant is $\tau = x_0/R$. Typical parameters are $x_0 = 1.5$–3.5 Å, while the sputtering rate R is typically about 0.1 Å/sec for power density 0.1 W/cm^2 when the O_2 pressure is on the order of 10^{-2} torr. These parameters give τ on the order of 1 min. Use of rf oxidation in making Nb Josephson junctions is described by Raider and Drake (1981). One of the useful variants of the above procedure, in which the vacuum chamber is filled with the oxygen or argon/oxygen mixture, is to keep the chamber pressure low and to introduce only a beam of the ionized gas mixture. This task can be done by introducing gas through a leak valve mounted on the flange of a commercial argon sputter gun, for example. The use of such a reactive ion–beam oxidation technique in fabrication of Nb Josephson juntions has been described by Kleinsasser and Buhrman (1980). One of the advantages of this method is that, since the ion beam here is focused on the sample alone, the oxidizing surface can be kept free of contamination from material which may be released from contaminated chamber walls, etc. if the glow discharge fills the entire chamber. Earlier comments on sputtering as a means for generating tunnel junctions on Nb have been given by Keith and Leslie (1978).

A.5 Artificial Barriers

The term *artificial* has become associated with a barrier not grown on the electrode material but deposited artificially in some way onto the electrode. Historically, there has been great skepticism (e.g., see Giaever, 1969) that such an approach can produce pinhole-free barriers in the relevant thickness range.

A.5.1 Totally Oxidized Metal Overlayers

At least one counterexample to Giaever's pessimistic view of 1969 is now known to exist (see Fig. A.7). This barrier, convincingly pinhole-free, was produced on sputtered Nb by subsequent, *in situ*, sputter deposit of a thin (~20-Å) layer of Al, which was then totally oxidized. While this Al overlayer technique was actually invented very early in the history of

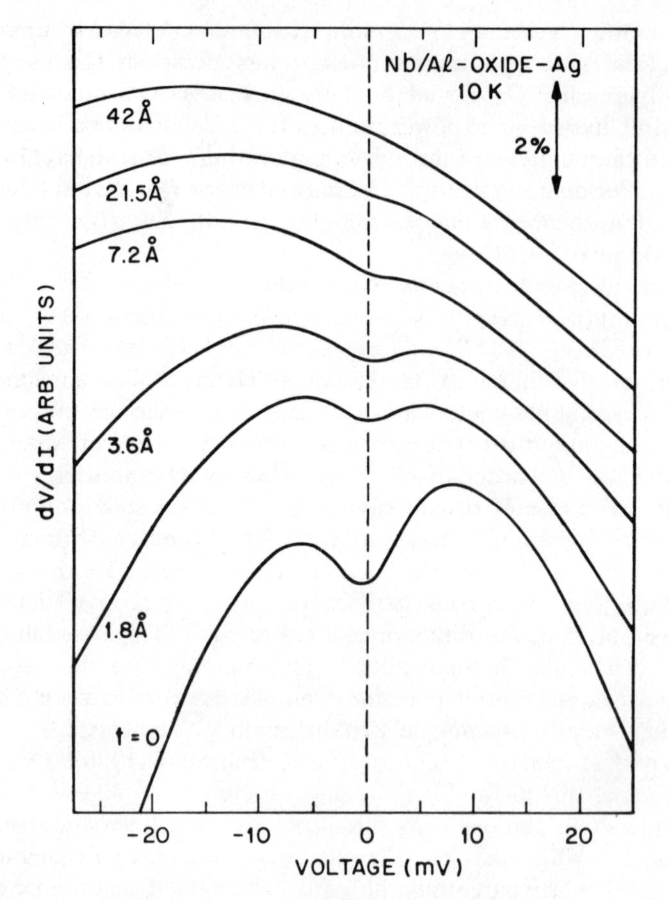

Fig. A.7. The dV/dI spectra of sputtered Ag-oxide-Al/Nb tunnel junctions measured at 10 K. Monolayer thicknesses t of Al sputtered on an Nb surface immediately after Nb deposition are seen to alter the subsequently formed oxide barrier, which implies uniform coverage of the Nb microcrystals by Al. (After Rowell, Gurvitch, and Geerk, 1981.)

superconductive tunneling spectroscopy by Adkins (1963) and used by Hauser, (1966) and by Hauser, Bacon, and Haemmerle (1966) (and probably by others), legitimate concern about definitive interpretation of the early results took two forms. First, one could question the continuity of the initial 20-Å deposit, leading to the real possibility of pinholes, or at least thickness fluctuations, in the resulting oxide. Secondly, the question of the influence on tunneling characteristics of any *metallic* Al left by incomplete oxidization of the Al metal was not answerable until recently. The answer to the first question—or, rather, demonstration of experimental conditions under which continuity is convincingly achieved—has come only recently, in the course of the work reported in Chap. 5. See, specifically, Wolf, Zasadzinski, Osmun, and Arnold (1980) for demonstration of foil-based "aluminum overlayer" samples, and the tunneling work of Rowell, Gurvitch, and Geerk (1981) and the XPS study of Kwo, Wertheim, Gurvitch, and Buchanan (1982) for sputtered films. The question of the influence of the remaining unoxidized metal has also been answered only relatively recently, again, in the course of the development of the PETS proximity tunneling methods (Chap. 5), originating in the paper of Arnold (1978).

Other metals which evidently will make continuous layers and hence continuous artificial barrier oxides on *clean* refractory metals are Mg (Burnell and Wolf, 1982) and rare earth metals Er and Lu (Umbach, Goldman, and Toth, 1982). General considerations and tabulated instances relevant to the "wetting" or epitaxy of the typically soft artificial barrier metal on the hard S-electrode metal are discussed by Biberian and Somorjai (1979). Thermodynamic considerations at the surface of a two phase system are carefully discussed by Miedema and den Broeder (1979). However, the understanding of the relevant conditions is regarded as still incomplete. Another question of practical importance, independent of the equilibrium arrangement of a thin layer of metal A on metal B, is the time and temperature regime such that inter-diffusion of the two metals is of no importance.

These considerations are also important when the objective is a metallic overlayer to form an NS bilayer for proximity tunneling.

A.5.2 Directly Deposited Artificial Barriers

A different type of artificial barrier, also depending upon oxidation of an initial deposit, is that obtained with ~20-Å deposit of amorphous silicon (Rudman and Beasley, 1980). This method has been used in quasiparticle tunneling studies on several A–15 superconductors (Moore, Zubeck, Rowell, and Beasley, 1979; Kwo and Geballe, 1981; Kihlstrom and Geballe, 1981). The use of hydrogenated amorphous Si barriers in Josephson junctions on Nb is reported by Kroger, Potter and Jillie (1979) and by Kroger, Smith, and Jillie (1981).

A variety of other materials have been used as artificial barriers. These include evaporated amorphous carbon (MacVicar, 1970), Ge (Giaever and Zeller, 1968*b*; 1969; Ladan and Zylbersztejn, 1972), CdS (Giaever and Zeller, 1968*b*; 1969) Lubberts, 1971), CdSe (Lubberts, 1971; Josefowicz and Smith, 1973; Rissman, 1973), and InSb, as well as Ge, by Keller and Nordman (1973). In addition, ZnO, CdO, and ZnS have been mentioned by Giaever and Zeller (1969) and Te barriers by Seto and Van Duzer (1971) and also by Cardinne, Marti, and Renard (1971), while cryogenically deposited Al_2O_3 has been used by Moodera, Meservey, and Tedrow (1982). The narrow band gap oxides In_2O_3 and Bi_2O_3 have been used by Aspen and Goldman (1976), while we have already mentioned the use of SiO and naphthalene by Sixl, Gromer, and Wolf (1974) and formvar by Mancini et al. (1979). Finally, wax is used by Beuermann and El Haffar (1981).

A.5.3 Polymerized Organic Films

The use of polymerized benzene barriers is described by Magno and Adler (1977*a*, 1977*b*). Earlier reports of organic barriers deposited by the Blodgett-Langmuir (1937) technique are given by Miles and McMahon (1961), Simpson and Reucroft (1970), Léger, Klein, Belin, and Defourneau (1971), and Mann and Kuhn (1971).

A.6 Point-Contact Barrier Tunneling Methods

The use of a preoxidized wire or Schottky barrier probe in physical contact with, or extremely closely spaced from, a counterelectrode of interest is an appealing idea which has received attention at several times, most recently in terms of the scanning tunneling microscope of Binnig, Rohrer, Gerber, and Weibel, (1982*a*, 1982*b*). The objective in these experiments is to observe tunneling, either through a Schottky barrier, oxide layer, or vacuum, and is different from the barrierless, metallic point-contact spectroscopic techniques (Yanson, 1974; Jansen, Mueller, and Wyder, 1977; Jansen, Von Gelder, and Wyder, 1980; Yanson, Kulik, and Batrak 1981).

A.6.1 Anodized Metal Probes

The early use of anodized (oxidized) tips of Nb, Ta, and Al to make tunneling measurements of the energy gaps of Nb_3Sn, V_3Si, and V_3Ge was reported by Levinstein and Kunzler (1966). The contact diameter of the junctions was estimated to be less than 10 μm. The tips were etched to a conical shape, heavily anodized, and then brought into mechanical contact with the counterelectrode under liquid helium. It is not clear whether tunneling occurred through a thick anodic oxide layer (presumably of low barrier height) or whether mechanical contacting was used to break away portions of the oxide, leaving a conventionally thin barrier. Levinstein and Kunzler reported that, within limits, the contact resistances could be varied by changing the pressure, controlled by an externally operated screw.

A.6.2 Schottky Barrier Probes

The idea of using the Schottky barrier layer of an etched degenerate semiconductor point (von Molnar, Thompson, and Edelstein, 1967) apparently was stimulated by the brief report of Rowell and Chynoweth (1962) of freezing Hg directly onto degenerate silicon and observing tunneling through the intrinsic Schottky barrier layer. [This scheme was revived, using clean In probes, by Wolf (1968) and also by others.]

The GaAs Schottky barrier probes were more fully described by Thompson and von Molnar (1970), and the technique was also independently assessed by Tsui (1970*c*). Single-crystal *p*-type GaAs with carrier concentration of $\sim 2 \times 10^{19}\ \mathrm{cm}^{-3}$ is typically ground to a conical point mechanically and then etched to remove surface damage, as verified by observation of sharp Laue spots (Tsui, 1970*c*). The Schottky barrier tunneling properties of *p*-GaAs, well known from the work of Conley, Duke, Mahan, and Tiemann (1966) and Conley and Mahan (1967), were verified in the probe configuration using gold as a counterelectrode (Thompson and von Molnar, 1970) and with Pb by Tsui (1970*c*). Tsui reported, however, that even with a freshly etched and polished Pb crystal with a mirrorlike finish, the "sharp point is needed to penetrate the contaminations of the freshly prepared Pb surface." Such an embedding procedure also is no doubt helpful in overcoming the sensitivity to vibration that is usually a problem with point-contact methods. While rather nice gap structures and repeatable second-derivative $\mathrm{d}^2I/\mathrm{d}V^2$ Pb phonon spectra were obtained, Tsui (1970*c*) found that the phonon spectra were systematically shifted to higher energy in a fashion consistent with the known effects of high pressure on the Pb phonon spectrum (e.g., Svistunov, D'yachenko, and Belogolovskii, 1981). Local damage in the region under the tip also seems likely.

An *in situ* sputter-cleaning procedure for the surface opposite the GaAs probe was devised by Thompson and von Molnar (1970). This consisted essentially of a local helium plasma discharge cleaning, arranged with the probe withdrawn slightly from the surface (all immersed in liquid helium), and connected to a small capacitor charged from 3 to 6 V. The probe was then moved back toward the sample until an arc was struck. Evidence is given by Thompson and von Molnar (1970) that this method is effective in sputter cleaning the surface to the extent of improving the gap characteristics, but few phonon spectra resulting from such methods have been published. One problem inherent in this scheme, of course, is that no annealing of the sputter-etched surface, as frequently found beneficial in UHV cleaning, is possible. However, interesting results using this method on intermediate-valence materials have been given recently by Güntherodt, Thompson, Holtzberg, and Fisk (1982).

Improved control over the motion of the tip is available with piezoelectric drives, although this technique is useful only if the level of microphonic vibrations is initially low enough to make this improvement noticeable. In this connection, the improved method of suspension of the point-contact tip (Binnig, Rohrer, Gerber, and Weibel, 1982*a*, 1982*b*), described in Sec. 11.6, is especially significant. An additional important advance in vacuum tunneling has been reported very recently by Moreland et al. (1983) and by Moreland and Hansma (1983).

A.6.3 Deformable Metal Vacuum Tunneling Probes

With improved stability of suspension and the sensitivity of the piezoelectric drive (~2 Å/V), Poppe (1981) has reported vacuum tunneling determination of superconducting gap characteristics of a cleaved

$ErRh_4B_4$ crystal. In this case, the presumably hard and clean $ErRh_4B_4$ cleaved surface is initially contacted with an electropolished Ag or Au tip, which has no inherent barrier and which is reportedly soft enough to deform to match the contour of the local $ErRh_4B_4$ surface. The tip is then retracted to a vacuum spacing of ~100 Å and tunneling spectra obtained. Presumably, wide voltage conductance spectra would help to discriminate between the vacuum tunneling mechanism and an alternative possibility, Schottky barrier tunneling through a conjectured charge compensation layer on the $ErRh_4B_4$ surface.

A.6.4 Analysis of Point-Contact Data

The problem of separating various tunneling and nontunneling current contributions at temperatures up to 300 K in point-contact measurements on nonsuperconducting organic metals such as $(CH)_x$ (Leo, Gusman, and Deltour, 1982) and TTF–TCNQ (Leo, 1981) has been approached by a mechanical modulation technique described by Thielemans, Leo, Deltour, and Mehbod (1979). However, the mechanical modulation seems not to have been utilized in a revised point-contact spectrometer design (Leo, 1982), which also seems inadequate in its approach to the important mechanical stability problem when compared with the design of Binnig, Rohrer, Gerber, and Weibel (1982*a*, 1982*b*). The useful analysis suggestion of Thielemans, Leo, Deltour, and Mehbod (1979) is to separate the contributions in "tunneling" point-contact currents arising, in fact, from tunneling, thermionic emission, and resistive shunts or microshorts by the markedly more rapid dependence, $\sim e^{-2\kappa t}$, of the tunneling current contribution upon barrier thickness, t. Thus, an approach is used in which the point-contact pressure is sinusoidally modulated by an ac voltage on a piezotransducer, with lock-in detection of the current. Stability of the local oxide sample surface configuration under such modulation would appear to require that a purely elastic mechanical operating point be maintained.

A.7 Characterization of Tunnel Junctions

We have already mentioned various techniques, including surface analysis, channelling, and others, which may be used to characterize the component films making up a tunnel junction. The purpose of the present section is to describe several primarily electrical measurements other than the tunneling derivative spectra which can be performed to characterize the completed structure. The techniques for obtaining dI/dV and d^2V/dI^2 spectra will be considered separately.

A.7.1 Initial Characterization of Junctions

Initial tests of a set of completed tunnel junctions to estimate their quality, the absence of shorts, approximate barrier thickness, and so forth are worth mention. A simple measurement of the zero-bias resistance using a low-voltage ohmmeter, to avoid damaging barriers, (typical of such devices available commercially is the Fluke 8024A), should give values varying inversely with the cross-sectional area of the junctions on a given substrate. Further, the resistance should increase slightly as the junctions are cooled to 77 K. The opposite behavior is a sure indication that shorts dominate the measured conductance.

Measurement of the junction capacitance will, in principle, determine the average barrier thickness if the area and dielectric constant are known. In practice, this measurement may be complicated by the typically high shunt tunnel conductance, so that an ac method at perhaps 100 kHz [which can be done by using the bridge type of tunneling measurement circuit, originally described by Adler and Jackson (1966), if care is taken to subtract the capacitance of cables, etc. in the measurement circuit] may be necessary. The other point is that, in the presence of any thickness fluctuations, the average barrier thickness inferred from the capacitance may well be larger than that typical at the thinner regions dominating the measured conductance. A much higher capacitance than expected may signal the presence of metallic conductivity in the nominal barrier insulator, with tunneling occurring through a Schottky barrier at its interface with one of the electrodes. Such a situation was diagnosed by Magerlein (1983) in Pb–In–An alloy Josephson tunnel junctions for which the oxide barrier grown on the Pb–In–An alloy consists largely of In_2O_3. Magerlein concluded that the In_2O_3 layer was likely a degenerately doped semiconductor, with a Schottky barrier occurring at its interface with the electrode metal. The variation of the capacitance with bias V is of interest in the Schottky barrier case, in which $1/C^2(V)$ vs. V ideally gives a linear plot, with the Schottky barrier height V_B occurring as the $1/C^2=0$ intercept and the donor-center density N_D given (see, e.g., Sze, 1969) by the slope through the relation $\mathrm{d}(C^{-2})/\mathrm{d}V=-2/(q\varepsilon N_D)$.

Estimates of the barrier heights in the more usual case of a "trapezoidal" barrier can be obtained by modeling the conductance $G(V)$, as discussed in Chap. 2, and also by examining the current-voltage $J(V)$ relation over a large voltage range, if this experiment can be done without danger of damaging the devices in question.

Two approaches to determination of barrier heights from the J–V data, where J is the current density, will be mentioned. In the method of McBride, Rochlin, and Hansma (1974), a plot of $\log(J/V^2)$ vs. $1/V$ (Fowler and Nordheim, 1928) is used initially to obtain the product $S\phi^{3/2}$, where S is the effective barrier thickness and ϕ is the effective barrier height. Then, a plot of the experimental $\log J$ vs. V data is compared with calculated J–V curves, the latter obtained assuming a trapezoidal model in which the product $S\phi^{3/2}$ is constrained to be constant but S and ϕ are individually varied. It was found by McBride, Rochlin, and Hansma (1974) that separate determinations of S and ϕ are possible in this fashion. This method, when applied by McBride, Rochlin, and Hansma

(1974) to Al–AlO_x barriers, gave $\phi = 2.62 \pm 0.12$ eV and $S = 20$ Å for a junction resistance of 1300 Ω/cm^2. A related approach is to plot $d(\log |J_{\pm}|)/dV$ vs. V (Gundlach and Hölzl, 1971; Gundlach, 1973), which yields peaks when $\pm eV = \phi \gtrless$, $\phi \gtrless$ being the higher (lower) of the barrier heights. A further variant is a plot of $(d^2J/dV^2)/J$ vs. V.

A second analysis used to determine barrier heights is based on the temperature-dependent ratio

$$J(V, T_1, T_2) = \frac{(J(V, T_1) - J(V, T_2)}{J(V, T_2)}$$

(Simmons, 1964), which again gives maxima vs. V at $\pm eV = \phi \gtrless$. This analysis is based on the fact that the temperature dependence of the tunnel current is greatest when $eV = \phi$, the barrier height.

If difficulties are experienced in deducing ϕ_1 and ϕ_2 from such methods, one possible recourse, useful if the barrier height is not too large (and is independent of barrier thickness), is to make the barrier sufficiently thick to eliminate the tunnel current, and to measure the thermionic (over the barrier) current as a function of T for fixed bias V. If this thermionic range is attainable, plots of $\log(J_{\pm})$ vs. $1/T$ can be analyzed to give the barrier heights $\phi \gtrless$ (for $\pm$ biases) as the usual activation energies from the slopes $d(\log J)d(1/T)$. An example of this method for a metal-semiconductor system is given by Lubberts, Burkey, Bücher, and Wolf (1974), who also confirmed their results by internal photoemission measurements (Fowler, 1931; Schermeyer, Young, and Blastinggame, 1968; Sze, 1969).

Barrier thickness uniformity over a large-area of oxidized electrode was investigated by Miles and Smith (1963) by evaporating an array of small tunnel contacts, measuring these individually, and plotting contours of equal junction resistance.

A rather sophisticated method of assessing barrier uniformity, based on the Fraunhofer-like dependence of the Josephson current $I_c(B)$ on magnetic field B, was described by Dynes and Fulton (1971). Treating a rectangular junction area of sides a, b, with $\vec{B}$ parallel to the b direction, one should obtain for an ideal junction

$$I_c(B) \propto \left|\frac{\sin \frac{1}{2}\beta a}{\frac{1}{2}\beta a}\right|, \qquad \text{where} \qquad \beta = \frac{2\pi(2\lambda + d)B}{\Phi_0}$$

where λ is the penetration depth, d is the barrier thickness, and Φ_0 is the flux quantum. Deviations from this behavior are analyzed in terms of a current profile

$$J(x) = \int_{-b/2}^{b/2} J(x, y)\, dy$$

obtained by a Fourier transform procedure. The even and odd parts of

$J(x)$ are given by Dynes and Fulton (1971) as

$$J_e(x) = \int_{-\infty}^{\infty} d\beta I_c(\beta)\cos[\theta(\beta) + \tfrac{1}{2}\beta a]\cos \beta x$$

$$J_0(x) = \int_{-\infty}^{\infty} d\beta I_c(\beta)\sin[\theta(\beta) + \tfrac{1}{2}\beta a]\sin \beta x$$

where

$$\theta(\beta) = \frac{\beta}{2\pi}\int_{-\infty}^{\infty} db \frac{\ln I_c(b) - \ln I_c(\beta)}{\beta^2 - b^2}$$

The resulting profiles from the data of Fig. A.8 from an Sn–SnO_2–Sn junction at 1.46 K have been shown in Fig. 2.19. The results indicate that, for junctions with plasma-grown oxides, the current density is greatest at the center of the barrier, but the variation, ~10%, is remarkably small considering the extreme sensitivity of $J(x, y)$ to barrier thickness.

In situ scanning methods are also possible to infer properties of tunnel junctions. Reports have included use of scanned optical illumination on superconducting metal films (Chi, Loy, and Cronemeyer, 1982) and on Al–I–Al junctions (Gilmartin, 1982) and of electron scanning with the tunnel junction mounted on a 4.2 K cold stage of an SEM (Epperlein, Seifert, and Huebener, 1982; Eichele et al., 1983) while monitoring a current or voltage.

Subtle changes in the microscopic condition of the electrode in a

Fig. A.8. Use of the Josephson critical current $I_c(H)$ dependence upon magnetic field to infer the uniformity of the current density in a tunnel junction. The $I_c(H)$ pattern at 1.46 K for an Sn–SnO_2–Sn junction is shown. The central peak is reduced by a factor of 5; the circles indicate the positions and heights for the maxima of a true Fraunhofer pattern. (After Dynes and Fulton, 1971.)

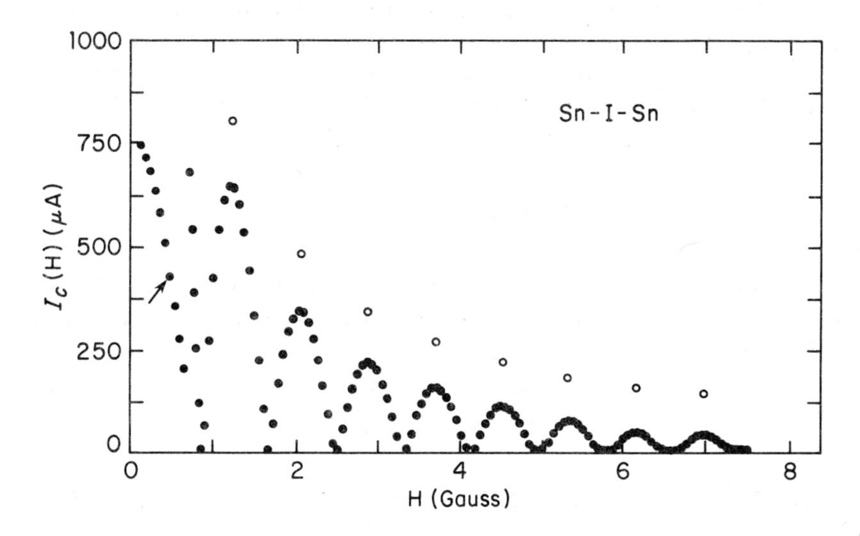

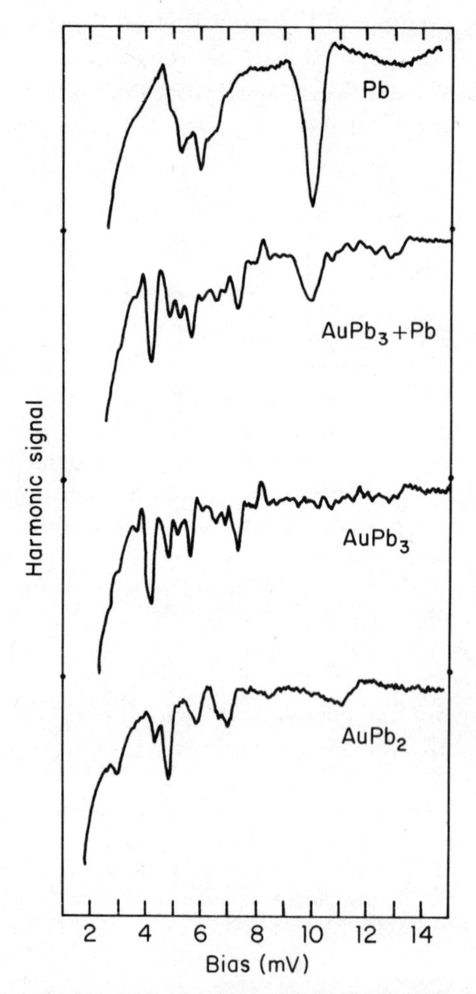

Fig. A.9. Second-derivative spectra obtained from Pb films overlayed with Au after degrees of aging (top to bottom) show changes indicative of alloying and, finally, formation of intermetallic compounds $AuPb_3$ and $AuPb_2$. (After Hebard, 1973.)

superconducting tunnel junction may lead to important effects. One example is illustrated in Fig. A.9 (Hebard, 1973), which shows pronounced changes in the phonon spectra of initially superimposed Au/Pb bilayers (due to interdiffusion and compound formation.) Demonstration of such measurements as a diagnostic for film interdiffusion has been given also by Donovan-Vojtovic, Dodds, Nasser, and Chaikin (1976), also dealing with soft metals in which such effects are possible at room temperature. If one is dealing with bilayers of such metals, it may be necessary to fabricate and measure the structures in situ at low temperature. Methods of doing so have been described, e.g., by Woolf and Reif (1965), Garno (1978), and Bermon and So (1978*a*, 1978*b*). In the case of transition metals, such interdiffusion effects are negligibly slow at room temperature (Geerk, Gurvitch, McWhan, and Rowell, 1982).

A more subtle property of a superconducting electrode which may noticeably affect tunneling derivative spectra is a state of nonuniform strain. This condition may result, in a polycrystalline evaporated film, from differential contraction in cooling to 4.2 K, if the expansion properties of the film and the substrate differ. One way to minimize such strains is to use the same material comprising the films of the junction also as the substrate. In this fashion, by using a single crystal of Pb as the substrate for an evaporated Pb–I–Pb junction, Banks and Blackford (1973) have obtained and displayed (Fig. A.10) better resolution of Pb phonon features (ω_2, ω_3 in the figure) in d^2V/dI^2 than can be obtained by using glass substrates. The evident sensitivity of phonon features to strain is also reflected in the well-known sensitivity of phonon features in transition metal tunnel junctions to disorder or contamination in the region of the electrode immediately beneath the tunnel barrier (or proximity N layer).

The distorting effect of voltage drops in thin *normal-state* electrodes on the energy of observation of gaps, phonon spectra, etc. has been emphasized and analyzed by Osmun (1980).

A.7.2 Derivative Measurement Circuitry

One can identify three distinct measurement tasks in connection with superconductive tunneling and IETS, for which the requirements, and hence the optimum experimental circuits, differ considerably.

First, measurements in the superconducting gap region $|eV| \leq \Delta_1 + \Delta_2$ of S–I–N and S–I–S junctions usually encounter large changes in conductance, including the possibility of negative values of dI/dV. In good junctions with large gaps at low temperatures, one can expect an increase of dI/dV in excess of 10^3 in crossing $eV = 2\Delta$. At the same time, resolution of the gap edge in SIS junctions may require voltage resolution on the order of 10 μV (see Fig. 7.7). For study of the negative-resistance region beyond the difference gap $eV \geq |\Delta_2 - \Delta_1|$ in S–I–S junctions, a low-impedance voltage-biasing source (e.g., Blackford, 1971) is required.

Secondly, quite different requirements apply to obtain accurate measurements of the tunneling density of states $(dV/dI)_N/(dV/dI)_S = \sigma$ in the phonon region. Since σ differs from unity by at most a few percent and itself must be accurate to a percent, one needs stability and resolution in measurements of dV/dI to one part in 10^4 or better, together with voltage resolution of the order of 100 μV, to preserve details of van Hove singularities in phonon spectra. Since the phonon information is present most directly in the second derivative d^2V/dI^2, one should preferably

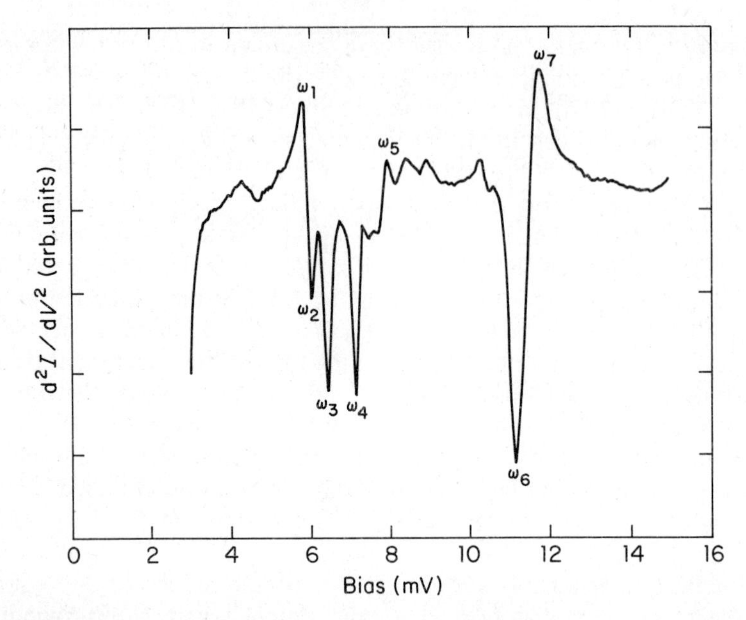

Fig. A.10. Effect of strain on phonon structure observed in tunneling. The peaks labeled ω_2 and ω_3 in this figure are rarely if ever resolved in films of lead deposited upon glass. In this case, improvement has been obtained by depositing the Pb–PbO–Pb junction on a single-crystal Pb substrate, thereby avoiding differential expansion effects in cooling to cryogenic temperature. (After Banks and Blackford, 1973.)

measure either this function or the closely related quantity $d^2I/dV^2 = -(d^2V/dI^2)/(dV/dI)^3$. Various forms of the Kelvin double-bridge circuit are in use for these precision measurements.

The third case, which does not differ greatly in its requirements from the case of superconducting phonon spectroscopy, is in IETS, where one needs primarily high sensitivity and high resolution in measurement of d^2I/dV^2 or d^2V/dI^2. The highly accurate $(1/10^4)$ measurement of the first derivative is usually not necessary in IETS, roughly because there is no theory of accuracy equivalent to Eliashberg theory to use in full analysis of IETS spectra. In this application, bridge circuits are usually considered unnecessary, and sharply tuned filters for rejection of the first harmoic and enhancement of the second harmonic are used to obtain the most detailed spectra. In IETS, then, sensible compromises on circuit design are those favoring a high signal-to-noise ratio and permitting better energy resolution at the expense of minor distortion of the scale of d^2I/dV^2 (but not of the voltage measurement).

A.7.2.1 Gap Region I–V Measurements. The I–V measurements of S–I–S and N–I–S junctions are accomplished with series connection of the (four-terminal) junction, a current-measuring resistance R_m, and a voltage source V of low internal resistance R_V. Circuitry for this purpose is described, e.g., by Adler and Jackson (1966). However straightforward this experiment may seem, great care must be exercised to faithfully reveal the Josephson current, possible Fiske mode structures, as well as the cusp and negative-resistance region at $eV \geqslant |\Delta_2 - \Delta|$ and the sum-gap edge $eV \simeq \Delta_1 + \Delta_2$. The features of interest have energy widths on the microvolt level; therefore, shielding and grounding must exclude pickup of extraneous signals on this level. A low source resistance R_V is needed to observe the negative-resistance regions. In order to observe the maximum Josephson current, one must rigorously exclude flux quanta from the junction. Magnetic shielding is necessary, and it may be desirable to provide for easy warming of the junction above T_c to expel trapped flux. On the other hand, if subgap structure is of interest, the availability of a small magnetic field may be desirable to simplify the spectra by quenching the Josephson current and possible Fiske modes. Such a magnetic field should be provided by an air core solenoid or equivalent near the sample, since it is essential to avoid residual magnetic fields and hysteresis. In this kind of adjustment and subsequent measurements, an oscilloscope display of the I–V curve is a great convenience. We will return to the question of gap region derivative measurements after a discussion of resistance and conductance bridge circuits.

A.7.2.2 Harmonic Detection and Bridge Circuits for Derivative Measurements. The most common approach to obtaining derivative spectra is the dV/dI measurement with sinusoidal current modulation. Constant-current conditions are usually approximated by applying ac and dc biases through series limiting resistors R_L whose values are large compared with the typical junction resistance R_J. The measured quantity is $V(I)$, obtained from the separate pair of potential terminals of the four-terminal cross-strip junction. Usually, the current I is swept slowly $[I_0(t)]$ with the addition of a sinusoidal component, $I(t) = I_0(t) + (\delta I)(\cos \omega t)$. The principle of harmonic detection is easily seen by Taylor series expansion of the $V(I)$ relationship about the value I_0 in powers of $\delta I = I - I_0$:

$$\begin{aligned} V(I) &= V(I_0) + \left(\frac{dV}{dI}\right)\bigg|_{I_0} \delta I \cos \omega t + \frac{1}{2}\left(\frac{d^2V}{dI^2}\right)\bigg|_{I_0} (\delta I \cos \omega t)^2 + \cdots \\ &= V(I_0) + \left(\frac{dV}{dI}\right)\bigg|_{I_0} \delta I \cos \omega t + \frac{1}{4}\left(\frac{d^2V}{dI^2}\right)\bigg|_{I_0} (\delta I)^2(1 + \cos 2\omega t) + \cdots \end{aligned}$$

Synchronous detection at ω and 2ω, respectively, thus provides signals proportional to dV/dI and d^2V/dI^2, evaluated at $V(I_0)$, with the modulation amplitude δI sufficiently small.

The problem of maintaining stability in measurement of dV/dI to one part in 10^4 or better is usually solved by incorporating the junction in a bridge circuit, matching the junction resistance R_J with a decade resistor R_S, balanced at $R_J = R_S$ for the bias region of interest. The signal to the lock-in amplifier (synchronous detector) is then the difference of the voltages across R_J and R_S. This balance condition also provides a null of the first harmonic signal, which is useful if the second harmonic is to be measured.

In Fig. A.11a is sketched the basic double–Kelvin bridge circuit of Rogers, Adler, and Woods (1964), adequate for precision dV/dI measurement of four-terminal tunnel junctions. In this circuit, the bridge unbalance voltage V_ω is given by

$$V_\omega = \left(\frac{\delta I}{\beta}\right)\left[R_S - \left(\frac{dV}{dI}\right)\bigg|_{I_0}\right]\cos \omega t$$

where the circuit parameter β has the form $\beta = 1 + R_S/R_B + 2R_S/Z$ and Z is the input impedance of the amplifier (or transformer, if this device precedes the amplifier). The important function of the large series bridge resistors $R_B \gg R_J$ in series with both current terminals of the junction is to make the results insensitive to changes in the junction lead resistances R_L. The latter typically effectively include the resistance of leads into a dewar, which may vary as the liquid helium level changes and, more importantly, as portions of the connecting leads and films may change from the superconducting to the normal state during the necessary measurement of the normal-state spectra. The improvement brought about by this circuit is best understood by thinking of the series current leads R_L to the junction as small parts of the large bridge resistors R_B. Rogers, Adler, and Woods (1964) show, in fact, that the change $\delta\sigma$ in $\sigma = (dV/dI)_N/(dV/dI)_S$ due to a change in lead resistance δR_L (for a balanced bridge) is given by $\delta\sigma = (\delta R_L/2)(R_P/R_B R_S)$, where R_P is the resistance of the potential lead to the junction. This value can be kept small with large values, 10^4 or $10^5\ \Omega$, of R_B. The circuit originally described by Rogers, Adler, and Woods (1964) was refined by Adler and Jackson (1966) and by Adler and Straus (1975) and also adapted to act as a conductance bridge by Rogers (1970).

The disadvantages of the resistance bridge are, first, its unsuitability for measuring in the gap region $|eV| < \Delta_1 + \Delta_2$ where R_J becomes very large and, second, the complication of having subsequently to numerically convert measurements of d^2V/dI^2 to the more fundamental current derivative d^2I/dV^2. These disadvantages would not be present if a true conductance bridge could be used. In such a circuit, in direct analogy to the resistance bridge, constant-voltage dc and ac biases would be required, and the current in each arm of the bridge would be measured as

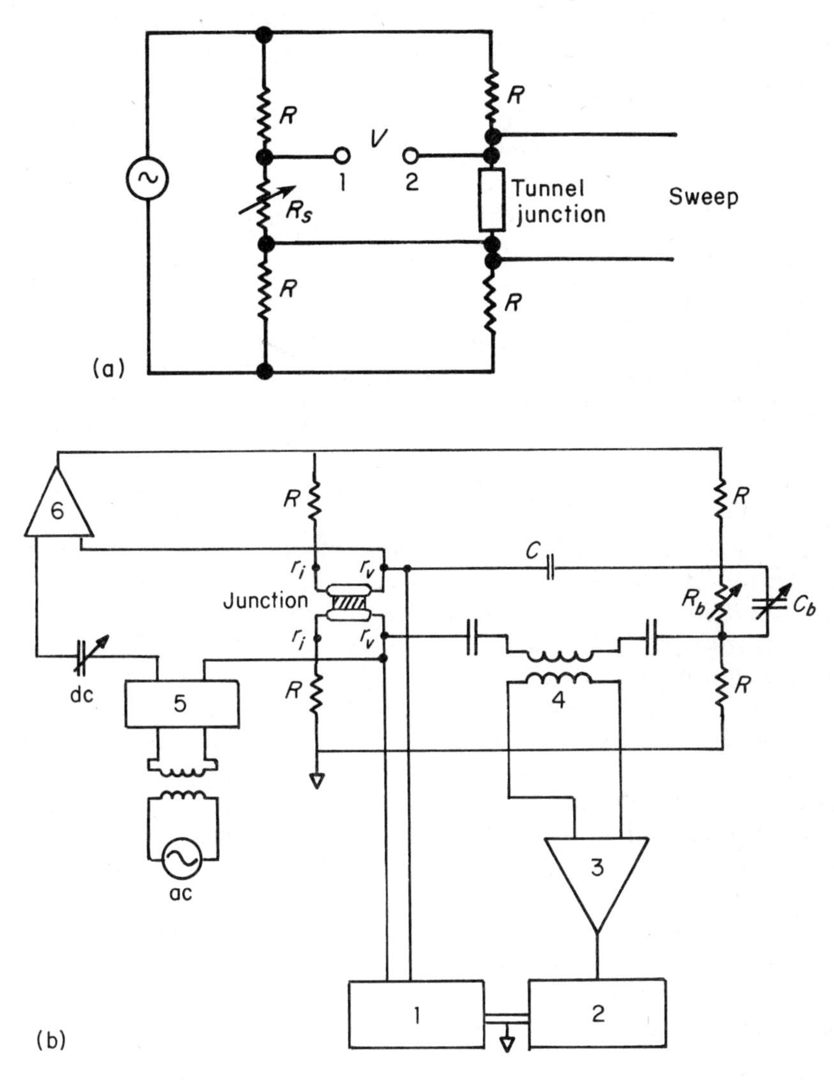

Fig. A.11. In bridge-type measurement circuits, choice of the balancing resistor R_S at any given bias voltage enables one to eliminate the first-harmonic contribution to the voltage V which goes to the lock-in detector. This facilitates determination of the second harmonic at high gain. A feature of the resistance bridge circuit (a) is also that large resistors R on both sides of the junction reduce the sensitivity of the measurement to changes in the resistance of the leads to the samples, which may occur in variation of temperature. (After Rogers, Adler, and Woods, 1964; Adler and Jackson, 1966.) (b) Conductance bridge circuit of Moody, Paterson, and Ciali (1979) described in the text. The role of the capacitor C is to avoid significant errors in the junction voltage measurements with the bridge substantially off balance.

the voltage across a measuring resistor R_m. In order to satisfy constant-voltage conditions, however, one would have to maintain $R_m \ll R_J$ and also $R_L \ll R_J$. Since these conditions are not easily met, the circuits in use for direct measurement of dI/dV and d^2I/dV^2 resort to feedback circuitry to maintain a constant voltage across the tunnel junction. This experiment was first attempted by Rogers (1970) (see also Blackford, 1971) in a resistance bridge circuit such as that of Fig. A.11a by connecting an operational amplifier input between the sample (R_J) and the balancing resistor R_S. The output of this amplifier was connected through a feedback impedance Z_F to the upper terminal of the sample R_J and used to maintain bridge balance by adjusting the current through the sample R_J. The in-phase output V_ω of the operational amplifier was monitored to record the conductance, from the equation (Rogers, 1970)

$$V_\omega = Z_F\left(\frac{dI}{dV} - R_S^{-1} + R_F^{-1}\right)\delta V \cos \omega t$$

Here, δV is the voltage modulation amplitude, R_S is the balancing resistance, and R_F is the resistive part of the feedback impedance Z_F. This analysis neglects the role of lead resistances R_L; and as later pointed out by Blackford (1971), these will enter the equation as additions to R_S. Hence, the bridge balance is potentially affected by the liquid helium levels, etc. This same feature is apparently retained in the subsequent and more completely analyzed conductance bridge circuit of Hebard and Shumate (1974) (cf. Eq. 8 ff, of the latter paper), which does achieve an important improvement in noise relative to the circuits of Rogers (1970) and of Blackford (1971) (which also avoids the lead-resistance drift problem). The fully evolved conductance bridge circuit shown in Fig. A.11b is taken from the paper of Moody, Paterson, and Ciali (1979), which is recommended also for a review and comparison of the several versions of the conductance bridge circuit. In this circuit of Moody, Paterson, and Ciali (1979), the first-harmonic voltage measured by the preamplifier is given by

$$V_\omega = \left(\frac{R_B}{R_B + R_S}\right)\delta V R_S\left(\frac{dI}{dV} - R_S^{-1}\right)$$

and is, hence, free of lead-resistance drift effects. The noise analysis presented by Moody, Paterson, and Ciali (1979) indicates a best–noise voltage value of $2nV\,\mathrm{Hz}^{-1/2}$, and the resulting resolution in conductance with 10% unbalance of the bridge is 4×10^{-5} using 35-μV rms modulation and a 1-sec time constant. This result appears to be only slightly inferior to the performance reported for the Hebard-Shumate circuit.

The second-derivative signals in all of the bridge circuits are available as the 2ω component of the bridge output and are free of ω content at balance.

A.7.2.3 Dedicated Second-Derivative Circuits Utilizing Filter Networks. The original electronic circuit of Thomas and Rowell (1965) solved the problem of measuring specifically the superconducting second derivative, at high resolution, by use of a rejection-selection filter network. This circuit borrowed from earlier designs of Hall, Racette, and Ehrenreich (1960) and of Thomas and Klein (1969), which used electromechanical servomechanisms; the first purely electronic circuit to be reported is apparently that of Patterson and Shewchun (1964). The Thomas-Rowell circuit contains two meshes: an input mesh feeds the fundamental signal ω to the junction and simultaneously suppresses the transmission of any second-harmonic content from the source, while the second mesh feeds the second-harmonic signal generated by the junction to the detector and simultaneously suppresses the first-harmonic output. The second mesh also transforms the impedance level from the low value of the junction to the high value of the preamplifier input. This circuit provides high resolution and low noise, but it has the disadvantages of being limited in resistance level of junctions to be measured and of making only two-terminal connections to junctions. This disadvantage leaves the possibility of errors in measurement of the electrode potential difference if the leads or film contacts should be appreciably resistive or nonlinear in their behavior. This basic approach has been carried forward in several circuits designed for IETS and described, e.g., by Lambe and Jaklevic (1968) and Weinberg (1978). One of the simplest and best of these circuits, judging from the resulting spectra (e.g., Figs. 1.9 and 10.3), is shown in Fig. A.12 (Walmsley, 1983; Walmsley et al., 1983) and will be described in some detail.

The main problems in obtaining excellent, high-sensitivity d^2V/dI^2 spectra are to avoid extraneous signals from pickup or ground loops; to avoid generation or feedthrough of the second harmonic from any source other than the nonlinearity of the tunnel junction I–V characteristic; to effectively filter out the first-harmonic signal before the amplifier input, and to obtain optimum impedance matching from the junction to the amplifier or preamplifier. The need for high sensitivity stems from the weakness of the nonlinearity due to molecular vibrations in the tunnel barrier ($\Delta G/G \simeq 10^{-2}$) and from the necessity of using small equivalent-voltage modulation ($\delta V \simeq 1$ mV rms in IETS) to preserve energy resolution of the vibrational structure. Since the second-harmonic signal is proportional to $(\delta V)^2$, this requirement is a serious one. The choice of 50 kHz as the fundamental frequency, with detection at 100 kHz, follows Lambe and Jaklevic (1968), who noted qualitatively that junction noise falls off with increasing frequency, while frequencies above about 100 kHz lead to capacitive currents which cause increasing difficulty. The 100-kHz range is also optimum from the point of view of noise figures in typical amplifiers.

The circuit of Walmsley (1983) (Fig. A.12) deals with these requirements in a simple and effective fashion. Ground loops and sources of

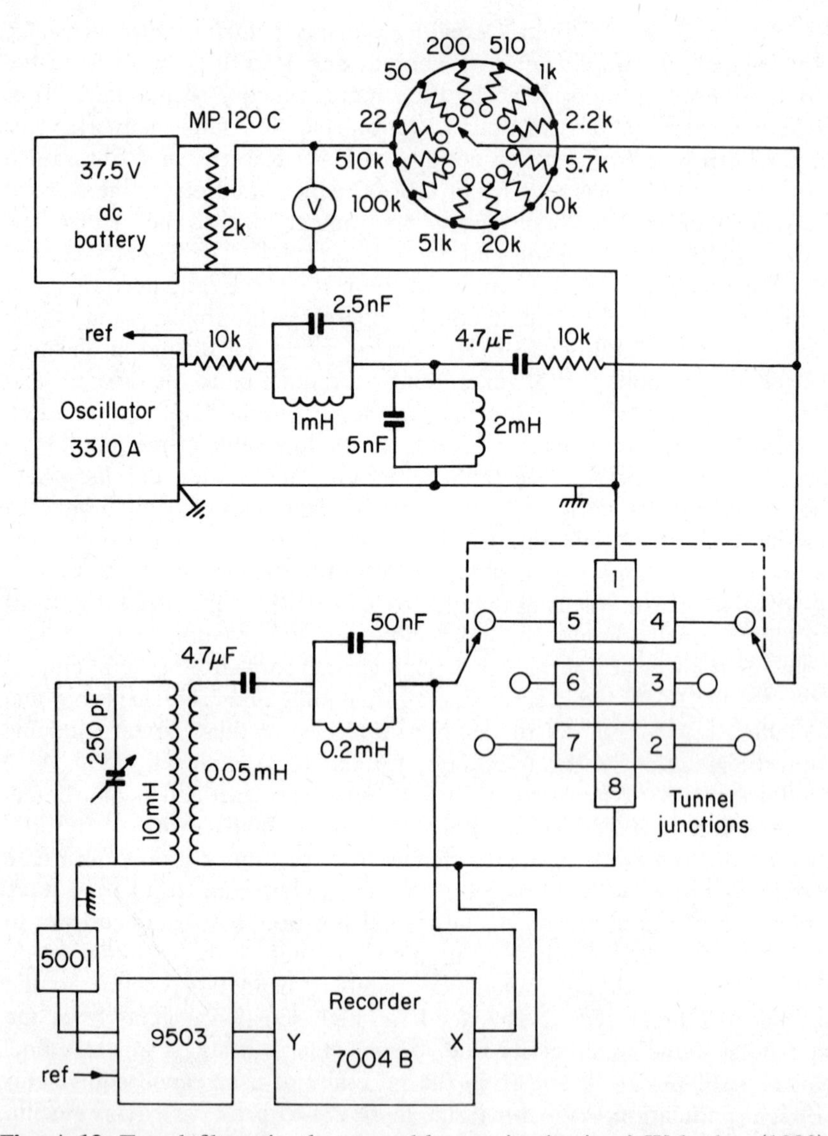

Fig. A.12. Tuned filters in the second-harmonic circuit of Walmsley (1983), designed for IETS measurements, block the second harmonic in the junction modulation input and also block the first harmonic in the junction output to the preamplifier. Earlier circuits of this type have been described by Thomas and Rowell (1965) and by Lambe and Jaklevic (1968). (After Walmsley et al., 1983.)

pickup related to the ac mains are minimized by making only one earth connection (no connections to the third pin of the usual ac plug) and by using batteries for the sweep voltage circuit. The second harmonic is filtered from the oscillator output by use of 100-kHz reject and 50-kHz pass filters in the line to the sample. The pickup of airborne signals (including 60-Hz magnetic signals) by the ferrite cores of these filters is simply but effectively dealt with by maintaining a high-level signal from the oscillator into the filters (the HP 3310A oscillator can produce a 30-V pp signal) sufficient to dwarf pickup signals, followed by a pick-up-free resistive potential divider to reduce the level to that appropriate for the junction modulation. Walmsley (1983) includes a 10-Ω resistor at the oscillator to avoid overload (harmonic) distortion of the oscillator at resonance; the 4.7-μF capacitor to block dc current from the sweep supply through the inductors (which could lower the resonant circuit Q values by polarizing the ferrite inductor cores); and the 10-Ω level-reducing divider resistor at the output of the filter, in series with the junctions.

Optimum transfer of the second-harmonic (2ω) signal generated by the junction to the high-impedance ($10^8\ \Omega$) preamplifier input (a Brookdeal-Ortec 5001 in this particular circuit) is accomplished by a step-up transformer whose secondary circuit is tuned to resonance (for each sample) at the second-harmonic frequency 2ω. Walmsley (1983) points out that tuning of the secondary circuit to resonance with the 250-pF capacitor (which is done by using the 5001 as a high-impedance input, rather than with an oscilloscope directly) is possible only if the cable to the preamplifier is short (perhaps 25 cm) so that its capacitance is low. The resonant filter in the transformer primary circuit blocks any 50-kHz signal from the preamplifier, so that it can be operated at high gain on the 2ω signal.

It may be useful to quote Professor Walmsley's (1983) comments on the construction of the high-Q filter and transformer elements which form the heart of this system and on the operation of the system:

> Every effort was made to optimize the quality factor of the coils in the resonant circuits. The three single coils were wound on ITT ferrite cores type C22 following the manufacturer's advice on wire diameter. Specifically, the 1 mH inductance has 60 turns of 26 SWG lacquer insulated copper wire, the 2 mH inductance has 85 turns of the same wire and the 0.2 mH inductance has 28 turns of 24 SWG wire. The cores are 2 mm in diameter and 13 mm high with an inductance factor, A_L, of 250 nH per turn. The transformer is on a type C26 core (26 mm diameter, 16 mm high) with an A_L of 400 nH per turn. The 0.05 mH primary has 10 turns of 24 SWG wire and the secondary has 155 turns of 34 SWG wire to give a self inductance of 10 mH. Typical quality factor values were 50 to 150. The ferrite cores have tuning rods but they do not

offer good resolution and it is convenient to have 100 pF trim capacitors on the 2.5 nF and 5 nF components. The circuits cannot be tuned with a standard 1M ohm input impedance oscilloscope; the 9001 preamplifier performs well as a buffer in this task.

In recording spectra, a lock-in time constant of 3 sec and sweep time of 40 minutes for 0 to 500 mV offers a fair compromise between requirements of noise reduction and signal lag as against overall time for data acquisition.

Switching between samples such as from terminals 4 and 5 to 3 and 6 should always be done with the dc bias set to zero as otherwise the samples may be destroyed.

Tests of performance of the circuit are given in the recent article by Walmsley et al. (1983).

APPENDIX B

Numerical Methods

Rather elaborate calculations are often required to obtain the desired information from tunneling measurements, especially in the case of superconductive tunneling when one extracts information on an $\alpha^2F(\omega)$ function of a superconductor. Even more involved calculations based on the Eliashberg equations are required for analysis of proximity (PETS) data. The objective of the present appendix is to briefly survey numerical methods that have been developed for these purposes, starting with elementary considerations in digitizing data, and such basic operations on data as smoothing and differentiation. Discussions of several of the basic numerical tasks in tunneling and some of the fairly standard programs which have come into use then follow.

B.1 Digitizing and Manipulating Data

The most common primary records of a tunneling measurement are the continuous but probably noisy plots of $\mathrm{d}V/\mathrm{d}I$ or $\mathrm{d}^2V/\mathrm{d}I^2$ vs. bias voltage obtained from an x–y recorder. A time-honored method of digitizing such records on an appropriate mesh of voltage points, typically spaced by 0.1 to 0.5 meV, is to smooth the curve by eye with a pencil and to read the best values into a table or directly into a computer file. While this method is laborious, it has at least two merits.

First, the initial data taking on the x–y chart is not hindered by providing appropriate settings for the additional instruments that may be necessary to record the data in an initially digitized form. In many cases, only a relatively small fraction of samples, or of x–y curves on a given sample, are sufficiently excellent in all respects to warrant complete numerical analysis, say to provide $\alpha^2F(\omega)$ in the case of conventional McMillan–Rowell spectroscopy. Since many forms of defect can be identified on the x–y plot, time and expense may be avoided by initially recording only in the traditional plot, carefully screening these records, and devoting full attention only to those curves that will lead, e.g., to definitive $\alpha^2F(\omega)$ functions.

Secondly, the smoothing by eye and pencil of a slightly noisy chart record is an efficient procedure for preserving the information obtained in the measurement. It is superior in this regard, e.g., to sampling over a short interval and digitally recording the output of a lock-in amplifier with a digital voltmeter at only the discrete bias voltages of the chosen mesh. The task of digitizing the x–y curve can be done also with the aid of commercial digitizing instruments such as those marketed by Summagraphics of Fairfield, Connecticut.

Having thus paid due respect to traditional methods, one can say that a superior method of recording data at the present state of the art is possible with a multichannel analyzer (MCA). Such a device will commonly have 4096 channels, each capable of storing a digit between 1 and 2^{20} or about 10^6. In the usual MCA, the input to a given channel is a train of voltage pulses which are counted. By using the internal time base of the MCA to control the bias voltage sweep in the tunneling experiment, a correspondence between channel address and bias voltage can easily be set up. An efficient approach to direct digital recording of data from a good tunnel junction would be to record dV/dI and d^2V/dI^2 simultaneously from separate lock-in amplifiers, breaking the MCA channels into four groups of 1024. This technique provides space for a 0.1-mV mesh up to 100 mV, more than is usually needed. The record might consist of dV/dI, d^2V/dI^2, and possibly also $I(V)$ and the bias voltage V. An efficient means of recording the information is by means of voltage-to-frequency converters; the MCA, in counting the pulsed output of these devices, in fact integrates the record from the lock-in amplifier (which should be set with a short time constant) over the chosen channel dwell time, which now provides the smoothing of the data. It is desirable to use a short integration time on the lock-in amplifier itself, as noted by McMillan and Rowell (1969), to minimize disruption of the data by very intense, but infrequent and short, noise pulses. The data can be quickly stored on an auxiliary disk file or directly read to a computer file over a telephone or direct line from the MCA.

Manipulation of data is facilitated by the use of the MCA acquisition scheme. In the first place, repeated sweeps through the bias region under study can easily be made and the cumulative record inspected on the oscilloscope display of the MCA. This method is preferable to the conventional scheme where one must guess at the optimum choices of sweep rate, modulation level, and lock-in amplifier time constant before doing a sweep and does not learn of errors in these choices until the sweep is substantially complete, a period which may vary up to an hour or so in difficult cases.

While some elementary operations on digitized data can be done directly in many of the commercially available multichannel analyzers, greater flexibility and control of such operations as smoothing and differentiating is possible when they are done in a programmable computer. Smoothing of a digitized trace, e.g., $\sigma(V_i) \equiv \sigma_i$ when $i = 1, \ldots, N$, can be accomplished by finite convolution with a weighting function W_j, normalized so that

$$\sum_{j=-n}^{n} W_j = 1.$$

The expression is

$$\bar{\sigma}(V_i) = \sum_{j=-n}^{n} \sigma_{i+j} W_j$$

An example of a suitable weighting function is the Gaussian $W_j = C_n e^{-\alpha j^2}$. Here, the choices of α and n depend upon the data but are coupled by a condition such as $e^{-\alpha n^2} = 0.01$: the constant C_n is determined by the normalization condition. An alternative would be a Lorentzian weighting function

$$W_j = C_n'(1 + \beta j^2)^{-1}$$

with similar conditions for β and C_n'.

Such operations may be necessary before one can numerically differentiate noisy data to get a reasonably smooth result. Another recommended approach to differentiation of noisy data is to use a least squares fitting of a parabola to a small number (e.g., five) of points and then to evaluate the derivative analytically from the parameters of the best parabola. In practical cases, such a numerical differentiation might be applied to data already smoothed by one of the schemes mentioned above. Data-taking schemes have in fact been reported (Dargis, 1981) in which numerical differentiation is the primary source of d^2I/dV^2 data. Integration of data numerically is ordinarily done by use of Simpson's rule. Other operations probably best done on the digitized data after reading into the computer are calibration, normalization, and (possibly) conversion of d^2V/dI^2 to d^2I/dV^2, making use of a measured dV/dI function and the relation

$$\frac{d^2I}{dV^2} = \frac{-(d^2V/dI^2)}{(dV/dI)^3}$$

Calibration of dV/dI data is usually accomplished by substituting a four-terminal decade resistance box for the sample, recording the deflection on the x–y chart for three or four values of resistance in the range of interest and using these values to establish a correspondence between the primary record (chart units or MCA values) and the recorded differential resistance values from the junction. A thorough approach would establish the correspondence between dV/dI and the primary record by least squares fit of several calibration points using a quadratic relationship. In ordinary circumstances, the coefficient of the quadratic term in the relation should be small. Calibration of d^2V/dI^2 data is not as direct but can be done by fitting an integrated d^2V/dI^2 to a calibrated dV/dI. For calculations on superconductive tunneling data, the normalized conductance $\sigma = (dV/dI)_N/(dV/dI)_S$ is required and is ordinarily calculated numerically at the beginning of any program dealing with superconductive results. An elegant method of combining $(d^2V/dI^2)_S$ and $(dV/dI)_{S,N}$ data to provide a detailed σ function will be briefly described in connection with the McMillan Stage I program. A summary of methods of data handling when a dedicated minicomputer is available has been given by Adler (1982).

B.2 Gap Region Programs

A primary role for numerical calculation in connection with NIS gap region data is to generate curves of $\sigma(V, \Delta_0, T)$ from the BCS $N_T(E)$, using Eq. (3.1), to determine the value of Δ_0 which most closely matches the measured normalized $\sigma(V)$. A calculation program for $\sigma(V, \Delta, T)$ has been discussed (and a numerical tabulation provided) by Bermon (1964). A similar program utilizing the Arnold (1978) gap region N-side proximity density of states has been used by Wolf, Zasadzinski, Osmun, and Arnold (1980). Finally, calculated curves of $\sigma(V, \Delta_0, T)$ using the N-side density of states of the McMillan tunneling model may be useful in some cases.

Numerical modeling of the SIS I–V and σ–V cases is also of interest. The earliest detailed calculations are those of Shapiro et al. (1962), and further sources are the thesis of Taylor (1963) and the article of Douglass and Falicov (1964) discussed in Chap. 3.

All that has been mentioned above is in the category of modeling, i.e., computing I or $\mathrm{d}I/\mathrm{d}V$ given the relevant gap parameter(s) and the temperature. The related inversion or deconvolution problem in the S_1IS_2 case to obtain $N_2(E)$ from measured $\mathrm{d}I/\mathrm{d}V$ or $\mathrm{d}^2I/\mathrm{d}V^2$ data, assuming that $N_1(E)$ is the BCS density of states and that thermal excitations are unimportant, has been solved by Vopat, Lee, and Tomasch (1976).

The central result of these authors, applicable at $T=0$, is

$$N_2(E) = -\frac{1}{eG_{NN}} \int_{\Delta_1}^{E+\Delta_1} \left(\frac{\mathrm{d}^2 I}{\mathrm{d}V^2}\right) \exp F\left(\frac{E-eV}{\Delta_1}\right) \mathrm{d}V. \tag{B.1}$$

The function $F(y)$ can be expressed as

$$F(y) = \begin{cases} 1 - G(y), & \mathrm{Re}(y) > -1 \\ 0, & \mathrm{Re}(y) \leqslant -1 \end{cases} \tag{B.2}$$

where

$$G(y) = \int_0^{x_0=\infty} \frac{\mathrm{d}x}{x^2} \frac{\exp(-yx) I_1(x)}{\{[K_1(x)]^2 + \pi^2 [I_1(x)]^2\}^2} \tag{B.3}$$

and where $K_1(x)$ and $I_1(x)$ are, respectively, modified Bessel functions of the first and second kinds. A plot of the function $F(y)$ is given in Fig. (B.1). In numerical work, slow convergence of $G(y)$, Eq. (B.3), near $y=-1$ necessitates large values of the upper limit $x_0 \simeq 1000$; a means of estimating truncation error is also presented by Vopat, Lee, and Tomasch (1976).

Demonstrations of the accuracy of the inversion method are also given in the original paper.

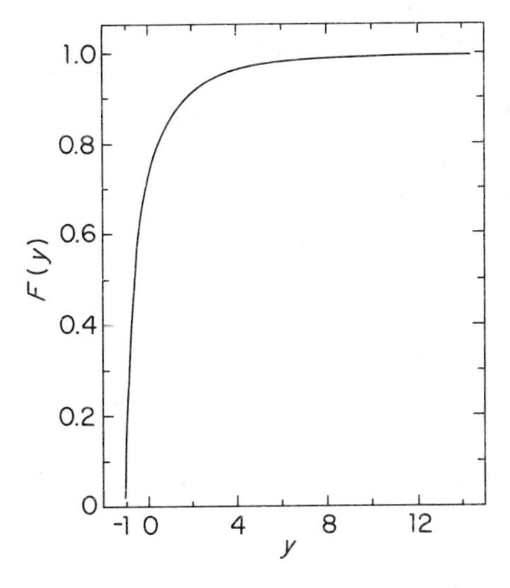

Fig. B.1. The function $F(y)$ defined in Eq. (B.2). (After Vopat, Lee, and Tomasch, 1976.)

B.3 Superconducting Phonon Inversion Methods

The classic numerical methods for inverting the tunneling conductance $\sigma(V)$ in the strong-coupling case for the superconducting phonon spectrum $\alpha^2F(\omega)$ were developed by McMillan and described by McMillan and Rowell (1969). The basic approach in the Stage III program for $\alpha^2F(\omega)$ has already been described in Chap. 4. A detailed report specifically on the three McMillan programs is that of Hubin (1970); more recent comments and extensions are described by Nedrud (1980). In view of these sources, we will only briefly outline the aspects of the McMillan programs not already described in Chap. 4.

The alternative numerical method for inverting phonon spectra for $\alpha^2F(\omega)$ is that Galkin, D'yachenko, and Svistunov (1974), the basic features of which have again already been described in Chap. 4.

B.3.1 McMillan Stage I Programs

The inputs for the Stage I program are the superconducting (S) and normal-state (N) derivative data: $(\mathrm{d}V/\mathrm{d}I)_{\mathrm{S,N}} \equiv V'_{\mathrm{S,N}}$ and $(\mathrm{d}^2V/\mathrm{d}I^2)_{\mathrm{S,N}} \equiv V''_{\mathrm{S,N}}$, measured on a mesh of voltage points with equal spacing H, starting from $V=0$. Additional inputs include the experimentally determined gap values for each electrode. The objective is to obtain, from both $V'_{\mathrm{S,N}}$ and $V''_{\mathrm{S,N}}$ data sets, the best BCS-reduced conductance $\sigma/\sigma_{\mathrm{BCS}}-1$, where $\sigma=\bar{V}'_{\mathrm{N}}/\bar{V}'_{\mathrm{S}}$, $\bar{V}''_{\mathrm{N,S}}$ are the smoothed derivatives discussed below, and $\sigma_{\mathrm{BCS}}=|E|/(E^2-\Delta_0^2)^{1/2}$, Δ_0 being the gap value in the electrode of interest. McMillan's Stage I program also provides interpolation to shift the zero of the energy scale from $V=0$ to $V=(\Delta_c+\Delta_0)/e$, so that energies are finally measured from the gap edge.

We will focus on the problem of combining V' and V'' data to preserve, in the smoothed derivatives $\bar{V}'$, both the detail obtained in V'' and the accuracy of calibration of V', and we will sketch a minor variation on the actual approach of McMillan's program. This problem is closely related to the problem of calibration of V'' data mentioned earlier and is accomplished by least squares fitting to the measured function V', via coefficients A, B, C, and D of an integrated function V'_{fitted}:

$$V'_{\text{fitted}}=A\int V''\frac{dV}{V'}+B\int V''\,dV+CV+D \tag{B.4}$$

In this equation, the coefficients A and C account for the scale factor and zero, respectively, of the d^2V/dI^2 signal, while B, which should be found to be small, accounts for the one type of distortion in the measurement of V''. The V' in the denominator of the first integral term is needed to change the variable of integration from I to V. Other integrand functions may be appropriate in (B.4), depending upon the actual measurement circuit employed. It seems most sensible to use this fitting procedure separately on the superconducting and normal data sets, taking a ratio of the form $V'_N/V'_S=\sigma$ only afterward. However, since relatively little detail is present in V'_N, it may be adequate merely to self-smooth V'_N, dispensing with measurement and subsequent integration of V''_N.

Before the ratio V'_N/V'_S is formed, however, it is necessary to provide a means of smoothing the connection between V'_{fitted} and V', in the usual case that the bias range over which V''_S measurements are made is smaller than the range for V'_S. For the details of McMillan's procedures for this purpose, the reader is referred to Hubin (1970). The final resulting σ is the ratio of the smoothed V' functions V'_N/V_S'. The output of a Stage I program is $\sigma/\sigma_{\text{BCS}}-1$ on a voltage mesh extending from the sum-gap voltage. The next step in the procedure is to subtract the influence of the counterelectrode gap Δ_c, in the event that the counterelectrode is a superconductor. This subtraction is done via the McMillan Stage II program.

B.3.2 *Stage II*

This program is applicable to S_1IS_2 data only in the case when S_1 is a weak-coupling material, with negligible phonon structure of its own, and S_2 is the strong-coupling superconductor of interest. This stage is described in detail by Hubin (1970).

B.3.3 *Stage III Programs*

The role of a Stage III program is to extract from the BCS-normalized conductance $n(E)=\sigma/\sigma_{\text{BCS}}-1$ the superconducting functions $\Delta(E)$, $Z(E)$, and $\alpha^2F(\omega)$. As has already been outlined in Chap. 4, this extraction is done by making use of the Eliashberg equations of strong-coupling superconductivity.

As an initial test of the quality of input BCS-reduced data $n(E)$, Ochik, Svistunov, Belogolovskii, and D'yachenko (1978) have shown that the relation

$$\int_{\Delta_0}^{E_c} n(E)\sigma_{\mathrm{BCS}}(E)\,\mathrm{d}E \simeq 0 \tag{B.5}$$

should be satisfied. Qualitatively, this expression says that the average deviation of the measured conductance data from the corresponding σ_{BCS} conductance must be zero, or that the BCS conductance should run through the center of phonon-induced structures in the measured (and predicted) conductances. Condition (B.5) follows from the basic normalization property of the spectral function:

$$\int_{-\infty}^{\infty} A(p, \omega)\,\mathrm{d}\omega = 1$$

which implies

$$\int_{-\infty}^{\infty} [A(p, \omega) - A_{\mathrm{BCS}}(p, \omega)]\,\mathrm{d}\omega = 0$$

The validity of this statement between the finite integration limits in (B.5) follows from the fact that A and A_{BCS} differ appreciably only in the energy range below the cutoff E_c. Hence,

$$\int_{\Delta_0}^{E_c} \sigma(\omega)\,\mathrm{d}\omega \simeq \int_{\Delta_0}^{E_c} \sigma_{\mathrm{BCS}}\,\mathrm{d}\omega$$

which is equivalent to (B.5). This expression is consistent with the assumptions of Arnold, Zasadzinski, and Wolf (1978), who also found means for correction of data in certain cases for failure in satisfying (B.5). The standard conventional Stage III programs of McMillan and Rowell (1969) and of Galkin, D'yachenko, and Svistunov (1974) have already been described in Sec. 4.4.

B.4 Proximity Electron Tunneling Programs

Several programs specifically useful for PET have been written by G. B. Arnold and briefly described by Arnold, Zasadzinski, Osmun, and Wolf (1980) and Burnell (1982).

B.4.1 MMR (Stage III)

A modified McMillan-Rowell Stage III (MMR) program for PETS data is used to determine $\alpha^2F_S(\omega)$ as in the analysis of conventional tunneling data. In the proximity case, σ_{prox} is calculated from Eq. (5.28), with $\Delta_S(E)=\Delta_{SO}$ and with input values for $\Delta_N(E)$, $Z_N(E)$, d, and l. From this equation, the "proximity-reduced conductance"

$$N_{prox} \equiv \frac{\sigma}{\sigma_{prox}} - 1$$

is obtained, separating structure arising from the proximity effect from deviations due to the structure in $\Delta_S(E)$.

The derivative of N_{prox} with respect to energy is then calculated and used subsequently as in the standard Stage III treatment. The linear response of the energy derivative of the PET density of states to changes in $\alpha^2F_S(\omega)$ is used to correct $\alpha^2F_S(\omega)$ to match the derivative of N_{prox} in successive iterations. After $\alpha^2F_S(\omega)$ and $\Delta_S(E)$ have been found to sufficient accuracy (determined by the least squares fit to dN_{prox}/dE), σ_{calc} is obtained from Eq. (5.28) and compared with the input BCS-reduced conductance.

Failure to correctly parametrize d and d/l has been shown by Arnold, Zasadzinski, and Wolf (1978) to produce an accurate fit for dN_T^{calc}/dE to the experimental derivative, while N_T^{calc} displays a positive upward displacement with respect to N_T^{exp}, thus violating (B.5). The source of this offset lies in underestimation of the electron-phonon coupling used in the determination of μ_S^*. In the program μ^* is adjusted at each stage of the iterative calculation so that the experimentally observed Δ_{SO} is obtained. Specifically, the observed gap Δ_{SO} must satisfy

$$\Delta_S(\Delta_{SO}) = \Delta_{SO}\frac{\alpha_S^2 I_1 - \mu_S^* I_2}{1-(\alpha_S^2/\Delta_{SO})I_3} \tag{B.6}$$

where α_S^2 is a number reflecting the average square of $\alpha_S(\omega)$, and I_1, I_2, and I_3 are integrals derived from the Eliashberg equations. Solving (B.6) for μ_S^* gives

$$\mu_S^* = \frac{\alpha_S^2(I_1+I_3)-\Delta_{SO}}{I_2} \tag{B.7}$$

For each successive $\alpha^2F_S(\omega)$ obtained in the iterations, μ_S^* is adjusted according to this equation. Now, for a fit of the amplitude of the phonon structure, the only free parameter available is α_S^2, since d, d/l, and Δ_{SO} are fixed through the calculation. If d/l is underestimated, a smaller α_S^2 than the true value is obtained, making $\Delta_S(\Delta_{SO})$ less than Δ_{SO}. So that agreement between the latter is obtained, μ_S^* is corrected *below* its true value, and possibly to a negative value. When $\Delta_S(E)$ is calculated with the new μ_S^*, it is shifted upward by a constant relative to the value for which σ_{calc} and σ_{exp} would agree. Thus, minimizing the displacement between σ_{calc} and σ_{exp} becomes a selection criterion for the d/l parameter, or, since d is experimentally determined, for the scattering parameter l.

B.4.2 DELN Program

The DELN program is designed to determine $\Delta_N(E)$ from the conductance data of a junction of intermediate thickness displaying moderately strong N-phonon characteristics yet thin enough to meet the PET criteria described in Chap. 5. The functions $\Delta_S(E)$, $Z_N(E)$, obtained from a thinner junction, and the relevant d and d/l parameters are the required inputs. Term $\Delta_N(E)$ is adjusted to bring the theoretical density of states into agreement with the junction conductance in the following manner.

The density of states given in Eq. (5.28) is proportional to the imaginary part of a Green's function. Since the imaginary part of a Green's function also determines its real part via a Hilbert transform, $\sigma(E)$ likewise determines its Hilbert transform $\tilde{\sigma}(E)$:

$$\tilde{\sigma}(E)=\frac{2}{\pi}\int_0^\infty \frac{\sigma(E')}{E^2-(E')^2} \tag{B.8}$$

Thus, when (5.28) is set equal to the experimental conductance, and the Hilbert transform of (5.28) is set equal to the Hilbert transform of the experimental conductance, two equations for Re Δ_N and Im Δ_N are established, directly incorporating the data. To ensure that the integrand of (B.8) goes to zero at a finite energy, one sets

$$\sigma(E)-\sigma_{\text{calc}}(E)=\frac{2E}{\pi}\int_0^\infty \frac{\sigma(E)-\sigma_{\text{calc}}(E')}{E^2-(E')^2}$$

This technique is preferable to a modified McMillan–Rowell approach in that it does not require the determination of μ_N^* in order to find $\Delta_N(E)$. To do so, the McMillan-Rowell approach would require $\Delta_N(E)$ to be directly determinable at a specific energy analogous to the energy gap for $\Delta_S(E)$; this determination is generally not possible. An additional advantage is that multiple iterations of the determined functions are eliminated.

Unlike the MMR program, the DELN program has no internal criterion for the appropriate selection of d/l. In practice, this parameter is selected to be consistent with the values indicated by the MMR inversion of the thinner-N data within the bounds of physical acceptability [i.e., Im $\Delta_N(E)$ cannot be negative at energies below the transverse phonon, as this result would imply a negative value for $\alpha^2F_N(\omega)$].

B.4.3 A2FN

The program A2FN determines the underlying $\alpha^2F_N(\omega)$ for a given $\Delta_N(E)$, making use of the corresponding $\Delta_S(E)$ responsible via the proximity effect for the N pair-potential enhancement. By taking the imaginary part of Eq. (5.45), one eliminates μ_N^* from the equation, leaving

$$\text{Im}\,\phi_N(E)=\int_{\Delta_{\text{SO}}}^{E} d\nu\ \text{Re}\frac{\Delta_S(\nu)}{[\nu^2-\Delta_S^2(\nu)]^{1/2}}\,\pi\alpha^2F_N(E-\nu)\theta(\omega_{0N}-E+\nu) \tag{B.9}$$

If one describes each function as an array over an energy grid $E(E) = I \cdot H$, where I is an integer and H is the energy spacing between sampled array points, (B.9) can be written

$$\Phi_{\mathrm{N}}(I) = \sum_{J=1}^{I-1} TP(J)^{**}\alpha^2 F(I-J+1) \qquad \text{(B.10)}$$

where

$$\mathrm{Re}\frac{\Delta_{\mathrm{S}}(\nu)}{[\nu^2 - \Delta_{\mathrm{S}}^2(\nu)]^{1/2}} \rightarrow TP(J)$$

and

$$\varphi_{\mathrm{N}}(E) = \mathrm{Im}[\Delta_{\mathrm{N}}(E) \cdot Z_{\mathrm{N}}(E)]$$

From Eq. (B.10), the $\alpha^2 F_{\mathrm{N}}(I)$ are successively determined in a bootstrap method, starting with $\alpha^2 F_{\mathrm{N}}(1) = 0.0$, $\alpha^2 F_{\mathrm{N}}(2) = \Phi(2)/\pi \cdot \mathrm{TP}(1)$, etc., until the entire function is determined for all $I \cdot H \leqslant \omega_{\mathrm{ON}}$.

APPENDIX C

Tabulated Results

Summaries of superconducting parameters for several classes of superconductors are contained in Tables C.1–C.5. Many of the T_c and θ_D values were obtained from the extensive compilation of Roberts (1976). Many of the values of $\omega_{\log} \equiv \ln\langle \exp \omega \rangle$, especially those in Table C.2, devoted largely to s–p-band alloys, are from the paper of Allen and Dynes (1975b). Additional sources, including tabulated values of superconducting functions [e.g., $\Delta(E)$, $Z(E)$] are the two compilations of strong-coupling tunneling results due to J. M. Rowell, W. L. McMillan, and R. C. Dynes and available from these authors. The estimated values of λ for many weak-coupling metals are obtained by use of McMillan's T_c formula,

$$T_c = \frac{\theta_D}{1.45} \exp \frac{1.04(1+\lambda)}{\lambda - \mu^*(1+0.62\lambda)} \tag{C.1}$$

(McMillan, 1968*a*). In general, the intent has been to provide the best present estimates of parameters rather than a full history of work on each superconductor. Excellent sources of tabulated information on superconductors, not restricted to tunneling results, in addition to the compilation of Roberts (1976), include the recent books by Grimvall (1981) and Vonsovsky, Izyamov, and Kurmaev (1982); as well as the earlier monograph by Savitskii et al. (1973).

Table C.6 lists primarily recent reports of metals used as an initial overlayer subsequently oxidized to produce a tunneling barrier. Table C.7 provides a partial list of references to tunneling work on thick N/S or S/N bilayers in which the oscillatory (Tomasch and McMillan-Rowell) phenomena were of interest. Excellent and more complete reviews of this area of research have been given by Nédellec, Dumoulin, and Guyon (1976) and by Nédellec (1977). A companion review to the latter (also in French) is that of Gilabert (1977) on the proximity effect.

Finally, Table C.8 gives a partial listing of tunneling results on superconductivity under hydrostatic pressure.

Table C.1. s, p Elements

Material	T_C (K)	θ_D (K)	ω_{log} (meV)	$\langle\omega\rangle$ (meV)	$\sqrt{\langle\omega^2\rangle}$ (meV)	λ	$2\Delta/k_BT_C$	Tunneling Reference
Be	0.026	1390				0.23[a]		
Mg	≤0.3 mK	400				0.29±0.03		Burnell and Wolf (1982)
Al	1.18	420				0.38[a]	3.53, 3.50	Blackford and March (1968) Hubin and Ginsberg (1969)
Cd	0.52	209				0.38[a]	3.2	Kumbhare, Tedrow, and Lee (1969)
Zn	0.85	310				0.38[a]		
Ga	1.08	325				0.40[a]	3.63	Yoshihiro and Sasaki (1968) Wühl, Jackson, and Briscoe (1968)
Am–Ga	8.56		4.74	6.63	8.70	1.62	4.5	Chen, Chen, Leslie, and Smith (1969) Leslie, Chen, and Chen (1970)
Sn	3.72	195	8.53	9.48	10.43	0.72	3.7	Rowell, McMillan, and Feldmann (1969) Zavaritskii (1965)
In	3.41	109	5.86	6.81	7.67	0.805	3.68	Dynes (1970)
Tl	2.38	79	4.48	5.0	5.51	0.795	3.6	Dynes (1970) Clark (1968)
Pb	7.196	105	4.83	5.20	5.55	1.55	4.67, 4.30	Townsend and Sutton (1962) McMillan and Rowell (1965, 1969) Dynes and Rowell (1975)
Am–Pb	7.2						4.65	Ziemba and Bergmann (1970)
Am–Hg	3.90						4.4	Ziemba and Bergmann (1970)
Hg(α)	4.15	72	2.5	3.27	4.22	1.6	4.61	Hubin and Ginsberg (1969)

[a] Value obtained using McMillan's T_c formula (McMillan, 1968*a*).

Table C.2. Alloys and Unusual Phases: s, p Elements

Material	T_C (K)	θ_D (K) $\omega_{\log}$ (meV)	$\langle\omega\rangle$ (meV)	$\sqrt{\langle\omega^2\rangle}$ (meV)	λ	$2\Delta/k_BT_C$	Tunneling Reference
$Tl_{0.9}Bi_{0.1}$	2.3	4.14	4.74	5.34	0.78	3.58	Dynes (1970) Allen and Dynes (1975*b*)
$In_{0.5}Tl_{0.5}$	2.52	4.57	5.51	6.29	0.83	—	Allen and Dynes (1975*b*)
$In_{0.9}Tl_{0.1}$	3.28	5.43	6.46	7.41	0.85	—	Allen and Dynes (1975*b*)
$In_{0.57}Tl_{0.43}$	2.60	4.57	5.51	6.38	0.85	—	Allen and Dynes (1975*b*)
$In_{0.67}Tl_{0.33}$	3.26	4.91	5.86	6.81	0.90	—	Allen and Dynes (1975*b*)
$In_{0.07}Tl_{0.93}$	2.77	4.22	4.83	5.43	0.89	—	Allen and Dynes (1975*b*)
$In_{0.73}Tl_{0.27}$	3.36	4.74	5.77	6.63	0.93	—	Dynes (1970) Allen and Dynes (1975*b*)
β–Ga	5.90	7.5	9.31	11.1	0.97		Allen and Dynes (1975*b*)
$In_{0.17}Tl_{0.83}$	3.19	3.88	4.74	5.43	0.98	—	Dynes (1970) Allen and Dynes (1975*b*)
$In_{0.27}Tl_{0.73}$	3.64	3.62	4.57	5.43	1.09	—	Allen and Dynes (1975*b*)
$Pb_{0.4}Tl_{0,6}$	4.60	4.14	4.79	5.31	1.15	4.06	Dynes and Rowell (1975)
$Pb_{0.6}Tl_{0.4}$	5.90	4.31	4.87	5.34	1.38	4.25	Dynes and Rowell (1975)
In_2Bi	5.6	3.96	4.91	5.77	1.40		Rowell, McMillan, and Dynes (1978)
Sb_2Tl_7	5.2	3.19	4.14	5.00	1.43		Rowell, McMillan, and Dynes (1978)
$Pb_{0.8}Tl_{0.2}$	6.8	4.31	4.84	5.27	1.53	4.37	Dynes and Rowell (1975)
Bi_2Tl	6.4	4.05	4.57	5.08	1.63		Rowell, McMillan, and Dynes (1978)
$Pb_{0.9}B_{0.1}$	7.65	4.31	4.80	5.20	1.66	4.67	Dynes and Rowell (1975)
$Pb_{0.6}Tl_{0.2}Bi_{0.2}$	7.26	4.14	4.57	4.96	1.81	4.80	Dynes and Rowell (1969)
$Pb_{0.8}Bi_{0.2}$	7.95	3.96	4.44	4.88	1.88	4.70	Allen and Dynes (1975*b*)
$Pb_{0.7}Bi_{0.3}$	8.45	4.05	4.48	4.87	2.01	4.86	Dynes and Rowell (1975)
$Pb_{0.65}Bi_{0.35}$	8.95	3.88	4.31	4.74	2.13	4.78	Dynes and Rowell (1975)
Am–$Pb_{0.45}Bi_{0.55}$	7.0	2.50	3.27	4.05	2.59		Allen and Dynes (1975*b*)

Table C.3. d-Band Elements

Material	T_C (K)	θ_D(K)	$\omega_{\log}$ (meV)	$\langle\omega\rangle$ (meV)	$\sqrt{\langle\omega^2\rangle}$ (meV)	λ	$2\Delta/k_BT_C$	Tunneling Reference
Ag	0					0.11	—	Wilson (1979)
Cu	>0			17.9		0.13	—	Wilson (1979)
W	0.015	390				0.28[a]		
Ir	0.1125	420				0.34[a]		
Ti	0.40	425				0.38[a]		
Zr	0.61	290				0.41[a]		
Ru	0.49	550				0.38[a]		
Os	0.66	500				0.39[a]		
Mo	0.92	460				0.41[a]		
Re	1.70	415				0.46[a]	3.59	Ochiai et al. (1971)
Ta	4.47	258	11.37	12.06	12.75	0.69	3.66	Shen (1970) Townsend and Sutton (1962)
	4.49		10.9			0.73	3.72	Wolf, Burnell, Khim, and Noer (1981)
V	5.40	383				0.60[a]	3.5	Noer (1975) Robinson and Rowell (1978)

	5.35		14.8	17.1	18.8	0.82[b]	3.51	Zasadzinski, Burnell, Wolf and Arnold (1982)
Tc	7.77							
Nb	9.25	276				0.82[a]	3.89	Townsend and Sutton (1962)
								Bostock et al. (1976)
								Robinson, Geballe, and Rowell (1976)
								Wolf and Zasadzinski (1978)
	9.2	276	12.7	14	15.3	1.04[c]	3.91	Arnold, Zasadzinski, Osmun, and Wolf (1980)
								Khim, Burnell, and Wolf (1981)
Am–Mo	8.8			12.0	14.0	0.9	3.7	Kimhi and Geballe (1980)
Am–Nb	5.2			10.8	12.5	0.8	3.7	Kimhi and Geballe (1980)

[a] Value obtained by using McMillan's T_c formula (McMillan, 1968*a*).
[b] A λ value of 0.88 is estimated after correction for spin fluctuations (Burnell et al., 1982).
[c] A λ value of 1.09 is estimated after correction for spin fluctuations (Burnell et al., 1982).

Table C.4. d-Band Alloys and Compounds

Material	T_C (K)	θ_D (K)	$\omega_{\log}$ (meV)	$\langle\omega\rangle$ (meV)	$\sqrt{\langle\omega^2\rangle}$ (meV)	λ	$2\Delta/k_BT_C$	Tunneling Reference
$Nb_{0.6}Ti_{0.4}$	9.8							Hulm and Blaugher (1961)
$Nb_{0.75}Zr_{0.25}$		253	9.4	11.3	12.8	1.31	4.1	Wolf, Noer, and Arnold (1980)
$Nb_{0.80}Zr_{0.20}$	11.0							Hulm and Blaugher (1961)
								Dietrich (1964)
$Mo_{0.6}Re_{0.4}$	12.2							Hulm and Blaugher (1961)
NbN	15							Horn and Saur (1968)
								Komenou, Yamashita, and Onodera (1968)
V_3Ga	15.9	310						Flükiger and Jorda (1977)
						1.12		Zasadzinski, Schubert, Wolf, and Arnold (1980)
V_3Si	17.0	530					3.8	Blaugher et al. (1969)
								Hauser, Bacon, and Haemmerle (1966)
$NbN_{0.65}C_{0.35}$	17.5							Pessall and Hülm (1966)

Nb_3Sn	18.3	290				~1.67	4.3	Matthias, Geballe, Geller, and Corenzwit (1954)
								Shen (1972*b*)
								Moore, Rowell, and Beasley (1976)
								Allen and Dynes (1975b)
	17.5		10.8	13.1	15.0	1.78	4.37	Wolf et al. (1980)
Nb_3Al	18.9							Sahm and Pruss (1969)
$NbAl_{0.23}$	16.4		9.5	11.4	13.5	1.7	4.3	Flükiger and Jorda (1977)
$NbTa_x$								Hertel, McWhan, and Rowell (1982)
$Nb_3Al_{0.8}Ge_{0.2}$	21							Matthias et al. (1967)
								Dew-Hughes (1975)
								Gregory, Bostock, MacVicar, and Rose (1973)
Nb_3Ge	23	378					4.2	Gavaler (1973)
								Testardi, Wernick, and Royer (1974)
								Rowell and Schmidt (1976)
$NbGe_{0.23}$	19.8				14	1.64 ±0.2	4.35	Kihlstrom and Geballe (1981)

Table C.5 f-Band Elements

Material	T_C (K)	θ_D (K)	$\omega_{\log}$ (meV)	$\langle\omega\rangle$ (meV)	$\sqrt{\langle\omega^2\rangle}$ (meV)	λ	$2\Delta/k_B T_C$	Tunneling Reference
Hf	0.128							
U(α)	0.68							
Th	1.38	165					3.47	Haskell, Keeler, and Finnemore (1972)
La(α)	4.88	151					3.8	Lou and Tomasch (1972)
La(β)	6.00	139						
La(dhcp)	4.95						3.75	Wühl, Eichler, and Wittig (1973)

Table C.6. Metal Overlayers for Barrier Formation

Metals	References
Al	Adkins (1963); Wolf, Zasadzinski, Osmun, and Arnold (1980); Rowell, Gurvitch, and Geerk (1981); Kwo, Wertheim, Gurvitch, and Buchanan (1982)
Mg	Burnell and Wolf (1982)
Y	Kwo, Wertheim, Gurvitch, and Buchanan (1983)
Eu, Lu	Umbach, Goldman, and Toth (1982)

Table C.7. Studies of Tomasch Oscillations in Thick Superconducting Films and of McMillan-Rowell Oscillations in Thick Normal Films

Metal(s)[a]	References
Zn/Pb	Rowell (1973); Wong, Shih, and Tomasch (1981)
Pb/Ag	Lykken, Geiger, Dy, and Mitchell (1971); Haywood and Mitchell (1974)
Pb/Cu	Nédellec (1977)
Ag/Pb	Rowell and McMillan (1966); Khim and Tomasch (1979)
Al/Pb	Khim and Tomasch (1978)
In/Ag	Nédellec (1977)
Sn	Tomasch (1966*b*); Nedellec (1977)
In/Pb	Tomasch (1966*a*)
Pb	Tomasch (1965*a*)
Cu	Nédellec, Dumoulin, and Guyon (1971)
Cd/Pb	Colucci, Tomasch, and Lee (1974)
Sn/Pb	Chaikin (1975)
In	Tomasch and Wolfram (1966)

[a] The first-mentioned metal faces the tunneling barrier.

Table C.8 Tunneling Studies of Superconductor Phonons Under Hydrostatic Pressure

Material	References
Pb	Franck, Keeler, and Wu (1969); Svistunov, Chernyak, Belogolovskii, and D'yachenko (1981)
Pb, Tl	Galkin, Svistunov, Dikii, and Taranenko (1971)
Nb, Ta	Revenko, D'yachenko, Svistunov, and Shonaikh (1980)
$PbIn_x$	Hansen, Pompi, and Wu (1973); Wright and Franck (1977); Svistunov, Chernyak, Belogolovskii, and D'yachenko (1981)
La	Wühl, Eichler, and Wittig (1973)

References

Abrahams, E., Anderson, P. W., Licciardello, D. C. and Ramakrishnan, T. V. (1979). Phys. Rev. Lett. *42*, 673.

Abrikosov, A. A. (1957). Zh. Eksp. Teor. Fiz. *32*, 1442 (Eng. transl. Sov. Phys.-JETP *5*, 1174).

Abrikosov, A. A. and Gor'kov, L. P. (1960). Zh. Eksp. Teor. Fiz. *39*, 1781 (Engl. trans. Sov. Phys.-JETP *12*, 1243).

Adams, A., Wyss, J. C. and Hansma, P. K. (1979). Phys. Rev. Lett. *42*, 912.

Adane, A., Fauconnet, A., Klein, J., Léger, A., Belin, M. and Defourneau, D. (1975). Solid St. Commun. *16*, 1071.

Adkins, C. J. (1963). Phil. Mag. *8*, 1051.

Adkins, C. J. and Kington, B. W. (1966). Phil. Mag. *13*, 971.

——— (1969). Phys. Rev. *177*, 777.

Adler, J. G. (1969*a*). Solid St. Commun. *7*, 1635.

——— (1969*b*). Phys. Lett. *29A*, 675.

——— (1982). In *Tunneling Spectroscopy: Capabilities, Applications, and New Techniques* (ed. P. K. Hansma), p. 423. Plenum Press, New York.

Adler, J. G. and Chen, T. T. (1971). Solid St. Commun. *9*. 1961.

Adler, J. G. and Jackson, J. E. (1966). Rev. Sci. Instr. *37*, 1049.

Adler, J. G. and Straus, J. (1975). Rev. Sci. Instr. *46*, 158.

Adler, J. G., Chen, T. T. and Straus, J. (1971). Rev. Sci. Instr. *42*, 362.

Adler, J. G., Kreuzer, H. J. and Straus, J. (1975). Phys. Rev. B*11*, 2812.

Adler, J. G., Kreuzer, H. J. and Wattamaniuk, W. J. (1971). Phys. Rev. Lett. *27*, 185.

Akoh, H. and Kajimura, K. (1981). Solid St. Commun. *38*, 1147.

Allen, P. B. (1972). Phys. Rev. B*6*, 2577.

Allen, P. B. and Cohen, M. L. (1969). Phys. Rev. *187*, 525.

——— (1972). Phys. Rev. Lett. *29*, 1593.

Allen, P. B. and Dynes, R. C. (1975*a*). Phys. Rev. B*11*, 1895.

——— (1975*b*). Phys. Rev. B*12*, 905.

Alexandrov, A. and Ranninger, J. (1981). Phys. Rev. B*24*, 1164.

Allender, D., Bray, J. and Bardeen, J. (1973). Phys. Rev. B*7*, 1020.

Al'tshuler, B. L. and Aronov, A. G. (1979*a*). Solid St. Commun. *30*, 115.

——— (1979*b*). Zh. Eksp. Teor. Fiz. *77*, 2028 (Engl. transl. Sov. Phys.-JETP *50*, 968).

Al'tshuler, B. L., Aronov, A. G. and Lee, P. A. (1980). Phys. Rev. Lett. *44*, 1288.

Ambegaokar, V. and Baratoff, A. (1963). Phys. Rev. Lett. *10*, 486; *11*, 104(E).

Ambegaokar, V. and Tewordt, L. (1964). Phys. Rev. *134*, A805.

Anacker, W. (1980). IBM J. Res. Dev. *24*, 107.

Anderson, P. W. (1958). Phys. Rev. *109*, 1492.

——— (1959). J. Phys. Chem. Solids *11*, 26.

——— (1961). In *Proc. of the VIIth Int. Conf. on Low Temp. Physics* (eds. G. M. Graham and A. C. H. Hallett). p. 298. University of Toronto Press, Toronto.

——— (1964). In *Lectures on the Many-Body Problem* (ed. E. R. Caianiello), p. 113. Academic Press, New York.

——— (1966). Phys. Rev. Lett. *17*, 95.

——— (1975). Phys. Rev. Lett. *34*, 953.

Anderson, P. W. and Dayem, A. H. (1964). Phys. Rev. Lett. *13*, 195.

Anderson, P. W. and Rowell, J. M. (1963). Phys. Rev. Lett. *10*, 230.

Andreev, A. F. (1964). Zh. Eksp. Teor. Fiz. *46*, 1823 (Engl. trans. Sov. Phys.-JETP *19*, 1228).

Andrei, N. (1980), Phys. Rev. Lett. *45*, 379.

Andrei, N. and Lowenstein, J. H. (1981). Phys. Rev. Lett. *46*, 356.

Andrews, A. M., Korb, H. W., Holonyak, N., Duke, C. B. and Kleiman, G. G. (1972). Phys. Rev. B*5*, 2273.

Appelbaum, J. A. (1967). Phys. Rev. *154*, 633.

Appelbaum, J. A. and Brinkman, W. F. (1969*a*). Phys. Rev. *183*, 553.

——— (1969*b*). Phys. Rev. *186*, 464.

——— (1970). Phys. Rev. B*2*, 907.

Appelbaum, J. A. and Shen, L. Y. L. (1972). Phys. Rev. B*5*, 544.

Arnold, G. B. (1978). Phys. Rev. B*18*, 1076.

——— (1981). Phys. Rev. B*23*, 1171.

——— (1982). Phys. Rev. B*25*, 5998.

Arnold, G. B. and Wolf, E. L. (1982). Phys. Rev. B25, 1541.

Arnold, G. B., Gallagher, W. J. and Wolf, E. L. (1982). In *Superconductivity in d- and f-Band Metals* (eds. W. Buckel and W. Weber), p. 305. Kernforschungszentrum Karlsruhe, Karlsruhe.

Arnold, G. B., Zasadzinski, J. and Wolf, E. L. (1978). Phys. Lett. *69A*, 136.

Arnold, G. B., Zasadzinski, J., Osmun, J. W. and Wolf, E. L. (1980). J. Low Temp. Phys. *40*, 225.

Artemenko, S. N., Volkov, A. F. and Zaitsev, A. V. (1979). Zh. Eksp. Teor. Fiz. *76*, 1816 (Eng. transl. Sov. Phys.-JETP *49*, 924).

Arthur, J. R. (1976). In *Critical Reviews of Solid State Science, Vol. 6* (eds. D. E. Schuele and R. W. Hoffman), p. 413. CRC Press, Cleveland.

Arzoumanian, C., Dumoulin, L., Nédellec, P. and Burger, J. P. (1979). Zeit. f Phys. Chem. *116*, 117.

Ashraf, M. and Swihart, J. C. (1982). Phys. Rev. B*25*, 2094.

Aslamazov, L. G. and Larkin, A. I. (1968). Phys. Lett. *26A*, 238.

Aspen, F. and Goldman, A. M. (1976). Cryogenics *16*, 721.

Baird, K. M. (1983). Physics Today *36*, 52.

Baker, J. M. and Blakely, J. M. (1972). Surf. Sci. *32*, 45.

Balkashin, O. P., Khotkevich, A. V. and Yanson, I. K. (1979). Fiz. Nizk. Temp. *5*, 28 (Sov. J. Low Temp. Phys. *5*, 12).

Balkashin, O. P., Yanson, I. K. and Khotkevich, A. V. (1977). Zh. Eksp. Teor. Fiz. *72*, 1182 (Sov. Phys.-JETP *45*, 618.)

Band, W. T. and Donaldson, G. B. (1973). Phys. Rev. Lett. *31*, 20.

Banks, D. E. and Blackford, B. L. (1973). Can. J. Phys. *51*, 2505.

Banninger, U., Busch, G., Campagna, M. and Siegmann, H. C. (1970). Phys. Rev. Lett. *25*, 585.

Baquero, R., Daams, J. M. and Carbotte, J. P. (1981). J. Low Temp. Phys. *42*, 585.

Baratoff, A., Binnig, G., Bednorz, J. G., Gervais, F. and Servoin, J. L. (1982). In *Superconductivity in d- and f-Band Metals* (eds. W. Buckel and W. Weber), p. 419. Kernforschungszentrum Karlsruhe, Karlsruhe.

Bardeen, J. (1961). Phys. Rev. Lett. *6*, 57.
Bardeen, J. and Pines, D. (1955). Phys. Rev. *99*, 1140.
Bardeen, J., Cooper, L. N. and Schrieffer, J. R. (1957). Phys. Rev. *108*, 1175.
Barnes, L. J. (1969). Phys. Rev. *184*, 434.
Barone, A. and Paterno, G. (1982). *Physics and Applications of the Josephson Effect.* Wiley-Interscience, New York.
Bar-Sagi, J. and Entin-Wohlman, O. (1977). Solid St. Commun. *22*, 29.
Batlogg, B., Remeika, J. P., Dynes, R. C., Barz, H., Cooper, A. S. and Garno, J. P. (1982). In *Superconductivity in d- and f-Band Metals* (eds. W. Buckel and W. Weber), p. 401. Kernforschungszentrum Karlsruhe, Karlsruhe.
Bauriedl, W., Ziemann, P. and Buckel, W. (1981). Phys. Rev. Lett. *47*, 1163.
Baym, G. (1969). *Lectures on Quantum Mechanics.* W. A. Benjamin, New York.
Bayman, A. and Hansma, P. K. (1980). Nature *285*, 97.
Bechgaard, K., Carneiro, K., Olsen, M., Rasmussen, F. B. and Jacobsen, C. S. (1981). Phys. Rev. Lett. *46*, 852.
Bellanger, D., Klein, J., Léger, A., Belin, M. and Defourneau, D. (1973). Phys. Lett. *42A*, 459.
BenDaniel, D. J. and Duke, C. B. (1967). Phys. Rev. *160*, 679.
Bennett, A. J., Duke, C. B. and Silverstein, S. D. (1968). Phys. Rev. *176*, 969.
Bergmann, G. (1971). Phys. Rev. B*3*, 3797.
Berk, N. F. and Schrieffer, J. R. (1966). Phys. Rev. Lett. *17*, 433.
Bermon, S. (1964). University of Illinois, Urbana, Tech. Rep. 1, NSF GP1100.
Bermon, S. and So, C. K. (1978*a*). Phys. Rev. Lett. *40*, 53.
——— (1978*b*). Phys. Rev. B*17*, 4256.
——— (1978*c*). Solid St. Commun. *27* 723.
Bermon, S. and Ware, M. (1971). Phys. Lett. *35A*, 226.
Bermon, S., Paraskevopoulos, D. E. and Tedrow, P. M. (1978). Phys. Rev. B*17*, 2110.
Beuermann, G. and El Haffar, H. (1981). Z. Phys. B*44*, 25.
Biberian, J. P. and Somorjai, G. A. (1979). J. Vac. Sci. Techn. *16*, 2073.
Binnig, G. and Gerber, Ch. (1979). IBM Techn. Discl. Bull. *22*, 2897.
Binnig, G., Baratoff, A., Hoenig, H. E. and Bednorz, J. G. (1980). Phys. Rev. Lett. *45*, 1352.
Binnig, G. and Rohrer, H. (1982). Helv. Phys. Acta. *55*, 726.
Binnig, G., Rohrer, H., Gerber, Ch. and Weibel, E. (1982*a*). Appl. Phys. Lett. *40*, 178.
——— (1982*b*). Phys. Rev. Lett. *49*, 57.
——— (1982*c*). Physica *109 & 110B*, 2075.
Bishop, M. F. and Overhauser, A. W. (1981). Phys. Rev. B*23*, 3627.
Blackford, B. L. (1971). Rev. Sci. Instr. *42*, 1198.
——— (1976). J. Low Temp. Phys. *23*, 43.
Blackford, B. L. and Hill, K. (1981). J. Low Temp. Phys. *43*, 25.
Blackford, B. L. and March, R. H. (1968). Can. J. Phys. *46*, 141.
——— (1969). Phys. Rev. *186*, 397.
Blanc, C., More, C., Roger, G. and Sorbier, J. P. (1981). Chemica Scripta *17*, 63.
Blaugher, R. D., Hein, R. A., Cox, J. E. and Waterstrat, R. M. (1969). J. Low Temp. Phys. *1*, 539.
Blodgett, K. and Langmuir, I. (1937). Phys. Rev. *51*, 964.
Blonder, G. E., Tinkham, M. and Klapwijk, T. M. (1982). Phys. Rev. B*25*, 4515.
Bloomfield, P. E. and Hamann, D. R. (1967). Phys. Rev. *164*, 856.
Bobetic, V. M. (1964). Phys. Rev. *136*, A1535.

Bogoliubov, N. N. (1958*a*). Nuovo Cimento *7*, 794.
——— (1958*b*). Zh. Eksp. Teor. Fiz. *34*, 58 (Engl. transl. Sov. Phys.-JETP 7, 41).
Bogoliubov, N. N., Tolmachev, V. V. and Shirkov, D. V. (1958). Academy of Sciences, Moscow. *A New Method in the Theory of Superconductivity* (Consultants Bureau, New York, 1959).
Bogoliubov, N. N. (1959). Uspekhi Fiz. Nauk *67*, 549 (Soviet Phys.-Uspekhi *67*, 236).
Bolz, J. and Pobell, F. (1975*a*). In *Proc. 14th Conf. on Low Temperature Physics* (eds. M. Krusius and M. Vuorio), Vol. 2, p. 425. North-Holland, Amsterdam.
——— (1975*b*). Z. Phys. B*20*, 95.
Bostock, J. L. and MacVicar, M. L. A. (1977). In *Anisotropy Effects in Superconductors* (ed. H. W. Weber), p. 213. Plenum, New York.
Bostock, J., Agyeman, K., Frommer, M. H. and MacVicar, M. L. A. (1973). J. Appl. Phys. *44*, 5567.
Bostock, J., Diadiuk, V., Cheung, W. N., Lo, K. H., Rose, R. M. and MacVicar, M. L. A. (1976), Phys. Rev. Lett. *36*, 603.
Bowser, W. M. and Weinberg, W. H. (1977). Surf. Sci. *64*, 377.
——— (1981). J. Am. Chem. Soc. *103*, 1453.
Brailsford, A. D. and Davis, L. C. (1970). Phys. Rev. B*2*, 1708.
Brinkman, W. F., Dynes, R. C. and Rowell, J. M. (1970). J. Appl. Phys. *41*, 1915.
Brockhouse, B. N. and Iyengar, P. K. (1958). Phys. Rev. *111*, 747.
Brockhouse, B. N., Arase, T., Caglioti, G., Rao, K. R. and Woods, A. D. B. (1962). Phys. Rev. *128*, 1099.
Broom, R. F. and Mohr, T. O. (1977). J. Vac. Sci. Technol. *15*, 1166.
Broom, R. F., Oosenbrug, A. and Walter, W. (1980). Appl. Phys. Lett. *37*, 237.
Brown, G. S., Testardi, L. R., Wernick, J. H., Hallak, A. B. and Geballe, T. H. (1977). Solid St. Commun. *23*, 875.
Brunner, M., Ekrut, H. and Hahn, A. (1982). J. Appl. Phys. *53*, 1596.
Bruno, R. C. and Schwartz, B. B. (1973). Phys. Rev. B*8*, 3161.
Burnell, D. M. (1982). Ph.D. Thesis, Department of Physics, Iowa State University (unpublished).
Burnell, D. M. and Wolf, E. L. (1982). Phys. Lett. *90A*, 471.
Burnell, D. M., Zasadzinski, J. and Wolf, E. L. (1981). Phys. Lett. *85A*, 383.
Burnell, D. M., Zasadzinski, J., Noer, R. J., Wolf, E. L. and Arnold, G. B. (1982). Solid St. Commun. *41*, 637.
Burstein, E., Langenberg, D. N. and Taylor, B. N. (1961). Phys. Rev. Lett. *6*, 92.
Burstein, E. and Lundqvist, S. (1969). *Tunneling Phenomena in Solids*, Plenum Press, New York.
Butler, W. H. (1977). Phys. Rev. B*15*, 5267.
——— (1980*a*). Phys. Rev. Lett. *44*, 1516.
——— (1980*b*). In *Superconductivity in d- and f-Band Metals* (eds. H. Suhl and M. B. Maple), p. 443. Academic, New York.
Butler, W. H., Pinski, F. J. and Allen, P. B. (1979). Phys. Rev. B*19*, 3708.
Butler, W. H., Smith, H. G. and Wakabayashi, N. (1977). Phys. Rev. Lett. *39*, 1004.
Cabrera, N. and Mott, N. F. (1949). Rept. Progr. Phys. *12*, 163.
Caputo, A. J. (1977). Thin Solid Films *40*, 49.
Carbotte, J. P. (1977). In *Anisotropy Effects in Superconductors* (ed. H. W. Weber), p. 183. Plenum, New York.
Carbotte, J. P. and Dynes, R. C. (1968). Phys. Rev. *172*, 476.

Cardinne, Ph., Marti, M. and Renard M. (1971). Rev. Phys. Appl. *6*, 547.
Caro, J., Coehoorn, R. and de Groot, D. G. (1981). Solid St. Commun. *39*, 267.
Caroli, C., Lederer, D. and Saint-James, D. (1972). Surf. Sci. *33*, 228.
Caroli, C., Combescot, R., Nozières, P. and Saint-James, D. (1971). J. Phys. C: Solid St. Phys. *4*, 916.
——— (1972). J. Phys. C: Solid St. Phys. *5*, 21.
Caroli, C., Combescot, R., Lederer, D., Nozières, P. and Saint-James, D. (1971). J. Phys. C: Solid St. Phys. *4*, 2598.
Caroli, C., Combescot, R., Lederer-Rozenblatt, D., Nozières, P. and Saint-James, D. (1975). Phys. Rev. B*12*, 3977.
Cederberg, A. A. (1981). Surf. Sci. *103*, 148.
Cederberg, A. A. and Kirtley (1979). Solid St. Commun. *30*, 381.
Chaikin, P. M. (1975). Solid St. Commun. *17*, 1471.
Chaikin, P. M. and Hansma, P. K. (1976). Phys. Rev. Lett. *36*, 1552.
Chaikin, P. M., Arnold, G. and Hansma, P. K. (1977). J. Low Temp. Phys. *26*, 229.
Chaikin, P. M., Hansma, P. K. and Greene, R. L. (1978). Phys. Rev. B*17*, 179.
Chang, C. C. (1971). Surf. Sci. *25*, 53.
Chang, J. J. (1978). Phys. Rev. B*17*, 2137.
Chang, J. J. and Scalapino, D. J. (1977). Phys. Rev. B*15*, 2651.
——— (1978). J. Low Temp. Phys. *31*, 1.
Chang, L. L., Esaki, L. and Tsu, R. (1974). Appl. Phys. Lett. *24*, 593.
Chazalviel, J.-N. and Yafet, Y. (1977). Phys. Rev. B*15*, 1062.
Chen, T. T. and Adler, J. G. (1970). Solid St. Commun. *8*, 1965.
Chen, T. T., Leslie, J. D. and Smith, H. J. T. (1971). Physica *55*, 439.
Chen, T. T., Chen, J. T., Leslie, J. D. and Smith, H. J. T. (1969). Phys. Rev. Lett. *22*, 526.
Chi, C. C. and Clarke, J. (1979). Phys, Rev. B*20*, 4465.
Chi, C. C., Loy, M. M. T. and Cronemeyer, D. C. (1982). Appl. Phys. Lett. *40*, 437.
Chiang, C. K., Fincher, C. R., Park, Y. M., Heeger, A. J., Shirakawa, H., Louis, E. J., Gau, S. C. and MacDiarmid, A. G. (1977). Phys. Rev. Lett. *39*, 1098.
Cho, A. Y. (1971). Appl. Phys. Lett. *19*, 467.
Chopra, K. L. (1969). *Thin Film Phenomena*. McGraw-Hill, New York.
Chow, C. K. (1963). J. Appl. Phys. *34*, 2490.
Christopher, J. E., Coleman, R. V., Isin, A. and Morris, R. C. (1968). Phys. Rev. *172*, 485.
Christopher, J. E., Darley, H. M., Lehman, G. W. and Tripathi, S. N. (1975). Phys. Rev. B*11*, 754.
Chu, H. T., Eib, N. K. and Henriksen, P. N. (1975). Phys. Rev. B*12*, 518.
Chynoweth, A. G., Logan, R. A. and Thomas, D. E. (1962). Phys. Rev. *125*, 877.
Clark, H. E. and Young, R. D. (1968). Surf. Sci. *12*, 385.
Clark, J. M. and Coleman, R. V. (1976). Proc. Natl. Acad. Sci. USA *73*, 1598.
Clark, T. D. (1968). J. Phys. C: Solid St. Phys. *1*, 732.
Clarke, J. (1966). Phil. Mag. *13*, 115.
——— (1968). Phys. Rev. Lett. *21*, 1566.
——— (1969). Proc. Roy. Soc. A*308*, 447.
——— (1972). Phys. Rev. Lett. *28*, 1363.
Clarke, J. and Paterson, J. L. (1974). J. Low Temp. Phys. *15*, 491.
Clarke, J., Goubau, W. M. and Ketchen, M. B. (1976). J. Low Temp. Phys. *25*, 99.

Clarke, J., Freake, S. M., Rappaport, M. L. and Thorp, T. L. (1972). Solid St. Commun. *11*, 689.
Clayman, B. P. (1972). Can. J. Phys. *50*, 3193.
Clem, J. R. (1966*a*). Phys. Rev. *148*, 392.
——— (1966*b*). Ann. Phys. *40*, 268.
——— (1974). In *Proc. 13th Conf. on Low Temperature Physics* (eds. K. D. Timmerhaus, W. J. O'Sullivan and E. F. Hammel), Vol. 3, p. 102. Plenum, New York.
Cohen, M. H., Falicov, L. M. and Phillips. J. C. (1962). Phys. Rev. Lett. *8*, 316.
Cohen, R. W., Abeles, B. and Fuselier, C. R. (1969). Phys. Rev. Lett. *23*, 377.
Coleman, R. V., Morris, R. C. and Christopher, J. E. (1974). *Methods of Experimental Physics*, Vol. 11, p. 123. Academic, New York.
Colley, S. and Hansma, P. K. (1977). Rev. Sci. Instr. *48*, 1192.
Collver, M. M. and Hammond, R. H. (1973). Phys. Rev. Lett. *30*, 92.
Colucci, S. L., Tomasch, W. J. and Lee, H. J. (1974). Phys. Rev. Lett. *32*, 590.
Colvard, C., Merlin, R., Klein, M. V. and Gossard, A. C. (1980). Phys. Rev. Lett. *45*, 298.
Comberg, A. and Ewert, S. (1976). Z. Phys. B*25*, 173.
Combescot, R. (1971). J. Phys. C: Solid St. Phys. *4*, 2611.
Combescot, R. and Schreder, G. (1973). J. Phys. C: Solid St. Phys. *6*, 1363.
——— (1974). J. Phys. C: Solid St. Phys. *7*, 1318.
Conley, J. W. and Mahan, G. D. (1967). Phys. Rev. *161*, 681.
Conley, J. W. and Tiemann, J. J. (1967). J. Appl. Phys. *38*, 2880.
Conley, J. W., Duke, C. B., Mahan, G. D. and Tiemann, J. J. (1966). Phys. Rev. *150*, 466.
Conwell, E. M. (1967). Solid St. Phys. *9*, 149.
Cook, C. F. and Everett, G. E. (1967). Phys. Rev. *159*, 374.
Coon, D. D. and Fiske, M. D. (1965). Phys. Rev. *138*, A744.
Cooper, J. R. and Wyatt, A. F. G. (1973). J. Phys. F*3*, L120.
Cooper, L. N. (1956). Phys. Rev. *104*, 1189.
——— (1960). Am. J. Phys. *28*, 91.
Corak, W. S., Goodman, B. B., Satterthwaite, C. B. and Wexler, A. (1956). Phys. Rev. *102*, 656.
Cox, D. E. and Sleight, A. W. (1976). In *Proc. Conf. on Neutron Scattering* (ed. R. M. Moon), p. 45. Gatlinburg, Tennessee.
Cullen, D. E., Wolf, E. L. and Compton, W. D. (1970). Phys. Rev. B*2*, 3157.
Cullen, G. W., Kaldis, E., Parker, R. L. and Rooymans, C. J. M., eds. (1975). *Vapor Growth and Epitaxy*. North-Holland, Amsterdam.
Daams, J. M. and Carbotte, J. P. (1980). J. Low Temp. Phys. *40*, 135.
——— (1981). J. Low Temp. Phys. *43*, 263.
Daams, J. M., Mitrović, B. and Carbotte, J. P. (1981). Phys. Rev. Lett. *46*, 65.
Dargis, A. B. (1981). Rev. Sci. Instr. *52*, 46.
Davenport, J. W., Ho, W. and Schrieffer, J. R. (1978), Phys. Rev. B*17*, 3115.
Davis, L. C. (1969). Phys. Rev. *187*, 1177.
——— (1970*a*). Phys. Rev. B*2*, 1714.
——— (1970*b*). Phys. Rev. B*2*, 4943.
——— (1977). Phys. Rev. B*16*, 2482.
Davis, L. C. and Duke, C. B. (1968). Solid St. Commun. *6*, 193.
——— (1969). Phys. Rev. *184*, 764.
Davis, L. C. and Steinrisser, F. (1970). Phys. Rev. B*1*, 614.

Davis, L. C., Jaklevic, R. C. and Lambe, J. (1975). Phys. Rev. B*12*, 798.
Dayan, M. (1978). J. Low Temp. Phys. *32*, 643.
Dayem, A. H. (1972). J. Phys. (Paris) Suppl. *33*, 15.
Dayem, A. H. and Martin, R. J. (1962). Phys. Rev. Lett. *8*, 246.
Deaver, B. S. and Fairbank, W. M. (1961). Phys. Rev. Lett. *7*, 43.
Deaver, B. S., Jr., Falco, C. M., Harris, J. H. and Wolf, S. A., eds. (1978). *Future Trends in Superconductive Electronics.* American Institute of Physics, New York.
de Cheveigné, S., Klein, J. and Léger, A. (1978). *Inelastic Electron Tunneling Spectroscopy* (ed. T. Wolfram), p. 202. Springer-Verlag, Berlin.
——— (1982). In *Tunneling Spectroscopy: Capabilities, Applications and New Techniques.* (ed. P. K. Hansma) p. 109. Plenum, New York.
de Cheveigné, S., Klein, J., Léger, A., Belin, M. and Defourneau, D. (1977). Phys. Rev. B*15*, 750.
Dees, J. W. (1966). Microwave J. *9*, 48.
de Gennes, P. G. (1964*a*). Rev. Mod. Phys. *36*, 225.
——— (1964*b*). Phys. Kond. Mat. *3*, 79.
——— (1966). *Superconductivity of Metals and Alloys.* W. A. Benjamin, New York.
de Gennes, P. G. and Saint-James, D. (1963). Phys. Lett. *4*, 151.
Demers, J. and Griffin, A. (1971). Canad. J. Phys. *49*, 285.
Denninger, G., Dose, V. and Bonzel, H. P. (1982). Phys. Rev. Lett. *48*, 279.
Deutscher, G. and de Gennes, P. G. (1969). In *Superconductivity.* (ed. R. D. Parks), p. 1005. Marcel Dekker, New York.
Dew-Hughes, D. (1975). Cryogenics *15*, 435.
Dietrich, I. (1964). Phys. Lett. *9*, 221.
Dietz, R. E., Parisot, G. I. and Meixner, A. E. (1971). Phys. Rev. B*4*, 2302.
Dingle, R., Gossard, A. C. and Weigmann, W. (1975). Phys. Rev. Lett. *34*, 1327.
Dmitrenko, I. M., Yanson, I. K. and Svistunov, V. M. (1965). Sov. Phys.-JETP Lett. *2*, 10.
Dodson, B. W., McMillan, W. L., Mochel, J. M. and Dynes, R. C. (1981). Phys. Rev. Lett. *46*, 46.
Doll, R. and Näbauer, M. (1961). Phys. Rev. Lett. *7*, 51.
Doniach, S. and Engelsberg, S. (1966). Phys. Rev. Lett. *17*, 750.
Donovan-Vojtovic, B. F., Dodds, S. A., Nasser, M. S. and Chaikin, P. M. (1976). Phil. Mag. *34*, 893.
Dose, V., Fauster, Th. and Scheidt, H. (1981). J. Phys. F*11*, 1801.
Douglass, D. H., Jr. and Falicov, L. M. (1964). In *Progress in Low Temperature Physics,* Vol. IV (ed. C. J. Gorter), p. 97 North-Holland, Amsterdam.
Dowman, J. E., MacVicar, M. L. A. and Waldram, J. R. (1969). Phys. Rev. *186*, 452.
Dragoset, R. A., Phillips, E. S. and Coleman, R. V. (1982). Phys. Rev. B*26*, 5333.
Drullinger, R. E., Evenson, K. M., Jennings, D. A., Peterson, F. R., Berquist, J. C., Burkins, L. and Daniel, H. U. (1983). Optics Lett. (to be published).
D'Silva, A. P., Rice, G. W. and Fassel, V. A. (1980). Appl. Spectroscopy *34*, 578.
Dubois, L. H. (1982). In *Tunneling Spectroscopy: Capabilities, Applications and New Techniques* (ed. P. K. Hansma), p. 153. Plenum, New York.
Dubois, L. H., Hansma, P. K. and Somorjai, G. A. (1980). Appl. Surf. Sci. *6*, 173.
Duke, C. B. (1969). *Tunneling in Solids.* Academic, New York.

Duke, C. B. and Alferieff, M. E. (1967). J. Chem. Phys. *46*, 923.
Duke, C. B., Rice, M. J. and Steinrisser, F. (1969). Phys. Rev. *181*, 733.
Duke, C. B., Silverstein, S. D. and Bennett, A. J. (1967). Phys. Rev. Lett. *19*, 315.
Dumoulin, L., Guyon, E. and Nédellec, P. (1970). Solid St. Commun. *8*, 885.
——— (1973). J. Phys. (Paris) *34*, 1021.
——— (1977). Phys. Rev. B*16*, 1086.
Dumoulin, L., Guyon, E. and Rochlin, G. I. (1970). Solid St. Commun. *8*, 287.
Dumoulin, L., Nédellec, P. and Chaikin, P. M. (1981). Phys. Rev. Lett. *47*, 208.
Dumoulin, L., Nédellec, P. and Guyon, E. (1972). Solid St. Commun. *11*, 1551.
Dumoulin, L., Nédellec, P., Burger, J. P. and Creppy, F. V. (1982). In *Superconductivity in d- and f-Band Metals* (eds. W. Buckel and W. Weber), p. 473. Kernforschungszentrum Karlsruhe, Karlsruhe.
Dumoulin, L., Nédellec, P., Chaumont, J., Gilbon, D., Lamoise, A.-M. and Bernas, H. (1976). C. R. Acad. Sci. Paris *283*, B285.
Durbin, S. (1982). (Unpublished.)
Durbin, S. M., Cunningham, J. E. and Flynn, C. P. (1982). J. Phys. F.*12*, L75.
Durbin, S. M., Cunningham, J. E., Mochel, M. E. and Flynn, C. P. (1981). J. Phys. F*11*, L223.
Dynes, R. C. (1970). Phys. Rev. B*2*, 644.
——— (1982). Physica *109 & 110B*, 1857.
Dynes, R. C. and Fulton, T. A. (1971). Phys. Rev. B*3*, 3015.
Dynes, R. C. and Garno, J. P. (1975). Bull. Am. Phys. Soc. (II) *20*, 422.
——— (1981). Phys. Rev. Lett. *46*, 137.
Dynes, R. C. and Rowell, J. M. (1969). Phys. Rev. *187*, 821.
——— (1975). Phys. Rev. B*11*, 1884.
Dynes, R. C., Narayanamurti, V. and Chin, M. (1971). Phys. Rev. Lett. *26*, 181.
Dynes, R. C., Narayanamurti, V. and Garno, J. P. (1977). Phys. Rev. Lett. *39*, 229.
——— (1978). Phys. Rev. Lett. *41*, 1509.
Dynes, R. C., Carbotte, J. P., Taylor, D. W. and Campbell, C. K. (1969). Phys. Rev. *178*, 713.
Eck, R. E., Scalapino, D. J. and Taylor, B. N. (1964). Phys. Rev. Lett. *13*, 15.
Eib, W. and Alvarado, S. F. (1976). Phys. Rev. Lett. *37*, 444.
Eichele, R. Freytag, L., Siefert, H., Huebener, R. P. and Clem, J. R. (1983). J. Low Temp. Phys. *52*, 449.
Eichler, A., Wühl, H. and Stritzker, B. (1975*a*). In *Proc. 14th Conf. on Low Temperature Physics* (eds M. Krusius and M. Vuorio), Vol. 2, p. 52. North-Holland, Amsterdam.
——— (1975*b*). Solid St. Commun. *17*, 213.
Eisenmenger, W. (1969). In *Tunneling Phenomena in Solids* (eds. E. Burstein and S. Lundqvist), p. 371. Plenum, New York.
Elchinger, G. M. Sanchez, A., Davis, C. F. and Javan, A. (1976). J. Appl. Phys. *47*, 591.
Eldridge, J. M. and Matisoo, J. (1971). In *Proc. 12th Int. Conf. on Low Temperature Physics* (ed. E. Kanda), p. 427. Keigaku, Tokyo.
Eliashberg, G. M. (1960). Zh. Eksp. Teor. Fiz. *38*, 966 (Engl. trans. Sov. Phys.-JETP *11*, 696).
——— (1970). Zh. Eksp. Teor. Fiz. Pis. Red. *11*, 186 (Engl. trans. Sov. Phys.-JETP Lett. *11*, 114).

Ellialtioglu, R. M., White, H. W., Godwin, L. M. and Wolfram, T. (1980). J. Chem. Phys. *72*, 5291.
——— (1981). J. Chem. Phys. *75*, 2432.
Elliott, R. J. and Taylor, D. W. (1967). Proc. Roy. Soc. (London) *A*296, 161. 867). Proc. Roy. Soc. (London) *A296*, 161.
El-Semary, M. A. and Rogers, J. S. (1972*a*). Solid St. Commun. *11*, 77.
——— (1972*b*). Phys. Lett. A*42*, 79.
El-Semary, M. A., Kaahwa, Y. and Rogers, J. S. (1973). Solid St. Commun. *12*, 593.
Engler, H. and Fulde, P. (1971). Z. Physik *247*, 1.
Enstrom, R. E., Hanak, J. J. and Cullen, G. W. (1970). RCA Review *31*, 702.
Entin-Wohlman, O. (1977). J. Low Temp. Phys. *27*, 777.
Entin-Wohlman, O. and Bar-Sagi, J. (1978). Phys. Rev. B*18*, 3174.
Epperlein, P. W. (1980). In *Squid '80* (eds. H. O. Hahlbohm and H. Lubbig), p. 131. de Gruyter, Berlin.
——— (1981). Physica *108B*, 999.
Epperlein, P. W., Seifert, H. and Huebener, R. P. (1982). Phys. Lett. *92A*, 146.
Esaki, L. (1958). Phys. Rev. *109*, 603.
——— (1974). Science *183*, 1149.
Esaki, L. and Chang, L. L. (1974). Phys. Rev. Lett. *33*, 495.
Esaki, L. and Miyahara, Y. (1960). Solid St. Electronics *1*, 13; and unpublished data.
Esaki, L. and Stiles, P. J. (1965). Phys. Rev. Lett. *14*, 902.
——— (1966). Phys. Rev. Lett. *16*, 574.
Esaki, L., Chang, L. L., Stiles, P. J., O'Kane, D. F. and Wiser, N. (1968). Phys. Rev. *167*, 637.
Evans, H. E. and Weinberg, W. H. (1980*a*, 1980*b*, 1980*c*). J. Am. Chem. Soc. *102*, 872; 2548; 2554.
Evans, H. E., Bowser, W. M. and Weinberg, W. H. (1974). Appl. Surf. Sci. *5*, 258.
Evans, H. E., Bowser, W. M. and Weinberg, W. H. (1979). Surf. Sci. *85*, L497.
——— (1980). Appl. Surf. Sci. *5*, 258.
Ewert, S. (1974). Z. Phys. *267*, 283.
Ewert, S. and Comberg, A. (1976). Solid St. Commun. *18*, 923.
Ewert, S., Comberg, A. and Wühl, H. (1976). Mater. Sci. Eng. *23*, 275.
Ewert, S., Comberg, A., Sander, W. and Wuhl, H. (1975). In *Proc. 14th Conf. on Low Temperature Physics* (eds. M. Krusius and M. Vuorio), Vol. 2, p. 407. North-Holland, Amsterdam.
Falco, C. M., Werner, T. R. and Schuller, I. K. (1980). Solid St. Commun. *34*, 535.
Falk, D. S. (1963). Phys. Rev. *132*, 1576.
Faris, S. M., Gustafson, T. K. and Wiesner, J. C. (1973). IEEE J. Quant. Electr. *QE–9*, 787.
Faris, S. M., Henkels, W. H., Valsamakis, E. A. and Zappe, H. H. (1980). IBM J. Res. Dev. *24*, 143.
Faris, S. M., Raider, S. I., Gallagher, W. J. and Drake, R. E. (1983). IEEE Trans. MAG-*19*, 1293.
Feher, G. and Gere, E. A. (1959). Phys. Rev. *114*, 1245.
Fehlner, F. P. and Mott, N. F. (1970). Oxidation of Metals *2*, 59.
Ferrell, R. A. and Prange, R. E. (1963). Phys. Rev. Lett. *10*, 479.

Fertig, W. A., Johnston, D. C., DeLong, L. E., McCallum, R. W., Maple, M. B. and Matthias, B. T. (1977). Phys. Rev. Lett. *38*, 987.
Feuchtwang, T. E. (1970). Phys. Rev. B*2*, 1863.
——— (1974*a*). Phys. Rev. B*10*, 4121.
——— (1974*b*). Phys. Rev. B*10*, 4135.
——— (1975). Phys. Rev. B*12*, 3979.
——— (1976). Phys. Rev. B*13*, 517.
Feuchtwang, T. E., Cutler, P. H. and Miskovsky, N. M. (1981). Phys. Rev. B*23*, 3563.
Feuchtwang, T. E., Cutler, P. H. and Nagy, D. (1978). Surf. Sci. *75*, 490.
Feuchtwang, T. E., Cutler, P. N. and Schmit, J. (1978). Surf. Sci. *75*, 401.
Feuerbacher, B. and Fitton, B. (1977). In *Electron Spectroscopy for Surface Analysis* (ed. H. Ibach), p. 151. Springer-Verlag, Berlin.
Finnegan, T. F., Denenstein, A. and Langenberg, D. N. (1970). Phys. Rev. Lett. *24*, 738.
Fisher, J. C. and Giaever, I. (1961). J. Appl. Phys. *32*, 172.
Fiske, M. D. (1964). Rev. Mod. Phys. *36*, 221.
Flesner, L. D. and Silver, A. H. (1980). Rev. Sci. Instr. *51*, 1411.
Flodström, S. A., Martinsson, C. W. B., Bachrach, R. F., Hagström, S. B. M. and Bauer, R. S. (1978). Phys. Rev. Lett. *40*, 907.
Floyd, R. B. and Walmsley, D. G. (1978). J. Phys. C*11*, 4601.
Flükiger, R. and Jorda, J. L. (1977). Solid St. Commun. *22*, 109.
Foner, S., McNiff, E. J., Jr., Gavaler, J. R. and Janocko, M. A. (1974). Phys. Lett. *47*A, 485.
Foner, S., McNiff, E. J., Jr., Matthias, B. T., Geballe, T. H., Willens, R. H. and Corenzwit, E. (1970). Phys. Lett. *31A*, 349.
Forest, G. and Erlbach, E. (1972). Solid St. Commun. *10*, 731.
Fowler, R. H. (1931). Phys. Rev. *38*, 45.
Fowler, R. H. and Nordheim, L. (1928). Proc. Roy. Soc. A*119*, 173.
Franck, J. P. and Keeler, W. J. (1967). Phys. Rev. *163*, 373.
Franck, J. P., Keeler, W. J. and Wu, T. M. (1969). Solid St. Commun. *7*, 483.
Franz, W. (1956). In *Handbuch der Physik*, Vol. 17. (ed. S. Flugge) p. 155. Springer-Verlag, Berlin.
Franz, W. (1969). In *Tunneling Phenomena in Solids* (eds. E. Burstein and S. Lundqvist), p. 13. Plenum, New York.
Freake, S. M. (1971). Phil. Mag. *24*, 319.
Fredkin, D. R. and Wannier, G. H. (1962). Phys. Rev. *128*, 2054.
Frenkel, J. (1930). Phys. Rev. *36*, 1604.
Frenkel, J. and Joffe, A. (1932). Phys. Z. Sowjetunion *1*, 60.
Frindt, R. F. (1972). Phys. Rev. Lett. *28*, 299.
Fröhlich, H. (1950). Phys. Rev. *79*, 845.
Fröhlich, H. (1952). Proc. Roy. Soc. (London) *A215*, 291.
Froitzheim, H. (1977). In *Electron Spectroscopy for Surface Analysis* (ed. H. Ibach), p. 205. Springer-Verlag, Berlin.
Fromhold, A. T., Jr. (1976). *Theory of Metal Oxidation*, Chap. 10. North-Holland, Amsterdam.
Fromhold, A. T., Jr. and Cook, E. L. (1967). Phys. Rev. *158*, 600.
Fuchs, J., Epperlein, P. W., Welte, M. and Eisenmenger, W. (1977). Phys. Rev. Lett. *38*, 919.
Fulde, P. (1965). Phys. Rev. *137*, A783.

——— (1969). In *Tunneling Phenomena in Solids* (eds. E. Burstein and S. Lundqvist), p. 427. Plenum, New York.

——— (1973). Adv. Phys. *22*, 667.

Fulton, T. A. and McCumber, D. E. (1968). Phys. Rev. *175*, 585.

Fulton, T. A., Hebard, A. F., Dunkleberger, L. N. and Eick, R. H. (1977). Solid St. Commun. *22*, 493.

Furukawa, T., Eib, N. K., Mittal, K. L. and Anderson, H. R., Jr. (1982). Surface and Interface Analysis *4*, 240.

Gadzuk, J. W. (1970). Phys. Rev. B*1*, 2110.

Galkin, A. A., D'yachenko, A. I. and Svistunov, V. M. (1974). Zh. Eksp. Teor. Fiz. *66*, 2262 (Engl. trans. Sov. Phys.-JETP *39*, 1115).

Galkin, A. A., Svistunov, V. M. and Dikii, A. P. (1969). Phys. Stat. Solidi *35*, 421.

Galkin, A. A., Svistunov, V. M., Dikii, A. P. and Taranenko, V. N. (1971). Zh. Eksp. Teor. Fiz. *59*, 77 (Engl. transl. Sov. Phys.-JETP *32*, 44).

Gallagher, W. J. (1980). Phys. Rev. B*22*, 1233.

——— (1982*a*). Bull. Am. Phys. Soc. (II) *27*, 215.

——— (1982*b*). (Unpublished notes.)

Gamow, G. (1928). Z. Phys. *51*, 204.

Garno, J. P. (1977). J. Appl. Phys. *48*, 4627.

——— (1978). Rev. Sci. Instr. *49*, 1218.

Gärtner, K. (1976). Z. Naturforschung *31a*, 858.

Gärtner, K. and Hahn, A. (1976). Z. Naturforschung *31a*, 861.

Gaspari, G. D. and Gyorffy, B. L. (1972). Phys. Rev. Lett. *28*, 801.

Gasparovic, R. F., Taylor, B. N. and Eck, R. E. (1966). Solid St. Commun. *4*, 59.

Gavaler, J. R. (1973). Appl. Phys. Lett. *23*, 480.

Geballe, T. H., Menth, A., Di Salvo, F. J. and Gamble, F. R. (1971). Phys. Rev. Lett. *27*, 314.

Geerk, J., Gurvitch, M., McWhan, D. B. and Rowell, J. M. (1982). Physica B & C *109B*, 1775.

Geerk, J., Glaser, W., Gompf, F., Reichardt, W. and Schneider, E. (1975). In *Proc. 14th Conf. on Low Temperature Physics* (eds. M. Krusius and M. Vuorio), Vol. 2, p. 411. North-Holland, Amsterdam.

Geiger, A. L., Chandrasekhar, B. S. and Adler, J. G. (1969). Phys. Rev. *188*, 1130.

Gheewala, T. R. (1980). IBM J. Res. Dev. *24*, 130.

Giaever, I. (1960*a*). Phys. Rev. Lett. *5*, 147.

——— (1960*b*). Phys. Rev. Lett. *5*, 464.

——— (1965). Phys. Rev. Lett. *14*, 904.

——— (1969). In *Tunneling Phenomena in Solids* (eds. E. Burstein and S. Lundqvist), p. 19. Plenum, New York.

——— (1974). Science *183*, 1253.

Giaever, I. and Megerle, K. (1961). Phys. Rev. *122*, 1101.

Giaever, I. and Zeller, H. R. (1968*a*). Phys. Rev. Lett. *20*, 1504.

——— (1968*b*). Phys. Rev. Lett. *21*, 1385.

——— (1969). J. Vac. Sci. Techn. *6*, 502.

——— (1970). Phys. Rev. B*1*, 4278.

Giaever, I., Hart, H. R. and Megerle, K. (1962). Phys. Rev. *126*, 941.

Gilabert, A. (1977). Ann. de Phys. *2*, 203.

Gilabert, A., Romagnon, J. P. and Guyon, E. (1971). Solid St. Commun. *9*, 1295.

Gilat, G. (1965). Solid St. Commun. *3*, 101.

Gilmartin, H. R. (1982). Ph.D. Thesis, University of Pennsylvania (unpublished).
Ginsberg, D. M. (1962). Phys. Rev. Lett. *8*, 204.
——— (1979). Phys. Rev. B*20*, 960.
Ginzburg, V. L. (1970). Usp. Fiz. Nauk. *101*, 185 (Engl. transl. Sov. Phys. Usp. *13*, 335).
Gleich, W., Regenfus, G. and Sizman, R. (1971). Phys. Rev. Lett. *27*, 1066.
Glotzl, D., Rainer, D. and Schober, H. R. (1975). (Unpublished.)
Glover, R. E. (1967). Phys. Lett. *25A*, 542.
Glover, R. E. and Tinkham, M. (1956). Phys. Rev. *104*, 844.
Godwin, L. M., White, H. W. and Ellialtioglu, R. (1981). Phys. Rev. B*23*, 5688.
Goetze, W. and Schlottmann, P. (1973). Solid St. Commun. *13*, 861.
Gollub, J. P., Beasley, M. R. and Tinkham, M. (1970). Phys. Rev. Lett. *25*, 1646.
Gomersall, I. R. and Gyorffy, B. L. (1974). J. Phys. F*4*, 1204.
Gompf, F. (1981). (Unpublished.)
Gompf, F., Richter, W., Scheerer, B. and Weber, W. (1981) Physica *108*B, 1337.
Good, R. H., Jr. and Müller, E. W. (1956). *Handbuch der Physik 21, p.* 176. Springer-Verlag, Berlin.
Goodman, W. L., Willis, W. D., Vincent, D. A. and Deaver, B. S. (1971). Phys. Rev. B*4*, 1530.
Gor'kov, L. P. (1958). Zh. Eksp. Teor. Fiz. *34*, 735 (Engl. transl. Sov. Phys.-JETP *7*, 505).
——— (1959). Zh. Eksp. Teor. Fiz. *36*, 1918 (Engl. transl. Sov. Phys.-JEPT *9*, 1364).
Granqvist, C. G. and Claeson, T. (1973). Phys. Rev. Lett. *31*, 456.
——— (1974*a*). Z. Phys. *269*, 23.
——— (1974*b*). Phys. Condens. Matter *18*, 99.
Grant, W. N., Barker, R. C. and Yelon A. (1969). *Programme and Abstracts of Symposium: "Electronic Density of States,"* p. 211 (abstr. only). National Bureau Standards, Gaithersburg, Maryland.
Gray, K. E. (1972). Phys. Rev. Lett. *28*, 959.
——— (1974). J. Low Temp. Phys. *15*, 335.
——— (1978*a*). Appl. Phys. Lett. *32*, 392.
——— (1978*b*). Solid St. Commun. *26*, 633.
——— (1980). In *Nonequilibrium Superconductivity, Phonons, and Kapitza Boundaries*, Proc. NATO Advanced Study Institute, p. 131. Plenum, New York, 1981.
Gray, K. E. and Willemsen, H. W. (1978). J. Low Temp. Phys. *31*, 911.
Gray, K. E., Long, A. R. and Adkins, C. J. (1969). Phil. Mag. *20*, 273.
Gregers-Hansen, P. E., Hendricks, E., Levinsen, M. T. and Pickett, G. R. (1973). Phys. Rev. Lett. *31*, 524.
Gregory, J. A., Bostock, J., MacVicar, M. L. A. and Rose, R. M. (1973). Phys. Lett. *46A*, 201.
Gregory, W. D., Averill, R. F. and Straus, L. S. (1971). Phys. Rev. Lett. *27*, 1503.
Greiner, J. H. (1971). J. Appl. Phys. *42*, 5151.
Griffin, A. and Demers, J. (1971). Phys. Rev. B*4*, 2202.
Grimes, C. C., Richards, P. L. and Shapiro, S. (1966). Phys. Rev. Lett. *17*, 431.
——— (1968). J. Appl. Phys. *39*, 3905.
Grimvall, G. (1981). *The Electron-phonon Interaction in Metals.* North-Holland, Amsterdam.
Gromer, J. and Sixl, H. (1974). Solid St. Commun. *14*, 353.

Grunewald, G. and Scharnberg, K. (1979). Solid St. Commun. *32*, 955.
Gubankov, V. M. and Margolin, N. M. (1979). JETP Lett. *29*, 673.
Guetin, P. and Schreder, G. (1970). Solid St. Commun. *8*, 291.
——— (1971*a*). Pnys. Rev. Lett. *27*, 326.
——— (1971*b*). Philips Tech. Rev. *32*, 211.
——— (1972*a*). Phys. Rev. B*5*, 3979.
——— (1972*b*). Phys. Rev. B*6*, 3816.
——— (1973). Phys. Rev. B*7*, 3697.
Gundlach, K. H. (1966). Solid St. Electronics *9*, 949.
——— (1973). J. Appl. Phys. *44*, 5005.
Gundlach, K. H. and Hölzl, J. (1971). Surf. Sci. *27*, 125.
Güntherodt, G., Thompson, W. A., Holtzberg, F. and Fisk, Z. (1982). Phys. Rev. Lett. *49*, 1030.
Gurney, R. W. and Condon, E. U. (1929). Phys. Rev. *33*, 127.
Guyon, E., Meunier, F. and Thompson, R. S. (1967). Phys. Rev. *156*, 452.
Guyon, E., Martinet, A., Matricon, J. and Pincus, P. (1965). Phys. Rev. *138*, A746.
Gyorgy, E. M., Dillon, J. F., Jr., McWhan, D. B., Rupp. L. W., Jr., Testardi, L. R. and Flanders, P. J. (1980). Phys. Rev. Lett. *45*, 57.
Hafstrom, J. W. and MacVicar, M. L. A. (1970). Phys. Rev. B*2*, 4511.
Halbohm, H. D. and Lubbig, H., eds. (1977). *SQUID, Superconducting Quantum Interference Devices and Their Applications.* De Gruyter, Berlin.
Hall, J. T. and Hansma, P. K. (1978). Surf. Sci. *71*, 1.
Hall, J. T., Holdeman, L. B. and Soulen, R. J., Jr. (1980). Phys. Rev. Lett. *45*, 1011.
Hall, R. N., Racette, J. H. and Ehrenreich, H. (1960). Phys. Rev. Lett. *4*, 456.
Hamann, D. R. (1967). Phys. Rev. *158*, 570.
Hamilton, C. A. and Shapiro, S. (1970). Phys. Rev. B*2*, 4494.
——— (1971). Phys. Rev. Lett. *26*, 426.
Hamilton, C. A., Lloyd, F. L. and Kautz, R. L. (1981). IEEE Trans. *MAG-17*, 577.
Hammond, R. H. (1975). IEEE Trans. *MAG-11*, 201.
Hanscom, D. H. (1970). Ph.D. Thesis, Case-Western Reserve University (unpublished).
Hansen, H. H., Pompi, R. L. and Wu, T. M. (1973). Phys. Rev. B*8*, 1042.
Hansma, P. K. (1977). Phys. Rep. *30*C, 145.
——— (1978). In *Inelastic Electron Tunneling Spectroscopy* (ed. T. Wolfram), p. 186. Springer-Verlag, Berlin.
——— (1982). *Tunneling Spectroscopy: Capabilities, Applications and New Techniques.* Plenum, New York.
Hansma, P. K. and Broida, H. P. (1978). Appl. Phys. Lett. *32*, 545.
Hansma, P. K. and Coleman, R. V. (1974). Science *184*, 1369.
Hansma, P. K. and Hansma, H. G. (1982). In *Tunneling Spectroscopy: Capabilities, Applications, and New Techniques.* (ed. P. K. Hansma) Plenum, New York, p. 475.
Hansma, P. K. and Kirtley, J. (1978). Accounts of Chem. Res. *11*, 440.
Hansma, P. K., Hickson, D. A. and Schwarz, J. A. (1977). J. Catal. *48*, 237.
Hansma, P. K., Kaska, W. C. and Laine, R. M. (1976). J. Am. Chem. Soc. *98*, 6064.
Hansma, P. K., Rochlin, G. I. and Sweet, J. N. (1971). Phys. Rev. B*4*, 3003.

Harris, R. E. (1974). Phys. Rev. B*10*, 84.
Harris, R. E., Dynes, R. C. and Ginsberg, D. M. (1976). Phys. Rev. B*14*, 993.
Harrison, W. A. (1961). Phys. Rev. *123*, 85.
Hartman, T. E. (1964). J. Appl. Phys. *35*, 3283.
Hartman, T. E. and Chivian, J. S. (1964). Phys. Rev. *134*, A1094.
Haskell, B. A., Keeler, W. J. and Finnemore, D. K. (1972). Phys. Rev. B*5*, 4364.
Hasselberg, L.-E., Levinsen, M. T. and Samuelson, M. R. (1974). Phys. Rev. B*9*, 3757.
——— (1975). J. Low Temp. Phys. *21*, 567.
Hauser, J. J. (1966). Physics *2*, 247.
——— (1967). Phys. Rev. *164*, 558.
Hauser, J. J. and Testardi, L. R. (1968). Phys. Rev. Lett. *20*, 12.
Hauser, J. J., Bacon, D. D. and Haemmerle, W. H. (1966). Phys. Rev. *151*, 296.
Havemann, R. H. (1978). J. Vac. Sci. Technol. *15*, 389.
Haywood, T. W. and Mitchell, E. N. (1974). Phys. Rev. B*10*, 876.
Heavens, O. S. (1965). *Optical Properties of Thin Films*, p. 156. Dover, New York.
Hebard, A. F. (1973). J. Vac. Sci. Techn. *10*, 606.
Hebard, A. F. and Arthur, J. R. (1977). Bull. Am. Phys. Soc. (II) *22*, 374.
Hebard, A. F. and Shumate, P. W. (1974). Rev. Sci. Instr. *45*, 529.
Hedin, L. and Lundqvist, S. (1969). Solid St. Phys. *23*, 1.
Heeger, A. J. and MacDiarmid, A. G. (1981). Mol. Cryst. Liq. Cryst. *77*, 1.
Heiblum, M., Wang, S. Whinnery, J. R. and Gustafson, T. K. (1978). IEEE J. Quant. El. *QE-14*, 159.
Heine, V. and Abarenkov, I. (1964). Phil. Mag. *9*, 451.
Heiniger, F., Bucher, E. and Muller, J. (1966). Phys. Kond. Mat. *5*, 243.
Herman, H. and Schmid, A. (1968). Z. Phys. *211*, 313.
Hertel, G., McWhan, D. B. and Rowell, J. M. (1982). In *Superconductivity in d- and f-Band Metals* (eds. W. Buckel and W. Weber), p. 299. Kernforschungszentrum Karlsruhe, Karlsruhe.
Hertz, J. A. and Aoi, K. (1973). Phys. Rev. B*8*, 3252.
Ho, K. M., Cohen, M. L. and Pickett, W. (1978). Phys. Rev. Lett. *41*, 815.
Hohenwarter, G. K. and Nordman, J. E. (1982). Appl. Phys. Lett. *40*, 436.
Holdeman, L. B., Hall, James, T., Van Vechten, D. and Soulen, R. B., Jr. (1981). Physica *108B*, 827.
Holland, L. (1956). *Vacuum Deposition of Thin Films.* Wiley, New York.
Holm, R. (1951). J. Appl. Phys. *22*, 569.
Holm, R. and Kirschstein, B. (1935). Z. Tech. Phys. *16*, 448.
——— (1939). Phys. Z. *36*, 882.
Holm, R. and Meissner, W. (1932). Z. Phys. *74*, 715.
——— (1933). Z. Phys. *86*, 787.
Holonyak, N., Lesk, I. A., Hall, R. N., Tiemann, J. T. and Ehrenreich, H. (1959). Phys. Rev. Lett. *3*, 167.
Hone, D., Mühlschlegel, B. and Scalapino, D. J. (1978). Appl. Phys. Lett. *33*, 203.
Hopkins, B. J., Leggett, M. and Watts, G. D. (1971). Surf. Sci. *28*, 581.
Horn, G. and Saur, E. (1968). Z. Phys. *210*, 70.
Horsch, P. and Rietschel, H. (1977). Z. Phys. B*27*, 153.
Howard, R. E., Hu, E. L., Jackel, L. D., Fetter, L. A. and Bosworth, R. H. (1979). Appl. Phys. Lett. *35*, 879.
Hsiang, T. Y. and Finnemore, D. K. (1980). Phys. Rev. B*22*, 154.

Hu, P., Dynes, R. C., Narayanamurti, V., Smith, H. and Brinkman, W. F. (1977). Phys. Rev. Lett. *38*, 361.
Hubin, W. N. (1970). University of Illinois, Urbana. Tech. Rep. 182, ARPA SD–131.
Hubin, W. N. and Ginsberg, D. M. (1969). Phys. Rev. *188*, 716.
Hulm, J. K. and Blaugher, R. D. (1961). Phys. Rev. *123*, 1569.
Hutchings, M. T. and Samuelson, E. J. (1971). Solid St. Commun. *9*, 1011.
Iguchi, I. and Langenberg, D. N. (1980). Phys. Rev. Lett. *44*, 486.
Iguchi, I., Kent, D., Gilmartin, H. and Langenberg, D. N. (1981). Phys. Rev. B*23*, 3240.
Iguchi, I., Kotani, S., Yamaki, Y., Suzuki, Y., Manabe, M. and Harada, K. (1981). Phys. Rev. B*24*, 1193.
Ivezić, T. (1975). J. Phys. C: Solid St. Phys. *8*, 3371.
Ivlev, B. I., Lisitsyn, S. G. and Eliashberg, G. M. (1973). J. Low Temp. Phys. *10*, 449.
Jablonski, D. G. and Waldram, J. R. (1980). In *Squid '80* (eds. H. O. Hahlbohm and H. Lubbig), p. 115. deGruyter, Berlin.
Jackel, L. D., Hu, E. L., Howard, R. E., Fetter, L. A. and Tennant, D. (1981). IEEE Trans. *MAG-17*, 295.
Jackson, J. E., Briscoe, C. V. and Wuhl, H. (1971). Physica *55*, 447.
Jacobsen, N. (1979). Z. Phys. B*34*, 149.
Jaklevic, R. C. (1980). Appl. Surf. Sci. *4*, 174.
——— (1982). In *Tunneling Spectroscopy: Capabilities, Applications and New Techniques* (ed. P. K. Hansma), p. 451. Plenum, New York.
Jaklevic, R. C. and Gaerttner, M. R. (1977). Appl. Phys. Lett. *30*, 646.
——— (1978). Appl. Surf. Sci. *1*, 479.
Jaklevic, R. C. and Lambe, J. (1966). Phys. Rev. Lett. *17*, 1139.
——— (1970). Phys. Rev. B*2*, 808.
——— (1973). Surf. Sci. *37*, 922.
——— (1975). Phys. Rev. B*12*, 4146.
Jaklevic, R. C., Lambe, J., Kirtley, J. and Hansma, P. K. (1977). Phys. Rev. B*15*, 4103.
Jaklevic, R. C., Lambe, J., Mercereau, J. E. and Silver, A. H. (1965). Phys. Rev. *140*, A1628.
Jaklevic, R. C., Lambe, J., Mikkor, M. and Vassell, W. C. (1971). Phys. Rev. Lett. *26*, 88.
——— (1972). Solid St. Commun. *10*, 199.
Jaklevic, R. C., Lambe, J., Silver, A. H. and Mercereau, J. E. (1964). Phys. Rev. Lett. *12*, 159.
Jansen, A. G. M., Mueller, F. M. and Wyder, P. (1977). Phys. Rev. B*16*, 1325.
Jansen, A. G. M., van Gelder, A. P. and Wyder, P. (1980). J. Phys. C*13*, 6073.
Jarlborg, T. (1979). J. Phys. F*9*, 283.
Jaworski, F. and Parker, W. H. (1979). Phys. Rev. B*20*, 945.
Jennings, R. J. and Merrill, J. R. (1972). J. Phys. Chem. Solids *33*, 1261.
Jérome, D. (1982*a*). Mol. Cryst. Liq. Cryst. *79*, 155.
——— (1982*b*). In *Superconductivity in d- and f-Band Metals* (ed. W. Buckel and W. Weber), p. 421. Kernforschungszentrum Karlsruhe, Karlsruhe.
Johnson, K. W. and Olson, D. H. (1971). Phys. Rev. B*3*, 1244.
Josefowicz, J. and Smith, H. J. T. (1973). J. Appl. Phys. *44*, 2813.
Joseph, A. S., Tomasch, W. J. and Fink, H. J. (1967). Phys. Rev. *157*, 315.

Josephson, B. D. (1962*a*). Phys. Lett. *1*, 251.
——— (1962*b*). Fellowship dissertation, Trinity College, University of Cambridge.
——— (1965). Adv. Phys. *14*, 419.
——— (1974). Science *184*, 527.
Joy, D. C., Newbury, D. E. and Davidson, D. L. (1982). J. Appl. Phys. *53*, R81.
Kaiser, A. B. (1977). J. Phys. F7, L339.
Kaiser, A. B. and Zuckermann, M. J. (1970). Phys. Rev. B*1*, 229.
Kalvey, V. R. and Gregory, W. D. (1979). Phys. Lett. *74A*, 256.
Kamitakahara, W. A., Smith, H. G. and Wakabayashi, N. (1977). Ferroelectrics *16*, 111.
Kane, E. O. (1959*a*). J. Phys. Chem. Solids *12*, 181.
——— (1959*b*). J. Appl. Phys. *32*, 83.
——— (1969). In *Tunneling Phenomena in Solids* (eds. E. Burstein and S. Lundqvist), p. 1. Plenum, New York.
Kaplan, S. B., Kirtley, J. R. and Langenberg, D. N. (1977). Phys. Rev. Lett. *39*, 291.
Kaplan, S. B., Chi, C. C., Langenberg, D. N., Chang, J. J., Jafarey, S. and Scalapino, D. J. (1976). Phys. Rev. B*14*, 4854.
Keck, B. and Schmid, A. (1976). J. Low Temp. Phys. *24*, 611.
Keil, R. G., Graham, T. P. and Roenker, K. P. (1976). Appl. Spectroscopy *30*, 1.
Keith, V. and Leslie, J. D. (1978). Phys. Rev. B*18*, 4739.
Keldysh, L. V. (1958*a*). Zh. Eksp. Teor. Fiz. *33*, 994 (Engl. transl. Sov. Phys.-JETP *6*, 763).
——— (1958*b*). Zh. Eksp. Teor. Fiz. *34*, 962 (Engl. transl. Sov. Phys.-JETP *7*, 665).
——— (1964). Zh. Eksp. Teor. Fiz. *47*, 1515 (Engl. Transl. Sov. Phys.-JETP *20*, 1018).
Keller, W. H. and Nordman, J. E. (1973). J. Appl. Phys. *44*, 4732.
Kerchner, H. R., Christen, D. K. and Sekula, S. T. (1981). Phys. Rev. B*24*, 1200.
Kern, W. and Ban, V. S. (1978). In *Thin Film Processes* (eds. J. L. Vossen and W. Kern), p. 257. Academic, New York.
Kerr, A. R., Pan, S.-K., Feldman, M. J. and Davidson, A. (1981). Physica *108B*, 1369.
Keskar, K. S., Yamashita, T. and Onadera, Y. (1971). Japan J. Appl. Phys. *10*. 370.
Khim, Z. G. (1979). Ph.D. Thesis, University of Notre Dame (unpublished).
Khim, Z. G. and Tomasch, W. J. (1978). Phys. Rev. *B18*, 4706.
——— (1979). Phys. Rev. Lett. *42*, 1227.
Khim, Z. G., Burnell, D. and Wolf, E. L. (1981). Solid St. Commun. *39*, 159.
Kieselmann, G. and Rietschel, H. (1982). J. Low Temp. Phys. *46*, 27.
Kihlstrom, K. E. and Geballe, T. H. (1981). Phys. Rev. B*24*, 4101.
Kimhi, D. B. and Geballe, T. H. (1980). Phys. Rev. Lett. *45*, 1039.
Kincaid, B. M. and Eisenberger, P. (1975). Phys. Rev. Lett. *34*, 1361.
Kirtley, J. R. (1980). In *Vibrational Spectroscopies of Adsorbed Species.* (eds. A. T. Bell and M. L. Hair), p. 217. American Chemical Society, Washington, D.C.
——— (1982). In *Tunneling Spectroscopy: Capabilities, Applications and New Techniques* (ed. P. K. Hansma), p. 43. Plenum, New York.
Kirtley, J. R. and Hall, J. T. (1980). Phys. Rev. B*22*, 848.
Kirtley, J. R. and Hansma, P. K. (1975). Phys. Rev. B*12*, 531.

——— (1976). Phys. Rev. B*13*, 2910.

Kirtley, J. R. and Soven, P. (1979). Phys. Rev. B*19*, 1812.

Kirtley, J. R., Scalapino, D. J. and Hansma, P. K. (1976). Phys. Rev. B*14*, 3177.

Kirtley, J. R., Theis, T. N. and Tsang, J. C. (1980). Appl. Phys. Lett. *37*, 435.

——— (1981). Phys. Rev. B*24*, 5650.

Kisker, E., Gudat, W., Campagna, M., Kuhlmann, E., Hopster, H. and Moore, I. D. (1979). Phys. Rev. Lett. *43*, 966.

Klapwijk, T. M., Blonder, G. E. and Tinkham, M. (1982). Physica B&C *109*, 1657.

Klein, B. M., Boyer, L. L., Papaconstantopoulos, D. A. and Mattheiss, L. F. (1978). Phys. Rev. B*18*, 6411.

Klein, J., Léger, A., Delmas, B. and de Cheveigné, S. (1976). Rev. Phys. Appl. *11*, 321.

Klein, J., Léger, A., Belin, M., Defourneau, D. and Sangster, M. J. L. (1973). Phys. Rev. B*7*, 2336.

Kleinsasser, A. W. and Buhrman, R. A. (1980). Appl. Phys. Lett. *37*, 841.

Knauss, H. P. and Breslow, R. A. (1962). Proc. IRE *50*, 1843.

Knorr, K. and Barth, N. (1970). Solid St. Commun. *8*, 1085.

Knorr, K. and Leslie, J. D. (1973). Solid St. Commun. *12*, 615.

Koehler, W. C. (1972). In *Magnetic Properties of Rare Earth Metals* (ed. R. J. Elliot), p. 83. Plenum, London.

Kofstad, P. (1966). *High-Temperature Oxidation of Metals*. Wiley, New York.

Komenou, K., Tanaka, K. and Onadera, Y. (1971). In *Proc. 12th Int. Conf. on Low Temperature Physics* (ed. E. Kanda), p. 510. Keigaku, Iokyo.

Komenou, K., Yamashita, T. and Onodera, Y. (1968). Phys. Lett. *28A*, 335.

Kommers, T. and Clarke, J. (1977). Phys. Rev. Lett. *38*, 1091.

Kondo, J. (1964). Prog. Theor. Phys. *32*, 37.

Korb, H. W., Andrews, A. M., Holonyak, N., Burnham, R. D., Duke, C. B. and Kleiman, G. G. (1971). Solid St. Commun. *9*, 1531.

Korringa, J. (1950). Physica *16*, 601.

Kotani, S., Suzuki, Y. and Iguchi, I. (1982). Phys. Rev. Lett. *49*, 391.

Krätzig, E. (1971). Solid St. Commun. *9*, 1205.

Krauss, G. and Buckel, W. (1975). Z. Phys. B*20*, 147.

Kroeker, R. M. (1982). In *Tunneling Spectroscopy: Capabilities, Applications and New Techniques* (ed. P. K. Hansma), p. 393. Plenum, New York.

Kroeker, R. M., Kaska, W. C. and Hansma, P. K. (1979). J. Catal. *57*, 72.

——— (1980), J. Catal. *61*, 87.

——— (1981). J. Chem. Phys. *74*, 732.

Kroger, H., Potter, C. N. and Jillie, D. W. (1979). IEEE Trans. *MAG-15*, 488.

Kroger, H., Smith, L. N. and Jillie, D. W. (1981). Appl. Phys. Lett. *39*, 280.

Kroo, N. and Szentirmay, Zs. (1970). Phys. Lett. A*32*, 543.

——— (1971). In *Proc. 12th. Int. Conf. on Low Temperature Physics* (ed. E. Kanda), p. 559. Keigaku, Tokyo.

Kroo, N., Szentirmay, Zs. and Félszerfalvi, J. (1980). Phys. Stat. Sol. (b) *102*, 227.

——— (1981). Phys. Lett. *81*A, 399.

——— (1982). Phys. Lett. *88A*, 90.

Kulik, I. O. and Yanson, I. K. (1970). *The Josephson Effect in Superconductive Tunneling Structures*. (Izdatel'stvo "Nauka," Moscow.) English trans. by Israel Program for Scientific Translations Ltd; 1972, published by Keter Press, Jerusalem.

Kumbhare, P., Tedrow, P. M. and Lee, D. M. (1969). Phys. Rev. *180*, 519.
Kunze, U. and Lautz, G. (1982). Surf. Sci. *113*, 55.
Kurtin, S. L., McGill, T. C. and Mead, C. A. (1970). Phys. Rev. Lett. *25*, 756.
——— (1971). Phys. Rev. B*3*, 3368.
Kuwasawa, Y., Rinderer, L. and Matthias, B. T. (1979). J. Low Temp. Phys. *37*, 179.
Kwak, J. F. (1982). Phys. Rev. B*26*, 4789.
Kwak, J. F. and Schirber, J. E. (1983) (to be published).
Kwo, J. and Geballe, T. H. (1981). Phys. Rev. B*23*, 3230.
Kwo, J., Wertheim, G. K., Gurvitch, M. and Buchanan, D. N. E. (1982). Appl. Phys. Lett. *40*, 675.
——— (1983). In *Proc. Conf. on Applied Superconductivity* (*Knoxville, 1982*). Paper SD–4.
Ladan, F. R. and Zylbersztejn, A. (1972). Phys. Rev. Lett. *28*, 1198.
Laks, B. and Mills, D. L. (1979). Phys. Rev. B*20*, 4962.
——— (1980). Phys. Rev. B*22*, 5723.
Lambe, J. and Jaklevic, R. C. (1968). Phys. Rev. *165*, 821.
Lambe, J. and McCarthy, S. L. (1976). Phys. Rev. Lett. *37*, 923.
Landau, L. D. and Lifshitz, E. M. (1958). *Quantum Mechanics*, p. 175. Addison-Wesley, Reading, Massachusetts.
Landolt, M. and Campagna, M. (1977). Phys. Rev. Lett. *38*, 663.
Landolt, M., Yafet, Y., Wilkens, B. and Campagna, M. (1978). Solid St. Commun. *25*, 1141.
Langan, J. D. and Hansma, P. K. (1975). Surf. Sci. *52*, 211.
Langenberg, D. N. (1974). Rev. Phys. Appl. *9*, 35.
——— (1975). In *Proc. 14th Conf. on Low Temperature Physics* (eds. M. Krusius and M. Vuorio), Vol. 5, p. 223. North-Holland, Amsterdam.
Langenberg, D. N., Scalapino, D. J. and Taylor, B. N. (1966). Proc. IEEE *54*, 560.
Langenberg, D. N., Scalapino, D. J., Taylor, B. N. and Eck, R. E. (1965). Phys. Rev. Lett. *15*, 294.
Langer, R. E. (1937). Phys. Rev. *51*, 669.
Langmuir, I. (1918). J. Am. Chem. Soc. *40*, 1361.
Larkin, A. I. (1965). Zh. Eksp. Teor. Fiz. *48*, 232 (Engl. transl. Soviet Phys.-JETP *21*, 153).
Larson, D. C. (1974). In *Methods of Experimental Physics*: *Volume 11, Solid State Physics* (ed. R. V. Coleman), p. 619. Academic, New York.
Lau, J. C. and Coleman, R. V. (1981). Phys. Rev. B*24*, 2985.
Leavens, C. R. (1974). Solid St. Commun. *15*, 1329.
Leavens, C. R. and Carbotte, J. P. (1972). Ann. Phys. *70*, 338.
Leavens, C. R. and MacDonald, A. H. (1983). Phys. Rev. B*27*, 2812.
Lee, P. A. and Payne, M. G. (1971). Phys. Rev. Lett. *26*, 1537.
Léger, A. and Klein, J. (1969). Phys. Lett. *28A*, 751.
Léger, A., Klein, J., Belin, M. and Defourneau, D. (1971). Thin Solid Films *8*, R51.
——— (1972). Solid St. Commun. *11*, 1331.
Léger, A., Klein, J., de Cheveigné, S., Belin, M. and Defourneau, D. (1975). J. Physique Lett. *36*, L301.
Leipold, W. C. and Feuchtwang, T. E. (1974). Phys. Rev. B*10*, 2195.
Leo, V. (1981). Solid St. Commun. *40*, 509.

——— (1982). Rev. Sci. Instr. *53*, 997.
Leo, V., Gusman, G. and Deltour, R. (1982). Phys. Rev. B*26*, 3285.
Leslie, J. D., Chen, J. T. and Chen, T. T. (1970). Can. J. Phys. *48*, 2783.
Leung, H. K., Carbotte, J. P. and Leavens, C. R. (1976). J. Low Temp. Phys. *24*, 25.
Levin, H., Selisky, H., Heim, G. and Buckel, W. (1976). Z. Phys. B*24*, 65.
Levinstein, H. J. and Kunzler, J. E. (1966). Phys. Lett. *20*, 581.
Lewicki, G. and Mead, C. A. (1966). Phys. Rev. Lett. *16*, 939.
Lewis, B. F., Mosesman, M. and Weinberg, W. H. (1974). Surf. Sci. *41*, 142.
Lie, S. G. and Carbotte, J. P. (1978). Solid St. Commun. *26*, 511.
Liehr, M. and Ewert, S. (1983). Z. Physik B*52*, 95.
Likharev, K. K. (1979). Rev. Mod. Phys. *51*, 101.
Lim, C. S., Leslie, J. D., Smith, H. J. T., Vashishta, P. and Carbotte, J. P. (1970). Phys. Rev. B*2*, 1651.
Liu, K. C., Davis, C. and Javan, A. (1979). Phys. Rev. Lett. *43*, 785.
Logan, R. A. and Rowell, J. M. (1964). Phys. Rev. Lett. *13*, 404.
London, F. and London, H. (1935*a*). Proc. Roy. Soc. A*149*, 71.
——— (1935*b*). Physica *2*, 341.
Losee, D. L. and Wolf, E. L. (1969). Phys. Rev. *187*, 925.
——— (1971). Phys. Rev. Lett. *26*, 1021.
Lou, L. F. and Tomasch, W. J. (1972). Phys. Rev. Lett. *29*, 858.
Lubberts, G. (1971). Phys. Rev. B*3*, 1965.
Lubberts, G., Burkey, B. C., Bücher, H. K. and Wolf, E. L. (1974). J. Appl. Phys. *45*, 2180.
Lüth, H., Roll, U. and Ewert, S. (1978). Phys. Rev. B*18*, 4241.
Lykken, G. I., Geiger, A. L., Dy, K. S. and Mitchell, E. N. (1971). Phys. Rev. B*4*, 1523.
Lykken, G. I. and Soonpaa, H. H. (1973). Phys. Rev. B*8*, 3186.
Lykken, G. I., Geiger, A. L. and Mitchell, E. N. (1970). Phys. Rev. Lett. *25*, 1578.
Lynn, K. G. (1980). Phys. Rev. Lett. *44*, 1330.
Lynton, E. A. (1962). *Superconductivity*. Methuen, London.
Lythall, D. J. and Wyatt, A. F. G. (1968). Phys. Rev. Lett. *20*, 1361.
McBride, D. E. and Hall, J. T. (1979). J. Catalysis *58*, 320.
McBride, D. E., Rochlin, G. and Hansma, P. K. (1974). J. Appl. Phys. *45*, 2305.
McColl, M., Millea, M. F. and Silver, A. H. (1973). Appl. Phys. Lett. *23*, 263.
McGrath, W. R., Richards, P. L., Smith, A. D., van Kempen, H., Batchelor, R. A., Prober, D. E. and Santhanam, P. (1981). Appl. Phys. Lett. *39*, 655.
Machida, K. (1977). J. Low Temp. Phys. *27*, 737.
Machida, K. and Dumoulin, L. (1978). J. Low Temp. Phys. *31*, 143.
McMillan, W. L. (1968*a*). Phys. Rev. *167*, 331.
——— (1968*b*). Phys. Rev. *175*, 537.
——— (1968*c*). Phys. Rev. *175*, 559.
——— (1981). Phys. Rev. B*24*, 2739.
McMillan, W. L. and Anderson, P. W. (1966). Phys. Rev. Lett. *16*, 85.
McMillan, W. L. and Mochel, J. (1981). Phys. Rev. Lett. *46*, 556.
McMillan, W. L. and Rowell, J. M. (1965). Phys. Rev. Lett. *14*, 108.
——— (1969). In *Superconductivity* (ed. R. D. Parks), Vol. 1, p. 561. Dekker, New York.
McMorris, I. W. N., Brown, N. M. D. and Walmsley, D. G. (1977). J. Chem. Phys. *66*, 3952.

MacVicar, M. L. A. (1970). J. Appl. Phys. *41*, 4765.
MacVicar, M. L. A. and Rose, R. M. (1968). J. Appl. Phys. *39*, 1721.
Magerlein, J. H. (1983). J. Appl. Phys. *54*, 2569.
Magno, R. and Adler, J. G. (1976). Phys. Rev. B*13*, 2262.
——— (1977*a*). Phys. Rev. B*15*, 1744.
——— (1977*b*). Thin Solid Films *42*, 237.
——— (1978). J. Appl. Phys. *49*, 5571.
Mahan, G. D. (1969). *Tunneling Phenomena in Solids*, (ed. E. Burstein and S. Lundqvist) p. 305. Plenum, New York.
Mahan, G. D. and Conley, J. W. (1967). Appl. Phys. Lett. *11*, 29.
Maissel, Leon I. and Glang, B., eds. (1970). *Handbook of Thin Film Technology*. McGraw-Hill, New York.
Maki, K. (1964). Prog. Theor. Phys. (Kyoto) *31*, 731.
——— (1969). In *Superconductivity* (ed. R. D. Parks), Vol. 2, p. 1035. Dekker, New York.
Mancini, N. A., Giaquinta, G., Burrafato, G., Di Mauro, C., Pennisi, A., Simone, F. and Troia, S. O. (1979). Il Nuovo Cimento *53b*, 435.
Mann, B. and Kuhn, H. (1971). J. Appl. Phys. *42*, 4398.
Manuel, P. and Veyssié, J. J. (1973). Solid St. Commun. *13*, 1819.
Markowitz, D. and Kadanoff, L. P. (1963). Phys. Rev. *131*, 563.
Mase, S. (1958). J. Phys. Soc. Japan *13*, 434.
——— (1959). J. Phys. Soc. Japan *14*, 584.
Maserjian, J. (1974). J. Vac. Sci. Techn. *11*, 996.
Masuda, Y., Nishioka, M. and Watanabe, N. (1967). J. Phys. Soc. Japan *22*, 238.
Mathews, J. W., ed. (1975). *Epitaxial Growth*, Part A. Academic, New York.
Matisoo, J. (1969). Phys. Lett. *29A*, 473.
——— (1980), IBM J. Res. Dev. *24*, 113.
Matsuzawa, M. (1968). J. Chem. Phys. *51*, 4705.
Mattheiss, L. F. and Hamann, D. R. (1982). Phys. Rev. B*26*, 2686.
Mattheiss, L. F. (1972). Phys. Rev. B*6*, 4740.
Matthias, B. T. (1955). Phys. Rev. *97*, 74.
——— (1960). J. Appl. Phys. Suppl. *31*, 235.
Matthias, B. T., Geballe, T. H. and Compton, V. B. (1963). Rev. Mod. Phys. *35*, 1.
Matthias, B. T., Geballe, T. H., Geller, S. and Corenzwit, E. (1954). Phys. Rev. *95*, 1435.
Matthias, B. T., Geballe, T. H., Longinotti, L. D., Corenzwit, E., Hull, G. W., Willens, R. H. and Maita, J. P. (1967). Science *156*, 645.
Maxwell, E. (1950). Phys. Rev. *78*, 477.
Mayer, H. (1955). *Physik Dünner Schichten*, p. 293. Wissenschaftliche Verlagsgesellschaft, Stuttgart.
Mayer, J. W. and Ziegler, J. F. (1974). In *Ion Beam Surface Layer Analysis* (ed. J. W. Mayer and J. F. Ziegler). Elsevier Sequoia, New York.
Meissner, H. (1960). Phys. Rev. *117*, 672.
Meissner, W. and Ochsenfeld, R. (1933). Naturwiss. *21*, 787.
Mele, E. J. and Rice, M. J. (1981). Phys. Rev. B*23*, 5397.
Meservey, R. (1978). *Inelastic Electron Tunneling Spectroscopy* (ed. T. Wolfram), p. 230. Springer-Verlag, Berlin.
Meservey, R. and Tedrow, P. M. (1971). J. Appl. Phys. *42*, 51.
Meservey, R., Paraskevopoulos, D. and Tedrow, P. M. (1978). *Proc. Conf. on Magnetism and Magnetic Materials*. J. Appl. Phys. *49*, 1405.

Meservey, R., Tedrow, P. M. and Fulde, P. (1970). Phys. Rev. Lett. *25*, 1270.
Meservey, R., Tedrow, P. M. and Kalvey, V. R. (1980). Solid St. Commun. *36*, 969.
Meservey, R., Tedrow, P. M., Kalvey, V. R. and Paraskevopoulos, D. (1979). J. Appl. Phys. *50*, 1935.
Methfessel, C. E., Stewart, G. R., Matthias, B. T. and Patel, C. K. N. (1980). Proc. Natl. Acad. Sci. USA *77*, 6307.
Mezei, F. (1967). Phys. Lett. *25A*, 534.
——— (1969). Solid St. Commun. *7*, 771.
Mezei, F. and Zawadowski, A. (1971*a*). Phys. Rev. B*3*, 167.
——— (1971*b*). Phys. Rev. B*3*, 3127.
Miedema, A. R. (1974). J. Phys. F*4*, 120.
Miedema, A. R. and den Broeder, F. J. A. (1979). Z. Metallkunde *70*, 14.
Migdal, A. B. (1958). Zh. Eksp. Teor. Fiz. *34*, 1438 (Engl. transl. Sov. Phys.-JETP *7*, 996].
Mikkor, M. and Vassell, W. C. (1970). Phys. Rev. B*2*, 1875.
Miles, J. L. and McMahon, H. O. (1961). J. Appl. Phys. *32*, 1176.
Miles, J. L. and Smith, P. (1963). J. Electrochem. Soc. *110*, 1240.
Miller, B. I. and Dayem, A. H. (1967). Phys. Rev. Lett. *18*, 1000.
Miller, J. C. P. (1965). In *Handbook of Mathematical Functions* (ed. M. Abramowitz and I. A. Stegun), p. 585ff. Dover, New York.
Miller, R. J. and Satterthwaite, C. B. (1975). Phys. Rev. Lett. *34*, 144.
Miller, S. C. Jr. and Good, R. H. Jr. (1953). Phys. Rev. *91*, 174.
Millstein, J. and Tinkham, M. (1967). Phys. Rev. *158*, 325.
Mitrović, B. and Carbotte, J. P. (1981*a*). Physica *108B*, 977.
——— (1981*b*). Solid St. Commun. *40*, 249.
Mohabir, S. and Nagi, A. D. S. (1979*a*). J. Low Temp. Phys. *35*, 671.
——— (1979*b*). J. Low Temp. Phys. *36*, 637.
Moodera, J. S., Meservey, R. and Tedrow, P. M. (1982). Appl. Phys. Lett. *41*, 488.
Moody, M. V., Paterson, J. L. and Ciali, R. L. (1979). Rev. Sci. Instr. *50*, 903.
Moore, D. F. (1978). Ph.D. Thesis, Stanford University (unpublished).
Moore, D. F. and Beasley, M. R. (1977). Appl. Phys. Lett. *30*, 494.
Moore, D. F., Beasley, M. R. and Rowell, J. M. (1978). J. Phys. *39*, C6-1390.
Moore, D. F., Rowell, J. M. and Beasley, M. R. (1976). Solid St. Commun. *20*, 305.
Moore, D. F., Zubeck, R. B., Rowell, J. M. and Beasley, M. R. (1979). Phys. Rev. B*20*, 2721.
Moore, E. L., Reed, R. W. and Brickwedde, F. G. (1977). Phys. Rev. B*15*, 187.
Moore, I. D. and Pendry, J. B. (1978). J. Phys. C*11*, 4615.
More, C., Roger, G., Sorbier, J. P., Jérome, D., Ribault, M. and Bechgaard, K. (1981). J. Phys. Lett. *42*, L313.
Morel, P. and Anderson, P. W. (1962). Phys. Rev. *125*, 1263.
Moreland, J. and Hansma, P. K. (1983). Rev. Sci. Instrum. *55*, 399.
Moreland, J., Alexander, S., Cox, M., Sonnenfeld, R., and Hansma, P. K. (1983). Appl. Phys. Lett. *43*, 387.
Morris, R. C. and Coleman, R. V. (1973). Phys. Lett. *43A*, 11.
Morris, R. C., Christopher, J. E. and Coleman, R. V. (1969). Phys. Lett. *A30*, 396.
Morse, R. W., Olsen, T. and Gavenda, J. D. (1959). Phys. Rev. Lett. *3*, 15.
Mott, N. F. (1939). Trans. Faraday Soc. *35*, 1175.
——— (1940). Trans. Faraday Soc. *36*, 472.

——— (1947). Trans. Faraday Soc. *43*, 429.
——— (1972). Adv. Phys. *21*, 785.
Mott, N. F. and Davis, E. A. (1979). *Electronic Processes in Non-Crystalline Materials*, 2nd Edition. Oxford Univ. Press (Clarendon), London and New York.
Mühlschlegel, B. (1959). Z. Physik *155*, 313.
Mukhopadhyay, P. (1979). J. Phys. F*9*, 903.
Muldoon, M. F., Dragoset, R. A. and Coleman, R. V. (1979). Phys. Rev. B*20*, 416.
Murphy, E. L. and Good, R. H., Jr. (1956). Phys. Rev. *102*, 1464.
Nam, S. B. (1967*a*). Phys. Rev. *156*, 470.
——— (1967*b*). Phys. Rev. *156*, 487.
Nambu, Y. (1960). Phys. Rev. *117*, 648.
Narayanamurti, V., Dynes, R. C., Hu, P., Smith, H. and Brinkman, W. F. (1978). Phys. Rev. B*18*, 6041.
Narayanamurti, V., Störmer, H. L., Chin, M. A., Gossard, A. C. and Wiegmann, W. (1979). Phys. Rev. Lett. *43*, 2012.
Nass, M. J., Levin, K. and Grest, G. S. (1980). Phys. Rev. Lett. *45*, 2070.
——— (1981). Phys. Rev. B*23*, 1111.
Nédellec, P. (1977). Ann. de Phys. *2*, 253.
Nédellec, P., Dumoulin, L. and Guyon, E. (1971). Solid St. Commun. *9*, 2013.
——— (1976). J. Low Temp. Phys. *24*, 663.
Nédellec, P., Dumoulin, L. and Noer, R. J. (1974). J. Phys. F: Metal Phys. *4*, L145.
Nedrud, B. W. (1980). *Analysis of Superconducting Tunneling Data by McMillan's Method.* Tech. Rep. 217. University of Illinois, Urbana-Champaign, Department of Physics.
Nedrud, B. W. and Ginsberg, D. M. (1981). Physica *108B*, 1175.
Neugebauer, C. A. and Ekvall, R. A. (1964). J. Appl. Phys. *35*, 547.
Newkirk, L. R., Valencia, F. A. and Wallace, T. C. (1976). J. Electrochem. Soc. *123*, 425.
Ng, H. K., Timusk, T., Delrieu, J. M., Jérome, D., Bechgaard, K. and Fabre, J. M. (1982). J. Phys. Lett. *43*, L513.
Ngai, K. L., Economou, E. N. and Cohen, M. H. (1969). Phys. Rev. Lett. *22*, 1375.
Nicklow, R. M., Gilat, G., Smith, H. G., Raubenheimer, L. J. and Wilkinson, M. K. (1967). Phys. Rev. *164*, 922.
Nicol, J., Shapiro, S. and Smith, P. H. (1960). Phys. Rev. Lett. *5*, 461.
Nicolet, M.-A., Mitchell, I. V. and Mayer, J. W. (1972). Science *177*, 841.
Nielsen, P. (1970). Phys. Rev. B*2*, 3819.
Niemeyer, J. and von Minnigerode, G. (1979). Z. Physik B*36*, 57.
Noer, R. J. (1975). Phys. Rev. B*12*, 4882.
Nordheim, L. (1932). Z. Phys. *75*, 434.
Ochiai, S. I., MacVicar, M. L. A. and Rose, R. M. (1971). Phys. Rev. B*4*, 2988.
Ochik, P., Svistunov, V. M., Belogolovskii, M. A. and D'yachenko, A. I. (1978). Sov. Phys. Solid St. *20*, 1101.
Onnes, H. K. (1911). Leiden Commun. *120b*, *120c*, *124c*.
Oppenheimer, J. R. (1928). Phys. Rev. *31*, 66.
Ord, J. L. (1969). Surf. Sci. *16*, 155.
Orlando, T. P. and Beasley, M. R. (1981). Phys. Rev. Lett. *46*, 1598.
Osmun, J. W. (1975). Phys. Rev. B*11*, 5008.

——— (1980). Phys. Rev. B*21*, 2829.
Ovadyahu, Z. and Entin-Wohlman, O. (1979). J. Phys. F*9*, 2091.
——— (1980). J. Phys. F*10*, 1525.
Owen, C. S. and Scalapino, D. J. (1972). Phys. Rev. Lett. *28*, 1559.
Oya, G., Onadera, Y. and Muto, Y. (1974). In *Low Temperature Physics-LT13, Volume 3*: *Superconductivity* (ed., K. D. Timmerhaus, W. J. O'Sullivan and E. F. Hammel), p. 399. Plenum, New York.
Palmberg, P. W. (1967). Rev. Sci. Instr. *38*, 834.
——— (1974). J. Electron Spec. Related Phenom. *5*, 691.
Pals, J. A. and van Haeringen, W. (1977). Physica *92*B, 360.
Pankove, J. I. (1966). Phys. Lett. *21*, 406.
Pannetier, B., Huet, D., Buechner, J. and Maneval, J. P. (1977). Phys. Rev. Lett. *39*, 646.
Paraskevopoulos, D., Meservey, R. and Tedrow, P. M. (1977). Phys. Rev. B*16*, 4907.
Park, R. L. (1975). Surf. Sci. *48*, 80.
Park, R. L., Houston, J. E. and Schreiner, B. G. (1970). Rev. Sci. Instr. *41*, 1810.
Parker, G. H. and Mead, C. A. (1968). Phys. Rev. Lett. *21*, 605.
——— (1969). Phys. Rev. *184*, 780.
Parker, W. H. (1974). Solid St. Commun. *15*, 1003.
——— (1975). Phys. Rev. B*12*, 3667.
Parker, W. H. and Williams, W. D. (1972). Phys. Rev. Lett. *29*, 924.
Parker, W. H., Taylor, B. N. and Langenberg, D. N. (1967). Phys. Rev. Lett. *18*, 287.
Parker, W. H., Langenberg, D. N., Denenstein, A. and Taylor, B. N. (1969). Phys. Rev. *177*, 639.
Parkin, S. S., Ribault, M., Jérome, D. and Bechgaard, K. (1981). J. Phys. C*14*, 5305.
Parks, R. D., ed. (1969). *Superconductivity*, Vols. 1 and 2. Marcel Dekker, New York.
Parmenter, R. H. (1961). Phys. Rev. Lett. *7*, 274.
Pashley, D. W. (1956). Adv. Phys. *5*, 173.
——— (1967). Adv. Phys. *14*, 327.
Passaglia, E., Stromberg, R. R. and Kruger, J., eds. (1964). *Ellipsometry in the Measurement of Surfaces and Thin Films.* National Bureau Standard Misc. Publ. 256, U.S. Government Printing Office, Washington, D.C.
Paterson, J. L. (1978). J. Low Temp. Phys. *33*, 285.
——— (1979). J. Low Temp. Phys. *35*, 371.
Patterson, W. R. and Shewchun J. (1964). Rev. Sci. Instr. *35*, 1704.
Payne, R. T. (1965). Phys. Rev. *139*, A570.
Pedan, A. G. and Kulik, I. O. (1982). Sov. J. Low Temp. Phys. *8*, 113.
Pedersen, N. F., Finnegan, T. F. and Langenberg, D. N. (1972). Phys. Rev. B*6*, 4151.
Penn, D. R. and Plummer, E. W. (1974). Phys. Rev. B*9*, 1216.
Pessall, N. and Hulm, J. K. (1966). Physics *2*, 311.
Phillips, N. E. (1964). Phys. Rev. *134*, 385.
Phillips, T. G. and Dolan, G. J. (1982). Physica *109 & 110B*, 2010.
Pickett, W. E. (1980). Phys. Rev. B*21*, 3897.
——— (1982). Phys. Rev. B*26*, 1186.
Phillips, T. G., Woody, D. P., Dolan, G. J., Miller, R. E. and Linke, R. A. (1981). IEEE Trans. *MAG-17*, 684.

Pickett, W. E. and Klein, B. M. (1981). Solid St. Commun. *38*, 95.
Pinski, F. J., Allen, P. B. and Butler, W. H. (1978). J. Phys. C6-472.
Plummer, E. W. (1975). In *Interactions on Metal Surfaces* (ed. R. Gomer), p. 144. Springer Verlag, New York.
Plummer, E. W. and Young, R. D. (1970). Phys. Rev. B*1*, 2088,
Pobell, F., Rainer, D. and Wühl, H. (1982). In *Superconductivity in Ternary Compounds I: Structural, Electronic and Lattice Properties* (eds. G. Fischer and M. B. Maple), p. 268. Springer-Verlag, Berlin, Heidelberg and New York.
Politzer, B. A. and Cutler, P. H. (1970*a*). Surf. Sci. *22*, 277.
——— (1970*b*). Mater. Res. Bull. *5*, 703 (1970).
Pollack, S. R. and Morris, C. E. (1965). Trans. AIME *233*, 497.
Pollock, C. R., Jennings, D. A., Petersen, F. R., Wells, J. S., Drullinger, R. E., Beaty, E. C. and Evenson, K. M. (1983). Optics Lett. *8*, 133.
Poon, S. J. (1980). Solid St. Commun. *34*, 659.
Poon, S. J. and Geballe, T. H. (1978). Phys. Rev. B*18*, 233.
Poppe, U. (1981). Physica *108B*, 805.
Poppe, U. and Wuhl, H. (1981). J. Low Temp. Phys. *43*, 371.
Prange, R. E. (1963). Phys. Rev. *131*, 1083.
——— (1970). Phys. Rev. B*1*, 2349.
Prothero, D. H. (1974). Phil. Mag. *29*, 829.
Pynn, R. and Squires, G. L. (1972). Proc. Roy. Soc. (London) *A326*, 347.
Rahman, A., Mandell, M. J. and McTague, J. P. (1976). J. Chem. Phys. *64*, 1564.
Raider, S. I. and Drake, R. E. (1981). IEEE Trans. *MAG-17*, 299.
Rasing, Th. H. M., Salemink, H. W. M., Wyder, P. and Strässler, S. (1981). Phys. Rev. B*23*, 4470.
Rath, J. and Wolfram, T. (1978). In *Inelastic Electron Tunneling Spectroscopy* (ed. T. Wolfram), p. 92. Springer-Verlag, Berlin.
Refai, T. F. and Wolf, E. L. (1983). Thin Solid Films *99*, 345.
Rehr, J. J. and Alben, R. (1977). Phys. Rev. B*16*, 2400.
Reif, F. and Woolf, M. A. (1962). Phys. Rev. Lett. *9*, 315.
Revenko, Y. F., D'yachenko, A. I., Svistunov, V. M. and Shonaikh, B. (1980). Sov. J. Low Temp. Phys. *6*, 635.
Reynolds, C. A., Serin, B., Wright, W. H. and Nesbitt, L. B. (1950). Phys. Rev. *78*, 487.
Rice, T. M. and Sneddon, L. (1981). Phys. Rev. Lett. *47*, 689.
Richards, P. L. (1977). In *Semiconductors and Semimetals* (eds. R. K. Willardson and A. C. Beer), Vol. 12, p. 395. Academic, New York.
——— (1978). In *Future Trends in Superconductive Electronics* (eds. B. S. Deaver, C. M. Falco, J. H. Harris and S. A. Wolf), p. 223. American Institute of Physics, New York.
Richards, P. L. and Shen, T.-M. (1980). IEEE Trans. *ED*-27, 1909.
Rickayzen, G. (1965). *Theory of Superconductivity*. Wiley-Interscience, New York.
Riedel, E. (1964). Z. Naturforschung *19a*, 1634.
Rietschel, H. (1978). Z. Phys. B*30*, 271.
Rietschel, H. and Winter, H. (1979). Phys. Rev. Lett. *43*, 1256.
Rietschel, H., Winter, H. and Reichardt, W. (1980). Phys. Rev. B*22*, 4284.
Rissman, P. (1973). J. Appl. Phys. *44*, 1893.
Roberts, B. W. (1976). J. Phys. Chem. Ref. Data *5*, 581.
Robinson, B. and Rowell, J. M. (1978). *Proc. Int. Conf. on Physics of Transition*

Metals (Inst. Phys. Conf. Ser. No. 39), p. 666. The Institute of Physics, Bristol.
Robinson, B., Geballe, T. H. and Rowell, J. M. (1976). In *Superconductivity in d- and f-Band Metals* (ed. D. H. Douglass), p. 381. Plenum, New York.
Rochlin, G. I. and Hansma, P. K. (1970). Phys. Rev. B2, 1460.
Rogers, J. S. (1970). Rev. Sci. Instr. *41*, 1184.
Rogers, J. S., Adler, J. G. and Woods, S. B. (1964). Rev. Sci. Instr. *35*, 208.
Rosenbaum, T. F., Andres, K., Thomas, G. A. and Bhatt, R. N. (1980). Phys. Rev. Lett. *45*, 1723.
Rothwarf, A. and Taylor, B. N. (1967). Phys. Rev. Lett. *19*, 27.
Rowe, J. M., Rush, J. J., Smith, H. G., Mostoller, M. and Flotow, H. E. (1974). Phys. Rev. Lett. *33*, 1297.
Rowell, J. M. (1963). Phys. Rev. Lett. *11*, 200.
——— (1969*a*). J. Appl. Phys. *40*, 1211.
——— (1969*b*). In *Tunneling Phenomena in Solids* (eds. E. Burstein and S. Lundqvist), p. 385. Plenum, New York.
——— (1973). Phys. Rev. Lett. *30*, 167.
——— (1980). Private communication.
Rowell, J. M. and Chynoweth, A. G. (1962). Bull. Am. Phys. Soc. (II) 7, 473.
Rowell, J. M. and Dynes, R. C. (1971). In *Phonons* (ed. M. A. Nusimovici), p. 150. Flammarion, Paris.
Rowell, J. M. and Feldmann, W. L. (1968). Phys. Rev. *172*, 393.
Rowell, J. M. and McMillan, W. L. (1966). Phys. Rev. Lett. *16*, 453.
Rowell, J. M. and Schmidt, P. H. (1976). Appl. Phys. Lett. *29*, 622.
Rowell, J. M. and Shen, L. Y. L. (1966). Phys. Rev. Lett. *17*, 15.
Rowell, J. M., Dynes, R. C. and Schmidt, P. H. (1980). In *Superconductivity in d- and f-Band Metals* (eds. H. Suhl and M. B. Maple), p. 409. Academic, New York.
Rowell, J. M., Gurvitch, M. and Geerk, J. (1981). Phys. Rev. B*24*, 2278.
Rowell, J. M., McMillan, W. L. and Anderson, P. W. (1965). Phys. Rev. Lett. *14*, 633.
Rowell, J. M., McMillan, W. L. and Dynes, R. C. (1978). (unpublished).
Rowell, J. M., McMillan, W. L. and Feldmann, W. L. (1969). Phys. Rev. *178*, 897.
Roy, A. P. and Brockhouse, B. N. (1970). Can. J. Phys. *48*, 1781.
Rudman, D. A. and Beasley, M. R. (1980). Appl. Phys. Lett. *36*, 1010.
Rudman, D. A., Howard, R. E., Moore, D. F., Zubeck, R. B. and Beasley, M. R. (1979). IEEE Trans. *MAG-15*, 582.
Rusinov, A. I. (1969). Sov. Phys.-JETP *29*, 1101.
Sahm, P. R. and Pruss, T. V. (1969). Phys. Lett. *28A*, 707.
Sai-Halasz, G. A., Chi, C. C., Denenstein, A. and Langenberg, D. N. (1974). Phys. Rev. Lett. *33*, 215.
Saint-James, D. (1964). J. Physique *25*, 899.
Samsonov, G. V. (1973). *The Oxide Handbook.* Plenum, New York.
Sanchez, A., Davis, C. F., Jr., Liu, K. C. and Javan, A. (1978). J. Appl. Phys. *49*, 5270.
Savitskii, E. M., Baron, V. V., Efimov, Yu. V., Bychkova, M. I. and Myzenkova, L. F. (1973). *Superconducting Materials.* Plenum, New York.
Scalapino, D. J. and Anderson, P. W. (1964). Phys. Rev. *133*, A921.
Scalapino, D. J. and Huberman, B. A. (1977). Phys. Rev. Lett. *39*, 1365.
Scalapino, D. J. and Marcus, S. M. (1967). Phys. Rev. Lett. *18*, 459.

Scalapino, D. J., Schrieffer, J. R. and Wilkins, J. W. (1966). Phys. Rev. *148*, 263.
Scalapino, D. J., Wada, Y. and Swihart, J. C. (1965). Phys. Rev. Lett. *14*, 102.
Schachinger, E., Daams, J. M. and Carbotte, J. P. (1980). Phys. Rev. B*22*, 3194.
Schein, L. B. and Compton, W. D. (1970). Appl. Phys. Lett. *17*, 236.
——— (1971). Phys. Rev. B*4*, 1128.
Schermeyer, F. L., Young, C. R. and Blastinggame, J. M. (1968). J. Appl. Phys. *39*, 1791.
Schiff, L. I. (1955). *Quantum Mechanics*, p. 184. McGraw-Hill, New York.
Schmid, A. and Schön, G. (1975). J. Low Temp. Phys. *20*, 207.
Schmidt, P. H., Spencer, E. G., Joy, D. C. and Rowell, J. M. (1976). In *Superconductivity in d- and f-Band Metals* (ed. D. H. Douglass), p. 431. Plenum, New York.
Schmidt, P. H., Castellano, R. N., Barz, H., Cooper, A. S. and Spencer, E. G. (1973). J. Appl. Phys. *44*, 1833.
Schnupp, P. (1967*a*). Phys. Stat. Sol. *21*, 567.
——— (1967b). Solid St. Electronics *10*, 785.
Schoneich, B., Elefant, D., Otschik, P. and Schumann, J. (1979). Phys. Stat. Sol. (b) *91*, 99.
Schopohl, N. and Scharnberg, K. (1977). Solid St. Commun. *22*, 371.
Schrieffer, J. R. (1964). *Theory of Superconductivity*. W. A. Benjamin, New York.
——— (1967). J. Appl. Phys. *38*, 1143.
——— (1968). J. Appl. Phys. *39*, 642.
Schrieffer, J. R. and Wilkins, J. W. (1963). Phys. Rev. Lett. *10*, 17.
Schrieffer, J. R. and Wolff, P. A. (1962). Phys. Rev. *149*, 491.
Schrieffer, J. R., Scalapino, D. J. and Wilkins, J. W. (1963). Phys. Rev. Lett. *10*, 336.
Schubert, W. K. and Wolf, E. L. (1979). Phys. Rev. B*20*, 1855.
Schuller, I. K. (1980). Phys. Rev. Lett. *44*, 1597.
Schuller, I. K., Orbach, R. and Chaikin, P. M. (1978). Phys. Rev. Lett. *41*, 1413.
Schulz, H. J., Jérome, D., Mazaud, A., Ribault, M. and Bechgaard, K. (1981). J. Physique *42*, 991.
Schweiss, B. P., Renker, B., Schneider, E. and Reichardt, W. (1976). In *Superconductivity in d- and f-Band Metals* (ed. D. H. Douglass), p. 189. Plenum, New York.
Schwidtal, K. (1960). Z. Phys. *158*, 563.
Scott, W. C. (1970). Appl. Phys. Lett. *17*, 166.
Seto, J. and Van Duzer, T. (1971). Appl. Phys. Lett. *19*, 488.
Shapiro, S. (1963). Phys. Rev. Lett. *11*, 80.
Shapiro, S., Smith, P. H., Nicol, J., Miles, J. L. and Strong, P. F. (1962). IBM J. Res. Dev. *6*, 34.
Shen, L. Y. L. (1969). J. Appl. Phys. *40*, 5171.
——— (1970). Phys. Rev. Lett. *24*, 1104.
——— (1972*a*). In *Superconductivity in d- and f-Band Metals* (ed. D. H. Douglass), p. 31. American Institute of Physics, New York.
——— (1972*b*). Phys. Rev. Lett. *29*, 1082.
——— (1972*c*). Rev. Sci. Instr. *43*, 1301.
Shen, L. Y. L. and Rowell, J. M. (1967). Solid St. Commun. *5*, 189.
——— (1968). Phys. Rev. *165*, 566.
Shen, T.-M., Richards, P. L., Harris, R. E. and Lloyd, F. L. (1980). Appl. Phys. Lett. *36*, 777.
Shiba, H. (1968). Prog. Theor. Phys. (Kyoto) *40*, 435.

Shih, S., Khim, Z. G., Arnold, G. B. and Tomasch, W. J. (1981). Phys. Rev. B*24*, 6440.

Shklyarevskii, O. I., Yanson, I. K. and Zaporozhskii, V. D. (1974). Solid St. Commun. *14*, 327.

Silver, A. H., Chase, A. B., McColl, M. and Millea, M. F. (1978). In *Future Trends in Superconductive Electronics* (eds. B. S. Deaver, C. M. Falco, J. H. Harris and S. A. Wolf), p. 364. American Institute of Physics, New York.

Silverman, P. J. and Briscoe, C. V. (1975). Phys. Lett. *53A*, 221.

Silvert, W. (1975). J. Low Temp. Phys. *20*, 439.

Simmons, J. G. (1963*a*, 1963*b*, 1963*c*). J. Appl. Phys. *34*, 238, 1793, 2581.

——— (1964). J. Appl. Phys. *35*, 2655.

Simons, A. L. and Varma, C. M. (1980). Solid St. Commun. *35*, 317.

Simonsen, M. G. and Coleman, R. V. (1973). Phys. Rev. B*8*, 5875.

Simonsen, M. G., Coleman, R. V. and Hansma, P. K. (1974). J. Chem. Phys. *61*, 3789.

Simpson, W. H. and Reucroft, P. J. (1970). Thin Solid Films *6*, 167.

Sixl, H., Gromer, J. and Wolf, H. C. (1974). Z. Naturforschung *29a*, 319.

Skalski, S., Betbeder-Matibet, O. and Weiss, P. R. (1964). Phys. Rev. *136*, A1500.

Slichter, C. P. (1963). *Principles of Magnetic Resonance*, p. 126. Harper and Row, New York.

Smith, A. D., McGrath, W. R., Richards, P. L., van Kemper, H., Prober, D. and Santhanam, P. (1981). Physica *108B*, 1367.

Smith, A. D., Skocpol, W. J. and Tinkham, M. (1980). Phys. Rev. B*21*, 3879.

Smith, H. G. and Glaser, W. (1971). *Proc. Int. Conf. on Phonons, Rennes* (ed. E. Nusimovici), p. 145. Flammarion, Paris.

Smith, L. N. and Mochel, J. M. (1975). Phys. Rev. Lett. *35*, 1597.

Smith, P. H. (1969). Surf. Sci. *16*, 34.

Smith, P. H., Shapiro, S., Miles, J. L. and Nicol, J. (1961). Phys. Rev. Lett. *6*, 686.

Solymar, L. (1972). *Superconductive Tunneling and Applications.* Chapman and Hall, London.

Solyom, J. and Zawadowski, A. (1968*a*) Phys. Kond. Mat. *7*, 325.

——— (1968*b*) Phys. Kond. Mat. *7*, 342.

Sommerfeld, A. and Bethe, H. (1933). *Handbuch der Physik*, (ed. S. Flügge) Vol. 24/2, p. 450. Springer-Verlag, Berlin.

Sood, B. R. (1972). Phys. Rev. B*6*, 136.

Sroubek, Z. (1969). Solid St. Commun. *7*, 1561.

Stearns, M. B. (1977). J. Magn. and Magn. Mater. *5*, 167.

Stedman, R., Almqvist, L. and Nilsson, G. (1967). Phys. Rev. *162*, 549.

Steinrisser, F., Davis L. C. and Duke, C. B. (1968). Phys. Rev. *176*, 912.

Stern, M. and Uhlig, H. H. (1952*a*, 1952*b*). J. Electrochem. Soc. *99*, 381; 389.

Strässler, S. and Wyder, P. (1967). Phys. Rev. *158*, 319.

Strässler, S. and Zeller, H. R. (1971). Phys. Rev. B*3*, 226.

Stratton, R. (1962). J. Phys. Chem. Solids *23*, 1177.

Straus, J. and Adler, J. G. (1974). Solid St. Commun. *15*, 1639.

Strieder, E. (1968). Ann. Phys. *22*, 15.

Stritzker, B. (1974). Z. Phys. *268*, 261.

Strongin, M. (1972). In *Superconductivity in d- and f-Band Metals* (ed. D. H. Douglass), p. 223. American Institute of Physics, New York.

Strongin, M., Thompson, R. S., Kammerer, O. F. and Crow, J. E. (1970). Phys. Rev. B*1*, 1078.
Su, W. P., Schrieffer, J. R. and Heeger, A. J. (1979). Phys. Rev. Lett. *42*, 1698.
Sutton, J. (1966). Proc. Phys. Soc. *87*, 791.
Svistunov, V. M., Chernyak, O. I., Belogolovskii, M. A. and D'yachenko, A. I. (1981). Phil. Mag. B*43*, 75.
Svistunov, V. M., D'yachenko, A. I. and Belogolovskii, M. A. (1978). J. Low Temp. Phys. *31*, 339.
Sweet, J. N. and Rochlin, G. I. (1970). Phys. Rev. B*2*, 656.
Swihart, J. C. (1961). J. Appl. Phys. *32*, 461.
Swihart, J. C., Scalapino, D. J. and Wada, Y. (1965). Phys. Rev. Lett. *14*, 106.
Sze, S. M. (1969). *Physics of Semiconductor Devices*, Chap. 8. Wiley-Interscience, New York.
Takada, Y. (1980). J. Phys. Soc. Japan *49*, 1267.
Taur, Y. (1980). IEEE Trans. *ED-27*, 1921.
Taur, Y. and Kerr, A. R. (1978). Appl. Phys. Lett. *32*, 775.
Taur, Y., Claasen, J. H. and Richards, P. L. (1974). Appl. Phys. Lett. *24*, 101.
Taylor, B. N. (1963). Ph.D. Thesis, University of Pennsylvania (unpublished).
Taylor, B. N. and Burstein, E. (1963). Phys. Rev. Lett. *10*, 14.
Taylor, B. N., Burstein, E. and Langenberg, D. N. (1962). Bull. Am. Phys. Soc. *7*, 190.
Taylor, B. N., Parker, W. H., Langenberg, D. N. and Denenstein, A. (1967). Metrologia *3*, 89.
Tedrow, P. M. and Meservey, R. (1971). Phys. Rev. Lett. *27*, 919.
——— (1973). Phys. Rev. B*7*, 318.
——— (1975). Phys. Lett. *51A*, 57.
——— (1979). Phys. Rev. Lett. *43*, 384.
Testardi, L. R. (1975). Rev. Mod. Phys. *47*, 637.
Testardi, L. R. and Mattheiss, L. F. (1978). Phys. Rev. Lett. *41*, 1612.
Testardi, L. R., Wernick, J. H. and Royer, W. A. (1974). Solid St. Commun. *15*. 1.
Testardi, L. R., Meek, R. L., Poate, J. M., Royer, W. A., Storm, A. R. and Wernick, J. H. (1975). Phys. Rev. B*11*, 4304.
Thielemans, M., Leo, V., Deltour, R. and Mehbod, M. (1979). J. Appl. Phys. *50*, 5841.
Thomas, D. E. and Klein, J. M. (1963). Rev. Sci. Instr. *34*, 920.
Thomas, D. E. and Rowell, J. M. (1965). Rev. Sci. Instr. *36*, 1301.
Thomas, P. and Queisser, H. J. (1968). Phys. *175*, 983.
Thompson, W. A. and von Molnar, S. (1970). J. Appl. Phys. *41*, 5218.
Thompson, W. A., Holtzberg, F. and McGuire, T. R. (1971). Phys. Rev. Lett. *26*, 1308.
Thompson, W. A., Holtzberg, F., McGuire, T. R. and Petrich, G. (1972). In *Magnetism and Magnetic Materials*, AIP Conf. Proc. No. 5, p. 827. American Institute of Physics, New York.
Thompson, W. A., Penney, T., Holtzberg, F. and Kirkpatrick, S. (1972). *Proc. 11th Int. Conf. Phys. Semicond.*, p. 1255. Elsevier, Amsterdam.
Thornton, J. A. and Penfold, A. S. (1978). In *Thin Film Processes* (eds. J. L. Vossen and W. Kern), p. 107. Academic, New York.
Thouless, D. J. (1960). Phys. Rev. *117*, 1256.
Tien, P. K. and Gordon, J. P. (1963). Phys. Rev. *129*, 647.

Tinkham, M. (1972). Phys. Rev. B*6*, 1747.
——— (1975). *Introduction to Superconductivity*. McGraw-Hill, New York.
——— (1979). In *Festkörperprobleme* (ed. J. Treusch), XIX, p. 363. Braunschweig, Vieweg.
Tinkham, M. and Clarke, J. (1972). Phys. Rev. Lett. *28*, 1366.
Tomasch, W. J. (1965*a*). Phys. Rev. Lett. *15*, 672.
——— (1965*b*). Phys. Rev. *139*, A746.
——— (1966*a*). Phys. Rev. Lett. *16*, 16.
——— (1966*b*). Phys. Lett. *23*, 204.
Tomasch, W. J. and Wolfram, T. (1966). Phys. Rev. Lett. *16*, 352.
Tomlinson, P. G. and Swihart, J. C. (1979). Phys. Rev. B*19*, 1867.
Toplicar, J. R. and Finnemore, D. K. (1977). Phys. Rev. B*16*, 2072.
Townsend, P. and Sutton, J. (1962). Phys. Rev. *128*, 591.
Trofimenkoff, P. N. and Carbotte, J. P. (1970). Phys. Rev. B*1*, 1136.
Trofimenkoff, P. N., Kreuzer, H. J., Wattamaniuk, W. J. and Adler, J. G. (1972). Phys. Rev. Lett. *29*, 597.
Truant, P. T. and Carbotte, J. P. (1972). Phys. Rev. B*6*, 3642.
Tsang, J. C., Kirtley, J. R. and Theis, T. N. (1980). Solid St. Commun. *35*, 667.
Tsang, J.-K. and Ginsberg, D. M. (1980*a*). Phys. Rev. B*21*, 132.
——— (1980*b*). Phys. Rev. B*22*, 4280.
Tsuei, C. C., Johnson, W. L., Laibowitz, R. B. and Viggiano, J. M. (1977). Solid St. Commun. *24*, 615.
Tsui, D. C. (1968). Phys. Rev. Lett. *21*, 994.
——— (1969). Phys. Rev. Lett. *22*, 293.
——— (1970*a*). Phys. Rev. Lett. *24*, 303.
——— (1970*b*). Solid St. Commun. *8*, 113.
——— (1970*c*). J. Appl. Phys. *41*, 2651.
——— (1971*a*). Phys. Rev. B*4*, 4438.
——— (1971*b*). Solid St. Commun. *9*, 1789.
——— (1972). *Proc. 11th Conf. on the Physics of Semiconductors*, p. 109. Elsevier, Amsterdam.
——— (1973). Phys. Rev. B*8*, 2657.
——— (1975*a*). Phys. Rev. B*12*, 5739.
——— (1975*b*). Phys. Rev. B*12*, 5853.
Tsui, D. C. and Barker, A. S. (1969). Phys. Rev. *186*, 590.
Tsui, D. C. and Kaminsky, G. (1973). Bull. Am. Phys. Soc. (II) *18*, 412.
Tsui, D. C., Dietz, R. E. and Walker, L. R. (1971). Phys. Rev. Lett. *27*, 1729.
Tsui, D. C., Kaminsky, G. and Schmidt, P. H. (1974). Phys. Rev. B*9*, 3524.
Tucker, J. R. (1979). IEEE J. Quant. Electron. *QE-15*, 1234.
——— (1980). Appl. Phys. Lett. *36*, 477.
Tucker, J. R. and Millea, M. F. (1978). Appl. Phys. Lett. *33*, 611.
Tüttö, I. and Ruvalds, J. (1979). Phys. Rev. B*19*, 5641.
Umbach, C. P. and Goldman, A. M. (1982). Phys. Rev. Lett. *48*, 1433.
Umbach, C. P., Goldman, A. M. and Toth, L. E. (1982). Appl. Phys. Lett. *40*, 81.
Umbach, C. P., Toth, L. E., Dahlberg, E. D. and Goldman, A. M. (1981). Physica *108B*, 803.
Valatin, J. G. (1958). Nuovo Cimento *7*, 843.
van Duzer, T. and Turner, C. W. (1981). *Principles of Superconductive Devices and Circuits*. Elsevier, New York.

van Gelder, A. P. (1969). Phys. Rev. *181*, 787.
Varma, C. M. and Weber, W. (1979). Phys. Rev. B*19*, 6142.
Vashishta, P. and Carbotte, J. P. (1970). Solid St. Commun. *8*, 161.
Vettiger, P., Moore, D. F. and Forster, T. (1981). IEEE Trans. *ED-28*, 1385.
von Baltz, R. and Urban, B. (1977). Phys. Stat. Solidi (b) *79*, 185.
von Molnar, S., Thompson, W. A. and Edelstein, A. S. (1967). Appl. Phys. Lett. *11*, 163.
Vonsovsky, S. V., Izyamov, Y. A. and Kurmaev, E. Z. (1982). *Superconductivity of Transition Metals.* Springer-Verlag, Berlin.
Vopat, F. E., Lee, H. J. and Tomasch, W. J. (1976). J. Appl. Phys. *47*, 329.
Vossen, J. L. and Kern, W., eds. (1978). *Thin Film Processes.* Academic, New York.
Vrba, J. and Woods, S. B. (1971*a*). Phys. Rev. B*3*, 2243.
——— (1971*b*). Phys. Rev. B*4*, 87.
Vujicić, G. M. (1979). J. Phys. C*12*, 1699.
Waldram, J. R. (1976). Rep. Prog. Phys. *39*, 751.
Wallace, P. R. and Stavn, M. J. (1965). Can. J. Phys. *43*, 411.
Wallis, R. H. and Wyatt, A. F. G. (1972). Phys. Rev. Lett. *29*, 479.
Walmsley, D. G. (1980). In *Vibrational Spectroscopy of Adsorbates,* Springer Series in Chemical Physics 15, Chap. 5. Springer-Verlag, Berlin.
——— (1983). Private communication.
Walmsley, D. G., Floyd, R. B. and Timms, W. E. (1977). Solid St. Commun. *22*, 497.
Walmsley, D. G., McMorris, I. W. N. and Brown, N. M. D. (1975). Solid St. Commun. *16*, 663.
Walmsley, D. G., Quinn, H. F. and Dawson, P. (1982). Phys. Rev. Lett. *49*, 892.
Walmsley, D. G., Timms, W. E. and Brown, N. M. D. (1976). Solid St. Commun. *20*, 627.
Walmsley, D. G., Wolf, E. L. and Osmun, J. W. (1979). Thin Solid Films *62*, 61.
Walmsley, D. G., Nelson, W. J., Brown, N. M. D. and Floyd, R. B. (1980). Appl. Surf. Sci. *5*, 107.
Walmsley, D. G., McMorris, I. W. N., Timms, W. E., Nelson, W. J., Tomlin, J. L., and Griffin, T. J. (1983). J. Phys. E *16*, 1052.
Wattamaniuk, W. J., Kreuzer, H. J. and Adler, J. G. (1971). Phys. Lett. *37A*, 7.
Weinberg, W. H. (1978). Ann. Rev. Phys. Chem. *29*, 115.
——— (1982). In *Tunneling Spectroscopy: Capabilities, Applications, and New Techniques* (ed. P. K. Hansma) p. 359. Plenum, New York.
Werthamer, N. R. (1966). Phys. Rev. *147*, 255.
White, H. W., Godwin, L. M. and Wolfram, T. (1978). In *Inelastic Electron Tunneling Spectroscopy* (ed. T. Wolfram), p. 70. Springer-Verlag, Berlin).
White, R. M. and Geballe, T. H. (1979). *Long Range Order in Solids.* Academic, New York.
Whitmer, M. D. (1981). Can. J. Phys. *59*, 309.
Wilkins, J. W. (1969). In *Tunneling Phenomena in Solids* (ed. E. Burstein and S. Lundqvist), p. 333. Plenum, New York.
Willemson, H. W. and Gray, K. E. (1978). Phys. Rev. Lett. *41*, 812.
Wilson, A. H. (1932). Proc. Roy. Soc. *A136*, 487.
Wilson, J. A. (1979). J. Low Temp. Phys. *35*, 135.
——— (1980). Ph.D. thesis, University of California, Los Angeles (unpublished).
Wilson, J. A. and Chaikin, P. M. (1980). J. Low Temp. Phys. *38*, 315.

Wilson, J. A., Simon, R. W., McGinnis, W. C. and Chaikin, P. M. (1980). Bull. Am. Phys. Soc. *25*, 168.
Wilson, K. G. (1975). Rev. Mod. Phys. *47*, 773.
Wimmers, J. T. and Christopher, J. E. (1981). J. Phys. C*14*, 977.
Wolf, E. L. (1968). Phys. Rev. Lett. *20*, 204.
——— (1974). Phys. Rev. B*10*, 784.
——— (1975). Solid St. Phys. *30*, 1.
——— (1978). Rep. Prog. Phys. *41*, 1439.
Wolf, E. L. and Arnold, G. B. (1982). Phys. Rpts. *91*, 31.
Wolf, E. L. and Compton, W. D. (1969). Rev. Sci. Instr. *40*, 1497.
Wolf, E. L. and Losee, D. L. (1969). Phys. Rev. Lett. *23*, 1457.
——— (1970). Phys. Rev. B*2*, 3660.
Wolf, E. L. and Noer, R. J. (1979). Solid St. Commun. *30*, 391.
Wolf, E. L. and Zasadzinski, J. (1977). Phys. Lett. *62A*, 165.
——— (1978). *Proc. Conf. on Physics of Transition Metals* (Inst. Phys. Conf. Ser. No. 39), p. 666. The Institute of Physics, Bristol.
Wolf, E. L., Noer, R. J. and Arnold, G. B. (1980). J. Low Temp. Phys. *40*, 419.
Wolf, E. L., Wallis, R. H. and Adkins, C. J. (1975). Phys. Rev. B*12*, 1603.
Wolf, E. L., Burnell, D. M., Khim, Z. G. and Noer, R. J. (1981). J. Low Temp. Phys. *44*, 89.
Wolf, E. L., Losee, D. L., Cullen, D. E. and Compton, W. Dale (1971). Phys. Rev. Lett. *26*, 438.
Wolf, E. L., Zasadzinski, J., Osmun, J. W. and Arnold, G. B. (1979). Solid St. Commun. *31*, 321.
——— (1980). J. Low Temp. Phys. *40*, 19.
Wolf, E. L., Noer, R. J., Burnell, D., Khim, Z. G. and Arnold, G. B. (1981). J. Phys. F*11*, L23.
Wolf, E. L., Zasadzinski, J., Arnold, G. B., Moore, D. F., Rowell, J. M. and Beasley, M. R. (1980). Phys. Rev. B*22*, 1214.
Wolfram, T. (1968). Phys. Rev. *170*, 481.
——— (1978). *Inelastic Electron Tunneling Spectroscopy*. Springer-Verlag, Berlin.
Wong, L., Shih, S. and Tomasch, W. J. (1981). Phys. Rev. B*23*, 5775.
Woodruff, D. P. and Smith, N. V. (1982). Phys. Rev. Lett. *48*, 283.
Woods, A. D. B. (1964). Phys. Rev. *136*, A781.
Woolf, M. A. and Reif, F. (1965). Phys. Rev. *137*, A557.
Wright, P. W. and Franck, J. P. (1977). J. Low Temp. Phys. *27*, 459.
Wu, T. M. (1967). Phys. Rev. Lett. *19*, 508.
Wühl, H., Eichler, A. and Wittig, J. (1973). Phys. Rev. Lett. *31*, 1393.
Wühl, H., Jackson, J. E. and Briscoe, C. V. (1968). Phys. Rev. Lett. *20*, 1496.
Wyatt, A. F. G. (1964). Phys. Rev. Lett. *13*, 401.
——— (1971). Phys. Lett. *37A*, 399.
——— (1973). J. Phys. C*6*, 673.
——— (1974). J. Phys. C*7*, 1303.
Wyatt, A. F. G. and Lythall, D. J. (1967). Phys. Lett. *25A*, 41.
Wyatt, P. W., Barker, R. C. and Yelon, A. (1972). Phys. Rev. B*6*, 4169.
Yanson, I. K. (1971). Zh. Eksp. Teor. Fiz. *60*, 1759 (Engl. transl. Sov. Phys.-JETP *33*, 951).
——— (1974). Zh. Eksp. Teor. Fiz. *66*, 1035 (Engl. transl. Sov. Phys.-JETP *39*, 506).
Yanson, I. K. and Batrak, A. G. (1978). Pis'ma Zh. Eksp. Teor. Fiz. *27*, 212 [JETP Lett. *27*, 197 (1978)].

Yanson, I. K. and Bogatina, N. I. (1971). Zh. Eksp. Teor. Fiz. *59*, 1509 (Engl. transl. Sov. Phys.-JETP *32*, 823).
Yanson, I. K., Kulik, I. O. and Batrak, A. G. (1981). J. Low Temp. Phys. *42*, 527.
Yanson, I. K., Svistunov, V. M. and Dmitrenko, I. M. (1965). Zh. Eksp. Teor. Fiz. *48*, 976 (Engl. transl. Sov. Phys.-JETP. *21*, 650).
Yanson, I. K., Bogatina, N. I., Verkin, B. I. and Shkylarevskii, O. I. (1972). Zh. Eksp. Teor. Fiz. *62*, 1023 (Engl. transl. Sov. Phys.-JETP *35*, 540).
Yasuoka, Y., Heiblum, M. and Gustafson, T. K. (1979). Appl. Phys. Lett. *34*, 823.
Yoshihiro, K. and Sasaki, W. (1968). J. Phys. Soc. Japan *24*, 426.
Yosida, K. (1957). Phys. Rev. *106*, 893.
Yu, C. C. and Anderson, P. W. (1984). Phys. Rev. B*29*, 6165.
Zaitsev, A. V. (1980). Zh. Eksp. Teor. Fiz. *78*, 221 (Engl. transl. Sov. Phys.-JETP *51*, 111).
Zasadzinski, J., Burnell, D. M., Wolf, E. L. and Arnold, G. B. (1982). Phys. Rev. B*25*, 1622.
Zasadzinski, J., Schubert, W. K., Wolf, E. L. and Arnold, G. B. (1980). *Superconductivity in d- and f-Band Metals* (eds. H. Suhl and B. Maple), p. 159. Academic, New York.
Zavaritskii, N. V. (1964). Zh. Eksp. Teor. Fiz. *45*, 1839 (Engl. transl. Sov. Phys.-JETP *18*, 1260).
——— (1965). Zh. Eksp. Teor. Fiz. *48*, 837 (Engl. transl. Sov. Phys.-JETP *21*, 557).
——— (1970). Zh. Eksp. Teor. Fiz. *57*, 752 (Engl. transl. Sov. Phys.-JETP *30*, 412).
Zawadowski, A. (1967). Phys. Rev. *163*, 341.
Zeller, H. R. (1972). Phys. Rev. B*5*, 1813.
Zeller, H. R. and Giaever, I. (1969). Phys. Rev. *181*, 789.
Zener, C. (1934). Proc. Roy. Soc. A*145*, 523.
Ziemba, G. and Bergmann, G. (1970). Z. Phys. *237*, 410.
Ziman, J. M. (1960). *Electrons and Phonons*. Oxford Univ. Press, Oxford.
Zimmerman, J. E. and Silver, A. H. (1966). Phys. Rev. *141*, 367.
Zimmerman, J. E., Thiene, P. and Harding, J. T. (1970). J. Appl. Phys. *41*, 1572.
Zittartz, J., Bringer, A. and Müller-Hartmann, E. (1972). Solid St. Commun. *10*, 513.
Zittartz, J. and Müller-Hartmann, E. (1970). Z. Phys. *232*, 11.
Zorin, A. B., Kulik, I. O., Likharev, K. K. and Schrieffer, J. R. (1979). Fiz. Nizk. Temp. *5*, 1138 [Engl. Transl. Sov. J. Low Temp. Phys. *5*, 537].
Zwerger, W. (1983). Solid St. Commun. *45*, 841.

Index